Power Generation Maintenance Electrician

Level 1

Trainee Guide
First Edition

Prentice Hall

Boston Columbus Indianapolis New York San Francisco Upper Saddle River
Amsterdam Cape Town Dubai London Madrid Milan Munich Paris Montreal Toronto
Delhi Mexico City Sao Paulo Sydney Hong Kong Seoul Singapore Taipei Tokyo

National Center for Construction Education and Research

President: Don Whyte
Director of Product Development: Daniele Stacey
Power Generation Maintenance Electrician Project Manager: Daniele Stacey
Production Manager: Tim Davis
Quality Assurance Coordinator: Debie Ness
Editors: Rob Richardson, Matt Tischler
Desktop Publishing Coordinator: James McKay
Production Assistant: Laura Wright

NCCER would like to acknowledge the contract service provider for this curriculum:
Topaz Publications, Syracuse, New York.

10 9 8 7 6 5 4 3 2 1

Prentice Hall
is an imprint of

PEARSON

www.pearsonhighered.com

ISBN 13: 978-0-13-215421-5

Preface

To the Trainee

Congratulations! If you are training under an NCCER Accredited Training Sponsor, you have successfully completed *Power Industry Fundamentals*. Now you are well on your way to more specific skills training in power generation maintenance. This is an area that combines the knowledge and skills of many different trade areas, including electrical, instrument and controls (I&C), mechanical, and welding.

As the first of four levels of *Power Generation Maintenance Electrician*, this curriculum provides a basic foundation for skill development in testing, diagnosing, and repairing power generation plant equipment. From reading plant construction drawings, to using test instruments, to hands-on experience in valve installation and shielded metal arc welding, this training is a rich source of knowledge and experience that will give you a competitive edge as you grow in your profession.

We invite you to visit the NCCER website at www.nccer.org for the latest releases, training information, newsletter, and much more. You can also reference the Contren® product catalog online at www.nccer.org.

Your feedback is welcome. You may email your comments to curriculum@nccer.org or send general comments and inquiries to info@nccer.org.

Opportunities for Multiple Credentials

Opportunities for multiple credentials now exist if your training program is being administered through an NCCER Accredited Training Program Sponsor. Successful completion of this training program could lead to a completion certificate for *Power Generation Maintenance Electrician Level One* as well as module completion credit for the original *Industrial Maintenance E&I Technician* and *Welding* modules. This is because the modules for *Power Generation Maintenance Electrician Level One* are presented as they originally appeared in these craft training programs. Example transcripts from the National Registry for these module completions appear on the following page.

If you are not training through an NCCER Accredited Training Program Sponsor, find out where one exists near you. Call NCCER at 1.888.622.3720 or visit us at www.nccer.org.

Contren® Learning Series

The National Center for Construction Education and Research (NCCER) is a not-for-profit 501(c)(3) education foundation established in 1995 by the world's largest and most progressive construction companies and national construction associations. It was founded to address the severe workforce shortage facing the industry and to develop a standardized training process and curricula. Today, NCCER is supported by hundreds of leading construction and maintenance companies, manufacturers, and national associations. The Contren® Learning Series was developed by NCCER in partnership with Pearson Education, Inc., the world's largest educational publisher.

Some features of NCCER's Contren® Learning Series are as follows:

- An industry-proven record of success
- Curricula developed by the industry for the industry
- National standardization providing portability of learned job skills and educational credits
- Compliance with Office of Apprenticeship requirements for related classroom training (CFR 29:29)
- Well-illustrated, up-to-date, and practical information

NCCER also maintains a National Registry that provides transcripts, certificates, and wallet cards to individuals who have successfully completed modules of NCCER's Contren® Learning Series. *Training programs must be delivered by an NCCER Accredited Training Sponsor in order to receive these credentials.*

Examples of credentials available on successful completion of Power Generation Maintenance Electrician Level One

NATIONAL CENTER FOR CONSTRUCTION EDUCATION AND RESEARCH
BUILDING TOMORROW'S WORKFORCE
3600 NW 43rd St, Bldg G • Gainesville, FL 32606
P 352.334.0911 • F 352.334.0932 • www.nccer.org
Affiliated with the University of Florida

May 25, 2010

Official Transcript

Student Sample24
Green Equipment Company
2563 Gravel Dr
Fort Worth. TX 76118

Current Employer/School:

Card #: 7691824

Course / Description	Instructor	Training Location	Date Compl.
00101-09 Basic Safety	Don E Whyte	Green Equipment Company	5/1/2010
00102-09 Introduction to Construction Math	Don E Whyte	Green Equipment Company	5/1/2010
00103-09 Introduction to Hand Tools	Don E Whyte	Green Equipment Company	5/1/2010
00104-09 Introduction to Power Tools	Don E Whyte	Green Equipment Company	5/1/2010
00105-09 Introduction to Construction Drawings	Don E Whyte	Green Equipment Company	5/1/2010
00106-09 Basic Rigging	Don E Whyte	Green Equipment Company	5/1/2010
00107-09 Basic Communication Skills	Don E Whyte	Green Equipment Company	5/1/2010
00108-09 Basic Employability Skills	Don E Whyte	Green Equipment Company	5/1/2010
00109-09 Introduction to Materials Handling	Don E Whyte	Green Equipment Company	5/1/2010
29107-09 SMAW-Equipment and Setup	Don E Whyte	Green Equipment Company	5/1/2010
40102-07 Tools of the Trade	Don E Whyte	Green Equipment Company	5/1/2010
40103-07 Fasteners and Anchors	Don E Whyte	Green Equipment Company	5/1/2010
40104-07 Oxyfuel Cutting	Don E Whyte	Green Equipment Company	5/1/2010
40105-07 Gaskets and Packing	Don E Whyte	Green Equipment Company	5/1/2010
40106-07 Craft-Related Mathematics	Don E Whyte	Green Equipment Company	5/1/2010
40107-07 Construction Drawings	Don E Whyte	Green Equipment Company	5/1/2010
40108-07 Pumps and Drivers	Don E Whyte	Green Equipment Company	5/1/2010

Page 1

Donald E. Whyte
President, NCCER

NATIONAL CENTER FOR CONSTRUCTION EDUCATION AND
BUILDING TOMORROW'S WORKFORCE
3600 NW 43rd St, Bldg G • Gainesville, FL 32606
P 352.334.0911 • F 352.334.0932 • www.nccer.org
Affiliated with the University of Florida

May 25, 2010

Official Transcript

Student Sample24
Green Equipment Company
2563 Gravel Dr
Fort Worth. TX 76118

Card #: 7691824

Course / Description	Instructor	Tra
40109-07 Introduction to Valves	Don E Whyte	Gre
40110-07 Introduction to Test Equipment	Don E Whyte	Gre
40111-07 Material Handling and Hand Rigging	Don E Whyte	Gre
40112-07 Mobile and Support Equipment	Don E Whyte	Gre
40113-07 Lubrication	Don E Whyte	Gre
49101-10 Introduction to the Power Industry	Don E Whyte	Gre

NO ENTRIES BELOW THIS LINE

Page 2

National Center for Construction Education and Research

This is to certify that

Student Sample24

has fulfilled the requirements for

Power Gen Maintenance Electrician Level One

in NCCER's standardized training curriculum
this First day of May, 2010

NATIONAL CENTER FOR CONSTRUCTION EDUCATION AND RESEARCH • FOUNDED 1996 •

Donald E. Whyte
Donald E. Whyte
President

Acknowledgments

This curriculum was revised as a result of the farsightedness and leadership of the following sponsors:

Arizona Public Service Company
– Cholla Power Plant
Constellation Energy
Dunellon High School
Fluor
Georgia Power Company
IMTI of NY and IMTI of CT
Mechanical Construction Services, Inc.
Progress Energy
Salt River Project – Coronado Generating Station
South Carolina Electric and Gas
Southern Company Service
TIC Industrial

This curriculum would not exist were it not for the dedication and unselfish energy of those volunteers who served on the Authoring Team. A sincere thanks is extended to the following:

Russell Battles
Allen Burlison
Randy Fabry
Ernie Griego
Phillip (Chip) Hammock, Jr.
Michael Hancock
Kenneth (Scott) Huggins
Keith Hunter
Chris Kelley
Darren Marvin
Yolanda McCloud
Jim Mitchem
Johnny Penrod
Martin Reece
Walter Ernie Tomlinson
Marcel Veronneau

NCCER Partners

American Fire Sprinkler Association
Associated Builders and Contractors, Inc.
Associated General Contractors of America
Association for Career and Technical Education
Association for Skilled and Technical Sciences
Carolinas AGC, Inc.
Carolinas Electrical Contractors Association
Center for the Improvement of Construction Management and Processes
Construction Industry Institute
Construction Users Roundtable
Design Build Institute of America
Merit Contractors Association of Canada
Metal Building Manufacturers Association
NACE International
National Association of Manufacturers
National Association of Minority Contractors
National Association of Women in Construction
National Insulation Association
National Ready Mixed Concrete Association
National Technical Honor Society
National Utility Contractors Association
NAWIC Education Foundation
North American Technician Excellence
Painting & Decorating Contractors of America
Portland Cement Association
SkillsUSA
Steel Erectors Association of America
U.S. Army Corps of Engineers
Women Construction Owners & Executives, USA
University of Florida, M.E. Rinker School of Building Construction

NCCER Business Partners

Contents

The modules for this curriculum are from *Industrial Maintenance E&I Technician Level One* and *Welding Level One,* and carry their original module numbering format. The accompanying course map illustrates the recommended training sequence for these modules in *Power Generation Maintenance Electrician Level One.*

40111-07
Material Handling and Hand Rigging

Introduces the equipment and techniques of material handling, and describes the procedures for rigging and communicating with riggers. (15 Hours)

40112-07
Mobile and Support Equipment

Introduces the safety procedures and methods of operation for motorized support equipment, including forklifts, manlifts, compressors, and generators. (10 Hours)

40113-07
Lubrication

Explains lubrication safety, storage, and classifications. Also explains selecting lubricants, additives, lubrication equipment, and lubricating charts. (12.5 Hours)

29107-09
SMAW Equipment and Setup

Describes SMAW welding and welding safety. Explains how to connect welding current and setup arc welding equipment. Identifies and explains using tools for cleaning welds. (5 Hours)

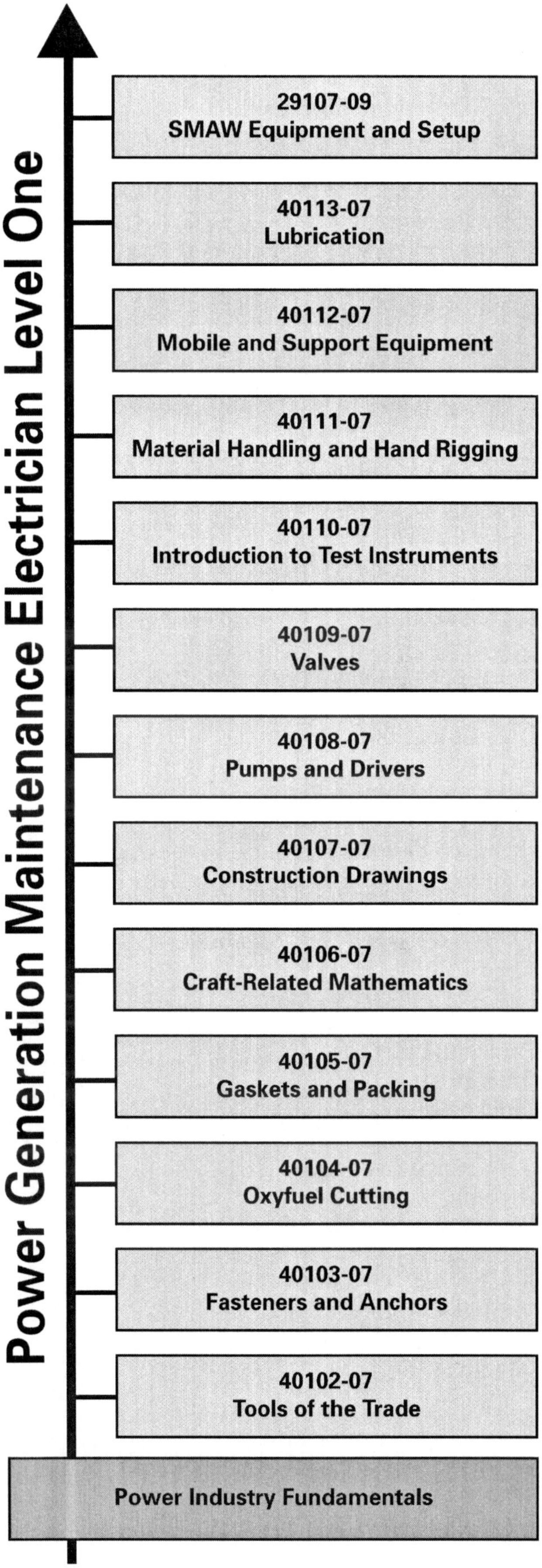

Contren® Curricula

NCCER's training programs comprise nearly 80 construction, maintenance, pipeline, and utility areas and include skills assessments, safety training, and management education.

Boilermaking
Cabinetmaking
Carpentry
Concrete Finishing
Construction Craft Laborer
Construction Technology
Core Curriculum:
  Introductory Craft Skills
Drywall
Electrical
Electronic Systems Technician
Heating, Ventilating, and
  Air Conditioning
Heavy Equipment Operations
Highway/Heavy Construction
Hydroblasting
Industrial Coating and Lining
  Application Specialist
Industrial Maintenance
  Electrical and
  Instrumentation Technician
Industrial Maintenance
  Mechanic
Instrumentation
Insulating
Ironworking
Masonry
Millwright
Mobile Crane Operations
Painting
Painting, Industrial
Pipefitting
Pipelayer
Plumbing
Reinforcing Ironwork
Rigging
Scaffolding
Sheet Metal
Site Layout
Sprinkler Fitting
Tower Crane Operator
Welding

Green/Sustainable Construction

Your Role in the Green
  Environment
Sustainable Construction
  Supervisor
Introduction to Weatherization
Weatherization Installer

Energy

Introduction to the Power
  Industry
Power Industry Fundamentals
Power Generation Maintenance
  Electrician
Power Generation I&C
  Maintenance Technician
Power Generation Maintenance
  Mechanic
Steam and Gas Turbine
  Technician
Introduction to Solar
  Photovoltaics
Introduction to Wind Energy

Pipeline

Control Center Operations,
  Liquid
Corrosion Control
Electrical and Instrumentation
Field Operations, Liquid
Field Operations, Gas
Maintenance
Mechanical

Safety

Field Safety
Safety Orientation
Safety Technology

Management

Introductory Skills for the
  Crew Leader
Project Management
Project Supervision

Supplemental Titles

Applied Construction Math
Careers in Construction
Tools for Success

Spanish Translations

Basic Rigging
  (Principios Básicos de
  Maniobras)
Carpentry Fundamentals
  (Introducción a la
  Carpintería, Nivel Uno)
Carpentry Forms
  (Formas para Carpintería,
  Nivel Trés)
Concete Finishing, Level One
  (Acabado de Concreto,
  Nivel Uno)
Core Curriculum:
  Introductory Craft Skills
  (Currículo Básico:
  Habilidades Introductorias
  del Oficio)
Drywall, Level One
  (Paneles de Yeso, Nivel Uno)
Electrical, Level One
  (Electricidad, Nivel Uno)
Field Safety
  (Seguridad de Campo)
Insulating, Level One
  (Aislamiento, Nivel Uno)
Masonry, Level One
  (Albañilería, Nivel Uno)
Pipefitting, Level One
  (Instalación de Tubería
  Industrial, Nivel Uno)
Reinforcing Ironwork,
  Level One
  (Herreria de Refuerzo,
  Nivel Uno)
Safety Orientation
  (Orientación de Seguridad)
Scaffolding
  (Andamios)
Sprinkler Fitting, Level One
  (Instalación de Rociadores,
  Nivel Uno)

Industrial Maintenance E & I Technician Level One

40102-07

Tools of the Trade

40102-07
Tools of the Trade

Topics to be presented in this module include:

Overview

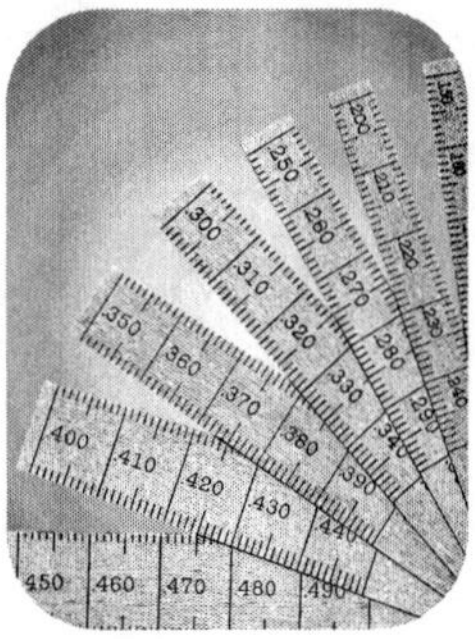

A craft is identified by the tools it uses. An industrial maintenance craftworker uses many very specialized tools ranging from heavy-duty devices to precision measuring and leveling instruments. Industrial maintenance work involves tolerances that are often measured in tiny fractions of inches; even a slight error in such an environment can have a large impact on the operation of equipment.

The industrial maintenance craftworker is required to lay out equipment locations, assemble and disassemble mechanical equipment, align mechanical equipment, and check for proper calibration of equipment. Each of these functions requires knowledge of specialized tools. The mechanic learns which tools to use in a given situation, as well as the proper application of the tool.

Objectives

When you have completed this module, you will be able to do the following:

1. Explain the purpose of each of the tools commonly used by industrial maintenance craftworkers.
2. Describe how to maintain each of the tools used by industrial maintenance craftworkers.
3. Demonstrate the proper use and basic maintenance of selected industrial maintenance tools.

Trade Terms

Abrasive
Align
Bevel
Burr
Chamfer
Chuck
Component
Contamination
Diameter
Die
Flanges
Galling
Horsepower (hp)
Pipe fitting
Pitch
Revolutions per minute (rpm)
Sheave
Shims
Tolerance

Required Trainee Materials

1. Pencil and paper
2. Appropriate personal protective equipment

Prerequisites

Before you begin this module, it is recommended that you successfully complete *Core Curriculum;* and *Industrial Maintenance E & I Technician Level One*, Module 40101-07.

This course map shows all of the modules in the first level of the *Industrial Maintenance E & I Technician* curriculum. The suggested training order begins at the bottom and proceeds up. Skill levels increase as you advance on the course map. The local Training Program Sponsor may adjust the training order.

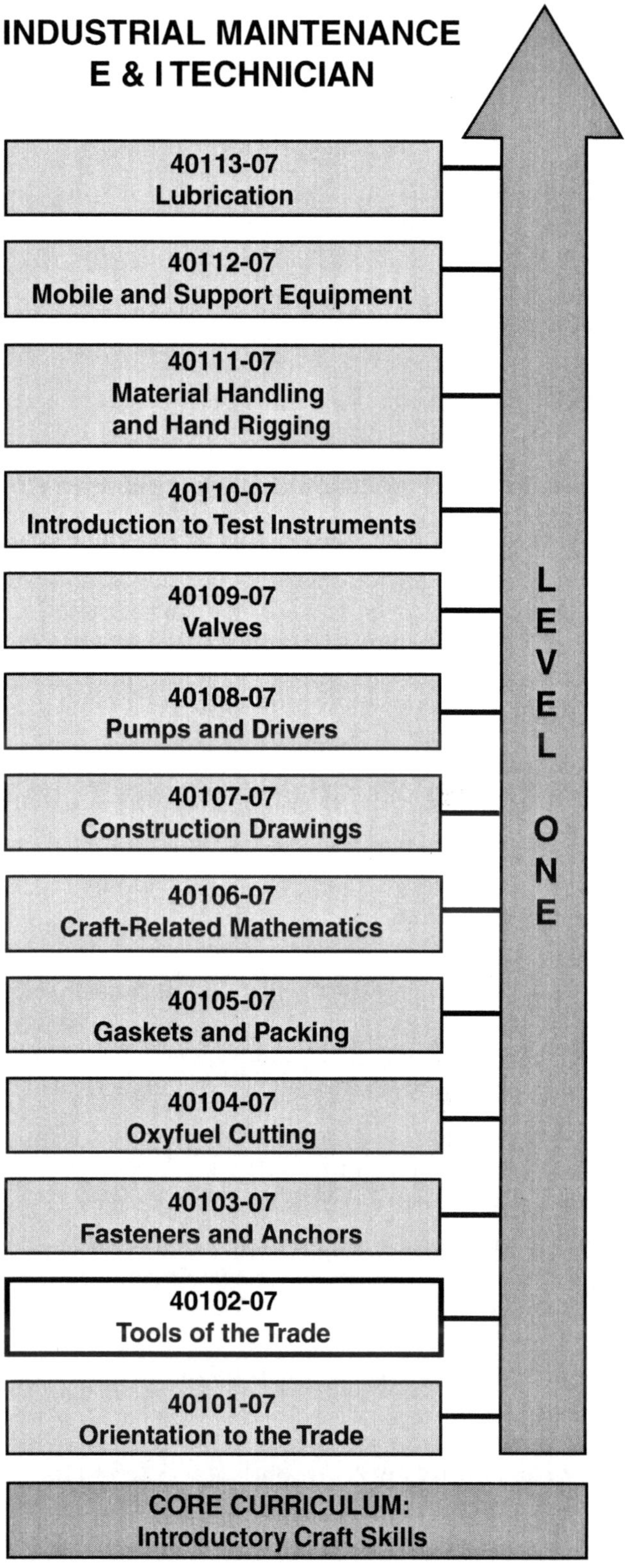

102CMAP.EPS

1.0.0 ◆ INTRODUCTION

Industrial maintenance workers use a variety of specialized hand tools. There are hand tools with multiple uses and those that are designed for a specific purpose. As a maintenance craftworker, you must be able to recognize, select, inspect, maintain, and safely use hand tools to perform a job safely and efficiently. This module describes the specialized hand tools normally used in industrial maintenance.

2.0.0 ◆ GENERAL HAND TOOL SAFETY

Although hand tools do not present the immediate hazards of power tools, improper use of hand tools can cause serious personal injury. Safely using hand tools requires following some basic safety rules. Some hand tools are more hazardous than others and require additional personal protection and care when using them. For safe use of hand tools, follow these basic rules:

- Always choose the proper tool for the job.
- Never use a tool for anything other than its intended purpose. For example, do not use a screwdriver as a prybar.
- Never use dull or broken tools. Dull tools require greater force to do the job. A sharp tool is a safe tool.
- Always keep your hands and fingers away from the sharp edges of cutting tools.
- Always work away from your body when using cutting tools.
- Always ensure that a tool is in good condition and handles are tightly fastened to the tool.
- Always keep tools clean and free of rust.
- Always wear eye protection, such as safety glasses or goggles, when using chisels, punches, or other tools that may produce flying particles and debris.
- Always wear gloves when using tools that require them.
- Always use tools in a safe and proper manner. For example, an improperly held wrench can slip, causing injury to the hand or knuckles.

3.0.0 ◆ USING AND CARING FOR INDUSTRIAL MAINTENANCE HAND TOOLS

Hand tools are a necessity for mechanics to perform their work. Good hand tools can be very expensive to buy. If you must provide your own hand tools, it can mean substantial cost out of

> **NOTE**
> Some industrial machinery and piping systems contain flammable liquids and gases. For that reason, it is recommended that you use non-sparking tools. Keep in mind, however, that non-sparking tools are made of metal and can produce sparks. They simply are less likely to spark than steel tools. Your company should have safety procedures that deal with working in flammable environments, if applicable. For your own safety and the safety of others, you must follow these procedures without fail.

your pocket. If your employer furnishes your hand tools, this expense adds to the cost of doing business and affects the contractor's ability to do work in a competitive manner. Regardless of who pays, you can see that properly using and caring for these tools is important. This module explains the use and care of the following tools:

- Strap/chain wrenches
- Spanner wrenches
- Taper gauges
- Pipe and tubing cutters
- Honing stones
- Putty knives/scrapers
- Drift pins
- Mallets
- Diagonal cutters
- Tin snips
- Taps and **dies**
- Thread gauges
- Scribers
- Tension meters
- **Sheave** gauges
- Cylinder hones
- Gear pullers
- Packing pullers
- Reamers
- Inspection mirrors
- Retaining ring pliers
- Extractors
- Feeler gauges
- Alignment bars
- Sleever bars
- Dial indicators
- Portaband
- End grinder
- Angle grinder

- Vertical grinder
- Pipe vise
- Threading machines

3.1.0 Strap and Chain Wrenches

Chain wrenches and strap wrenches (*Figure 1*) are types of pipe wrenches. They are used for many of the same purposes as pipe wrenches. They may also be used in situations in which a pipe wrench will not work. For instance, when working on a pipe that is close to a wall, there need only be enough room between the pipe and the wall to pass the chain or strap through. Chain wrenches come in various sizes to handle pipe up to 18 inches in **diameter**. Strap wrench sizes go up to 12 inches. Strap wrenches are often used on pipe or round material that is too hard for pipe or chain wrenches to grip and on finished surfaces that pipe wrenches or chain wrenches would mar.

Chain wrenches and strap wrenches are both used in much the same manner. Follow these steps to select, inspect, use, and maintain chain wrenches:

Step 1 Select a chain wrench of adequate size to fit around the pipe.

Step 2 Inspect the wrench to ensure that the chain is not broken or excessively worn and that the handle is not cracked or broken.

Step 3 Position the wrench on the pipe; wrap the chain around the pipe, and hook the chain into the hooks on the handle. The chain will fit snugly, but not tightly, around the pipe. The chain will tighten on the pipe when pressure is applied.

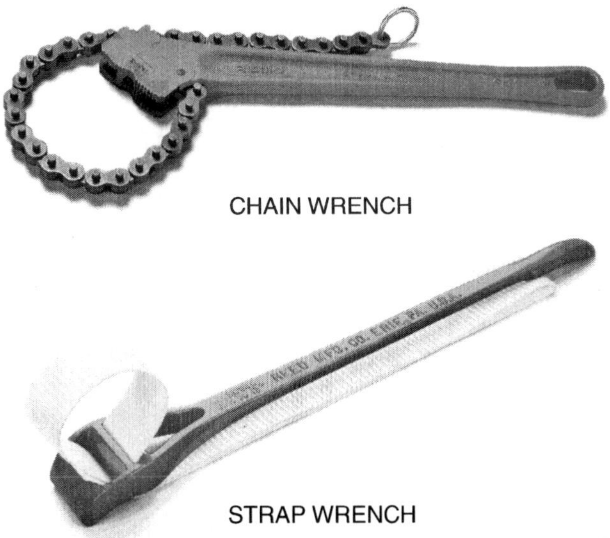

102F01.EPS

Figure 1 ◆ Chain wrench and strap wrench.

Step 4 Apply pressure on the wrench handle in the direction the pipe is intended to turn until the chain tightens and the wrench grips the pipe.

Step 5 Rotate the wrench as far as clearances will permit or as far as is convenient.

Step 6 Release pressure; return the wrench to the starting position; establish a new grip; and turn the pipe again. Repeat this process until the pipe has been given as many turns as needed.

Step 7 Clean the wrench thoroughly.

Step 8 Oil the wrench lightly to prevent rust.

> **NOTE**
> Do not oil the strap of a strap wrench. This will make the strap slick and defeat its gripping ability.

Step 9 Store the wrench in its proper place.

3.2.0 Spanner Wrenches

Spanner wrenches are specially made wrenches with specific purposes. They are usually either furnished with the tool or equipment that they are used on or are offered as service items for the equipment. Refer to a parts list or service bulletin to identify a spanner for a particular application. *Figure 2* shows some common spanner wrenches.

Spanner wrenches are relatively easy to use, and all spanners are used in a similar manner. Follow these steps to select, inspect, use, and maintain a face spanner wrench:

Step 1 Select a spanner wrench that is suitable for the job to be done. It must be the correct size or adjustable.

Step 2 Inspect the wrench to ensure that the pins are not bent, excessively worn, or otherwise damaged, and that the handle is not bent or damaged.

Step 3 Clean the holes of the tool or equipment part being worked on thoroughly to remove dirt, grease, or other obstructions that would prevent the wrench from fitting into the holes.

Step 4 Insert the pins of the wrench into the holes provided in the tool or equipment part being worked on. Be sure that the pins fit snugly into the holes and that the holes are not disfigured beyond use.

Step 5 Hold the wrench in the holes with one hand, and apply pressure to the wrench with the other.

> **NOTE**
> Be sure that you are turning the wrench in the proper direction. Although most nuts are loosened by turning them in a counterclockwise direction, some nuts have left-hand threads and are loosened by turning them clockwise.

Step 6 Turn the wrench as far as clearances will allow or as far as is convenient.

Step 7 Remove the wrench from the holes; return to the starting position, and reinsert the wrench. Continue turning the nut until it has been rotated as many turns as needed.

Step 8 Clean the wrench thoroughly, using a rag and solvent if needed.

Step 9 Oil the wrench lightly to prevent rust.

Step 10 Store the wrench in its proper place.

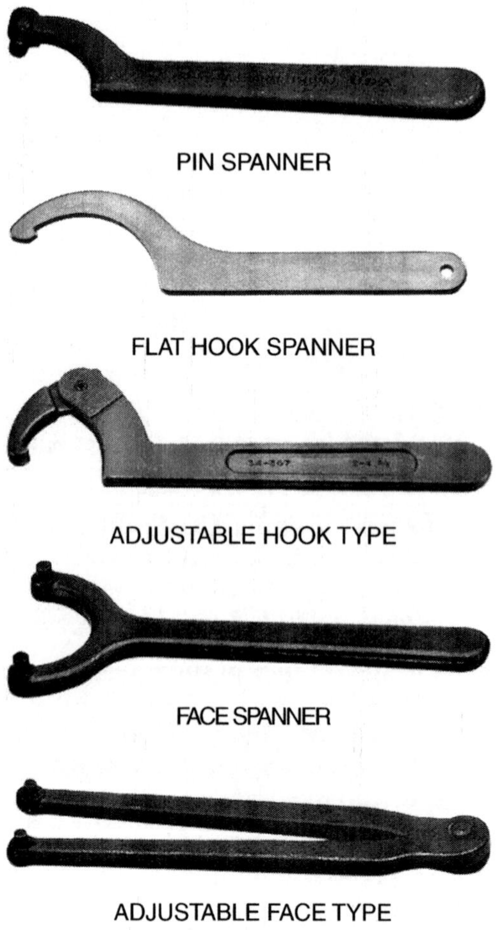

102F02.EPS

Figure 2 ◆ Spanner wrenches.

3.3.0 Taper Gauges

A taper gauge consists of a solid shaft-shaped bar with a tapered section that is used to measure the inside diameters of tapered holes. Taper gauges come in various sizes to measure different size holes. There is a gauge line on the tapered part of the gauge. To use a taper gauge, the gauge is inserted in a tapered hole and read at the gauge line. The hole is the right diameter if the gauge fits exactly up to the gauge line. If the gauge does not go all the way up to the gauge line, the hole is too small. If the gauge goes in past the gauge line, the hole is too large. Taper gauges are precision instruments and are very fragile. When not in use, the gauge should be wrapped in an oily cloth to prevent rust and stored in a place where other tools and materials will not be thrown or dropped on it. Once a gauge is damaged, it is of no further use. *Figure 3* shows a taper gauge.

3.4.0 Pipe and Tubing Cutters

Pipe cutters are used to cut steel, brass, copper, and iron pipe. A pipe cutter has a special alloy-steel cutting wheel and two pressure rollers. The cutting wheel is mounted so that it is stationary in the cutter frame. The pressure rollers are mounted on a sliding frame and are adjusted and tightened by turning the handle. To cut a pipe, the cutter is placed on the pipe; pressure is applied to the pressure rollers, and the cutter is rotated around the pipe until it is cut.

Tubing cutters closely resemble pipe cutters and are used in the same way. Tubing cutters are used to cut thin-wall iron, brass, steel, copper, and aluminum tubing. There are various sizes of pipe and tubing cutters for different sizes of pipe and tubing. *Figure 4* shows some different sizes and types of pipe and tubing cutters.

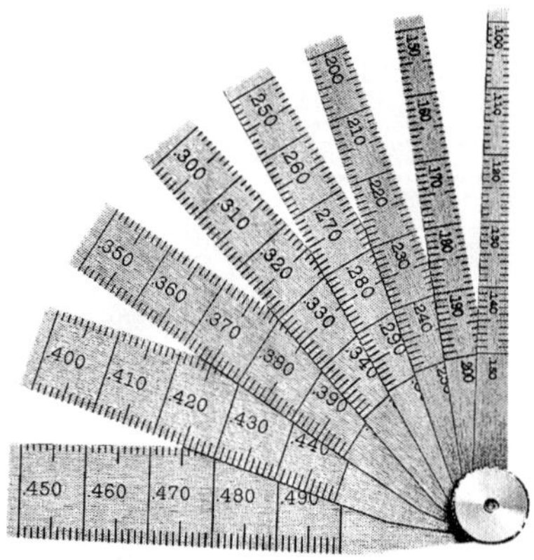

102F03.EPS

Figure 3 ◆ Taper gauge.

Pipe cutters and tubing cutters are used in essentially the same way. Follow these steps to select, inspect, use, and maintain pipe and tubing cutters:

Step 1 Select the proper type and size cutter for the pipe to be cut.

> **NOTE**
> The cutter must be of the proper size to fit over the pipe and heavy enough to cut the wall thickness of the pipe being cut. Heavy-wall pipe cannot be cut with this type of cutter.

Step 2 Inspect the cutter wheel to ensure that it is not nicked, bent, or dulled.

Step 3 Inspect the pressure rollers to ensure that they are not damaged and that there is no buildup of foreign materials on the rollers.

Step 4 Inspect the cutter frame to ensure that it is not bent, cracked, or otherwise damaged.

Step 5 Place the cutter over the pipe at the point where it is to be cut, then tighten the pressure rollers until there is light pressure on the pipe.

Step 6 Grip the cutter by the handle and rotate it around the pipe one complete turn.

Step 7 Tighten the pressure rollers a small amount, about one to one and one-half turns of the handle, and rotate the cutter another turn around the pipe.

> **CAUTION**
> Do not overtighten the cutter. The cutting wheel could be damaged or the cutter broken. Be sure that the cutter follows in the same groove cut by the first turn on the second and all consecutive turns of the cutter. If the cutter is moving out of the groove, you may be applying too much pressure, or the cutter may be defective.

Step 8 Continue this procedure until the cut is finished.

Step 9 Clean the cutter thoroughly to remove any dirt, grease, or buildup.

Step 10 Oil the cutter lightly to prevent rust.

Step 11 Store the cutter in its proper place.

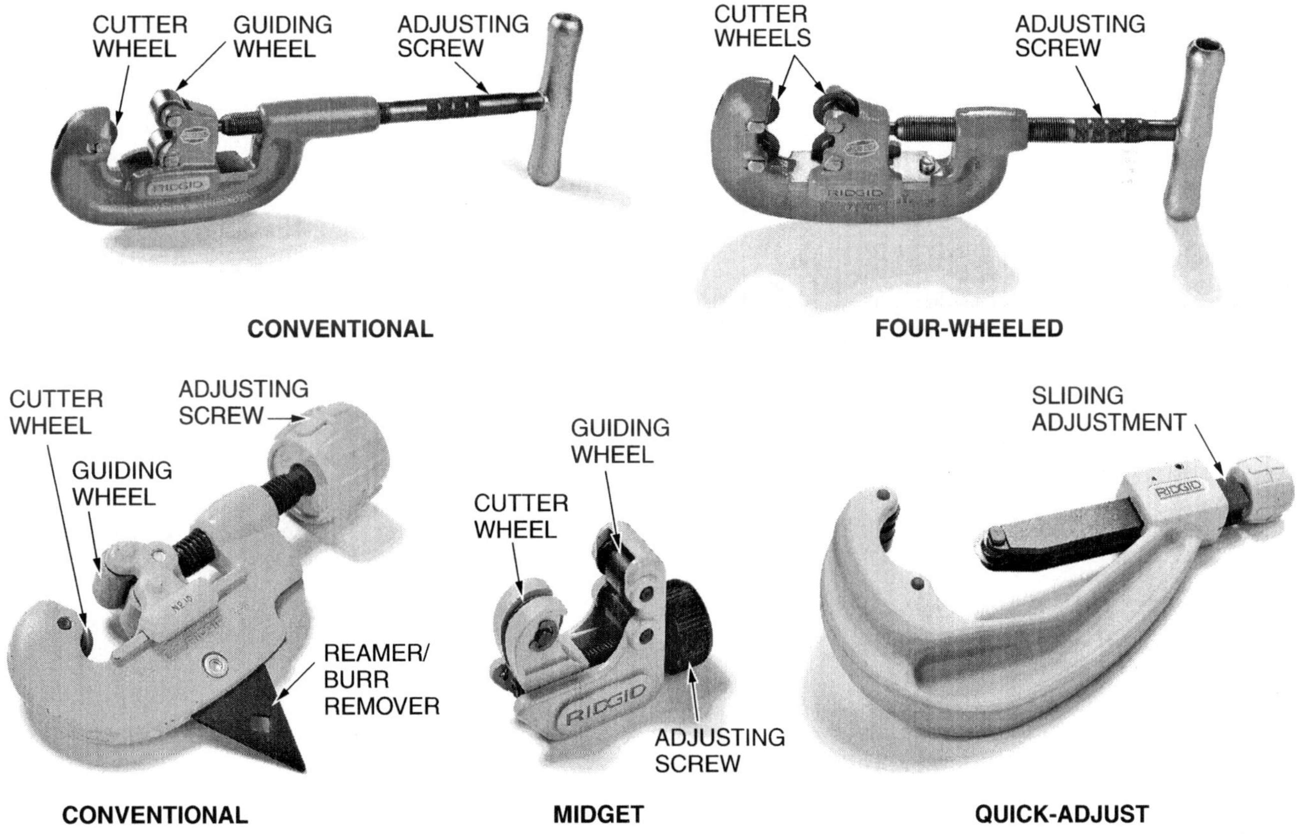

102F04.EPS

Figure 4 ◆ Pipe and tubing cutters.

3.5.0 Putty Knives/Scrapers

Putty knives are used to scrape off gaskets, glue, sealer, and other substances when cleaning pipe **flanges** and other machinery and equipment. Putty knives have a wide, square blade and are available in many different lengths and widths. *Figure 5* shows a common putty knife.

This procedure explains how to scrape off a gasket. Follow these steps to select, inspect, use, and maintain a putty knife:

Step 1 Select a putty knife of the proper length and width to do the job.

> **NOTE**
> Do not choose a knife that is larger than needed and cannot be handled safely.

Step 2 Inspect the putty knife to ensure that the blade is not bent or dull and that the handle is in good condition and firmly attached to the blade.

Step 3 Hold the knife in one hand at an effective angle to the work, and scrape the gasket off using smooth, even strokes.

> **WARNING!**
> Do not try to chip the gasket with the putty knife. This could nick the surface being scraped and dull the knife. Keep the knife blade flat on the surface. Do not try to scrape with the side edge of the blade. Keep your hands away from the cutting edge of the blade, and work away from your body to prevent personal injury.

Step 4 Continue to scrape the surface until all the gasket is removed.

Step 5 Clean the knife thoroughly with a rag, using solvent if needed.

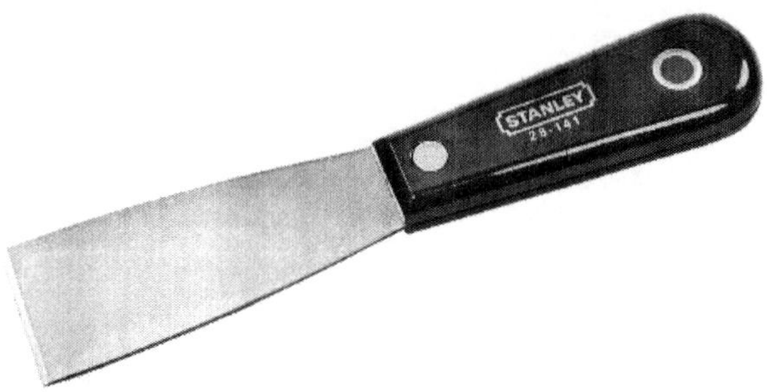

102F05.EPS

Figure 5 ◆ Putty knife.

Step 6 Oil the knife lightly to prevent rust.

Step 7 Store the knife in its proper place.

3.6.0 Drift Pins

A drift pin is a round steel piece that is tapered on one end. A barrel pin is a drift pin that is tapered on both ends. Drift pins and barrel pins are used to **align** bolt holes when connecting flanges, structural steel, and other holes in mating surfaces. Drift pins and barrel pins come in various sizes to fit different size bolt holes. *Figure 6* shows a common drift pin and a barrel pin.

Follow these steps to select, inspect, use, and maintain a drift pin:

Step 1 Select a pin of the proper size for the holes being aligned.

> **NOTE**
> The proper size drift barrel pin will fit all the way through both holes being aligned and will still be large enough and strong enough to pull the mating pieces together.

Step 2 Inspect the pin to ensure that it is not bent and that it does not have **burrs** that could cut your hands.

Step 3 Insert the pin into the holes of the mating pieces and use it to pull the holes into alignment.

Step 4 Clamp the mating pieces together using a C-clamp to hold the parts in alignment while the pin is removed and a bolt installed in its place.

Step 5 Clean the pin using a rag and solvent if needed.

Step 6 Cover the pin with a light coat of oil to prevent rust.

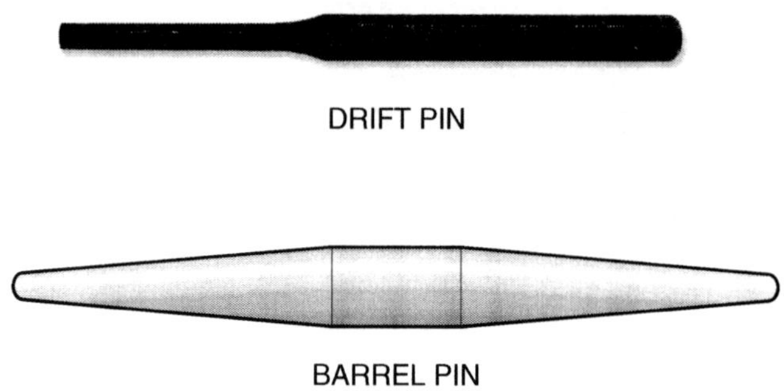

102F06.EPS

Figure 6 ◆ Drift pin and barrel pin.

> **NOTE**
> On pieces with more than one hole, the drift pin can be left in until bolts can be installed in the other holes. Then the drift pin is removed, and a bolt is installed in its place.

Step 7 Store the pin in its proper place.

3.7.0 Mallets

Dead-blow mallets (*Figure 7*) and rawhide mallets are used whenever using steel hammers would deface the work. Other types of mallets are made of lead, brass, wood, leather, plastic, or rubber. Never use your mallet to drive nails. It is not designed for this purpose. Mallets used to drive nails won't work properly for delicate forming tasks.

Mallet faces are designed to shape, bend, or force metal parts together and form light-gauge metals without marring or damaging them.

> **NOTE**
> The head of the dead-blow mallet is encased in plastic. Hidden inside the head is a small piece of steel, which provides the striking power. Unlike other mallets, the dead-blow mallet will not rebound after striking an object; that is how it got its name.

102F07.EPS

Figure 7 ◆ Mallets.

3.8.0 Diagonal Cutters

Diagonal cutters are small pliers used to cut wire, cotter pins, and other items made of light material that are inaccessible with other types of cutters. Since the cutting edges are diagonally offset approximately 15 degrees, they are also used to cut small objects close to a surface. *Figure 8* shows a diagonal cutter.

Follow these steps to select, inspect, use, and maintain a diagonal cutter:

Step 1 Select a diagonal cutter of the proper size for the material to be cut.

Step 2 Inspect the cutter to ensure that the cutting edges are not nicked or dull.

Step 3 Position the cutter to cut the material in the desired location.

Step 4 Squeeze the handles of the cutter to cut the material.

Step 5 Clean the cutter.

Step 6 Store the cutter in its proper place.

3.9.0 Tin Snips

Tin snips work like scissors. They are used to cut light-gauge sheet metal where the shape, position, or location prevents the use of power metal-cutting tools. Tin snips are made in a variety of configurations for specific jobs.

There are several types of snips, including those that make straight and circular cuts. Both are available in right-hand and left-hand styles.

Some of the most commonly used snips are aviation snips. The handles of aviation snips are color-coded to indicate whether they cut a curve to the right, a curve to left, or a straight line. Red-handled aviation snips cut a curve to the left. Green-handled aviation snips cut a curve to the

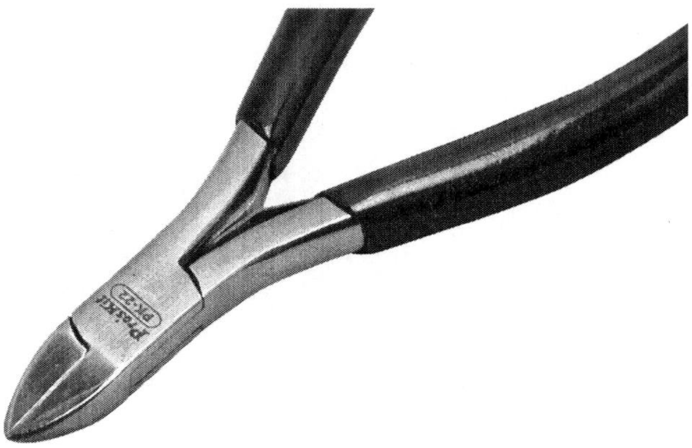

102F08.EPS

Figure 8 ◆ Diagonal cutter.

right. Yellow-handled aviation snips cut a straight line. *Figure 9* shows some common tin snips.

Tin snips are used to cut many shapes from sheet metal. This procedure explains cutting a hole in a piece of sheet metal. Follow these steps to select, inspect, use, and maintain tin snips:

Step 1 Select a snip that is of adequate capacity to cut the sheet metal and of the style needed to make the cut (right cut or left cut).

Step 2 Inspect the snips to ensure that they are not bent or damaged, that the blades are not sprung, gaped, or dull, and that the handles are in good condition.

CAUTION

The blades of tin snips can be easily damaged if they are used improperly. They should never be used to cut wire, rod, or material that is beyond their capacity. The blades must be sharp to produce a good cut.

Step 3 Ensure that the cut line on the sheet metal is distinct and easily seen.

Step 4 Drill or cut a hole using a hammer and chisel to make a starter hole in the center of the intended cut.

Step 5 Slip the blade of the snips into the starter hole and start the cut, using the snips like a pair of scissors and allowing the cut to curve as you go.

WARNING!

You should wear gloves when cutting sheet metal. Cutting the metal leaves a sharp edge and may also create burrs that can seriously cut your hands.

Step 6 Continue to cut in a circular motion from the starter hole until you reach the cut line. *Figure 10* shows starting a circle cut.

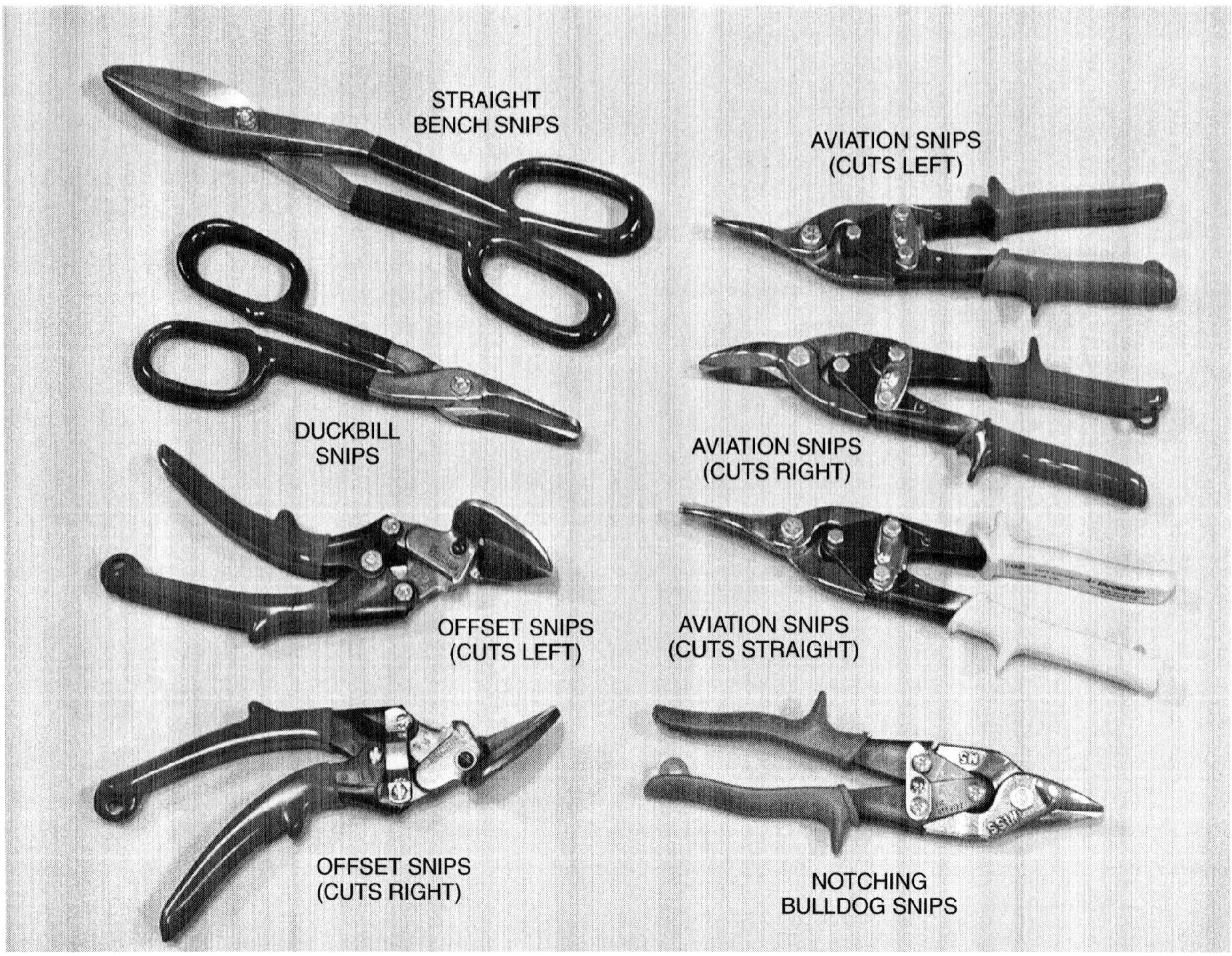

102F09.EPS

Figure 9 ◆ Tin snips.

Step 7 Cut along the cut line around the circle until the cut is complete.

NOTE
You should not completely close the snips during the cut. Close the snips approximately halfway on each cut; slip the snips further into the cut, and cut halfway again. Closing the snips completely will cause burrs on the metal that must be filed off later.

Step 8 Clean the snips using a rag.

Step 9 Coat the blades of the snips lightly with oil.

Step 10 Store the snips in their proper place.

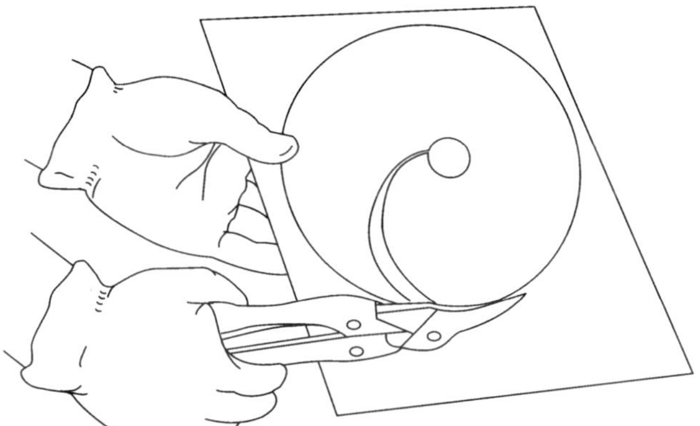

Figure 10 ◆ Starting a circle cut.

3.10.0 Taps and Dies

Taps and dies are used to cut threads in materials such as metal, plastic, and hard rubber. Taps are used to cut internal threads in materials; dies are used to cut external threads on bolts and rods. The most common types of taps are the following:

- *Taper taps* – Taper taps are used when starting a tapping operation or when tapping through holes. They have a **chamfer** length of 8 to 10 threads.
- *Plug taps* – Plug taps are designed to be used after the taper tap. They have a chamfer length of 3 to 5 threads.
- *Bottoming taps* – Bottoming taps are always used after the taper and plug taps have already been used. They are used for threading the bottom of a blind hole and have a chamfer length of only 1 to 1½ threads.
- *Pipe taps* – Pipe taps are used where extremely tight fits are necessary, as with **pipe fittings**. The diameter of the tap from end to end of the threaded portion is tapered at the rate of ¾ inch per foot.

Figure 11 shows the different types of taps.

Dies are made in a variety of shapes and sizes and are solid or adjustable. The following are common types of dies:

- Pipe dies will cut American Standard Pipe Threads or other pipe (such as metric) threads.
- Rethreading dies (die nuts) are used for repairing damaged or rusty threads on bolts or

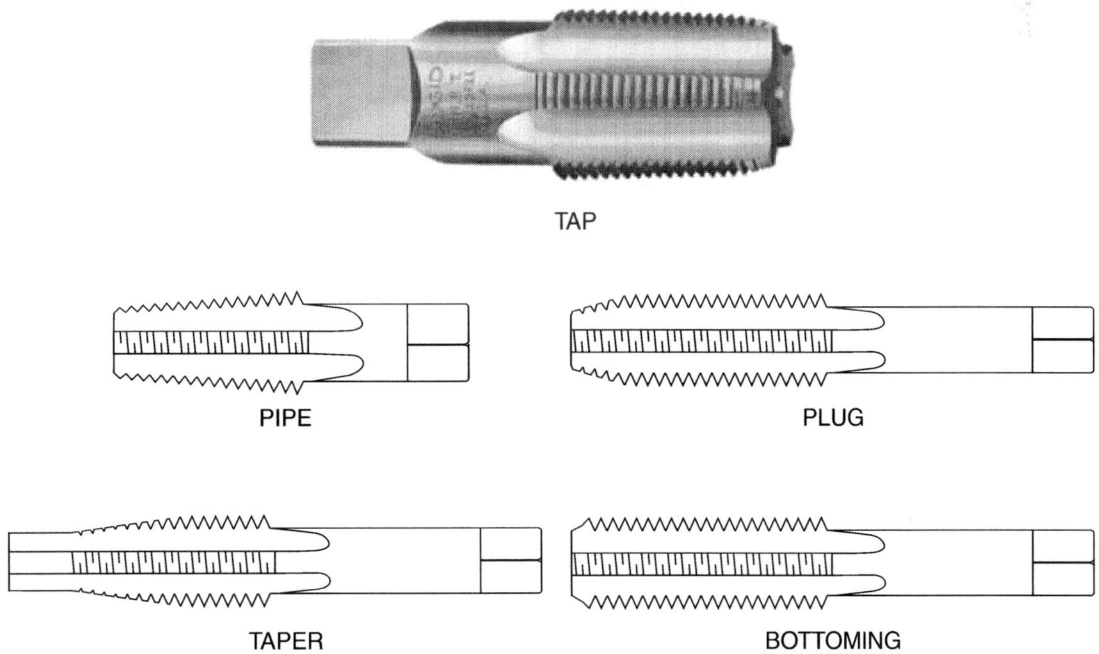

Figure 11 ◆ Taps.

screws. They are usually hexagon-shaped and can be turned using any type of wrench that will fit.

- Screw adjusting dies are adjusted by a screw in the die.
- Open adjusting dies are adjusted by three screws in the die holder.
- Two-piece rectangular dies are adjusted by setscrews and held in ordinary or ratchet die-stocks.

Taps and dies can be obtained in complete sets that include various taps and dies, as well as die-stocks, tap wrenches, guides, and the screw-drivers and wrenches necessary to loosen and tighten adjusting screws. *Figure 12* shows a common tap and die set.

Taps are used to thread holes that go all the way through a piece of material and holes that end inside the material (blind holes). This procedure explains threading a blind hole. Follow these steps to select, inspect, use, and maintain a tap:

Step 1 Select a tap of the proper size for the hole being tapped and the proper type for the tapping operation (*Table 1*).

CAUTION

All three types of dies are used when tapping a blind hole. A taper die is used to start the tapping operation.

Step 2 Inspect the tap to ensure that there are no broken or nicked teeth and that the tap is sharp and clean of any shavings or chips.

102F12.EPS

Figure 12 ◆ Tap and die set.

Step 3 Place the end of the tap in the hole, and turn it while applying light pressure on the tap wrench to start the tap.

Step 4 Check the tap, using a square to ensure that it is square to the surface of the work.

CAUTION

The tap must be square to the surface of the work to ensure that the hole is tapped straight. Check the tap in several places, using the square.

Step 5 Continue to turn the tap, applying light pressure until the tap feeds itself. The pressure can then be relaxed. Lubricate the tap regularly during the entire tapping process. The tap must be kept lubricated during the tapping process to prevent the tap from **galling** or breaking.

Step 6 Stop the forward cutting, and back the tap out one-quarter turn to clear away chips.

Step 7 Continue the tapping process of turning in one-half turn and then backing out one-quarter turn until the tap touches the bottom of the hole.

Step 8 Back the tap out of the hole.

Step 9 Select a plug tap for the second phase of the tapping process.

Step 10 Tap the hole, using the plug tap, following the same procedure as with the taper tap.

Step 11 Back the plug tap out of the hole.

Step 12 Select a bottom tap for the final phase of the tapping process.

Step 13 Tap the hole, using the bottom tap.

Table 1 Drill and Tap Sizes

Pipe Tap Size (inches)	Drill Size (inches)	Pipe Tap Size (inches)	Drill Size (inches)
1/8	11/32	2	2 3/16
1/4	7/16	2 1/2	2 5/8
3/8	16/32	3	3 1/4
1/2	23/32	3 1/2	3 3/4
3/4	15/16	4	4 1/4
1	1 5/32	4 1/2	4 3/4
1 1/4	1 1/2	5	5 5/16
1 1/2	1 23/32	6	6 3/8

102T01.EPS

> **NOTE**
> The procedure for the bottom tap is the same as for the taper and plug taps, except that the tap must be backed all the way out every two or three turns and the chips cleaned out to prevent them from collecting in the bottom of the hole.

Step 14 Continue the process of turning the tap forward two or three turns, backing it out, and cleaning the hole until the bottom tap touches the bottom of the hole.

Step 15 Clean the taps thoroughly to remove all chips and shavings.

Step 16 Coat the tools lightly with oil to prevent rust.

Step 17 Return the tools to their protective case and store them in their proper place.

3.11.0 Thread Gauges

Thread gauges, also known as screw pitch gauges, are used to measure the **pitch** of threaded fasteners. Thread gauges can be used to measure both external and internal threads. A thread gauge is a thin, flat metal leaf that has teeth that correspond to standard thread sections. Thread gauges usually come in sets of several sizes, held together by a screw. *Figure 13* shows a thread gauge that measures internal and external threads.

Follow these steps to select, inspect, use, and maintain thread gauges:

Step 1 Select a complete set of gauges.

Step 2 Inspect the gauges to ensure that they are not damaged, dirty, or rusty.

Step 3 Hold one of the gauges against the threads being measured, and compare the fit.

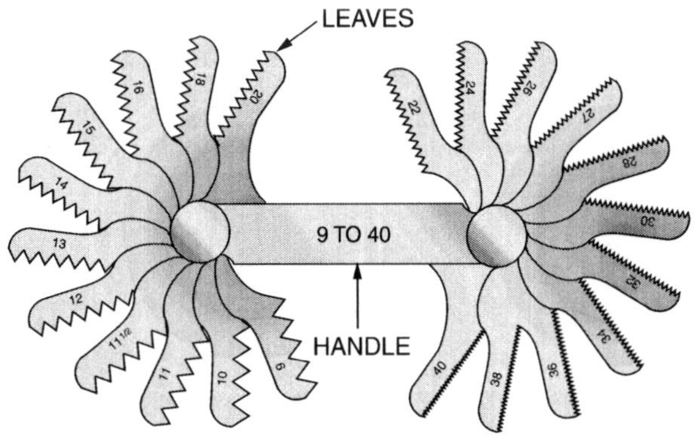

Figure 13 ◆ Thread gauge.

Step 4 Try various sizes of gauges until you find an exact fit. The correct size will fit the threads perfectly.

> **NOTE**
> The number on the gauge indicates the number of threads per inch. Once this value is determined, you can use it to select nuts, bolts, taps, and dies of the proper pitch.

Step 5 Clean the gauges thoroughly.

Step 6 Coat the gauges lightly with oil to prevent rust.

Step 7 Store the gauges in their proper place.

3.12.0 Scribers

Scribers are sharp, pointed, pencil-like tools used to mark lines on sheet or plate metal. They are usually made of 3/16-inch round tool steel and vary in length from 6 to 12 inches, depending on the handle style. A blue or purple dye, called layout dye, is frequently used to make scriber marks more visible. Layout dye is painted on the surface and allowed to dry before marking. *Figure 14* shows two styles of scribers.

Follow these steps to select, inspect, use, and maintain a scriber:

Step 1 Select a scriber of a convenient length for the job.

Step 2 Inspect the scriber to ensure that the points are not bent or dull. Sharpen the points of the scriber if needed.

> **NOTE**
> If the points of the scriber are not sharp, it will not produce a distinct mark. Sharpen the points on a honing stone if needed.

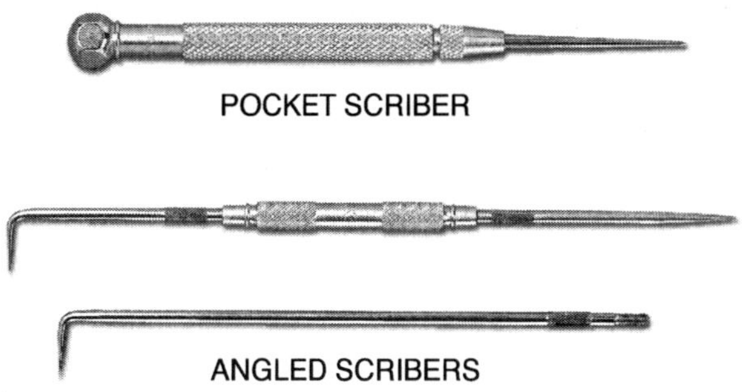

Figure 14 ◆ Scribers.

Step 3 Hold the scriber at a 45-degree angle to the surface of the material. Apply moderate pressure and pull it toward you, following a straightedge or pattern to mark the surface.

CAUTION

Hold the point of the scribe tightly against the straightedge or pattern so that your marks will be accurate.

Step 4 Clean the scriber.

Step 5 Store the scriber in its proper place.

3.13.0 Tension Meters

A tension meter is used to measure the tension on belts. It is a precision instrument and must be cared for as you would any precision instrument. It should be kept clean and should be stored in a place where it will not get damaged by other tools or materials being thrown on top of it. To use a tension meter, hold it in one hand and press it against the belt about midway between the pulleys. Read the tension on the scale on the side of the meter. *Figure 15* shows using a tension meter.

3.14.0 Strobe Lights

Any cyclic movement can be analyzed through the use of stroboscopic light. A stroboscopic light, commonly known as a strobe light, is a light that flashes at a certain controlled frequency. If the strobe light is set to the frequency of an industrial motor, the light will always show the same point on the motor's shaft. If that point appears to move from side to side, the shaft is vibrating. If a belt on a shaft appears to move when seen with a strobe light, the belt is traveling. If the strobe is set to slightly less or more frequent flashing than the rpm of the motor, the process of the motor turning can be seen in slow motion because a slightly different point will be seen on each flash.

3.15.0 Sheave Gauges

A sheave gauge is used to measure the side wall wear of a belt sheave. It is used when changing V-belts or when trouble is detected in a V-belt drive. When measuring sheave grooves, ensure that the proper gauge is used. The gauge must be correct for the type and size of belt used and for the pitch diameter of the sheave being checked. *Figure 16* shows a sheave being measured with a sheave gauge.

Follow these steps to select, inspect, use, and maintain a sheave gauge:

Step 1 Select a gauge that is correct for the type and size of V-belts being used and for the correct pitch diameter of the sheave being checked.

Step 2 Inspect the gauge to ensure that it is not bent, worn, or otherwise damaged.

Step 3 Insert the gauge into the groove of the sheave, and observe the fit. It is important that the correct groove dimensions and

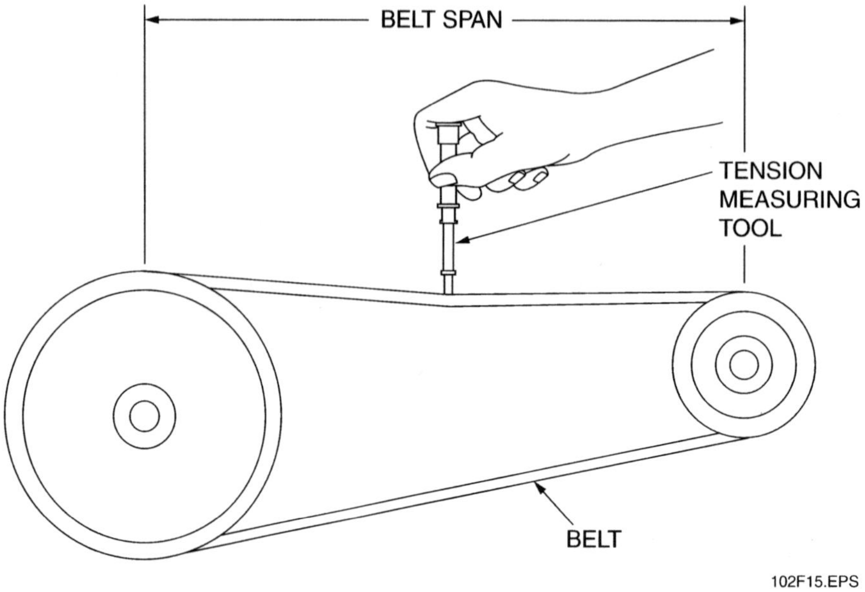

Figure 15 ◆ Using a tension meter.

angles be maintained on sheaves. Refer to the manufacturer's dimension data to ensure that the sheave is within **tolerances**.

Step 4 Clean the gauge thoroughly.

Step 5 Coat the gauge lightly with oil to prevent rust.

Step 6 Store the gauge in its proper place.

3.16.0 Cylinder Hones

A cylinder hone (*Figure 17*) is used to dress the walls of cylinders when rebuilding and repairing engines, pumps, compressors, and other equipment with cylinders. To hone a cylinder, the hone is turned in the cylinder by an electric drill or motor. The hone is equipped with **abrasive** stones that cut small amounts of metal away from the cylinder wall as the hone turns. If large amounts of metal must be removed to true the cylinder, the cylinder must be bored. The amount of cylinder wear and the type of rings to be installed usually determine whether the cylinder should be honed or bored. Honing is a precise procedure, and the hone must be held rigid and square to the cylinder to produce a good job. *Figure 18* shows a hone set up to hone a cylinder block.

Follow these steps to select, inspect, use, and maintain a cylinder hone:

Step 1 Select a hone of the proper size for the cylinder being honed.

Step 2 Inspect the hone to ensure that it is not bent and that the stones are not chipped, broken, or excessively worn.

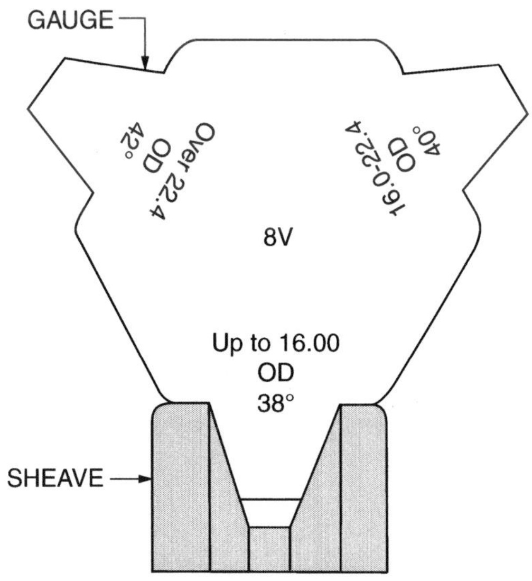

102F16.EPS

Figure 16 ◆ Measuring a sheave using a sheave gauge.

Step 3 Install the hone in a drill, and set up the honing operation.

> **CAUTION**
>
> On cylinders that are small enough, it is best to set up the honing operation on a drill press or in a lathe. On larger equipment, a jig or a magnetic drill must be set up to hold the drill and hone rigid and true to the cylinder. If the hone is not held true to the cylinder, the cylinder can be ruined.

Step 4 Move the hone to the center of the cylinder.

Step 5 Tighten the adjusting knob by hand or using a small screwdriver to apply light pressure to the stones.

Step 6 Lubricate the stones according to the manufacturer's recommendations.

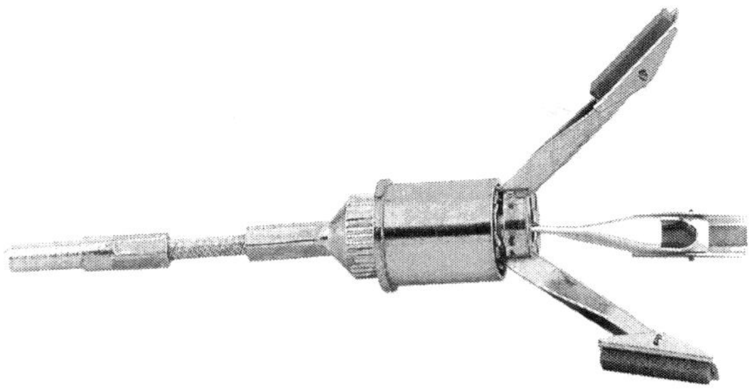

102F17.EPS

Figure 17 ◆ Cylinder hone.

102F18.EPS

Figure 18 ◆ Using a cylinder hone.

Step 7 Adjust the speed of the drill. The hone should operate at 300 to 700 **revolutions per minute (rpm)** to avoid overheating and damaging the stones or the cylinder.

> **WARNING!**
> Be sure to wear eye protection to prevent the possibility of flying debris injuring your eyes.

Step 8 Start the drill and move the hone up and down a few strokes in the lower half of the cylinder until the bottom of the cylinder begins to increase in diameter.

Step 9 Gradually increase the stroke toward the top of the cylinder until the hone travels the full length of the cylinder bore.

> **CAUTION**
> Do not extend the hone more than ¾ inch past either end of the cylinder to prevent damaging the stones.

Step 10 Stop the drill and readjust the pressure on the stones as the diameter of the cylinder increases and the pressure on the stones decreases.

Step 11 Change to a finishing stone if coarse stones were used on the first part of this procedure. Hone the cylinder again to remove all rough hone marks.

Step 12 Extract the hone from the drill when honing is complete, and clean the drill thoroughly.

Step 13 Coat the hone and stones lightly with oil and store them in their proper place.

Step 14 Take down the setup; store all tools, and clean up the work area.

3.17.0 Gear Pullers

Many precision gears and bearings are press-fitted in place and are difficult to remove. Attempting to remove them with a hammer or by prying them off may damage the gear or bearing and the shaft. Gear pullers are made to grip the gears or bearings in a way that will not damage them. Gear pullers apply force to the gear or bearing in a straight and even motion. Pullers are usually made of tool steel and have two or more interchangeable, reversible jaws and a pressure bolt mounted in the middle of a yoke. A substitute for some of the mechanical pullers is a Porta Power® kit. The Porta Power® system of hydraulic tools includes gear pullers, clamps, bending tools, and expanders. These tools come in sizes rated for the tonnage of power to be applied. The tools are portable, and can be brought to the work site. Gear pullers come in various styles and sizes for different applications. *Figure 19* shows a gear puller.

Follow these steps to select, inspect, use, and maintain gear pullers:

Step 1 Select a gear puller with the correct jaw length and spread width for the gear or bearing being pulled.

Step 2 Inspect the puller to ensure that the jaws are not bent or excessively worn and that the pressure bolt threads are not stripped.

Step 3 Ensure that all setscrews or locking screws have been loosened or removed from the gear or bearing.

Step 4 Inspect the shaft to ensure that there are no nicks, burrs, or foreign objects that would interfere with the gear or bearing sliding off the shaft.

Step 5 Position the puller on the gear or bearing, and screw the pressure screw in until it presses against the shaft.

Step 6 Check the puller to ensure that the jaws are properly positioned on the gear or bearing.

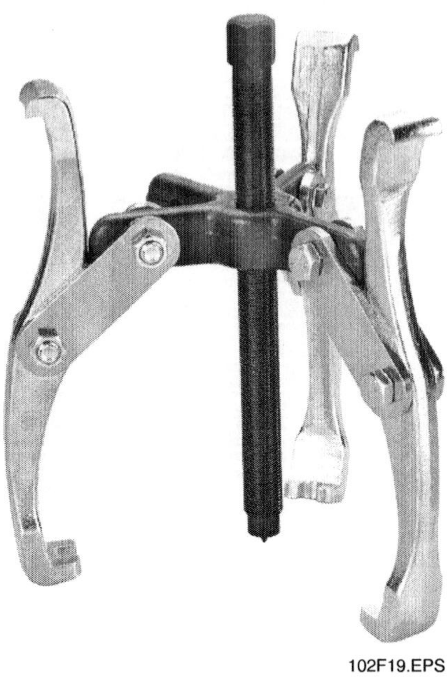

102F19.EPS

Figure 19 ◆ Gear puller.

> **CAUTION**
> The puller jaws must be positioned so that they will provide a straight, even pull on the gear or bearing and not damage it.

Step 7 Turn the pressure bolt slowly clockwise to begin applying pressure to the gear or bearing.

> **WARNING!**
> Always wear eye protection when using a puller, because particles could break off and injure your eyes.

Step 8 Continue to tighten the pressure bolt until the gear or bearing is removed from the shaft.

Step 9 Clean the puller thoroughly.

Step 10 Coat the puller lightly with oil to prevent rust.

Step 11 Store the puller in its proper place.

3.18.0 Packing Pullers

A packing puller is used to remove packing material from valves, pumps, and some motors. There are several types of packing pullers. Some pullers resemble a pick with a curved point. Other types are flexible and have a corkscrew-like end on them. Special care must be taken when using a packing puller to prevent scarring the valve stem, pump shaft, or stuffing box. *Figure 20* shows two types of packing pullers.

Follow these steps to remove packing:

Step 1 Select a corkscrew-type puller of the proper size and length for the job.

Step 2 Inspect the puller to ensure that it is sharp and not bent out of shape.

Step 3 Insert the puller into the stuffing box, turning it one or two turns clockwise to screw it into the packing.

Step 4 Pull on the puller carefully to remove the packing from the box. The packing will usually break up into small pieces when it is removed.

Step 5 Repeat step 4 until all the packing has been removed from the stuffing box.

Step 6 Clean the puller thoroughly.

> **NOTE**
> After the packing has been removed, there will usually be some packing pieces left that the corkscrew puller cannot pick up. Choose a pick-type puller and remove the packing pieces.

Step 7 Coat the puller lightly with oil to prevent rust.

Step 8 Store the puller in its proper place.

3.19.0 Reamers

Reamers are used to enlarge and true holes. A reamer consists of three parts: the body, the shank, and the cutting blades. The shank has a square end that allows the reamer to be turned with a wrench. The blades are made of hardened steel and are supported by the body of the reamer. Reamers are available in standard sizes, as well as in size variations of 0.001 inch for special work.

The hardened blades of a reamer are brittle and fragile. Special care should be taken when using and storing the reamer to prevent chipping the blades. *Figure 21* shows some common reamers.

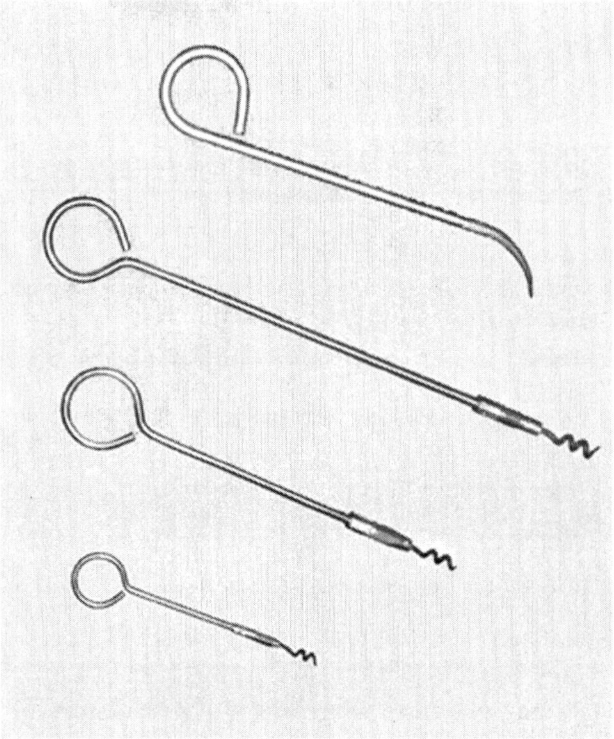

102F20.EPS

Figure 20 ◆ Packing pullers.

Follow these steps to select, inspect, use, and maintain a reamer:

Step 1 Select a reamer of the proper size for the hole to be reamed.

Step 2 Inspect the reamer to ensure that it is not bent and that the blades are not chipped or nicked.

Step 3 Select a wrench of the proper size to turn the reamer.

Step 4 Position the reamer in the hole to be reamed.

Step 5 Squirt a small amount of oil into the hole around the reamer.

Step 6 Hold the reamer square to the surface of the piece being worked on and turn the reamer clockwise to start reaming.

> **CAUTION**
> Always turn the reamer in the cutting direction (clockwise). Never turn it backwards because this could chip or dull the blades.

Step 7 Continue to turn the reamer until the hole is properly reamed.

Step 8 Clean the reamer thoroughly to remove shavings and debris.

Step 9 Wrap the reamer in an oily cloth, and store it in its proper place.

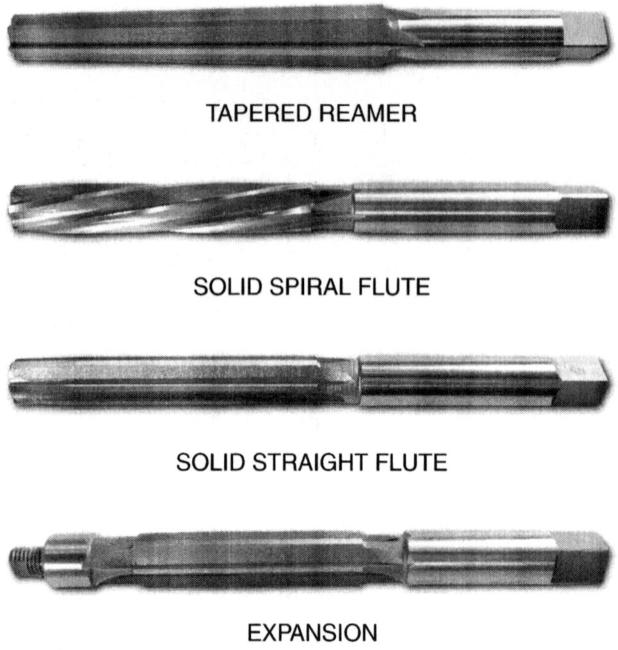

102F21.EPS

Figure 21 ◆ Reamers.

> **CAUTION**
> The reamer should be stored in a place where other tools will not be thrown or dropped on it because this can damage the blades.

3.20.0 Inspection Mirrors

An inspection mirror is used for close inspection of areas that cannot be seen directly. Inspection mirrors come in a variety of sizes and may be rectangular or round. The mirror is attached to the end of a rod and may be fixed or adjustable. By using a flashlight and holding the mirror in the proper position, you can inspect most areas. Some of the newer mirrors have built-in lights for more convenience.

Inspection mirrors should be kept clean and must be stored in a place where they will not be damaged. *Figure 22* shows an inspection mirror set.

3.21.0 Retaining Ring Pliers

Retaining ring pliers are used to remove and install retaining rings from shafts, rods, and spindles. They come in various sizes and shapes for different uses. Some retaining ring pliers come with interchangeable jaws. Retaining ring pliers are made to use on retaining rings only and should not be used for any other purpose. They are fragile and should be used with care and stored in a safe place to prevent damage. To use retaining ring pliers, the points of the pliers are inserted into the holes of the ring. The handle is squeezed to open the ring, and the ring is removed or installed. *Figure 23* shows three types of retaining ring pliers.

3.22.0 Extractors

If a screw or tap breaks off in a workpiece, it is necessary to remove it. If the screw or tap protrudes

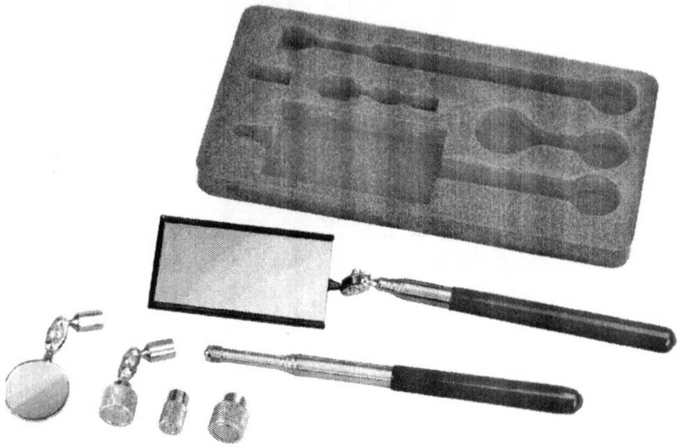

102F22.EPS

Figure 22 ◆ Adjustable inspection mirrors.

above the surface of the work, removing it may be as simple as using a pair of locking pliers to grip and remove it. If the screw or tap is broken near or below the surface of the work, you will need an extractor to remove it. Two common types of extractors are the spiral screw extractor and tap extractor.

3.22.1 Spiral Screw Extractors

Spiral screw extractors, or easy-outs, are reverse-threaded to remove right-hand threaded screws or bolts. They are made in various sizes to use on many size screws. The easy-out manufacturer provides recommendations as to what size easy-out should be used to remove a given size screw and what size drill should be used with the easy-out. You should follow the manufacturer's recommendations, because an easy-out that is too small may not be strong enough to turn the screw. The easy-out may break off in the screw, compounding the problem of removing the screw. Follow these steps to select, inspect, use, and maintain an easy-out:

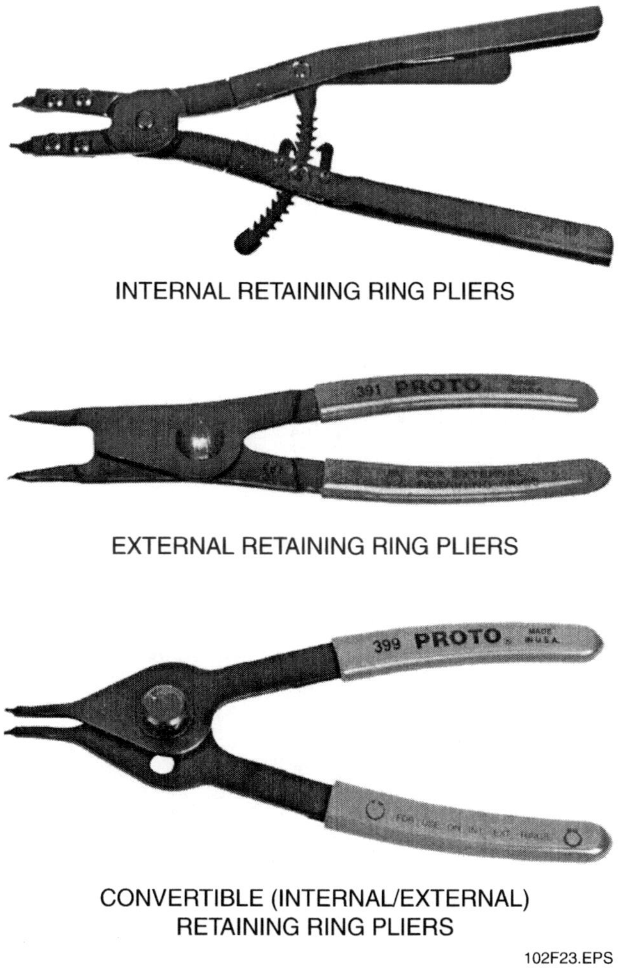

Figure 23 ◆ Retaining ring pliers.

Step 1 Select an easy-out of the proper size, according to the manufacturer's recommendations.

Step 2 Inspect the easy-out to ensure that the threads are not damaged.

Step 3 Select a drill that is the proper size for the easy-out, according to the manufacturer's recommendations.

Step 4 Apply penetrating oil to the broken screw, and allow it to soak.

NOTE

It is best to allow the penetrating oil to soak overnight if possible, adding oil periodically. Allow it to soak as long as time will allow.

Step 5 Prick the exact center of the screw using a center punch. Pricking the center of the screw will help to ensure that the hole will be drilled in the center of the screw.

Step 6 Drill a hole in the center of the screw.

CAUTION

Take care not to drill all the way through the screw and into the workpiece, since this could damage the workpiece.

Step 7 Insert the easy-out into the drilled hole, and begin turning it in a counterclockwise direction until it gets a bite.

Step 8 Continue to slowly turn the easy-out until the screw is removed. *Figure 24* shows the use of an easy-out and an extractor set.

Step 9 Clean the easy-out and the workpiece thoroughly to remove the penetrating oil.

Step 10 Oil the easy-out lightly, and store it in its proper place.

3.22.2 Tap Extractors

Tap extractors are specially made to remove broken taps from a workpiece. The tap extractor has four prongs that fit into the flutes of the broken tap to turn it. There is a special size tap extractor for each size tap. Follow these steps to select, inspect, use, and maintain tap extractors:

Step 1 Select an extractor of the proper size for the tap being removed.

Step 2 Inspect the tap to ensure that it is not damaged.

Step 3 Insert the prongs of the extractor into the flutes of the tap.

Step 4 Push the lower collar of the extractor down against the surface of the workpiece to hold the prongs securely against the body of the tap.

Step 5 Turn the tap back and forth slowly, using a tap handle on the extractor to loosen the tap.

Step 6 Turn the tap counterclockwise to screw the tap out of the hole.

Step 7 Clean the extractor thoroughly.

Step 8 Oil the extractor lightly to prevent rust and store it in its proper place.

3.23.0 Feeler Gauge

Required clearances between parts are usually specified by the equipment manufacturer. A feeler gauge (*Figure 25*) is used to measure this clearance. The feeler gauge contains several blades, which are lengths of metal of different thicknesses. The measurement is marked on each blade. In some cases, the blades will be marked with both standard and metric sizes.

Plastic gauges, including thickness gauges, are made by several companies, allowing the mechanic to use plastic where surfaces are vulnerable to scratching, such as polytetrafluoroethylene (PTFE)-coated or epoxy-coated surfaces. The plastic gauges also are non-sparking, and will not conduct electricity. Most sets are color-coded for thickness, since they can lose lettering from abrasion.

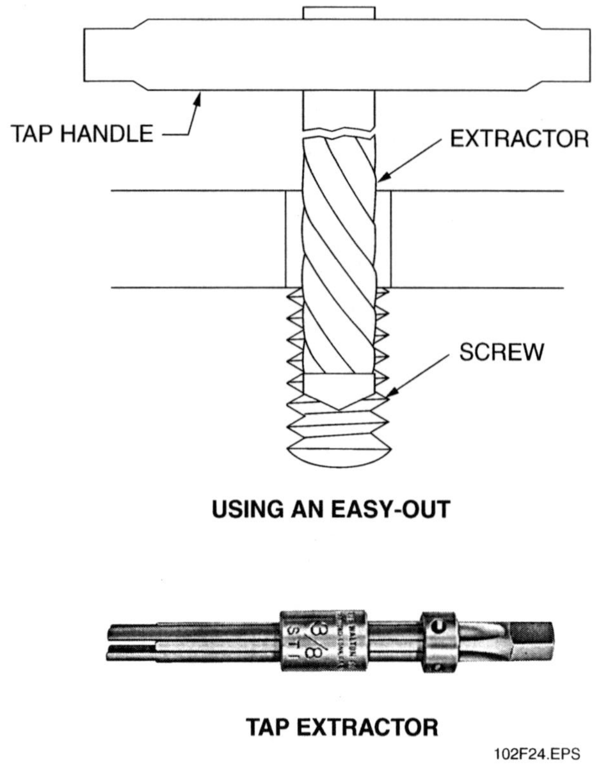

Figure 24 ◆ Using an easy-out.

3.24.0 Dial Indicators

A dial indicator is a direct-reading instrument used to measure machined parts for accuracy or surfaces of machinery to determine runout or accuracy of alignment. Dial indicators are also used when installing bearings and seals, setting up lathes and milling machines, and checking the concentricity of a diameter.

A dial indicator consists of a graduated dial with an indicator hand and a contact point attached to a spring-loaded plunger. Any movement of the plunger causes the pointer on the dial to move. Dial indicators range in size from 1 to 4½ inches in face diameter. The dials are usually divided into one of the following increments:

- Hundredths of a millimeter (0.01 millimeter)
- Two-thousandths of a millimeter (0.002 millimeter)
- Thousandths of an inch (0.001 inch)
- Ten-thousandths of an inch (0.0001 inch)

The two most common types of dial indicators are the balanced type and the continuous-reading

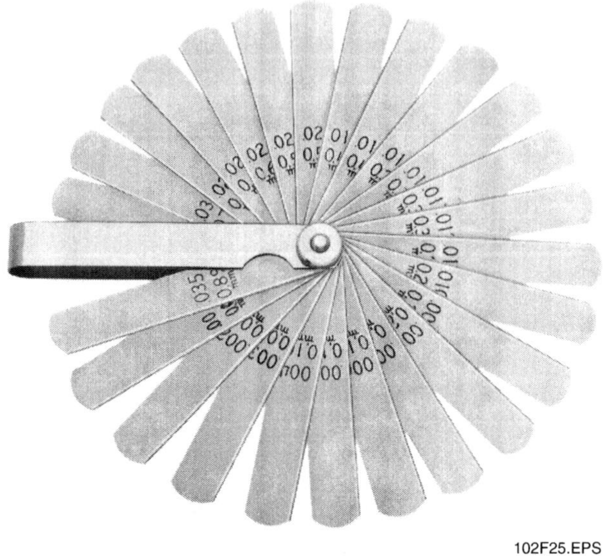

Figure 25 ◆ Feeler gauge.

type (see *Figure 26*). The numbers on the balanced type start at zero and increase in both directions. The numbers on the continuous-reading type start at zero and continue around the dial clockwise. The dials can be reset to zero by rotating the entire dial face. Follow these steps to use a dial indicator:

Step 1 Clean the surfaces that the indicator will touch.

Step 2 Determine where to attach the base; clamp the indicator base onto the workpiece or machine using a C-clamp or a magnetic base.

> **NOTE**
> For quick setup, dial indicators can be placed in a magnetic base holder. A magnet in the base holds to any flat or round steel or iron surface, eliminating the time necessary to clamp the dial indicator to a machine. The magnetic force is turned on or off by pressing a push button.

Step 3 Mount the indicator to the indicator base using the attachments on the indicator holding rod, and tighten all locknuts securely (see *Figure 27*).

Step 4 Move the workpiece or the indicator, depending on what is being measured.

Step 5 Loosen the clamp on the base post and push the indicator into the workpiece by at least half the total travel of the indicator to preload the indicator.

Step 6 Tighten the locknuts securely.

Step 7 Loosen the dial face lock screw and rotate the face to read absolute zero to zero the indicator.

Step 8 Tighten the dial face lock screw.

Step 9 Check all other adjustments to ensure that they are tight and secure.

> **NOTE**
> Take care not to bottom the indicator plunger because that will damage the indicator. If the indicator is too close, reposition it with less preload.

Step 10 Rotate the shaft or move the indicator slowly to obtain a total runout. This records the most extreme position of measurement in both directions of travel.

Step 11 Position the indicator base and take new measurements to get at least three points of measurement for an average.

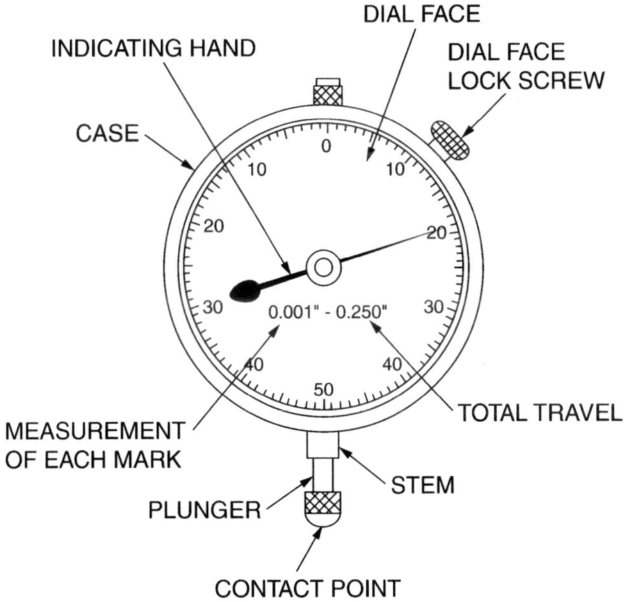

BALANCED TYPE
ONE REVOLUTION = 0.100"
EACH MARK = 0.001

CONTINUOUS-READING TYPE
ONE REVOLUTION = 0.010"
EACH MARK = 0.0001

102F26.EPS

Figure 26 ◆ Two types of dial indicators.

3.25.0 Bars

Equipment manufacturers make a variety of bars to help in moving and aligning heavy equipment. *Figure 28* shows a sleever bar and an alignment bar. One end of each bar is pointed so it can be used in aligning flanges. The sleever bar can be up to 36 inches long.

Technicians use alignment bars to roll up equipment so that **shims** can be placed under it.

> **NOTE**
> Maintenance technicians will often add a lock washer and rubber grommet to a round sleever bar so it can be carried in a holster.

4.0.0 ◆ INTRODUCTION TO POWER TOOLS

The widespread use of power tools in the industrial maintenance trade has greatly increased the individual production of the craftworker. Power tools are available for cutting, grinding, and shaping all types of piping material. Industrial maintenance craftworkers must know how to safely operate these power tools for the specific jobs for which the tools are designed. This rest of this module introduces the trainee to the different types of power tools and procedures for selecting, using, caring for, and maintaining these tools.

5.0.0 ◆ POWER TOOL SAFETY

If not used properly, power tools can cause severe injuries and even death. The following safety precautions should be followed when using power tools:

- Always keep the work area clean. Many accidents are caused by materials or tools carelessly spread about the work area.
- Do not use electrical power tools in wet or damp locations. There is always a danger of electrical shock, even if the tool is properly grounded.
- Store tools when they are not in use. Allowing power tools to lie around the work area when they are not in use increases the possibility that the tools may be damaged and become unsafe for use.

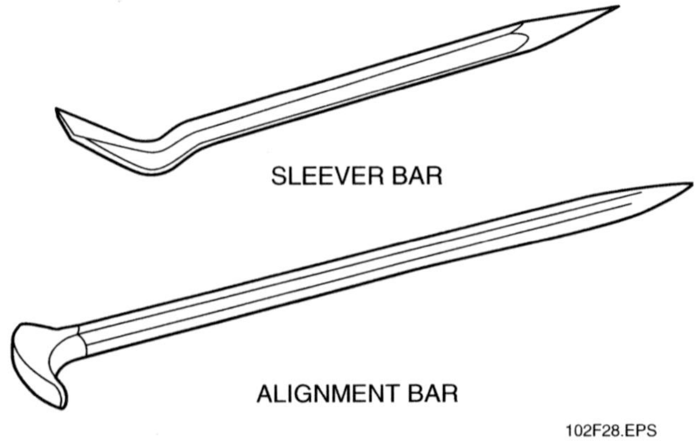

Figure 28 ◆ Alignment bars.

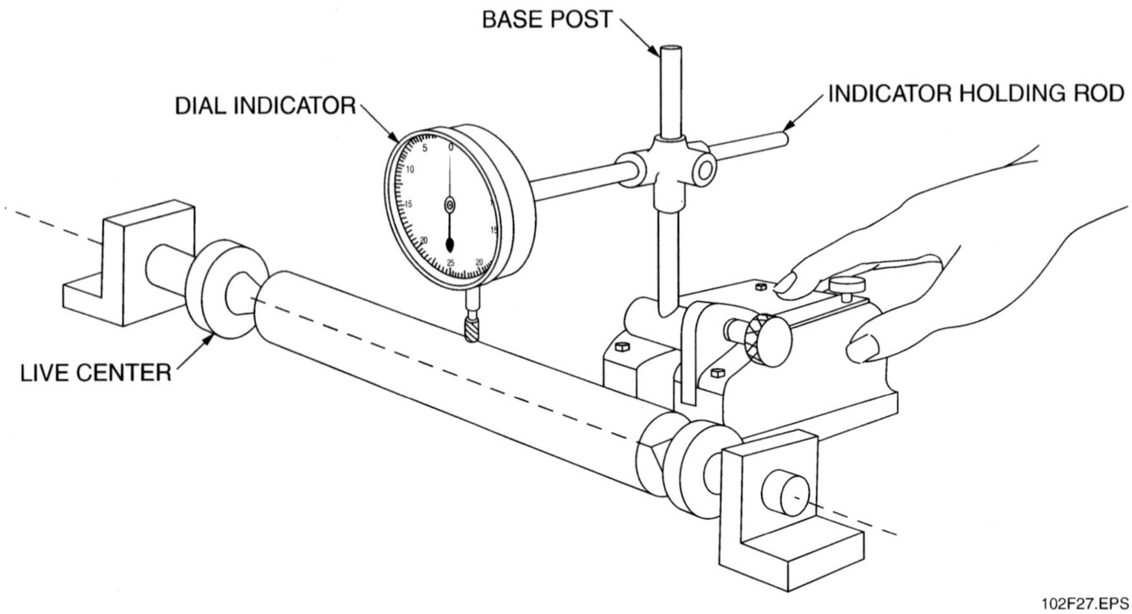

Figure 27 ◆ Indicator mounted on the base.

- Use tools only for their intended purposes, and use the proper tool for each job. Know the capabilities of the tools, and work within those limits. If a tool is not suited for a particular job, select a tool that can perform the job.
- Never force a tool beyond its capabilities. Forcing a tool to work beyond the limits of its design not only wears out the tool prematurely, but also places you in an unsafe situation.
- Always wear proper clothing. Do not wear loose-fitting clothing when working with or around power tools. There is always the danger of loose clothing, especially shirt tails, getting caught in the moving parts of the tool.
- Wear safety glasses when using power tools. Most power tools cause material to be ejected, which could cause severe damage to the eyes.
- Always maintain proper footing and balance when using a power tool. Do not overreach. Any fall can be dangerous, but a fall while using a power tool can be much worse.
- Respect your tools and always follow the recommended operating and maintenance procedures. A well-maintained tool is a safe tool.
- Disconnect the power source before performing maintenance on the tool or changing bits, cutting tips, or other accessories.
- Allow only competent technicians to repair power tools.
- Do not alter or change a power tool in any way.
- Inspect power tools before each use. Make sure that there is no damage to the tool or power cord and that all the safety guards are in place.
- Never carry a power tool by its power cord or an air tool by its air line.
- Avoid accidental starts. Make sure that the power switch is in the OFF position before plugging in the power cord.
- Be aware of the torque or kickback of the power tool being used. Maintain good balance and properly brace yourself when using power tools.

5.1.0 Electric Power Tool Safety

Electrical power tools are extremely dangerous and can cause severe injury and even death if misused. The following safety precautions apply to electric power tools:

- Do not use electric power tools in wet or damp locations. There is always a danger of electric shock, even if the tool is properly grounded.
- Respect power cords. Keep power cords out of situations where they can be frayed, cut, or damaged. Always disconnect a power cord by grasping the plug and pulling. Never yank on the cord to disconnect a power cord.
- The color-code system is used to ensure proper grounding and operation of the power cord. At most job sites, power cords are periodically inspected, and the color-code strips are changed to indicate that the cord has been recently inspected. Make sure that all power cords that you use have the current color-code strip attached to them.
- Ensure that the current available is the same current for which the tool is designed. This will be indicated on the identification plate attached to the tool.
- All power cords must be tied off overhead, at least 7 feet off the floor, to prevent tripping hazards.

5.2.0 Pneumatic Power Tool Safety

Pneumatic power tools can also be dangerous and cause severe personal injury if misused. The following safety precautions apply to pneumatic power tools:

- All air hoses must be properly drained or blown clear before the hose is attached to the tool.
- Line pressure must be checked to make sure that it is applicable to the tool.
- Ensure that the oil reservoir is at the proper level before using the tool.
- Never direct the air flow from an air nozzle toward your body or anyone else.
- Secure all hose connections to avoid accidental disconnection.
- Do not turn air on until you trace the hose and ensure that the end is securely attached to an air tool.
- Do not crimp the hose to turn off the air to the tool or to change a tool.
- Inspect the air line for damage before using the tool.
- When connecting into an air line, make sure that it is an air line and not an inert gas line, such as nitrogen or argon.
- Do not attempt to repair a damaged air line. Always replace it.
- Make sure that the air line is rated for the air pressure with which it will be used. The air line should have a pressure rating printed on the hose.
- Air lines must be tied off at least 7 feet above the ground or floor to prevent tripping hazards.

6.0.0 ◆ CUTTING PIPE USING PORTABLE BAND SAWS

In addition to structural members, portable band saws are used to cut carbon steel pipe, galvanized pipe, and pipe made from other materials. These pipes have varying diameters. The larger diameter pipes, usually 4 inches and larger, require a special procedure to ensure that the pipe is cut square throughout. The following sections explain how to select the proper blades for different types of materials and cut pipe square using the portable band saw.

6.1.0 Selecting Band Saw Blades

Band saw blades are selected depending on the type and wall thickness of material to be cut. Blades are classified by the material of the blade and the number of teeth per inch. A coarse blade is a blade with fewer teeth per inch and should be used on softer materials. A fine blade has more teeth per inch and should be used on harder materials. *Table 2* shows a blade selection chart and recommended usages.

6.2.0 Replacing Portable Band Saw Blades

The portable band saw blade must be replaced if it becomes worn and dull or if you are cutting a material that the blade in the saw was not designed to cut. Follow these steps to replace a band saw blade:

> **WARNING!**
> Make sure that the portable band saw is unplugged before replacing the blade.

Step 1 Rotate the blade adjust knob 180 degrees to release the tension on the blade. *Figure 29* shows portable band saw **components**.

Step 2 Turn the saw upside down on your work table.

Step 3 Remove the blade from the blade pulleys underneath the saw.

Table 2 Blade Selection Chart and Recommended Usages

Band Saw Speed		Use Single-Speed Band Saw or Higher Speeds on Two-Speed and Variable-Speed Band Saws									Use Lower Speeds on Two-Speed or Variable-Speed Band Saws							
Material to be cut		**Aluminum-Brass-Copper-Bronze-Mild Steel**					**Angle Iron-Cast Iron-Galvanized Pipe**				**Stainless Steel**				**Fiber Glass Asbestos-Plastics**			
Wall or Material Thickness	**Teeth Per Inch**	3/32 to 1/8	1/8 to 1/4	5/32 to 1/2	3/16 to 3/4	11/32 & Over	3/32 to 1/8	1/8 to 1/4	5/32 to 1/2	3/16 & Over	3/32 to 1/8	1/8 to 1/4	5/32 to 1/2	3/16 & Over	3/32 to 1/8	1/8 to 1/4	5/32 to 1/2	3/16 & Over
Carbon Steel Blades	6					X												
	8					X												
	10				X					X								
	14			X					X									
	18		X					X										
	24	X					X											
Alloy Steel Blades	10				X					X								X
	14			X					X								X	
	18		X					X								X		
	24	X					X								X			
High-Speed Steel Blades	10				X					X				X				X
	14			X					X				X				X	
	18		X					X				X				X		
	24										X							

102T02.EPS

> **WARNING!**
> Be careful when removing the blade, because the blade has a tendency to spring out when you take it off the blade pulleys.

Step 4 Slip a new blade around the blade pulleys.

Step 5 Slip the new blade into the blade guide and the back stop.

Step 6 Turn the saw over, and rotate the blade adjust knob to put tension on the blade.

6.3.0 Cutting Pipe

When cutting pipe 4 inches in diameter or larger, it is important not to cut straight through the pipe because the blade may bend and may not produce a straight cut. The procedure for cutting pipe using a portable band saw consists of cutting around the pipe instead of straight through it. Follow these steps to cut pipe using a portable band saw.

Step 1 Select the pipe to be cut, and determine what kind of blade you need to cut the pipe.

Step 2 Replace the blade in the portable band saw if necessary.

Step 3 Measure and mark the desired length on the pipe.

Step 4 Mark a straight line around the pipe at the cut mark, using a wraparound.

Step 5 Start the band saw and cut the top of the pipe, keeping the blade on the cut line and the pipe snug against the back stop of the saw. *Figure 30* shows starting the cut.

> **NOTE**
> Keep the blade at a right angle to the surface of the pipe at all times.

Step 6 Continue cutting on the cut line by rotating the saw over the top of the pipe. *Figure 31* shows rotating the saw over the pipe.

Step 7 Stop cutting and remove the blade from the pipe when approximately ¾ of the blade between the front blade guide and back stop has cut into the pipe.

Step 8 Reposition the pipe in the vise so that the next section of uncut pipe is facing you.

Step 9 Repeat steps 5 through 8 until the pipe is cut completely through.

> **WARNING!**
> Avoid allowing the pipe fall to the ground. It can fall on your foot or bounce back into your leg, causing injury. The freshly cut pipe end is extremely sharp and may be hot. Use care to avoid cutting yourself on the pipe end.

Step 10 Remove the pipe from the vise.

7.0.0 ◆ PORTABLE GRINDERS

Maintenance craftworkers use portable grinders for a variety of tasks, such as cleaning scale from pipe before welding, **beveling** pipe ends, and grinding away welding splatter. If not used properly, the portable grinder can cause severe injury. Follow these safety guidelines when using a portable grinder:

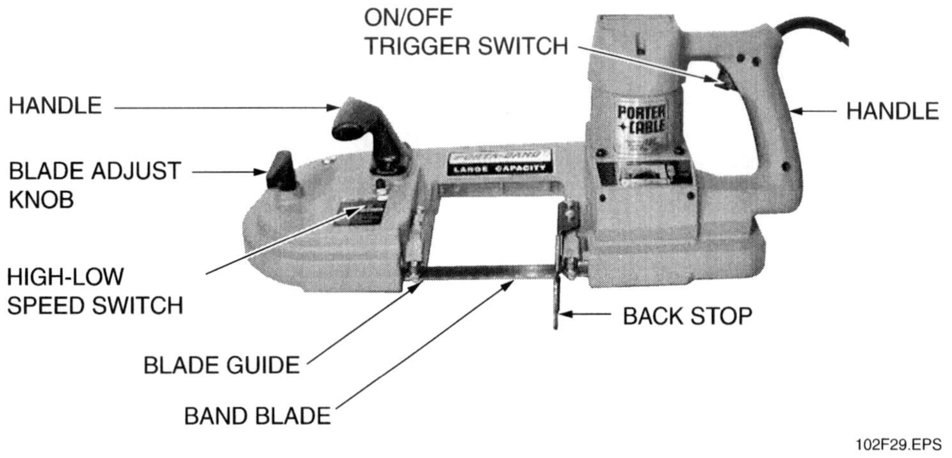

Figure 29 ◆ Portable band saw components.

> **WARNING!**
> Compare the grinder wheel speed rating to the grinder operating speed. The wheel speed rating must meet or exceed the grinder operating speed.

- Inspect the grinding wheel visually before mounting it to the grinder. Do not mount a damaged wheel.
- Examine the grinder wheel guard for damage. It must be in place before starting the grinder.
- Inspect the power cord or air line for damage.
- Wear gloves and a safety shield to protect yourself from flying sparks and metal fragments when using a grinder.
- Make sure the moveable side handle is in place.
- Run the grinder with the safety guard in place for a moment before grinding.

7.1.0 Types of Portable Grinders

Portable grinders can be either electrically or pneumatically powered, rated up to 5 **horsepower (hp)**, and rotate up to 20,000 rpm at no load. Portable grinders are usually identified by the angle of the grinding wheel in relation to the motor housing. The three basic types of portable grinders are end, angle, and vertical grinders.

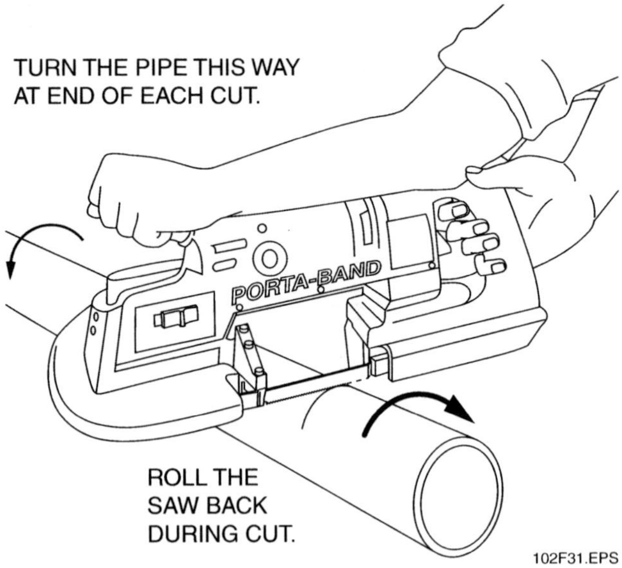

Figure 31 ◆ Rotating saw over pipe.

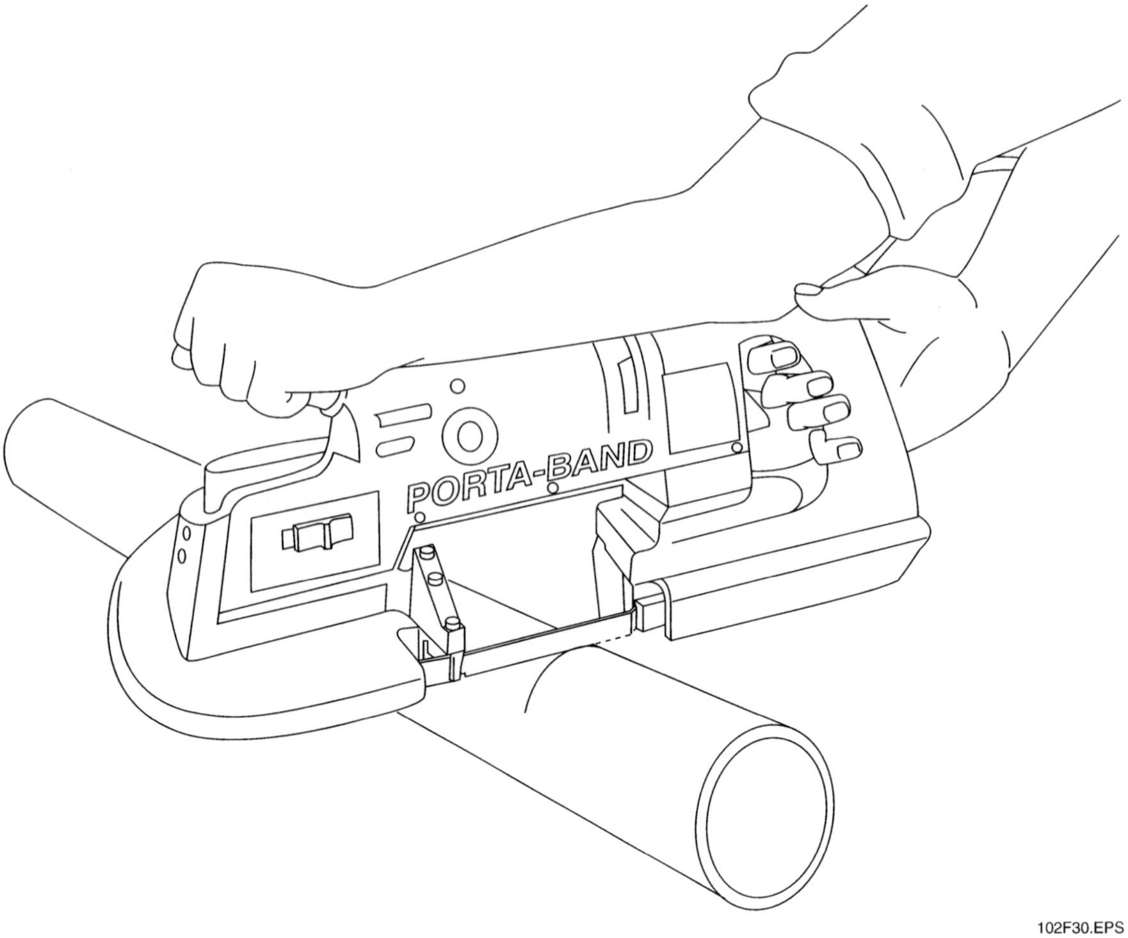

Figure 30 ◆ Starting cut.

7.1.1 End Grinders

End grinders, or horizontal grinders, have the grinder spindle, motor, and handle all in one line. They are designed to reach into small areas with limited working space. Smaller versions of the end grinders are the pencil grinder and the die grinders, which are the least powerful of the portable grinders but which operate at a much higher speed than the other grinders to produce a fine finish. *Figure 32* shows an end grinder.

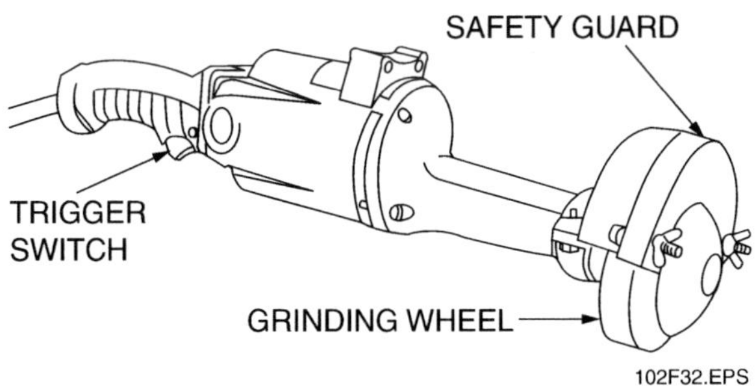

Figure 32 ◆ End grinder.

7.1.2 Angle Grinders

Angle grinders, also known as side grinders, are designed with the spindle at a 90-degree angle to the motor axis. Angle grinders remove material very rapidly and leave a medium to rough finish. An auxiliary handle that fits either side of the casing gives the operator more control of the tool and allows whatever downward pressure is required to do the job. *Figure 33* shows an electric angle grinder.

7.1.3 Vertical Grinders

Vertical grinders are very similar to angle grinders. The only difference is that the motor and spindle are in line with each other but are set at a 90-degree angle from the handle. They also have an auxiliary side handle. *Figure 34* shows a pneumatic vertical grinder.

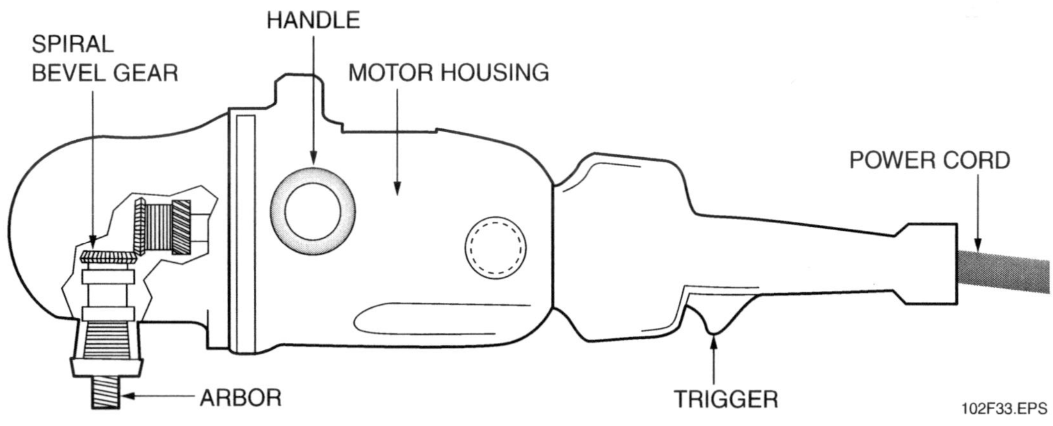

Figure 33 ◆ Electric angle grinder.

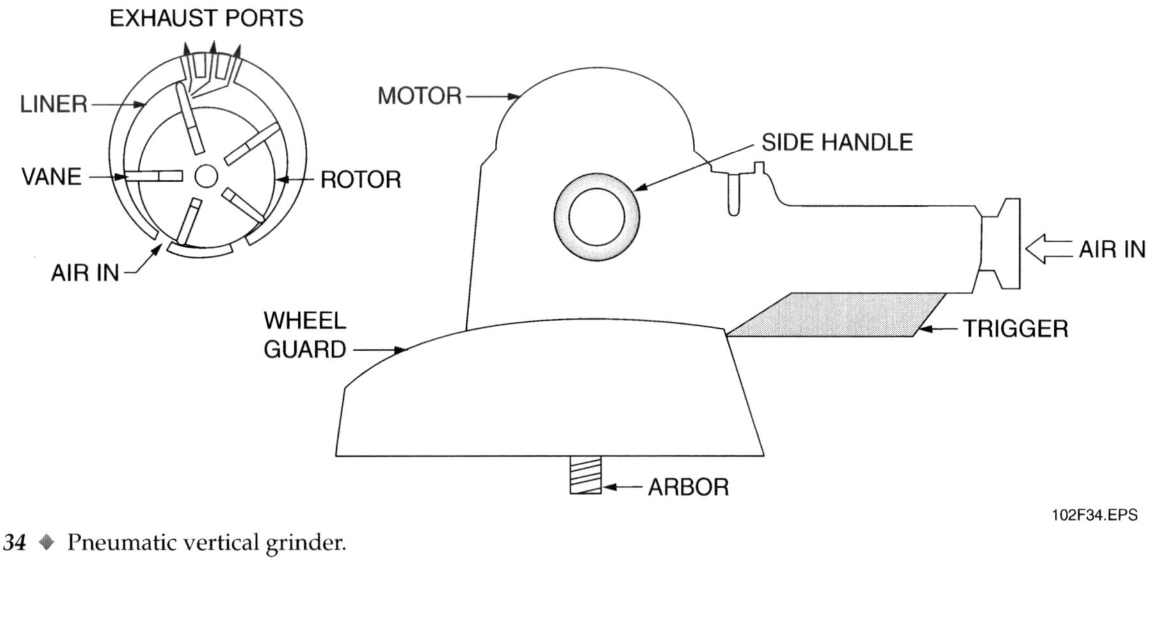

Figure 34 ◆ Pneumatic vertical grinder.

7.2.0 Grinder Accessories

A wide variety of accessories are available for grinders. Grinder accessories are usually called wheels, whether they are in the shape of wheels, discs, cones, or wire brushes. There are aluminum oxide wheels for metal grinding, silicon carbide wheels for cutting stone and concrete, wire cup brushes or buffing wheels for rust and scale removal, flared-cup grinding wheels, and abrasive sanding wheels. No matter what type of portable grinder is used, proper selection of the accessory is necessary to get the job done efficiently and safely. Consider the type of application, the type of material, and the appropriate grinder rpm speed while selecting the accessories. If the wrong wheel is chosen for a given job, the wheel may break apart and send pieces flying at high speeds. *Figure 35* shows different types of grinding wheels. *Figure 36* shows a buffing wheel.

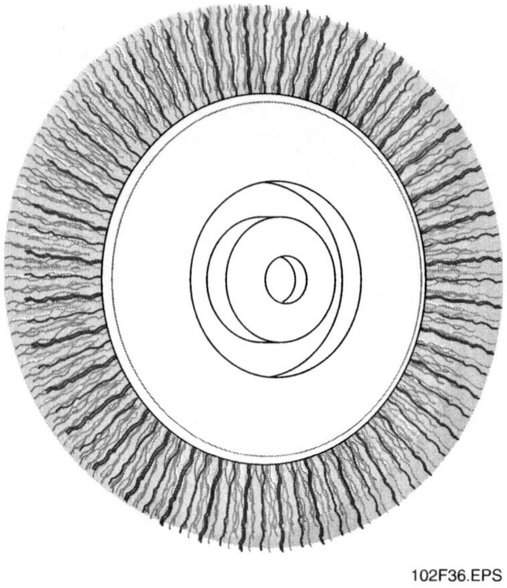

Figure 36 ◆ Buffing wheel.

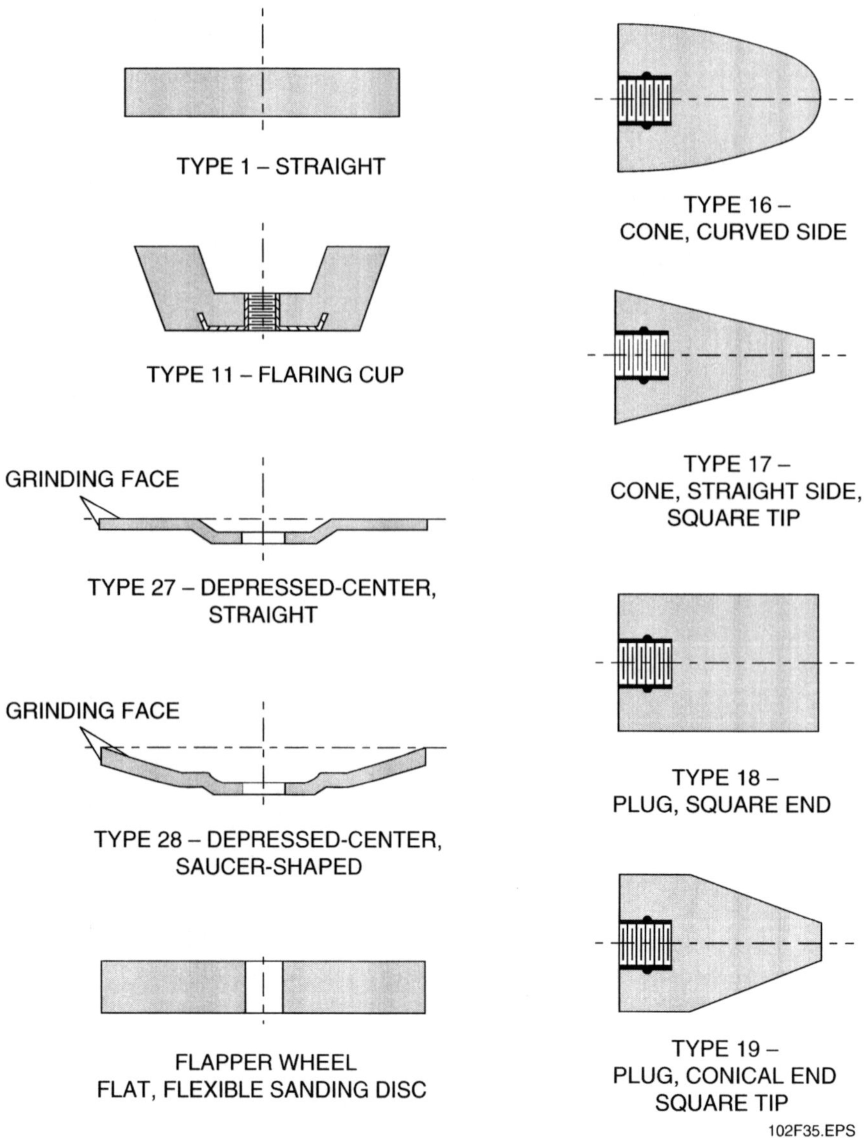

Figure 35 ◆ Different types of grinding wheels.

When selecting a grinding wheel for a given job, two factors must be considered. The wheel must be suited to the material being ground, and the maximum rotational speed of the wheel must be rated higher than the rotational speed of the grinder. The correct wheel can be identified by referring to the manufacturer code stamped or printed on the wheel. The code describes the contents and characteristics of the wheel. The maximum rpm of the wheel is also stamped or printed on the wheel. *Figure 37* shows the standard marking system for grinding wheels.

Whatever you need to do with a grinder of any kind, including changing accessories, you must always wear the personal protective equipment for that job. Safety goggles, hearing protection, and gloves are a minimum, and you may find that your company requires a great deal more, such as

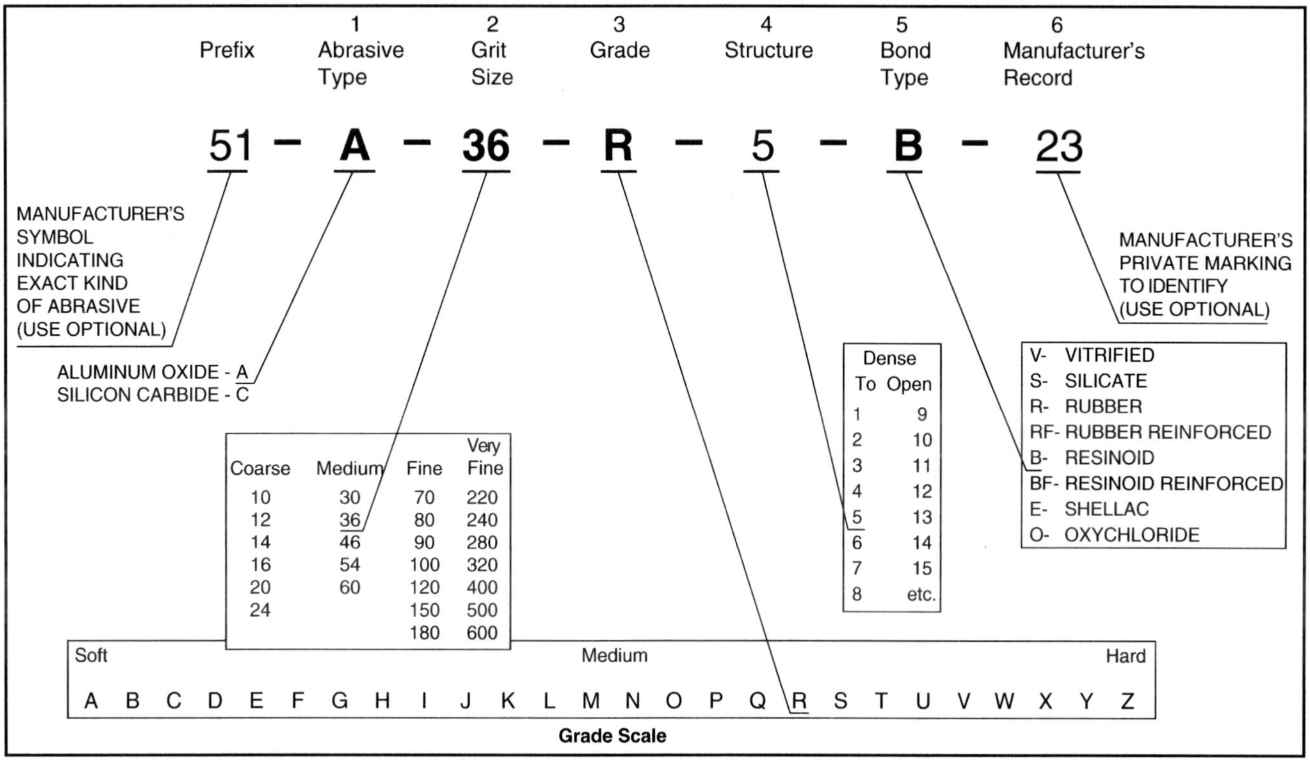

BOND MATERIAL CHARACTERISTICS

Type of Bond	Characteristics
V-vitrified bond	A tough, glass-like ceramic bond used in 75 percent of the grinding wheels.
B-resinoid bond	The second most popular bond, it is made from resin (phenol formaldehyde).
R-rubber bond	This bond is used for very fine abrasives used in polishing metals.
E-shellac bond	This bond is used for very high polishing and is not commonly used.
S-silicate bond	A very special (but rarely used) bond (sodium silicate) where heat must be kept to a minimum.

WHEEL SPECIFICATIONS FOR PORTABLE GRINDING

Workpiece	Abrasive	Grit Size	Grade	Bond
Aluminum				
to 6,500 ft/min	C	24	0	B
to 9,500 ft/min	C	30	0	V
Cast iron	C	20	Q	B
Ductile iron	A	20	P	B
Steel				
Low-carbon	A	20	Q	B
Stainless, free machining	A	20	Q	B
Welds				
Carbon-alloy steels				
5,000-6,500 ft/min	A	24	Q	V
7,000-9,500 ft/min	A	16	R	B
9,501-12,500 ft/min	A	16	R	B
Stainless steel				
7,000-9,500 ft/min	A	16	R	B
9,501-12,500 ft/min	A	16	R	B

USED WITH PERMISSION OF WELDING DESIGN AND FABRICATION

102F37.EPS

Figure 37 ◆ Standard marking system for grinding wheels.

full-face shields over goggles or fireproof aprons. Grinders are easily capable of doing a lot of damage; don't take chances.

7.2.1 Changing Accessories on Grinders

The grinder wheel must be changed if the wheel is not the correct type for the job or if the wheel is worn or cracked. The thin flat type of wheel can be tested by striking the wheel lightly with a screwdriver handle while holding the wheel by the center hole. If the wheel produces a clear note, it is not cracked. This does not apply to other wheel types. All others must be visually inspected and used with caution against the possibility of hidden cracks. Follow these steps to change a grinding wheel:

> **CAUTION**
> Always disconnect the power source before removing the disc.

Step 1 If the grinder has a spindle lock button behind the guard housing, depress and hold the spindle lock. If there is not a button, place an open-end wrench on the adapter nut located behind the grinder wheel to hold the arbor while removing the lock nut. *Figure 38* shows the grinding wheel hardware.

> **NOTE**
> On some grinders you have to use special wrenches provided by the manufacturer.

Step 2 Place a wrench on the lock nut in front of the wheel.

> **NOTE**
> Some lock nuts use an Allen wrench; others use an open-end wrench.

Step 3 Turn the lock nut clockwise to remove it.

> **NOTE**
> The spindles on grinders use left-hand threads because they rotate clockwise.

Step 4 Remove the grinding wheel from the grinder.

Step 5 Repeat this procedure in reverse to put a new grinding wheel on the grinder.

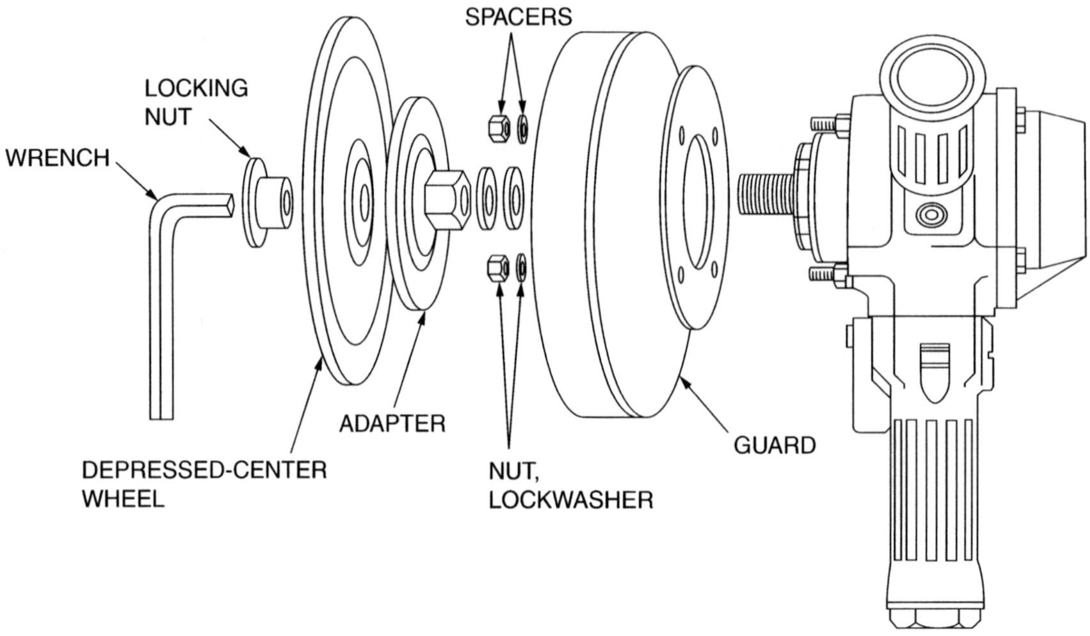

Figure 38 ◆ Grinding wheel hardware.

> **NOTE**
> If a washer was located between the lock nut and the wheel, be sure to install it back into place when putting on the new wheel. Make certain that the grinding wheel is tightened firmly to the arbor. The wheel can come off if carelessly mounted, and it can injure you or others.

7.2.2 Inspecting Grinders Before Use

Whether the portable grinder being used is electric or pneumatic, a thorough inspection of the equipment is required before the grinder is used. Follow these steps to inspect a grinder:

Step 1 Inspect the air inlet and the air line of a pneumatic grinder to ensure there are no signs of damage that could cause a bad connection or loss of air.

Step 2 Inspect the power cord and plug on electric models to ensure there are no signs of damage.

Step 3 Inspect the handle to make sure it is not loose, which could cause a loss of control.

Step 4 Inspect the grinder housing and body for defects.

Step 5 Ensure that the trigger switch works properly and does not stick in the ON position.

Step 6 Ensure that the safety guard is in good condition and securely attached to the grinder.

Step 7 Check the oil level in pneumatic grinders.

Step 8 Ensure that the maximum rotating speed of the grinding wheel is higher than the maximum rotating speed of the grinder.

Step 9 Start the grinder and allow it to run for a moment while checking for visual abnormalities, excessive vibration, extreme temperature changes, or noisy operation.

Step 10 Inspect the work area to ensure the safety of yourself and others and to make sure that the heat and sparks generated by the grinder cannot start any fires.

7.2.3 Operating Grinders

When operating grinders, you must pay full attention to the grinder, the work being performed, and the flow of sparks and metal bits coming off the wheel. Always remember that each grinding accessory is designed to be used in only one way, and that the grinding wheel rotates clockwise. When a flat grinding disc is being used, only put the flat surface of the disc against the work. If a cutoff wheel is being used, only use the edge of the wheel to do the cutting. Follow these steps to operate a grinder:

Step 1 Perform the inspection procedures outlined in this module to inspect the grinder.

Step 2 Ensure that no combustible material will be struck by sparks.

Step 3 Put on personal protective equipment, including gloves, a face shield, ear plugs, and safety shoes.

Step 4 Tuck in all loose clothing.

Step 5 Secure the object to be worked on in a vise or clamps to ensure that it does not move.

Step 6 Obtain any hot work permits required. It may also be required that you place barriers or screens to contain sparks.

Step 7 Attach the grinder to the power source.

Step 8 Position yourself with good footing and balance, and establish a firm hold on the grinder to avoid kickback.

Step 9 Verbally warn bystanders before pulling the trigger. Then, pull the trigger to start the grinder.

Step 10 Apply the grinder to the work, and direct the sparks to the ground whenever possible and away from any hazards in the area, such as combustible debris or acetylene tanks. Protect nearby alloy metals, such as stainless steel tanks, from cross **contamination** of other metals being ground in the area.

WARNING!
Special precautions must be taken to avoid kick-back while prepping fittings and pipe. Always use a grinding wheel rather than a flapper wheel or a buffing wheel when cleaning a bevel on a pipe end. Flapper wheels and buffing brushes have more of a tendency to grab into the pipe and kick back than grinding wheels do. Remember that the wheel rotates clockwise, and you must position yourself and the wheel on the pipe end correctly to avoid the wheel grabbing the pipe end and kicking back toward you. Always grind with the direction of the bevel and position the wheel to the bevel, so that if it kicks off, it will kick off away from you. Be especially careful if you are grinding an inside angle, as the grinder wheel can catch in the angle and come off at unpredictable angles. The wheel is properly positioned on the bevel when the torque pulls the wheel in the direction that the wheel would leave the pipe end if it grabbed. Reposition the wheel on the pipe end when moving from the top half to the bottom half of the pipe.

Step 11 Apply proper force to the grinder on the grinding surface. If you are applying too much force on the grinder, you can hear the motor strain.

Step 12 Periodically stop grinding and inspect the work and the grinding wheel. Replace the wheel if necessary.

Step 13 Turn off and disconnect the grinder from the power source when grinding is complete.

Step 14 Inspect the grinder for any signs of damage.

Step 15 Return the grinder and any accessories to the storage area.

7.3.0 Pipe Vise

The pipe vise (*Figure 39*) is used to hold pipe being threaded, cut, or worked with portable machinery. The pipe is held in place with a chain whose latch is adjustable.

8.0.0 ◆ PIPE THREADING MACHINES

Pipe threading machines are multipurpose power tools that center, hold, and rotate pipe, conduit, or bolt stock while cutting, reaming, and threading operations are performed. All pipe threading machines are designed with a gear drive and **chucks**, which hold and turn the pipe, and a three-position power switch consisting of forward, off, and reverse drive positions. Depending on the size and complexity of the machine, pipe threading machines can be mounted on tripods, tables, or special stands.

8.1.0 Types of Pipe Threading Machines

Pipe threading machines come in a variety of models, from the basic gear-drive type with front and rear pipe-holding chucks to the all-purpose model with built-in cutters, oilers, reamers, and a die for threading pipe. These pipe threading machines are covered in the following sections:

- Ridgid® Power Drive 300
- Ridgid® Threading Machine 535

8.1.1 Ridgid® Power Drive 300

The Model 300 power drive unit consists of a ½-hp reversible motor, a locking speed chuck, a rear centering device, and two sliding heavy-duty support bars. *Figure 40* shows a Ridgid® Power Drive 300.

A universal die head, pipe cutter, pipe reamer, moving carriage, and a tripod stand can be attached to the power drive to convert it into a portable pipe threading machine. The threading machine can thread pipe from ⅛ to 2 inches in diameter. A geared threader, used to thread pipe up to 6 inches in diameter, or a power-grooving machine, used to cut or roll grooves into the pipe, can also be

102F39.EPS

Figure 39 ◆ Pipe vise.

secured on the support bars. *Figure 41* shows a power drive converted into a threading machine.

8.1.2 Ridgid Threading Machine 535

The Ridgid® Threading Machine 535 is a complete pipe threading machine capable of threading pipe ranging from ⅛ to 2 inches in diameter. It can also be used to drive a geared threader to thread pipe up to 6 inches in diameter. It has a ½-hp reversible motor, a speed chuck with replaceable jaws, a rear centering device, a pipe cutter, a reamer, a universal die head, and a safety foot switch. A handwheel located on the side of the machine allows the operator to move the tools into the rotating pipe and measure the distance that the tools travel. A built-in lubrication system automatically filters and delivers cutting oil to the pipe. *Figure 42* shows the Ridgid® Threading Machine 535.

8.2.0 Loading Pipe Into Threading Machine

On most threading machines, the pipe can be inserted from either end of the chuck. Some models, however, must be loaded from the rear. Either way, the operator must position the pipe so that the end to be threaded extends out away from the chuck. On most threading machines, there is a stop to help position the pipe properly. Follow these steps to load pipe into the threading machine:

Step 1 Measure and mark the length of pipe to be worked.

Step 2 Insert the pipe into the threading machine.

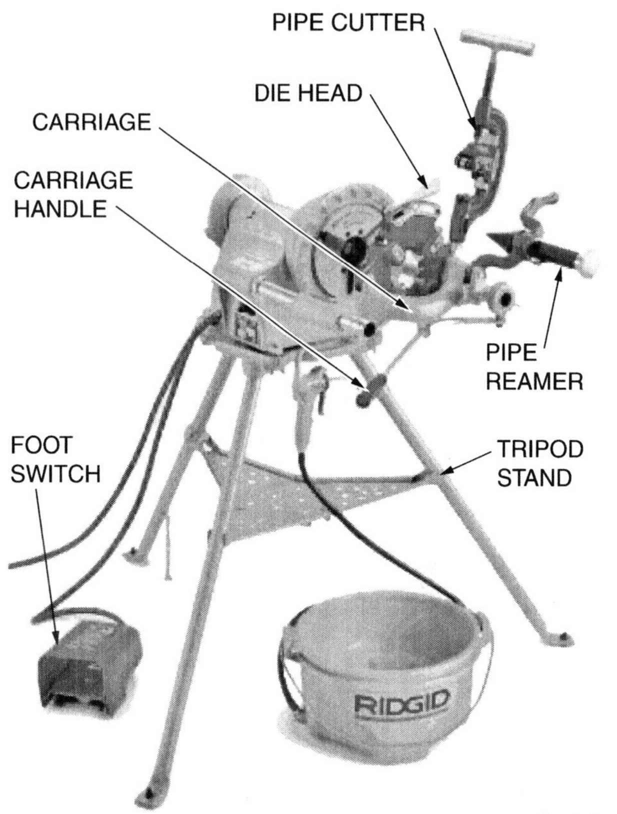

Figure 41 ◆ Power drive converted into threading machine.

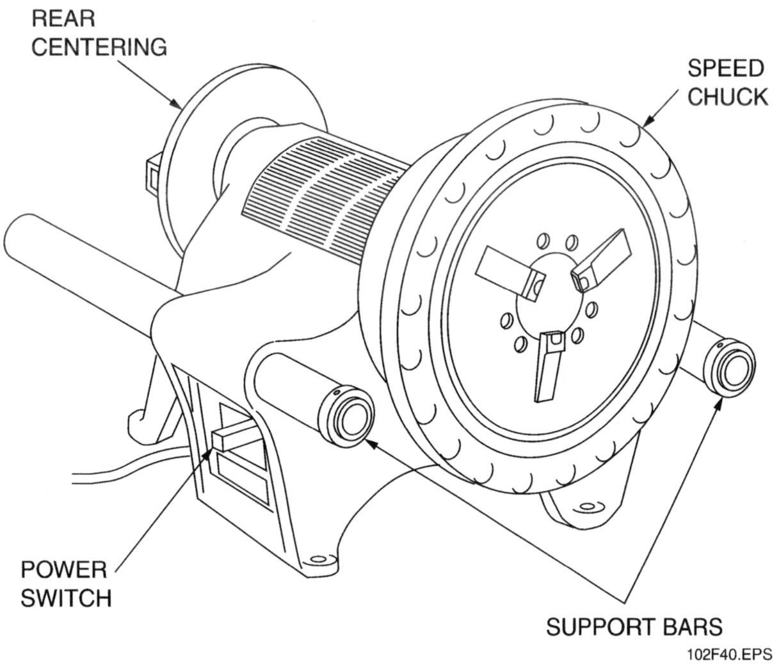

Figure 40 ◆ Ridgid Power Drive 300.

> **NOTE**
> If the pipe is long enough to be held by the centering device, insert the pipe through the front or rear of the machine. If the pipe is short, insert the pipe through the front of the machine. For long pipe, support the end of the pipe with an adjustable support.

Step 3 Make sure that the pipe is centered in the centering device.

Step 4 Tighten the centering device.

Step 5 Spin the handwheel counterclockwise to tighten the chuck jaws. This tightens the jaws on the pipe. A clockwise spin releases the jaws.

> **WARNING!**
> Do not wear loose clothing or jewelry. Remove safety gloves and keep hair pulled back while operating a threading machine.

8.3.0 Cutting and Reaming Pipe

Follow these steps to cut and ream pipe:

Step 1 Determine the length of pipe needed.

Step 2 Measure the pipe to determine how much must be cut off.

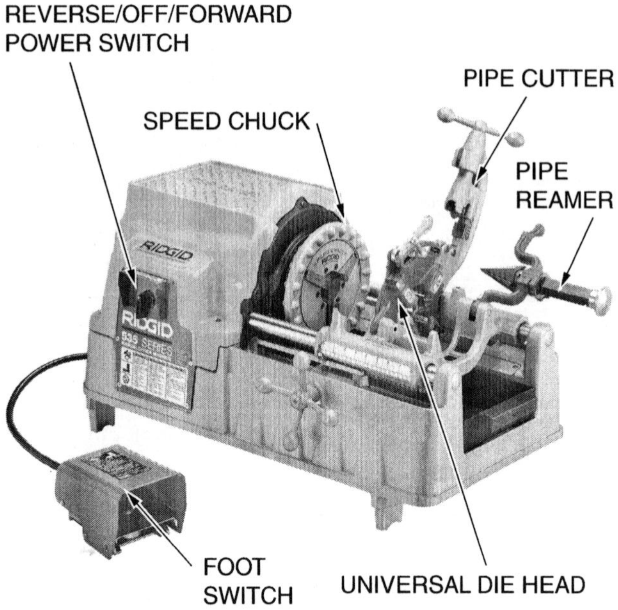

102F42.EPS

Figure 42 ◆ Ridgid Threading Machine 535.

Step 3 Mark the pipe at the point where it is to be cut.

Step 4 Swing the cutter, threader, and reamer up and out of the way.

Step 5 Load the pipe into the threading machine.

Step 6 Place pipe stands as needed.

Step 7 Turn the centering device to center the pipe in the machine.

Step 8 Spin the chuck handwheel counterclockwise to lock the pipe in place.

Step 9 Loosen the cutter until it fits over the pipe.

Step 10 Lower the cutter onto the pipe. *Figure 43* shows the cutter over the pipe.

Step 11 Rotate the handwheel to move the cutter wheel to the cutting mark on the pipe.

Step 12 Rotate the cutter handle to tighten the cutter on the pipe.

Step 13 Direct the cutting oil to the cutting area.

Step 14 Turn the pipe machine switch to the FORWARD position.

Step 15 Step on the foot switch to start the machine and turn the pipe.

Step 16 Rotate the cutter handle to apply pressure to the cutter until the cut is complete.

Step 17 Release the foot switch.

Step 18 Raise the cutter to an out-of-the-way position.

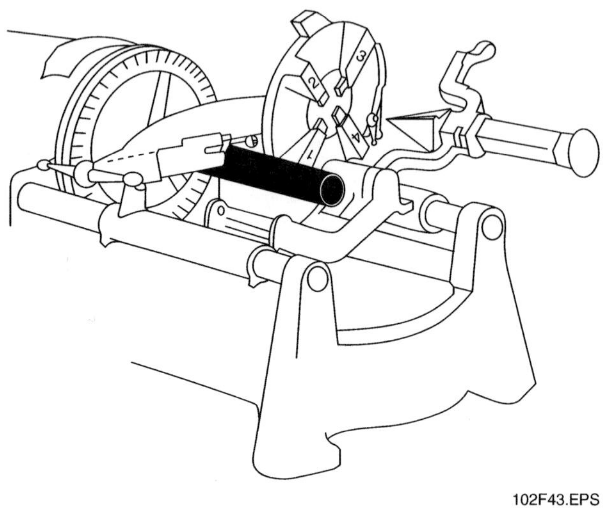

102F43.EPS

Figure 43 ◆ Cutter over pipe.

Step 19 Lower the reamer into place. *Figure 44* shows the reamer in place.

Step 20 Press the latch on the reamer, and slide the reamer bar toward the pipe until the latch catches.

Step 21 Step on the foot switch to start the machine.

Step 22 Rotate the handwheel to move the reamer into the end of the pipe.

Step 23 Ream the end of the pipe.

> **CAUTION**
>
> Do not overream the pipe. Only ream enough to remove the burrs.

Step 24 Rotate the handwheel to move the reamer out of the pipe.

Step 25 Release the foot switch.

Step 26 Slide the power switch to the OFF position.

Step 27 Raise the reamer to an out-of-the-way position.

Step 28 Loosen the chuck handwheel and the centering device.

Step 29 Remove the pipe from the machine.

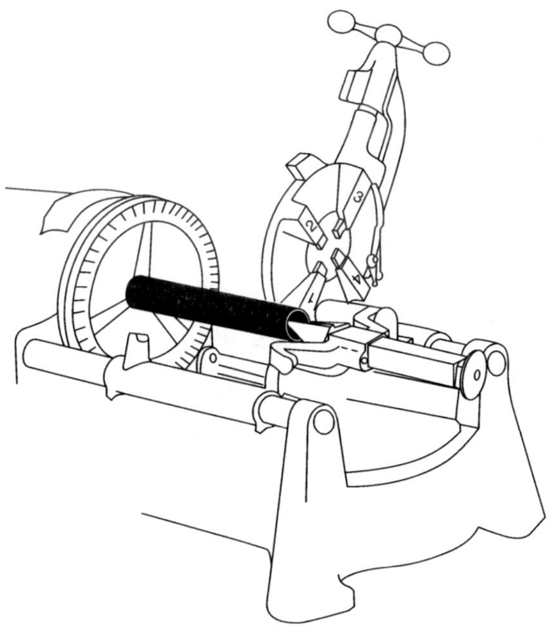

102F44.EPS

Figure 44 ◆ Reamer in place.

8.4.0 Replacing Dies in Threading Machine

Four sets of matched dies are used with pipe threading machines that thread pipe up to 2 inches in diameter, and another set is used for pipe 2½ inches and larger. Sets of dies are matched; be sure to choose a matched set or the thread will be unusable. Dies are classified according to the number of threads per inch. *Table 3* shows proper die selection.

Once selected, the pipe thread dies must be installed and adjusted for different sizes of pipe. Pipe threading machines have universal die heads that adjust the dies for different sizes of pipe. *Figure 45* shows the die head details.

Each set of dies, except those used on ⅛-inch pipe, can be used for several sizes of pipe. To adjust the die head for different sizes of pipe, loosen the clamp lever and move the size bar until the line underneath the desired pipe size lines up with the index line. Tighten the clamp lever after the adjustment has been made. If oversized or undersized threads are required, set the index line in the direction of the over or under mark on the size bar.

The throw-out lever, which is marked OPEN and CLOSE, is a quick-release handle used to close the dies into the threading position and open the dies to retract the dies away from the pipe.

Table 3 Proper Die Selection

Nominal Pipe Size (inches)	Die Threads (per inch)	Thread Length (inches)	Number of Threads
⅛	27	⅜	10
¼	18	⅝	11
⅜	18	⅝	11
½	14	¾	10
¾	14	¾	10
1	11½	⅞	10
1¼	11½	1	11
1½	11½	1	11
2	11½	1	11
2½	8	1½	12
3	8	1½	12
3½	8	1⅝	13
4	8	1⅝	13
5	8	1¾	14
6	8	1¾	14
8	8	1⅞	15
10	8	2	16
12	8	2⅛	17

102T03.EPS

After threading a pipe, the operator can lift this handle to the open position and quickly remove the dies from the pipe. Before cutting threads with the pipe threading machine, the proper dies must be installed in the die head. Follow these steps to replace the dies in a threading machine:

Step 1 Remove the die head from the machine and place it on a workbench with the numbers facing up.

Step 2 Open the throw-out lever.

Step 3 Loosen the clamp lever about three turns.

Step 4 Lift the tongue of the clamp lever washer up and out of the slot under the size bar.

Step 5 Slide the throw-out lever all the way to the end of the slot in the OVER direction of the slide bar.

Step 6 Remove the old dies from the die head.

Step 7 Select a matched set of replacement dies, specified by serial numbers on the dies.

Step 8 Clean and inspect the new dies.

Step 9 Insert the new dies in the die head.

NOTE
The numbers 1 through 4 on the dies must match the numbers on the slots of the die head.

Step 10 Slide the throw-out lever back to where the tongue of the clamp lever washer drops in the slot under the size bar.

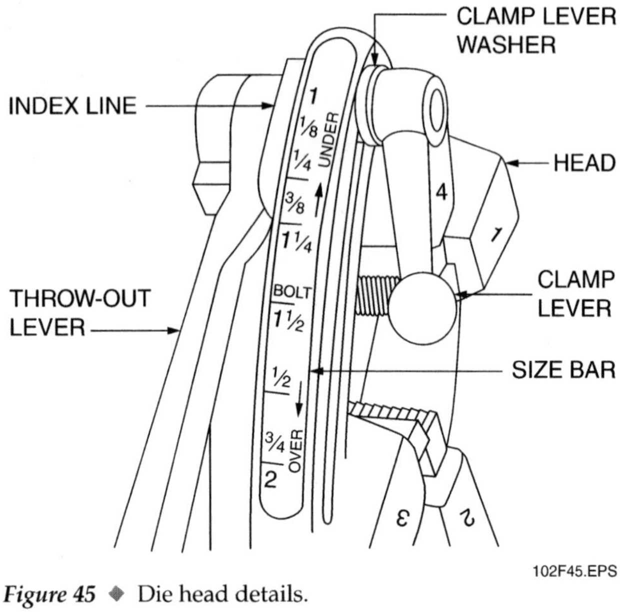

Figure 45 ◆ Die head details.

Step 11 Adjust the die head size bar until the index line is lined up with the proper size mark on the size bar.

Step 12 Tighten the clamp lever.

Step 13 Install the die head into the machine.

8.5.0 Performing Threading Operations

This section explains the threading operations that need to be performed to thread pipe using the Ridgid model 535 pipe threading machine. Before actually threading the pipe, the operator should check the thread cutting oil level and oil pump operation and prime the oil pump if necessary.

8.5.1 Checking Thread Cutting Oil Level

Before any threading operations are performed, it is extremely important to check the level of the special thread cutting oil. Follow these steps to check the thread cutting oil level:

Step 1 Slide the chip pan out from the base of the threading machine.

Step 2 Check to see if the oil is up to the fill-level line in the reservoir. Fill the reservoir with cutting oil as necessary.

Step 3 Lower the lubrication arm over the open reservoir. Oil cannot flow from the lubrication arm in its upright position.

Step 4 Turn the power selector switch to the FORWARD position.

Step 5 Step on the foot switch.

NOTE
Oil should flow from the lubrication arm. If it does not, it may be necessary to prime the pump.

Step 6 Turn the power selector switch to the OFF position to stop the threading machine.

Step 7 Raise the lubrication arm and slide the chip pan back into the base of the threading machine.

8.5.2 Priming Oil Pump

If the oil does not flow from the lubrication arm, it may be necessary to prime the pump. Follow these steps to prime the oil pump.

Step 1 Remove the button plug on the machine cover.

Step 2 Remove the primer screw through the opening.

Step 3 Fill the pump with cutting oil.

Step 4 Replace the primer screw and the button plug.

> **CAUTION**
> Failure to replace the primer screw and button plug causes the pump to drain itself immediately when the power is turned on.

Step 5 Turn the power on in the reverse direction, and check the flow of oil from the lubrication arm. Running the machine in reverse primes the pump more efficiently than running the machine forward.

8.5.3 Threading Pipe

After checking the cutting oil level and priming the pump, the operator is ready to thread the pipe. Follow these steps to thread pipe:

Step 1 Load the pipe into the threading machine.

Step 2 Place pipe stands under the pipe as needed.

> **WARNING!**
> Do not wear loose clothing or jewelry. Remove safety gloves and keep hair pulled back while operating threading machine.

Step 3 Cut and ream the pipe to the required length as explained earlier in this module.

Step 4 Make sure that the proper dies are in the die head.

Step 5 Loosen the clamp lever.

Step 6 Move the size bar to select the proper die setting for the size of pipe being threaded.

Step 7 Lock the clamp lever.

Step 8 Swing the die head down to the working position.

Step 9 Close the throw-out lever.

Step 10 Lower the lubrication arm and direct the oil supply onto the die.

Step 11 Turn the machine switch to the FORWARD position.

Step 12 Step on the foot switch to start the machine.

Step 13 Turn the carriage to bring the die against the end of the pipe.

Step 14 Apply light pressure on the handwheel to start the die.

Step 15 Release the handwheel once the dies have started to thread the pipe. The dies feed onto the pipe automatically as they follow the newly cut threads.

> **CAUTION**
> Make sure that the die is flooded with oil at all times while the die is cutting to prevent overheating the die and the pipe. Overheating can cause damage to the die and to the pipe threads.

Step 16 Open the throw-out lever as soon as one full thread extends from the back of the dies.

Step 17 Release the foot switch to stop the machine.

Step 18 Turn the carriage handwheel to back the die off the pipe.

Step 19 Swing the die and the oil spout up and out of the way.

Step 20 Screw a fitting that is the same size as the pipe you are threading onto the end of the pipe to check the threads. If the threads are correct, you should be able to screw a fitting three and a half revolutions by hand onto the new threads. If the threads are not deep enough, adjust the die using the clamp lever and repeat the threading operation.

Step 21 Turn the machine switch to the OFF position.

Step 22 Open the chuck and remove the pipe.

8.6.0 Threading Machine Maintenance

By following a few simple guidelines, you can expect a long life of service from the threading machine. This section describes the procedures to keep the threading machine in proper working condition.

- Never reverse the rotation of the machine while the machine is running. Always turn off the power and wait until the drive has stopped rotating before changing the direction of the drive mechanism.
- At the start of each day, clean the chuck jaws with a stiff brush to remove rust, scale, chips, pipe coating, or other foreign matter. Apply lubricating oil to the machine bedways, cutter rollers, and feed screw.
- Always use sharp dies, which produce smoother threads and require less motor power than dull dies. Threading stainless steel pipe requires special dies, usually made of high-speed steels and special lubricants.
- Refer to and follow the manufacturer's instructions for lubrication schedules for bearings and gears. The threading machine should be lubricated at least once for every 40 hours of running time.
- Ensure that there is always plenty of cutting oil in the machine. The cutting oil should be periodically cleaned to remove accumulated sludge, chips, and other foreign matter from the oil. Replace the cutting oil when it becomes dirty or contaminated.
- Remove the oil filter periodically and clean it using solvent. Blow the filter clean using compressed air. Do not operate the machine without the oil filter.
- Keep the chuck jaws in good condition, and replace the jaw inserts as necessary when they become worn.

9.0.0 ◆ SPECIAL THREADING APPLICATIONS

The special threading applications that can be performed with the threading machine include threading short nipples using a nipple chuck and threading pipe larger than 2 inches in diameter using a geared threader or mule.

9.1.0 Cutting and Threading Nipples

A nipple is a piece of pipe less than 12 inches long that is threaded on each end. Threaded pipe longer than 12 inches is considered cut pipe. The nipple chuck is a useful tool for holding nipples in the threading machine when threading. A nipple chuck with inserts and adapters will thread nipples from ⅛ inch to 2 inches in diameter. *Figure 46* shows a nipple chuck kit.

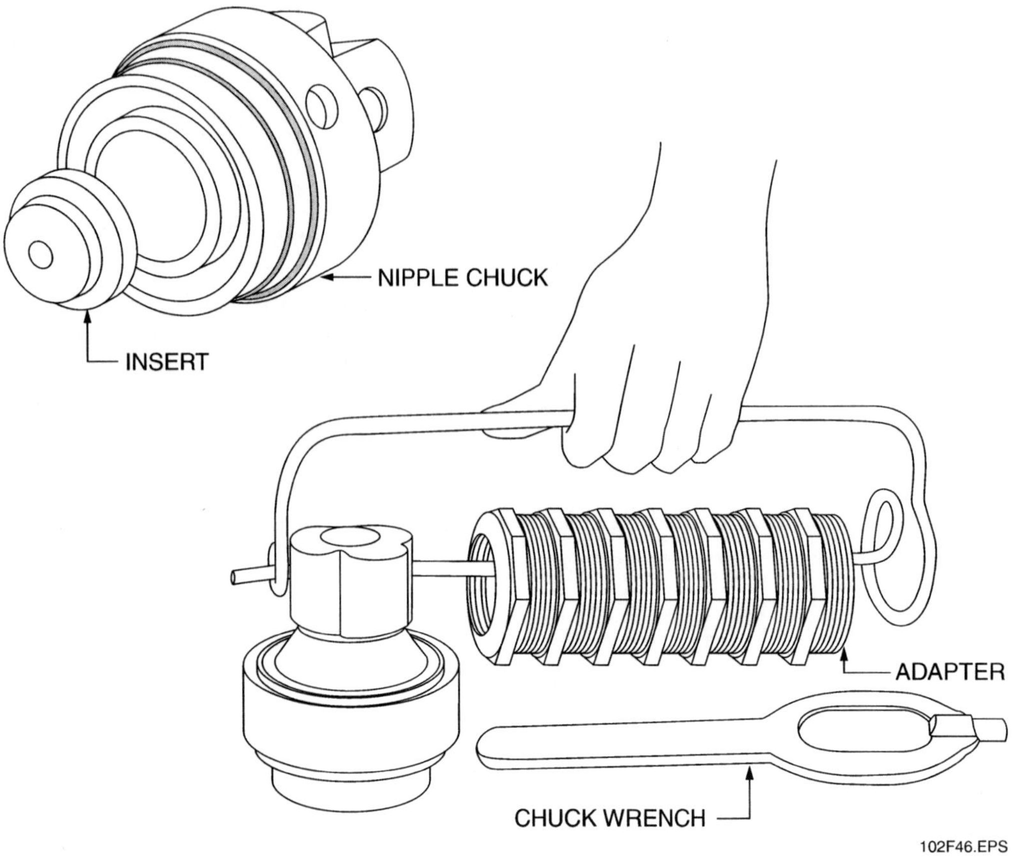

Figure 46 ◆ Nipple chuck kit.

Follow these steps to cut and thread nipples:

Step 1 Load the pipe into the threading machine.

Step 2 Ream the end of the pipe.

Step 3 Thread the pipe according to the pipe threading procedures outlined in this module.

Step 4 Measure from the threaded end the desired length of the nipple and mark a line on the pipe at this point.

Step 5 Cut the pipe at the mark.

Step 6 Remove the pipe from the power drive chuck.

Step 7 Place the nipple chuck into the power drive chuck and tighten the jaws on the nipple chuck.

Step 8 Place the insert inside the end of the nipple chuck.

NOTE

Place the small end toward the outside if threading ⅛- to ¾-inch pipe. Place the large end toward the outside if threading 1-inch pipe. Use no insert with pipe that is 1¼ inch and larger.

Step 9 Select the proper size nipple chuck adapter depending on the size pipe you are threading.

NOTE

If you are threading 2-inch pipe, you do not need a nipple chuck adapter.

Step 10 Screw the nipple chuck adapter into the nipple chuck.

Step 11 Tighten the nipple chuck adapter into the nipple chuck using a chuck wrench.

WARNING!

Do not start the threading machine with the chuck wrench over the nipple chuck adapter.

Step 12 Screw the nipple into the end of the nipple chuck.

Step 13 Tighten the nipple chuck release collar using the chuck wrench. *Figure 47* shows tightening the nipple chuck.

Step 14 Ream the end of the nipple.

Step 15 Thread the nipple according to the pipe threading procedures outlined in this module.

Step 16 Insert the chuck wrench into the release collar and turn the release collar to loosen the nipple from the nipple chuck.

Step 17 Unscrew the nipple from the nipple chuck.

WARNING!

Use a rag to grab the threads on the freshly cut nipple. These threads are extremely sharp and can cut you.

Step 18 Place the chuck wrench over the nipple chuck adapter and unscrew the adapter from the nipple chuck.

Step 19 Wipe off all excess cutting oil from the nipple chuck adapter and the insert, and store them with the nipple chuck kit.

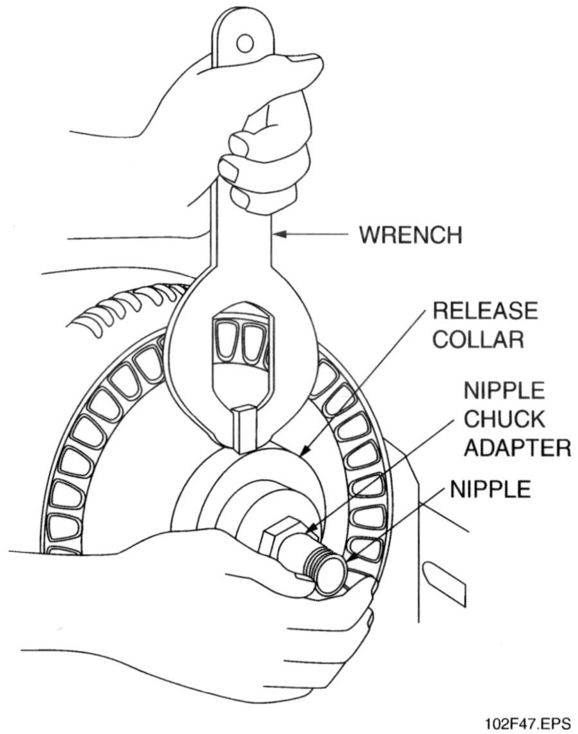

Figure 47 ◆ Tightening nipple chuck.

Step 20 Release the power drive chuck to remove the nipple chuck from the threading machine.

Step 21 Wipe off the nipple chuck, and store it with the rest of the nipple chuck kit.

9.2.0 Threading Pipe Using Geared Threader

A geared threader, or mule, is used to thread pipe from 2½ to 6 inches in diameter. Two sets of dies are used with the geared threaders. One set is used for pipe that is 2½ to 4 inches in diameter, and the other set is used for pipe that is 4 to 6 inches in diameter. Each die set contains five dies that are numbered just like the dies for the universal die head. The geared threader is connected to the pipe threading machine by a universal drive shaft. *Figure 48* shows a geared threader and a universal drive shaft. *Figure 49* shows the proper setup for a geared threader.

Follow these steps to thread pipe using a geared threader:

> **WARNING!**
> A geared threader produces an enormous amount of torque while threading. It can easily tip over the threading machine if the machine is not properly secured. If possible, bolt the threading machine to a steel table and secure the pipe in a vise that is bolted to a steel table.

Step 1 Make sure the correct die head is in place in the threader and that the die has been backed the right distance from the chuck by turning the drive shaft of the threader backwards.

Step 2 Make certain that the pipe to be threaded or cut-grooved is securely fastened to a fixed stand, such as a steel plate table or fixed bench, with approximately 18 inches of pipe clear of the table.

Step 3 Open the locking chuck at the back of the threader by turning the cam.

Step 4 Slip the threader over the pipe until the die head is centered on the pipe and the dies are just resting against the end of the pipe all the way around.

Step 5 Clamp the chuck at the rear of the threader onto the pipe, by spinning the cam wheel.

Step 6 Attach the universal drive shaft from the drive motor to the threader, and tighten the setscrew firmly.

> **CAUTION**
> Make sure all connections are firm and will not come loose while the machine is running. The drive motor is very powerful, and the drive shaft could either come loose or pull the workpiece loose.

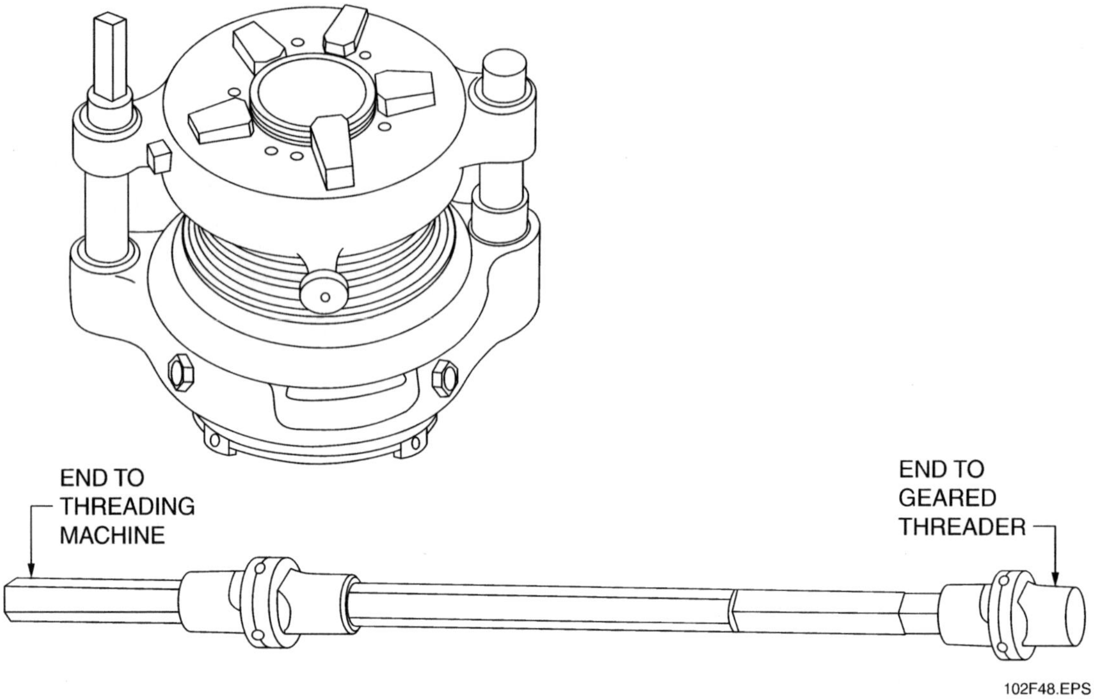

Figure 48 ◆ Geared threader and universal drive shaft.

Step 7 Turn on the automatic oiler or oil the dies manually, and turn on the drive motor in forward gear.

Step 8 Keep the work and dies oiled until the correct length of thread has been cut; then cut the drive motor off and release the universal drive shaft. Open the die and chuck, and remove the work piece.

Step 9 Clean the threader, and return it to storage.

10.0.0 ◆ PORTABLE POWER DRIVES

A portable power drive is a hand-held power unit that can be used for various jobs, such as cutting threads, driving hoists and winches, and operating large valves. An electric motor provides the power to rotate a ring in the end of the tool. Threading dies and other accessories can be inserted into the circular ring, which has grooved spines to lock these accessories into place. *Figure 50* shows portable power drives.

Threading die heads for pipe 2 inches and smaller either fit directly into the power drive or fit into a die head adapter and then into the power drive.

Follow these steps to thread pipe using a portable power drive:

Step 1 Secure the pipe to be threaded in a vise.

Step 2 Identify the size of the pipe to be threaded and insert the appropriate die into the portable power drive.

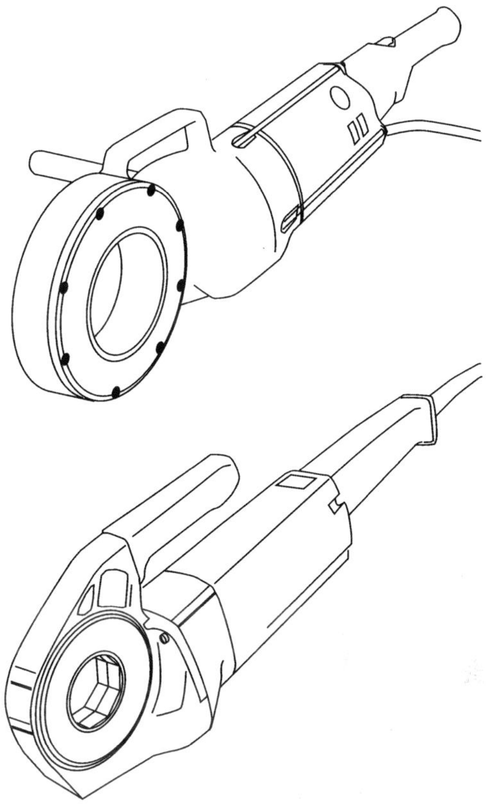

Figure 50 ◆ Portable power drives.

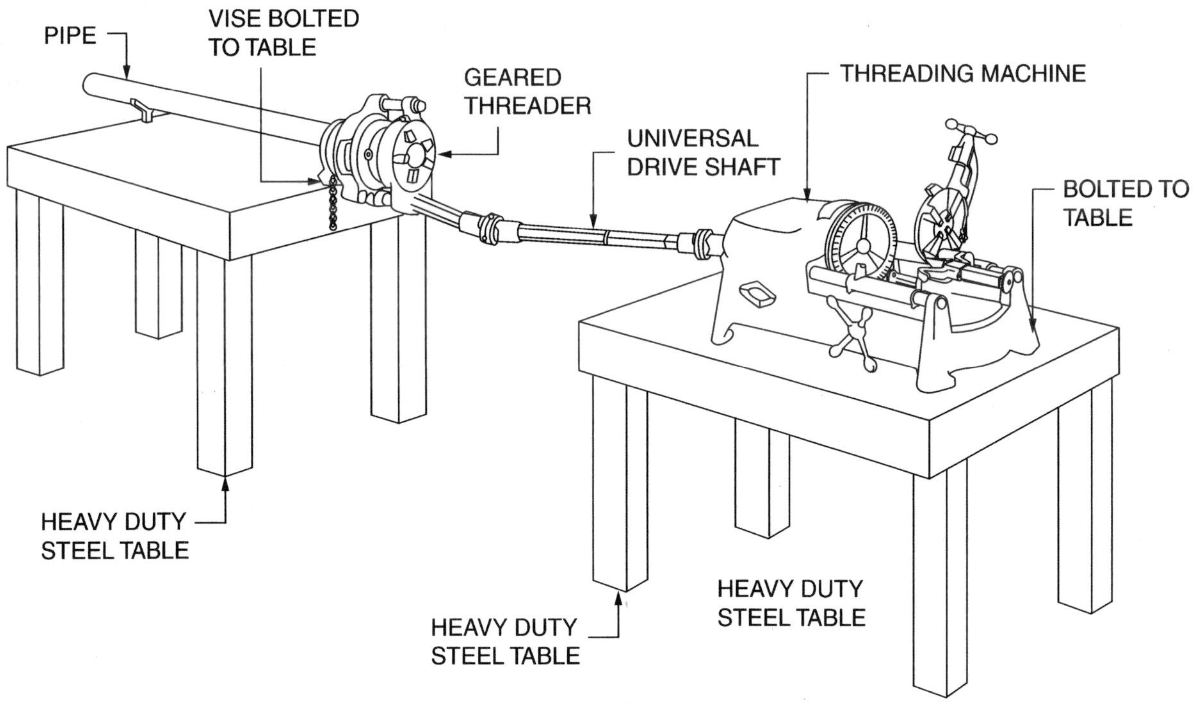

Figure 49 ◆ Proper setup for geared threader.

Step 3 Position the die head onto the end of the pipe with the face of the die stock facing away from the pipe.

> **NOTE**
> The drive unit should be positioned on the end of the pipe with the handle of the drive unit in a nearly vertical position above the pipe.

Step 4 Hold the handle of the drive unit with your right hand and center the die head onto the end of the pipe with your left hand.

Step 5 Pull the handle of the drive unit down sharply with your right hand while pushing the die stock against the end of the pipe with your left hand to make the dies bite into the end of the pipe.

Step 6 Continue to move the handle down to a position where you can straighten your elbow to provide a firm, strong grip on the drive unit. You should lower your right shoulder and balance your weight above the drive unit to resist the threading torque. If you are threading pipe larger than 1 inch, attach a torque arm to support the drive unit at this point. *Figure 51* shows the power drive support arm.

Step 7 Start the power drive, and cut the thread. Be sure to keep an adequate supply of cutting oil on the thread as it is being cut.

Step 8 Stop the power drive as soon as the thread has been cut to the required length.

Step 9 Turn the power switch to the reverse position, and back the dies off the thread.

Step 10 Inspect the threads to make sure they have been cut properly.

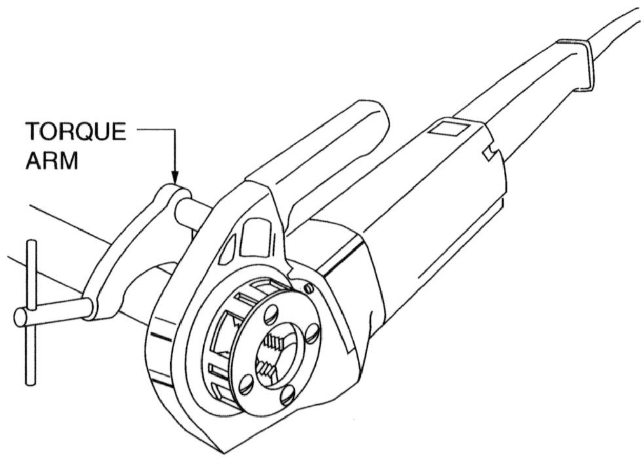

Figure 51 ◆ Power drive support arm.

Review Questions

1. Which of the following tools would be used on a polished steel assembly?
   a. Chain wrench
   b. Pipe wrench
   c. Strap wrench
   d. Spanner

2. The type of wrench designed for use on a specific part is the _____ wrench.
   a. strap
   b. chain
   c. spanner
   d. pipe

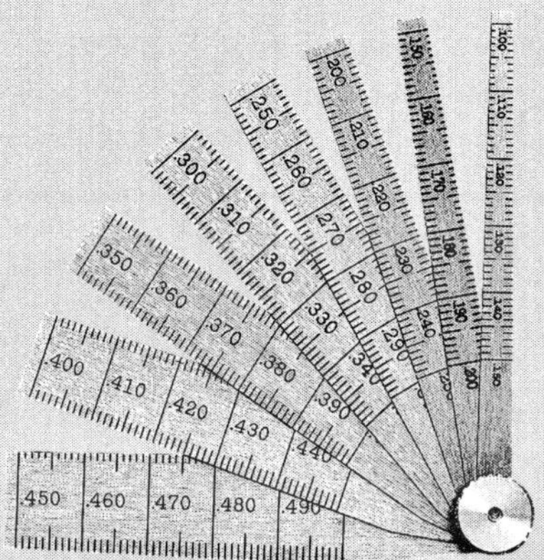

102RQ01.EPS

Figure 1

3. The tool shown in *Figure 1* is a _____.
   a. taper gauge
   b spanner
   c. thread gauge
   d. feeler gauge

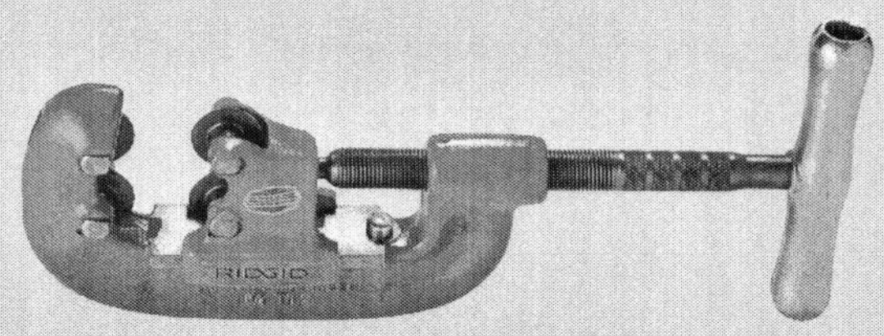

102RQ02.EPS

Figure 2

4. The tool shown in *Figure 2* is commonly used to cut steel pipe.
   a. True
   b. False

5. A putty knife is used to _____.
   a. sharpen tools
   b. remove gaskets, glue, or sealer
   c. measure small openings
   d. measure thread pitch

6. The tool used to line up holes in flanges is a _____.
   a. mallet
   b. sleever bar
   c. feeler gauge
   d. drift pin

7. The primary purpose for a mallet is to drive nails.
   a. True
   b. False

8. The handle of an aviation snip used to cut to the left is _____.
   a. blue
   b. black
   c. green
   d. red

9. A tap is used to _____.
   a. locate hollow spots
   b. cut internal threads
   c. cut external threads
   d. punch holes in metal

10. The tool used to repair damaged threads on bolts or screws is a _____.
    a. plug tap
    b. open adjusting die
    c. rethreading die
    d. reamer

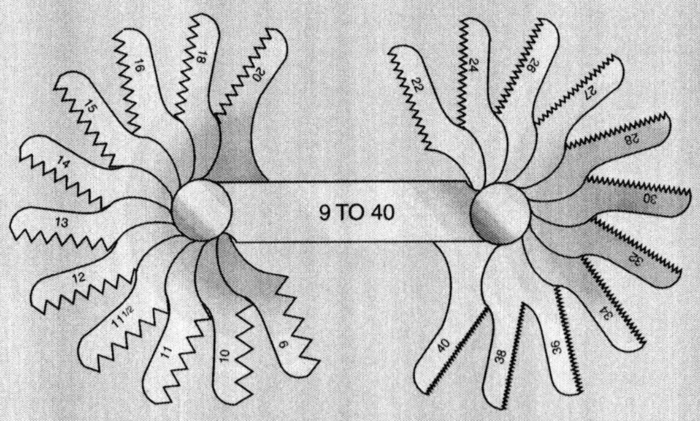

102RQ03.EPS

Figure 3

11. The tool shown in *Figure 3* is a _____.
    a. feeler gauge
    b. thread gauge
    c. packing puller
    d. cylinder hone

12. The tool used to mark lines on metal is a _____.
    a. mechanic's pencil
    b. barrel pin
    c. sheave gauge
    d. scriber

13. A sheave gauge is used to check belt tension.
    a. True
    b. False

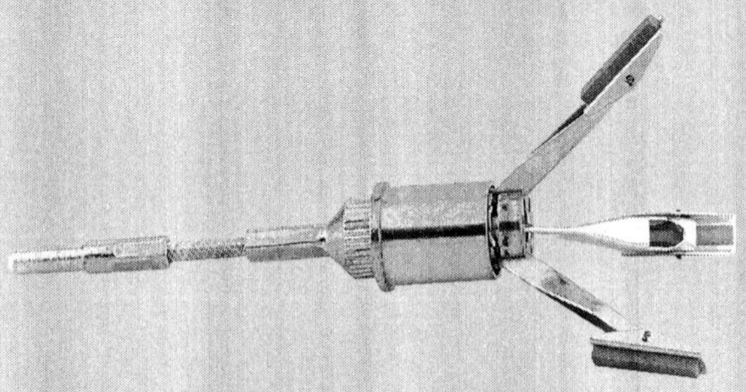

102RQ04.EPS

Figure 4

14. The tool shown in *Figure 4* is used to _____.
    a. pull gears
    b. hone cylinders
    c. remove retaining rings
    d. pull packings

15. When using a cylinder hone, it should be operated at _____ rpm to avoid overheating.
    a. 200 – 400
    b. 300 – 700
    c 500 – 900
    d. 900 – 1,500

16. The tool shown in *Figure 5* is used to _____.
    a. hone cylinders
    b. install bearings
    c. remove retaining rings
    d. pull gears and bearings

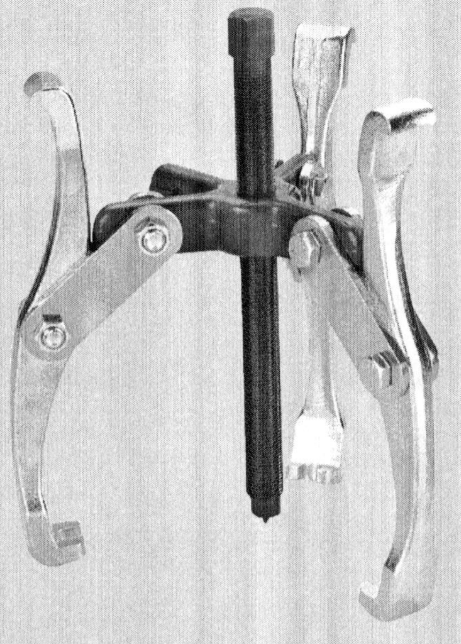

102RQ05.EPS

Figure 5

17. Reamers are used to enlarge holes.
    a. True
    b. False

18. The tool shown in *Figure 6* is used to _____.
    a. remove packings
    b. install or remove retaining rings
    c. cut wire
    d. remove bearings

Review Questions

102RQ06.EPS

Figure 6

19. An easy-out is used to remove bearings.
    a. True
    b. False

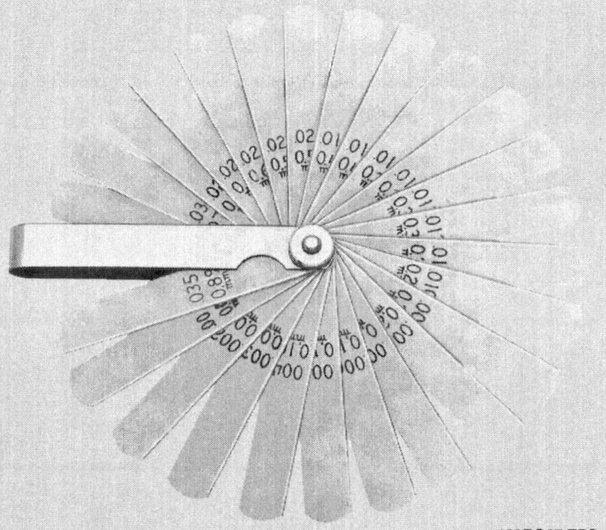

102RQ07.EPS

Figure 7

20. The tool shown in *Figure 7* is used to _____.
    a. measure thread depth
    b. remove packings
    c. check belt tension
    d. measure clearance between parts

21. Do *not* use electric tools in _____ conditions.
    a. dark
    b. noisy
    c. wet
    d. dry

22. Before using an air line check the _____.
    a. pressure rating of the line
    b. color of the line
    c. number of threads on the hose connector
    d. voltage on the line

23. Band saw blades are chosen on the basis of the _____ of the material to be cut.
    a. number of pieces
    b. amount of rust
    c. type and wall thickness
    d. height from the ground

24. When cutting pipe more than 4 inches in diameter with a porta-band, you should _____ the pipe.
    a. cut around
    b. cut straight through
    c. not clamp
    d. cut along the centerline of

25. The operating speed of a grinder should be higher than the wheel speed rating.
    a. True
    b. False

26. A grinding wheel rotates _____.
    a. counterclockwise
    b. clockwise
    c. away from you
    d. toward you

27. A pipe threading machine is used to _____ pipe.
    a. bend, weld, and thread
    b. straighten, weld, and thread
    c. cut, ream, and thread
    d. grind, bend, and thread

28. Threading dies are classified according to _____.
    a. length of the die
    b. number of threads per inch
    c. type of material to be threaded
    d. length of pipe to be threaded

29. A portable power drive can be used to thread pipe 6 inches or more in diameter.
    a. True
    b. False

30. A portable power drive does *not* require cutting oil.
    a. True
    b. False

Summary

An industrial maintenance craftworker must learn how to use a large number of specialized hand and power tools. Your work will include precise equipment layouts, alignments, and assembly and disassembly of many kinds of equipment. Where some crafts may work in inches, industrial maintenance work must be accurate to very small fractions of an inch, and inaccuracies can cause equipment damage and danger to workers. Hand and power tools must be maintained carefully and used correctly, because defective tools can cause a very large problem. In addition, your tools will frequently be quite expensive, and you would do well to take care of them. Power tools, in addition to their expense, can injure you or others very easily if misused or used carelessly.

Notes

Trade Terms Introduced in This Module

Abrasive: A rough material used for sanding, grinding, sharpening, or cutting.

Align: To line up two or more parts.

Bevel: An angle cut or ground on the end of a piece of solid material.

Burr: A sharp, ragged edge produced by cutting sheet metal.

Chamfer: An angle cut or ground only on the edge of a piece of material.

Chuck: The part of a machine that holds a piece of work tightly in the machine. A chuck is normally used only when the pipe or cutter will be rotated.

Component: A single part in a system.

Contamination: To make impure by contact or mixture.

Diameter: The width of a circle. The measurement from side to side.

Die: A tool used to make male threads on a pipe or a bolt.

Flanges: Projecting rims or collars on pipes used to hold them in place, give them strength, or attach them to something else.

Galling: Surface damage on mating, moving metal parts that is caused by friction

Horsepower (hp): A unit of power equal to 745.7 watts or 33,000 foot-pounds per minute.

Pipe fitting: A unit attached to a piping system.

Pitch: The number of threads per inch on bolts, screws, and threaded rods.

Revolutions per minute (rpm): The number of complete revolutions an object will make in one minute.

Sheave: A grooved wheel used as a belt pulley.

Shims: A thin strip of wood or metal used to align parts.

Tolerance: The difference between the allowed maximum and minimum limits of size.

Resources & Acknowledgments

Additional Resources

This module is intended to be a thorough resource for task training. The following reference work is suggested for further study. This is optional material for continued education rather than for task training.

Tools and Their Uses, Latest Edition. Naval Education and Training Program and Development Center. Washington, DC: US Government Printing Offices.

Figure Credits

Ridge Tool Co. (Ridgid®), 102F01 (chain wrench), 102F04, 102F11 (photo), 102F12, 102F39, 102F41, 102F42

Reed Manufacturing Company, 102F01 (strap wrench)

Danaher Tool Group, 102F02 (pin spanner, adjustable hook, face spanner, adjustable face), 102F17, 102F19

KW Automotive America, Inc., 102F02 (flat hook spanner)

The L.S. Starrett Company, 102F03, 102F14

Stanley Proto, 102F05, 102F07 (bottom), 102F23

Topaz Publications, Inc., 102F06 (drift pin), 102F09

Hartville Tool Company, 102F07 (top)

Eclipse Tools, 102F08

Dresser-Rand Company, 102F18

American Seal & Packing, 102F20

Alvord-Polk Incorporated, 102F21

SPX Corp, 102F22

Brownells, 102F24 (tap extractor)

Kastar Hand Tools, 102F25

Porter-Cable Corporation, 102F29

Used with permission of Welding Design & Fabrication, 102F37

CONTREN® LEARNING SERIES — USER UPDATE

NCCER makes every effort to keep these textbooks up-to-date and free of technical errors. We appreciate your help in this process. If you have an idea for improving this textbook, or if you find an error, a typographical mistake, or an inaccuracy in NCCER's Contren® textbooks, please write us, using this form or a photocopy. Be sure to include the exact module number, page number, a detailed description, and the correction, if applicable. Your input will be brought to the attention of the Technical Review Committee. Thank you for your assistance.

Instructors – If you found that additional materials were necessary in order to teach this module effectively, please let us know so that we may include them in the Equipment/Materials list in the Annotated Instructor's Guide.

Write: Product Development and Revision
National Center for Construction Education and Research
3600 NW 43rd St. Bldg G, Gainesville, FL 32606

Fax: 352-334-0932

E-mail: curriculum@nccer.org

Craft ________________ Module Name ________________

Copyright Date ________ Module Number ________ Page Number(s) ________

Description

____

____

____

____

(Optional) Correction

____

____

____

(Optional) Your Name and Address

____

____

____

40103-07

Fasteners and Anchors

40103-07
Fasteners and Anchors

Topics to be presented in this module include:

Overview

Many types of fasteners are available for installing, repairing, and connecting equipment and components. Each is designed for a specific purpose. The industrial maintenance craftworker must be able to recognize various fasteners, identify their intended uses, and install them properly.

You must match the fastener to both the object being secured and the surface or material receiving the fastener. For example, many fasteners can be used for hollow wall anchoring, but not all of these fasteners can secure heavy objects. Using the wrong fastener or installing the right fastener incorrectly can lead to property damage or injury. The project specifications and the manufacturer's installations often provide information on the types of fasteners to use and how they are to be installed. Other times, you will be expected to select the right type of fastener to use in a particular application. Knowing how to do this is an important skill.

Objectives

When you have completed this module, you will be able to do the following:

1. Identify and explain the use of threaded fasteners.
2. Identify and explain the use of non-threaded fasteners.
3. Identify and explain the use of anchors.
4. Select the correct fasteners and anchors for given applications.
5. Install fasteners and anchors.

Trade Terms

American Society for Testing and Materials (ASTM) International
Clearance
Foot pounds (ft lbs)
Inch pounds (in lbs)
Key
Keyway
Nominal size
Society of Automotive Engineers (SAE)
Thread classes
Thread identification
Thread standards
Tolerance
Torque
Unified National Coarse (UNC) thread
Unified National Extra Fine (UNEF) thread
Unified National Fine (UNF) thread

Required Trainee Materials

1. Pencil and paper
2. Appropriate personal protective equipment

Prerequisites

Before you begin this module, it is recommended that you successfully complete *Core Curriculum*; and *Industrial Maintenance E & I Technician Level One*, Modules 40101-07 and 40102-07.

This course map shows all of the modules in the first level of the *Industrial Maintenance E & I Technician* curriculum. The suggested training order begins at the bottom and proceeds up. Skill levels increase as you advance on the course map. The local Training Program Sponsor may adjust the training order.

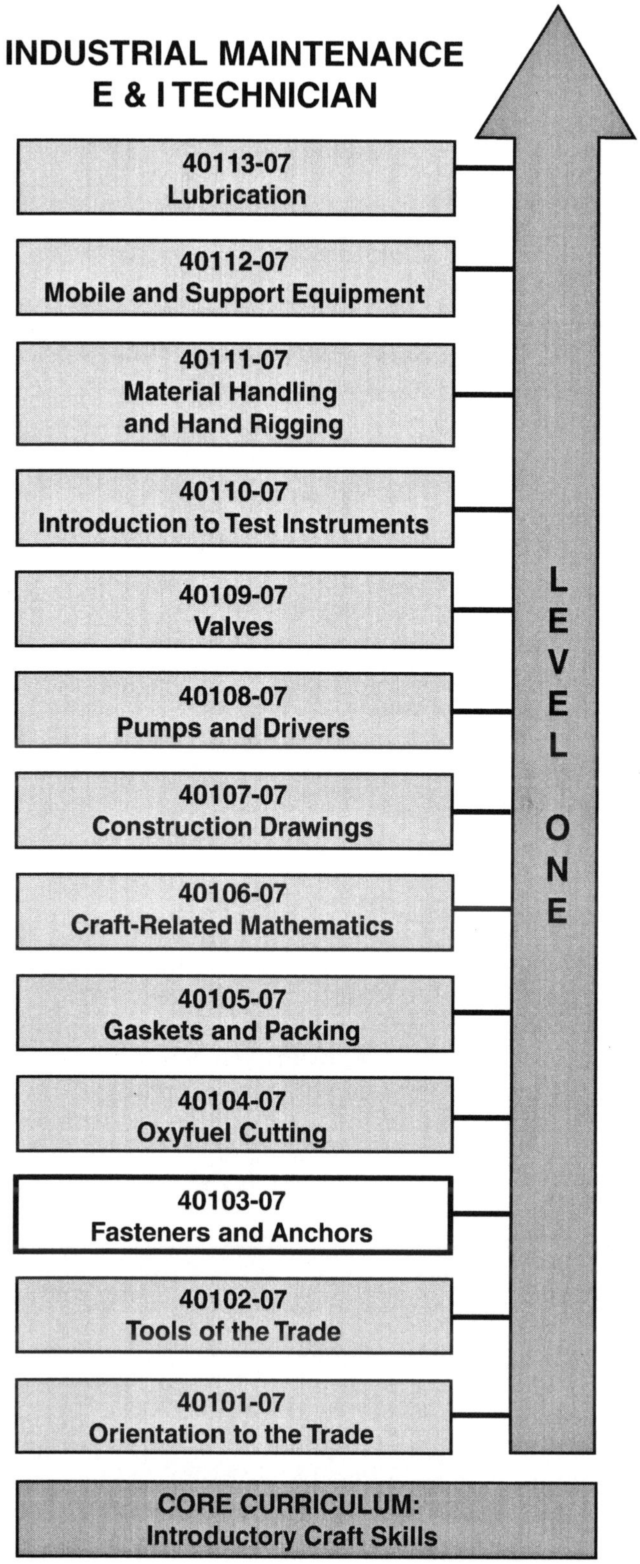

103CMAP.EPS

1.0.0 ◆ INTRODUCTION

Fasteners are used to assemble and install many different types of equipment, parts, and materials. Fasteners include screws, bolts, nuts, pins, clamps, retainers, tie wraps, rivets, and **keys**. Fasteners are used extensively in the industrial maintenance craft. You need to be familiar with the many different types of fasteners in order to identify, select, and properly install the correct fastener for a specific application.

The two primary categories of fasteners are threaded fasteners and non-threaded fasteners. Within each of these two categories, there are numerous different types and sizes of fasteners. Each type of fastener is designed for a specific application. The kind of fastener used for a job may be listed in the project specifications, or you may have to select an appropriate fastener.

Failure of fasteners can result in a number of different problems. To perform quality work, it is important to use the correct type and size of fastener for the particular job. It is equally important that the fastener be installed properly.

In this module, you will be introduced to fasteners commonly used in industrial maintenance work.

2.0.0 ◆ THREADED FASTENERS

Threaded fasteners are the most commonly used type of fastener. Many threaded fasteners are assembled with nuts and washers. The following sections describe standard threads used on threaded fasteners, as well as different types of bolts, screws, nuts, and washers. *Figure 1* shows several types of threaded fasteners.

2.1.0 Thread Standards

There are many different types of threads used for manufacturing fasteners. The different types of threads are designed to be used for different jobs. Threads used on fasteners are manufactured to industry-established standards for uniformity. The most common **thread standard** is the Unified standard, sometimes known as the American standard. Unified standards are used to establish thread series and **thread classes**.

2.1.1 Thread Series

Unified standards are established for three series of threads, depending on the number of threads per inch for a certain diameter of fastener. These three series are as follows:

- ***Unified National Coarse (UNC) thread*** – Used for bolts, screws, nuts, and other general purposes. Fasteners with UNC threads are commonly used for rapid assembly or disassembly of parts and where corrosion or slight damage may occur.
- ***Unified National Fine (UNF) thread*** – Used for bolts, screws, nuts, and other applications where a finer thread for a tighter fit is desired.
- ***Unified National Extra Fine (UNEF) thread*** – Used on thin-walled tubes, nuts, ferrules, and couplings.

2.1.2 Thread Classes

The Unified standards also establish thread classes. Classes 1A, 2A, and 3A apply to external threads only. Classes 1B, 2B, and 3B apply to internal

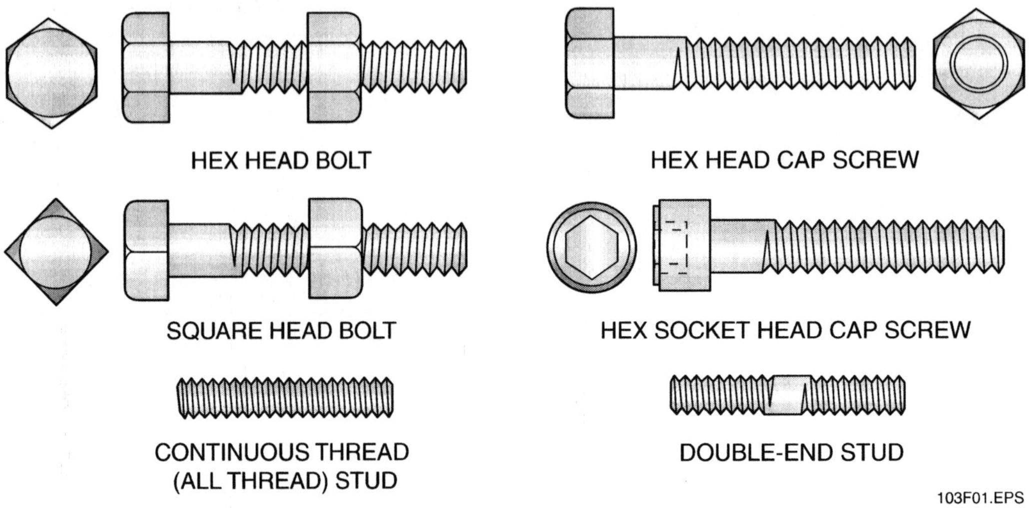

Figure 1 ◆ Threaded fasteners.

threads only. Thread classes are distinguished from each other by the degrees of **tolerance** permitted. Classes 3A and 3B allow a minimum **clearance** and Classes 1A and 1B allow a maximum clearance.

Classes 2A and 2B are the most commonly used. Classes 3A and 3B are used when close tolerances are needed. Classes 1A and 1B are used where quick and easy assembly is needed and a large tolerance is acceptable.

2.1.3 Thread Identification

Thread identification is done using a standard method. *Figure 2* shows how screw threads are designated for a common fastener.

- ***Nominal size*** – The nominal size is the approximate diameter of the fastener.
- *Number of threads per inch (TPI)* – The TPI is standard for all diameters.
- *Thread series symbol* – This symbol indicates the Unified standard thread type (UNC, UNF, or UNEF).
- *Thread class symbol* – This symbol indicates the closeness of fit between the bolt threads and nut threads.
- *Left-hand thread symbol* – This symbol is indicated by the letters LH. Unless threads are specified with the LH symbol, the threads are right-hand threads.

2.1.4 Thread Design

There are several types of thread design. The thread design used depends on the purpose of the fastener. Power transmission threads are special threads that are used to move machine parts for adjusting, setting, and transmitting power. Three common transmission threads are the square thread, acme thread, and buttress thread. *Figure 3* shows an enlargement of the three types.

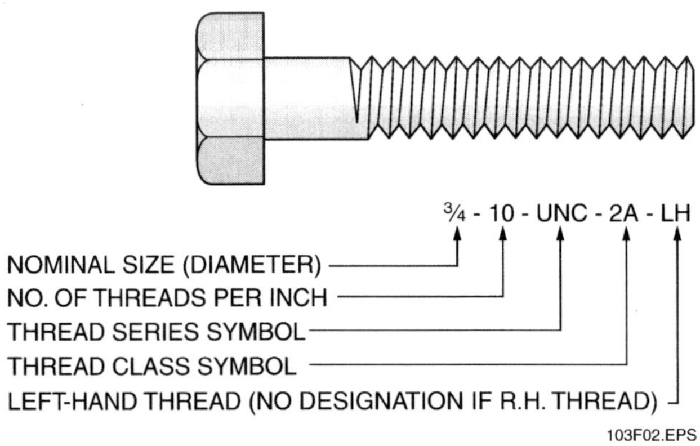

Figure 2 ◆ Screw thread designations.

The buttress thread has one side of the thread cut square and the other side cut at a slant. It has great strength along the thread axis in one direction only. This thread form is used to screw thin, tubular parts together. *Figure 3A* shows an enlargement of a buttress thread.

The square thread is the strongest and most useful of all thread forms. It is also the most difficult to make because its parallel sides must be accurately machined. *Figure 3B* shows an enlargement of a square thread.

The acme thread has replaced the square thread in most cases because it is easier to machine. *Figure 3C* shows an enlargement of an acme thread.

2.1.5 Grade Markings

Special markings on the head of a bolt or screw can be used to determine the quality of the fastener. The **Society of Automotive Engineers (SAE)** and the **American Society for Testing and**

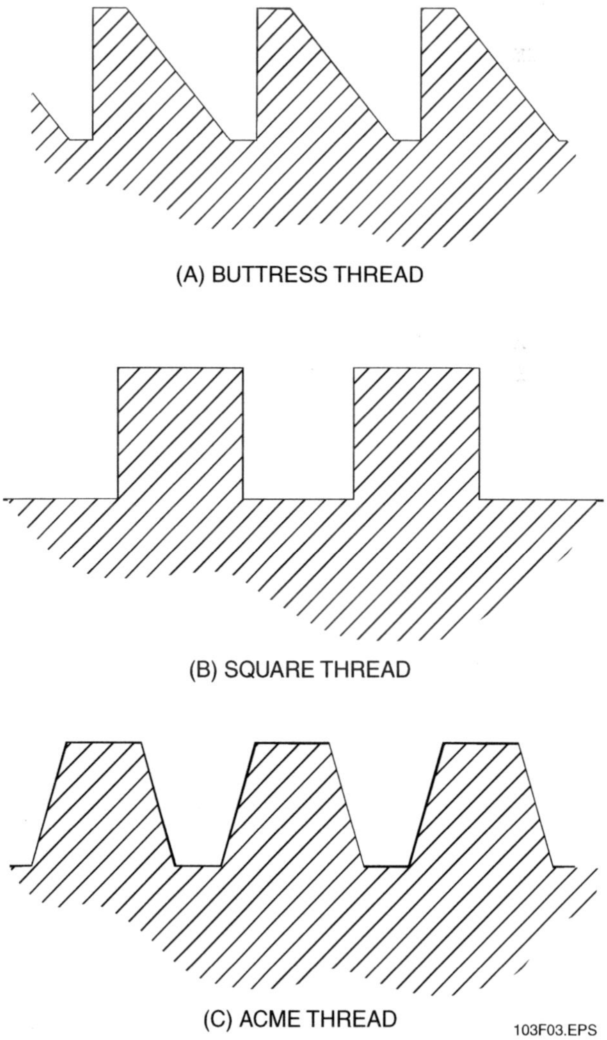

Figure 3 ◆ Thread types.

Materials (ASTM) International have developed the standards for these markings. These grade or line markings for steel bolts and screws are shown in *Figure 4*.

Generally, the higher-quality steel fasteners have a greater number of marks on the head. If the head is unmarked, the fastener is usually considered to be made of mild steel, which is steel with a low carbon content.

2.2.0 Bolt and Screw Types

Bolts and screws are made in many different sizes and shapes and from a variety of materials. They are usually identified by the head type or other special characteristics. The following sections describe several different types of bolts and screws.

ASTM AND SAE GRADE MARKINGS FOR STEEL BOLTS & SCREWS

GRADE MARKING	SPECIFICATION	MATERIAL
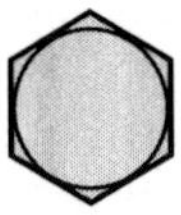	SAE-GRADE 0	STEEL
	SAE-GRADE 1 ASTM-A 307	LOW CARBON STEEL
	SAE-GRADE 2	LOW CARBON STEEL
	SAE-GRADE 3	MEDIUM CARBON STEEL, COLD WORKED
	SAE-GRADE 5	MEDIUM CARBON STEEL, QUENCHED AND TEMPERED
	ASTM-A 449	
	ASTM-A 325	MEDIUM CARBON STEEL, QUENCHED AND TEMPERED
	ASTM-A 354 GRADE BB	LOW ALLOY STEEL, QUENCHED AND TEMPERED
	ASTM-A 354 GRADE BC	LOW ALLOY STEEL, QUENCHED AND TEMPERED
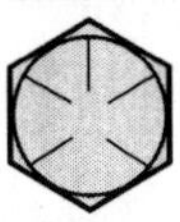	SAE-GRADE 7	MEDIUM CARBON ALLOY STEEL, QUENCHED AND TEMPERED ROLL THREADED AFTER HEAT TREATMENT
	SAE-GRADE 8	MEDIUM CARBON ALLOY STEEL, QUENCHED AND TEMPERED
	ASTM-A 354 GRADE BD	ALLOY STEEL, QUENCHED AND TEMPERED
	ASTM-A 490	ALLOY STEEL, QUENCHED AND TEMPERED

ASTM SPECIFICATIONS

A 307 – LOW CARBON STEEL EXTERNALLY AND INTERNALLY THREADED STANDARD FASTENERS.
A 325 – HIGH STRENGTH STEEL BOLTS FOR STRUCTURAL STEEL JOINTS, INCLUDING SUITABLE NUTS AND PLAIN HARDENED WASHERS.
A 449 – QUENCHED AND TEMPERED STEEL BOLTS AND STUDS.
A 354 – QUENCHED AND TEMPERED ALLOY STEEL BOLTS AND STUDS WITH SUITABLE NUTS.
A 490 – HIGH STRENGTH ALLOY STEEL BOLTS FOR STRUCTURAL STEEL JOINTS, INCLUDING SUITABLE NUTS AND PLAIN HARDENED WASHERS.

SAE SPECIFICATION

J 429 – MECHANICAL AND QUALITY REQUIREMENTS FOR THREADED FASTENERS.

103F04.EPS

Figure 4 ◆ Grade markings for steel bolts and screws.

2.2.1 Machine Screws

Machine screws (*Figure 5*) are used for general assembly work. They come in a variety of types, with slotted or recessed heads. Machine screws are generally available in diameters ranging from 0 (0.060 inch) to ½ inch (0.500 inch). The length of machine screws typically varies from ⅛ inch to 3 inches. Machine screws are also manufactured in metric sizes.

As shown, the heads of machine screws are made in different shapes and with slots made to fit various kinds of manual and powered screwdrivers. Flat-head screws are used in a countersunk hole and are tightened so that the head is flush with the surface. Oval-head screws are also used in a countersunk hole in applications where a more decorative finish is desired. Pan and round-head screws are general-use fastening screws. Fillister, hex socket, and TORX® socket screws are typically used in confined space applications on machined assemblies that need a finished appearance. They are often installed in a recessed hole. Truss screws are low-profile screws generally used without a washer. To prevent damage when tightening and removing machine screws, regardless of head type, make sure to use a screwdriver or power tool bit with the proper tip to drive them.

2.2.2 Machine Bolts

Machine bolts are generally used to assemble parts where close tolerances are not required. Machine bolts have square or hexagonal heads and are generally available in diameters ranging from ¼ inch to 3 inches. The length of machine bolts varies from ½ inch to 30 inches. Nuts used with machine bolts are similar in shape to the bolt heads. The nuts are usually purchased at the same time as the bolts. *Figure 6* shows two different types of machine bolts.

2.2.3 Cap Screws

Cap screws are often used on high-quality assemblies requiring a finished appearance. The cap screw passes through a clearance hole in one of the assembly parts and screws into a threaded hole in the other part. The clamping action occurs by tightening the cap screw.

Cap screws are made to close tolerances and are provided with a machined or semi-finished bearing surface under the head. They are normally made in coarse and fine thread series and in diameters from ¼ inch to 2 inches. Lengths may range from ⅜ inch to 10 inches. Metric sizes are also available. *Figure 7* shows typical cap screws.

2.2.4 Setscrews

Heat-treated steel is normally used to make setscrews. Common uses of setscrews include preventing pulleys from slipping on shafts, holding collars in place on shafts, and holding shafts in place. The head style and point style are typically used to classify setscrews. *Figure 8* shows several types of screw heads and point styles.

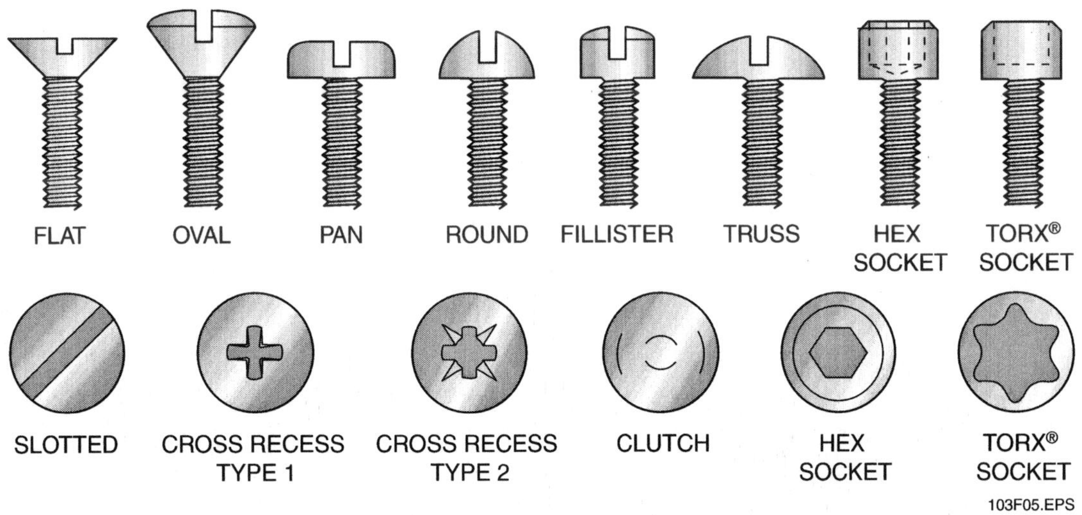

Figure 5 ◆ Machine screws.

correction has been made to Quark file and final PDF on 7/22/10 by LW; will need to upload corrected page for next reprint

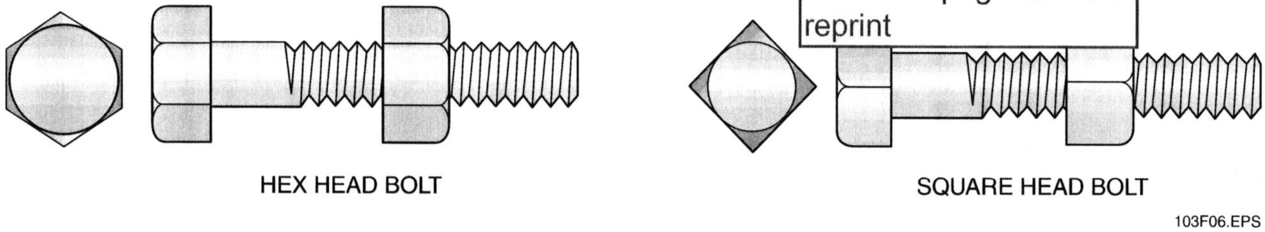

Figure 6 ◆ Machine bolts.

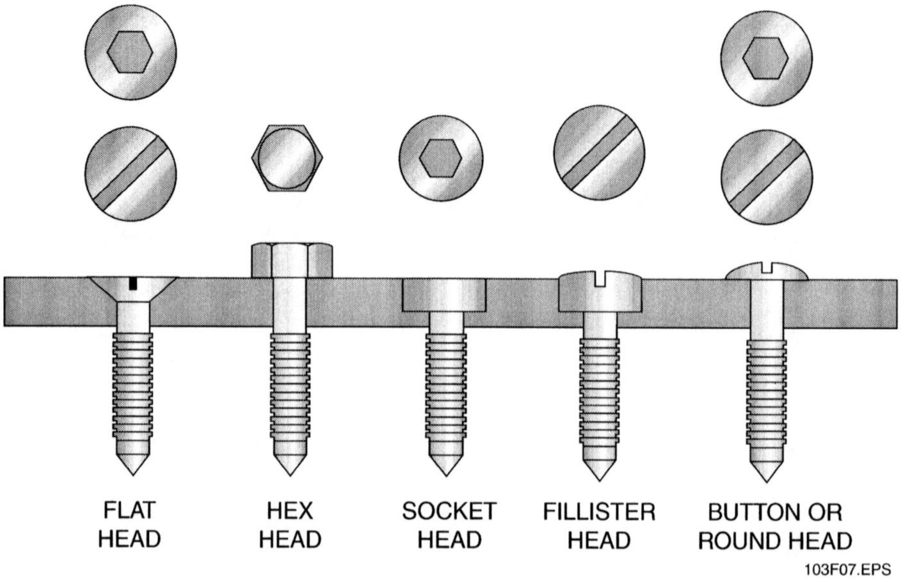

Figure 7 ◆ Cap screws.

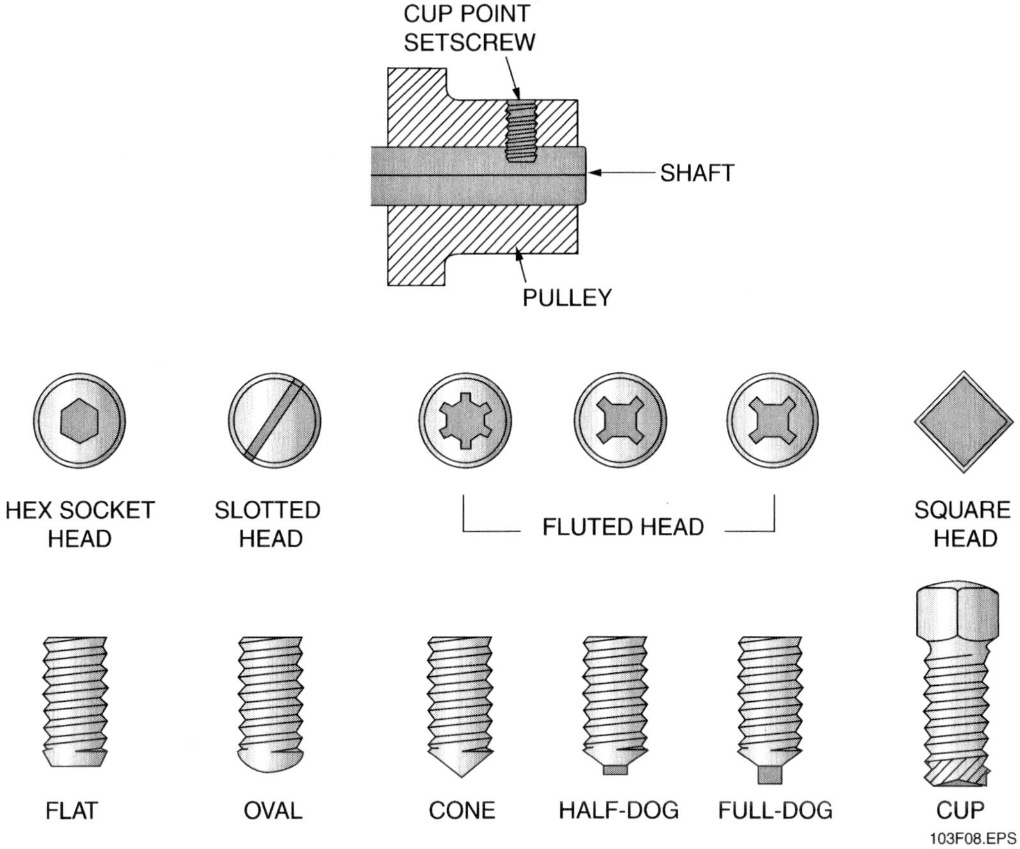

Figure 8 ◆ Setscrews.

2.2.5 Stud Bolts

Stud bolts, also known as threaded rods (*Figure 9*), are headless bolts that are threaded over the entire length of the bolt or for a length on both ends of the bolt. One end of the stud bolt is screwed into a tapped hole. The part to be clamped is placed over the remaining portion of the stud, and a nut and washer are screwed on to clamp the two parts together. Other stud bolts have machine-screw threads on one end and lag-screw threads on the other so that they can be screwed into wood.

Stud bolts are used for several purposes, including holding together inspection covers on equipment and bearing caps.

CAUTION

When installing stud bolts, the thread engagement must be equal to or greater than the stud diameter. Check the equipment manufacturer's literature for stud penetration and projection requirements.

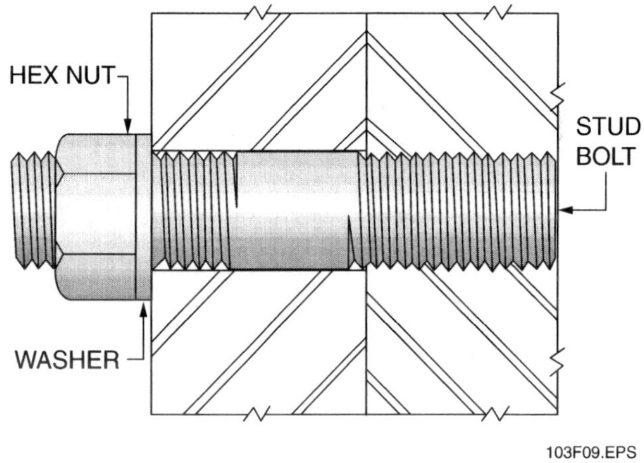

Figure 9 ◆ Stud bolt.

2.3.0 Nuts

Most nuts used with threaded fasteners are hexagonal or square. They are generally used with bolts having the same shape head. *Figure 10* shows several different types of nuts that are used with threaded fasteners.

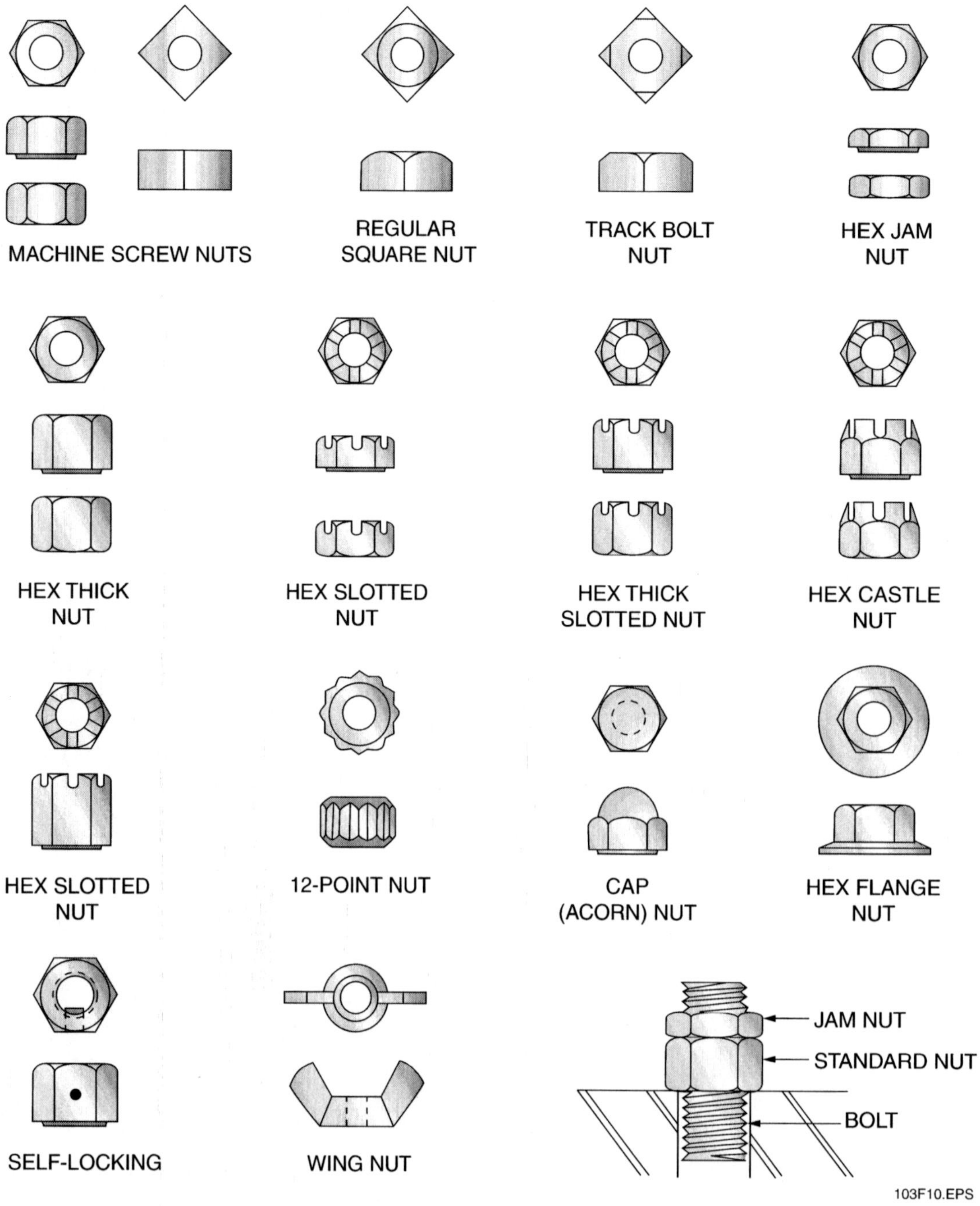

Figure 10 ◆ Nuts.

Nuts are typically classified as regular, semi-finished, or finished. The only machining done on regular nuts is to the threads. In addition to the threads, semi-finished nuts are also machined on the bearing face. Machining the bearing face makes a truer surface for fitting the washer. The only difference between semi-finished and finished nuts is that finished nuts are made to closer tolerances.

The standard machine screw nut has a regular finish. Regular and semi-finished nuts are shown in *Figure 11*.

2.3.1 Jam Nuts

A jam nut is used to lock a standard nut in place. A jam nut is a thin nut installed on top of the standard nut. *Figure 12* shows an example of a jam nut installation. Note that a regular nut can also be used as a jam nut.

2.3.2 Castellated, Slotted, and Self-Locking Nuts

Castellated (castle) and slotted nuts are slotted across the flat part of the nut. They are used with specially manufactured bolts in applications where little or no loosening of the fastener can be tolerated. After the nut has been tightened, a cotter pin is fitted in through a hole in the bolt and one set of slots in the nut. The cotter pin keeps the nut from loosening under working conditions.

Self-locking nuts are also used in many applications where loosening of the fastener cannot be tolerated. Self-locking nuts are designed with nylon inserts, or they are deliberately deformed in such a manner that they cannot work loose. An advantage of self-locking nuts is that no hole in the bolt is needed. *Figure 13* shows typical castellated, slotted, and self-locking nuts.

2.3.3 Acorn Nuts

When appearance is important, or exposed sharp thread edges on the fastener must be avoided, acorn (cap) nuts are used. The acorn nut tightens on the bolt and covers the ends of the threads. The tightening capability of an acorn nut is limited by the depth of the nut. *Figure 14* shows a typical acorn nut.

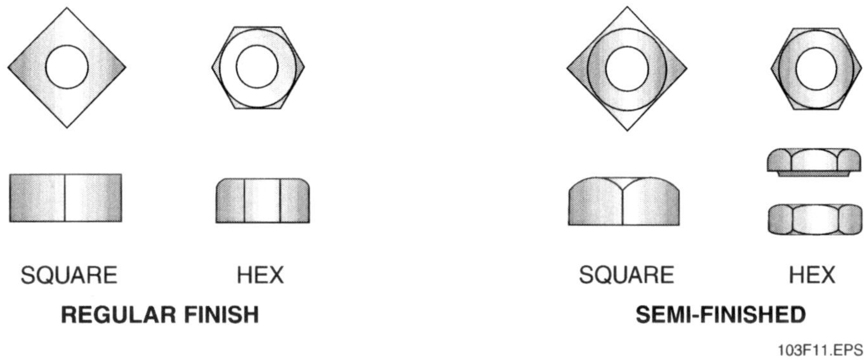

Figure 11 ◆ Nut finishes.

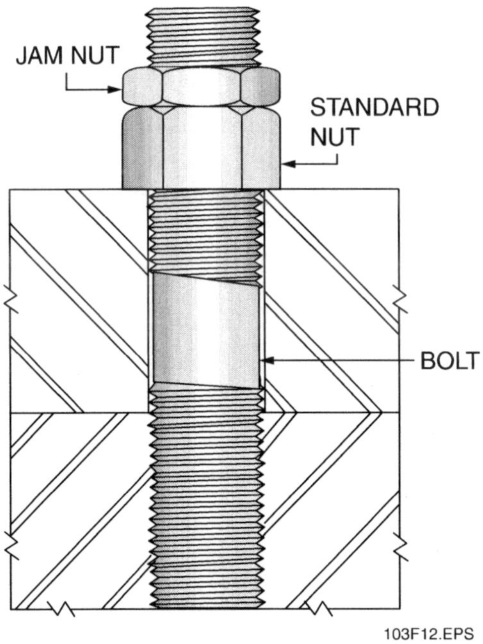

Figure 12 ◆ Jam nut.

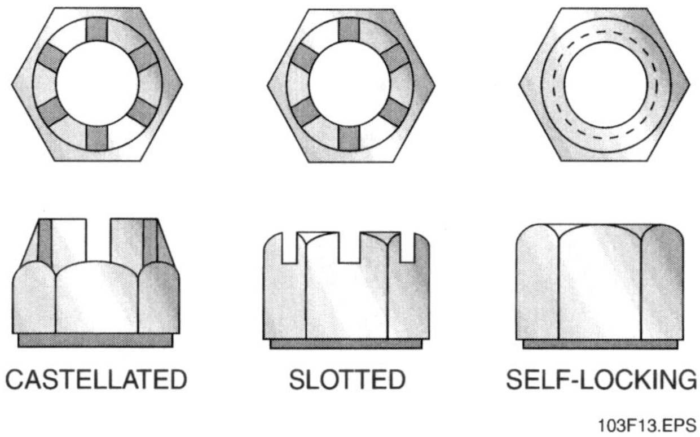

Figure 13 ◆ Castellated, slotted, and self-locking nuts.

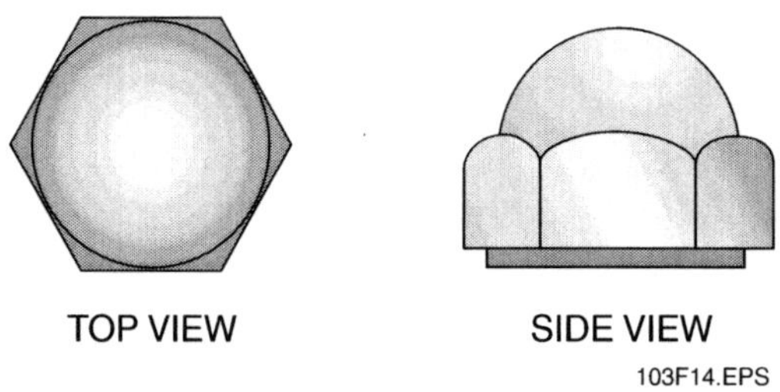

Figure 14 ◆ Acorn nut.

2.3.4 Wing Nuts

Wing nuts are designed to allow rapid loosening and tightening of the fastener without the need for a wrench. They are used in applications where limited **torque** is required and where frequent adjustments and service are necessary. *Figure 15* shows a typical wing nut.

> **NOTE**
> Wing nuts should be used for applications where hand tightening is sufficient.

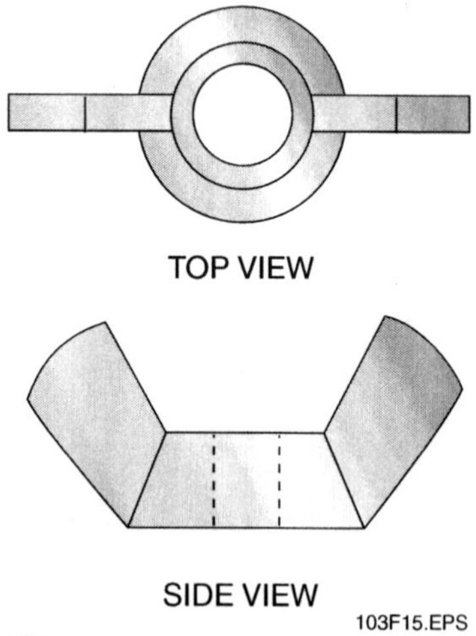

Figure 15 ◆ Wing nut.

2.4.0 Washers

There are several different types and sizes of washers. They fit over a bolt or screw to provide an enlarged surface for bolt heads and nuts. Washers also serve to distribute the fastener load over a larger area and to prevent marring of the surfaces. Standard washers are made in light, medium, heavy-duty, and extra heavy-duty series. *Figure 16* shows different types of washers.

2.4.1 Lock Washers

Lock washers are designed to keep bolts or nuts from working loose. There are various types of lock washers for different applications:

- *Split-ring* – Commonly used with bolts and cap screws.
- *External* – Used for the greatest resistance.
- *Internal* – Used with small screws.
- *Internal-external* – Used for oversized mounting holes.
- *Countersunk* – Used with flat or oval-head screws.

2.4.2 Flat and Fender Washers

Flat washers are used under bolts or nuts to spread the load over a larger area and protect the surface. Common flat washers are made to fit bolt or screw sizes ranging from No. 6 up to 1 inch with outside diameters ranging from ⅜ inch to 2 inches, respectively.

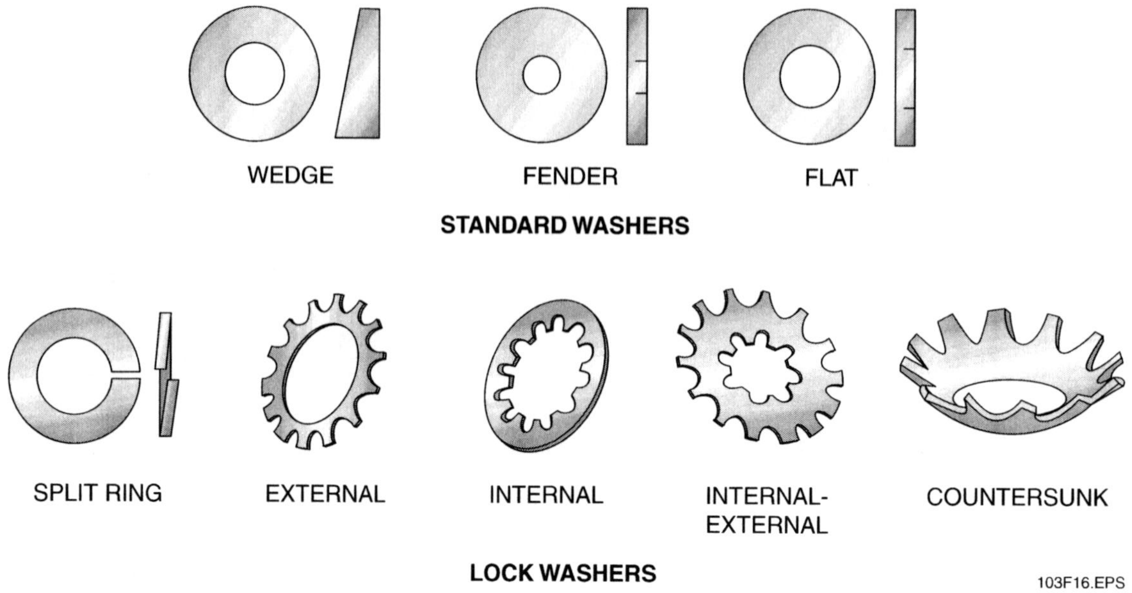

Figure 16 ◆ Washers.

NOTE
The threads of the bolt or screw should have minimal clearance from the hole in the washer.

Fender washers are wide-surfaced washers made to bridge oversized holes or other wide clearances to keep bolts or nuts from pulling through the material being fastened. They are flat washers that have a larger diameter and surface area than regular washers. They may also be thinner than regular washers. Fender washers are typically made to fit bolt or screw sizes ranging from 3⁄16 inch to ½ inch, with outside diameters ranging from ¾ inch to 2 inches, respectively.

NOTE
A nail apron is handy for carrying a large amount of fasteners.

2.5.0 Installing Threaded Fasteners

Different types of fasteners require different installation techniques. However, all installations require knowing the proper installation methods, tightening sequence, and torque specifications for the type of fastener being used. Some bolts and nuts require that special safety wires or pins be installed to keep them from working loose.

Most fastener manufacturers provide charts that specify the size hole that should be drilled into the base material for use with each of their products (*Table 1*). The charts typically show the proper size drill bit to use if it is necessary to first drill and tap holes for use with machine bolts, screws, or other threaded fasteners. They also show the proper size drill to use for drilling pilot holes used with metal and wood screws. (Various kinds of screws are described in detail later in this module.)

To properly tighten a threaded fastener, two primary factors must be considered:

- The strength of the fastener material
- The degree to which the fastener is tightened

A torque wrench is used to control the degree of tightness. The torque wrench measures how much a fastener is being tightened. Torque is the turning force applied to the fastener. Torque is normally expressed in **inch pounds (in lbs)** or **foot pounds (ft lbs)**. A one-pound force applied to a wrench that is 1 foot long exerts 1 foot pound, or 12 inch pounds, of torque. The torque reading is shown on the indicator on the torque wrench as the fastener is being tightened.

NOTE
Torquing screwdrivers are calibrated in inch pounds.

Different types of bolts, nuts, and screws are torqued to different values, depending on the application. Always check the project specifications and the manufacturer's manual to determine the proper torque for a particular type of fastener. *Figure 17* shows selected torque values for various graded steel bolts.

NOTE
Steel fasteners are often tightened using a torque wrench. Whenever possible, apply force to the torque wrench by pulling rather than pushing. This reduces the chance of injury to the fingers or knuckles should the wrench slip. Be sure to grip the handle. Do not use extensions.

The following general procedure can be used to install threaded fasteners in a variety of applications.

CAUTION
When installing threaded fasteners for a specific job, make sure to check all installation requirements. Except in special cases, do not use lubricants such as anti-seize compounds on bolts that will be torqued, as the oil may affect the torque. Manufacturer's specifications will tell you whether to use lubricants or not.

Table 1 Fastener Hole Guide Chart

DRILL THIS SIZE HOLE		To Tap for This Size Bolt or Screw	For this Size Wood Screw Pilot in Hard Wood
Drill Size	Dec. Equiv.		
60	0.0400		
59	0.0410		
58	0.0420		
57	0.0430		
56	0.0465	0 × 80	
3/64	0.0469		
55	0.0520		
54	0.0550	1 × 56	No. 3
53	0.0595	1 × 64-72	
1/16	0.0625		
52	0.0635		No. 4
51	0.0670		
50	0.0700	2 × 56-64	
49	0.0730		No. 5
48	0.0760		
5/64	0.0781		
47	0.0785	3 × 48	No. 6
46	0.0810		
45	0.0820	3 × 56	
44	0.0860	4 × 36	No. 7
43	0.0890	4 × 40	
42	0.0935	4 × 48	
3/32	0.0937		
41	0.0960		
40	0.0980	5 × 36	No. 8
39	0.0995		
38	0.1015	5 × 40	
37	0.1040	5 × 44	No. 9
36	0.1069		
7/64	0.1094		
35	0.1100	6 × 32	
34	0.1110	6 × 36	
33	0.1130	6 × 40	No. 10
32	0.1160		
31	0.1200		No. 11
1/8	0.1250	7 × 36	
30	0.1285	8 × 30	No. 12
29	0.1360	8 × 32-36	
28	0.1405	8 × 40	
27	0.1440	9 × 30	
26	0.1470	3/16 × 24	
25	0.1495	10 × 24	No. 14
24	0.1520		
23	0.1540	10 × 28	
5/32	0.1562		
22	0.1570	10 × 30	
21	0.1590	10 × 32	
20	0.1610	3/16 × 32	
19	0.1660		
18	0.1695		No. 16
11/64	0.1719		
17	0.1730		
16	0.1770	12 × 24	
15	0.1800		
14	0.1820	12 × 28	
13	0.1850	12 × 32	No. 18
3/16	0.1875		
12	0.1890		

DRILL THIS SIZE HOLE		To Tap for This Size Bolt or Screw	For this Size Wood Screw Pilot in Hard Wood
Drill Size	Dec. Equiv.		
11	0.1910		
10	0.1935	15 × 20	
9	0.1960		
8	0.1990		
7	0.2010	1/4 × 20	
13/64	0.2031		
6	0.2040		
5	0.2055		
4	0.2090	1/4 × 24	No. 20
3	0.2130	1/4 × 28	
7/32	0.2187	1/4 × 32	
2	0.2210		
1	0.2280		No. 24
A	0.2340		
15/64	0.2344		
B	0.2380		
C	0.2420		
D	0.2460		
1/4	0.2500		

DRILL THIS SIZE HOLE		To Tap for This Size Bolt or Screw
Drill Size	Dec. Equiv.	
E	0.2500	
F	0.2570	5/16 × 18
G	0.2610	
17/64	0.2656	5/16 × 18
H	0.2660	
I	0.2720	
J	0.2770	5/16 × 24-32*
K	0.2810	
9/32	0.2812	5/16 × 24-32*
L	0.2900	
M	0.2950	
19/64	0.2969	
N	0.3020	
5/16	0.3125	3/8* × 16-1/8* P
O	0.3160	
P	0.3230	
21/64	0.3281	3/8 × 20-24
Q	0.3332	
R	0.3390	
11/32	0.3437	
S	0.3480	
T	0.3580	
23/64	0.3594	
U	0.3680	
3/8	0.3750	7/16 × 14
V	0.3770	
W	0.3860	
25/64	0.3906	7/16 × 14
X	0.3970	
Y	0.4040	
13/32	0.4062	
Z	0.4130	
27/64	0.4219	1/2 × 12-13
7/16	0.4375	1/4* Pipe
29/64	0.4531	1/2 × 20-24
15/32	0.4687	1/2 × 27
31/64	0.4844	9/16 × 12
1/2	0.5000	

* All tap drill sizes are for 75% full thread except asterisked sizes which are 60% full thread.

103T01.EPS

Step 1 Select the proper bolts or screws for the job.

Step 2 Check for damaged or dirty internal and external threads. Replace the fastener if the threads are damaged.

Step 3 Clean the bolt or screw threads. If the fastener is being installed in a threaded hole, inspect the hole threads for damage. If the threads are damaged, the hole should be retapped. Be sure to use the correct size tap for the fastener being used. Tap drill size charts are used to select the proper size drill and tap. *Table 2* shows an example of a tap drill size chart.

Step 4 Insert the bolts through the predrilled holes and tighten the nuts by hand. Or, insert the screws through the holes and start the threads by hand.

> **NOTE**
> Turn the nuts or screws several turns by hand and check for cross threading.

Step 5 Tighten the fasteners in increments as specified by the manufacturer. Follow the proper tightening sequence for the bolt pattern and fastener type before reaching the final torque. *Figure 18* shows the typical tightening sequence for flanges.

> **CAUTION**
> Failure to follow the specified sequence, or over-tightening the bolts, can damage the fasteners or distort the connection. This requirement applies to steel, and not to soft metal.

TORQUE IN FOOT POUNDS

FASTENER DIAMETER	THREADS PER INCH	MILD STEEL	STAINLESS STEEL 18-8	ALLOY STEEL
1/4	20	4	6	8
5/16	18	8	11	16
3/8	16	12	18	24
7/16	14	20	32	40
1/2	13	30	43	60
5/8	11	60	92	120
3/4	10	100	128	200
7/8	9	160	180	320
1	8	245	285	490

SUGGESTED TORQUE VALUES FOR GRADED STEEL BOLTS

GRADE		SAE 1 OR 2	SAE 5	SAE 6	SAE 8
TENSILE STRENGTH		64,000 PSI	105,000 PSI	130,000 PSI	150,000 PSI
GRADE MARK		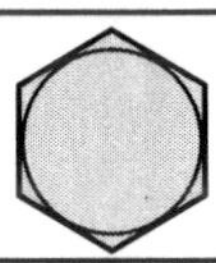		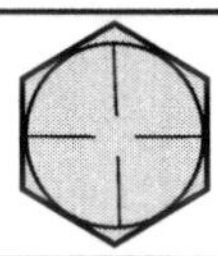	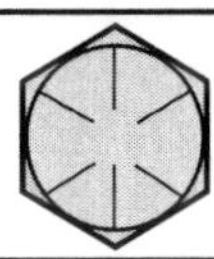
BOLT DIAMETER	THREADS PER INCH	FOOT POUNDS TORQUE			
1/4	20	5	7	10	10
5/16	18	9	14	19	22
3/8	16	15	25	34	37
7/16	14	24	40	55	60
1/2	13	37	60	85	92
9/16	12	53	88	120	132
5/8	11	74	120	169	180
3/4	10	120	200	280	296
7/8	9	190	302	440	473
1	8	282	466	660	714

103F17.EPS

Figure 17 ◆ Torque value chart.

Table 2 Tap Drill Size Chart

SCREW THREAD			COMMERCIAL TAP DRILLS		SCREW THREAD			COMMERCIAL TAP DRILLS	
Outside Diameter	Pitch	Root Diam.	Size or Number	Decimal Equiv.	Outside Diameter	Pitch	Root Diam.	Size or Number	Decimal Equiv.
1/16	64	0.0422	3/64	0.0469		27	0.4519	15/32	0.4687
	72	0.0445	3/64	0.0469	9/16	12	0.4542	31/64	0.4844
5/64	60	0.0563	1/16	0.0625		18	0.4903	33/64	0.5156
	72	0.0601	52	0.0635		27	0.5144	17/32	0.5312
3/32	48	0.0667	49	0.0730	5/8	11	0.5069	17/32	0.5312
	50	0.0678	49	0.0730		12	0.5168	35/64	0.5469
7/64	48	0.0823	43	0.0890		18	0.5528	37/64	0.5781
1/8	32	0.0844	3/32	0.0937		27	0.5769	19/32	0.5937
	40	0.0925	38	0.1015	11/16	11	0.5694	19/32	0.5937
9/64	40	0.1081	32	0.1160		16	0.6063	5/8	0.6250
5/32	32	0.1157	1/8	0.1250	3/4	10	0.6201	21/32	0.6562
	36	0.1202	30	0.1285		12	0.6418	43/64	0.6719
11/64	32	0.1313	9/64	0.1406		16	0.6688	11/16	0.6875
3/16	24	0.1334	26	0.1470		27	0.7019	23/32	0.7187
	32	0.1469	22	0.1570	13/16	10	0.6826	23/32	0.7187
15/64	24	0.1490	20	0.1610	7/8	9	0.7307	49/64	0.7656
7/32	24	0.1646	16	0.1770		12	0.7668	51/64	0.7969
	32	0.1782	12	0.1890		14	0.7822	13/16	0.8125
15/64	24	0.1806	10	0.1935		18	0.8028	53/64	0.8281
1/4	20	0.1850	7	0.2010		27	0.8269	27/32	0.8437
	24	0.1959	4	0.2090	15/16	9	0.7932	53/64	0.8281
	27	0.2019	3	0.2130	1	8	0.8376	7/8	0.8750
	28	0.2036	3	0.2130		12	0.8918	59/64	0.9219
	32	0.2094	7/32	0.2187		14	0.9072	15/16	0.9375
5/16	18	0.2403	F	0.2570		27	0.9519	31/32	0.9687
	20	0.2476	17/64	0.2656	1 1/8	7	0.9394	63/64	0.9844
	24	0.2584	I	0.2720		12	1.0168	1 3/64	1.0469
	27	0.2644	J	0.2770	1 1/4	7	1.0644	1 7/64	1.1094
	32	0.2719	9/32	0.2812		12	1.1418	1 11/64	1.2187
3/8	16	0.2938	5/16	0.3125	1 3/8	6	1.1585	1 7/32	1.1719
	20	0.3100	21/64	0.3281		12	1.2668	1 19/64	1.2969
	24	0.3209	Q	0.3320	1 1/2	6	1.2835	1 11/32	1.3437
	27	0.3269	R	0.3390		12	1.3918	1 27/64	1.4219
7/16	14	0.3477	U	0.3680	1 5/8	5 1/2	1.3888	1 29/64	1.4531
	20	0.3726	25/64	0.3906	1 3/4	5	1.4902	1 9/16	1.5625
	24	0.3834	X	0.3970	1 7/8	5	1.6152	1 11/16	1.6875
	27	0.3894	Y	0.4040	2	4 1/2	1.7113	1 25/32	1.7812
1/2	12	0.3918	27/64	0.4219	2 1/8	4 1/2	1.8363	1 29/32	1.9062
	13	0.4001	27/64	0.4219	2 1/4	4 1/2	1.9613	2 1/32	2.0312
	20	0.4351	29/64	0.4531	2 3/8	4	2.0502	2 1/8	2.1250
	24	0.4459	29/64	0.4531	2 1/2	4	2.1752	2 1/4	2.2500

103T02.EPS

Step 6 Check the torque specification and torque each bolt in the specified sequence.

> **NOTE**
> It is not acceptable to use an impact wrench to torque bolts.

Step 7 If required to keep the bolts or nuts from working loose, install jam nuts, cotter pins, or safety wire. *Figure 19* shows fasteners with a safety wire installed. Safety wire should be flexible without fatiguing easily. Do not use mig welding wire, as it is too stiff, and will break too easily. Wire is specially manufactured for safety wire in many different materials, to be used in different work environments. Such wire is sold as safety wire, airplane safety wire, or lockwire. In addition to the crenellations or notches in nuts, sometimes the bolt itself is also drilled through, and the wire passed through.

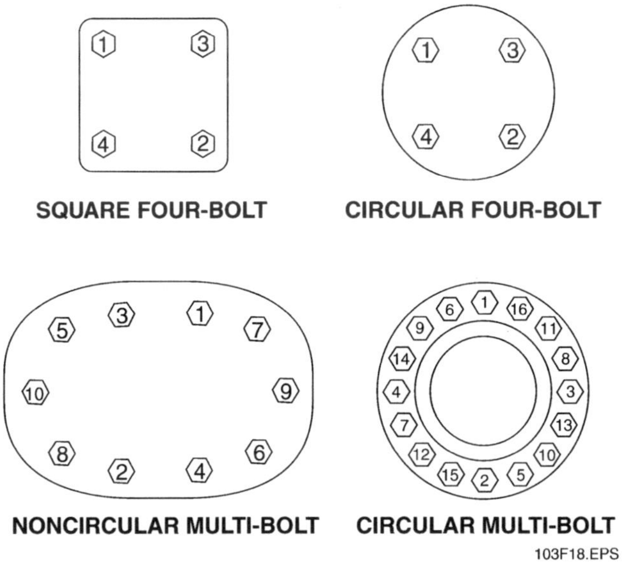

Figure 18 ◆ Proper tightening sequence.

> **NOTE**
> Hydraulic bolt tightening machines are often used to tighten bolts in critical applications and to tighten large bolts. A typical hydraulic bolt tightening machine consists of a hydraulic pump unit and one or more hydraulic torque wrenches. The hydraulic wrench is placed over the bolt and the pump supplies high-pressure hydraulic fluid to drive the wrench in small increments and tighten the bolt. In addition to providing precise bolt tightening, a major advantage of the hydraulic equipment is that multiple bolts can be tightened at once. This allows for even tightening and can significantly reduce the time required for a major tightening operation. Another advantage is that the hydraulic wrenches can be placed in locations that are difficult to access with hand tools.

3.0.0 ◆ NON-THREADED FASTENERS

Non-threaded fasteners have many uses. Different types of non-threaded fasteners include retainers, keys, pins, clamps, washers, rivets, and tie wraps.

3.1.0 Retainer Fasteners

Retainer fasteners, also called retaining rings, are used for both internal and external applications. Some retaining rings are seated in grooves in the fastener. Other types of retainer fasteners are self-locking and do not require a groove. To easily remove internal and external retainer rings

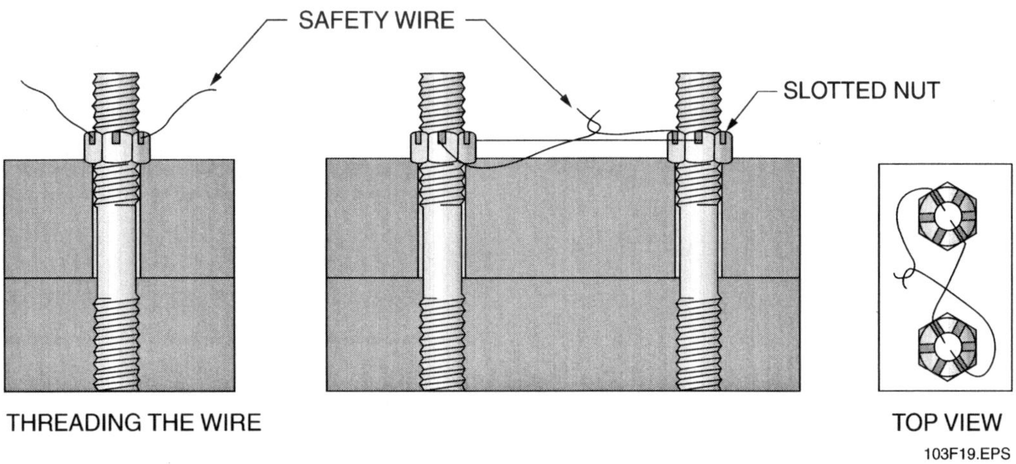

Figure 19 ◆ Safety-wired fasteners.

without damaging the ring or the fastener, special pliers are used. *Figure 20* shows several types of retainer fasteners.

> **NOTE**
> External retainer fasteners are sometimes called clips.

3.2.0 Keys

Keys are inserted in a shaft to prevent a gear or pulley from rotating on the shaft. Half of the key fits into a key seat on the shaft. The other half fits into a **keyway** in the hub of the gear or pulley. The key fastens the two parts together, stopping the gear or pulley from turning on the shaft. *Figure 21* shows different types of keys and keyways and their uses.

Some different types of keys include the following:

- *Square key* – Usually one-quarter of the shaft diameter. It may be slightly tapered on the top for easier fitting.
- *Pratt and Whitney key* – Similar to the square key, but rounded at both ends. It fits into a keyseat of the same shape.

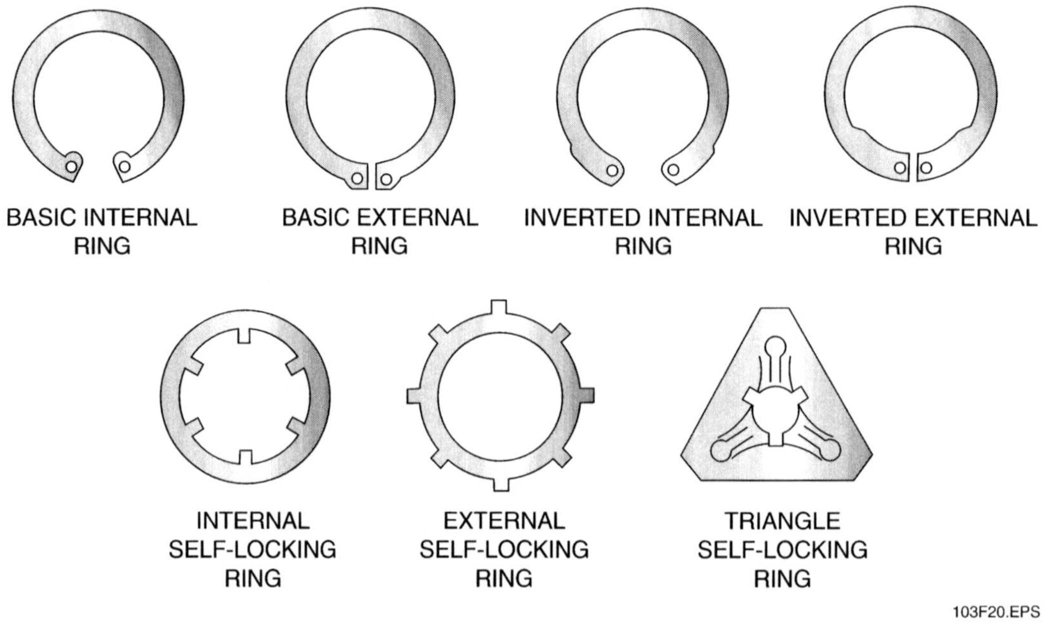

Figure 20 ◆ Retainer fasteners (rings).

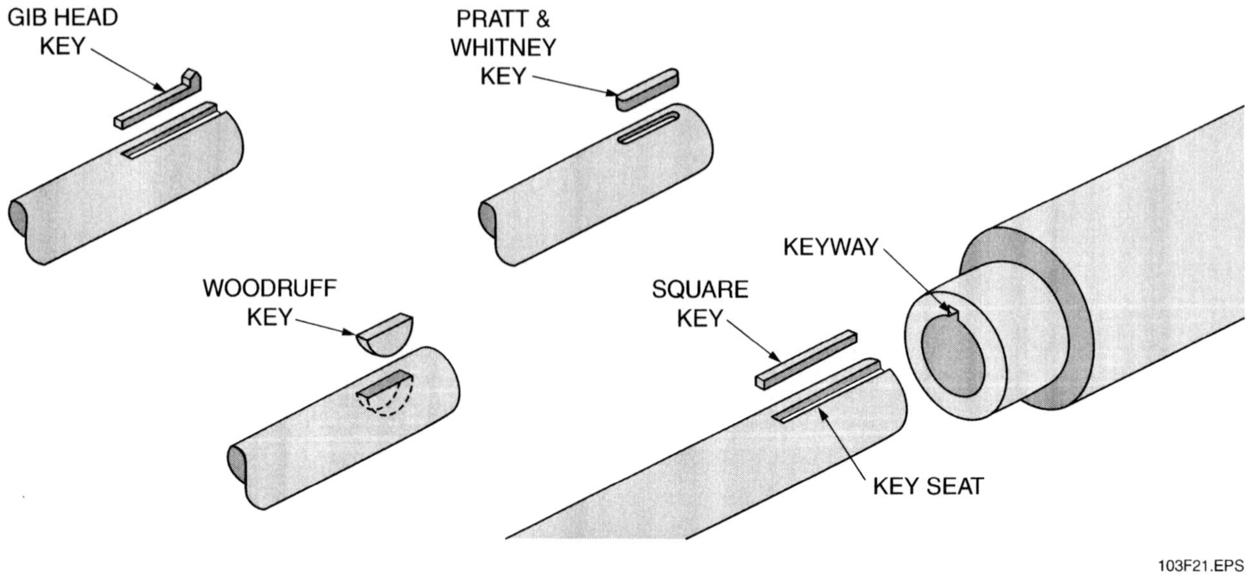

Figure 21 ◆ Keys and keyways.

- *Gib head key* – Interchangeable with the square key. The head design allows easy removal from the assembly.
- *Woodruff key* – Semicircular shape that fits into a keyseat of the same shape. The top of the key fits into the keyway of the mating part.

3.3.0 Pin Fasteners

Pin fasteners come in several types and sizes. They have a variety of applications. Common uses of pin fasteners include holding moving parts together, aligning mating parts, fastening hinges, holding gears and pulleys on shafts, and securing slotted nuts. *Figure 22* shows several pin fasteners.

3.3.1 Dowel Pins

Dowel pins fit into holes to position mating parts. They may also support a portion of the load placed on the parts. *Figure 23* shows an application of dowel pins used to position mating parts.

3.3.2 Taper and Spring Pins

Taper and spring pins are used to fasten gears, pulleys, and collars to a shaft. *Figure 24* shows how taper and spring pins are used to attach a component to a shaft. The groove in a spring pin allows it to compress against the walls in a spring-like fashion.

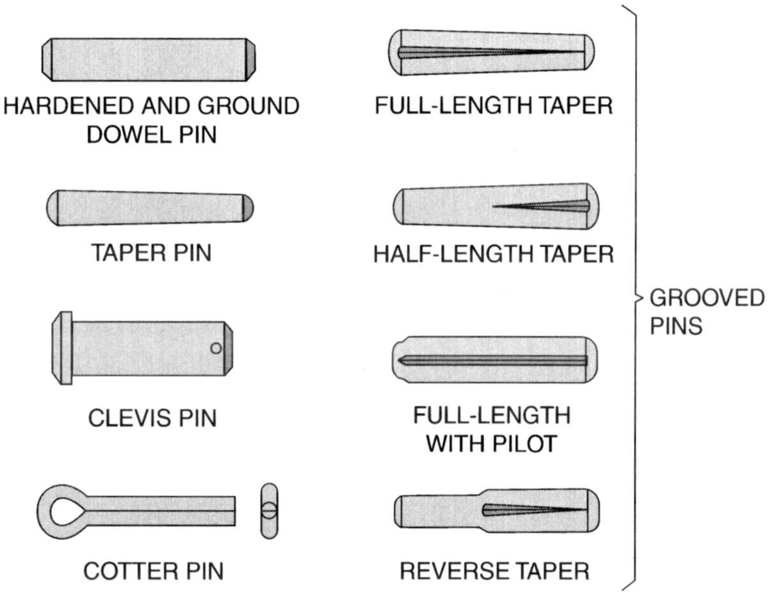

Figure 22 ◆ Pin fasteners.

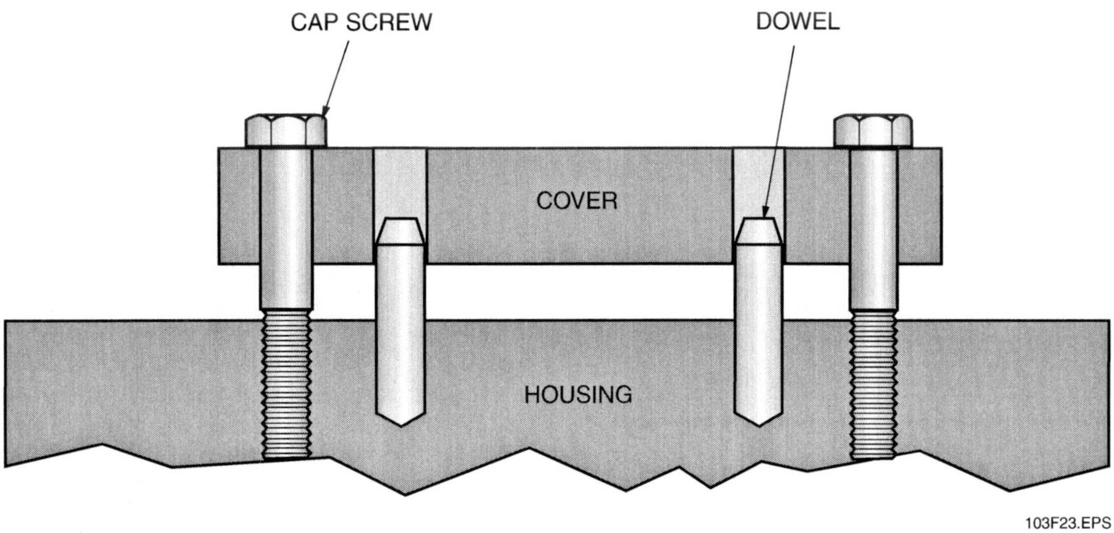

Figure 23 ◆ Dowel pins.

3.3.3 Cotter Pins

There are several different types of cotter pins. They are used as a locking device for a variety of applications. Cotter pins are often inserted through a hole drilled crosswise through a shaft to prevent parts from slipping on or off the shaft. They are also used to keep slotted nuts from working loose. Standard cotter pins are general-use pins. When installed, the extended prong is normally bent back over the nut to provide the locking action. If it is ever removed, throw it away and replace it with a new one. The humped, cinch, and hitch-type cotter pins are self-locking pins. The humped and cinch type should also be thrown away and replaced with a new one if removed. The hitch pin, also called a hair pin, is a reusable pin made to be installed and removed quickly. *Figure 25* shows several common types of cotter pins.

3.4.0 Blind/Pop Rivets

When only one side of a joint can be reached, blind rivets can be used to fasten the parts together. Some applications of blind rivets include fastening light to heavy gauge sheet metal, fiberglass, plastics, and belting. Blind rivets are made of a variety of materials and come in several sizes and lengths. They are installed using special riveting tools. *Figure 26* shows a typical blind rivet installation.

Blind rivets are installed through drilled or punched holes using a special blind (pop) rivet gun (*Figure 27*).

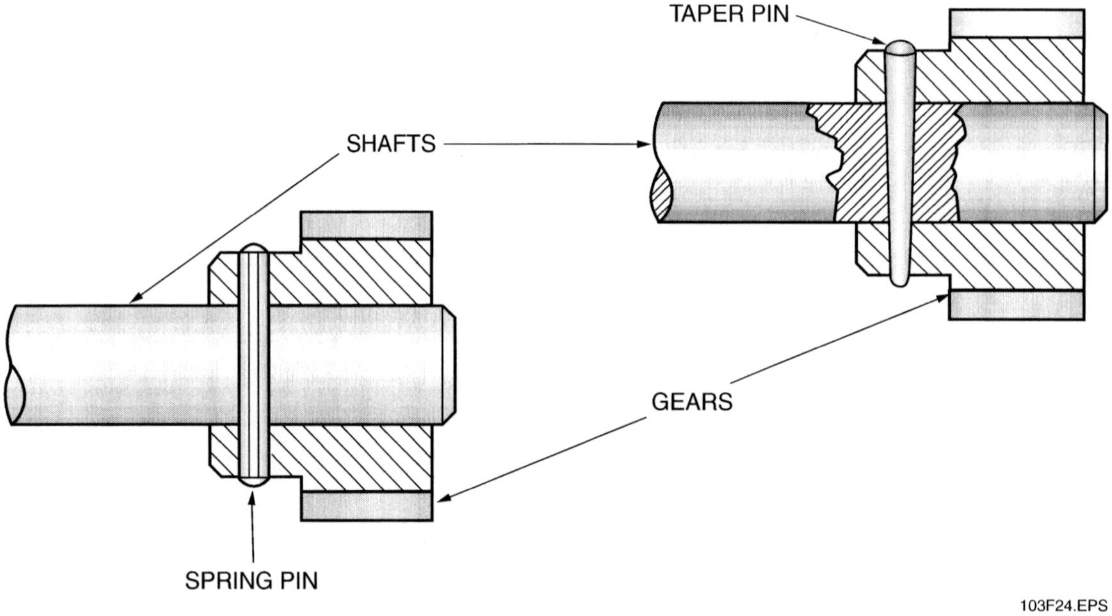

Figure 24 ◆ Taper and spring pins.

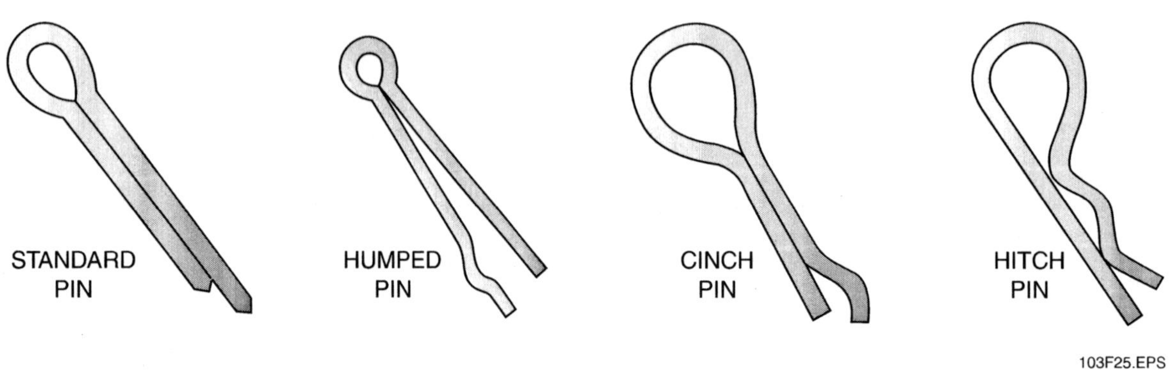

Figure 25 ◆ Cotter pins.

NOTE

To remove rivets, chisel or grind off the heads of tinner's rivets or steel blind rivets; remove the shaft by striking through with a punch. Drill off blind or countersunk rivet heads using a high-speed electric drill, then punch out the shaft.

Use the following general procedure to install blind rivets.

WARNING!

Follow all safety precautions. Make sure to wear proper eye and face protection when riveting.

Step 1 Select the correct length and diameter of blind rivet to be used.

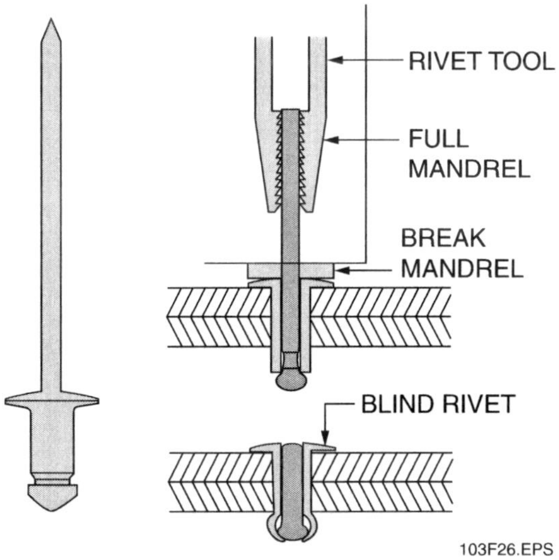

Figure 26 ◆ Blind rivet installation.

Step 2 Select the appropriate drill bit for the size of rivet being used.

Step 3 Drill a hole through both parts being connected.

Step 4 Inspect the rivet gun for any defects that might make it unsafe for use.

Step 5 Place the rivet mandrel into the proper size setting tool.

Step 6 Insert the rivet end into the predrilled hole.

Step 7 Install the rivet by squeezing the handle of the rivet gun, causing the jaws in the setting tool to grip the mandrel. The mandrel is pulled up, expanding the rivet until it breaks at the shear point. *Figure 28* shows the rivet and tool positioned for joining parts together.

Step 8 Inspect the rivet to make sure the pieces are firmly riveted together and that the rivet is properly installed. *Figure 29* shows a properly installed blind rivet.

NOTE

Before you attach materials using a rivet gun and possibly do something you'll regret later, make a spacing template. Hold the template up to your material to ensure that all rivets fall into the correct locations, then secure your pieces. When using the template for actual riveting, hold it in position with an awl or drift pin inserted into your predrilled rivet holes. Note that rivets may pull through certain materials or when joining two pieces of flexible material together when the hole is slightly larger than required.

A manual pop rivet tool can be used to install blind rivets in a shop or at the job site.

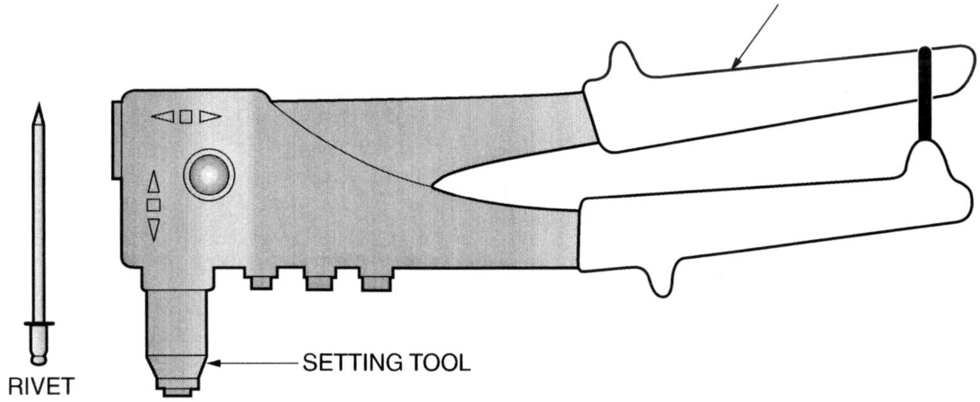

Figure 27 ◆ Rivet gun.

3.5.0 Tie Wraps

A tie wrap is a one-piece, self-locking cable tie, usually made of nylon, that is used to fasten a bundle of wires and cables together. Tie wraps can be quickly installed either manually or using a special installation tool. Black tie wraps resist ultraviolet light and are recommended for outdoor use.

Tie wraps are made in standard, cable strap and clamp, and identification configurations (*Figure 30*). All types function to clamp bundled wires or cable together. In addition, the cable strap and clamp has a molded mounting hole in the head used to secure the tie with a rivet, screw, or bolt after the tie wrap has been installed around the wires or cable. Identification tie wraps have a large flat area provided for imprinting or writing cable identification information. There is also a releasable version available. It is a non-permanent tie used for bundling wires or cable that may require frequent additions or deletions. Cable ties are made in various lengths ranging from about 3 to 30 inches, allowing them to be used for fastening wires and cable into bundles with diameters ranging from about ½ to 9 inches, respectively. Tie wraps can also be attached to a variety of adhesive mounting bases made for that purpose.

4.0.0 ◆ SPECIAL THREADED FASTENERS

Special threaded fasteners consist of hardware manufactured in several shapes and sizes and designed to perform specific jobs. Certain types of nuts may be considered special threaded fasteners if they are designed especially for a particular application. In the electrical craft, special threaded fasteners are used on a number of different jobs.

The types of special threaded fasteners described in this section are eye bolts, inserts, and panel and electrical mounts.

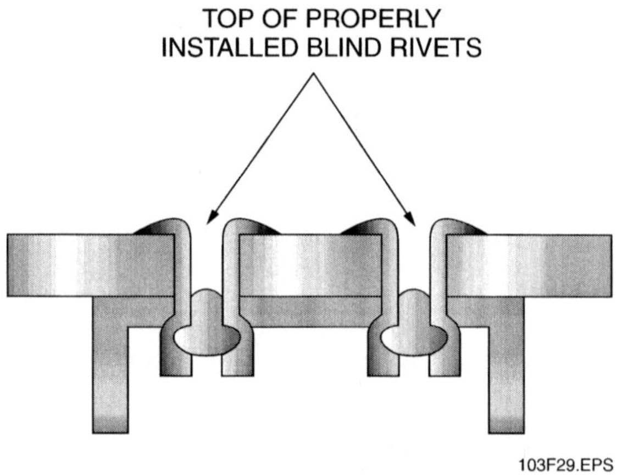

Figure 29 ◆ Properly installed blind rivets.

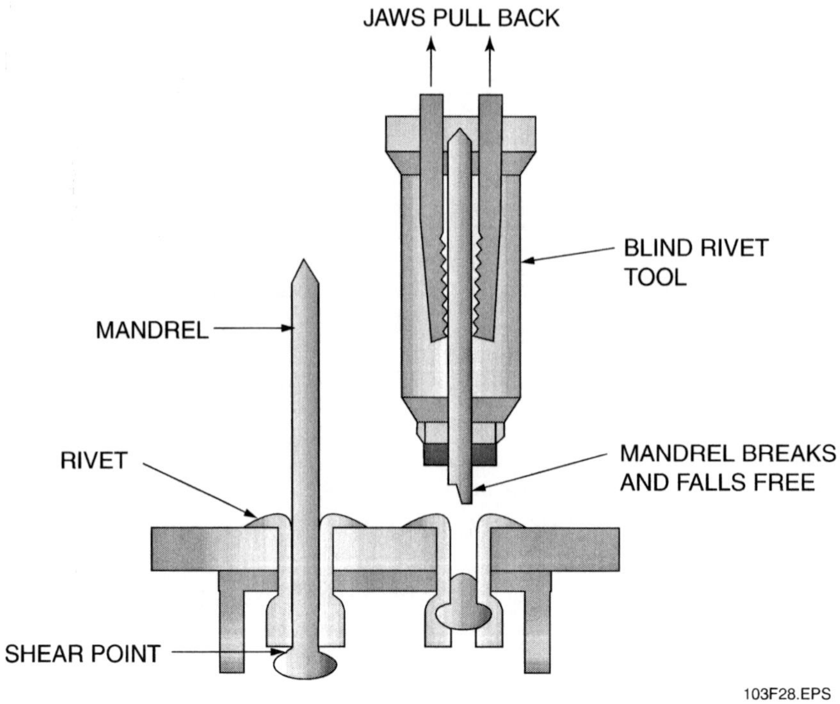

Figure 28 ◆ Joining parts.

4.1.0 Eye Bolts

Eye bolts get their name from the eye or loop at one end. The other end of an eye bolt is threaded. There are many types of eye bolts. The eye on some eye bolts is formed and welded, while the eye on other types is forged. Shoulder-forged eye bolts are commonly used as lifting devices and guides for wires, cables, and cords. *Figure 31* shows some typical eye bolts.

4.2.0 Thread Inserts

Inserts, also called heli-coils, are a special kind of nut used to provide high-strength threads in soft metals and plastics. They are also used to replace damaged or stripped threads in a tapped hole. Internal threads are made in standard sizes and forms. *Figure 32* shows a thread insert.

4.3.0 Panel and Electrical Mounts

Some specialized hardware is used to fasten instruments to panels and to attach electrical conduit safely. Two types mentioned here are J-nuts and cage nuts.

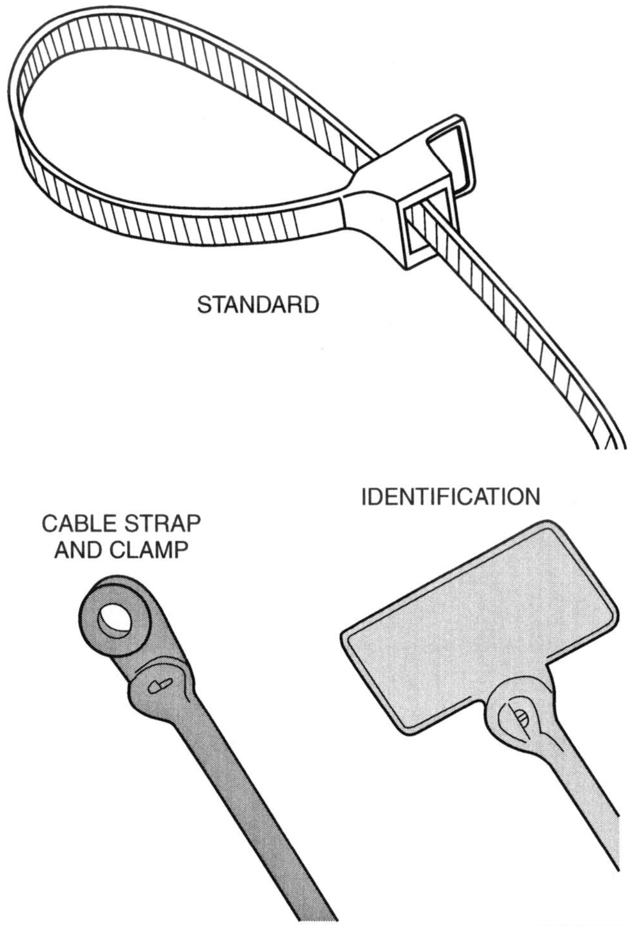

Figure 30 ◆ Tie wraps.

J-nuts and U-nuts are sheetmetal stampings, threaded and designed to clip on to the edge of panel boxes and automobile doors to hold paneling in place. *Figure 33* shows U-nuts and J-nuts.

Cage nuts are designed to mount electronics and computer hardware on racks; they consist of

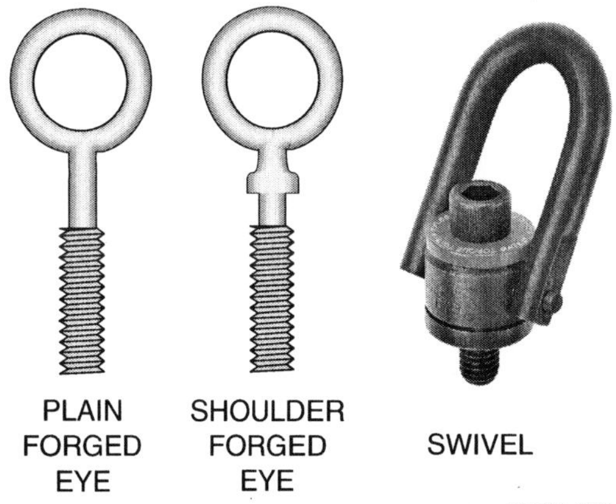

Figure 31 ◆ Eye bolts.

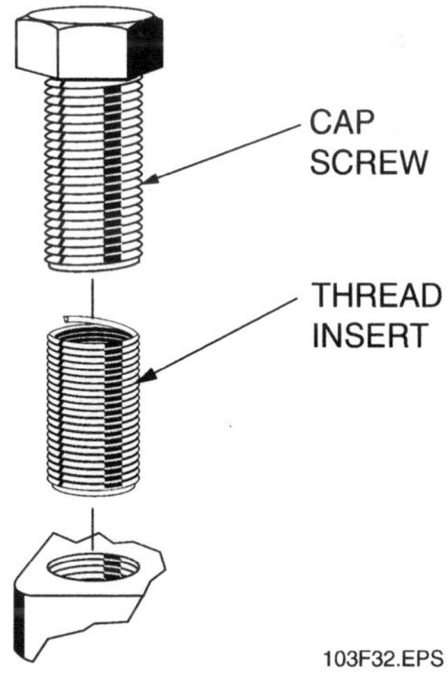

Figure 32 ◆ Thread insert.

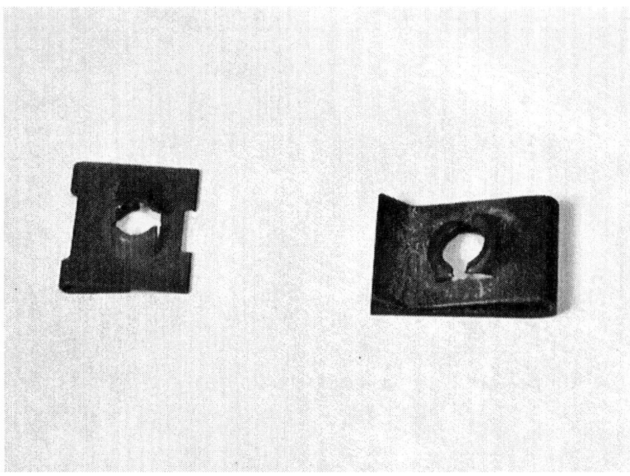

Figure 33 ◆ U-nuts and J-nuts.

a nut mounted in a spring-steel cage. The spring steel cage allows the nut to be mounted on a square hole on a rack. *Figure 34* shows cage nuts.

5.0.0 ◆ MECHANICAL ANCHORS

Mechanical anchors are devices used to give fasteners a firm grip in a variety of materials, where the fasteners by themselves would otherwise have a tendency to pull out. Anchors have been classified in many ways by different manufacturers. In this module, anchors have been divided into four broad categories:

- One-step anchors
- Bolt anchors
- Screw anchors
- Self-drilling anchors

5.1.0 One-Step Anchors

One-step anchors are designed so that they can be installed through the mounting holes in the component to be fastened. This is because the anchor and the drilled hole into which it is installed have the same diameter. They come in various diameters ranging from ¼ inch to 1¼ inches, with lengths ranging from 1¾ inches to 12 inches. Wedge, stud, sleeve, one-piece, screw, nail, and threaded rod anchors (*Figure 35*) are common types of one-step anchors.

5.1.1 Wedge Anchors

Wedge anchors are heavy-duty anchors supplied with nuts and washers. The drill bit size used to drill the hole is the same diameter as the anchor. The depth of the hole is not critical, as long as the minimum length recommended by the manufacturer is drilled. After the hole is blown clean of dust and other material, the anchor is inserted into the hole and driven with a hammer far enough so that at least six threads are below the top surface of the component. Then, the component is fastened by tightening the anchor nut to expand the anchor and tighten it in the hole.

5.1.2 Stud Bolt Anchors

Stud bolt anchors are heavy-duty threaded anchors. Because this type of anchor is made to bottom in its mounting hole, it is a good choice to use when jacking or leveling of the fastened component is needed. The depth of the hole drilled in the masonry must be as specified by the manufacturer in order to achieve proper expansion. After the hole is blown clean of dust and other material, the anchor is inserted in the hole with the expander plug end down. Following this, the anchor is driven into the hole with a hammer (or setting tool) to expand the anchor and tighten it in the hole. The anchor is fully set when it can no longer be driven into the hole. The component is fastened using the correct size and thread bolt for use with the anchor stud.

5.1.3 Sleeve Anchors

Sleeve anchors are multi-purpose anchors. The depth of the anchor hole is not critical, as long as the minimum length recommended by the manufacturer is drilled. After the hole is blown clean of dust and other material, the anchor is inserted into the hole and tapped until flush with the component. Then, the anchor nut or screw is tightened to expand the anchor and tighten it in the hole.

Figure 34 ◆ Cage nuts.

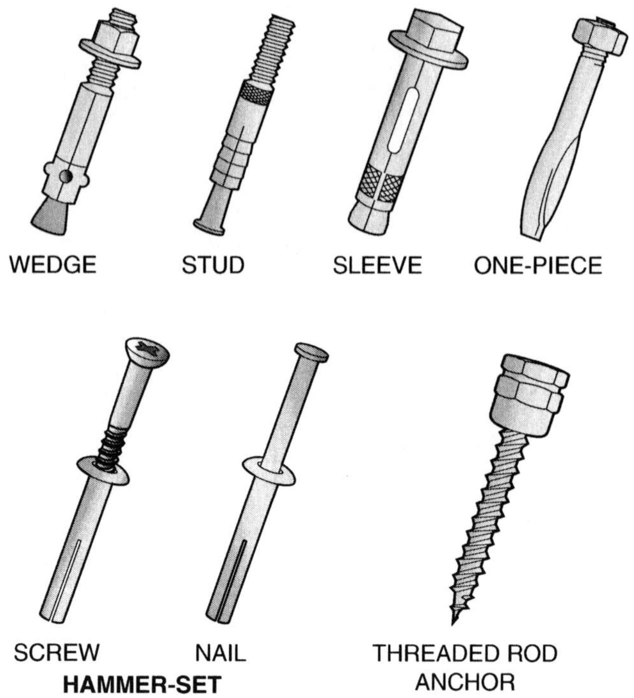

Figure 35 ◆ One-step anchors.

5.1.4 One-Piece Anchors

One-piece anchors are multi-purpose anchors. They work on the principle that as the anchor is driven into the hole, the spring force of the expansion mechanism is compressed and flexes to fit the size of the hole. Once set, it tries to regain its original shape. The depth of the hole drilled in the masonry must be at least ½ inch deeper than the required embedment. The proper depth is crucial. Refer to the manufacturer's recommendation for the correct depth. After the hole is blown clean of dust and other material, the anchor is inserted through the component and driven with a hammer into the hole until the head is firmly seated against the component. Make sure that the anchor is driven to the proper embedment depth. Manufacturers also make specially designed drivers and manual tools that are used instead of a hammer to drive one-piece anchors. These tools allow the anchors to be installed in confined spaces and help prevent damage to the component from stray hammer blows.

5.1.5 Hammer-Set Anchors

Hammer-set anchors are made for use in concrete and masonry. There are two types: nail and screw. An advantage of the screw-type anchors is that they are removable. Both types have a diameter the same size as the anchoring hole. For both types, the anchor hole must be drilled to the diameter of the anchor and to a depth of at least ¼ inch deeper than that required for embedment. After the hole is blown clean of dust and other material, the anchor is inserted into the hole through the mounting holes in the component to be fastened; then the screw or nail is driven into the anchor body to expand it. Make sure that the head is seated firmly against the component and is at the proper embedment.

5.1.6 Threaded Rod Anchors

Threaded rod anchors are available for installation in concrete, steel, or wood. The anchor is designed to support a threaded rod, which is screwed into the head of the anchor after the anchor is installed. A special nut driver is available for installing the screws.

5.2.0 Bolt Anchors

Bolt anchors are designed to be installed flush with the surface of the base material. They are used in conjunction with threaded machine bolts or screws. In some types, they can be used with threaded rod. Drop-in, single- and double-expansion, and caulk-in anchors (*Figure 36*) are commonly used types of bolt anchors.

5.2.1 Drop-In Anchors

Drop-in anchors are typically used as heavy-duty anchors. There are two types of drop-in anchors. The first type, made for use in solid concrete and masonry, has an internally threaded expansion anchor with a preassembled internal expander plug. The anchor hole must be drilled to the specific diameter and depth specified by the manufacturer. After the hole is blown clean of dust and other material, the anchor is inserted into the hole and tapped until it is flush with the surface. Following this, a setting tool supplied with the anchor is driven into the anchor to expand it. The component to be fastened is positioned in place and is fastened by threading and tightening the correct size machine bolt or screw into the anchor.

The second type, called a hollow-set drop-in anchor, is made for use in hollow concrete and masonry base materials. Hollow-set drop-in anchors have a slotted, tapered expansion sleeve and a serrated expansion cone. They come in various lengths compatible with the outer wall thicknesses of most hollow base materials. They can also be used in solid concrete and masonry. The anchor hole must be drilled to the specific diameter specified by the manufacturer. When installed in hollow base materials, the hole is drilled into the cell or void. After the hole is blown clean of dust and other material, the anchor is inserted

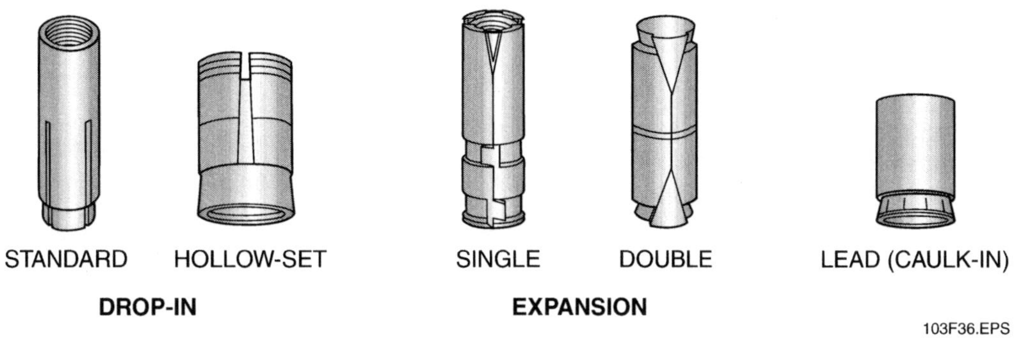

Figure 36 ◆ Bolt anchors.

into the hole and tapped until it is flush with the surface. Following this, the component to be fastened is positioned in place; then the proper size machine bolt or screw is threaded into the anchor and tightened to expand the anchor in the hole.

5.2.2 Single- and Double-Expansion Anchors

Single- and double-expansion anchors are both made for use in concrete and other masonry. The double-expansion anchor is used mainly when fastening into concrete or masonry of questionable strength. For both types, the anchor hole must be drilled to the specific diameter and depth specified by the manufacturer. After the hole is blown clean of dust and other material, the anchor is inserted into the hole, threaded cone end first. It is then tapped until it is flush with the surface. Following this, the component to be fastened is positioned in place; then the proper size machine bolt or screw is threaded into the anchor and tightened to expand the anchor in the hole.

5.2.3 Lead (Caulk-In) Anchors

Lead (caulk-in) anchors are a cast-type anchor used in concrete and masonry. They consist of an internally threaded expander cone with a series of vertical internal ribs and a lead sleeve. The vertical internal ribs prevent the cone from turning in the sleeve as the anchor is tightened. The anchor hole must be drilled to the specific diameter and depth specified by the manufacturer. However, in weak or soft masonry, a slightly deeper hole can be drilled to countersink the anchor below the surface. After the hole is blown clean of dust and other material, the anchor is inserted into the hole, threaded cone end first. Following this, a setting tool supplied with the anchor is driven into the anchor to expand it. The component to be fastened is positioned in place and fastened by threading and tightening the correct size machine bolt or screw into the anchor.

5.3.0 Screw Anchors

There are a number of options for installing screws in relatively fragile materials such as plaster. One type is the wallboard anchor, which has a plastic socket that is expanded in a hole in the wall. Another type screws into the wall material, and the screw is inserted into the anchor. Most of these systems are designed to be loaded in shear, that is, at right angles to the length of the screw and anchor.

Tapcons® (*Figure 37*) are a light- to medium-strength screw that cuts threads in masonry. They are supplied with a drill that is the correct size to install the screw. After drilling the hole, the Tapcon® is screwed into the hole until it seats. Do not install the screw too deeply; the threads can easily be stripped out if the screw is tightened too far. The Tapcon® should not be relied on for high-strength applications, as the threads are too easily stripped in brittle masonry.

5.4.0 Self-Drilling Anchors

Some anchors made for use in masonry are self-drilling anchors. *Figure 38* is typical of those in common use. This fastener has a cutting sleeve that is first used as a drill bit and later becomes the expandable fastener itself. A rotary hammer is used to drill the hole in the concrete using the anchor sleeve as the drill bit. After the hole is drilled, the anchor is pulled out and the hole cleaned. This is followed by inserting the anchor's expander plug into the cutting end of the sleeve. The anchor sleeve and expander plug are driven back into the hole with the rotary hammer until they are flush with the surface of the concrete. As the fastener is hammered down, it hits the bottom, where the tapered expander causes the fastener to expand and lock into the hole. The anchor is then snapped off at the shear point with a quick lateral movement of the hammer. The component to be fastened can then be attached to the anchor using the proper size bolt.

103F37.EPS

Figure 37 ◆ Tapcons®.

5.5.0 Guidelines for Drilling Anchor Holes in Hardened Concrete or Masonry

When selecting masonry anchors, regardless of the type, always take into consideration and follow the manufacturer's recommendations pertaining to hole diameter and depth, minimum embedment in concrete, maximum thickness of material to be fastened, and the pullout and shear load capacities.

When installing anchors and/or anchor bolts in hardened concrete, make sure the area where the equipment or component is to be fastened is smooth so that it will have solid footing. Uneven footing might cause the equipment to twist, warp, not tighten properly, or vibrate when in operation. Before starting, carefully inspect the rotary hammer or hammer drill and the drill bit(s) to ensure they are in good operating condition. Be sure to use the type of carbide-tipped masonry or percussion drill bits recommended by the drill/hammer or anchor manufacturer because these bits are made to take the higher impact of the masonry materials. Also, it is recommended that the drill or hammer tool depth gauge be set to the depth of the hole needed. The trick to using masonry drill bits is not to force them into the material by pushing down hard on the drill. Use a little pressure and let the drill do the work. For large holes, start with a smaller bit, then change to a larger bit.

The methods for installing the different types of anchors in hardened concrete or masonry were briefly described in the sections above. Always install the selected anchors according to the manufacturer's directions. Here is an example of a typical procedure used to install many types of expansion anchors in hardened concrete or masonry.

Refer to *Figure 39* as you study the procedure.

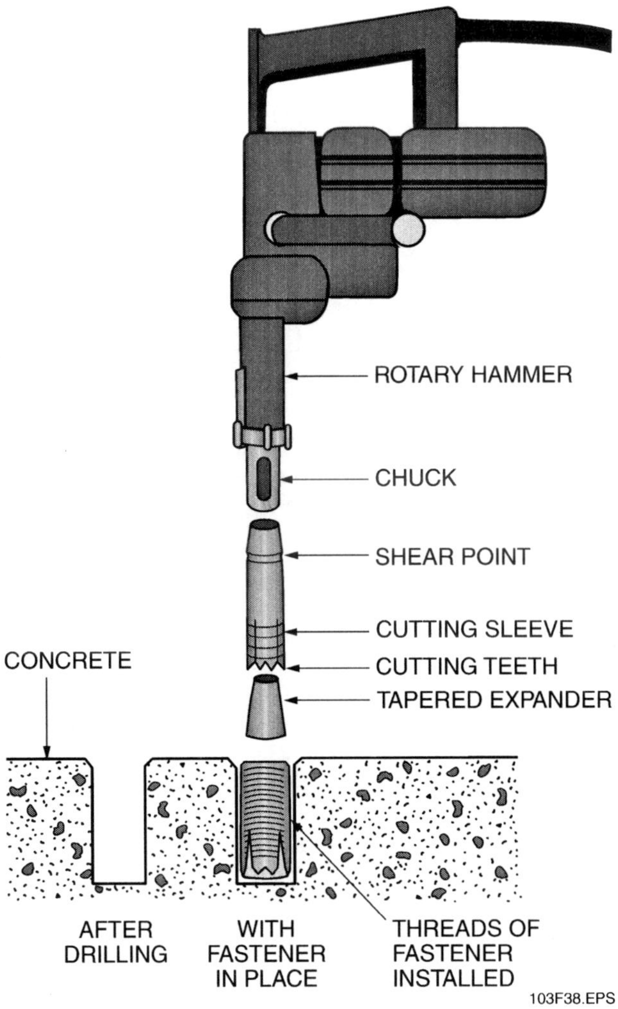

Figure 38 ◆ Self-drilling anchor.

> **WARNING!**
> Drilling in concrete generates noise, dust, and flying particles. Always wear safety goggles, ear protectors, and gloves. Make sure other workers in the area also wear protective equipment.

Step 1 Drill the anchor bolt hole the same size as the anchor bolt. The hole must be deep enough for six threads of the bolt to be below the surface of the concrete (see *Figure 39*, Step 1). Clean out the hole using a squeeze bulb.

Step 2 Drive the anchor bolt into the hole using a hammer (*Figure 39*, Step 2). Protect the threads of the bolt with a nut that does not allow any threads to be exposed.

Step 3 Put a washer and nut on the bolt, and tighten the nut with a wrench until the anchor is secure in the concrete (*Figure 39*, Step 3).

Although it is possibly more accurate to drill through the base hole where the bolt and anchor are to be placed, it is not always an available option. Most commonly, it will be necessary to place the base, use a centerpunch to mark the hole locations, and then move the base and drill. Once the holes have been drilled, place the anchors, move the base into place, and, if necessary, use washers under the base to keep the bolts from pulling the anchors through the base holes.

> **NOTE**
> Be sure to wear safety goggles whenever you tackle any fastening project, regardless of how small the job may seem. Remember, you can never replace lost eyesight.

6.0.0 ◆ EPOXY ANCHORING SYSTEMS

Epoxy resin compounds can be used to anchor threaded rods, dowels, and similar fasteners in solid concrete, hollow walls, and brick. For one manufacturer's product, a two-part epoxy is packaged in a two-chamber cartridge that keeps the resin and hardener ingredients separated until use. This cartridge is placed into a special tool similar to a caulking gun. When the gun handle is pumped, the epoxy resin and hardener components are mixed within the gun; then the epoxy is ejected from the gun nozzle.

To use the epoxy to install an anchor in solid concrete (*Figure 40*), a hole of the proper size is drilled in the concrete and cleaned using a nylon (not metal) brush. Following this, a small amount of epoxy is dispensed from the gun to make sure that the resin and hardener have mixed properly. This is indicated by the epoxy being of a uniform color. The gun nozzle is then placed into the hole, and the epoxy is injected into the hole until half the depth of the hole is filled. Following this, the selected fastener is pushed into the hole with a slow twisting motion to make sure that the epoxy fills all voids and crevices, then is set to the required plumb (or level) position. After the recommended cure time for the epoxy has elapsed, the fastener nut can be tightened to secure the component or fixture in place.

Another technique used for individual holes employs capsules of epoxy and hardener, prepackaged. The two forms in use are packaged in either glass capsules or foil. The hardener is in a separate chamber inside the resin chamber. When the hole has been drilled, the capsule is inserted, and the bolt is inserted to break the chambers. The bolt is then turned so as to mix the resin and hardener thoroughly, and the epoxy is allowed to set (*Figure 41*). Manufacturers supply instructions as to how long the bolt must be turned to completely mix the adhesive. With the glass capsule, the glass fragments will be brought to the surface, so the worker must take care not to wipe the top off empty-handed.

In this application, the washer is not necessary, and the hole is much smaller without sacrificing strength. In either case, follow manufacturer's directions so that the anchor will set completely. Adhesive anchors should not be used in concrete that has not set for at least seven days, nor should they be used below 40°F. Adhesive anchors must be allowed to set completely before any stress or movement is applied to the bolt.

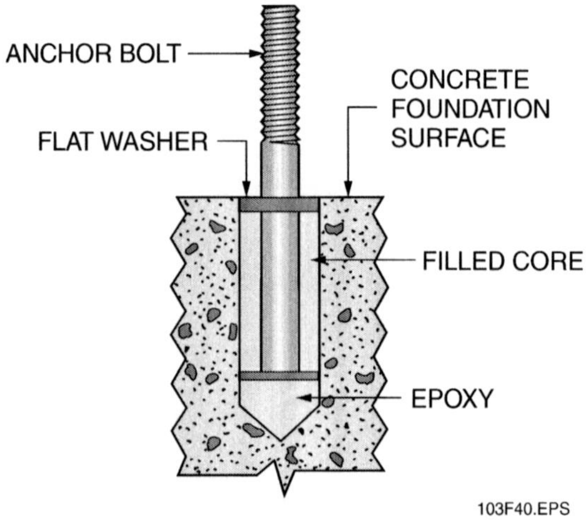

Figure 40 ◆ Fastener anchored in epoxy.

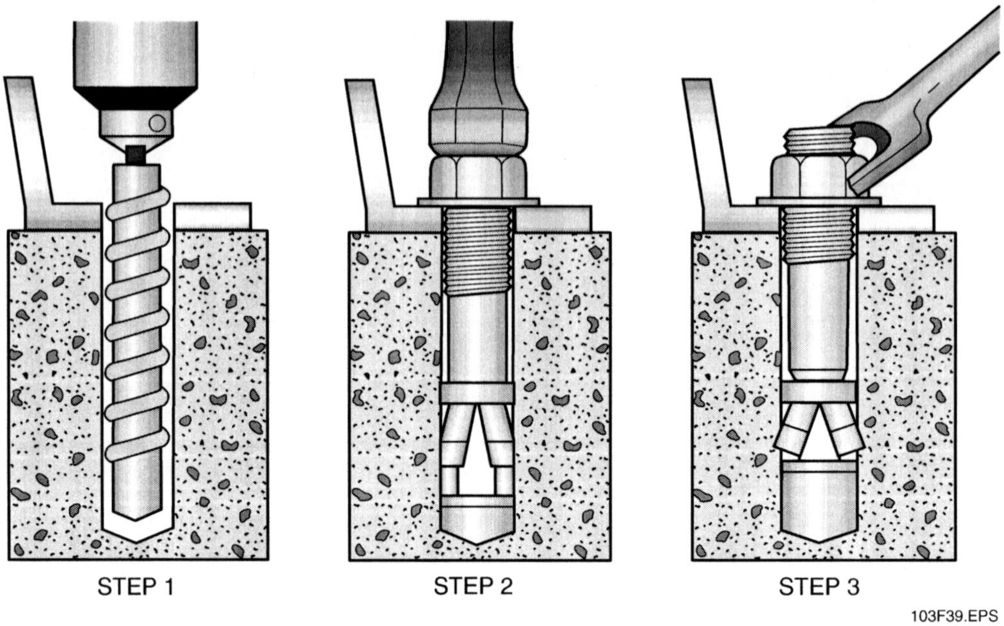

Figure 39 ◆ Installing an anchor bolt in hardened concrete.

The procedure for installing a fastener in a hollow wall or brick using epoxy is basically the same as previously described. The difference is that the epoxy is first injected into an anchor screen to fill the screen, then the anchor screen is installed into the drilled hole. Use of the anchor screen is necessary to hold the epoxy intact in the hole until the anchor is inserted into the epoxy.

> **NOTE**
>
> Once mixed, epoxy has a limited working time. Therefore, mix exactly what you need and work quickly. After the working time is up, epoxy requires a specific curing time. Always give epoxy its recommended curing time. Because epoxy is so strong and sets so quickly, you'll be tempted to stress the bond before it's fully cured.

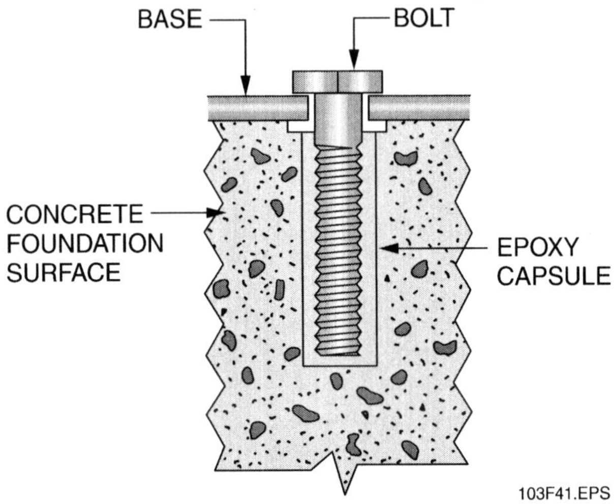

Figure 41 ◆ Fastener installed in epoxy capsule.

Review Questions

1. The thread class that would be used when close tolerances are required is _____.
   a. 1A
   b. 2A
   c. 2B
   d. 3A

2. The quality of some fasteners can be determined by the _____ on the head of a bolt or screw.
   a. number of grooves cut
   b. number of sides
   c. length of the lines
   d. grade markings

3. The type of bolt that is threaded over its entire length is the machine bolt.
   a. True
   b. False

4. The purpose of a jam nut is to _____.
   a. hold a piece of material stationary
   b. lock a standard nut in place
   c. stop rotation of a machine quickly for safety reasons
   d. compress a lock washer in place

5. The type of nut that is used to prevent contact with sharp edges of threads on a fastener is the _____ nut.
   a. castellated
   b. jam
   c. self-locking
   d. acorn

6. Washers are used to _____.
   a. distribute the load over a larger area
   b. attach an item to a hollow surface
   c. anchor materials that expand due to temperature changes
   d. allow the bolts to expand with temperature changes

7. Torque is normally expressed in _____.
   a. pounds per square inch (psi)
   b. gallons per minute (gpm)
   c. foot pounds (ft lbs)
   d. cubic feet per minute (cpm)

8. The fastener that is used to keep a gear or pulley from rotating on a shaft is the _____.
   a. key
   b. retaining ring
   c. throw bolt
   d. safety wire

9. What type of fasteners are commonly used as lifting devices?
   a. Toggle bolts
   b. Anchor bolts
   c. J-bolts
   d. Eye bolts

10. Wedge and sleeve anchors are classified as _____ anchors.
    a. bolt
    b. hollow-wall
    c. one-step
    d. screw

Summary

Fasteners and anchors are used for a variety of tasks in the industrial maintenance craft. In this module, you learned about various types of fasteners and anchors and their uses. Basic installation procedures were also included. Selecting the correct fastener or anchor for a particular job is required to perform high-quality work. It is important to be familiar with the correct terms used to describe fasteners and anchors. Using the proper technical terms helps avoid confusion and improper selection. Installation techniques for fasteners and anchors may vary depending on the job. Make sure to check the project specifications and manufacturer's information when installing any fastener or anchor. New fasteners and anchors are being developed every day. Your local distributor/manufacturer is an excellent source of information about anchors.

Notes

Trade Terms Introduced in This Module

American Society for Testing and Materials (ASTM) International: An organization that publishes specifications and standards relating to fasteners.

Clearance: The amount of space between the threads of bolts and their nuts.

Foot pounds (ft lbs): The normal method used for measuring the amount of torque being applied to bolts or nuts.

Inch pounds (in lbs): A method of measuring the amount of torque applied to small bolts or nuts that require measurement in smaller increments than foot pounds.

Key: A machined metal part that fits into a keyway and prevents parts such as gears or pulleys from rotating on a shaft.

Keyway: A machined slot in a shaft and on parts such as gears and pulleys that accepts a key.

Nominal size: A means of expressing the size of a bolt or screw. It is the approximate diameter of a bolt or screw.

Society of Automotive Engineers (SAE): An organization that publishes specifications and standards relating to fasteners.

Thread classes: Threads are distinguished by three classifications according to the amount of tolerance the threads provide between the bolt and nut.

Thread identification: Standard symbols used to identify threads.

Thread standards: An established set of standards for machining threads.

Tolerance: The amount of difference allowed from a standard.

Torque: The turning force applied to a fastener.

Unified National Coarse (UNC) thread: A standard type of coarse thread.

Unified National Extra Fine (UNEF) thread: A standard type of extra-fine thread.

Unified National Fine (UNF) thread: A standard type of fine thread.

Appendix

Mechanical Anchors and Their Uses

Anchor Type	Typically Used In	Use With Fastener	Typical Working Load Range*
One-Step Anchors			
Wedge	Concrete **Stone	None	Light, medium, and heavy duty
Stud	Concrete **Stone, solid brick and block	None	Light, medium, and heavy duty
Sleeve	Concrete, solid brick and block **Stone, hollow brick and block	None	Light and medium duty
One-piece	Concrete, solid block **Stone, solid and hollow brick, hollow block	None	Light and medium duty
Hammer-set	Concrete, solid block **Stone, solid and hollow brick, hollow block	None	Light duty
Bolt Anchors			
Drop-in	Concrete **Stone, solid brick	Machine screw or bolt	Light, medium, and heavy duty
Hollow-set drop-in	Concrete, solid brick and block **Stone, hollow brick and block	Machine screw or bolt	Light and medium duty
Single-expansion	Concrete, solid brick and block **Stone, hollow brick and block	Machine screw or bolt	Light and medium duty
Double-expansion	Concrete, solid brick and block **Stone, hollow brick and block	Machine screw or bolt	Light and medium duty
Lead (caulk-in)	Concrete, solid brick and block **Stone, hollow brick and block	Machine screw or bolt	Light and medium duty
Screw Anchors			
Lag shield	Concrete **Stone, solid and hollow brick and block	Lag screw	Light and medium duty
Fiber	Concrete, stone, solid brick and block **Hollow brick and block, wallboard	Wood, sheet metal, or lag screw	Light and medium duty
Lead	Concrete, solid brick and block **Hollow brick and block, wallboard	Wood or sheet metal screw	Light duty
Plastic	Concrete, stone, solid brick and block **Hollow brick and block, wallboard	Wood or sheet metal screw	Light duty
Hollow-Wall Anchors			
Toggle bolts	Concrete, plank, hollow block, wallboard, plywood/paneling	None	Light and medium duty
Plastic toggle bolts	Wallboard, plywood/paneling **Hollow block, structural tile	Wood or sheet metal screw	Light duty
Sleeve-type wall	Wallboard, plywood/paneling **Hollow block, structural tile	None	Light duty
Wallboard	Wallboard	Sheet metal screw	Light duty
Metal drive-in	Wallboard	Sheet metal screw	Light duty

*Anchor working loads given in the table are defined below. These are approximate loads only. Actual allowable loads depend on such factors as the anchor style and size, base material strength, spacing and edge distance, and the type of service load applied. Always consult the anchor manufacturer's product literature to determine the correct type of anchor and size to use for a specific application.

- Light duty—Less than 400 lbs.
- Medium duty—400 to 4,000 lbs.
- Heavy duty—Above 4,000 lbs.

**Indicates use may be suitable depending on the application.

Resources & Acknowledgments

Additional Resources

This module is intended to be a thorough resource for task training. The following reference works are suggested for further study. These are optional materials for continued education rather than for task training.

http://www.Thomasglobal.com
http://www.confast.com
http://www.boltdepot.com/fastener-information

Figure Credits

The Crosby Group, 103F31

Topaz Publications, Inc., 103F33

Middle Atlantic Products, Inc., 103F34

Tapcon® is a registered trademark of ITW Buildex and Illinois Tool Works Inc., 103F37

CONTREN® LEARNING SERIES — USER UPDATE

NCCER makes every effort to keep these textbooks up-to-date and free of technical errors. We appreciate your help in this process. If you have an idea for improving this textbook, or if you find an error, a typographical mistake, or an inaccuracy in NCCER's Contren® textbooks, please write us, using this form or a photocopy. Be sure to include the exact module number, page number, a detailed description, and the correction, if applicable. Your input will be brought to the attention of the Technical Review Committee. Thank you for your assistance.

Instructors – If you found that additional materials were necessary in order to teach this module effectively, please let us know so that we may include them in the Equipment/Materials list in the Annotated Instructor's Guide.

Write: Product Development and Revision
National Center for Construction Education and Research
3600 NW 43rd St. Bldg G, Gainesville, FL 32606

Fax: 352-334-0932

E-mail: curriculum@nccer.org

Craft ______________________ Module Name ______________________

Copyright Date ____________ Module Number ____________ Page Number(s) ____________

Description

____

____

____

(Optional) Correction

____

____

(Optional) Your Name and Address

____

____

Industrial Maintenance E & I Technician Level One

40104-07

Oxyfuel Cutting

40104-06
Oxyfuel Cutting

Topics to be presented in this module include:

Overview

Industrial maintenance technicians frequently fabricate or modify mounting plates or other hardware. Larger items cannot be shaped with hand or power tools. Oxyfuel cutting can be used to quickly cut, trim, and shape ferrous metals.

Oxyfuel cutting combines flame and oxygen to heat the metal and oxidize it, cutting even the hardest steel rapidly. However, particular attention must be paid to safety. Special protective clothing must be worn to protect your skin and eyes against flame and molten metal. The equipment and work area must be set up properly to minimize fire and explosion hazards. But with these precautions, oxyfuel cutting will become a useful skill to aid you in your craft.

Objectives

1. Identify and explain the use of oxyfuel cutting equipment.
2. State the safety precautions for using oxyfuel equipment.
3. Set up oxyfuel cutting equipment.
4. Light and adjust an oxyfuel torch.
5. Shut down oxyfuel cutting equipment.
6. Disassemble oxyfuel cutting equipment.
7. Change empty cylinders.
8. Perform oxyfuel cutting:
   - Straight line and square shapes
   - Piercing and slot cutting
   - Bevels
   - Washing
9. Apply a rosebud flame to remove frozen components (also for preheat and expanding larger fittings).
10. Operate a motorized, portable oxyfuel gas cutting machine.

Trade Terms

Backfire
Carburizing flame
Drag lines
Dross
Ferrous metals
Flashback
Kerf
Neutral flame
Oxidizing flame
Pierce
Soapstone

Required Trainee Materials

1. Pencil and paper
2. Appropriate personal protective equipment

Prerequisites

Before you begin this module, it is recommended that you successfully complete *Core Curriculum;* and *Industrial Maintenance E & I Technician Level One,* Modules 40101-07 through 40103-07.

This course map shows all of the modules in the first level of the *Industrial Maintenance E & I Technician* curriculum. The suggested training order begins at the bottom and proceeds up. Skill levels increase as you advance on the course map. The local Training Program Sponsor may adjust the training order.

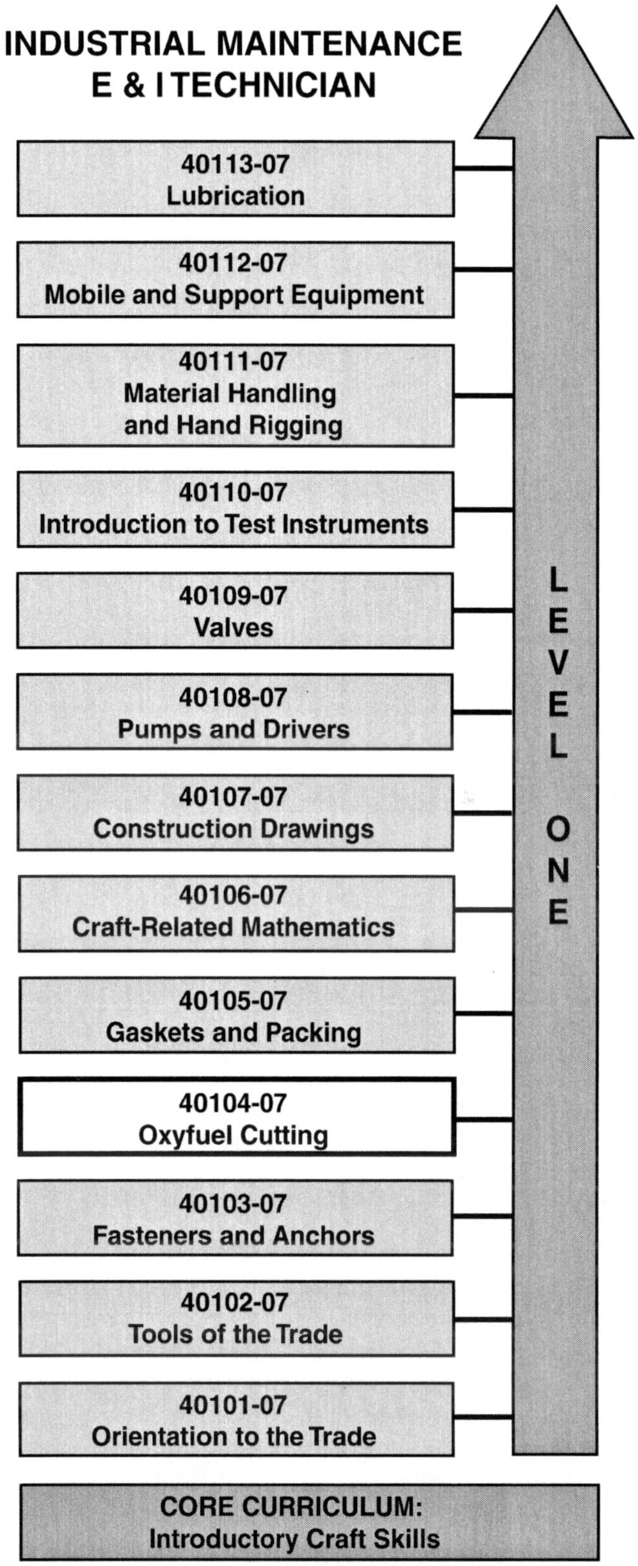

104CMAP.EPS

1.0.0 ◆ INTRODUCTION

Oxyfuel cutting (OFC), also called flame cutting or burning, is a process that uses the flame and oxygen from a cutting torch to cut **ferrous metals**. The flame is produced by burning a fuel gas mixed with pure oxygen. The flame heats the metal to be cut to the kindling temperature (a cherry-red color); then a stream of high-pressure pure oxygen is directed from the torch at the metal's surface. This causes the metal to instantaneously oxidize or burn. The cutting process results in oxides that mix with molten iron and produce **dross**, which is blown from the cut by the jet of cutting oxygen.

The oxyfuel cutting process (*Figure 1*) is usually used only on ferrous metals such as straight carbon steels, which oxidize rapidly. This process can be used to quickly cut, trim, and shape ferrous metals, including the hardest steel. An industrial maintenance technician may use a cutting torch to disassemble damaged machinery, cut pipe, loosen stuck parts, or fabricate components such as hangers and brackets.

Oxyfuel cutting can be used for certain metal alloys, such as stainless steel; however, the process requires higher preheat temperatures (white heat) and about 20 percent more oxygen for cutting. In addition, sacrificial steel plate or rod may have to be placed on top of the cut to help maintain the burning process. Other methods, such as carbon arc cutting, powder cutting, inert gas cutting, and plasma arc cutting, are much more practical for cutting steel alloys and nonferrous metals.

2.0.0 ◆ OXYFUEL CUTTING SAFETY

The proper safety equipment and precautions must be used when working with oxyfuel equipment because of the potential danger from the high-pressure flammable gases and high temperatures used. The following is a summary of safety procedures and practices that must be observed while cutting or welding. Keep in mind that this is just a summary. Above all, be sure to wear appropriate protective clothing and equipment when welding or cutting.

2.1.0 Protective Clothing and Equipment

- Always use safety goggles with a full face shield or a helmet. The goggles, face shield, or helmet lens must have the proper light-reducing tint for the type of welding or cutting to be performed. Never directly or indirectly view an electric arc without using a properly tinted lens (*Figure 2*).
- Wear proper protective leather and/or flame retardant clothing along with welding gloves that will protect you from flying sparks and molten metal, as well as heat.
- Wear 8-inch or taller high-top safety shoes or boots. Make sure that the tongue and lace area of the footwear will be covered by a pant leg. If the tongue and lace area is exposed or the footwear must be protected from burn marks, wear leather spats under the pants or chaps and over the top of the footwear.
- Wear a solid material (nonmesh) hat with a bill pointing to the rear or, if much overhead cutting or welding is required, a full leather hood with a welding face plate and the correctly tinted lens. If a hard hat is required, use a hard hat that allows the attachment of rear deflector material and a face shield.
- If a full leather hood is not worn, wear a face shield and snugly fitting welding goggles over safety glasses for gas welding or cutting. Either the face shield or the lenses of the welding goggles must be an approved shade 5 or 6 filter. Depending on the method used for electric arc cutting, wear safety goggles and a welding hood with the correctly tinted lens (shade 5 to 14).
- If a full leather hood is not worn, wear earmuffs, or at least earplugs, to protect your ear canals from sparks.

2.1.1 Ear Protection

Welding areas can be very noisy. In addition, if overhead work is being performed, hot sparks can cause burns to the ears and ear canals unless a leather hood is used. For maximum protection, earmuff-type hearing protectors (*Figure 3*) should also be used. They are available in varying degrees of protection from all noise, including low frequencies. Most earmuffs have adjustable headbands that can be worn over the head, behind the neck, or under the chin. To use earmuffs, adjust the tension on the headband and ear cushion pads to obtain the best possible seal. Check the earmuff

104F01.EPS

Figure 1 ◆ Oxyfuel cutting.

shell for cracks and the ear cushion pads for tears before each use. Any damaged, cracked, or torn part must be repaired or replaced. As minimal protection, earplugs (*Figure 3*) can be used. Disposable earplugs are the most common form of hearing protection used in the industry. These devices usually have an outer layer of pliable foam and a core layer of acoustical fiber that filters out harmful noise yet allows you to hear normal conversation. To use disposable earplugs, simply roll each plug into a cone and insert the tapered end into the ear canal while pulling up on the upper portion of your ear. The earplugs will expand, filling the ear canal and creating a proper fit. Reusable earplugs that can be cleaned and worn repeatedly are also commonly used. These are typically cleaned with boiling water or alcohol. Plain cotton placed in the ear is not an acceptable protective device.

2.1.2 Eye, Face, and Head Protection

The heat and light produced by cutting or welding operations can damage the skin and eyes. Injury to the eyes may result in permanent loss of vision. Oxyfuel cutting and welding can cause eye fatigue and mild burns to the skin because of the infrared heat radiated by the process. Welding or cutting operations involving an electric arc of any kind produce UV radiation, which can cause severe burns to the eyes and exposed skin and permanent damage to the retina. A flash burn can harm unprotected eyes in just seconds. Welders should never view an electric arc directly or indirectly without wearing a properly tinted lens designed for electric arc use. If electric arc operations are

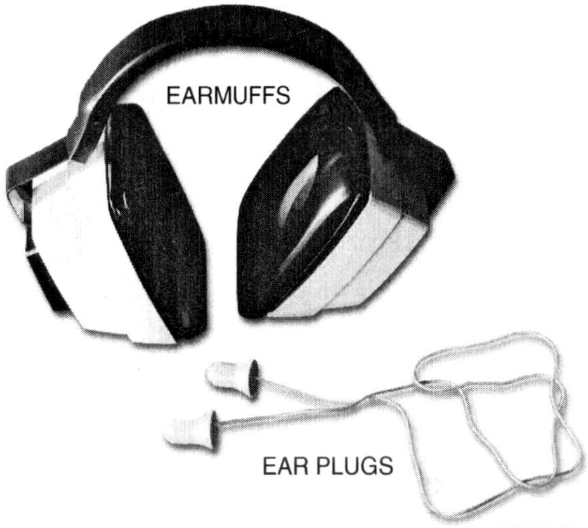

Figure 3 ◆ Typical ear protection.

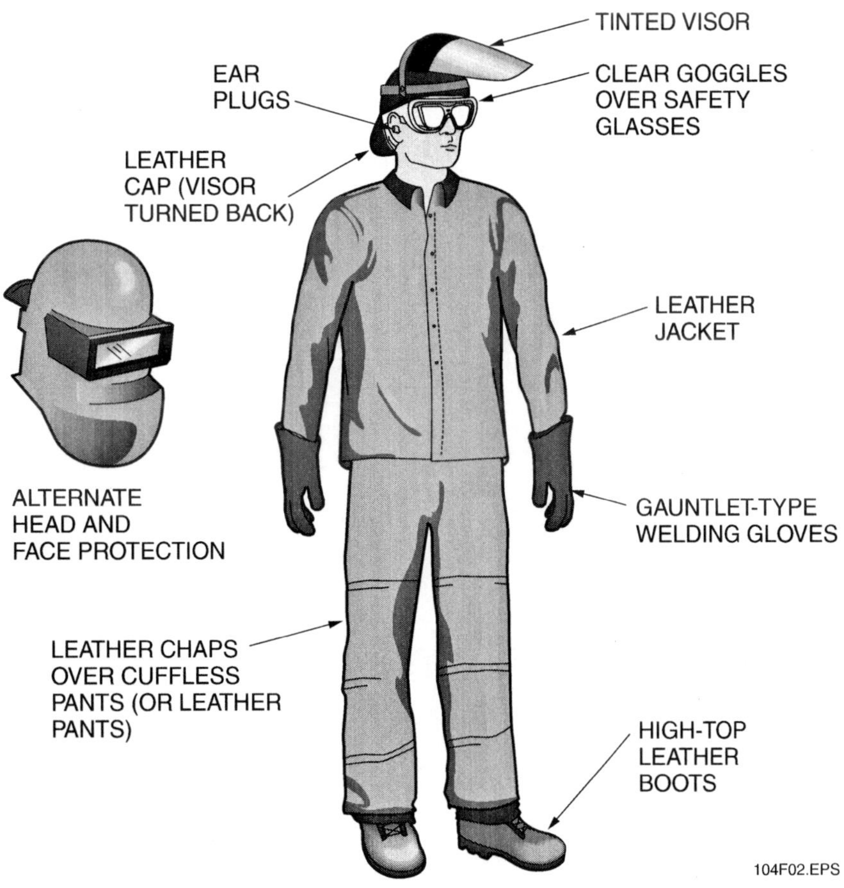

Figure 2 ◆ Typical personal protective equipment.

occurring in the vicinity, safety goggles with a tinted lens (shades 3 to 5) and tinted side shields must be worn at all times.

For oxyfuel welding and cutting, wear tinted welding goggles (shades 4 to 6) over safety glasses, and wear a clear face shield. Clear safety glasses and goggles with a tinted face shield can also be used. Most oxyfuel welders prefer the latter combination because, for clear vision, only the face shield has to be flipped up. For overhead oxyfuel operations, a leather hood may be used in place of the face shield to obtain protection from sparks and molten metal (*Figure 4*).

For electric arc operations, a leather hood or welding helmet with a properly tinted lens (shades 9 to 14) must be worn over safety goggles to provide proper protection. Many varieties of helmets are available; typical styles are shown in *Figure 5*. Some of the helmets are available with additional side-view lenses in a lighter tint so that welders can sense, by peripheral vision, any activities occurring beside them.

Most welding and cutting tasks require the use of safety goggles, chemical-resistant goggles, dust goggles, or face shields. Always check the material safety data sheet (MSDS) for the welding or cutting product being used to find out what type of eye protection is needed.

2.2.0 Ventilation

Adequate mechanical ventilation must be provided to remove fumes that are produced by welding or cutting processes. *ANSI Z49.1-1999* on welding safety covers such ventilation procedures. The gases, dust, and fumes caused by welding or cutting operations can be hazardous if the appropriate safety precautions are not observed. The following general rules can be used to determine if there is adequate ventilation:

- The welding area must contain at least 10,000 cubic feet of air for each welder.
- There must be air circulation.
- Partitions, structural barriers, or equipment must not block air circulation.

Even when there is adequate ventilation, avoid inhaling welding or cutting fumes and smoke. The heated fumes and smoke generally rise straight up. Observe the column of smoke and position yourself to avoid it. A small fan may also be used to divert the smoke, but take care to keep the fan from blowing directly on the work area; the fumes and gases must be present at an electric arc in order to protect the molten metal from the air.

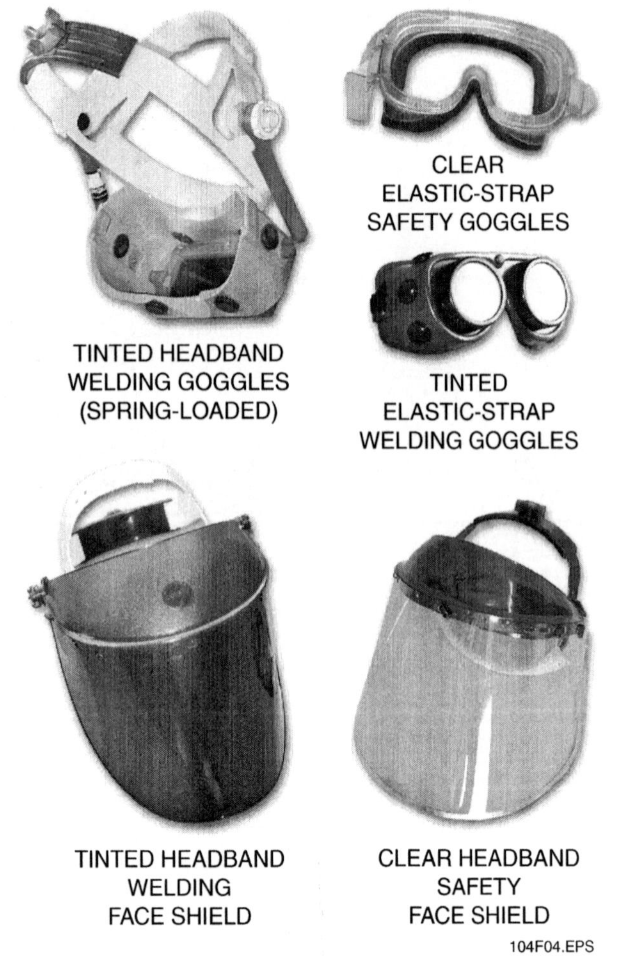

Figure 4 ◆ Oxyfuel welding/cutting goggles and face shield combinations.

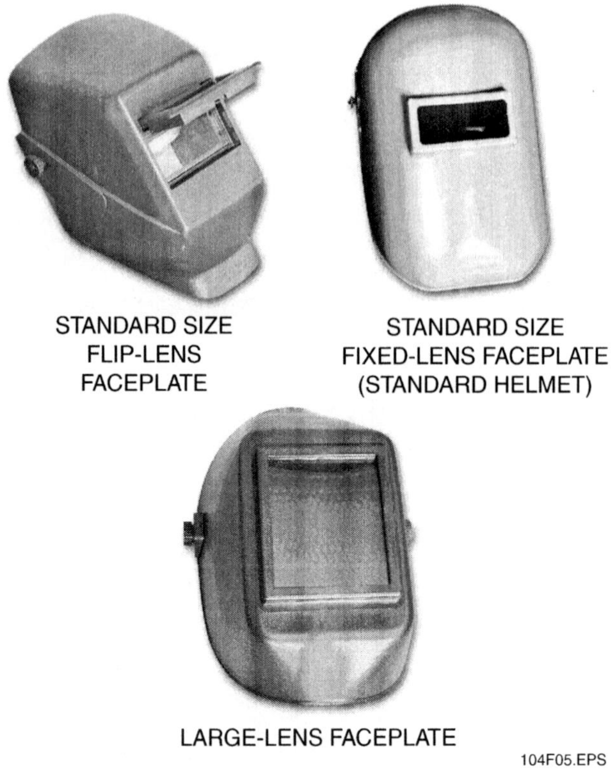

Figure 5 ◆ Typical electric arc welding helmets.

> **NOTE**
> Cutting or welding operations involving materials, coatings, or electrodes containing cadmium, mercury, lead, zinc, chromium, and beryllium result in toxic fumes. For cutting or welding such materials, always use proper area ventilation and wear an approved full-face, supplied-air respirator (SAR) that uses breathing air from outside the work area. For occasional, very short-term exposure to fumes from zinc- or copper-coated materials, a high-efficiency particulate arresting (HEPA)-rated or metal-fume filter may be used on a standard respirator.

2.2.1 Fume Hazards

Welding or cutting processes create fumes. Fumes are solid particles consisting of the base metal, electrodes or welding wire, and any coatings applied to them. Most fumes are not considered dangerous as long as there is adequate ventilation. If ventilation is questionable, use air sampling to determine the need for corrective measures. Adequate ventilation can be a problem in tight or cramped working quarters. To ensure adequate room ventilation, local exhaust ventilation should be used to capture fumes (*Figure 6*).

104F06.EPS

Figure 6 ◆ A flexible exhaust pickup.

The exhaust hood should be kept four to six inches away from the source of the fumes. Welders should recognize that fumes of any type, regardless of their source, should not be inhaled. The best way to avoid problems is to provide adequate ventilation. If this is not possible, breathing protection must be used. Protective devices for use in poorly ventilated or confined spaces are shown in *Figures 7* and *8*.

If respirators are used, your employer should offer worker training on respirator fitting and usage. Medical screenings should also be given.

> **WARNING!**
> Fumes and gases can be dangerous. Overexposure can cause nausea, headaches, dizziness, metal fume fever, and severe toxic effects that can be fatal. Studies have demonstrated irritation to eyes, skin, and the respiratory system and even more severe complications. In confined spaces, fumes and gases may cause asphyxiation.

2.3.0 Respirators

Special metals require the use of respirators to provide protection from harmful fumes. Respirators are grouped into three main types based on how they work to protect the wearer from contaminants. The types are:

- Air-purifying respirators
- Supplied-air respirators (SARs)
- Self-contained breathing apparatus (SCBA)

A respirator must be clean and in good condition, and all of its parts must be in place for it to give you proper protection. Respirators must be cleaned every day. Failure to do so will limit their effectiveness and offer little or no protection. For example, suppose you wore the respirator yesterday and did not clean it. The bacteria from breathing into the

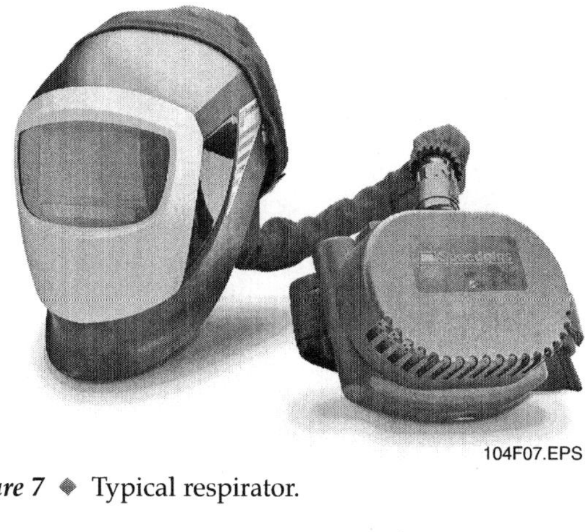

104F07.EPS

Figure 7 ◆ Typical respirator.

respirator, plus the airborne contaminants that managed to enter the facepiece, will have made the inside of your respirator very unsanitary. Continued use may cause you more harm than good. Remember, only a clean and complete respirator will provide you with the necessary protection. Follow these guidelines:

- Inspect the condition of your respirator before and after each use.
- Do not wear a respirator if the facepiece is distorted or if it is worn and cracked. You will not be able to get a proper face seal.
- Do not wear a respirator if any part of it is missing. Replace worn straps or missing parts before using.
- Do not expose respirators to excessive heat or cold, chemicals, or sunlight.
- Clean and wash your respirator after every time you use it. Remove the cartridge and filter, hand wash the respirator using mild soap and a soft brush, and let it air dry overnight.
- Sanitize your respirator each week. Remove the cartridge and filter, then soak the respirator in a sanitizing solution for at least two minutes. Thoroughly rinse with warm water and let it air dry overnight.
- Store the clean and sanitized respirator in its resealable plastic bag. Do not store the respirator face down. This will cause distortion of the facepiece.

104F08.EPS

Figure 8 ◆ A belt-mounted respirator.

2.3.1 Air-Purifying Respirators

Air-purifying respirators provide the lowest level of protection. They are made for use only in atmospheres that have enough oxygen to sustain life (at least 19.5 percent). Air-purifying respirators use special filters and cartridges to remove specific gases, vapors, and particles from the air. The respirator cartridges contain charcoal, which absorbs certain toxic vapors and gases. When the wearer detects any taste or smell, the charcoal's absorption capacity has been reached and the cartridge can no longer remove the contaminant. The respirator filters remove particles such as dust, mists, and metal fumes by trapping them within the filter material. Filters should be changed when it becomes difficult to breathe. Depending on the contaminants, cartridges can be used alone or in combination with a filter/pre-filter and filter cover. Air-purifying respirators should be used for protection only against the types of contaminants listed on the filters and cartridges and on the National Institute for Occupational Safety and Health (NIOSH) approval label affixed to each respirator carton and replacement filter/cartridge carton. Respirator manufacturers typically classify air-purifying respirators into four groups:

- No maintenance
- Low maintenance
- Reusable
- Powered air-purifying respirators (PAPRs)

No-maintenance and low-maintenance respirators are typically used for residential or light commercial work that does not call for constant and heavy respirator use. No-maintenance respirators are typically half-mask respirators with permanently attached cartridges or filters. The entire respirator is discarded when the cartridges or filters are spent. Low-maintenance respirators generally are also half-mask respirators that use replaceable cartridges and filters. However, they are not designed for constant use.

Reusable respirators (*Figure 9*) are made in half-mask and full facepiece styles. These respirators require the replacement of cartridges, filters, and respirator parts. Their use also requires a complete respirator maintenance program. Air respirator maintenance is discussed later.

Powered air-purifying respirators (PAPRs) are made in half-mask, full facepiece, and hood styles. They use battery-operated blowers to pull outside air through the cartridges and filters attached to the respirator. The blower motors can be either mask- or belt-mounted. Depending on the cartridges used, they can filter particulates, dusts, fumes, and mists along with certain gases and vapors. PAPRs like the one shown in *Figure 10* have a belt-mounted, powered air-purifier unit connected to the mask by a breathing tube. Many models also have an audible and visual alarm that is activated when airflow falls below the required minimum level. This feature gives an immediate indication of a loaded filter or low battery charge condition. Units with the blower mounted in the mask do not use a belt-mounted powered air purifier connected to a breathing tube.

2.3.2 Supplied-Air Respirators

Supplied-air respirators (*Figure 11*) provide a supply of air for extended periods of time via a high-pressure hose that is connected to an external source of air, such as a compressor, compressed-air cylinder, or pump. They provide a higher level of protection in atmospheres where air-purifying respirators are not adequate. Supplied-air respirators are typically used in toxic atmospheres. Some can be used in atmospheres that are immediately dangerous to life and health (IDLH) as long as they are equipped with an air cylinder for emergency escape. An atmosphere is considered IDLH if it poses an immediate hazard to life or produces immediate, irreversible, and debilitating effects on health. There are two types of supplied-air respirators: continuous-flow and pressure-demand.

The continuous-flow supplied-air respirator provides air to the user in a constant stream. One or two hoses are used to deliver the air from the air source to the facepiece. Unless the compressor or pump is especially designed to filter the air or a portable air-filtering system is used, the unit must be located where there is breathable air (grade D or better as described in *Compressed Gas Association*

104F09.EPS

Figure 9 ◆ Reusable half-mask air-purifying respirator.

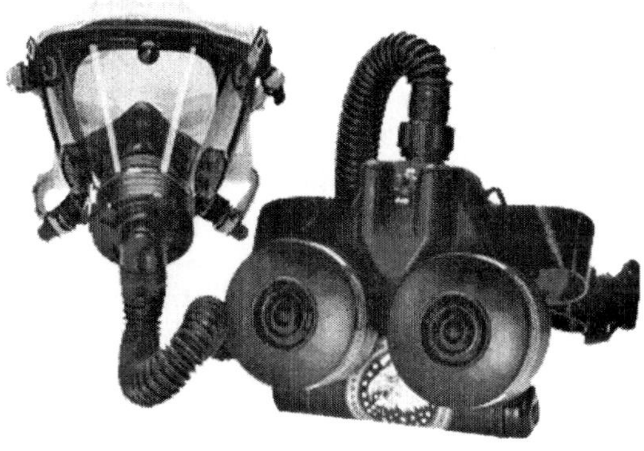

104F10.EPS

Figure 10 ◆ Typical powered air-purifying respirator (PAPR).

104F11.EPS

Figure 11 ◆ Supplied-air respirator.

[CGA] Commodity Specification G-7.1). Continuous-flow respirators are made with tight-fitting half-masks or full facepieces. They are also made with hoods. The flow of air to the user may be adjusted either at the air source (fixed flow) or on the unit's regulator (adjustable flow). Pressure-demand supplied-air respirators are similar to the continuous-flow type except that they supply air to the user's facepiece via a pressure-demand valve as the user inhales and fresh air is required. They typically have a two-position exhalation valve that allows the worker to switch between pressure-demand and negative-pressure modes to facilitate entry into, movement within, and exit from a work area.

2.3.3 Self-Contained Breathing Apparatus (SCBA)

SCBAs (*Figure 12*) provide the highest level of respiratory protection. They can be used in oxygen-deficient atmospheres (below 19.5 percent oxygen), in poorly ventilated or confined spaces, and in IDLH atmospheres. These respirators provide a supply of air for 30 to 60 minutes from a compressed-air cylinder worn on the user's back. Note that the emergency escape breathing apparatus (EEBA) is a smaller version of an SCBA cylinder. EEBA units are used for escape from hazardous environments and generally provide a five- to ten-minute supply of air.

2.3.4 Respiratory Program

Local and OSHA procedures must be followed when selecting the proper type of respirator for a particular job (*Figure 13*). A respirator must be properly selected (based on the contaminant present and its concentration level), properly fitted, and used in accordance with the manufacturer's instructions. It must be worn during all times of exposure. Regardless of the kind of respirator needed, OSHA regulations require employers to have a respirator protection program consisting of:

- Standard operating procedures for selection and use
- Employee training
- Regular cleaning and disinfecting
- Sanitary storage
- Regular inspection
- Annual fit testing
- Pulmonary function testing

As an employee, you are responsible for wearing respiratory protection when needed. When it comes to vapors or fumes, both can be eliminated in certain concentrations by the use of air-purifying devices as long as oxygen levels are acceptable. Examples of fumes are smoke billowing from a fire or the fumes generated when welding. Always check the cartridge on your respirator to make sure it is the correct type to use for the air conditions and contaminants found on the job site.

When selecting a respirator to wear while working with specific materials, you must first determine the hazardous ingredients contained in the material and their exposure levels, then

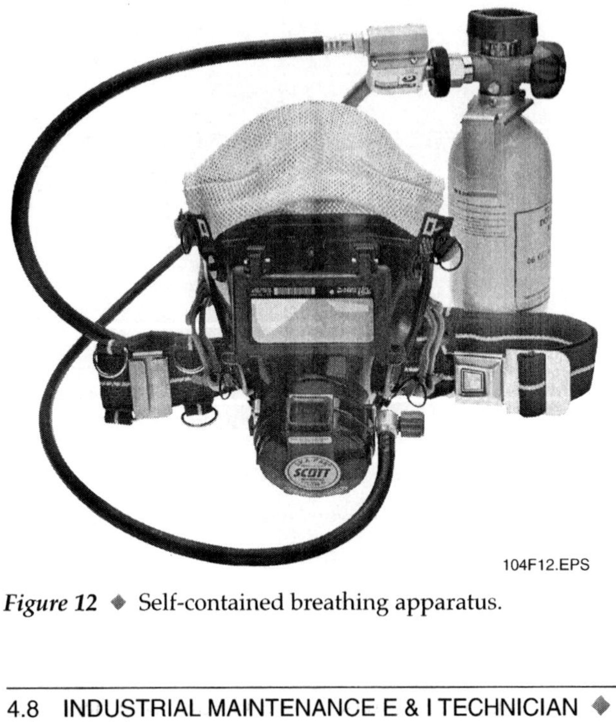

104F12.EPS

Figure 12 ◆ Self-contained breathing apparatus.

104F13.EPS

Figure 13 ◆ Use the right respirator for the job.

choose the proper respirator to protect yourself at these levels. Always read the product's MSDS. It identifies the hazardous ingredients and should list the type of respirator and cartridge recommended for use with the product.

Limitations that apply to all half-mask (air-purifying) respirators are as follows:

- These respirators do not completely eliminate exposure to contaminants, but they will reduce the level of exposure to below hazardous levels.
- These respirators do not supply oxygen and must not be used in areas where the oxygen level is below 19.5 percent.
- These respirators must not be used in areas where chemicals have poor warning signs, such as no taste or odor.

If your breathing becomes difficult, if you become dizzy or nauseated, if you smell or taste the chemical, or if you have other noticeable effects, leave the area immediately, return to a fresh air area, and seek any necessary assistance.

2.3.5 Positive and Negative Fit Checks

All respirators are useless unless properly fit-tested to each individual. To obtain the best protection from your respirator, you must perform positive and negative fit checks each time you wear it. These fit checks must be done until you have obtained a good face seal. To perform the positive fit check, do the following:

Step 1 Adjust the facepiece for the best fit, then adjust the head and neck straps to ensure good fit and comfort.

WARNING!

Do not overtighten the head and neck straps. Tighten them only enough to stop leakage. Overtightening can cause facepiece distortion and dangerous leaks.

Step 2 Block the exhalation valve with your hand or other material.

Step 3 Breathe out into the mask.

Step 4 Check for air leakage around the edges of the facepiece.

Step 5 If the facepiece puffs out slightly for a few seconds, a good face seal has been obtained.

To perform a negative fit check, do the following:

Step 1 Block the inhalation valve with your hand or other material.

Step 2 Attempt to inhale.

Step 3 Check for air leakage around the edges of the facepiece.

Step 4 If the facepiece caves in slightly for a few seconds, a good face seal has been obtained.

2.4.0 Confined Space Permits

A confined space refers to a relatively small or restricted space, such as a storage tank, boiler, or pressure vessel or small compartments, such as underground utility vaults, small rooms, or the unventilated corners of a room.

OSHA 29 CFR 1910.146 defines a confined space (*Figure 14*) as a space that:

- Is large enough and so configured that an employee can bodily enter and perform assigned work
- Has a limited or restricted means of entry or exit; for example, tanks, vessels, silos, storage bins, hoppers, vaults, and pits
- Is not designed for continuous employee occupancy.

OSHA 29 CFR 1910.146 further defines a permit-required confined space as a space that:

- Contains or has the potential to contain a hazardous atmosphere
- Contains a material that has the potential for engulfing an entrant

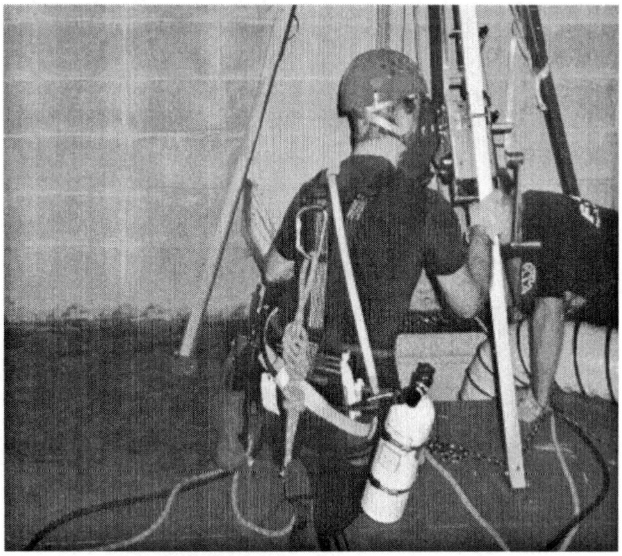

104F14.EPS

Figure 14 ◆ Worker entering a confined space with a restricted opening for entry and exit.

- Has an internal configuration such that an entrant could be trapped or asphyxiated by inwardly converging walls or by a floor that slopes downward and tapers to a smaller cross section
- Contains any other recognized serious safety or health hazard

For safe working conditions, the oxygen level in a confined space atmosphere must range between 19.5 and 21.5 percent by volume as measured with an oxygen analyzer, with 21 percent being considered the normal level. Oxygen concentrations below 19.5 percent by volume are considered deficient; those above 23.5 percent by volume are considered enriched (*Table 1*).

Table 1 indicates the effect of increases and decreases in oxygen levels in a confined space. If too much oxygen is ventilated into a confined space, it can be absorbed by the welder's clothing and ignite. If too little oxygen is present, it can lead to the welder's death in minutes. For this reason, the following precautions apply:

- Make sure confined spaces are ventilated properly for cutting or welding purposes.
- Never use oxygen in confined spaces for ventilation purposes.

When welding or cutting is being performed in any confined space, the gas cylinders and welding machines are left on the outside. Before operations are started, the wheels for heavy portable equipment are securely blocked to prevent accidental movement. Where a welder must enter a confined space through a manhole or other opening, all means are provided for quickly removing the worker in case of emergency. When safety harnesses and lifelines are used for this purpose, they are attached to the welder's body so that he or she cannot be jammed in a small exit opening. An attendant with a pre-planned rescue procedure is stationed outside to observe the welder at all times and must be capable of putting the rescue operations into effect.

Table 1 Effects of an Increase or Decrease in Oxygen Levels

Oxygen Level	Effects
> 21.5 percent	Easy ignition of flammable material such as clothes
19.5 – 21.5 percent	Normal
17 percent	Deterioration of night vision, increased breathing volume, accelerated heartbeat
14 – 16 percent	Very poor muscular coordination, rapid fatigue, intermittent respiration
6 – 10 percent	Nausea, vomiting, inability to perform, unconsciousness
< 6 percent	Spasmodic breathing, convulsive movements, and death in minutes

When welding or cutting operations are suspended for any substantial period of time, such as during lunch or overnight, all electrodes are removed from the holders, and the holders are carefully located so that accidental contact cannot occur. The welding machines are also disconnected from the power source.

In order to eliminate the possibility of gas escaping through leaks or improperly closed valves when gas welding or cutting, the gas and oxygen supply valves must be closed, the regulators released, the gas and oxygen lines bled, and the valves on the torch shut off when the equipment will not be used for a substantial period of time. Where practical, the torch and hose are also removed from the confined space. After welding operations are completed, the welder must mark the hot metal or provide some other means of warning other workers.

2.5.0 Area Safety

An important factor in area safety is good housekeeping. The work area should be picked up and swept clean. The floors and workbenches should be free of dirt, scrap metal, grease, oil, and anything that is not essential to accomplishing the given task. Collections of steel, welding electrode studs, wire, hoses, and cables are difficult to work around and easy to trip over. An electrode caddy can be used to hold the electrodes. Hooks can be made to hold hoses and cables, and scrap steel should be thrown into scrap bins.

The ideal welding shop should have bare concrete floors and bare metal walls and ceilings to reduce the possibility of fire. Never weld or cut over wood floors, as this increases the possibility of fire.It is important to keep flammable liquids as well as rags, wood scraps, piles of paper, and other combustibles out of the welding area.

If you must weld in an enclosed building, make every effort to eliminate anything that could trap a spark. Sparks can smolder for hours and then burst into flames. Regardless of where you're welding, be sure to have a fire extinguisher nearby. Also, keep a five-gallon bucket of water handy to cool off hot metal and quickly douse small fires. If a piece of hot metal must be left unattended, use **soapstone** to write the word HOT on it before leaving. This procedure can also be used to warn people of hot tables, vises, firebricks, and tools.

Never use a cutting torch inside your workshop unless a proper cutting area is available. Take whatever you're going to cut outside, away from flammables. Also be aware that welding sparks can ignite gasoline fumes in a confined space.

Whenever welding must be done outside a welding booth, use portable screens to protect other personnel from the arc or reflected glare (see *Figure 15*). The portable screen also prevents drafts of air from interfering with the stability of the arc.

The most common welding accident is burned hands and arms. Keep first-aid equipment nearby to treat burns in the work area. Eye injuries can also occur if you are careless. Post emergency phone numbers in a prominent location.

The following are some work-area reminders:

- Eliminate tripping hazards by coiling cables and keeping clamps and other tools off the floor.
- Don't get entangled in cables, loose wires, or clothing while you work. This allows you to move freely, especially should your clothing ignite or some other accident occur.
- Clean up oil, grease, or other agents that may ignite and splatter off surfaces while welding.
- Shut off the welder and disconnect the power plug before performing any service or maintenance.
- Keep the floor free of electrodes once you begin to weld. They could cause a slip or fall.
- Work in a dry area, booth, or other shielded area whenever possible.
- Make sure there are no open doors or windows through which sparks may travel to flammable materials.

104F15.EPS

Figure 15 ◆ A typical welding screen.

2.6.0 Hot Work Permits and Fire Watches

A hot work permit (*Figure 16*) is an official authorization from the site manager to perform work that may pose a fire hazard. The permit includes information such as the time, location, and type of work being done. The hot work permit system promotes the development of standard fire safety guidelines. Permits also help managers keep records of who is working where and at what time. This information is essential in the event of an emergency or at other times when personnel need to be evacuated.

During a fire watch, a person other than the welder or cutting operator must constantly scan the work area for fires. Fire watch personnel must have ready access to fire extinguishers and alarms and know how to use them. Cutting operations must never be performed without a fire watch. Whenever oxyfuel cutting equipment is used, there is a great danger of fire. Hot work permits and fire watches are used to minimize this danger. Most sites require the use of hot work permits and fire watches. When they are violated, severe penalties are imposed.

Localities often require longer fire watches. Be sure to find out the local fire watch requirements.

> **WARNING!**
> Never perform any type of heating, cutting, or welding until you have obtained a hot work permit and established a fire watch. If you are unsure of the procedure, check with your supervisor. Violation of hot work permit and fire watch procedures can result in serious injury or death.

2.7.0 Cutting Containers

Cutting and welding activities present unique hazards depending upon the material being cut or welded and the fuel used to power the equipment. All cutting and welding should be done in designated areas of the shop if possible. These areas should be made safe for welding and cutting operations with concrete floors, arc filter screens, protective drapes, curtains or blankets (*Figure 17*), and fire extinguishers. No combustibles should be stored nearby.

HOT WORK PERMIT

FOR CUTTING, WELDING, OR SOLDERING WITH PORTABLE GAS OR ARC EQUIPMENT

Job Date ___________ Start Time ___________ Expiration ___________ W/O# ___________

Applicant Name ______________________ Company / Dep't. ______________ Phone _________

Supervisor ______________________________________ Phone ___________________________

Location / Description of work____

____

IS FIRE WATCH REQUIRED? (ref. NFPA 51-B 3-3)

1.______(yes or no) Are combustible materials in building construction closer that 35 feet to the point of operation?

2.______(yes or no) Are combustibles more than 35 feet away but would be easily ignited by sparks?

3.______(yes or no) Are wall or floor openings within a 35 foot radius exposing combustible material in adjacent areas, including concealed spaces in floors or walls?

4.______(yes or no) Are combustible materials adjacent to the other side of metal partitions, walls, ceilings, or roofs which could be ignited by conduction or radiation?

5.______(yes or no) Does the work necessitate disabling a fire detection, suppression, or alarm system component?

YES to any of the above indicates that a qualified fire watch is required.

Fire Watcher Name(s) ____ Phone __________

NOTIFICATIONS

NOTIFY THE FOLLOWING GROUPS AT LEAST 72 HOURS PRIOR TO WORK AND 30 MINUTES AFTER WORK IS COMPLETED. Write in names of persons contacted.

*Facilities Management Service Desk (492-5522) ______________________________________

**Facilities Management Fire Alarm Supervisor (492-0633) ______________________________

*Facilities Management Fire Protection Group (FPG) (492-5681, 492-4042)___________________

*Environmental Health and Safety Industrial Hygiene Group (492-6025)_____________________

* Notify by phone or in person. If by phone, write down name of person and send them a completed copy of this permit.

**Notify in person.

SIGNATURES REQUIRED

University Project Manager __________________________ Date __________ Phone ________

I understand and will abide by the conditions described in this permit. I will implement the necessary precautions which are outlined on both sides of this permit form. Thirty minutes after each hot work session, I will reinspect work areas and adjacent areas to which spark and heat might have spread to verify that they are fire safe and contact Facilities Management Alarm Technicians to have any disabled fire protection systems reactivated.

___________________________ , ___________________ Date ___________ Phone __________ Permit Applicant Company or Dep't

104F16.EPS

Figure 16 ◆ Hot work permit.

WARNING!
Welding or cutting must never be performed on drums, barrels, tanks, vessels, or other containers until they have been emptied and cleaned thoroughly, eliminating all flammable materials and all substances (such as detergents, solvents, greases, tars, or acids) that might produce flammable, toxic, or explosive vapors when heated.

Containers must be cleaned by steam cleaning, flushing with water, or washing with detergent until all traces of the material have been removed.

WARNING!
Clean containers only in well-ventilated areas. Vapors can accumulate during cleaning, causing explosions or injury.

After cleaning the container (*Figure 18*), fill it with water or an inert gas such as argon or carbon dioxide (CO_2) for additional safety. Air, which contains oxygen, is displaced from inside the container by the water or inert gas. Without oxygen, combustion cannot take place. When using water, position the container to minimize the air space. When using an inert gas, provide a vent hole so the inert gas can purge the air.

3,000°F INTERMITTENT, 1,500°F CONTINUOUS SILICON DIOXIDE CLOTH

104F17.EPS

Figure 17 ◆ A typical welding blanket.

WARNING!
Do not assume that a container that has held combustibles is clean and safe until proven so by proper tests. Do not weld in places where dust or other combustible particles are suspended in air or where explosive vapors are present. Removal of flammable materials from vessels/ containers may be done by steaming or boiling.

Proper procedures for cutting or welding hazardous containers are described in the American Welding Society (AWS) *F4.1-1999, Recommended Safe Practices for the Preparation for Welding and*

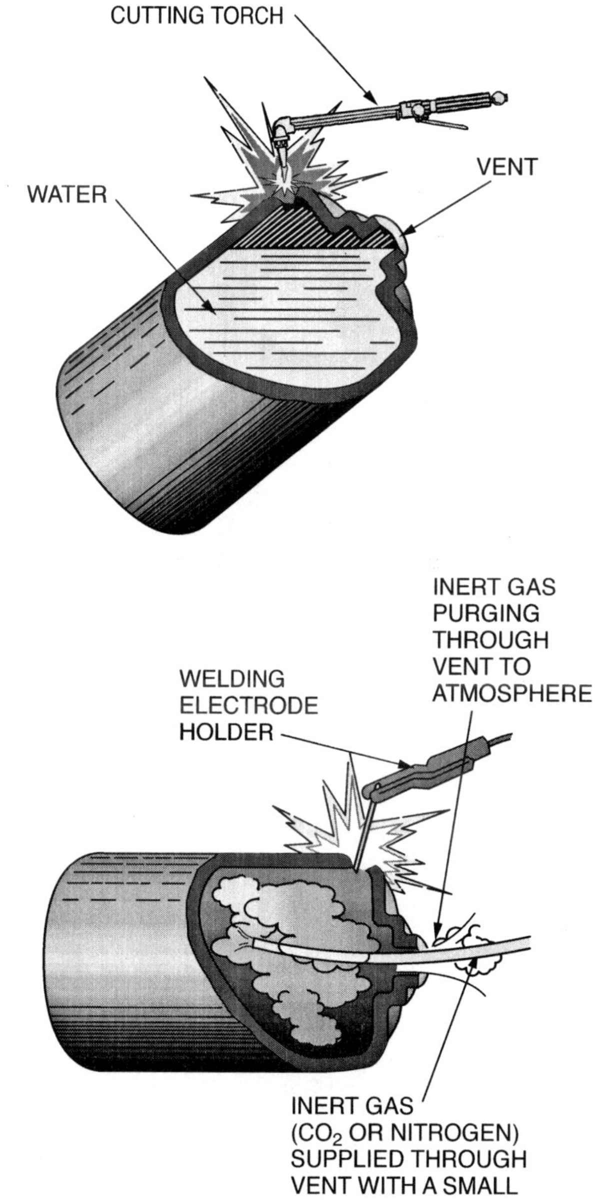

104F18.EPS

Figure 18 ◆ Purging containers of potential health hazards.

Cutting of Containers and Piping. You should also consult with the local fire marshal before welding or cutting such containers.

2.8.0 Cylinder Storage and Handling

Oxygen and fuel gas cylinders or other flammable materials must be stored separately. The storage areas must be separated by 20 feet or by a wall 5 feet high with at least a 30-minute burn rating. The purpose of the distance or wall is to keep the heat of a small fire from causing the oxygen cylinder safety valve to release. If the safety valve releases the oxygen, a small fire would become a raging inferno.

Inert gas cylinders may be stored separately or with either fuel cylinders or oxygen cylinders. Empty cylinders must be stored separately from full cylinders, although they may be stored in the same room or area. All cylinders must be stored vertically and have the protective caps screwed on firmly.

> **WARNING!**
> An accumulation of an oxygen and fuel-gas mixture can result in a dangerous explosion when ignited. A 2-inch balloon filled with an oxygen and acetylene mixture has the explosive power of an M80 firecracker (one-quarter stick of dynamite). The mixed gases from 100 feet of ¼-inch twin hose will fill a 9-inch balloon. A 9-inch balloon filled with mixed oxygen and acetylene gases has an explosive power that is 90 times more powerful than an M80 firecracker.

2.8.1 Securing Gas Cylinders

Cylinders must be secured with a chain or other device so that they cannot be knocked over accidentally. Even though they are more stable, cylinders attached to a manifold should be chained, as should cylinders stored in a special room used only for cylinder storage.

2.8.2 Storage Areas

Cylinder storage areas must be located away from halls, stairwells, and exits so that in case of an emergency they will not block an escape route. Storage areas should also be located away from heat, radiators, furnaces, and welding sparks. The location of storage areas should be where unauthorized people cannot tamper with the cylinders. A warning sign that reads Danger—No Smoking, Matches, or Open Lights, or similar wording, should be posted in the storage area.

2.8.3 Cylinders with Valve Protection Caps

Cylinders equipped with a valve protection cap must have the cap in place unless the cylinder is in use. The protection cap prevents the valve from being broken off if the cylinder is knocked over. If the valve of a full high-pressure cylinder (argon, oxygen, CO_2, or mixed gases) is broken off, the cylinder can fly around the shop like a missile if it has not been secured properly. Never lift a cylinder by the safety cap or valve. The valve can easily break off or be damaged. When moving cylinders, the valve protection cap must be replaced (*Figure 19*), especially if the cylinders are mounted on a truck or trailer. Cylinders must never be dropped or handled roughly.

High-pressure cylinders can also be equipped with a clamshell cap that can be closed to protect the cylinder valve with or without a regulator installed on the valve. This enables safe movement of the cylinder after the cylinder valve is closed. This type of cap is usually secured to the cylinder body cap threads when it is installed so that it cannot be removed. When the clamshell is closed, it can also be padlocked to prevent unauthorized operation of the cylinder valve.

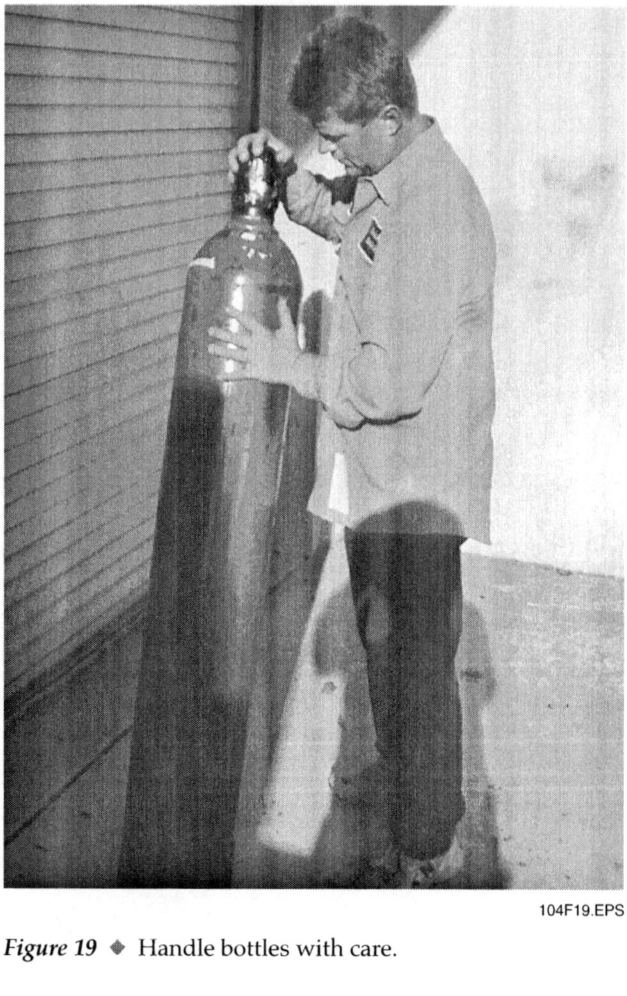

104F19.EPS

Figure 19 ◆ Handle bottles with care.

2.8.4 General Precautions

Use warm water (not boiling) to loosen cylinders that are frozen to the ground. Any cylinder that leaks, has a bad valve, or has gas-damaged threads must be identified and reported to the supplier. Use a piece of soapstone to write the problem on the cylinder. If closing the cylinder valve cannot stop the leak, move the cylinder outdoors to a safe location, away from any source of ignition, and notify the supplier. Post a warning sign, then slowly release the pressure.

In its gaseous form, acetylene is extremely unstable and explodes easily. For this reason it must remain at pressures below 15 pounds per square inch (psi). If an acetylene cylinder is tipped over, stand it upright and wait at least 30 minutes before using it. If liquid acetone is withdrawn from a cylinder, it will gum up the safety check valves and regulators and decrease the stability of the acetylene stored in the cylinder. For this reason, acetylene must never be withdrawn at a per-hour rate that exceeds 10 percent of the volume of the cylinder(s) in use. Acetylene cylinders in use should be opened no more than one and one-half turns and, preferably, no more than three-fourths of a turn. Other precautions include:

- Use only compressed gas cylinders containing the correct gas for the process used and properly operating regulators designed for the gas and pressure used.
- Make sure all hoses, fittings, and other parts are suitable and maintained in good condition.
- Keep cylinders in the upright position and securely chained to an undercarriage or fixed support.
- Keep combustible cylinders in one area of the building for safety. Cylinders must be at a safe distance from arc welding or cutting operations and any other source of heat, sparks, or flame.
- Never allow the electrode, electrode holder, or any other electrically hot parts to come in contact with the cylinder.
- When opening a cylinder valve, keep your head and face clear of the valve outlet.
- Always use valve protection caps on cylinders when they are not in use or are being transported.

3.0.0 ◆ OXYFUEL CUTTING EQUIPMENT

The equipment used to perform oxyfuel cutting includes oxygen and fuel gas cylinders, oxygen and fuel gas regulators, hoses, and a cutting torch. A typical movable oxyfuel (oxyacetylene) cutting outfit is shown in *Figure 20*.

If the torch cart in *Figure 20* is to be used only to transport the rig to the worksite, and to hold the tanks while the work is being done, it is OSHA compliant. If it is to be used to store the tanks, the cart shown is not compliant. In order to store the two tanks together on a cart, a dividing wall 5 feet tall and capable of withstanding a fire for 30 minutes must be placed in between the tanks. Such carts are now available from vendors, or can be retrofitted on the standard tank cart, by adding a half-inch steel plate, one foot wide and at least five feet tall, between the tanks.

3.1.0 Oxygen

Oxygen (O_2) is a colorless, odorless, tasteless gas that supports combustion. Combined with burning material, pure oxygen causes a fire to flare and burn out of control. When mixed with fuel gases, oxygen produces the high-temperature flame required in order to flame cut metals.

3.1.1 Oxygen Cylinders

Oxygen is stored at more than 2,000 pounds per square inch (psi) in hollow steel cylinders. The cylinders come in a variety of sizes based on the cubic feet of oxygen they hold. The smallest standard cylinder holds about 85 cubic feet of oxygen, and the largest ultra-high-pressure cylinder holds about 485 cubic feet. The most common size oxygen cylinder used for welding and cutting operations is the 227 cubic foot cylinder. It is more than

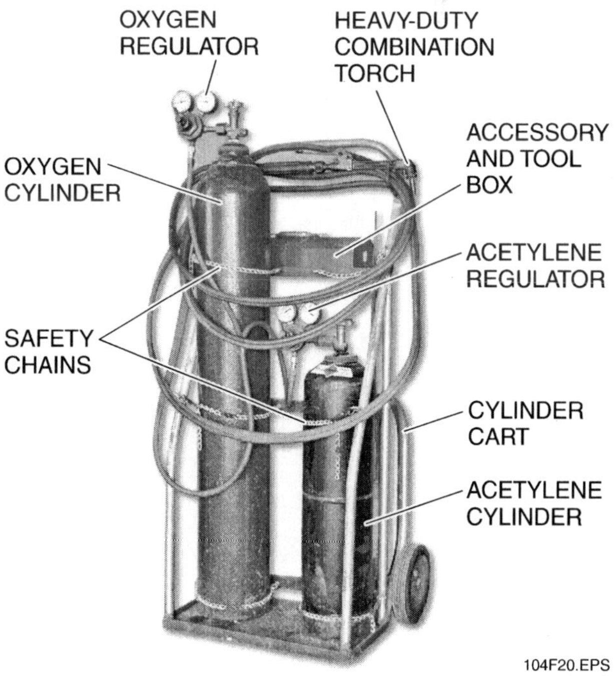

Figure 20 ◆ Typical oxyacetylene welding/cutting outfit.

4 feet tall and 9 inches in diameter. The shoulder of the oxygen cylinder has the name of the gas and / or the supplier stamped, labeled, or stenciled on it. *Figure 21* shows standard high-pressure oxygen cylinder markings and sizes. The cylinders must be tested every 10 years.

Oxygen cylinders have bronze cylinder valves on top (*Figure 22*). Turning the cylinder valve handwheel controls the flow of oxygen out of the cylinder. A safety plug on the side of the cylinder valve allows oxygen in the cylinder to escape if the pressure in the cylinder rises too high. Oxygen cylinders are usually equipped with Compressed Gas Association (CGA) 540 valves for service up to 3,000 per square inch gauge (psig). Some cylinders are equipped with CGA 577 valves for up to 4,000 psig service or CGA 701 valves for up to 5,500 psig service.

Use care when handling oxygen cylinders because oxygen is stored at such high pressures. When it is not in use, always cover the cylinder valve with the protective steel safety cap (*Figure 23*).

> **WARNING!**
> Do not remove the protective cap unless the cylinder is secured. If the cylinder falls over and the nozzle breaks off, the cylinder will be propelled like a rocket, causing severe injury or death to anyone in its way.

3.2.0 Acetylene

Acetylene gas (C_2H_2), a compound of carbon and hydrogen, is lighter than air. It is formed by dissolving calcium carbide in water. It has a strong, distinctive, garlic-like odor. In its gaseous form, acetylene is extremely unstable and explodes easily. Because of this instability, it cannot be compressed at pressures of more than 15 psi when in its gaseous form. At higher pressures, acetylene gas breaks down chemically, producing heat and pressure that could result in a violent explosion. When combined with oxygen, acetylene creates a flame that burns hotter than 5,500°F, one of the hottest gas flames. Acetylene can be used for flame cutting, welding, heating, flame hardening, and stress relieving.

3.2.1 Acetylene Cylinders

Because of the explosive nature of acetylene gas, it cannot be stored above 15 psi in a hollow cylinder. To solve this problem, acetylene cylinders are specially constructed to store acetylene at higher pressures. The acetylene cylinder is filled with a porous material that creates a solid, instead of a hollow, cylinder. The porous material is soaked with acetone, which absorbs the acetylene, stabilizing it for storage at higher pressures. Because of the liquid acetone inside the cylinder, acetylene cylinders must always be used in an upright position. If the cylinder is tipped over, stand the cylinder upright and wait at least 30 minutes before using it. If liquid acetone is withdrawn from a cylinder, it will gum up the safety check valves and regulators. Always take care to withdraw acetylene gas from a cylinder at pressures less than 15 psig and at hourly rates that do not exceed one-tenth of the cylinder capacity. Higher rates

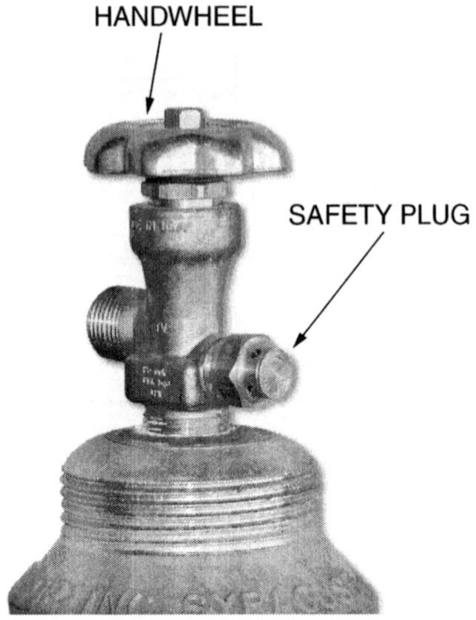

Figure 22 ◆ Oxygen cylinder valve.

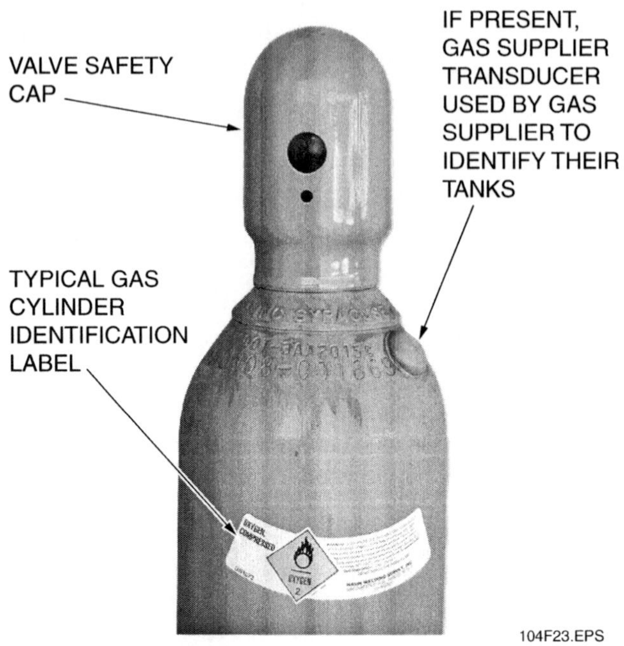

Figure 23 ◆ Oxygen cylinder with standard safety cap.

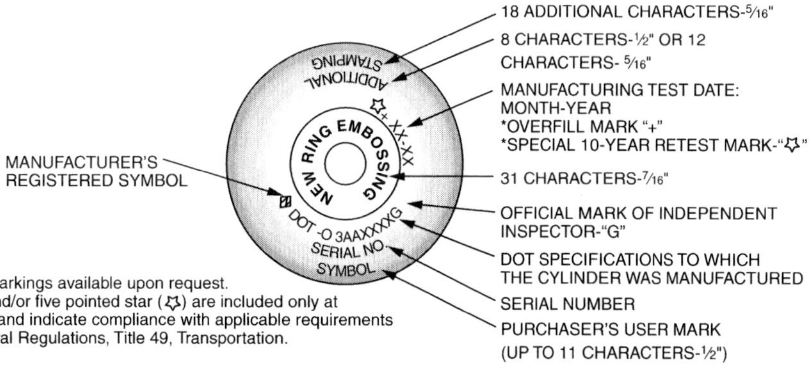

Transport Canada Markings available upon request.
*The plus sign (+) and/or five pointed star (☆) are included only at customer's request, and indicate compliance with applicable requirements of the Code of Federal Regulations, Title 49, Transportation.

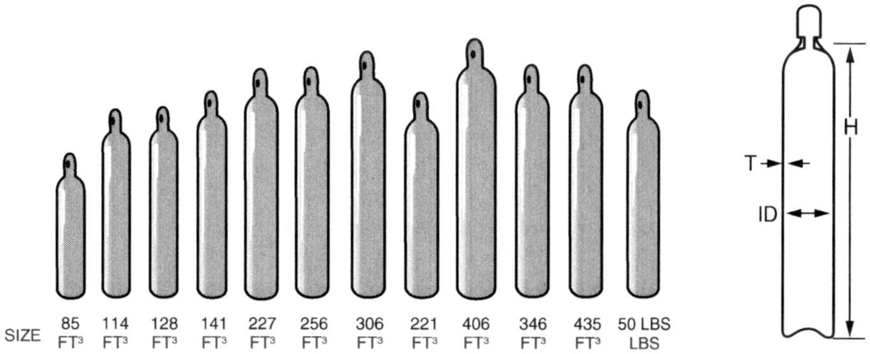

HIGH PRESSURE CYLINDER MARKINGS

DOT SPECIFICATIONS	O_2 CAPACITY (FT^3) AT RATED SERVICE PRESSURE	O_2 CAPACITY (FT^3) AT 10% OVERCHARGE	WATER CAPACITY (IN^3) MINIMUM	WATER CAPACITY (IN^3) MAXIMUM	NOMINAL DIMENSIONS (IN) AVG. INSIDE DIAMETER "ID"	NOMINAL DIMENSIONS (IN) HEIGHT "H"	NOMINAL DIMENSIONS (IN) MINIMUM WALL "T"	NOMINAL WEIGHT (LB)	PRESSURE (PSI) SERVICE	PRESSURE (PSI) TEST
STANDARD HIGH PRESSURE CYLINDERS[1]										
3AA2015	85	93	960	1040	6.625	32.50	0.144	48	2015	3360
3AA2015	114	125	1320	1355	6.625	43.00	0.144	61	2015	3360
3AA2265	128	140	1320	1355	6.625	43.00	0.162	62	2265	3775
3AA2015	141	155	1630	1690	7.000	46.00	0.150	70	2015	3360
3AA2015	227	250	2640	2710	8.625	51.00	0.184	116	2015	3360
3AA2265	256	281	2640	2710	8.625	51.00	0.208	117	2265	3775
3AA2400	306	336	2995	3060	8.813	55.00	0.226	140	2400	4000
3AA2400	405	444	3960	4040	10.060	56.00	0.258	181	2400	4000
ULTRALIGHT® HIGH PRESSURE CYLINDERS[1]										
E-9370-3280	365	NA	2640	2710	8.625	51.00	0.211	122	3280	4920
E-9370-3330	442	NA	3181	3220	8.813	57.50	0.219	147	3330	4995
ULTRA HIGH PRESSURE CYLINDERS[2]										
3AA3600	347[3]	374	2640	2690	8.500	51.00	0.336	170	3600	6000
3AA6000	434[3]	458	2285	2360	8.147	51.00	0.568	267	6000	10000
E-10869-4500	435[3]	NA	2750	2890	8.813	51.00	0.260	148	4500	6750
E-10869-4500	485[3]	NA	3058	3210	8.813	56.00	0.260	158	4500	6750

1. Regulators normally permit filling these cylinders with 10% overcharge, provided certain other requirements are met.
2. Under no circumstances are these cylinders to be filled to a pressure exceeding the marked service pressure at 70°F.
3. Nitrogen capacity at 70°F.

All cylinders normally furnished with 3/4" NGT internal threads, unless otherwise specified.
Nominal weights include neck ring but exclude valve and cap, add 2 lbs. (.91 kg) for cap and 1 1/2 lb. (.8 kg) for valve.
Cap adds approximately 5 in. (127 mm) to height.
Cylinder capacities are approximately 5 in. (127 mm) to height.
Cylinder capacities are approximately at 70°F. (21°C).

104F21.EPS

Figure 21 ◆ High-pressure oxygen cylinder markings and sizes.

may cause liquid acetone to be withdrawn along with the acetylene.

Acetylene cylinders have safety fuse plugs in the top and bottom of the cylinder that melt at 220°F (*Figure 24*). In the event of a fire, the fuse plugs will release the acetylene gas, preventing the cylinder from exploding.

Acetylene cylinders are available in a variety of sizes based on the cubic feet of acetylene that they can hold. The smallest standard cylinder holds about 10 cubic feet of gas. The largest standard cylinder holds about 420 cubic feet of gas. A cylinder that holds about 850 cubic feet is also available. *Figure 25* shows standard acetylene cylinder markings and sizes. Like oxygen cylinders, acetylene cylinders must be tested every 10 years.

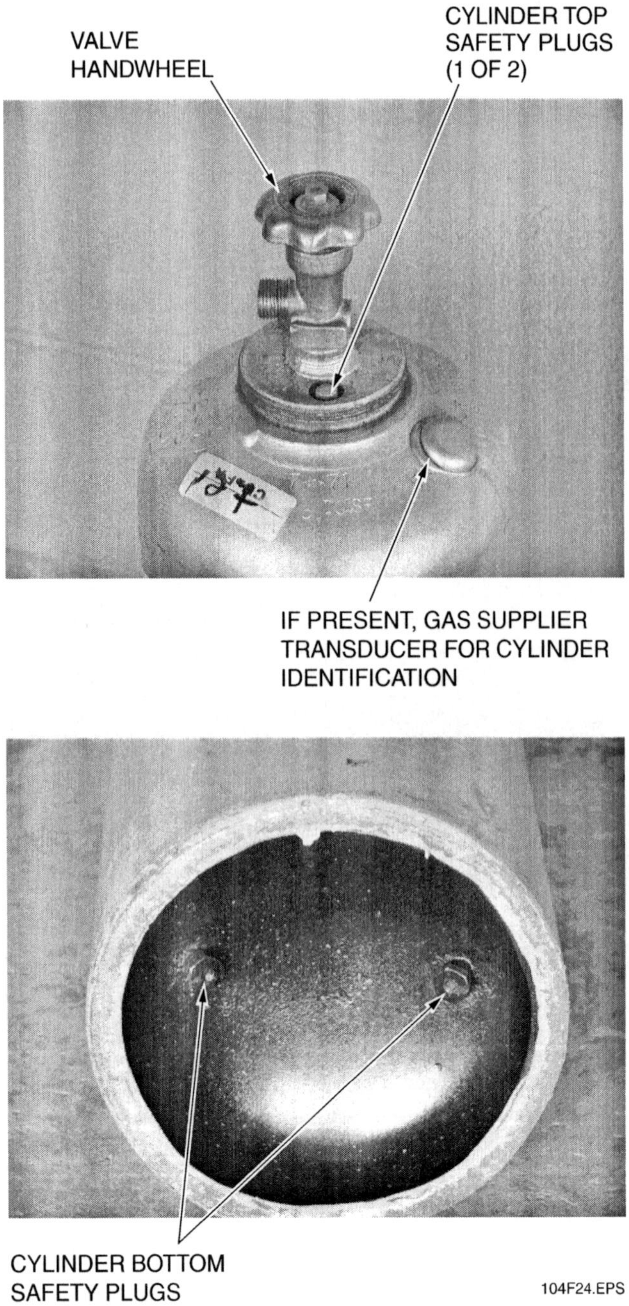

Figure 24 ◆ Acetylene cylinder valve and safety plugs.

Acetylene cylinders are usually equipped with a standard CGA 510 brass cylinder valve (see *Figure 24*). The handwheel of the valve controls the flow of acetylene from the cylinder to a regulator.

Some acetylene cylinders are equipped with an alternate standard CGA 300 valve. Some obsolete valves still in use require a special long-handled wrench with a square socket end to operate the valve.

The smallest standard acetylene cylinder, which holds 10 cubic feet, is equipped with a CGA 200 small series valve, and 40 cubic foot cylinders use a CGA 520 small series valve. As with oxygen cylinders, place a protective valve cap on the acetylene cylinders during transport (*Figure 26*).

NOTE

Acetylene cylinders can be equipped with a ring guard cap that protects the cylinder valve with or without a regulator installed on the valve. This enables safe movement of the cylinder after the cylinder valve is closed. This type of cap is usually secured to the cylinder body cap threads when it is installed so that it cannot be removed.

WARNING!

Do not remove the protective cap unless the cylinder is secured. If the cylinder falls over and the nozzle breaks off, the cylinder will release highly explosive gas.

3.3.0 Liquefied Fuel Gases

Many fuel gases other than acetylene are used for cutting. They include natural gas and liquefied fuel gases such as methylacetylene propadiene (MAPP®), propylene, and propane. Their flames are not as hot as acetylene, but they have higher British thermal unit (Btu) ratings and are cheaper and safer to use. The supervisor at your job site will determine which fuel gas to use.

Table 2 compares the flame temperatures of oxygen mixed with various fuel gases.

MAPP® is a Dow Chemical Company product that is a chemical combination of acetylene and propane gases. MAPP® gas burns at temperatures almost as high as acetylene and has the stability of propane. Because of this stability, it can be used at pressures over 15 psi and is not as likely as

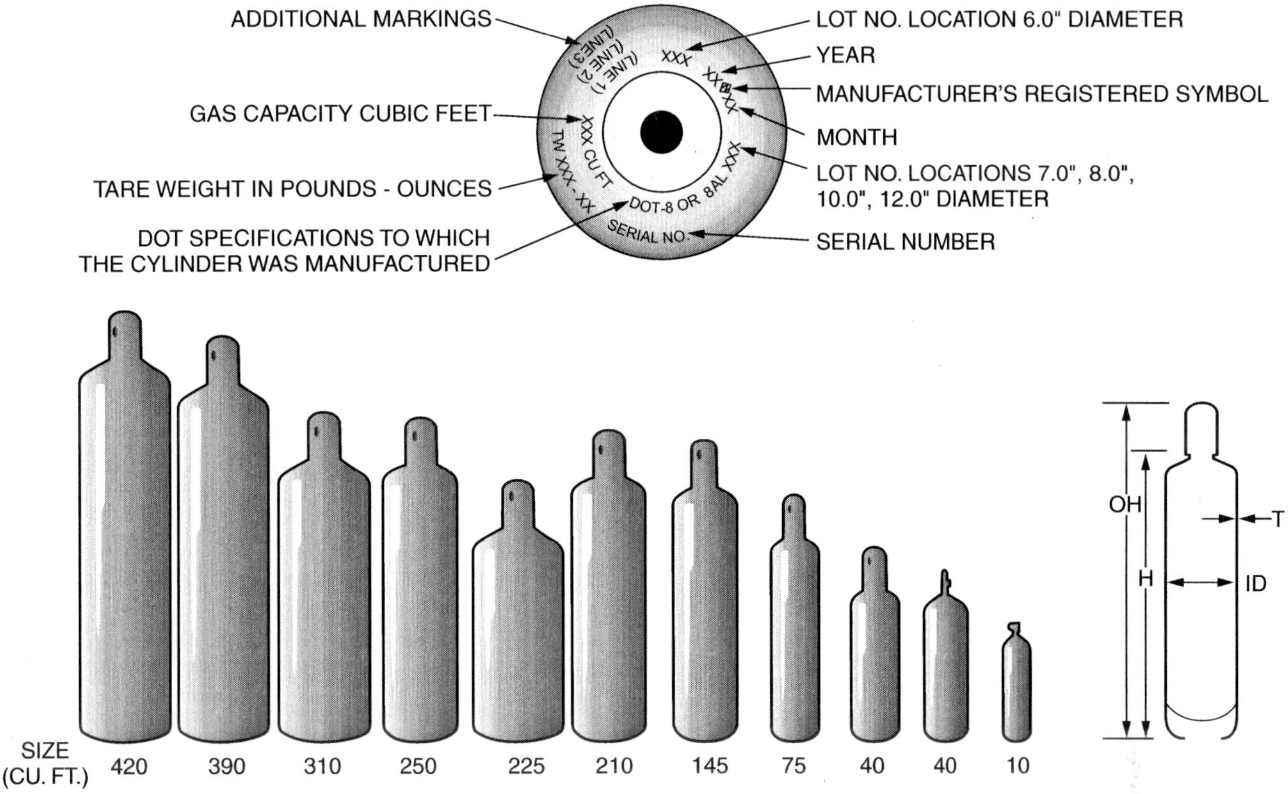

ACETYLENE CYLINDER MARKINGS

DOT SPECIFICATIONS	CAPACITY			NOMINAL DIMENSIONS (IN.)				ACETONE (LB. - OZ.)	APPROXIATE TARE WEIGHT WITH VALVE WITHOUT CAP (LB.)
	ACETYLENE	MIN. WATER							
	(FT.³)	(IN.³)	(LB.)	AVG. INSIDE DIAMETER "ID"	HEIGHT W/OUT VALVE OR CAP "H"	HEIGHT W/VALVE AND CAP "OH"	MINIMUM WALL "T"		
8 AL[1]	10	125	4.5	3.83	13.1375	14.75	0.0650	1-6	8
8[1]	40	466	16.8	6.00	19.8000	23.31	0.0870	5-7	25
8[2]	40	466	16.8	6.00	19.8000	28.30	0.0870	5-7	28
8[3]	75	855	30.8	7.00	25.5000	31.25	0.0890	9-8	45
8	100	1055	38.0	7.00	30.7500	36.50	0.0890	12-2	55
8	145	1527	55.0	8.00	34.2500	40.00	0.1020	18-10	76
8	210	2194	79.0	10.00	32.2500	38.00	0.0940	25-13	105
8AL	225	2630	94.7	12.00	27.5000	32.75	0.1280	29-6	110
8	250	2606	93.8	10.00	38.0000	43.75	0.0940	30-12	115
8AL	310	3240	116.7	12.00	32.7500	38.50	0.1120	39-5	140
8AL	390	4151	150.0	12.00	41.0000	46.75	0.1120	49-14	170
8AL	420	4375	157.5	12.00	43.2500	49.00	0.1120	51-14	187
8	60	666	24.0	7.00	25.79 OH		0.0890	7-11	40
8	130	1480	53.3	8.00	36.00 OH		0.1020	17-2	75
8AL	390	4215	151.8	12.00	46.00 OH		0.1120	49-14	180

1. Tapped for 3/8" valve but are not equipped with valve protection caps.
2. Includes Valve protection cap.
3. Can be tared to hold 60 ft³ (1.7 m³) of acetylene gas.

Standard tapping (except cylinders tapped for 3/8") 3/4"-14 NGT.

Weight includes saturation gas, filler, paint, solvent, valve, fuse plugs. Does not include cap of 2 lb. (91 kg.)

Cylinder capacities are based upon commercially pure acetylene gas at 250 psi (17.5 kg/cm²), and 70°F (15°C).

104F25.EPS

Figure 25 ◆ Acetylene cylinder markings and sizes.

acetylene to **backfire** or **flashback**. MAPP® also has an offensive odor that can be detected easily. MAPP® gas can be used for flame cutting, heating, stress relieving, brazing, soldering, and scarfing (cleaning cutting dross or other material from the workpiece).

Propylene mixtures are hydrocarbon-based gases that are stable and shock-resistant, making them relatively safe to use. They are purchased under trade names such as High Purity Gas (HPG™), Apachi™, and Prestolene™. These gases and others have distinctive odors to make leak detection easier. They burn at temperatures around 5,193°F, hotter than natural gas and propane. Propylene gases are used for flame cutting, scarfing, heating, stress relieving, brazing, and soldering.

Propane is also known as liquefied petroleum (LP) gas. It is stable and shock-resistant, and it has a distinctive odor for easy leak detection. It burns at 4,580°F, which is the lowest temperature of any fuel gas. It has a slight tendency toward backfire and flashback and is used quite extensively for cutting procedures.

Natural gas is delivered by pipeline rather than by cylinders. It burns at about 4,600°F. Natural gas is relatively stable and shock-resistant and has a slight tendency toward backfire and flashback. Because of its recognizable odor, leaks are easily detectable. Natural gas is used primarily for cutting on job sites with permanent cutting stations.

104F26.EPS

Figure 26 ◆ Acetylene cylinder with standard valve safety cap.

Table 2 Flame Temperatures of Oxygen with Various Fuel Gases

Type of Gas	Flame Temperature
Acetylene	More than 5,500°F
MAPP®	5,300°F
Propylene	5,190°F
Natural gas	4,600°F
Propane	4,580°F

3.3.1 Liquefied Fuel Gas Cylinders

Liquefied fuel gases are shipped in hollow steel cylinders (*Figure 27*). When empty, they are much lighter than acetylene cylinders.

Liquefied fuel gas is stored in hollow steel cylinders of various sizes. They can hold from 30 to 225 pounds of fuel gas. As the cylinder valve is opened, the vaporized gas is withdrawn from the cylinder. The remaining liquefied gas absorbs heat and releases additional vaporized gas. The pressure of the vaporized gas varies with the outside temperature. The colder the outside temperature, the lower the vaporized gas pressure will be. If high volumes of gas are removed from a liquefied fuel gas cylinder, the pressure drops, and the temperature of the cylinder will also drop. A ring of frost can form around the base of the cylinder. If high withdrawal rates continue, the regulator may also start to ice up. If high withdrawal rates are required, special regulators with electric heaters should be used.

WARNING!

Never apply heat directly to a cylinder or regulator. This can cause excessive pressures, resulting in an explosion.

The pressure inside a liquefied fuel gas cylinder is not an indicator of how full or empty the cylinder is. The weight of a cylinder determines how much liquefied gas is left. Liquefied fuel gas cylinders are equipped with CGA 510, 350, or 695 valves, depending on the fuel and storage pressures.

> **WARNING!**
> Do not remove the protective cap on liquefied fuel gas cylinders unless the cylinder is secured. If the cylinder falls over and the nozzle breaks off, the cylinder will release highly explosive gas. Cylinders containing a liquid such as propane must be kept in an upright position. If the valve is broken or is opened with the cylinder horizontal, the fuel can emerge as a liquid that will shoot a long distance before it vaporizes. If it is ignited, it will act like a flamethrower.

Liquefied fuel gas cylinders have a safety valve built into the valve at the top of the cylinder. The safety valve releases gas if the pressure begins to rise. Use care when handling fuel gas cylinders because the gas in cylinders is stored at such high pressures. Cylinders should never be dropped or hit with heavy objects, and they should always be stored in an upright position. When not in use, the cylinder valve must always be covered with the protective steel cap.

3.4.0 Regulators

Regulators (*Figure 28*) are attached to the cylinder valve. They reduce the high cylinder pressures to the required lower working pressures and maintain a steady flow of gas from the cylinder.

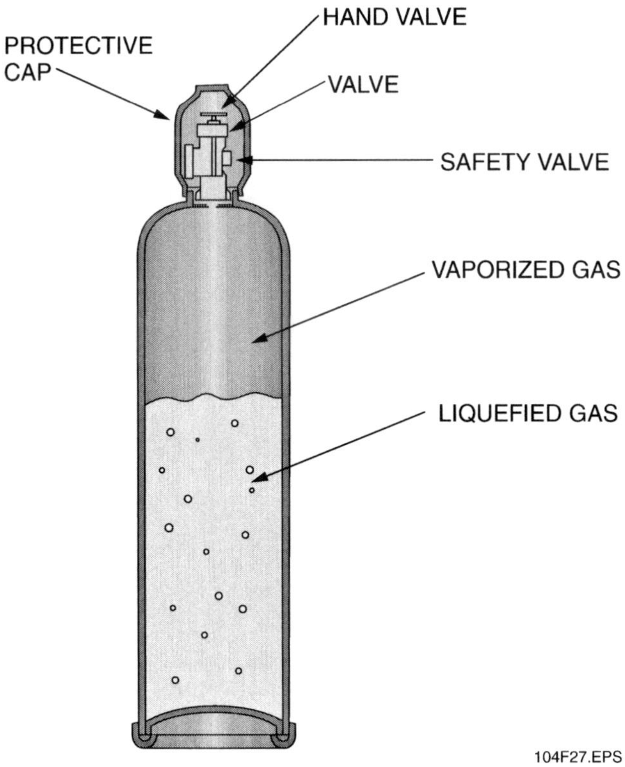

Figure 27 ◆ Liquefied fuel gas cylinder.

The pressure adjusting screw controls the gas pressure. Turned clockwise, it increases the flow of gas. Turned counterclockwise, it reduces the flow of gas. When turned counterclockwise until loose (released), it stops the flow of gas.

Most regulators contain two gauges. The high-pressure or cylinder-pressure gauge indicates the actual cylinder pressure; the low-pressure or working-pressure gauge indicates the pressure of the gas leaving the regulator.

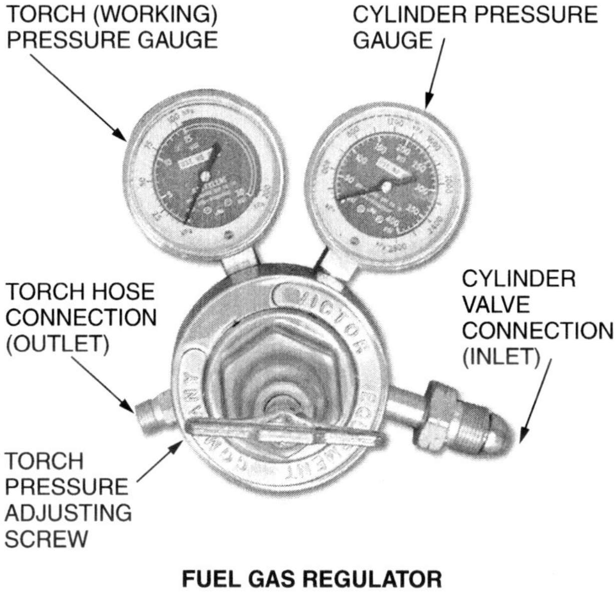

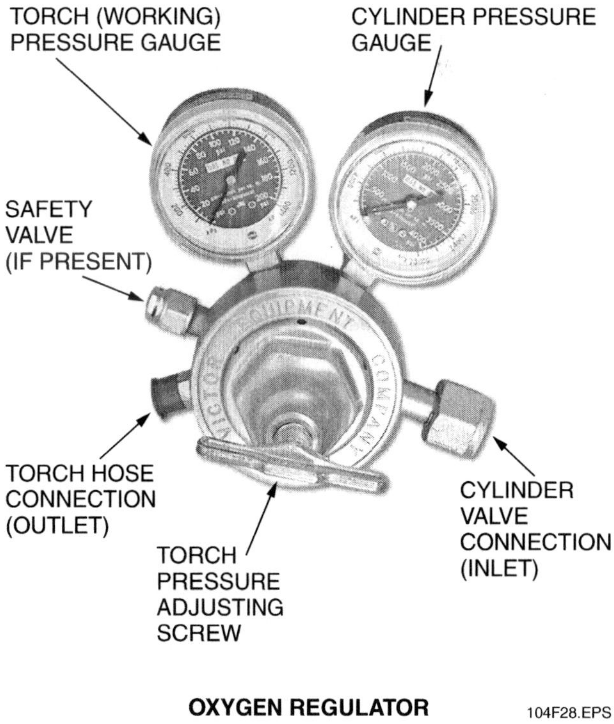

Figure 28 ◆ Oxygen and acetylene regulators.

Oxygen regulators differ from fuel gas regulators. Oxygen regulators are often painted green and always have right-hand threads on all connections. The oxygen regulator's high-pressure gauge generally reads up to 3,000 psi and includes a second scale that shows the amount of oxygen in the cylinder in terms of cubic feet. The low-pressure or working-pressure gauge may read 100 psi or higher.

Fuel gas regulators are often painted red and always have left-hand threads on all the connections. As a reminder that the regulator has left-hand threads, a V-notch may be cut around the nut. The fuel gas regulator's high-pressure gauge usually reads up to 400 psi. The low-pressure or working-pressure gauge may read up to 40 psi. Acetylene gauges, however, are always red-lined at 15 psi as a reminder that acetylene pressure should not be increased over 15 psi. There are two types of regulators: single-stage and two-stage.

> **WARNING!**
> To prevent injury and damage to regulators, always follow these guidelines:
>
> - Never subject regulators to jarring or shaking, as this can damage the equipment beyond repair.
> - Always check that the adjusting screw is fully released before the cylinder valve is turned on and when the welding has been completed.
> - Always open cylinder valves slowly and stand on the side of the cylinder opposite the regulator.
> - Never use oil to lubricate a regulator. This can result in an explosion when the regulator is in use.
> - Never use fuel gas regulators on oxygen cylinders or oxygen regulators on fuel gas cylinders.
> - Never work with a defective regulator. If it is not working properly, shut off the gas supply and have the regulator repaired by someone who is qualified to work on it.
> - Never use large wrenches, pipe wrenches, pliers, or slipjoint pliers to install or remove regulators.

3.4.1 Single-Stage Regulators

Single-stage, spring-compensated regulators reduce pressure in one step. As gas is drawn from the cylinder, the internal pressure of the cylinder decreases. A single-stage, spring-compensated regulator is unable to automatically adjust for this decrease in internal cylinder pressure. Therefore, it becomes necessary to adjust the spring pressure to periodically raise the output gas pressure as the gas in the cylinder is consumed. These regulators are the most commonly used because of their low cost and high flow rates.

3.4.2 Two-Stage Regulators

The two-stage, pressure-compensated regulator reduces pressure in two steps. It first reduces the input pressure from the cylinder to a predetermined intermediate pressure. The intermediate pressure is then adjusted by the pressure-adjusting screw. With this type of regulator, the delivery pressure to the torch remains constant, and no readjustment is necessary as the gas in the cylinder is consumed. Standard two-stage regulators are more expensive than single-stage regulators and have lower flow rates. There are also heavy-duty types with higher flow rates that are usually preferred for thick material and/or continuous-duty cutting operations.

3.4.3 Check Valves and Flashback Arrestors

Check valves and flashback arrestors (*Figure 29*) are safety devices for regulators, hoses, and torches. Check valves allow gas to flow in one direction only. Flashback arrestors stop fire.

Check valves consist of a ball and spring that open inside a cylinder. The valve allows gas to move in one direction but closes if the gas attempts to flow in the opposite direction. When a torch is first pressurized or when it is being shut off, back-pressure check valves prevent the entry

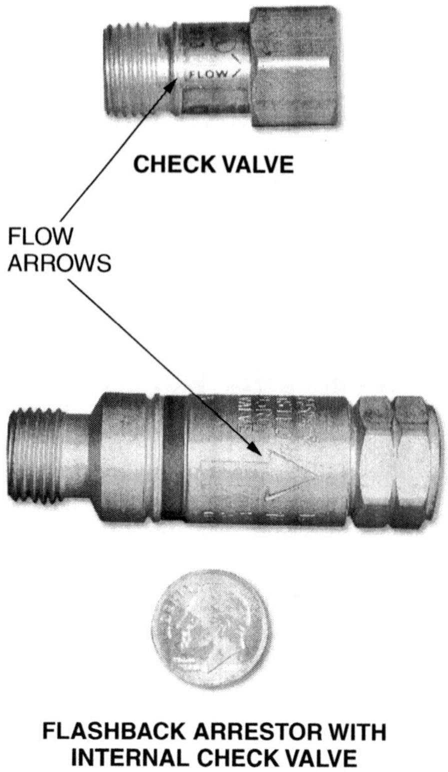

104F29.EPS

Figure 29 ◆ Torch handle add-on check valve or flashback arrestor.

and mixing of acetylene with oxygen in the oxygen hose or the entry and mixing of oxygen with acetylene in the acetylene hose.

Flashback arrestors prevent flashbacks from reaching the hoses and/or regulator. They have a flame-retarding filter that will allow heat, but not flames, to pass through. Most flashback arrestors also contain a check valve.

It is highly recommended that add-on flashback arrestors with check valves, rather than just check valves, be installed at both the torch handle and regulator connections to prevent serious injury or property damage. If the flame front of a flashback, sometimes called the backburn, gets by the check valves into the hoses or regulators, the hoses could burn through, and a large, very dangerous, uncontrolled fire could start. Newer torches are available with check valves and flashback arrestors built into the torch handle.

Add-on check valves and flashback arrestors are designed to be attached either to the torch handle connections or to the regulator outlets. As a minimum, flashback arrestors with check valves should be attached to the torch handle connections. Both devices have arrows on them to indicate flow direction. When installing add-on check valves and flashback arrestors, be sure the arrow matches the desired gas flow direction.

NOTE

Newer one-piece hand cutting torches and combination torches with built-in check valves and/or flashback arrestors are available from most manufacturers.

3.5.0 Hoses

Hoses transport gases from the regulators to the torch. Oxygen hoses are usually green or black with right-hand threaded connections. Hoses for fuel gas are usually red and have left-hand threaded connections. The fuel gas connections may also be grooved as a reminder that they have left-hand threads.

Proper care and maintenance of the hose is important for maintaining a safe, efficient work area. Remember the following guidelines for hoses:

- Protect the hose from molten dross or sparks, which will burn the exterior. Although some hoses are flame retardant, they will burn.
- Remove the hoses from under the metal being cut. If the hot metal falls on the hose, the hose will be damaged.
- Frequently inspect and replace hoses that show signs of cuts, burns, worn areas, cracks, or damaged fittings.
- Never use pipe-fitting compounds or lubricants around hose connections. These compounds often contain oil or grease, which ignite and burn or explode in the presence of oxygen.

3.6.0 Cutting Torches

Cutting torches mix oxygen and fuel gas for the torch flame and control the stream of oxygen necessary for the cutting jet. Depending on the job site, you may use either a one-piece or a combination cutting torch.

3.6.1 One-Piece Hand Cutting Torch

The one-piece hand cutting torch, sometimes called a demolition torch, contains the fuel gas and oxygen valves that allow the gases to enter the chambers and then flow into the tip where they are mixed. The main body of the torch is called the handle. The torch valves control the fuel gas and oxygen used for preheating the metal to be cut. The cutting oxygen lever, which is spring-loaded, controls the jet of cutting oxygen. Hose connections are located at the end of the torch body behind the valves.

Figure 30 shows a three-tube one-piece positive-pressure hand cutting torch in which the preheat fuel and oxygen are mixed in the tip. These torches are designed for heavy-duty cutting. They have long supply tubes from the torch handle to the torch head to reduce radiated heat to the operator's hands. The torches are generally available in capacities for cutting steel up to 12 inches thick. Larger-capacity torches, with the ability to cut steel up to 36 inches thick, can also be obtained.

Two different types of oxyfuel cutting torches are in general use. The positive-pressure torch is designed for use with fuel supplied through a regulator from pressurized fuel storage cylinders. The injector torch is designed to use a vacuum created by oxygen flow to draw the necessary amount of fuel from a very low-pressure fuel source, such as a natural gas line or acetylene generator. The injector torch, when used, is most often found in continuous-duty high-volume manufacturing applications. Both types may employ one of two different fuel-mixing methods:

- Torch-handle or supply-tube mixing
- Torch-head or tip mixing

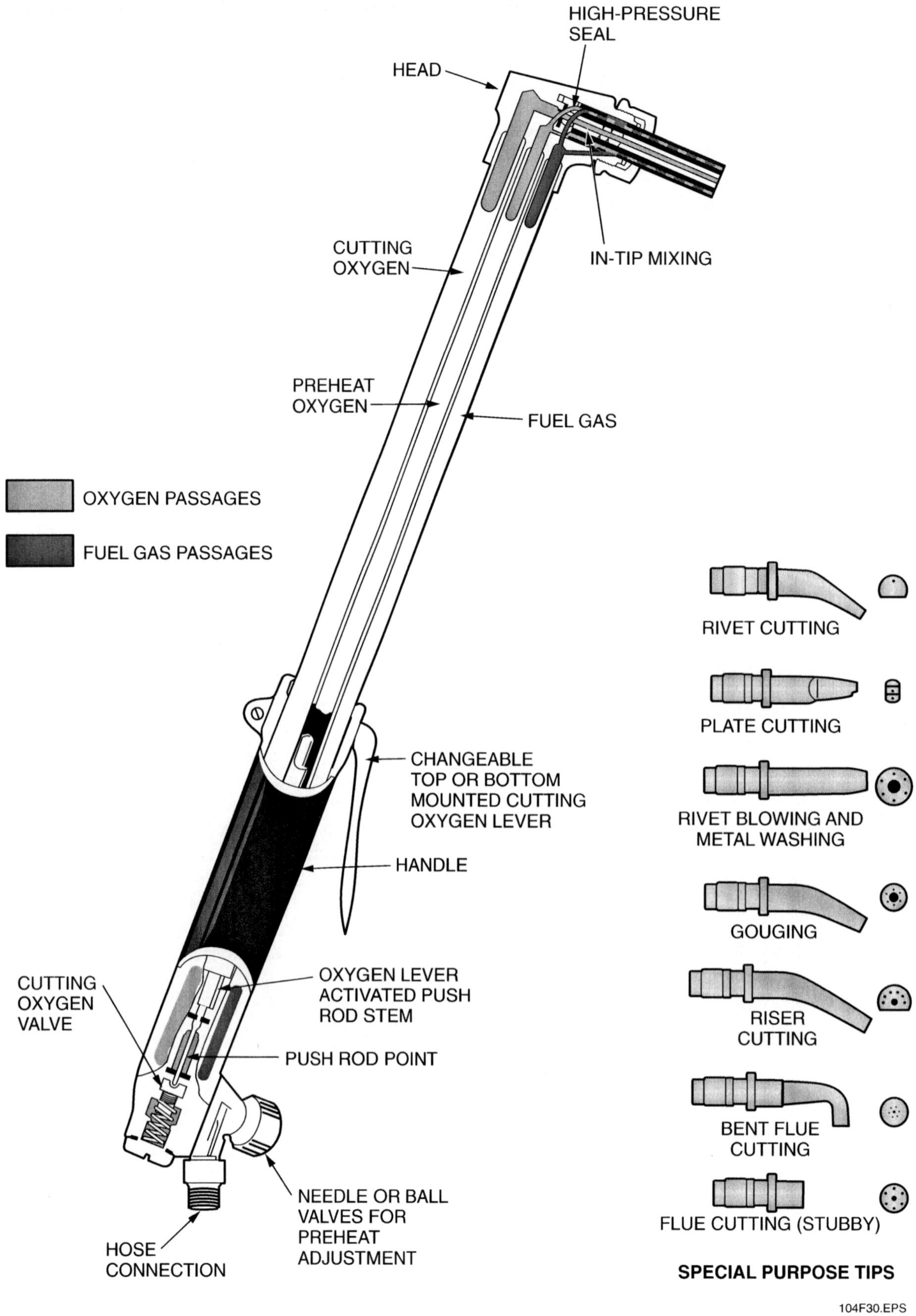

Figure 30 ◆ Heavy-duty three-tube one-piece positive-pressure hand cutting torch.

The two methods can normally be distinguished by the number of supply tubes from the torch handle to the torch head. Torches that use three tubes (see *Figure 30*) from the handle to the head mix the preheat fuel and oxygen at the torch head or tip. This method tends to help eliminate any flashback damage to the torch head supply tubes and torch handle. Torches with two tubes usually mix the preheat fuel and oxygen in a mixing chamber in the torch body or in one of the supply tubes. Injector torches usually have the injector located in one of the supply tubes, and the mixing occurs in that tube from the injector to the torch head. Some older torches that have only two visible tubes are actually three-tube torches that mix the preheat fuel and oxygen in the torch head or tip. This is accomplished by using a separate preheat fuel tube inside a larger preheat oxygen tube.

3.6.2 Combination Torch

The combination torch consists of a cutting torch attachment that fits onto a welding torch handle. These torches are normally used in light-duty or medium-duty applications. Fuel gas and oxygen valves are on the torch handle. The cutting attachment has a cutting oxygen lever and another oxygen valve to control the preheat flame. When the cutting attachment is screwed onto the torch handle, the torch handle oxygen valve is opened all the way, and the preheat oxygen is controlled by an oxygen valve on the cutting attachment. When the cutting attachment is removed, welding and heating tips can be screwed onto the torch handle. *Figure 31* shows a two-tube combination torch in which the preheat mixing is accomplished in a supply tube. These torches are usually positive-pressure torches with mixing occurring in the attachment body, supply tube, head, or tip. These torches are also equipped with built-in flashback arrestors and check valves.

3.7.0 Cutting Torch Tips

Cutting torch tips, or nozzles, fit into the cutting torch and are either screwed in or secured with a tip nut. There are one- and two-piece cutting tips (*Figure 32*).

One-piece cutting tips are made from a solid piece of copper. Two-piece cutting tips have a separate external sleeve and internal section.

Torch manufacturers supply literature explaining the appropriate torch tips and gas pressures to be used for various applications. *Table 3* shows a sample cutting tip chart that lists recommended tip sizes and gas pressures for use with acetylene fuel gas and a specific manufacturer's torch and tips.

The cutting torch tip to be used depends on the base metal thickness and fuel gas being used. Special-purpose tips are also available for use in such operations as gouging and grooving.

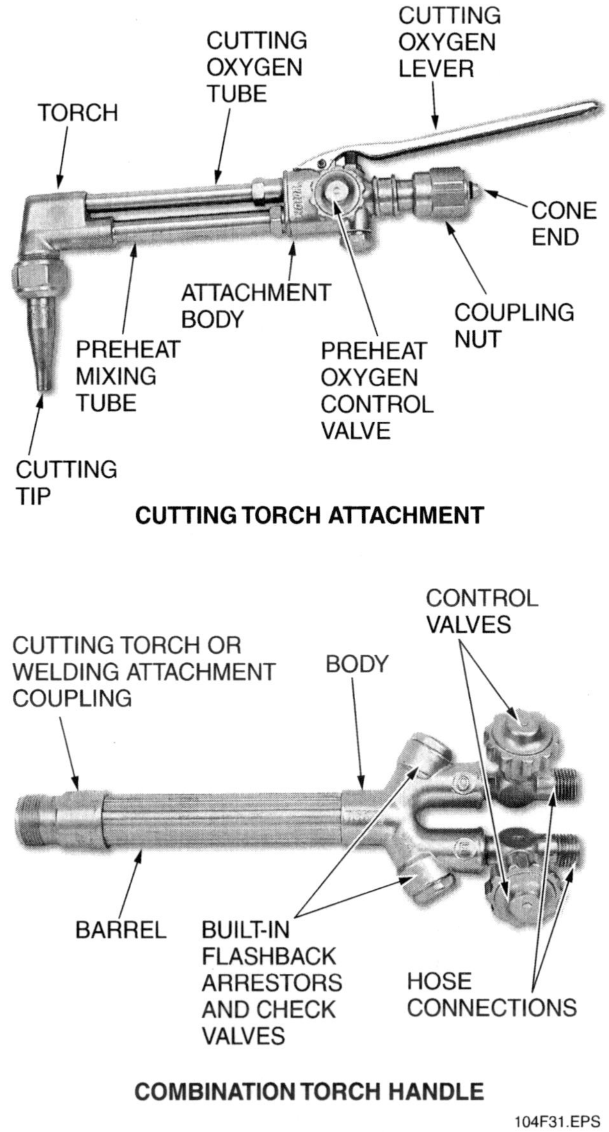

Figure 31 ◆ Typical combination torch.

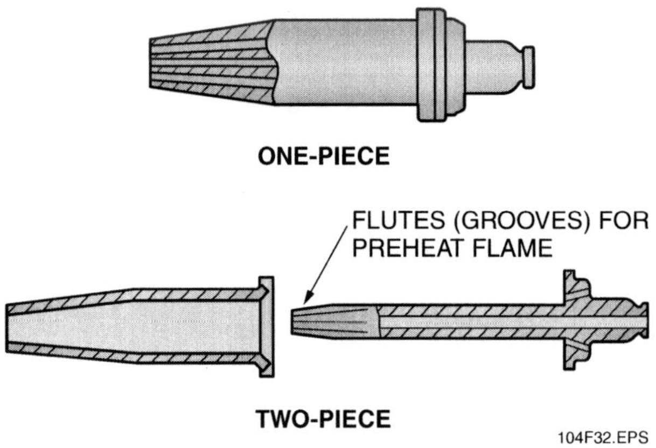

Figure 32 ◆ One- and two-piece cutting tips.

> **WARNING!**
> Do not use the cutting tip chart from one manufacturer for the cutting tips of another manufacturer. The gas flow rate of the tips may be different, resulting in excessive flow rates. Different gas pressures may also be required.

Nearly all manufacturers use different tip-to-torch mounting designs, sealing surfaces, and diameters. In addition, tip sizes and flow rates are usually not the same between manufacturers even though the number designations may be the same. This makes it impossible to safely interchange cutting tips between torches from different manufacturers. Even though some tips from different manufacturers may appear to be the same, do not interchange them. The sealing surfaces are very precise, and serious leaks may occur that could result in a dangerous fire or flashback.

3.7.1 Cutting Tips for Acetylene

One-piece torch tips are usually used with acetylene cutting because of the high temperatures involved. They can have four, six, or eight preheat holes in addition to the single cutting hole. *Figure 33* shows typical acetylene torch cutting tips.

Manufacturers provide listings of the maximum fuel flow rate for each acetylene tip size in addition to recommended acetylene pressures. When selecting a tip, make sure that its maximum flow rate (in cubic feet per hour) does not exceed one-tenth of the total fuel capacity (in cubic feet) for the acetylene cylinder in use. Multiple cylinders must be manifolded together if the flow rate exceeds the cylinder(s) in use in order to prevent withdrawal of acetone along with acetylene.

3.7.2 Cutting Tips for Liquefied Fuel Gases

Tips used with liquefied fuel gases must have at least six preheat holes. Because fuel gases burn at lower temperatures than acetylene, more holes are necessary for preheating. Tips used with liquefied fuel gases can be one- or two-piece cutting tips. *Figure 34* shows typical cutting tips used with liquefied fuel gases.

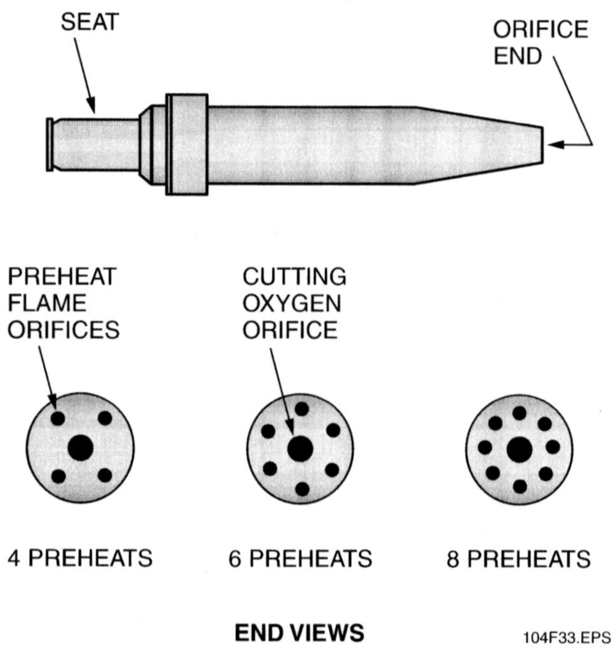

Figure 33 ◆ Typical acetylene torch cutting tips.

Table 3 Sample Acetylene Cutting Tip Chart

Cutting Tip Series 1-101, 3-101, and 5-101						
Metal Thickness (in)	**Tip Size**	**Cutting Oxygen Pressure (psig)**	**Preheat Oxygen (psig)**	**Acetylene Pressure (psig)**	**Speed (in/min)**	**Kerf Width**
⅛	000	20/25	3/5	3/5	20/30	.04
¼	00	20/25	3/5	3/5	20/28	.05
⅜	0	25/30	3/5	3/5	18/26	.06
½	0	30/35	3/6	3/5	16/22	.06
¾	1	30/35	4/7	3/5	15/20	.07
1	2	35/40	4/8	3/6	13/18	.09
2	3	40/45	5/10	4/8	10/12	.11
3	4	40/50	5/10	5/11	8/10	.12
4	5	45/55	6/12	6/13	6/9	.15
6	6	45/55	6/15	8/14	4/7	.15
10	7	45/55	6/20	10/15	3/5	.34
12	8	45/55	7/25	10/15	3/4	.41

3.7.3 Special-Purpose Tips

Special-purpose tips are available for special cutting jobs. These jobs include cutting sheet metal, rivets, risers, and flues, as well as washing and gouging. *Figure 35* shows special-purpose torch cutting tips.

- The sheet metal cutting tip has only one preheat hole. This minimizes the heat and prevents distortion in the sheet metal. These tips are normally used with a motorized carriage but can also be used for hand cutting.
- Rivet cutting tips are used to cut off rivet heads, bolt heads, and nuts.
- Rivet blowing and metal washing tips are heavy-duty tips designed to withstand high heat. They are used for coarse cutting and for removing such items as clips, angles, and brackets.
- Rosebuds are used to provide large amounts of heat to loosen frozen parts or to expand parts prior to pressing them together for a machine fit. Care must be taken when using a rosebud as they produce much more heat than a standard cutting or welding tip, and as such, the metal being worked will heat up much faster.
- Riser cutting tips are similar to rivet cutting tips and can also be used to cut off rivet heads, bolt heads, and nuts. They have extra preheat holes to cut risers, flanges, or angle legs faster. They can be used for any operation that requires a cut close to and parallel to another surface, such as in removing a metal backing.
- Flue cutting tips are designed to cut flues inside boilers. They also can be used for any cutting operation in tight quarters where it is difficult to get a conventional tip into position.

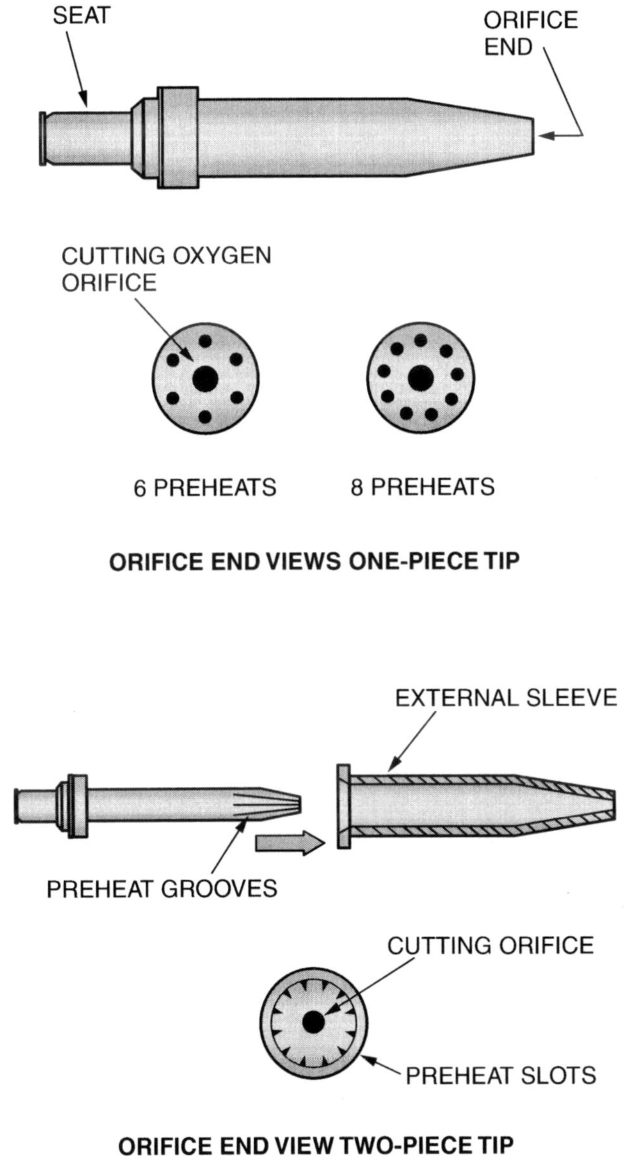

Figure 34 ◆ Typical cutting tips for liquefied fuel gases.

3.8.0 Tip Cleaners and Tip Drills

With use, cutting tips become dirty. Carbon and other impurities build up inside the holes, and molten metal often sprays and sticks onto the surface of the tip. A dirty tip will result in a poor-quality cut with an uneven **kerf** and excessive dross buildup. To ensure good cuts with straight kerfs and minimal dross buildup, clean cutting tips with tip cleaners or tip drills (*Figure 36*).

Tip cleaners are small round files. They usually come in a set with files to match the diameters of the various tip holes. In addition, each set usually includes a file that can be used to lightly recondition the face of the cutting tip. Tip cleaners are inserted into the tip hole and moved back and forth a few times to remove deposits from the hole.

Tip drills are used for major cleaning and for holes that are plugged. Tip drills are tiny drill bits that are sized to match the diameters of tip holes.

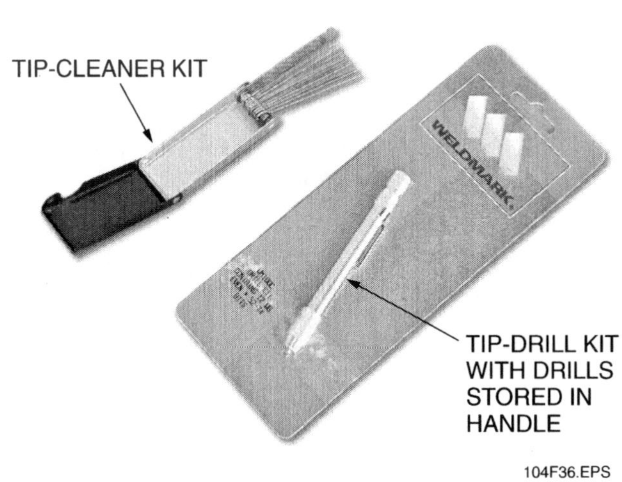

Figure 36 ◆ Tip cleaner and drill kits.

The drill fits into a drill handle for use. The handle is held, and the drill bit is turned carefully inside the hole to remove debris. They are more brittle than tip cleaners, making them more difficult to use.

> **CAUTION**
> Tip cleaners and tip drills are brittle. If you are not careful, they may break off inside a hole. Broken tip cleaners are difficult to remove. Improper use of tip cleaners or tip drills can enlarge the tip, causing improper burning of gases. If this occurs, tips must be discarded. If the end of the tip has been partially melted or deeply gouged, do not attempt to cut it off or file it flat. The tip should be discarded and replaced with a new tip. This is because some tips have tapered preheat holes, and if a significant amount of metal is removed from the end of the tip, the preheat holes will become too large.

3.9.0 Friction Lighters

Always use a friction lighter (*Figure 37*), also known as a striker or spark-lighter, to ignite the cutting torch. The friction lighter works by rubbing a piece of flint on a steel surface to create sparks.

> **WARNING!**
> Do not use a match or a gas-filled lighter to light a torch. This could result in severe burns and/or could cause the lighter to explode.

When using a cup-type striker to ignite a welding torch, hold the cup of the striker slightly below and to the side of the tip, parallel with the fuel gas stream from the tip. This prevents the ignited gas from deflecting back toward you from the cup and reduces the amount of carbon soot in the cup. Note that the flint in a striker can be replaced.

3.10.0 Cylinder Cart

The cylinder cart, or bottle cart, is a modified hand truck that has been equipped with seats and chains to hold cylinders firmly in place. Bottle carts help ensure the safe transportation of gas cylinders. *Figure 38* shows a typical cylinder cart for a welding/cutting rig. Some carts are equipped with tool/accessory trays or boxes as well as rod holders.

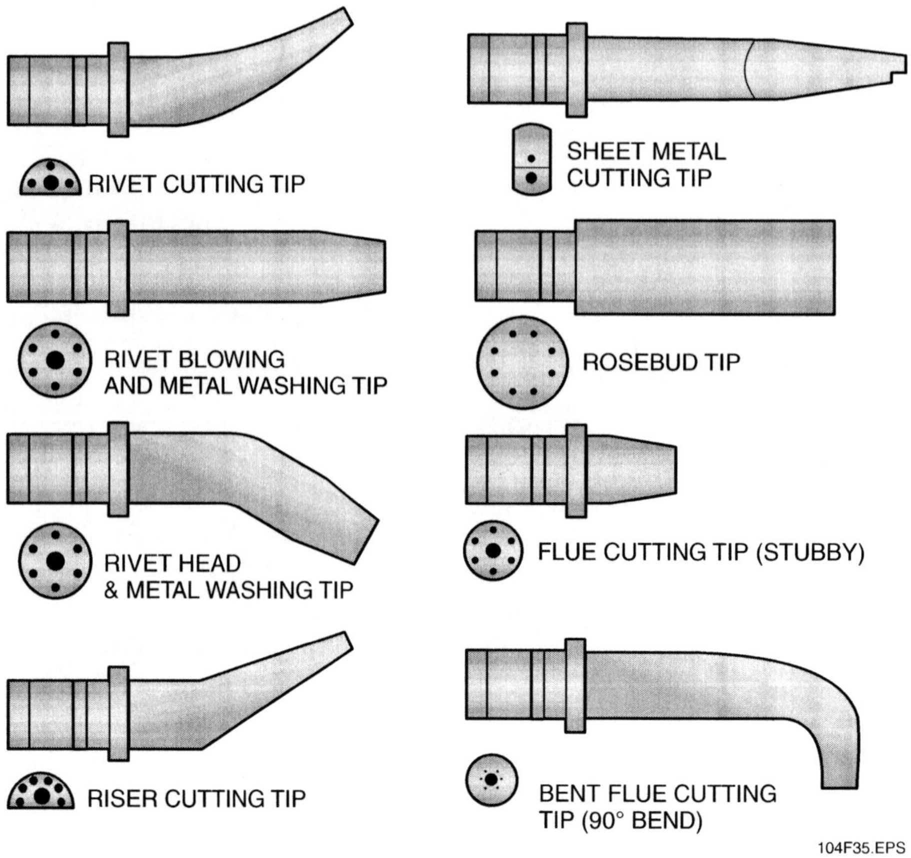

Figure 35 ◆ Special-purpose torch cutting tips.

3.11.0 Soapstone Markers

Because of the heat involved in welding and cutting operations, along with the tinted lenses that are required, ordinary pen or pencil marking for cutting lines or welding locations is not effective. The oldest and most common material used for marking is soapstone in the form of sticks or cylinders (*Figure 39*). Soapstone is soft and feels greasy and slippery. It is actually steatite, a dense, impure form of talc that is heat resistant. It also shows up well through a tinted lens under the illumination of an electric arc or gas welding/cutting flame. Some welders prefer to use silver-graphite pencils for marking dark materials and red-graphite pencils for aluminum or other bright metals. Graphite is also highly heat resistant (*Figure 39*). A few manufacturers also market heat-resistant paint/dye markers for welding.

The most effective way to sharpen a soapstone stick marker is to shave it on one side with a penknife. By leaving one side flat, accurate lines can be drawn very close to a straightedge or a pattern.

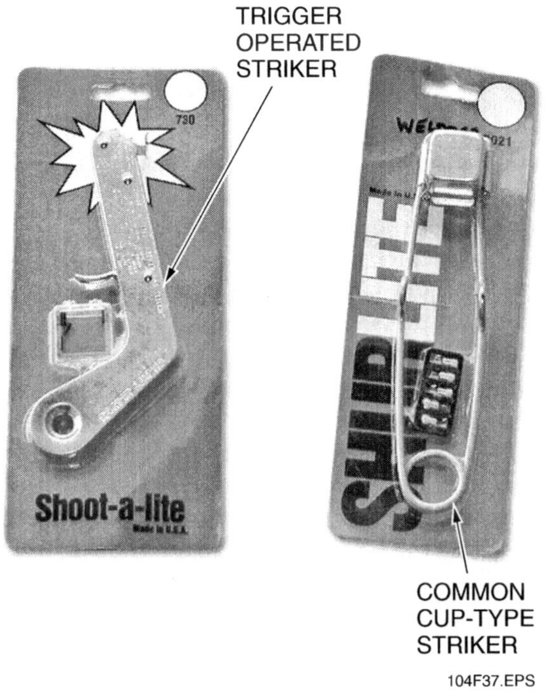

Figure 37 ◆ Typical friction lighters.

Figure 38 ◆ Typical cutting/welding rig cylinder cart.

3.12.0 Specialized Cutting Equipment

In addition to the common hand cutting torches, other types of equipment are used in oxyfuel cutting applications. This equipment includes mechanical guides used with a hand cutting torch, various types of motorized cutting machines, and oxygen lances. All of the motorized units use special straight body machine cutting or welding torches with a gear rack attached to the torch body to set the tip distance from the work.

3.12.1 Mechanical Guides

On long, circular, or irregular cuts, it is very difficult to control and maintain an even kerf with a hand cutting torch. Mechanical guides can help maintain an accurate and smooth kerf along the cutting line. For straight line or curved cuts, use a one- or two-wheeled accessory that clamps on the torch tip in a fixed position. The wheeled accessory

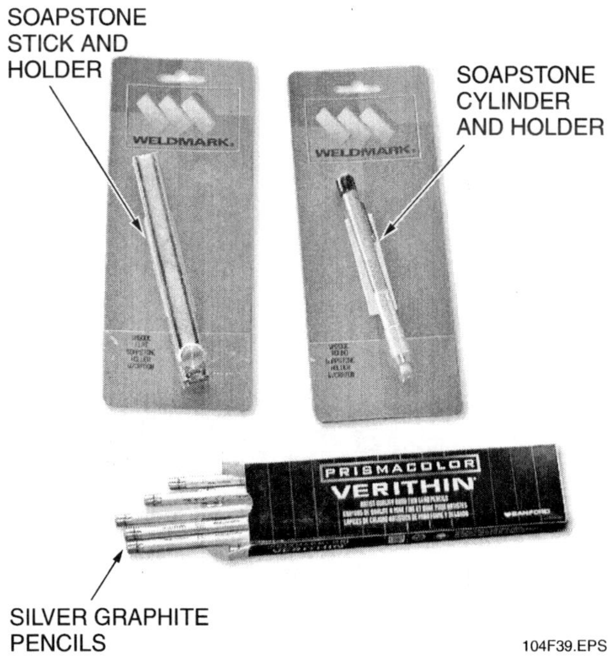

Figure 39 ◆ Typical soapstone and graphite markers.

maintains the proper tip distance while the tip is guided by hand along the cutting line. The fixed, two-wheeled accessory is similar to the rotating-mount wheeled unit used for a circle cutter but without the radius bar (*Figure 40*).

Perform arc or circular cuts with the circle cutting accessory shown in *Figure 40*. The torch tip fits through and is secured to a rotating mount between the two small metal wheels. The wheel heights are adjustable so that the tip distance from the work can be set. The radius of the circle is set by moving the pivot point on a radius bar. After a starting hole is cut (if needed), the torch tip is placed through and secured to the circle cutter rotating mount. Then the pivot point is placed in a drilled hole or a magnetic holder at the center of the circle. When the cut is restarted, the torch can be moved in a circle around the cut, guided by the circle cutter.

When large work with an irregular pattern must be cut, a template is often used. The torch is drawn around the edges of the template to trace the pattern as the cut is made. If multiple copies must be cut, a metal pattern, held in place and spaced for tip distance from the work by stacked magnets, is usually used. For a one- or two-time copy, a heavily weighted Masonite or aluminum template that is spaced off the workpiece could be carefully used and discarded.

A simple solution for straight line cutting is to clamp a piece of angle iron to the work and use a band clamp around the cutting torch tip to maintain the cutting tip distance from the work. When the cut is started, the band clamp rests on the top of the vertical leg of the angle iron, and the torch is drawn along the length of the angle iron at the correct cutting speed.

3.12.2 Motor-Driven Equipment

A variety of fixed and portable motorized cutting equipment is available for straight and curved cutting/welding. The computer-controlled gantry cutting machine (*Figure 41*) and the optical pattern-tracing machine (*Figure 42*) are fixed-location machines used in industrial manufacturing applications. The computer-controlled machine can be programmed to **pierce** and then cut any pattern from flat metal stock. The optical pattern-tracing machine follows lines on a drawing using a light beam and an optical detector. They both can be rigged to cut multiple items using multiple torches operated in parallel. Both units have a motor-driven gantry that travels the length of a table and a transverse motor-driven beam or torch head that moves back and forth across the table. Both units are also equipped to use both oxyfuel cutting and plasma cutting torches. The size of the patterns can also be adjusted.

Other types of pattern-tracing machines use metal templates that are clamped in the machine. A follower wheel traces the pattern from the template. The pattern size can be increased or decreased by electrical or mechanical linkage to a moveable arm holding one or more cutting torches that cut the pattern from flat metal stock.

Portable track cutting machines (track burners) can be used in the field for straight or curved cutting and beveling. *Figures 43* and *44* show units

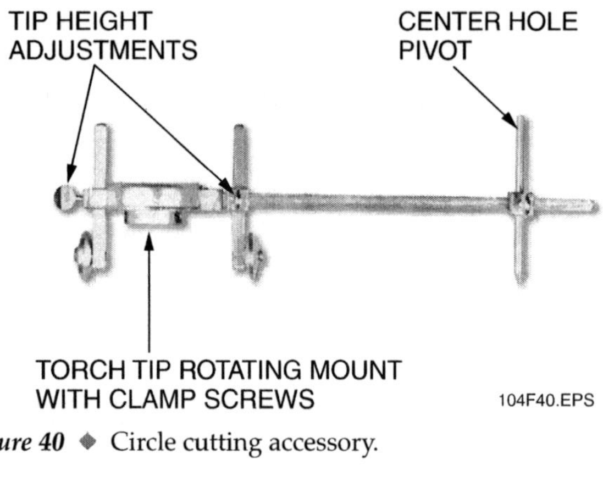

Figure 40 ◆ Circle cutting accessory.

Figure 41 ◆ Computer-controlled gantry cutting machine.

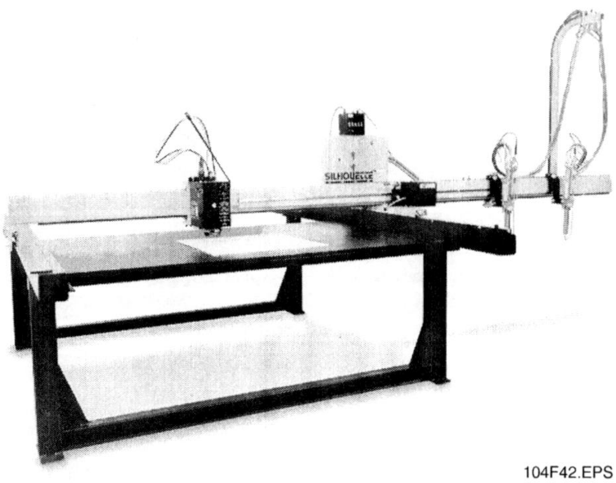

Figure 42 ◆ Optical pattern-tracing machine.

driven by a variable-speed motor. The unit shown in *Figure 43* is available with track extensions for any length of straight cutting or beveling, along with a circle cutting attachment. The unit shown in *Figure 44* uses a somewhat flexible magnetic track for both flat straight-line or large-diameter object cutting or beveling. It is shown equipped with an optional plasma machine torch. Both units can be adapted to metal inert gas (MIG) or tungsten inert gas (TIG) welding.

A portable, motor-driven band track or hand-cranked ring gear cutter/beveler can be set up in the field for cutting and beveling pipe with oxy-fuel or plasma machine torches (*Figures 45* and *46*). The stainless steel band track cutter uses a chain and motor sprocket drive to rotate the machine cutting torch around the pipe a full 360 degrees. The all-aluminum ring gear type of cutter/ beveler is positioned on the pipe, and then the saddle is clamped in place. In operation, the ring gear and the cutting torch rotate at different rates around the saddle for a full 360-degree cut.

A hand-guided oxyfuel cutting torch with an integral-precision, variable-speed motor drive (*Figure 47*), which can be used for straight line and curved cutting to achieve machine-quality cuts, is available. A circle cutting accessory is also available for the unit.

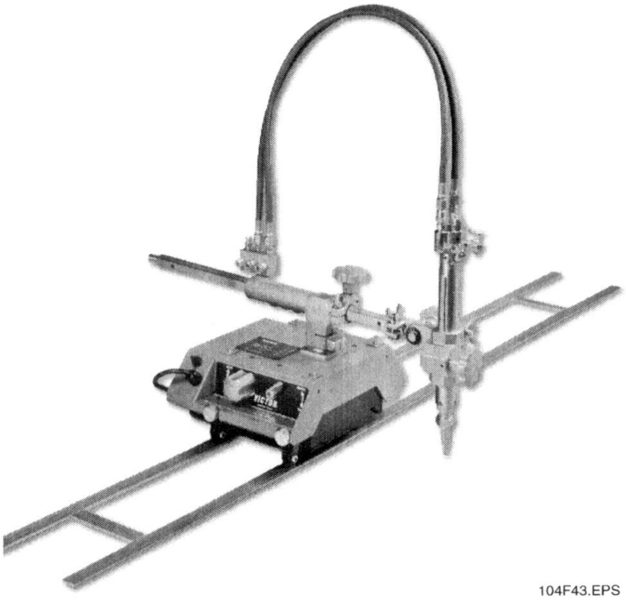

104F43.EPS

Figure 43 ◆ Track burner with an oxyfuel machine torch.

104F45.EPS

Figure 45 ◆ Band track pipe cutter/beveler.

104F44.EPS

Figure 44 ◆ Track burner with a plasma machine torch.

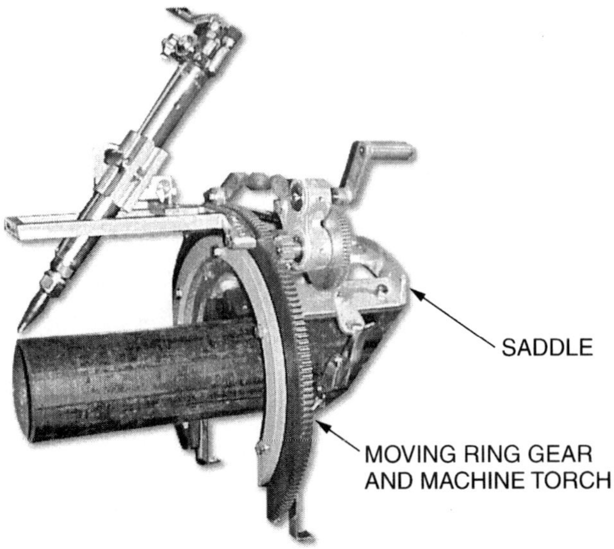

104F46.EPS

Figure 46 ◆ Ring gear pipe cutter/beveler.

3.12.3 Exothermic Oxygen Lances

Exothermic (combustible) oxygen lances are a special oxyfuel cutting tool usually used in heavy industrial applications and demolition work. The lance is a steel pipe that contains magnesium- and aluminum-cored powder or rods (fuel). In operation, the lance is clamped into a holder (*Figure 48*) that seals the lance to a fitting that supplies oxygen to the lance through a hose at pressures of 75 to 80 psi. With the oxygen turned on, the end of the lance is ignited with an acetylene torch or flare. As long as the oxygen is applied, the lance will burn and consume itself. The oxygen-fed flame of the burning magnesium, aluminum, and steel pipe creates temperatures approaching 10,000°F. At this temperature, the lance will rapidly cut or pierce any material, including steel, metal alloys, and cast iron, even under water.

The lances for the holder shown in *Figure 48* are 10 feet long and range in size from ⅜ to 1 inch in diameter. The larger sizes can be coupled to obtain a longer lance. A small pistol-grip heat-shielded unit that can, if desired, be used with an electric welder is also available. The arc, combined with an oxygen lance, can create temperatures exceeding 10,000°F. This small unit uses lances from ¼ to ⅜ inch in diameter that are 22 to 36 inches long and that cut very rapidly at a maximum burning time of 60 to 70 seconds. The small unit is primarily used to burn out large frozen pins and frozen headless bolts or rivets. Like a large lance, it can be used to cut any material, including concrete-lined pipe. It also can be used to remove hard-surfacing material that has been applied to wear surfaces. Both units are relatively inexpensive and can be set up in the field with only an oxygen cylinder, hose, and ignition device.

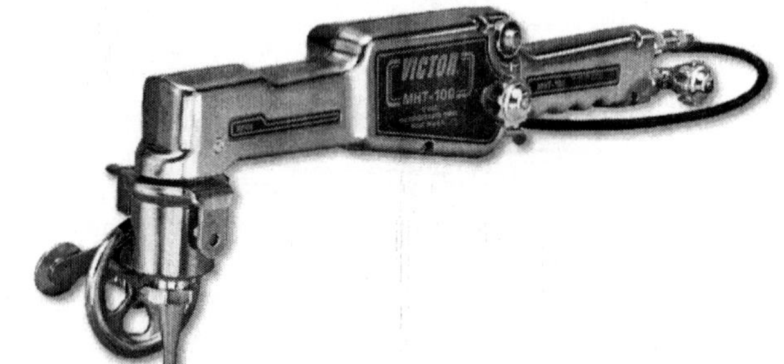

104F47.EPS

Figure 47 ◆ Hand-guided motorized oxyfuel cutting torch.

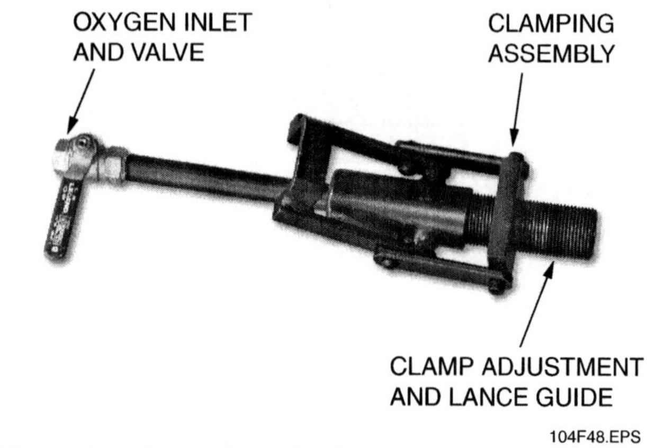

104F48.EPS

Figure 48 ◆ Oxygen lance holder.

> **NOTE**
>
> Fire brick, reinforced concrete, and large, thick iron castings or steel objects are some of the many materials that can quickly be cut or pierced with an exothermic oxygen lance. With the larger lances, use specially insulated heat-reflective suits and full head shields because of the amount of heat generated.

4.0.0 ◆ SETTING UP OXYFUEL EQUIPMENT

When setting up oxyfuel equipment, you must follow certain procedures to ensure that the equipment operates properly and safely. The following sections explain the procedures for setting up oxyfuel equipment.

> **NOTE**
>
> For fixed installations involving one or more cylinders coupled to a manifold, fuel and oxygen cylinders must be separated by at least 20 feet or be divided by a wall 5 feet or more high (*American National Standards Institute Z49.1*).

4.1.0 Transporting and Securing Cylinders

Follow these steps to transport and secure cylinders:

> **WARNING!**
>
> Always handle cylinders with care. They are under high pressure and should never be dropped, knocked over, or exposed to excessive heat. When moving cylinders, always be certain that the valve caps are in place. Use a cylinder cage to lift cylinders. Never use a sling or electromagnet.

Step 1 Transport cylinders to the workstation in the upright position on a hand truck or bottle cart (*Figure 49*).

Step 2 Secure the cylinders at the workstation.

Step 3 Remove the protective cap from each cylinder and inspect the outlet nozzles to ensure that the seat and threads are not damaged. Place the protective caps where they will not be lost and where they will be available when the cylinders are empty.

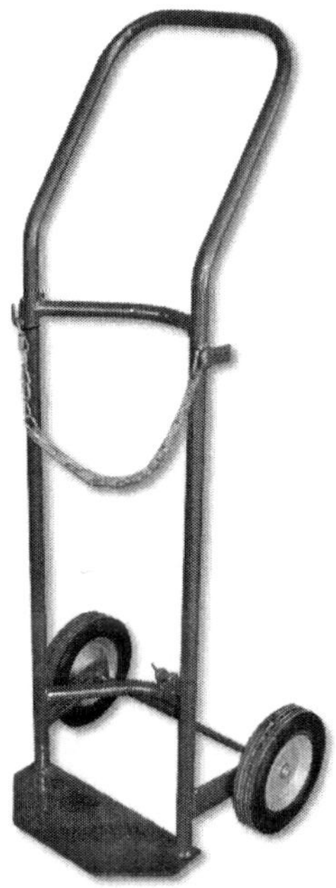

SINGLE-CYLINDER HAND CART

HEAVY-DUTY TWIN-CYLINDER HAND CART

104F49.EPS

Figure 49 ◆ Carts for transporting cylinders.

CAUTION

Do not transport or immediately use an acetylene cylinder found resting on its side. Stand it upright and wait at least 30 minutes to allow the acetone to settle before using it.

Never attempt to lift a cylinder using the holes in a safety cap. Always use a lifting cage. Make sure that the cylinder is secured in the cage. Various size cages are available for high-pressure cylinders and cylinders containing liquids.

4.2.0 Cracking Cylinder Valves

Follow these steps to crack cylinder valves:

Step 1 Crack open the cylinder valve momentarily to remove any dirt from the valves (*Figure 50*).

WARNING!

Always stand to one side of the valves when opening them to avoid injury from dirt that may be lodged in the valve.

Step 2 Wipe out the connection seat of the valves with a clean cloth. Dirt frequently collects in the outlet nozzle of a cylinder valve and must be cleaned out to keep it from entering the regulator when pressure is turned on.

WARNING!

Be sure the cloth used does not have any oil or grease on it. Oil or grease mixed with compressed oxygen will cause an explosion.

4.3.0 Attaching Regulators

Follow these steps to attach the regulators:

Step 1 Check that the regulator is closed (adjustment screw loose).

Step 2 Check the regulator fittings to ensure that they are free of oil and grease (*Figure 51*).

WARNING!

Do not work with a regulator that shows signs of damage, such as cracked gauges, bent thumbscrews, or worn threads. Set it aside for repairs. Never attempt to repair regulators yourself. When tightening the connections, always use a torch wrench.

Step 3 Connect and tighten the oxygen regulator to the oxygen cylinder using a torch wrench (*Figure 52*).

Step 4 Connect and tighten the fuel gas regulator to the fuel gas cylinder. Remember that all fuel gas fittings have left-hand threads.

Figure 50 ◆ Cracking a cylinder valve.

Figure 51 ◆ Checking connection fittings.

Step 5 Crack the cylinder valve slightly and open the regulator to expel any debris from the outlet. Shut the cylinder valve and close the regulator (*Figure 53*).

4.4.0 Installing Flashback Arrestors or Check Valves

Follow these steps to install flashback arrestors or check valves:

WARNING!

At least one flashback arrestor must be used with each hose. The absence of a flashback arrestor could result in flashback. Flashback arrestors can be attached either to the regulator, the torch, or both; however, flashback arrestors installed at the torch handle are preferred if only one is being used.

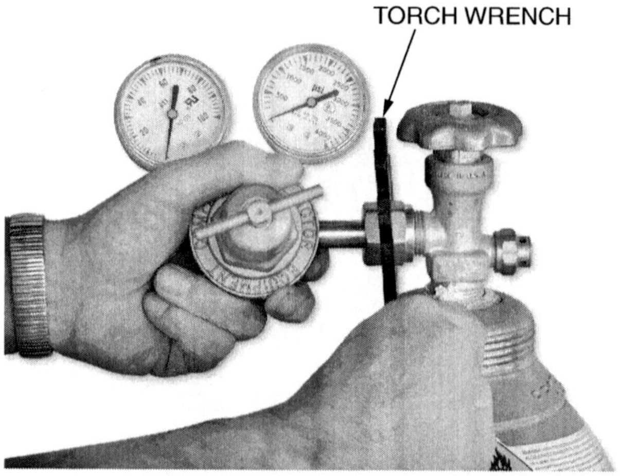

Figure 52 ◆ Tightening regulator connection.

Figure 53 ◆ Cleaning the regulator.

Step 1 Attach a flashback arrestor or check valve to the hose connection on the oxygen regulator (*Figure 54*) and tighten with a torch wrench.

Step 2 Attach and tighten a flashback arrestor or check valve to the hose connection on the fuel gas regulator. Keep in mind that all fuel gas fittings have left-hand threads.

4.5.0 Connecting Hoses to Regulators

New hoses contain talc and loose bits of rubber. These materials must be blown out of the hoses using an inert gas such as nitrogen or argon before the torch is connected. If they are not blown out, they will clog the torch needle valves.

WARNING!

Never blow out hoses with compressed air, fuel gas, or oxygen. Compressed air often contains some oil that could explode or cause a fire when compressed in the hose with oxygen. Using fuel gas or oxygen creates a fire and explosion hazard.

Check that used hoses are not cracked, cut, damaged, or contaminated with oil or grease. Replace the hoses if these conditions exist.

Follow these steps to connect the hoses to the regulators:

Step 1 Inspect both the oxygen and fuel gas hoses for any damage, burns, cuts, or fraying.

Step 2 Replace any damaged hoses.

Step 3 Connect the green oxygen hose to the oxygen regulator flashback arrestor or check valve (*Figure 55*).

Step 4 Connect the red or black fuel gas hose to the fuel gas regulator flashback arrestor or check valve. Keep in mind that all fuel gas fittings have left-hand threads.

4.6.0 Attaching Hoses to the Torch

Follow these steps to attach the hoses to the torch:

Step 1 Attach flashback arrestors to the oxygen and fuel gas hose connections on the torch body unless the torch has built-in flashback arrestors and check valves. Keep in mind that all fuel gas fittings have left-hand threads.

Step 2 Attach and tighten the green oxygen hose to the oxygen fitting on the flashback arrestor or torch (*Figure 56*).

Step 3 Attach and tighten the red or black hose to the fuel gas fitting on the flashback arrestor or torch. Remember that all fuel gas fittings have left-hand threads.

4.7.0 Connecting Cutting Attachments (Combination Torch Only)

If a combination torch is being used, connect cutting attachments as follows:

Step 1 Check the torch manufacturer's instructions for the correct method of installing the attachment.

Step 2 Connect the attachment and tighten by hand as required (*Figure 57*).

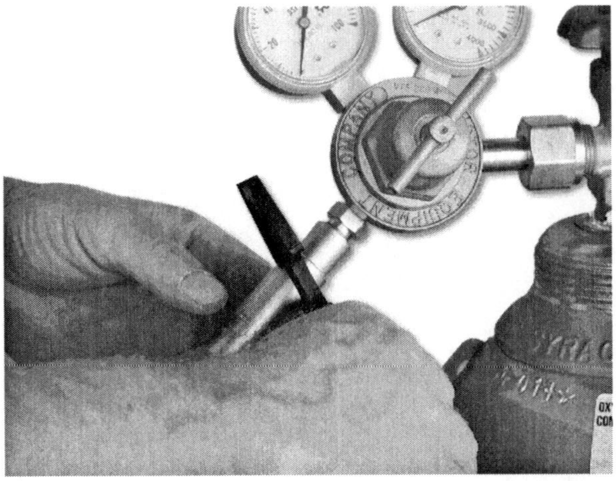

104F54.EPS

Figure 54 ◆ Attaching a flashback arrestor.

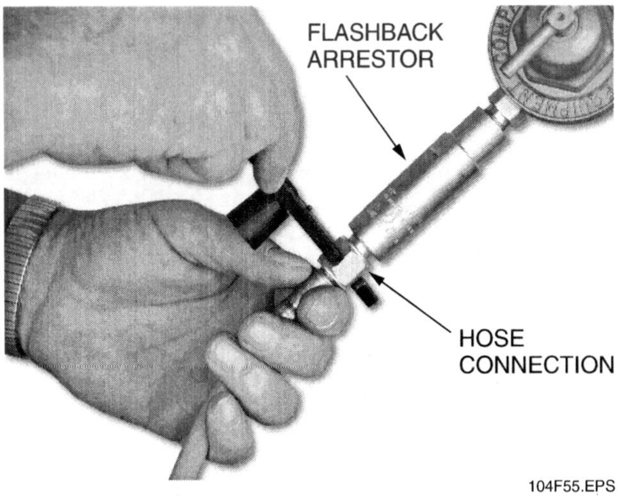

104F55.EPS

Figure 55 ◆ Connecting hose to regulator flashback arrestor.

4.8.0 Installing Cutting Tips

Follow these steps to install a cutting tip in the cutting torch:

Step 1 Identify the thickness of the material to be cut.

> **WARNING!**
>
> If acetylene fuel is being used, make sure that the maximum fuel flow rate per hour of the tip does not exceed one-tenth of the fuel cylinder capacity. If a purplish flame is observed when the torch is operating, the fuel rate is too high and acetone is being withdrawn from the acetylene cylinder along with the acetylene gas.

Step 2 Identify the proper size cutting tip from the manufacturer's recommended tip size chart for the fuel being used.

Step 3 Inspect the cutting tip sealing surfaces and orifices for damage or plugged holes. If the sealing surfaces are damaged, discard the tip. If the orifices are plugged, clean them with a tip cleaner or drill.

Step 4 Check the torch manufacturer's instructions for the correct method of installing cutting tips.

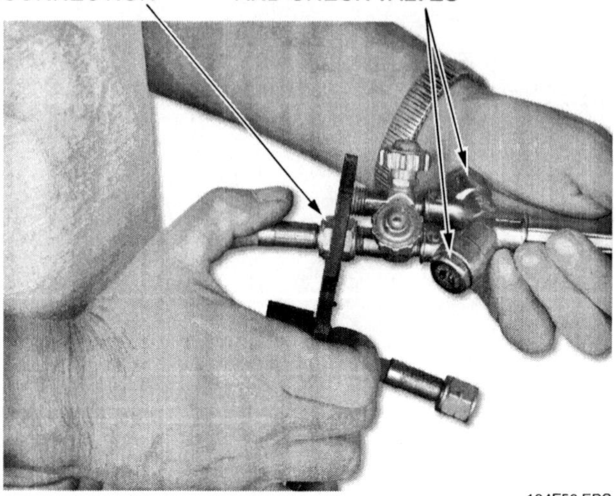

Figure 56 ◆ Connecting hoses to torch body.

Step 5 Install the cutting tip, securing it with a torch wrench or by hand as required (*Figure 58*).

4.9.0 Closing Torch Valves and Loosening Regulator Adjusting Screws

Follow these steps to close the torch valves and loosen the regulator adjusting screws (*Figure 59*):

Step 1 Check the fuel and oxygen valves on the torch to be sure they are closed. Closing the torch gas valves prevents gases from backing up inside the torch.

> **CAUTION**
>
> Loosening regulator adjusting screws closes the regulators and prevents damage to the regulator diaphragms when the cylinder valves are opened.

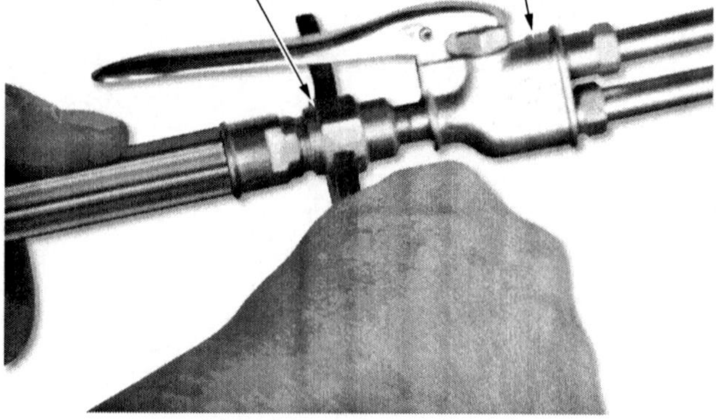

Figure 57 ◆ Connecting a cutting attachment.

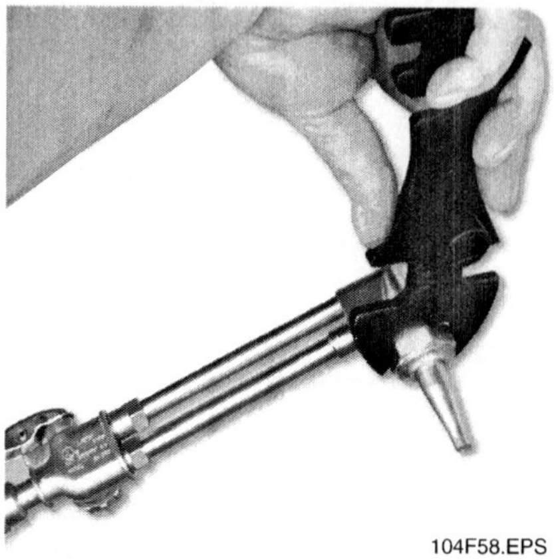

Figure 58 ◆ Installing a cutting tip.

Step 2 Check the oxygen regulator adjusting screw to be sure it is loose (backed out).

Step 3 Check the fuel gas regulator adjusting screw to be sure it is loose.

4.10.0 Opening Cylinder Valves

Follow these steps to open cylinder valves (*Figure 60*):

> **WARNING!**
> Never stand directly in front of or behind a regulator. The regulator adjusting screw can blow out, causing serious injury. Always open the cylinder valve gradually. Quick openings can damage a regulator or gauge or even cause a gauge to explode.
>
> Oxygen cylinder valves must be opened all the way until the valve seats at the top. Seating the valve at the fully open position prevents high-pressure leaks at the valve stem.

Step 1 Standing on the cylinder valve side of the oxygen regulator, slowly open the oxygen cylinder valve all the way, allowing the pressure in the cylinder pressure gauge to rise gradually until the gauge indicates the oxygen cylinder pressure.

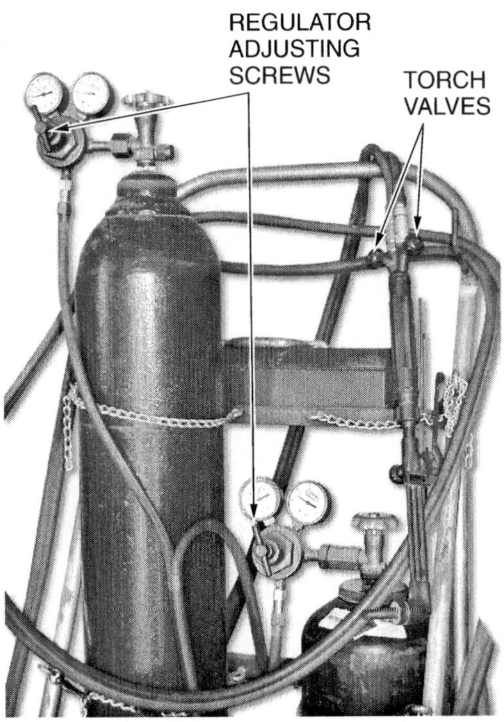

Figure 59 ◆ Torch valves and regulator adjusting screws.

Step 2 Standing on the cylinder valve side of the fuel gas regulator, slowly open the fuel gas cylinder valve a quarter turn or until the cylinder pressure gauge indicates the cylinder pressure. Opening the cylinder valve a quarter turn allows it to be quickly closed in case of a fire.

> **NOTE**
> If the fuel cylinder is equipped with a valve requiring a T-wrench, always leave the wrench in place on the valve so that the fuel can be quickly turned off. This type of valve is obsolete but still in use

4.11.0 Purging the Torch and Setting the Working Pressures

To reduce the chances of a flashback, the hoses and torch should always be purged and the working pressures checked each time the torch is to be ignited. Follow these steps to purge the torch and set the working pressures (*Figure 61*):

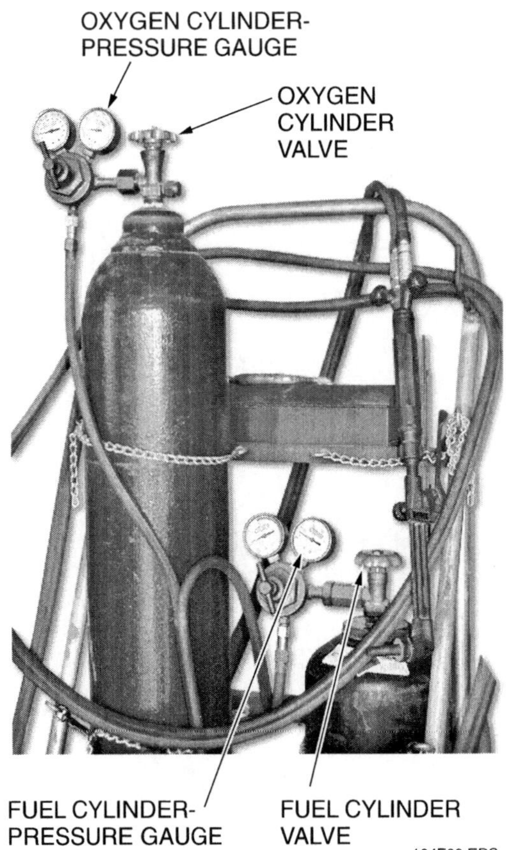

Figure 60 ◆ Cylinder valves and gauges.

> **WARNING!**
> The working pressure gauge readings on single-stage regulators will rise after the torch valves are turned off. This is normal. However, if acetylene is being used as the fuel gas, ensure that the static pressure does not rise to 15 psig. Make sure that equipment is purged and leak tested in a well-ventilated area to avoid creating an explosive concentration of gases.

Step 1 Fully open the oxygen valve on the torch. Then depress and hold or lock open the cutting oxygen lever.

Step 2 Tighten the oxygen regulator adjusting screw until the working pressure gauge shows the correct oxygen gas working pressure with the gas flowing. Allow the gas to flow for five to ten seconds to purge the torch and hoses of air or fuel gas.

Step 3 At the torch, release the cutting lever and close the oxygen valve.

Step 4 Open the fuel valve on the torch about ⅛ of a turn.

Step 5 Tighten the fuel regulator adjusting screw until the working pressure gauge shows the correct fuel gas working pressure with the gas flowing. Allow the gas to flow for five to ten seconds to purge the hoses and torch of air.

Step 6 At the torch, close the fuel valve. If acetylene is used, check that the acetylene static pressure does not rise to 15 psig. If it does, immediately open the torch fuel valve and reduce the regulator output pressure as required.

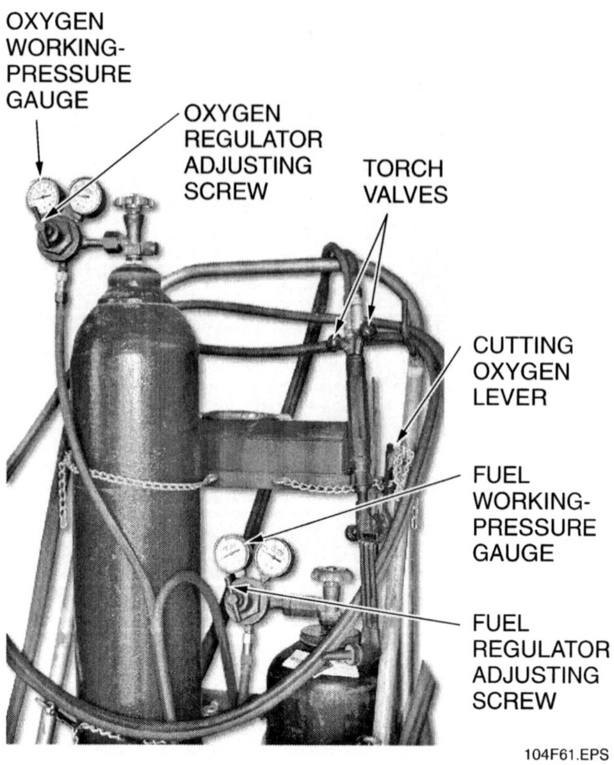

Figure 62 ◆ Typical points for purging the equipment and setting the working pressures

4.12.0 Testing for Leaks

Equipment must be tested for leaks immediately after it is set up and periodically thereafter. The torch should be checked for leaks each time before use. Leaks could cause a fire or explosion if undetected. To test for leaks, brush a commercially prepared leak-testing formula or a solution of detergent and water on the following points. If bubbles form, a leak is present.

> **WARNING!**
> Make sure the detergent contains no oil. In the presence of oxygen, oil can cause fires or explosions.

Leak points include the following:

- Oxygen cylinder valve
- Fuel gas cylinder valve
- Oxygen regulator and regulator inlet and outlet connections
- Fuel gas regulator and regulator inlet and outlet connections
- Hose connections at the regulators, check valves/flashback arrestors, and torch
- Torch valves and cutting oxygen lever valve
- Cutting attachment connection (if used)
- Cutting tip

If there is a leak at the fuel gas cylinder valve stem, attempt to stop it by tightening the packing gland. If this does not stop the leak, mark and remove the cylinder and notify the supplier. For other leaks, tighten the connections slightly with a wrench. If this does not stop the leak, turn off the gas pressure, open all connections, and inspect the screw threads.

> **CAUTION**
>
> Use care not to overtighten connections. The brass connections used will strip if overtightened.
>
> Repair or replace equipment that does not seal properly. If the torch valves leak, try lightly tightening the packing nut at the valves. If a cutting attachment or torch tip leaks, disassemble and check the sealing surfaces and/or O-rings for damage. If the equipment or torch tip is new and the sealing surfaces do not appear damaged, overtighten the connection slightly to seat the sealing surfaces; then loosen and retighten normally. If the leaks persist or the equipment is visibly damaged, replace the torch, cutting attachment (if used), or cutting tip as necessary.

> **WARNING!**
>
> Make sure that equipment is purged and leak tested in a well-ventilated area to avoid creating an explosive concentration of gases.

4.12.1 Initial and Periodic Leak Testing

Perform the following steps at initial equipment setup and periodically thereafter:

Step 1 Set the equipment to the correct working pressures with the torch valves turned off.

Step 2 Using a leak-test solution, check for leaks at the cylinder valves, regulator relief ports, and regulator gauge connections (*Figure 62*). Also, check for leaks at hose connections, regulator connections, and check valve/flame arrestor connections up to the torch.

4.12.2 Leak-Down Testing of Regulators, Hoses, and Torch

Perform the following steps to quickly leak test the regulators, hoses, and torch before the torch is ignited.

Step 1 Set the equipment to the correct working pressures with the torch valves turned off. Then loosen both regulator adjusting screws. Check the working pressure gauges after a minute or two to see if the pressure drops. If the pressure drops, check the hose connection and regulators for leaks; otherwise, proceed to Step 2.

Step 2 Place a thumb over the cutting tip orifices and press to block the orifices.

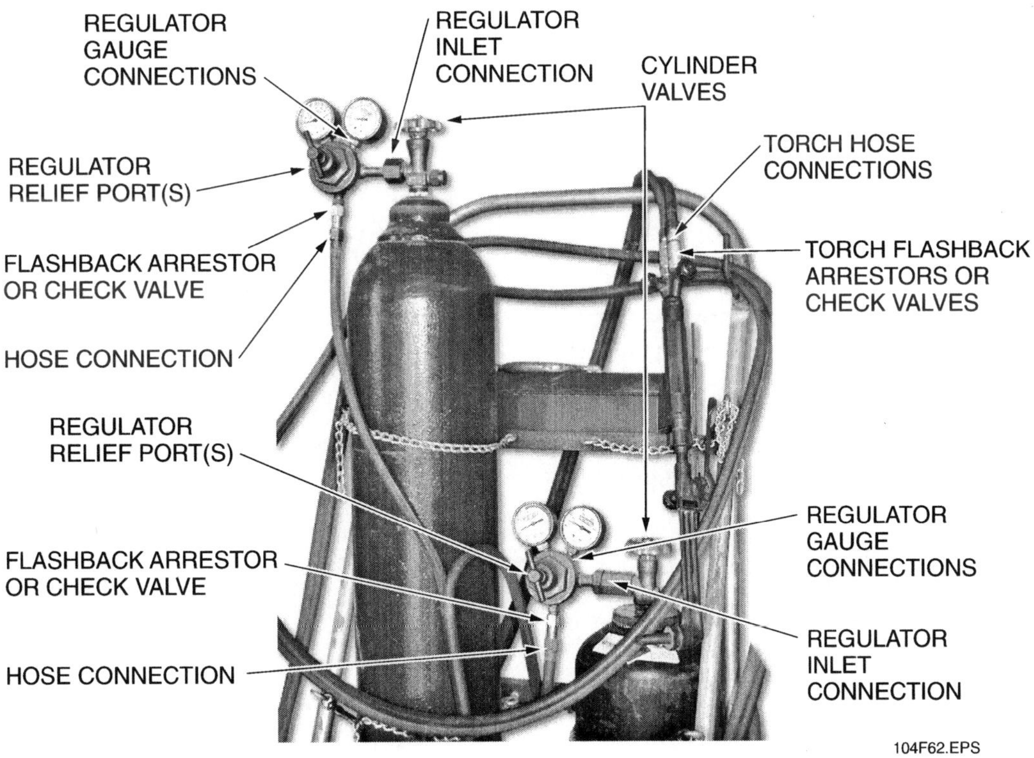

Step 3 Turn on the torch oxygen valve and then depress and hold the cutting oxygen lever down.

Step 4 After the gauge pressure drops slightly, observe the oxygen working pressure gauge for a minute to see if the pressure continues to drop. If the pressure keeps dropping, perform the leak test described in the following section to determine the source of the leak. If the pressure does not change, close the torch oxygen valve and release the pressure at the cutting tip.

Step 5 With the tip blocked, turn on the torch fuel valve. After the gauge pressure drops slightly, carefully observe the fuel working pressure gauge for a minute. If the pressure continues to drop, perform the leak test described in the following section to determine the source of the leak. If the pressure does not change, close the torch fuel valve and release the pressure at the cutting tip.

Step 6 If no leaks are apparent, set the equipment to the correct working pressures.

4.12.3 Full Leak Testing of a Torch

Always take the time to perform a leak test on regulators, hoses, and the torch before you use them the first time that day. Always perform a leak test on a torch after you change tips or if you convert the torch from welding to cutting or vice versa. Leaks, especially fuel leaks, can cause a fire or explosion after a cutting or welding operation begins. A dangerous fire occurring at or near the torch may not be immediately noticed by the welder due to limited visibility caused by the tinted lenses in the welding goggles or face shield.

Perform the following steps to test for and isolate torch leaks (*Figure 63*).

Step 1 Set the equipment to the correct working pressures with the torch valves turned off.

Step 2 Block the flow to the cutting tip orifices.

Step 3 Turn on the torch oxygen valve and then depress and lock the cutting oxygen lever down.

Step 4 With the cutting tip blocked, check for leaks using a leak-test solution at the torch oxygen valve, cutting oxygen lever valve, cutting attachment connection (if used), preheat oxygen valve (if used), and cutting tip seal at the torch head.

Step 5 Release the cutting oxygen lever and close the torch oxygen valve. Release the pressure at the cutting tip.

Step 6 With the cutting tip blocked, open the torch fuel valve.

Step 7 Using a leak-test solution, check for leaks at the torch fuel valve, cutting attachment (if used), and cutting tip seal at the torch head.

Step 8 Close the torch fuel valve and remove thumb from the cutting tip.

5.0.0 ◆ CONTROLLING THE OXYFUEL TORCH FLAME

To be able to safely use a cutting torch, the operator must understand the flame and be able to adjust it and react to unsafe conditions. The following sections will explain the oxyfuel flame and how to control it safely.

5.1.0 Oxyfuel Flames

There are three types of oxyfuel flames: **neutral flame**, **carburizing flame**, and **oxidizing flame**.

- *Neutral flame* – A neutral flame burns proper proportions of oxygen and fuel gas. The inner cones will be light blue in color, surrounded by a darker blue outer flame envelope that results when the oxygen in the air combines with the super-heated gases from the inner cone. A neutral flame is used for all but special cutting applications.

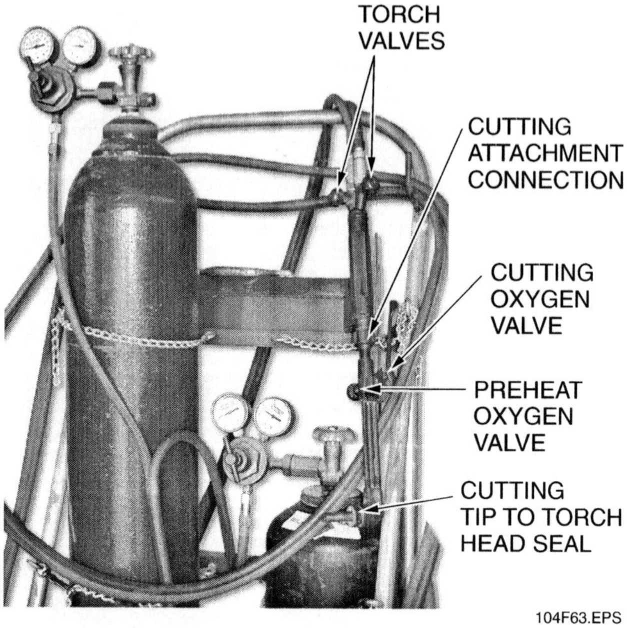

Figure 63 ◆ Torch leak-test points.

- *Carburizing flame* – A carburizing flame has a white feather created by excess fuel. The length of the feather depends on the amount of excess fuel present in the flame. The outer flame envelope is longer than that of the neutral flame, and it is much brighter in color. The excess fuel in the carburizing flame (especially acetylene) produces large amounts of carbon. The carbon will combine with red-hot or molten metal, making the metal hard and brittle. The carburizing flame is cooler than a neutral flame and is never used for cutting. It is used for some special heating applications.
- *Oxidizing flame* – An oxidizing flame has an excess of oxygen. The inner cones are shorter, much bluer in color, and more pointed than a neutral flame. The outer flame envelope is very short and often fans out at the ends. An oxidizing flame is the hottest flame. A slightly oxidizing flame is recommended with some special fuel gases, but in most cases it is not used. The excess oxygen in the flame can combine with many metals, forming a hard, brittle, low-strength oxide. However, the preheat flames of a properly adjusted cutting torch will be slightly oxidizing when the cutting oxygen is shut off.

Figure 64 shows the various flames that occur at a cutting tip for both acetylene and LP gas.

NOTE

Increasing the fuel flow until the flame pulls away from the tip and then decreasing the flow until the flame returns to the tip sets the maximum fuel flow for the tip size in use.

5.2.0 Backfires and Flashbacks

When the torch flame goes out with a loud pop or snap, a backfire has occurred. Backfires are usually caused when the tip or nozzle touches the work surface or when a bit of hot dross briefly interrupts the flame. When a backfire occurs, you can relight the torch immediately. Sometimes the torch even relights itself. If a backfire recurs without the tip making contact with the base metal, shut off the torch and find the cause. Possible causes are:

- Improper operating pressures
- A loose torch tip
- Dirt in the torch tip seat or a bad seat

When the flame goes out and burns back inside the torch with a hissing or whistling sound, a flashback is occurring. Immediately shut off the oxygen valve on the torch; the flame is burning inside the torch. If the flame is not extinguished quickly, the end of the torch will melt off. The flashback will stop as soon as the oxygen valve is closed. Therefore, quick action is crucial. Flashbacks can cause fires and explosions within the cutting rig and, therefore, are very dangerous. Flashbacks can be caused by:

- Equipment failure
- Overheated torch tip
- Dross or spatter hitting and sticking to the torch tip
- Oversized tip (tip is too large for the gas flow rate being used)

After a flashback has occurred, wait until the torch has cooled. Then, blow oxygen (not fuel gas) through the torch for several seconds to remove soot that may have built up in the torch during the flashback before relighting it. If you hear the hissing or whistling after the torch is reignited or if the

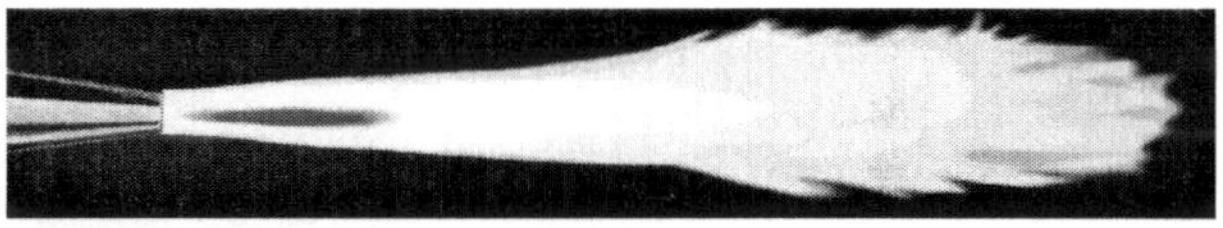

Acetylene Burning in Atmosphere
Open fuel gas valve until smoke clears from flame.

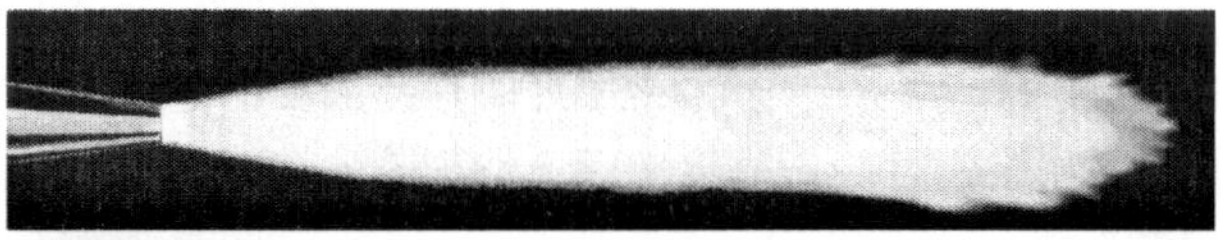

Carburizing Flame
(Excess acetylene with oxygen) Preheat flames require more oxygen.

Neutral Flame
(Acetylene with oxygen) Temperature 5589°F (3087°C). Proper preheat adjustment when cutting.

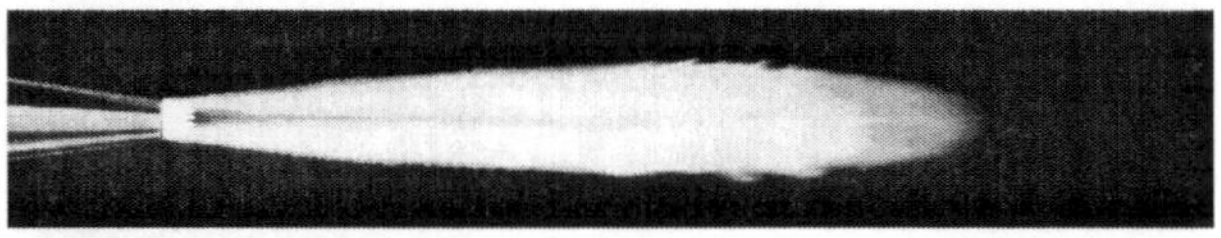

Neutral Flame with Cutting Jet Open
Cutting jet must be straight and clean. If it flares, the pressure is too high for the tip size.

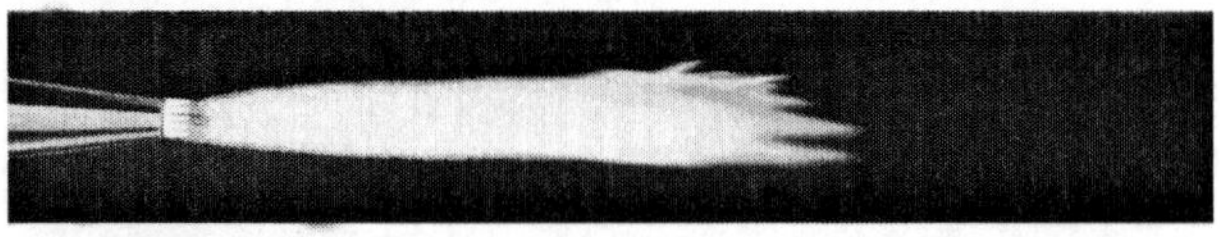

Oxidizing Flame
(Acetylene with excess oxygen) Not recommended for average cutting. However, if the preheat flame is adjusted for neutral with the cutting oxygen on, then this flame is normal after the cutting oxygen is off.

104F64A.EPS

Figure 64 ◆ Acetylene and LP gas flames. (1 of 2)

flame does not appear normal, shut off the torch immediately and have the torch serviced by a qualified technician. If the torch or tip has been damaged, replace the flashback arrestors.

5.3.0 Igniting the Torch and Adjusting the Flame

After the cutting equipment has been properly set up and purged, the torch can be ignited and the flame adjusted for cutting. Follow these steps to ignite the torch:

Step 1 Choose the appropriate cutting torch tip according to the base metal thickness you will be cutting and fuel gas you are using.

NOTE

Refer to the manufacturer's charts. You may have to readjust the oxygen and fuel gas pressure depending on the tip selected.

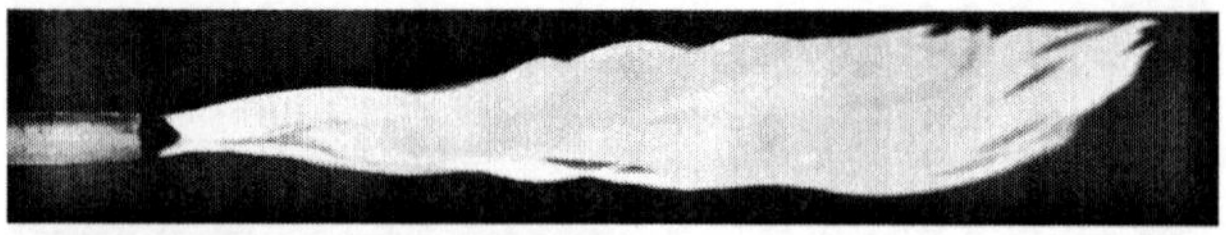

LP Gas Burning in Atmosphere
Open fuel gas valve until flame begins to leave tip end.

Reducing Flame
(Excess LP-gas with oxygen) Not hot enough for cutting.

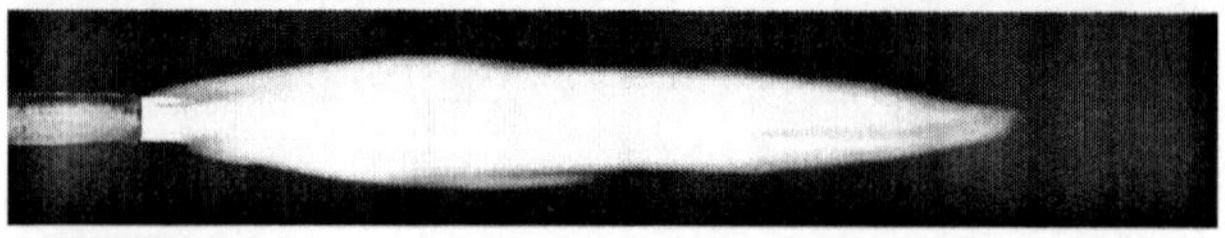

Neutral Flame
(LP-gas with oxygen) For preheating prior to cutting.

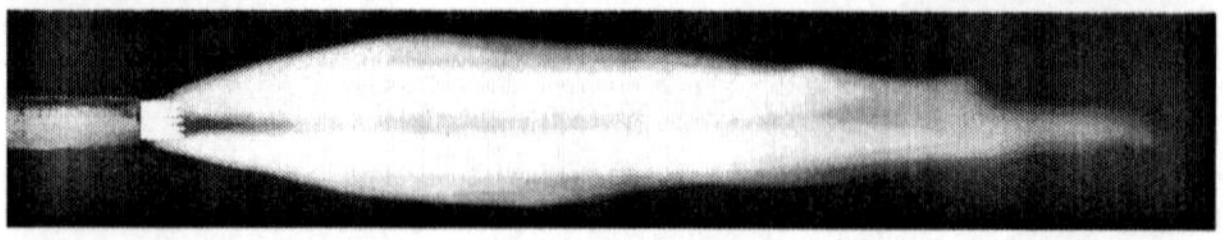

Oxidizing Flame with Cutting Jet Open
Cutting jet stream must be straight and clean.

Oxidizing Flame without Cutting Jet Open
(LP-gas with excess oxygen) The highest temperature flame for fast starts and high cutting speeds.

104F64B.EPS

Figure 64 ◆ Acetylene and LP gas flames. (2 of 2)

Step 2 Inspect the tip sealing surfaces and orifices. Attach the tip to the cutting torch or cutting attachment by placing it on the end of the torch and tightening the nut.

NOTE

Some manufacturers recommend tightening the nut with a torch wrench. Others recommend tightening the nut by hand. Check the tip manual to see if the manufacturer recommends that the nut be tightened manually or with a torch wrench.

Step 3 Put on proper protective clothing, gloves, and eye/face protection.

Step 4 Raise tinted eye protection and/or face shield.

Step 5 Release the oxygen cutting lever. If present, close the preheat oxygen valve and open the torch oxygen valve fully.

Step 6 Open the fuel gas valve on the torch handle about one-quarter turn.

Step 7 Holding the friction lighter near the side and to the front of the torch tip, ignite the torch.

WARNING!

Hold the friction lighter near the side of the tip when igniting the torch to prevent deflecting the ignited gas back toward you. Always use a friction lighter. Never use matches or cigarette lighters to light the torch because this could result in severe burns and/or could cause the lighter to explode. Always point the torch away from yourself, other people, equipment, and flammable material.

Step 8 Once the torch is lit, adjust the torch fuel gas flame by adjusting the flow of fuel gas with the fuel gas valve. Increase the flow of fuel gas until the flame stops smoking or pulls slightly away from the tip. Decrease the flow until the flame returns to the tip.

Step 9 Open the preheat oxygen valve (if present) or the oxygen torch valve very slowly and adjust the torch flame to a neutral flame.

Step 10 Press the cutting oxygen lever all the way down and observe the flame. It should have a long, thin, high-pressure oxygen cutting jet up to 8 inches long extending from the cutting oxygen hole in the center of the tip. If it does not:

- Check that the working pressures are set as recommended on the manufacturer's chart.
- Clean the cutting tip. If this does not clear up the problem, change the cutting tip.

Step 11 With the cutting oxygen on, observe the preheat flame. If it has changed slightly to a carburizing flame, increase the preheat oxygen until the flame is neutral. After this adjustment, the preheat flame will change slightly to an acceptable oxidizing flame when the cutting oxygen is shut off.

5.4.0 Shutting Off the Torch

Follow these steps to shut off the torch after a cutting operation is completed:

> **WARNING!**
> Always turn off the oxygen flow first to prevent a possible flashback into the torch.

Step 1 Release the cutting oxygen lever. Then close the torch or preheat oxygen valves.

Step 2 Close the torch fuel gas valve quickly to extinguish the flame.

6.0.0 ◆ SHUTTING DOWN OXYFUEL CUTTING EQUIPMENT

When a cutting job is completed and the oxyfuel equipment is no longer needed, it must be shut down. Follow these steps to shut down oxyfuel cutting equipment (*Figure 65*):

Step 1 Close the fuel gas and oxygen cylinder valves.

Step 2 Open the fuel gas and oxygen torch valves to allow gas to escape. Do not proceed to Step 3 until all pressure is released and all regulator gauges read zero.

Step 3 Back out the fuel gas and oxygen regulator adjusting screws until they are loose.

Step 4 Close the fuel gas and oxygen torch valves.

Step 5 Coil up the hose and secure the torch to prevent damage.

7.0.0 ◆ DISASSEMBLING OXYFUEL EQUIPMENT

Follow these steps if the oxyfuel equipment must be disassembled after use:

Step 1 Check to be sure the equipment has been properly shut down. This includes checking that:

- The cylinder valves are closed.
- All pressure gauges read zero.

Step 2 Remove both hoses from the torch.

Step 3 Remove both hoses from the regulators.

Step 4 Remove both regulators from the cylinder valves.

Step 5 Replace the protective caps on the cylinders.

Step 6 Return the oxygen cylinder to its proper storage place.

> **WARNING!**
> Always transport and store gas cylinders in the upright position. Be sure they are properly secured (chained) and capped.

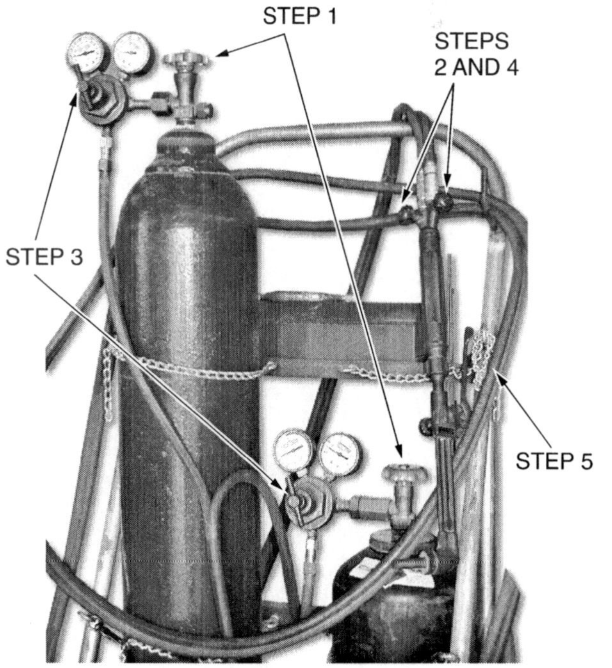

104F65.EPS

Figure 65 ◆ Shutting down oxyfuel cutting equipment.

Step 7 Return the fuel gas cylinder to its proper storage place.

> **WARNING!**
> Regardless if the cylinders are empty or full, never store fuel gas cylinders and oxygen cylinders together. Storing cylinders together is a violation of OSHA and local fire regulations and could result in a fire and explosion.

8.0.0 ◆ CHANGING EMPTY CYLINDERS

Follow these procedures to change a cylinder when it is empty:

> **WARNING!**
> When moving cylinders, always be certain that they are in the upright position and the valve caps are secured in place. Never use a sling or electromagnet to lift cylinders. To lift cylinders, use a cylinder cage.

Step 1 Check to be sure equipment has been properly shut down. This includes checking that:

- The cylinder valves are closed.
- All pressure gauges read zero.

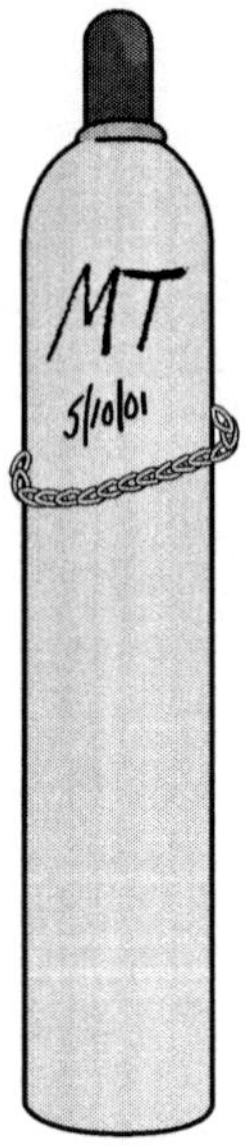

104F66.EPS

Figure 66 ◆ Typical empty cylinder marking.

Step 2 Remove the regulator from the empty cylinder.

Step 3 Replace the protective cap on the empty cylinder.

Step 4 Transport the empty cylinder from the workstation to the storage area.

Step 5 Mark MT (empty) and the date (or the accepted site notation for indicating an empty cylinder) near the top of the cylinder using soapstone (*Figure 66*).

Step 6 Place the empty cylinder in the empty cylinder section of the cylinder storage area for the type of gas in the cylinder.

> **NOTE**
> Do not use permanent markers on cylinders; use soapstone or another temporary marker. If a cylinder is defective, place a warning tag on it.

9.0.0 ◆ PERFORMING CUTTING PROCEDURES

The following sections explain how to recognize good and bad cuts, how to prepare for cutting operations, and how to perform straight-line cutting, piercing, bevel cutting, washing, and gouging.

9.1.0 Inspecting the Cut

Before attempting to make a cut, you must be able to recognize good and bad cuts and know what causes bad cuts. This is explained in the following list and illustrated in *Figure 67*:

- A good cut features a square top edge that is sharp and straight, not ragged. The bottom edge can have some dross adhering to it but not an excessive amount. What dross there is should be easily removable with a chipping hammer. The **drag lines** should be near vertical and not very pronounced.
- When preheat is insufficient, bad gouging results at the bottom of the cut because of slow travel speed.
- Too much preheat will result in the top surface melting over the cut, an irregular cut edge, and an excessive amount of dross.
- When the cutting oxygen pressure is too low, the top edge will melt over because of the resulting slow cutting speed.
- Using cutting oxygen pressure that is too high will cause the operator to lose control of the cut, resulting in an uneven kerf.

- A travel speed that is too slow results in bad gouging at the bottom of the cut and irregular drag lines.
- When the travel speed is too fast, there will be gouging at the bottom of the cut, a pronounced break in the drag line, and an irregular kerf.
- A torch that is held or moved unsteadily across the metal being cut can result in a wavy and irregular kerf.
- When a cut is lost and then not restarted carefully, bad gouges will result at the point where the cut is restarted.

> **NOTE**
>
> The following tasks are designed to develop your skills with a cutting torch. Practice each task until you are thoroughly familiar with the procedure. As you complete each task, take it to your instructor for evaluation. Do not proceed to the next task until your instructor tells you to continue.

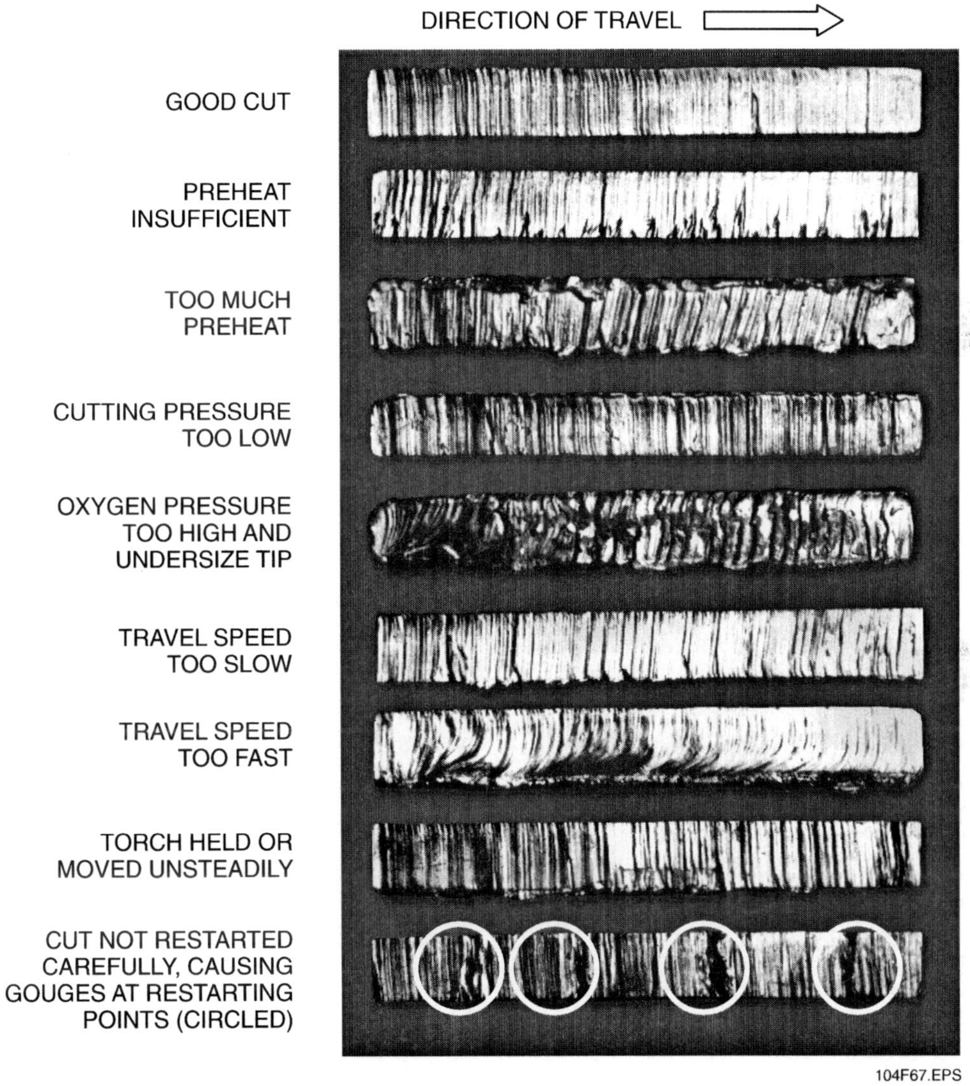

Figure 67 ◆ Examples of good and bad cuts.

9.2.0 Preparing for Oxyfuel Cutting with a Hand Cutting Torch

Before metal can be cut, the equipment must be set up and the metal prepared. One important step is to properly lay out the cut by marking it with soapstone or punch marks. The few minutes this takes will result in a quality job, reflecting craftsmanship and pride in your work. Follow these steps to prepare to make a cut:

Step 1 Prepare the metal to be cut by cleaning any rust, scale, or other foreign matter from the surface.

Step 2 If possible, position the work so you will be comfortable when cutting.

Step 3 Mark the lines to be cut with soapstone or a punch.

Step 4 Select the correct cutting torch tip according to the thickness of the metal to be cut, the type of cut to be made, the amount of preheat needed, and the type of fuel gas to be used.

Step 5 Ignite the torch.

Step 6 Use the procedures outlined in the following sections for performing particular types of cutting operations.

9.3.0 Cutting Thin Steel

Thin steel is 3/16 inch thick or less. A major concern when cutting thin steel is distortion caused by the heat of the torch and the cutting process. To minimize distortion, move as quickly as you can without losing the cut. Follow these steps to cut thin steel:

Step 1 Prepare the metal surface.

Step 2 Light the torch.

Step 3 Hold the torch so that the tip is pointing in the direction the torch is traveling at a 15- to 20-degree angle. Make sure that a preheat orifice and the cutting orifice are centered on the line of travel next to the metal (*Figure 68*).

CAUTION

Holding the tip upright when cutting thin steel will overheat the metal, causing distortion.

Step 4 Preheat the metal to a dull red. Use care not to overheat thin steel because this will cause distortion.

NOTE

The edge of the tip can be rested on the surface of the metal being cut and then slid along the surface when making the cut.

Step 5 Press the cutting oxygen lever to start the cut, and then move quickly along the line to be cut. To minimize distortion, move as quickly as you can without losing the cut.

9.4.0 Cutting Thick Steel

Most oxyfuel cutting will be on steel that is more than 3/16 inch thick. Whenever heat is applied to metal, distortion is a problem, but as the steel gets thicker, it becomes less of a problem. Follow these steps to cut thick steel with a hand cutting torch:

Step 1 Prepare the metal surface and torch. Light the torch.

Step 2 Ignite and adjust the torch flame.

Step 3 Follow the number sequence shown in *Figure 69* to perform the cut.

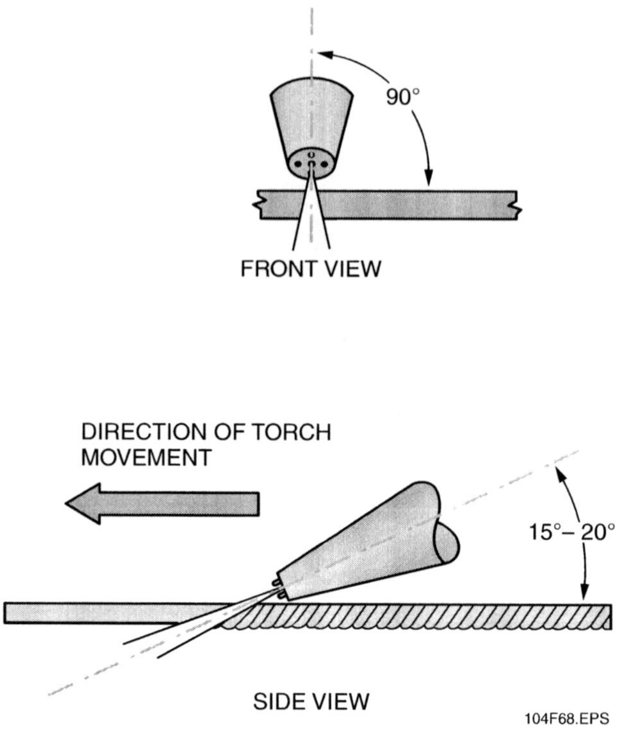

Figure 68 ◆ Cutting thin steel.

> **NOTE**
> The torch can be moved from either right to left or left to right. Choose the direction that is the most comfortable for you.

9.5.0 Piercing a Plate

Before holes or slots can be cut in a plate, the plate must be pierced. Piercing puts a small hole through the metal where the cut can be started.

> **NOTE**
> When cutting begins, the tips of the preheat flame should be held 1⁄16 to 1⁄8 inch above the workpiece. For steel up to 3⁄8 inch thick, the first and third procedures can usually be omitted.

Because more preheat is necessary on the surface of a plate than at the edge, choose the next-larger cutting tip than is recommended for the thickness to be pierced. When piercing steel that is more than 3 inches thick, it may help to preheat the bottom side of the plate directly under the spot to be pierced. Follow these steps to pierce a plate for cutting:

Step 1 Prepare the metal surface and torch.

Step 2 Ignite the torch and adjust the flame.

Step 3 Hold the torch tip 1⁄4 to 5⁄16 inch above the spot to be pierced until the surface is a bright cherry red (*Figure 70*).

Step 4 Slowly press the cutting oxygen lever. As the cut starts, raise the tip about 1⁄2 inch above the metal surface and tilt the torch slightly so that molten metal does not blow back onto the tip. The tip should be raised and tipped before the cutting oxygen lever is fully depressed.

Step 5 Maintain the tipped position until a hole burns through the plate. Then rotate the tip vertically.

Step 6 Lower the torch tip to about 3⁄16 inch above the metal surface and continue to cut outward from the original hole to the edge of the line to be cut.

9.6.0 Cutting Bevels

Bevel cutting is often performed to prepare the edge of steel plate for welding. Follow these steps to perform bevel cutting (*Figure 71*):

Step 1 Prepare the metal surface and the torch.

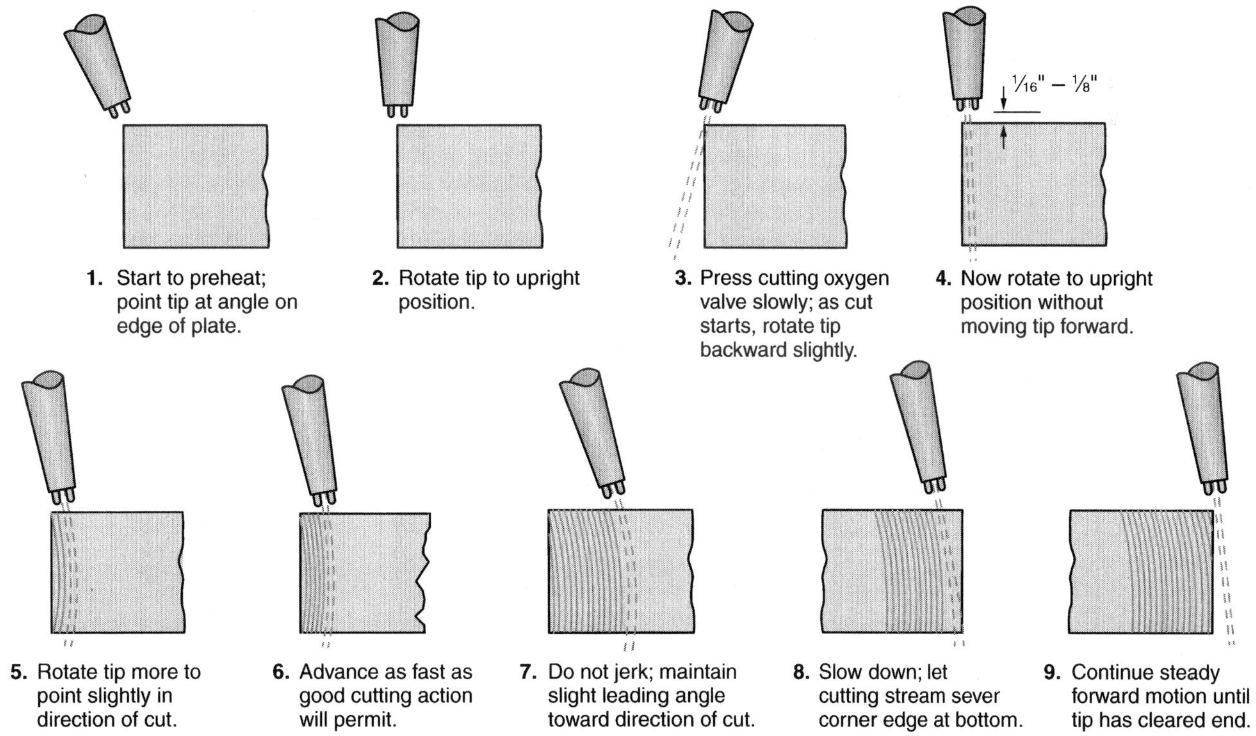

Figure 69 ◆ Cutting thick steel.

Step 2 Ignite the torch and adjust the flame.

Step 3 Hold the torch so that the tip faces the metal at the desired bevel angle.

> **NOTE**
> An angle iron can be used as a guide.

Step 4 Preheat the edge to a bright cherry red.

Step 5 Press the cutting oxygen lever to start the cut.

Step 6 As cutting begins, move the torch tip at a steady rate along the line to be cut. Pay particular attention to the torch angle to ensure it is uniform along the entire length of the cut.

9.7.0 Washing

Washing is a term used to describe the process of cutting out bolts or rivets. Washing operations use a special tip with a large cutting hole that produces a low-velocity stream of oxygen. The low-velocity oxygen stream helps prevent cutting into the surrounding base metal. Washing tips can also be used to remove items such as blocks, angles, or channels that are welded onto a surface. Follow these steps to perform washing (*Figure 72*):

Step 1 Prepare the metal surface and torch.

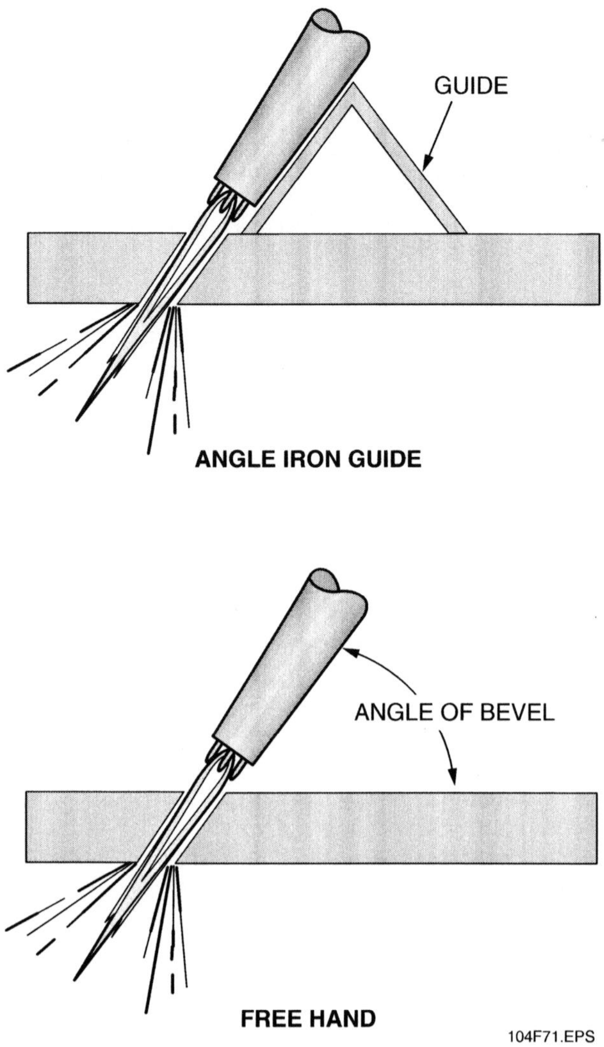

Figure 71 ◆ Cutting a bevel.

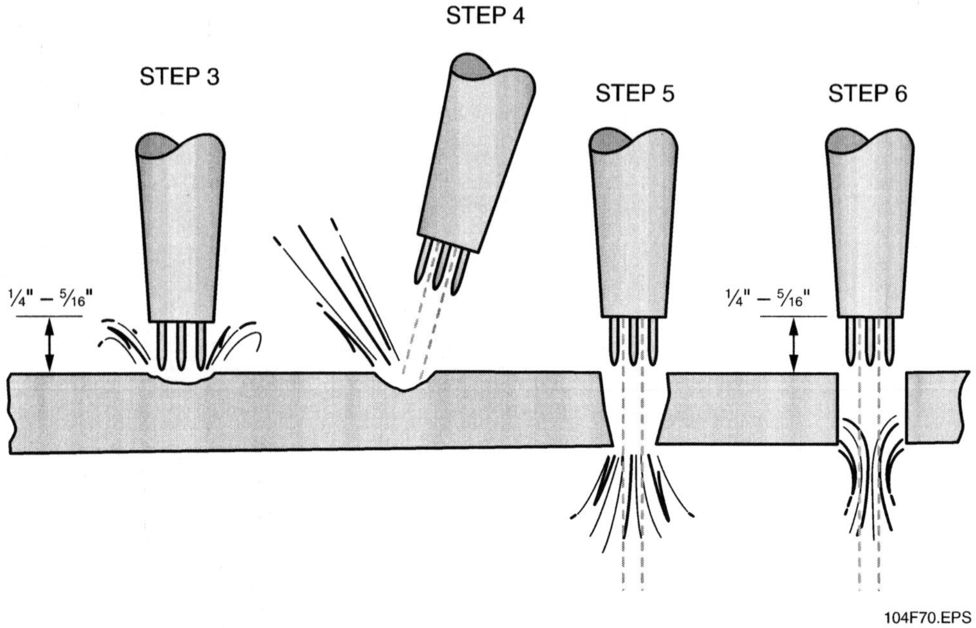

Figure 70 ◆ Steps for piercing steel.

Step 2 Ignite the torch and adjust the flame.

Step 3 Preheat the metal to be cut until it is a bright cherry red.

Step 4 Move the cutting torch at a 55-degree angle to the metal surface.

Step 5 At the top of the material, press the cutting oxygen lever to cut the material to be removed. Continue moving back and forth across the material while rotating the tip to a position parallel with the material. Move the tip back and forth and down to the surrounding metal. Use care not to cut into the surrounding metal.

> **CAUTION**
> As the surrounding metal heats up, there is a greater danger of cutting into it. Try to complete the washing operation as quickly as possible. If the surrounding metal gets too hot, stop and let it cool down.

10.0.0 ◆ PORTABLE OXYFUEL CUTTING MACHINE OPERATION

As explained previously, machine oxyfuel gas cutters or track burners are basic guidance systems driven by a variable speed electric motor to enable the operator to cut or bevel straight lines at any desired speed. The device (*Figure 73*) is usually mounted on a track or used with a circle-cutting attachment to enable the operator to cut various diameters from 4 to 96 inches. It consists of a heavy-duty tractor unit fitted with an adjustable torch mount and gas hose attachments. It is also equipped with an ON/OFF switch, a reversing switch, a clutch, and a speed-adjusting dial calibrated in feet/meters per minute.

The device shown in *Figure 73* offers the following operational features:

- Makes straight-line cuts of any length
- Makes circle cuts up to 96 inches in diameter
- Makes bevel or chamfer cuts
- Has an infinitely variable cutting speed from 1 to 60 inches per minute
- Has dual speed and clutch controls to enable operation of the machine from either end

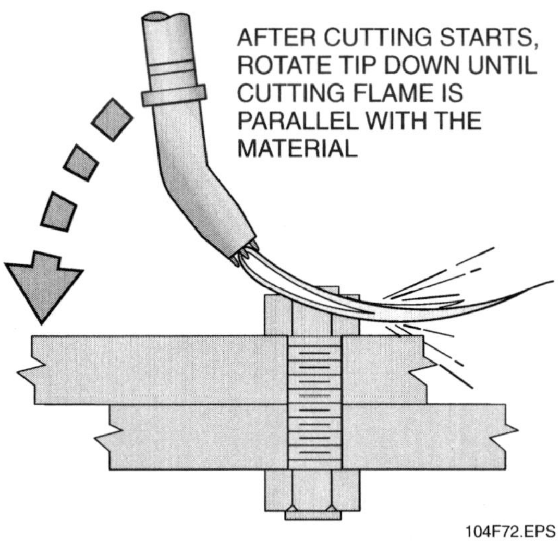

Figure 72 ◆ Washing.

10.1.0 Machine Controls

Figure 74 shows the location of the following controls:

- *Directional control* – Set the machine direction by toggling the FWD-OFF-REV toggle switch located next to the power cord.
- *Speed control* – Turn the large knob on either end of the machine to position the speed indicator at the desired cutting speed.
- *Clutch operation* – Engage the clutch by rotating one of the two clutch levers, located on either end of the machine, to the DRIVE position. Place the clutch lever in the FREE position to permit easy manual positioning of the machine prior to or after the actual cutting operation.

10.2.0 Torch Adjustment

The rack assembly permits the torch holder assembly to move toward or away from the tractor unit. The torch holder allows vertical positioning of the torch. The torch bevel adjustment allows torch positioning at any angle from 190 to 290 degrees in a plane perpendicular to the track. After adjusting the torch to the desired position, tighten all clamping screws to prevent the torch from making any unexpected movements.

10.3.0 Straight Line Cutting

Cut straight lines using the following procedure:

> **WARNING!**
> Most cutting machines are not designed to detect the end of their track or workpiece. Take care that an unattended machine does not fall from an elevated workpiece.

Step 1 Place the machine track on the workpiece and line it up before placing the machine on the track.

Step 2 Be sure the track is long enough for the cut to be made. If not, install additional track. Connect track sections carefully. Extend the track on both sides of the cut and support the track. When properly connected, the machine should travel smoothly from one track section to the next. If the cut is long, the track may have to be clamped at both ends beyond the cut to keep the track from moving during the cut.

Step 3 Place the machine on the track. Place the clutch lever in the FREE position. Be sure that the supply gas hoses and the power lines are long enough and free to move with the machine so that it can complete the cut properly.

Step 4 Move the machine to the approximate point where the cut will start. Set the drive speed control to the desired cutting speed. Set the FWD-OFF-REV switch to the OFF position. Plug the power cord into a 115 alternating current (AC), 60 Hertz (Hz) power outlet.

Step 5 Ensure that all clamping screws are properly tightened. Ignite and properly adjust the torch, then preheat the start of the cut. Set the FWD-OFF-REV switch to the desired direction of travel. Simultaneously turn on the cutting oxygen and set the clutch lever to the DRIVE position.

Step 6 When the cut is completed, stop the machine and shut off the torch.

10.4.0 Bevel Cutting

Perform bevel cutting operations using the following procedure:

WARNING!

Most cutting machines are not designed to detect the end of their track or workpiece. Take care that an unattended machine does not fall from an elevated workpiece.

Step 1 Place the machine track on the workpiece and line it up before placing the machine on the track.

Step 2 Be sure the track is long enough for the cut to be made. If not, install additional track. Connect track sections carefully. Extend the track on both sides of the cut and support the track. When properly connected, the machine should travel smoothly from one track section to the next. If the cut is long, the track may have to be clamped at both ends beyond the cut to keep the track from moving during the cut.

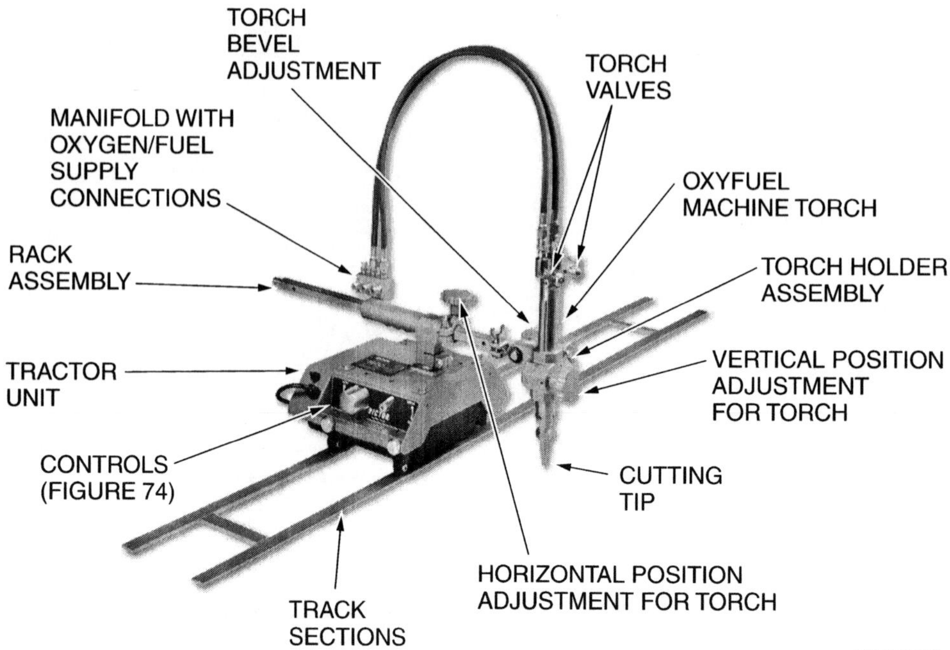

Figure 73 ◆ Victor track burner.

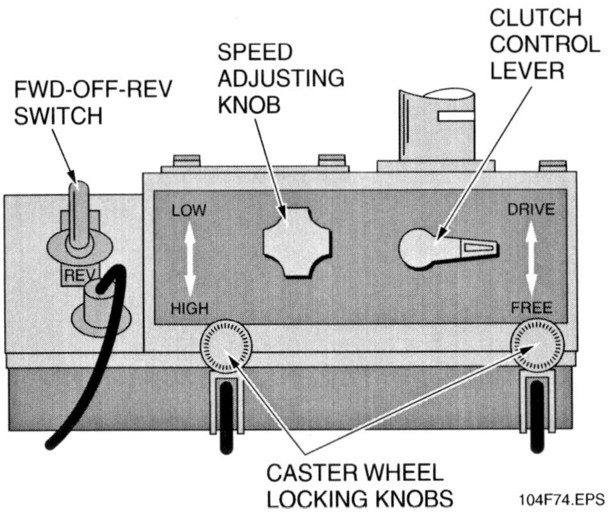

Figure 74 ◆ Victor track burner controls.

Step 3 Place the machine on the track. Place the clutch lever in the FREE position. Be sure that the supply gas hoses and the power lines are long enough and free to move with the machine so that it can complete the cut properly.

Step 4 Loosen the bevel adjusting knob, set the torch angle to the desired bevel angle, and then tighten the bevel adjusting knob.

Step 5 Move the machine to the approximate point where the cut will start. Set the drive speed control to the desired cutting speed. Set the FWD-OFF-REV switch to the OFF position. Plug the power cord into a 115AC, 60Hz power outlet.

Step 6 Ensure that all clamping screws are properly tightened. Ignite and properly adjust the torch, then preheat the start of the cut. Set the FWD-OFF-REV switch to the desired direction of travel. Simultaneously turn on the cutting oxygen and set the clutch lever to the DRIVE position.

Step 7 When the cut is completed, stop the machine and shut off the torch.

Review Questions

1. To prevent fume hazards, there must be at least _____ cubic feet of air for each welder.
   a. 2,000
   b. 4,000
   c. 5,000
   d. 10,000

2. A good safety practice is _____.
   a. welding near an open window
   b. welding over wooden floors
   c. to use cardboard to deflect welding/grinding sparks away from others
   d. writing HOT on hot metal before leaving it unattended

3. Cutting operations should never be performed without a _____ in the area.
   a. bucket of sand
   b. bucket of water
   c. fire watch
   d. fire hose

4. When pure oxygen is combined with fuel gases, the pure oxygen produces _____.
   a. argon
   b. hydrogen
   c. a colorless, odorless, and tasteless gas
   d. a high-temperature flame needed for flame cutting

5. The smallest standard oxygen cylinder holds about _____ cubic feet of oxygen.
   a. 40
   b. 65
   c. 80
   d. 85

6. Acetylene gas must be withdrawn from a cylinder at an hourly rate that does not exceed _____ of the cylinder capacity.
   a. 110
   b. 15
   c. 13
   d. 12

7. Safety fuse plugs in the top and bottom of an acetylene cylinder are designed to _____.
   a. release acetylene gas in the event of a fire
   b. prevent the release of acetylene gas
   c. prevent the withdrawal of acetone
   d. release acetone from the cylinder

8. Methylacetylene propadiene (MAPP®) gas, a liquefied fuel used in oxyfuel cutting, burns at temperatures almost as high as acetylene and has the stability of _____.
   a. natural gas
   b. propylene
   c. propane
   d. oxygen

9. The amount of liquefied gas remaining in a cylinder is determined by the _____ of the cylinder.
   a. color
   b. heat
   c. weight
   d. pressure

10. The regulators used on fuel gas cylinders are often painted red and always have _____ on all connections.
    a. right-hand threads
    b. left-hand threads
    c. metric threads
    d. safety latches

11. The attachment on the top of a fuel gas cylinder that allows the gas to flow only in one direction is called a _____.
    a. single-stage regulator
    b. two-stage regulator
    c. flashback arrestor
    d. check valve

12. The cutting tips used with liquefied fuel gases must have at least _____ preheat holes.
    a. four
    b. five
    c. six
    d. seven

13. When lifting oxyfuel cutting cylinders, always use a(n) _____.
    a. sling cable
    b. sling strap
    c. cylinder cage
    d. electromagnet

Review Questions

14. To avoid injury from dirt that may be lodged in the valve and regulator seat of a gas cylinder, always stand _____ the valve when opening the valve to clear the regulator seat.
    a. to the side of
    b. in front of
    c. behind
    d. above

15. When clearing debris from new oxyfuel cutting equipment hoses, blow the hoses out with _____.
    a. nitrogen
    b. propane
    c. propylene
    d. compressed air

16. The first step in installing a cutting tip is to _____.
    a. inspect the cutting tip sealing surfaces and orifices for damage
    b. determine the size of cutting tip to use
    c. determine the kind of gas being used
    d. identify the thickness of the material to be cut

17. Before opening cylinder valves, verify that the adjusting screws on the oxygen and fuel gas regulators are _____.
    a. tight
    b. fully clockwise
    c. loose
    d. fully counterclockwise

18. When a cutting flame has an excess of fuel, the flame is called a(n) _____ flame.
    a. cold
    b. neutral
    c. oxidizing
    d. carburizing

19. When disassembling oxyfuel equipment, verify that all pressure gauges read _____ before starting to take the equipment apart.
    a. –1.0
    b. 0
    c. within 0.3 of 0
    d. within 0.2 of 0

20. When inspecting a completed cut made with oxyfuel cutting equipment, the drag lines of the cut should be near _____ and not very pronounced.
    a. 3 degrees
    b. 45 degrees
    c. horizontal
    d. vertical

Summary

Oxyfuel cutting has many uses on job sites. It can be used to cut plate and shapes to size, prepare joints for welding, clean metals or welds, and disassemble structures. Because of the high pressures and flammable gases involved, there is a danger of fire and explosion when using oxyfuel equipment. However, these risks can be minimized when the oxyfuel cutting operator is well trained and knowledgeable. Be sure you know and understand the safety precautions and equipment presented in this module before using oxyfuel equipment.

Notes

Trade Terms Introduced in This Module

Backfire: A loud snap or pop as a torch flame is extinguished.

Carburizing flame: A flame burning with an excess amount of fuel; also called a reducing flame.

Drag lines: The lines on the kerf that result from the travel of the cutting oxygen stream into, through, and out of the metal.

Dross: The material (oxidized and molten metal) that is expelled from the kerf when cutting using a thermal process.

Ferrous metals: Metals containing iron.

Flashback: The flame burning back into the tip, torch, hose, or regulator, causing a high-pitched whistling or hissing sound.

Kerf: The edge of the cut.

Neutral flame: A flame burning with correct proportions of fuel gas and oxygen.

Oxidizing flame: A flame burning with an excess amount of oxygen.

Pierce: To penetrate through metal plate with an oxyfuel cutting torch.

Soapstone: Soft, white stone used to mark metal.

Resources & Acknowledgments

Additional Resources

This module is intended to be a thorough resource for task training. The following reference work is suggested for further study. This is optional material for continued education rather than for task training.

Safety in Welding, Cutting, and Allied Processes, ANSI Z49.1-99, 1999. Miami, FL: American Welding Society.

Figure Credits

Controls Corporation of America, 104F01

Topaz Publications, Inc., 104F03–104F05, 104F17, 104F20, 104F22–104F24, 104F26, 104F28, 104F29, 104F36–104F39, 104F48–104F63, 104F65

Nederman, Inc., 104F06

3M, 104F07, 104F08, 104F11

Scott Health & Safety, 104F09, 104F10, 104F12

Brad Krauel, 104F14

Sellstrom Manufacturing, 104F15

John Yochum, 104F19

Thermadyne, Inc., 104F31, 104F43, 104F47, 104F69, 104F73, 104F74

Lenco d/b/a NLC, Inc., 104F40

Koike Aronson, 104F41

ESAB Cutting and Welding Products, 104F42

Bug-O Systems, Inc., 104F44

H&M Pipe Beveling Machine Company, Inc., 104F45

Magnatech Limited Partnership, 104F46

Smith Equipment, 104F64

American Welding Society, 104F67

CONTREN® LEARNING SERIES — USER UPDATE

The NCCER makes every effort to keep these textbooks up-to-date and free of technical errors. We appreciate your help in this process. If you have an idea for improving this textbook, or if you find an error, a typographical mistake, or an inaccuracy in NCCER's Contren® textbooks, please write us, using this form or a photocopy. Be sure to include the exact module number, page number, a detailed description, and the correction, if applicable. Your input will be brought to the attention of the Technical Review Committee. Thank you for your assistance.

Instructors – If you found that additional materials were necessary in order to teach this module effectively, please let us know so that we may include them in the Equipment/Materials list in the Annotated Instructor's Guide.

Write: Product Development and Revision
National Center for Construction Education and Research
3600 NW 43rd St. Bldg G, Gainesville, FL 32606

Fax: 352-334-0932

E-mail: curriculum@nccer.org

Craft ______ Module Name ______

Copyright Date ______ Module Number ______ Page Number(s) ______

Description ______

(Optional) Correction ______

(Optional) Your Name and Address ______

Industrial Maintenance E & I Technician Level One

40105-07

Gaskets and Packing

40105-07
Gaskets and Packing

Topics to be presented in this module include:

Overview

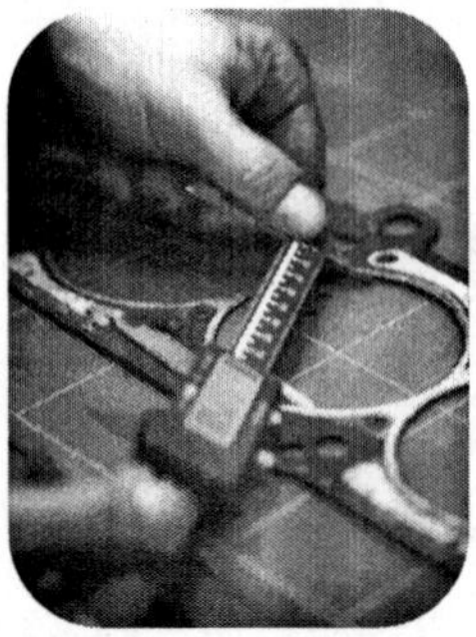

Whenever you work on machinery, you will have to learn to handle gaskets, packing, and O-rings. These are the ways of sealing fluids and gases in and out, and in some cases keeping lubricants and other fluids separate. Gaskets fill the spaces between faces of parts; packing is used to seal and hold lubrication around shafts; and O-rings seal stems in their housings. You will learn about the different materials and applications for each, as well as how to install them and how to make a gasket.

Objectives

When you have completed this module, you will be able to do the following:

1. Identify the various types of gaskets and explain their uses.
2. Identify the various types of gasket materials and explain their applications.
3. Lay out, cut, and install a flange gasket.
4. Describe the use of O-rings.
5. Explain the importance of selecting the correct O-ring for an application.
6. Select an O-ring for a given application and install it.
7. Describe the uses and methods of packing.

Trade Terms

Compressed
Concentric circles
Creep
Ethylene propylene dieneterpolymer (EPDM)
Extrusion
Impervious
Inert
Ozone
Plastic flow
Resilience
Resistance
Tetrafluoroethylene (TFE) (Teflon®)

Required Trainee Materials

1. Pencil and paper
2. Appropriate personal protective equipment

Prerequisites

Before you begin this module, it is recommended that you successfully complete *Core Curriculum;* and *Industrial Maintenance E & I Technician Level One,* Modules 40101-07 through 40104-07.

This course map shows all of the modules in the first level of the *Industrial Maintenance E & I Technician* curriculum. The suggested training order begins at the bottom and proceeds up. Skill levels increase as you advance on the course map. The local Training Program Sponsor may adjust the training order.

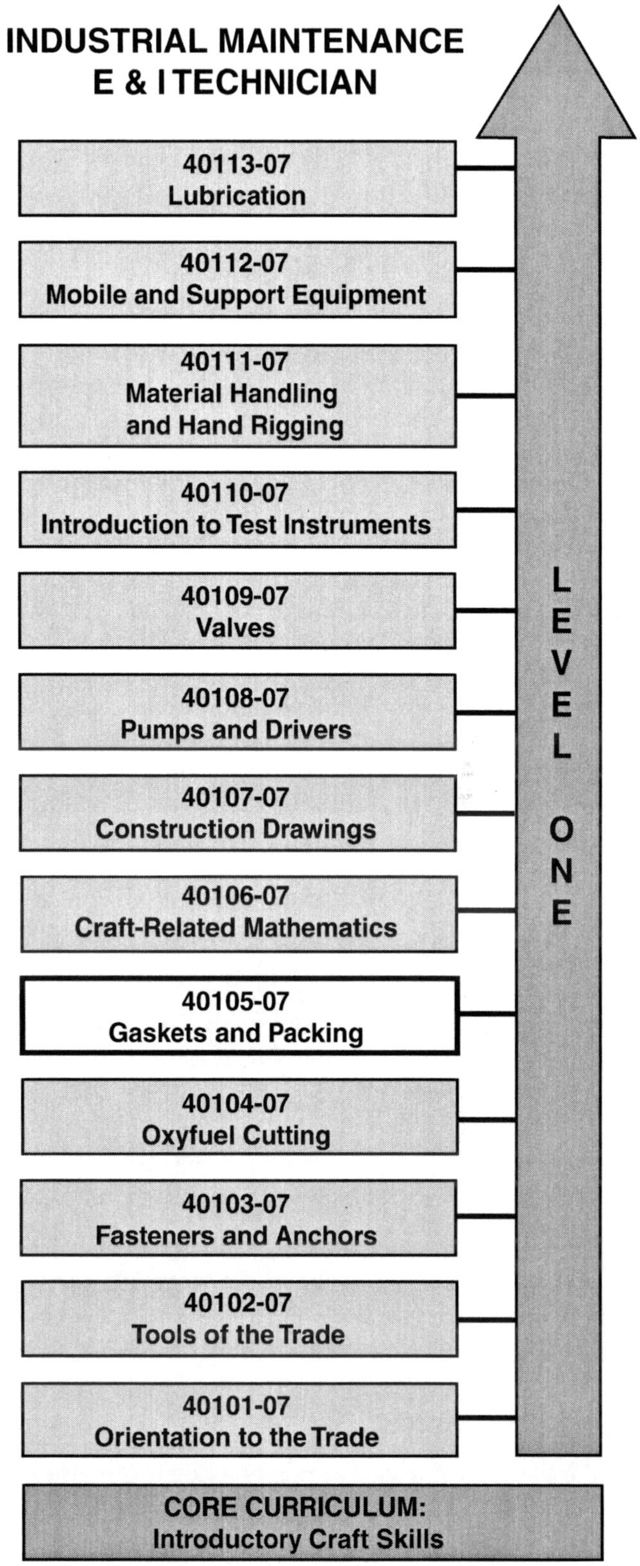

105CMAP.EPS

1.0.0 ◆ INTRODUCTION

Gaskets seal joints in pipe systems, gear boxes, pumps, motors, and other types of equipment. There are many types of gaskets and gasket materials for different applications. The type of gasket used in an application depends on the operating temperature of the system or equipment, the type of fluid the system or equipment handles, the type of connection being made, and the system or equipment pressure range. Many specialty gaskets, such as pipe flange gaskets and equipment assembly gaskets, are manufactured for specific uses. These gaskets are already cut to the required pattern, and the holes are already punched. When specialty gaskets are not available, you must identify and select the proper gasket material, cut the gasket, and punch the holes to match the application. This unit explains how to identify and select gasket types and gasket materials and how to lay out, cut, and install gaskets.

Pump shafts and valve stems are sealed with packing made from materials that will allow the stem to turn freely and not leak. As is the case with gaskets and O-rings, the synthetic materials have created many new options, and you will need to keep up with new developments in the field.

Different types of gaskets, packing, and O-rings are used extensively in maintenance work. Gaskets and O-rings are used to form a seal at the joint of two parts. Gaskets are made of material that can be compressed to fill irregularities in the mating surfaces of the parts. The gasket material conforms to the shapes of the surfaces and makes a seal between them.

There is one major difference between the way that gaskets and O-rings form seals. A gasket usually forms a seal between two flat surfaces. An O-ring fits into a groove on a shaft or surface and is made to protrude slightly from the groove. When another surface is brought in contact with the O-ring, the O-ring compresses to make a seal. Packing is **compressed** by being pushed into the area tightly.

1.1.0 Compatibility

The information given in this module is general in nature. Selecting gaskets, packing, and O-rings for a particular application involves more than just finding one that is the right size and shape. The properties of the material from which a particular gasket or O-ring is made must also be compatible with the process and environment to which the gasket or O-ring will be exposed. Use of the wrong gasket can cause leaks to occur. It can also result in a hazardous condition for people, equipment, or both, depending on the type of fluid or other material that is leaking from a damaged seal into the environment.

Gaskets or O-rings made of natural rubber have a limited **resistance** to petroleum products. Some others made of synthetic rubber compounds will degrade when exposed to specific kinds of chemicals. This degradation can cause the gaskets or O-rings to either shrink or swell, causing a leak to occur at the seal. O-ring, packing, and gasket materials can also be affected by the process temperature and ambient temperature to which they are exposed. Subjecting gaskets, packing, or O-rings to the wrong range of temperatures can cause them to shrink or swell. These temperatures can also cause the associated flange bolts to elongate or shrink. One or more of these conditions can result in a leaking seal.

In general, the higher the operating temperature of the process or environment, the more attention should be given to the selection of the gasket, packing, or O-ring material. Fluids are generally easier to seal than gases. However, ambient or process temperatures can cause some fluids to have an increased deteriorating effect on polymers, a common material used in many gaskets and O-rings. Some tests have shown that concentrations of oxygen at relatively high temperatures above 200°F (93°C) have a deteriorating effect on graphite as well as some elastomers and other materials used in the construction of gaskets or O-rings.

Manufacturers of seals, O-rings, and gaskets typically provide compatibility tables in their catalogs and product literature that list the composite materials from which their gaskets or O-rings are made and their compatibility or noncompatibility with certain application processes. Always check these compatibility charts whenever replacing or installing a gasket, packing, or O-ring in any process.

WARNING!

Gasket materials must never be arbitrarily substituted unless authorized by an engineer. Substituting gasket material may cause equipment failure and even personal injury. Always refer to engineering specifications before installing a gasket into a system.

2.0.0 ◆ TYPES OF GASKETS

Gaskets are available in a variety of types and materials. The type of gasket used must be matched to the operating conditions to which it will be exposed. For example, different gasket materials are capable of withstanding different temperature and pressure ranges. Certain gasket materials are compatible with different process fluids.

When replacing a gasket, it is considered good work practice to use a replacement gasket that is identical in properties to the gasket that has been removed. Replacement gaskets are generally supplied by the manufacturer of the equipment. However, in some cases, the replacement gasket will have to be fabricated.

In piping systems, gaskets are placed between two flanges to make the joint leakproof. There are different types of gaskets that must be matched to different types of flanges. The general rule is to use a full-face gasket with a flat-face flange and a ring-joint gasket with a raised-face flange. Gaskets may also be regularly or irregularly shaped. Irregularly shaped gaskets are often cut to fit around openings and fasteners. Gaskets are normally sized by the thickness of the gasket material in fractions of an inch, such as 1⁄16 inch or 1⁄8 inch. Some of the more common types of gaskets are shown in *Figure 1*.

The following types of gaskets are explained in this section:

- Ring
- Spiral wound
- Full face
- Jacketed
- Envelope
- Split ring
- Strip
- Ring-type joint
- Graphite-impregnated

2.1.0 Ring Gaskets

Ring gaskets, which are commonly used, are flat flange gaskets made of various materials for different applications. They are made to fit inside the bolt circle of a flange and they have no bolt holes. They usually have a pressure range of 150 to 200 psi.

2.2.0 Spiral Wound Gaskets

Spiral wound gaskets are flat flange gaskets, commonly made of stainless steel with a graphite insert. They are used on very high-temperature, high-pressure systems. A spiral wound gasket is a crushable gasket that has high **resilience** that allows it to adjust automatically to changes in line pressure, thermal shocks, vibration, and minor flange separation.

CAUTION

When tightening flange bolts it is important to tighten the bolts to an even torque. Overtightening can deform the flange, the gasket, or both, as well as possibly damaging the bolt itself. Do not exceed the torque recommendations given for the flange and gasket.

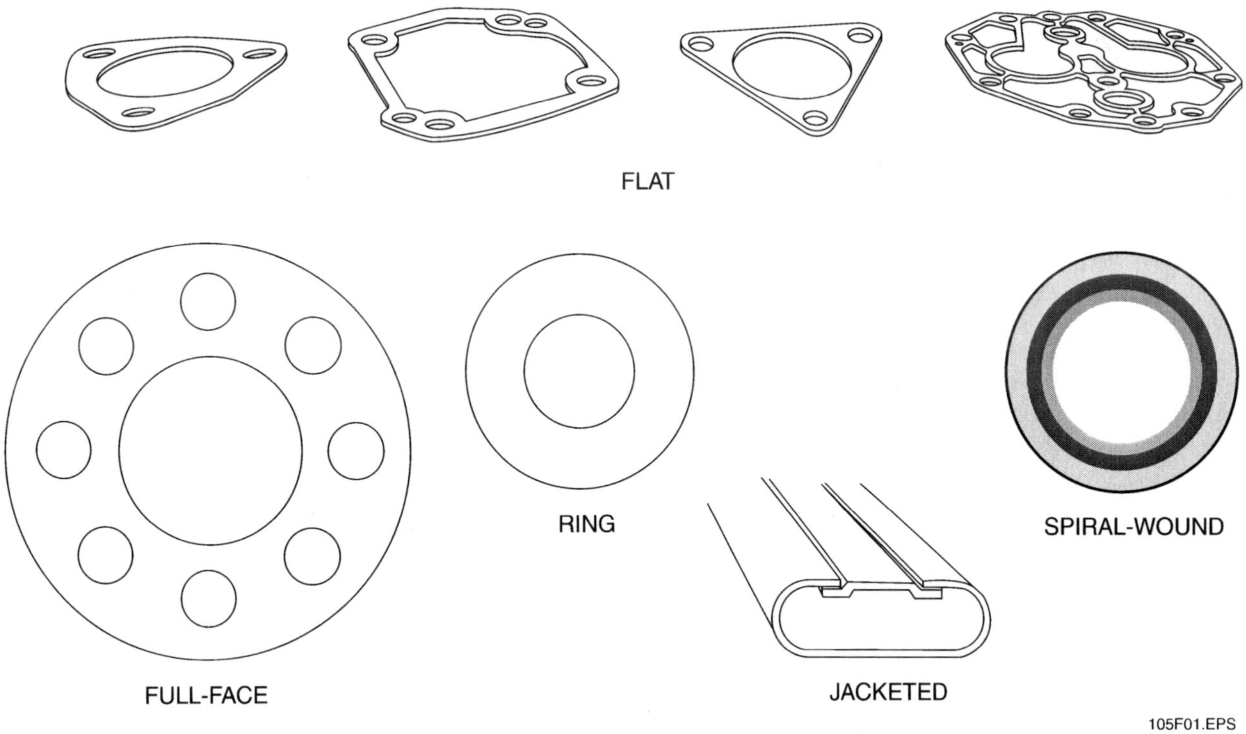

Figure 1 ◆ Common gasket types.

2.3.0 Full Face Gaskets

The full face gasket, another commonly used gasket, is a flat flange gasket. These gaskets are very similar to ring gaskets, but are made to fit the flange and have holes for the flange bolts. They are made from various materials for different applications.

2.4.0 Jacketed Gaskets

Jacketed gaskets have a metal exterior cover with an internal filler. The filler can be of various materials. The outer jacket prevents the fluid in a system from contacting the inner filler material and prevents the filler material from contaminating the system fluid. These gaskets are commonly made with a stainless steel jacket and an asbestos or other filler.

> **WARNING!**
> Asbestos is classified as a hazardous material. For this reason, removal of existing asbestos gaskets must be performed by authorized personnel only and under the supervision of the proper authority. All standards and safety regulations set forth by all jurisdictions having authority in the matter of the proper handling and disposal of asbestos products must be followed.

2.5.0 Envelope Gaskets

An envelope gasket is made using the same principle as a jacketed gasket. The outer cover is usually **tetrafluoroethylene (TFE) (Teflon®)**, with various types of filler. Envelope gaskets have a limited temperature range and are commonly used where flange imperfections exist because of their superior sealing ability.

2.6.0 Split Ring Gaskets

Split ring gaskets, which are not commonly used, are made in two pieces and are used where space is limited. They can be slipped into a flange without removing all the bolts.

2.7.0 Strip Gasket

Strip gasket material, usually made of Teflon®, comes in a roll. The gasket is pulled from the roll and placed on the flange in one or more strips. This gasket is only good for medium-pressure, low-temperature systems. Strip gaskets are considered a band-aid fix and are not commonly used for permanent service.

3.0.0 ◆ GASKET MATERIALS

Gaskets are made from a wide range of materials, including both natural and man-made substances. Some materials, called composite materials, contain both. The material chosen to make a gasket is very important. If the wrong material is chosen, the gasket will fail. There are many things to consider when choosing gasket material. The following are some of the criteria that must be considered when selecting gasket material:

- The material must be **impervious** to the fluid that the system handles.
- The gasket must have adequate chemical resistance at the fluid contact areas to prevent damage to the gasket.
- The material must not contaminate the fluid in the system.
- The material must have adequate elasticity to maintain an adequate seal in case there is joint movement.
- The material must develop sufficient friction in contact with the flange surface to resist excessive **creep** or **extrusion**.
- The material must not promote corrosion of the flange with which it comes in contact.
- The material must be able to withstand the required bolt torque without crushing or undergoing excessive **plastic flow**.
- In order to compensate for normal irregularities, the material must be sufficiently deformable to distribute the applied pressure evenly over the flange surfaces.
- The material must be rated to handle the maximum system pressure.
- The material must be rated to handle the maximum system temperature.

In addition to the above criteria, the pressure rating of a gasket must be matched to the system. The pressure rating of a gasket varies with the grade of material used and the type of gasket.

> **WARNING!**
> Always replace a gasket with one specified by the equipment manufacturer, or an approved equivalent. Substituting the wrong gasket or material may cause equipment failure and possible personal injury.

Many gasket suppliers provide a physical properties chart to be used to match gasket material to the application. You should use a chart of this type or contact the system engineer before installing a gasket in a system. You must be sure that the gasket is compatible with the system. The *Appendix* contains a sample of a physical properties chart.

Different gasket materials have very different properties. This section explains some of the properties of the following common gasket materials:

- Natural rubber
- **Ethylene propylene dieneterpolymer (EPDM)**
- Neoprene
- Nitrile
- Silicone
- Viton®
- Gylon
- Graphite-impregnated
- Ring-type joint
- Soft metal

3.1.0 Natural Rubber

Natural rubber is seldom used in installations today. It has low solvent and oil resistance, but is excellent when used with water. Natural rubber has excellent resilience and may be used at temperatures up to 175°F.

3.2.0 EPDM

EPDM is very good in an oxygen environment. It has good **ozone** resistance. EPDM has low resistance to solvent and oil. It is a good material to be used with water and chemicals and has a maximum temperature rating of 350°F. It also has excellent flame resistance.

3.3.0 Neoprene

Neoprene, which has good resilience, is only used in noncritical conditions. Neoprene has fair oil and solvent resistance, but has poor chemical resistance and is not recommended for water service. It has a maximum temperature rating of 250°F.

3.4.0 Nitrile

Nitrile is used with medium-pressure oil and solvent services. Nitrile cannot be used with acetone or methyl ethyl ketone (MEK). Nitrile has a maximum temperature rating of 250°F. Nitrile has poor ozone resistance, but is an excellent material to use with water or alcohol. Nitrile also has good chemical resistance but poor flame resistance.

3.5.0 Silicone

Silicone is a rubber-like material that is widely used. There are more grades of silicone material than any other rubber-type material. It has a maximum temperature rating of 550°F. Silicone is not recommended for use with oil or solvents. It has fair to good chemical resistance, excellent ozone resistance, and fair to good flame resistance. It has poor to excellent resilience depending on the grade used. It is not recommended for use on water systems.

3.6.0 Viton®

Viton® is a fluoroelastomer material. It has excellent chemical resistance and is the most commonly used chemical-resistant material. It has a maximum temperature rating of 400°F. Viton®, a hard material, has excellent oil and solvent resistance and also has excellent ozone resistance. Viton® is a relatively inexpensive gasket material.

3.7.0 Gylon® or Amerilon®

Under proprietary names like Gylon® and Amerilon®, there are also some new materials that combine Kevlar and PTFE to produce a very tough and resistant low temperature gasket. It is chemically **inert**, has a high resilience (45 percent recovery), and a maximum temperature rating of 500°F. Gylon® 3510 has one of the best sealability factors of any gasket material in the industry. It is one of the most commonly used gasket materials. It is used heavily in the chemical and paper industries.

3.8.0 Graphite-Impregnated Gaskets

These gaskets are made of fiberglass and impregnated with graphite particles. They are used to provide a tight seal in some high-temperature applications. However, they are reactive to oxidation.

3.9.0 Ring-Type Joint (RTJ) Gaskets

The RTJ gasket is used in high-pressure applications in excess of 2,000 psi. In special-purpose applications, it may be used up to 50,000 psi. This gasket seals by deforming a metal ring into concentric grooves machined into the flange faces. One concern with the RTJ gasket is that it is vulnerable to corrosion, especially when it is used in contact with corrosive fluids.

3.10.0 Soft Metal Gaskets

Soft metal and aluminum gaskets are designed to compress into the concentric rings machined into the surface of carbon steel and stainless steel flanges. Metal gaskets are used for oil and gas pipelines, where their non-reactive materials and high wear resistance keep them working over long periods without replacement.

3.11.0 Gasket Color Codes

Manufacturers use color codes to help identify gaskets made of different materials. These color codes will vary from one manufacturer to another. Some manufacturers use an additional color-coding scheme on the outside diameter of some gaskets so the material can be identified without breaking the seal. In such cases, the gasket body may be the same color as other types of gaskets, but the OD color will be different. The important thing is to check the color-coding scheme used by the given manufacturer before replacing a gasket.

> **NOTE**
> The use of a commercial liquid gasket remover can make removing old gasket material easier. It reduces the amount of scraping and sanding needed, thus preventing possible damage to flange surfaces.

4.0.0 ◆ FABRICATING GASKETS

The easiest way to make a gasket is to use the existing flange or the old gasket as a template to cut out the new gasket. If this is not possible, the following section describes the procedures for laying out and cutting a new gasket.

4.1.0 Laying Out a New Gasket

The following procedure describes how to lay out and cut a new gasket for a pipe flange.

> **WARNING!**
> Approved eye protection should be worn to protect the eyes from airborne gasket fibers or metal shards when fabricating gaskets. Hand protection should also be worn to protect the hands from injury when sharp gasket cutter blades are being used and/or sharp metal gasket edges are present.

Step 1 Select the proper gasket material for the conditions and the process.

Step 2 Take the following three measurements and draw them as **concentric circles**:

- Diameter of pipe opening
- Outside diameter of flange
- Diameter of the bolt hole circle

The diameter of the bolt circle is found by measuring from the edges of opposite holes, as shown in *Figure 2*. To get a center measurement, measure from the inside edge of one hole to the outside edge of the opposite hole. The easiest way to do this may be to put the tip of one arm of a pair of dividers on the inside of one bolt hole, and pull the other arm out to place the tip on the outside of the opposite bolt hole.

Step 3 Find the radius of the bolt circle as shown in *Figure 3*. The radius of the bolt circle is equal to half the diameter.

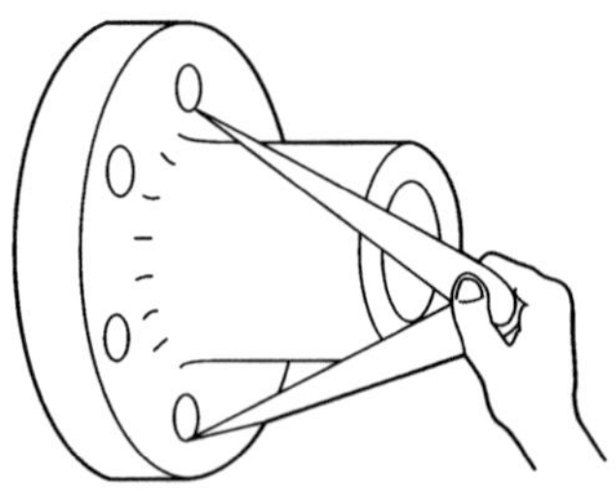

105F02.EPS

Figure 2 ◆ Measuring the diameter of a bolt circle.

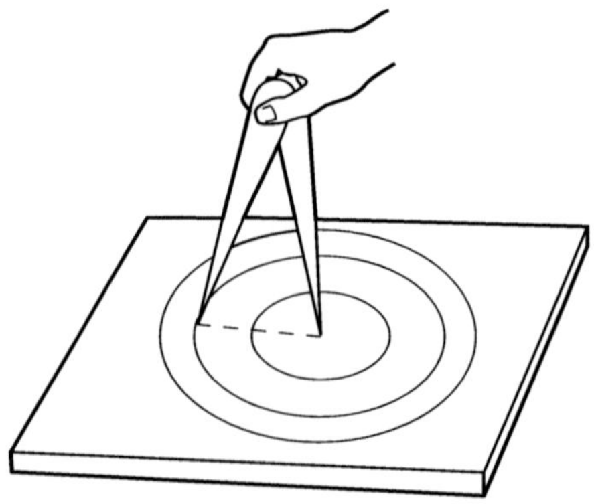

105F03.EPS

Figure 3 ◆ Radius of bolt circle.

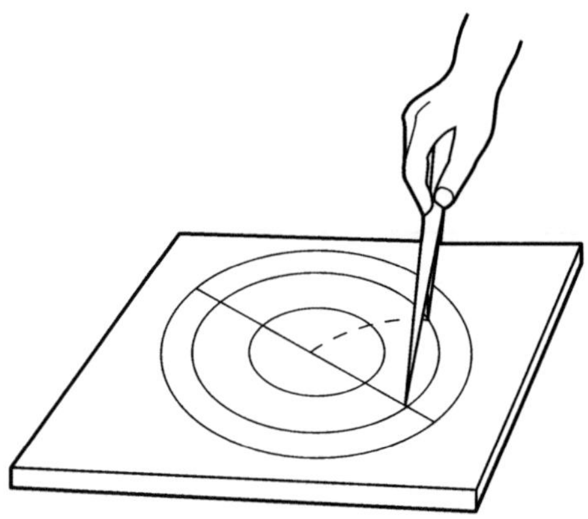

105F04.EPS

Figure 4 ◆ Bolt hole layout.

Step 4 Draw a line through the circle's center for opposite holes.

Step 5 On a six-hole flange, the radius is equal to the distance between bolt holes, as shown in *Figure 4*. Lay out holes a little larger than their actual size, as shown in *Figure 5*. Gaskets with an even number of holes (4, 8, 16) can also be laid out using the swing arc method (*Figure 6*). This method uses the divider to bisect angles. To draw an even number of holes, use a tape measure to mark the radius on the center line at each end, so as to define the centers of the two holes on the center line. Then, take a set of dividers or a compass and set the distance between the points of the divider at the diameter of the circle; that is, from one hole center to the opposite hole center. Set one point on one of the two hole centers, draw an arc above the center of the center line, and do the same thing below the center of the center line. Repeat these arcs, with the point of the

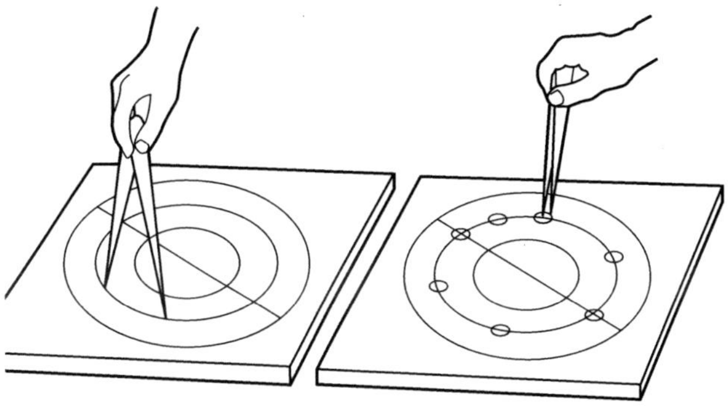

105F05.EPS

Figure 5 ◆ Layout bolt hole slightly larger than actual size.

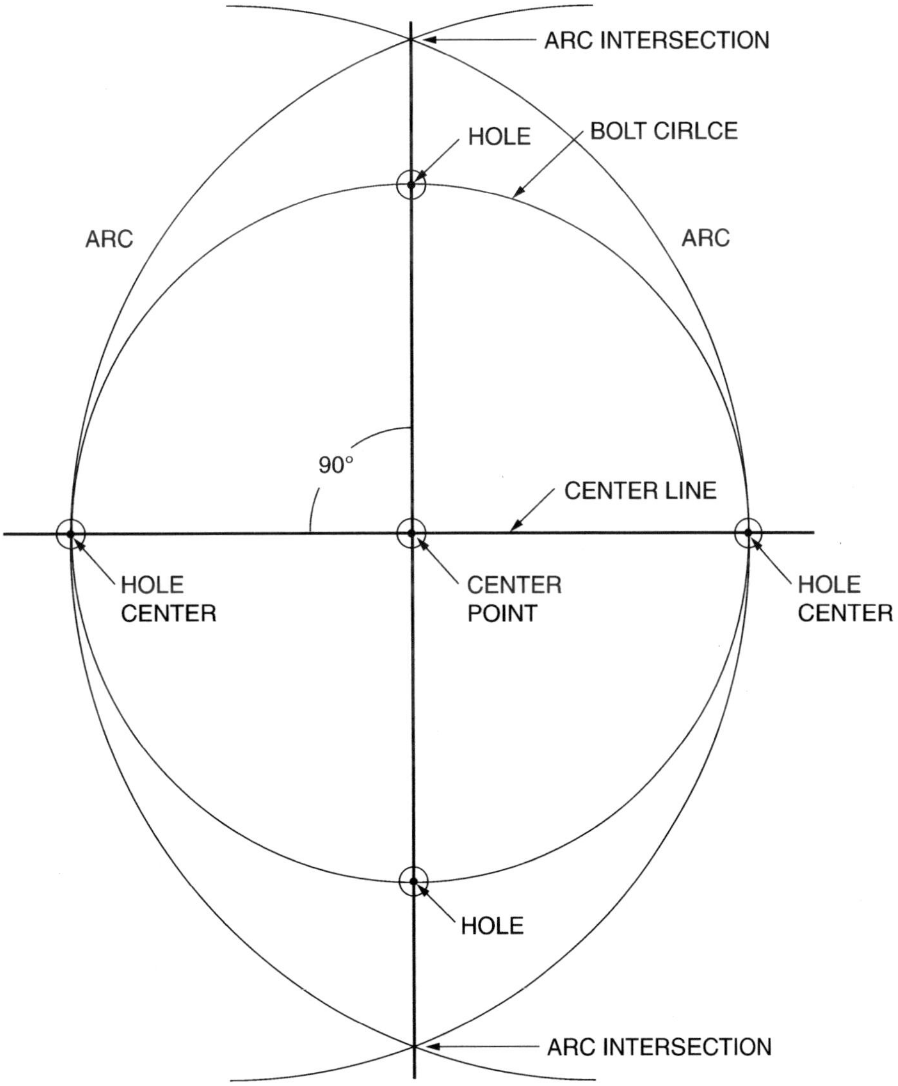

105F06.EPS

Figure 6 ◆ Swing-arc method of gasket layout.

dividers on the opposite hole center. Draw a line though the points where these arcs intersect above and below the center line. The line will be perpendicular to the center line and will pass through the middle of the circle. Any angle can be bisected (divided into two equal angles) by this process. Flanges with an odd number of holes can be laid out using a protractor by simply dividing 360 degrees by the number of holes.

Step 6 To check the distance, walk the dividers around the circle.

Step 7 Cut out the inside circle and outside circle of the gasket using a gasket cutter, as shown in *Figure 7*. The gasket cutter blade should not protrude more than 1/32 inch more than the thickness of the gasket material. Never hammer the gasket, as hammering may cause lumps in the gasket.

Step 8 Place the gasket material on hard wood to protect the punch edge. Then, punch out the holes using the proper size hole punch, as shown in *Figure 8*.

4.2.0 Tracing a New Gasket

The following procedure describes how to trace and cut a new metal gasket for a pipe flange.

Step 1 Spread bluing ink on the pipe flange face, as shown in *Figure 9*.

Step 2 Place the gasket material on the flange face and make an impression, as shown in *Figure 10*.

Step 3 Lift the gasket material off the flange face. You should have an impression of the gasket, as shown in *Figure 11*.

Step 4 Cut out the gasket using tin snips, as shown in *Figure 12*.

CAUTION

Never hammer the gasket. Hammering may cause lumps in the gasket that might result in failure of the gasket.

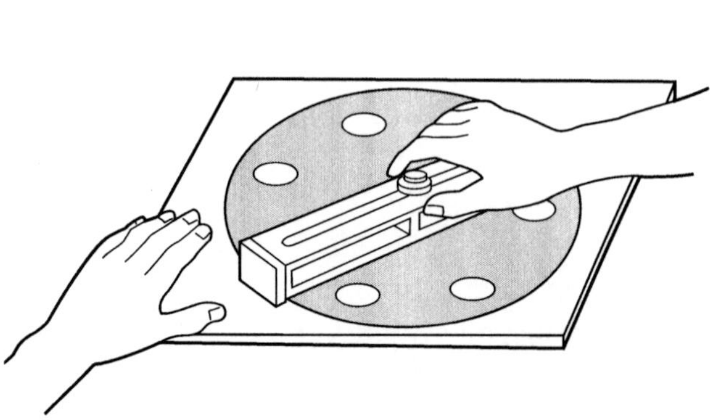
105F07.EPS

Figure 7 ◆ Using a gasket cutter.

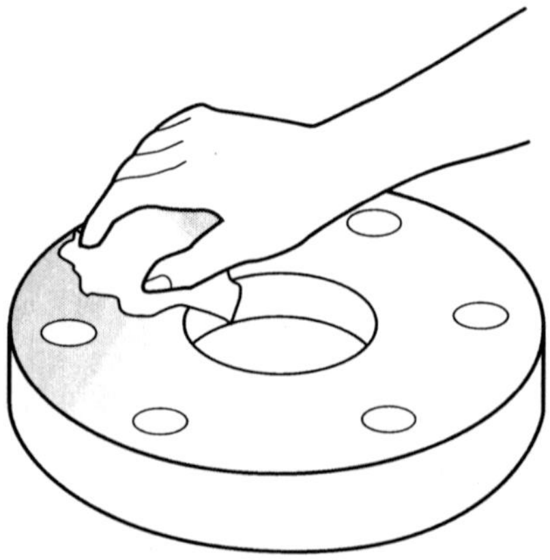
105F09.EPS

Figure 9 ◆ Bluing ink tracing.

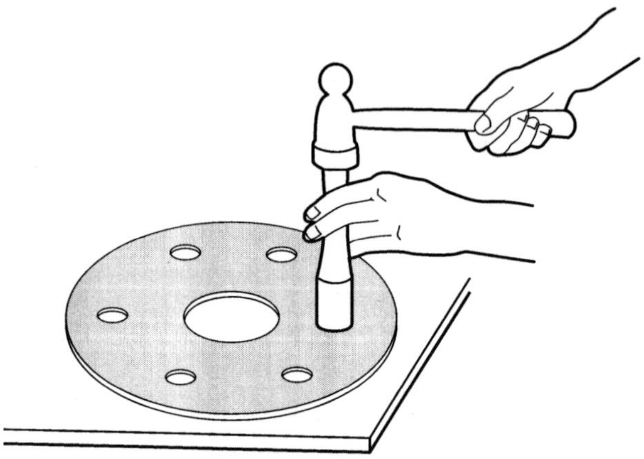
105F08 .EPS

Figure 8 ◆ Using a hole punch.

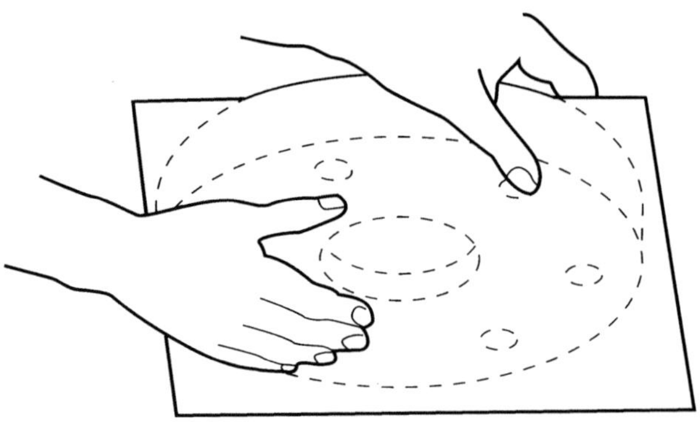
105F10.EPS

Figure 10 ◆ Making a bluing ink impression.

Step 5 Place the gasket material on hard wood to protect the punch edge, and then punch out the holes with the proper size hole punch.

Step 6 Clean any loose particles or wet ink off the gasket using the appropriate cleaner. It is not necessary to clean the dried bluing ink from the surfaces.

Step 7 Check the gasket against the flange for proper fit. If it does not fit correctly, discard the gasket and start over.

4.3.0 Machine Gaskets

Machine gaskets come in many configurations. It is not practical to try to lay out a machine gasket in the same manner as a flange gasket. You must transfer a pattern from the machine to the gasket. This is done using bluing ink. Follow these steps to lay out and cut a machine gasket:

Step 1 Thoroughly clean the machine flat (surface where the gasket fits) to remove the old gasket and any foreign matter. Use solvent if needed and prescribed.

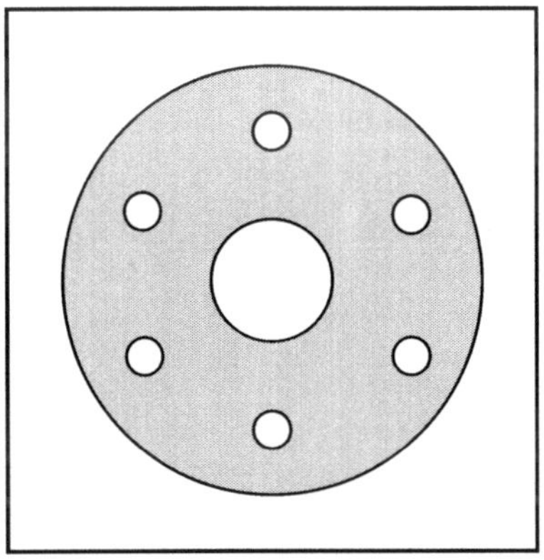

105F11.EPS

Figure 11 ◆ Bluing ink impression

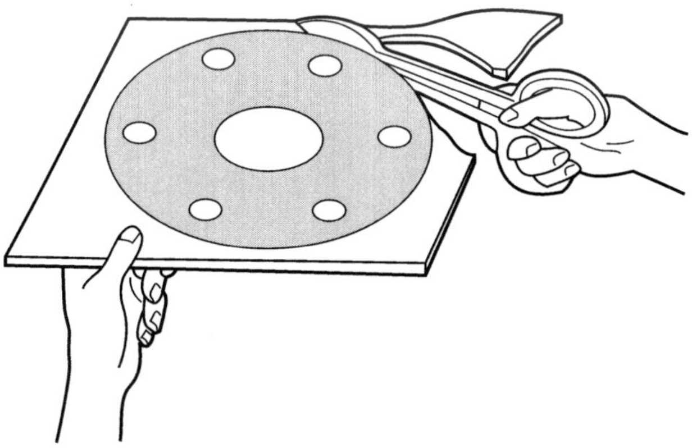

105F12.EPS

Figure 12 ◆ Cutting out the gasket with tin snips.

Step 2 Ensure that the machine surface is completely dry.

Step 3 Cut a piece of gasket material a little larger than the area to be covered by the gasket.

Step 4 Apply an even coat of bluing ink to the machine flat.

Step 5 Lay the gasket on the machine flat.

CAUTION

Use extra care when placing the gasket on the machine flat to prevent smearing the gasket with ink. The ink should make a distinct impression on the gasket.

Step 6 Rub the gasket with your finger along the machine flat to ensure that the ink contacts the gasket at all points.

Step 7 Carefully remove the gasket from the flat.

Step 8 Allow the ink to dry.

Step 9 Cut out the gasket using scissors or a utility knife.

Step 10 Punch the bolt holes using the proper size hole punch.

Step 11 Place the gasket on the machine flat and check the fit.

5.0.0 ◆ INSTALLING GASKETS

Once the gasket has been laid out and cut, it is ready to be installed on the flange or machine. The steps for installing a gasket may vary depending on the type of gasket material and the type of joint. Normally, the basic steps for installing gaskets are much the same. Follow these steps to install a gasket:

NOTE

When replacing gaskets, the old gasket must be removed and the flanges cleaned. The flanges and old gaskets should be inspected for irregularities. Any irregularities should be reported to your supervisor.

Step 1 Clean the surface where the gasket is to be installed, using solvent if needed.

Step 2 Apply any adhesive, sealer, or lubricant to the gasket as specified.

CAUTION

Normally, adhesives, sealers, or lubricants should be used only when specified. In some cases, adhesives can be used to hold the gasket in place during installation. If any of these materials is used, it must be compatible with the gasket material and the fluid in the system.

Step 3 Place the gasket in position on the flange or machine flat.

Step 4 Install the mating flange or machine part.

CAUTION

Make sure to leave approximately ¼" of the gasket protruding past any intersecting mating surfaces and trim the gasket after the mating flange has been bolted down.

Step 5 Install the hold-down bolts.

Step 6 Tighten the hold-down bolts, following the proper tightening sequence (*Figure 13*).

CAUTION

Tighten the fasteners in increments as specified by the manufacturer. Follow the proper tightening sequence for the bolt pattern and fastener type before reaching the final torque. Failure to follow this sequence, or overtightening the bolts, can damage the fasteners or distort the connection. This requirement applies to steel, and not to soft metal.

Step 7 Tighten the bolts to the specified torque, using a calibrated torque wrench and following the proper tightening sequence.

CAUTION

Do not overtighten the bolts because this could damage the gasket, the flange, the part, or the bolts. The required torque for an application depends on the type of gasket used, the size of the flange, and the grade bolts used. Contact the design engineer if you are not sure of the torque requirements.

Step 8 Retighten the bolts until there is no movement.

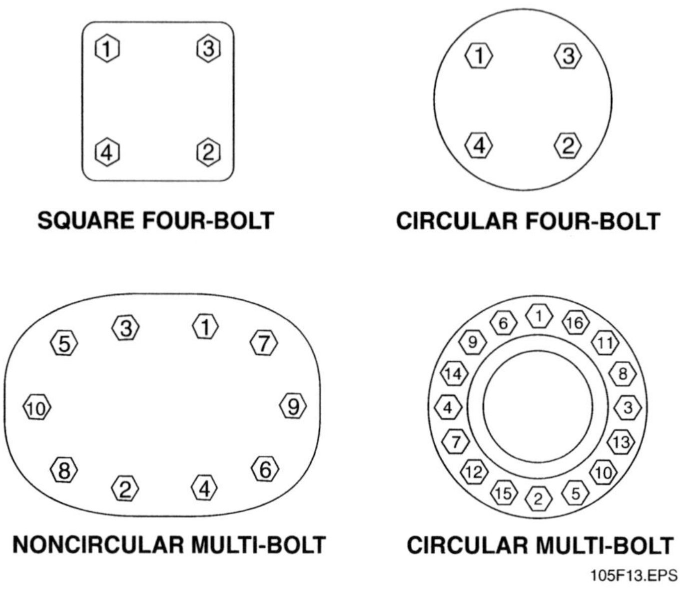

Figure 13 ◆ Proper tightening sequence.

6.0.0 ◆ PACKING

Packing is commonly used to form a seal around shafts, stems, or mandrels that either rotate or turn, yet require a sealing point. Uses in revolving equipment or devices include valve stems, faucets, pumps, compressors, and other equipment. Packing is sometimes used as gaskets for piping couplings, flange facings, and gear cases, with the packing installed in a packing groove. Various types of packing are shown in *Figure 14*.

Commonly used types of packing include Teflon® yarn and filament, lubricated graphite yarn, lubricated carbon yarn (graphite impregnated), and tetrafluorethylene (TFE)/synthetic fiber.

6.1.0 Teflon® Yarn Packing

Teflon® yarn packing is a cross-braided yarn impregnated with Teflon®. Teflon®-impregnated packing resists concentrated acids, such as sulfuric acid and nitric acid, sodium hydroxide, gases, alkalis, and most solvents. It is good for use in applications of up to approximately 550°F.

6.2.0 Teflon® Filament Packing

Teflon® filament (cord) packing is a braided packing made from TFE filament. It is impregnated with Teflon® and an inert softener lubricant. It is often used on rotating pumps, mixers, agitators, kettles, and other equipment.

6.3.0 Lubricated Graphite Yarn Packing

Lubricated graphite yarn packing is an intertwined braid of pure graphite yarn impregnated with inorganic graphite particles. The graphite particles dissipate heat. Lubricated graphite yarn packing also contains a special lubricant that provides a film to prevent wicking and reduce friction. It is good for use in high-temperature applications. It is reactive to oxygen atmospheres.

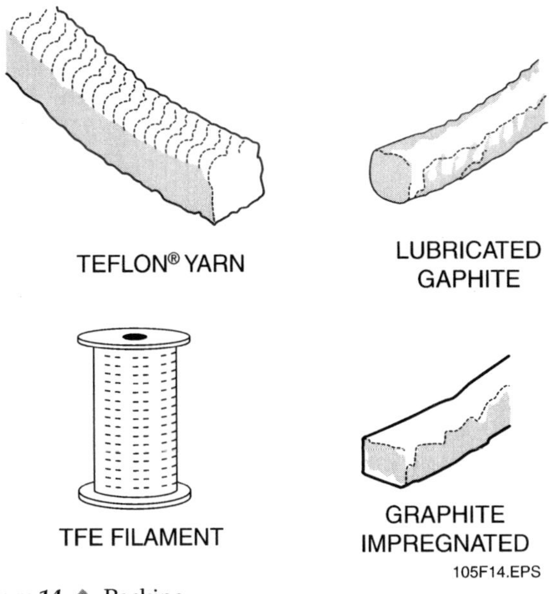

Figure 14 ◆ Packing.

6.4.0 Lubricated Carbon Yarn Packing

Lubricated carbon yarn (graphite impregnated) packing is made from an intertwined braid of carbon fibers impregnated with graphite particles and lubricants to fill voids and block leakage. It is used in systems containing water, steam, and solutions of acids and alkalis. When used for packing pumps, it is capable of handling shaft speeds up to approximately 3,000 rpm. It is considered suitable for steam application with temperatures up to approximately 1,200°F, and up to 600°F where oxygen is present. It is reactive to oxygen atmospheres.

6.5.0 TFE/Synthetic Fiber Packing

TFE/synthetic fiber packing is made from braided yarn fibers saturated and sealed with TFE particles before being woven into a multi-lock braided packing. TFE/synthetic fiber packing protects against a variety of chemical actions. It is used in applications where caustics, mild acids, gases, and many chemicals and solvents are present. It is often used in general service for rotating and reciprocating pumps, agitators, and valves.

6.6.0 Removing and Installing Packings

Packing rings used on revolving equipment shafts are usually installed in a packing gland similar to the one shown in *Figure 15*. In this example, the packing gland is shown with four packing rings installed. The number of rings used normally depends on the pressure of the process that needs to be contained by the packing rings. The following procedure describes how to remove and replace packing rings in a packing gland on a pump.

WARNING!

If repacking an operational process-connected piece of equipment, follow all safety procedures according to the standards, practices, and rules that apply to the specific location, equipment, and situation. Always wear approved eye protection when repacking a pump.

Step 1 De-energize/de-activate and lockout/tagout the equipment if operational or if connected to a process.

Step 2 Remove the packing follower plate, then use a packing extractor tool (*Figure 16*) to hook and pull out the old packing rings from inside the packing gland cavity.

Step 3 Inspect and clean the packing gland cavity and examine the pump shaft for any damage, burrs, or pitting. If any of these conditions exist, the equipment must be repaired; otherwise, the new packings can become damaged and will not properly seal.

Step 4 Size the new packing rings according to the old rings and cut new rings from a roll of matching packing material. Cut each piece at a 45-degree angle so that the ends overlap each other instead of butting up.

Step 5 Place the rings into the packing gland one at a time until the gland contains the correct number of rings. The last ring installed should extend slightly beyond the outer edge of the packing gland. When installing each ring, make sure to stagger its cut end 180 degrees from the cut end of the ring installed before it. Staggering the packing cut ends is important to prevent leaks. Tap each length of packing into place as you insert it. Unless instructions tell you otherwise, you will want to lubricate the packing with light mineral oil. In case of food-grade or potable water application, you may be advised to use light vegetable oil instead. The lubricant helps keep the packing from being affected by friction, and helps to produce a water- and air-tight seal.

Step 6 Secure the packing follower plate in place with the plate bolts and tighten just enough to apply a slight pressure on the packing rings.

Step 7 Momentarily operate the pump at its normal speed while checking for any leaks around the shaft. If no leaks are visible, slightly loosen the packing plate bolts until minimal leakage is visible around the pump shaft, then slightly tighten the bolts again until the leakage stops. This gradual but slight tightening procedure prevents undue pressure from being applied to the shaft by excessive compression of the packing rings, while still maintaining proper sealing.

7.0.0 ◆ O-RINGS

O-rings (*Figure 17*) are rugged and extremely dependable. They are used to seal against conditions ranging from strong vacuum to high pressure. O-rings are made from a variety of materials for different applications. Like gaskets, O-rings are often used to seal a mechanical connection between two parts of an instrument. O-rings that are used in instruments are often made of rubber

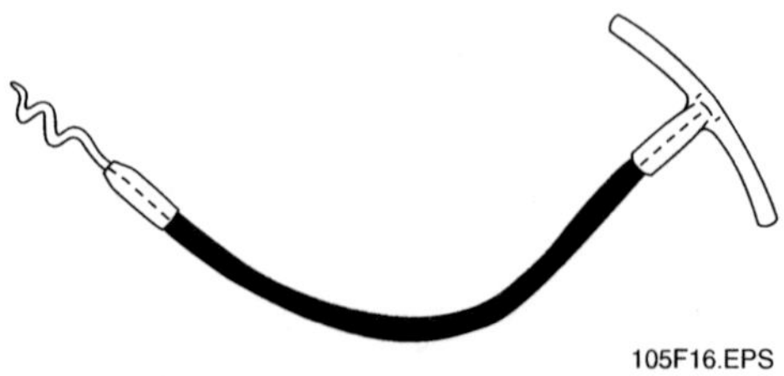

Figure 16 ◆ Packing extractor tool.

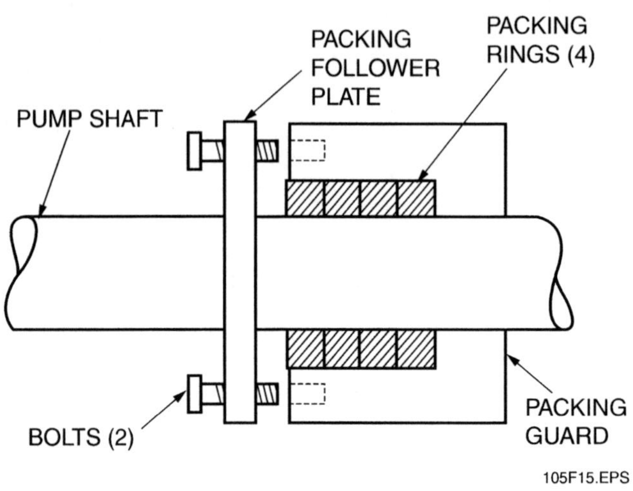

Figure 15 ◆ Packing gland.

or of a synthetic material. Occasionally, high-temperature or pressure applications may require the use of a metal O-ring.

Some of the more common O-rings are made from Buna-N (Nitrile), ethylene propylene, Viton®, Teflon®, silicone, Teflon®-encapsulated silicone, and polyurethane. O-rings are used in both static and dynamic seals. A static seal is not subjected to flow, but may have system pressure. A dynamic seal has both flow and pressure.

The sizes of O-rings are set by *Aerospace Standard AS568A* published by the Society of Automotive Engineers (SAE). Sizes are designated by dash numbers, such as AS568-006 and AS568-216.

7.1.0 Buna-N O-Rings

Buna-N O-rings are widely used. They are made of an elastomeric-sealing material. They are used with a variety of petroleum and silicone fluids, hydraulic and nonaromatic fuels, and solvents. Buna-N O-rings are not compatible with phosphate esters, ketones, brake fluids, strong acids, or ozone. They have an approximate temperature range of –65°F to 275°F. Buna-N O-rings do not weather well, especially in direct sunlight.

7.2.0 Ethylene Propylene O-Rings

Ethylene propylene O-rings resist automotive brake fluids, hot water, steam to approximately 400°F, silicone fluids, dilute acids, and phosphate esters. They are not compatible with petroleum fluids and diester lubricants. Ethylene O-rings have a good compression set plus high abrasion resistance, and they are weather resistant. They have an approximate temperature range of –70°F to 250°F.

7.3.0 Viton® O-Rings

Viton® O-rings offer excellent resistance to petroleum products, diester lubricants, silicone fluids, phosphate esters, solvents, and acids, except fuming nitric acid. They have a low compression set and low gas permeability. Viton® O-rings are often used for hard vacuum service. They should never be used with acetates, methyl alcohol, ketones, brake fluids, hot water, or steam. The approximate temperature range of Viton® O-rings is –31°F to 400°F.

105F17.EPS

Figure 17 ◆ O-rings.

7.4.0 Teflon® O-Rings

Teflon® O-rings lubricate well and have excellent chemical and temperature resistance. They make fine static seals, but need mechanical loading when used as dynamic seals. They have an approximate temperature range of –300°F to 500°F.

NOTE

All O-rings require lubrication. Keep in mind, however, that the lubricant and the O-ring material must be compatible. If they are not properly matched, the seal could be damaged. Because new lubricants are constantly being introduced, it is increasingly important to check the manufacturer's requirements to ensure a match between the O-ring and the lubricant.

7.5.0 Silicone O-Rings

Silicone O-rings are used where long-term exposure to dry heat is expected. Due to poor abrasion resistance, silicone O-rings perform best in static sealing applications. Silicone O-rings resist brake fluids and high aniline point oil. They are not recommended for use with ketones and most petroleum oils. They have an approximate temperature range of –80°F to 400°F.

7.6.0 Teflon®-Encapsulated Silicone O-Rings

Teflon®-encapsulated silicone O-rings are used in most of the same applications as silicone O-rings. The Teflon® coating makes them resistant to most solvents and chemicals. These O-rings have an extremely low coefficient of friction and low compression set. They are primarily used as seals in static applications. They have an approximate temperature range of –75°F to 400°F.

7.7.0 Polyurethane O-Rings

Polyurethane O-rings have high tensile strength, are abrasion resistant, and have excellent tear strength. Polyurethane is the toughest of the elastomers. Polyurethane O-rings can be used with petroleum fluids, ozone, and solvents, except ketones. Polyurethane O-rings are non-compatible with hot water, brake fluids, acids, and high temperature. They have poor compression set. Polyurethane O-rings have an approximate temperature range of –40°F to 200°F.

7.8.0 Removing and Installing O-Rings

The removal and installation of an O-ring seal is a relatively simple procedure. To achieve a proper seal and protect the new O-ring from damage, follow these guidelines:

- The equipment must be properly prepared to receive the new O-ring. The groove(s) that the O-ring fits into must be clean and free of any sharp edges.
- The type and size of the O-ring must be right for the application.
- As applicable, lubricate the shaft, O-ring, and O-ring groove(s) using a lubricant specified by the O-ring manufacturer.
- During assembly, protect the O-ring from being damaged by the sharp edges of threads, keyways, or the end of a shaft.
- Follow the proper bolt-tightening procedure.

CAUTION

Do not overtighten O-ring face seals. Overtightening can distort the O-ring, resulting in a leak.

Review Questions

1. Once the size of a gasket or O-ring has been determined, _____ factors must be considered before choosing a gasket or O-ring.
   a. cost
   b. availability
   c. compatibility
   d. maintenance

2. The gasket that is made of stainless steel with a graphite insert is the _____ gasket.
   a. strip
   b. ring
   c. spiral wound
   d. envelope

3. Natural rubber has limited chemical resistance to _____.
   a. petroleum products
   b. water
   c. organic substances
   d. steam

4. The gasket material that is *not* recommended for water service and has poor chemical resistance is _____.
   a. Teflon®
   b. neoprene
   c. natural rubber
   d. EPDM

5. Which of the following substances would be compatible with a nitrile gasket?
   a. Methyl ethyl ketone
   b. Ozone
   c. Alcohol
   d. Acetone

6. RTJ gaskets are used in _____ applications.
   a. low-pressure
   b. high-pressure
   c. medium-pressure
   d. corrosive

7. The radius of a bolt circle is equal to _____.
   a. one-fourth the diameter
   b. half the diameter
   c. the diameter
   d. twice the diameter

8. In gasket making, bluing ink is used to _____.
   a. color-code the gasket
   b. transfer the flange impression to the gasket material
   c. document the gasket measurement for future use
   d. mark the centers of the bolt holes

9. When tightening the bolts on a circular multi-bolt flange, you should start at 12 o'clock and proceed clockwise around the flange.
   a. True
   b. False

10. Graphite packings should *not* be used for applications where _____ is present.
    a. a caustic substance
    b. acidic material
    c. oxygen
    d. heat

11. TFE/synthetic fiber packing is used in _____.
    a. caustic chemical applications
    b. hot and cold water and antifreeze applications
    c. very low-temperature applications
    d. high-pressure piping

Review Questions

12. Packing is usually installed in a _____ on revolving equipment.
    a. motor housing
    b. seal unit
    c. packing gland
    d. groove

13. When installing new packing, stagger the new cut end _____ from the last ring emplaced.
    a. 30 degrees
    b. 90 degrees
    c. 180 degrees
    d. 220 degrees

14. The type of O-ring that can be used with most petroleum and silicone fluids, hydraulic fluids, and solvents is _____.
    a. Buna-N
    b. silicone
    c. ethylene propylene
    d. neoprene

15. The appropriate temperature range for _____ O-rings is –31°F to 400°F.
    a. Teflon®
    b. ethylene propylene
    c. Viton®
    d. silicone

Summary

In this training module, you learned about various types of gaskets, packing, and O-rings, and how to lay out and fabricate a gasket. Gaskets, packing, and O-rings are used with a variety of equipment and systems. Different types of gaskets, packing, and O-rings are designed to accommodate various pressure and temperature conditions and process characteristics.

The purpose of a gasket is to create a positive seal between two relatively stationary parts. The thinnest gasket that accomplishes this purpose is the best to use. The gasket used must be matched to the application. Gaskets are chosen for their pressure rating, their temperature rating, and their resistance to the process on which they are used. Preformed, factory-made gaskets need not be cut before installation. When preformed gaskets are not available, gaskets must be laid out and cut as needed. Gaskets must be cut to close tolerances, and torquing specifications must be followed when tightening the gasketed joint.

The knowledge and skills covered in this module provide basic and fundamental competencies required to identify gaskets, packing, and O-rings, and to effectively work with different gasket, packing, and O-ring materials. In your day-to-day work activities, take the time to practice the skills learned in this module. It takes a lot of practice, study, and on-the-job experience to develop the skills needed to be expert in working with gaskets, packing, and O-rings.

Notes

Trade Terms Introduced in This Module

Compressed: Pressed or squeezed together.

Concentric circles: Circles having a common center point.

Creep: The loss of thickness of a gasket, which results in bolt torque loss and leakage.

Ethylene propylene dieneterpolymer (EPDM): A gasket material and general-purpose polymer that is heat, ozone, and weather resistant. EPDM is not oil resistant.

Extrusion: A gasket protruding out of a flange.

Impervious: Cannot be penetrated.

Inert: Unreactive. Exhibiting no chemical activity.

Ozone: A form of oxygen, usually created when electricity passes through the oxygen.

Plastic flow: The flowing of gasket material under stress.

Resilience: The ability of a gasket to return to its original shape after being compressed.

Resistance: The ability of a gasket to withstand the effects of chemicals and other substances without damage or change.

Tetrafluoroethylene (TFE) (Teflon®): Used for many high-pressure, high-temperature applications.

Appendix

Physical Properties Chart

COMMON NAME			NATURAL RUBBER	SBR	BUTYL	EPDM	NEOPRENE	NITRILE
WEIGHT OF BASE ELASTOMER		LB/CU IN.	0.033	0.034	0.033	0.031	0.044	0.036
		SPEC. GR.	0.93	0.94	0.92	0.85	1.23	1.00
PHYSICAL PROPERTIES FOR ELASTOMER COMPOUNDS	DUROMETER, RANGE		30-100	40-100	30-100	30-90	45-95	20-90
	RESILIENCE		EXCELLENT	GOOD	FAIR	GOOD	EXCELLENT	GOOD
	TENSILE STRENGTH, PSI (REINFORCED)		4000+	2000+	2000+	2000-3000	3000+	1000-3500+
	ELONGATION, % (REINFORCED)		500	450	300-800	500	650-850	400-600
	DRIFT, ROOM TEMP		EXCELLENT	EXCELLENT	FAIR	FAIR	FAIR TO GOOD	GOOD
	COMPRESSION SET		GOOD	GOOD	FAIR	FAIR	FAIR TO GOOD	GOOD
	ELECTRICAL RESISTIVITY		EXCELLENT	EXCELLENT	EXCELLENT	EXCELLENT	FAIR	POOR
	IMPERMEABILITY, GAS		GOOD	FAIR	EXCELLENT	GOOD	GOOD	GOOD
RESISTANCE PROPERTIES — MECHANICAL	RESISTANCE TO	IMPACT	EXCELLENT	EXCELLENT	GOOD	GOOD	GOOD	FAIR
		ABRASION	EXCELLENT	EXCELLENT	GOOD	GOOD	GOOD TO EXCELLENT	EXCELLENT
		TEAR	EXCELLENT	FAIR	GOOD	POOR	GOOD	GOOD
		CUT GROWTH	EXCELLENT	GOOD	EXCELLENT	GOOD	GOOD	GOOD
TEMPERATURE	TENSILE STRENGTH; PSI,AT	250F	1800	1200	1000	2000	1500	700
		400F	125	170	350	400	180	130
	ELONGATION %, AT	250F	500	250	250	300-500	350	120
		400F	80	60	80	0-120	0-100	20
	DRIFT AT 212F		GOOD	GOOD	FAIR	FAIR	FAIR TO GOOD	EXCELLENT
	HEAT AGING AT 212F		GOOD	GOOD	EXCELLENT	EXCELLENT	GOOD	GOOD
	FLAME RESISTANCE		POOR	POOR	POOR	POOR	GOOD	POOR TO FAIR
	TEMPERATURE: MAXIMUM (¡F)		200	275	325	350	250	250
	LOW TEMPERATURE	STIFFENING, F	-20 TO -50	0 TO 50	10 TO 40	-20 TO -50	+10 TO -20	+30 TO -20
		BRITTLE POINT, F	-80	-80	-80	-90	-45	-65
ENVIRONMENTAL	WEATHER		FAIR	FAIR	EXCELLENT	EXCELLENT	EXCELLENT	GOOD
	OXIDATION		GOOD	GOOD	EXCELLENT	GOOD	GOOD	FAIR TO GOOD
	OZONE		POOR	POOR	EXCELLENT	EXCELLENT	EXCELLENT	POOR
	RADIATION		FAIR TO GOOD	GOOD	POOR	POOR	FAIR TO GOOD	FAIR TO GOOD
	WATER		EXCELLENT	EXCELLENT	EXCELLENT	GOOD TO EXCELLENT	GOOD	EXCELLENT
	ACID		FAIR TO GOOD	FAIR TO GOOD	EXCELLENT	GOOD TO EXCELLENT	GOOD	GOOD
	ALKALI		FAIR TO GOOD	FAIR TO GOOD	EXCELLENT	GOOD TO EXCELLENT	GOOD	FAIR TO GOOD
	GASOLINE, KEROSENE, ETC. (ALIPHATIC HYDOCARBONS)		POOR	POOR	POOR	POOR	GOOD	EXCELLENT
	BENZOL, TOLUOL, ETC. (AROMATIC HYDROCARBONS)		POOR	POOR	FAIR TO GOOD	FAIR	POOR	GOOD
	DEGREASER SOLVENTS (HALOGENATED HYDROCARBONS)		POOR	POOR	POOR	POOR	POOR	POOR
	ALCOHOL		GOOD	FAIR	EXCELLENT	POOR	FAIR	EXCELLENT
	SYNTHETIC LUBRICANTS (DIESTER)		POOR TO FAIR	POOR	FAIR	POOR TO FAIR	POOR	FAIR TO GOOD
	HYDRAULIC FLUIDS	SILICATES	POOR	POOR TO FAIR	FAIR	FAIR TO GOOD	GOOD	FAIR
		PHOSPATES	POOR TO FAIR	POOR	GOOD	GOOD TO EXCELLENT	POOR	POOR
SUBJECTIVE PROPERTIES	TASTE		FAIR TO GOOD	FAIR TO GOOD	FAIR TO GOOD	GOOD	FAIR TO GOOD	FAIR TO GOOD
	ODOR		FAIR TO GOOD	GOOD	GOOD	GOOD	FAIR TO GOOD	GOOD
	NONSTAINING		POOR TO GOOD	POOR TO GOOD	GOOD	GOOD	GOOD TO EXCELLENT	POOR TO GOOD
BONDED TO RIGID MATERIALS			EXCELLENT	EXCELLENT	FAIR TO EXCELLENT	POOR	GOOD TO EXCELLENT	GOOD TO EXCELLENT

105A01.EPS

Resources & Acknowledgments

Additional Resources

This module is intended to be a thorough resource for task training. The following reference works are suggested for further study. These are optional materials for continued education rather than for task training.

Specifications for Gaskets, O-Rings, and Packing. Washington, DC: American National Standards Institute (ANSI).

Specifications for Gaskets, O-Rings, and Packing. West Conshohoken, PA: American Society for Testing and Materials.

Specifications for Gaskets, O-Rings, and Packing. Warrendale, PA: Society of Automotive Engineers.

CONTREN® LEARNING SERIES — USER UPDATE

NCCER makes every effort to keep these textbooks up-to-date and free of technical errors. We appreciate your help in this process. If you have an idea for improving this textbook, or if you find an error, a typographical mistake, or an inaccuracy in NCCER's Contren® textbooks, please write us, using this form or a photocopy. Be sure to include the exact module number, page number, a detailed description, and the correction, if applicable. Your input will be brought to the attention of the Technical Review Committee. Thank you for your assistance.

Instructors – If you found that additional materials were necessary in order to teach this module effectively, please let us know so that we may include them in the Equipment/Materials list in the Annotated Instructor's Guide.

Write: Product Development and Revision
National Center for Construction Education and Research
3600 NW 43rd St. Bldg G, Gainesville, FL 32606

Fax: 352-334-0932

E-mail: curriculum@nccer.org

Craft ______ Module Name ______

Copyright Date ______ Module Number ______ Page Number(s) ______

Description ______

(Optional) Correction ______

(Optional) Your Name and Address ______

40106-07

Craft-Related Mathematics

40106-07
Craft-Related Mathematics

Topics to be presented in this module include:

Overview

Industrial maintenance craftspersons need specialized mathematical skills to do their work. In this module, you will learn how to use formulas, determine ratios and proportions, calculate areas and volumes, and use scales and tables of comparative values.

Objectives

When you have completed this module, you will be able to do the following:

1. Identify and explain the use of special measuring devices.
2. Use tables of weights and measurements.
3. Use formulas to solve basic problems.
4. Solve area problems.
5. Solve volume problems.
6. Solve circumference problems.
7. Solve right triangles using the Pythagorean theorem.

Trade Terms

Adjacent side
Apex
Arithmetic numbers
Circle
Circumference
Cubic
Cylinder
Exponent
Factors
Formula
Hypotenuse
Literal numbers
Opposite side
Perpendicular
Pi
Pyramid
Radius
Rectangular
Run
Set
Solid
Sphere
Travel
Volume

Required Trainee Materials

1. Pencil and paper
2. Scientific calculator

Prerequisites

Before you begin this module, it is recommended that you successfully complete *Core Curriculum; Industrial Maintenance E & I Technician Level One*, Modules 40101-07 through 40105-07.

This course map shows all of the modules in the first level of the *Industrial Maintenance E & I Technician* curriculum. The suggested training order begins at the bottom and proceeds up. Skill levels increase as you advance on the course map. The local Training Program Sponsor may adjust the training order.

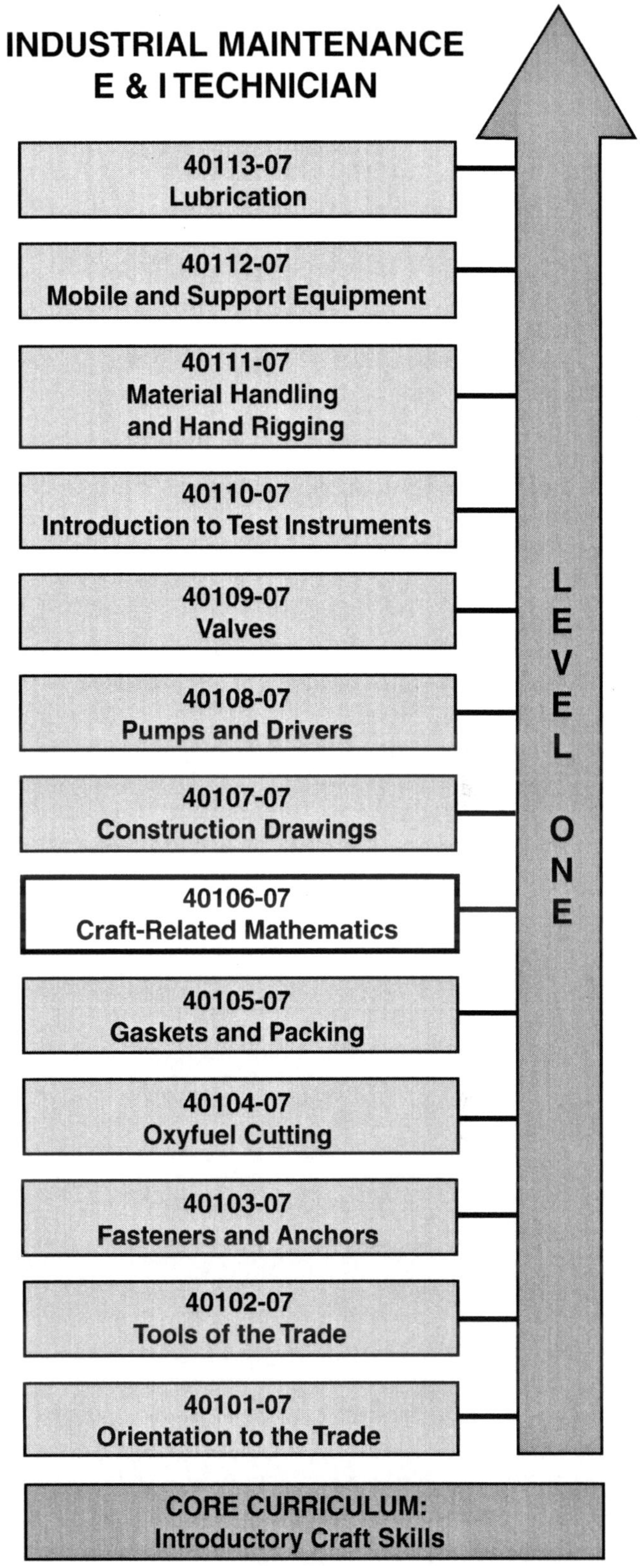

106CMAP.EPS

1.0.0 ◆ INTRODUCTION

In addition to the basic math skills you may have learned in school, a craftsperson needs advanced mathematical skills. You will use these skills to determine the **volume** of pipes and tanks, and to determine simple and rolling offsets. This module explains how to use an architect's scale and an engineer's scale; how to use tables of weights and measurements; use ratios; solve basic problems using **formulas**; use basic geometry to solve area, volume, perimeter, and **circumference** problems; and use the Pythagorean theorem. All of these skills are applied to your everyday duties as a maintenance craftsperson, and the use of mathematics can become a valuable tool that will make your job easier.

2.0.0 ◆ SPECIAL MEASURING DEVICES

Special measuring devices that an industrial maintenance craftsperson must be familiar with include the architect's scale and the engineer's scale *(Figure 1)*. These tools are useful when reading blueprints and orthographic drawings and when converting scale drawings to actual size.

2.1.0 Architect's Scale

The architect's scale is an all-purpose scale that has many uses. It has a full-size scale of inches, divided into sixteenths, and a number of reduced-size scales in which inches or fractions of an inch represent feet. An architect's scale has 11 separate scales divided into different increments (*Table 1*).

Each scale has a number located at the end of the scale. The numbers at each end of the scale designate the size of the increments on the scale. If the numbers on the left end designate the scale you are using, the scale is read from left to right. If the numbers on the right end of the scale designate the scale you are using, the scale is read from right to left. For example, on the quarter-inch scale, each quarter of an inch designates one foot. *Figure 2* shows an enlarged view of the quarter-inch scale.

Since the ¼-inch mark is at the right end of the scale, the ¼-inch scale is read from right to left. The scale has long vertical marks that represent feet. The even-numbered feet are labeled, and the odd-numbered feet are not. The fully-divided scale to the right of the zero represents inches.

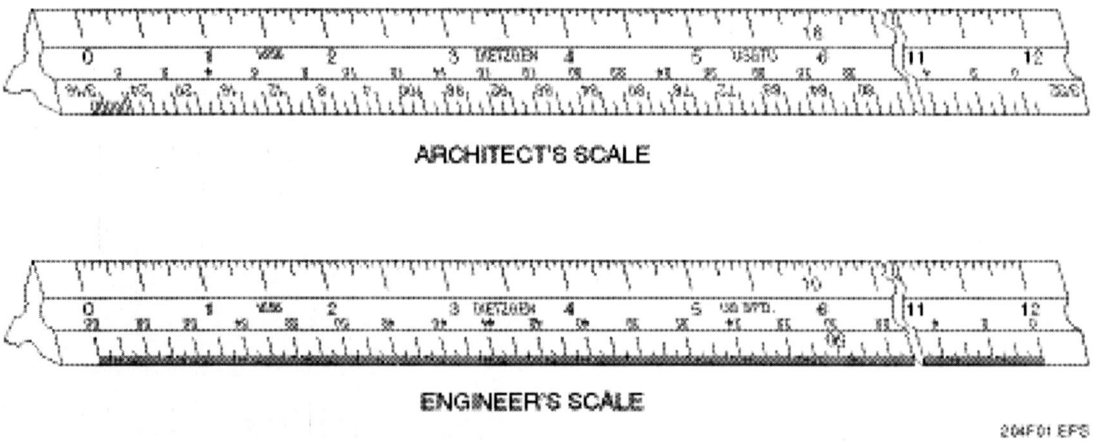

Figure 1 ◆ Architect's and engineer's scales.

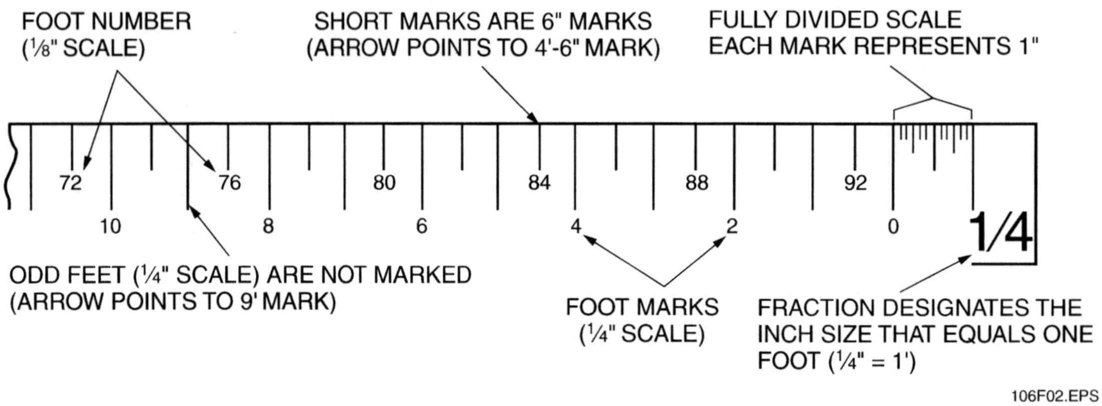

Figure 2 ◆ Enlarged view of quarter-inch scale.

2.2.0 Engineer's Scale

The engineer's scale has several scales, each of which is divided into 10, 20, 30, 40, 50, or 60 parts. These scales can be used to check drawings. If a blueprint is drawn to the 50 scale, meaning that 1 inch equals 50 feet or 50 meters, you would use the scale marked 50. These scales make it easier to determine quick measurements without having to use math to calculate distances. *Figure 3* shows the scales on the engineer's scale.

It is okay to check a given measurement with a scale rule, but you must not rely on scaled dimensions for actually building a structure. Scaled dimensions are only approximations and are not detailed enough for construction. The printing of blueprints can slightly shrink the drawing, causing inaccuracy, or the drawing may not be drawn to the indicated scale. This is highly likely when changes have been made to the drawing. Because of this, always use a written dimension rather than a scaled dimension.

3.0.0 ◆ USING TABLES

Tables consist of two or more parallel columns of data. They can be read quickly and can present large amounts of data clearly and concisely. Handbooks of tables are frequently useful as information references, and for solving mathematical problems. While tables vary in form, they are read following the same basic steps. The following sections explain comparative value tables and mathematical tables.

3.1.0 Comparative Value Tables

The simplest types of tables provide comparative values of related quantities. These values come from the definitions of quantities in the tables. Comparative value tables include the following:

- Tables of measure
- Multiplication tables
- Tables of weight
- Tables of money

One type of comparative value table is a table of linear measures. *Table 2* lists sample linear measures.

When mathematical problems are being solved, it is sometimes necessary to know the decimal equivalent of a fraction. In the following sample table, the first column lists fractions of an inch; the second column lists the decimal equivalents in inches, and the third column lists decimal equivalents in millimeters. If any of these values are known, the others can be found quickly and easily. *Table 3* lists decimal equivalents of some common fractions.

Table 1 Scales on Typical Architect's Scale

Scale	Relation of Scale to Object
16	Full Scale
3	3" = 1'
1½	1½" = 1'
1	1" = 1'
¾	¾" = 1'
½	½" = 1'
3/8	3/8" = 1'
¼	¼" = 1'
3/16	3/16" = 1'
1/8	1/8" = 1'
3/32	3/32" = 1'

106T01.EPS

Table 2 Sample Linear Measures

12 inches = 1 foot
3 feet = 1 yard
1,760 yards = 1 mile

106T02.EPS

Table 3 Decimal Equivalents of Common Fractions

Fraction (inches)	Decimal Equivalent	
	English (inches)	Metric (millimeters)
1/64	0.015625	0.3969
1/32	0.03125	0.7938
3/64	0.046875	1.1906
1/16	0.0625	1.5875
5/64	0.078125	1.9844
3/32	0.09375	2.3813
7/64	0.109375	2.7781
1/8	0.1250	3.1750
9/64	0.140625	3.5719
5/32	0.15625	3.9688
11/64	0.171875	4.3656
3/16	0.875	4.7625
13/64	0.203125	5.1594
7/32	0.21875	5.5563
15/64	0.234375	5.9531
1/4	0.250	6.3500

106T03.EPS

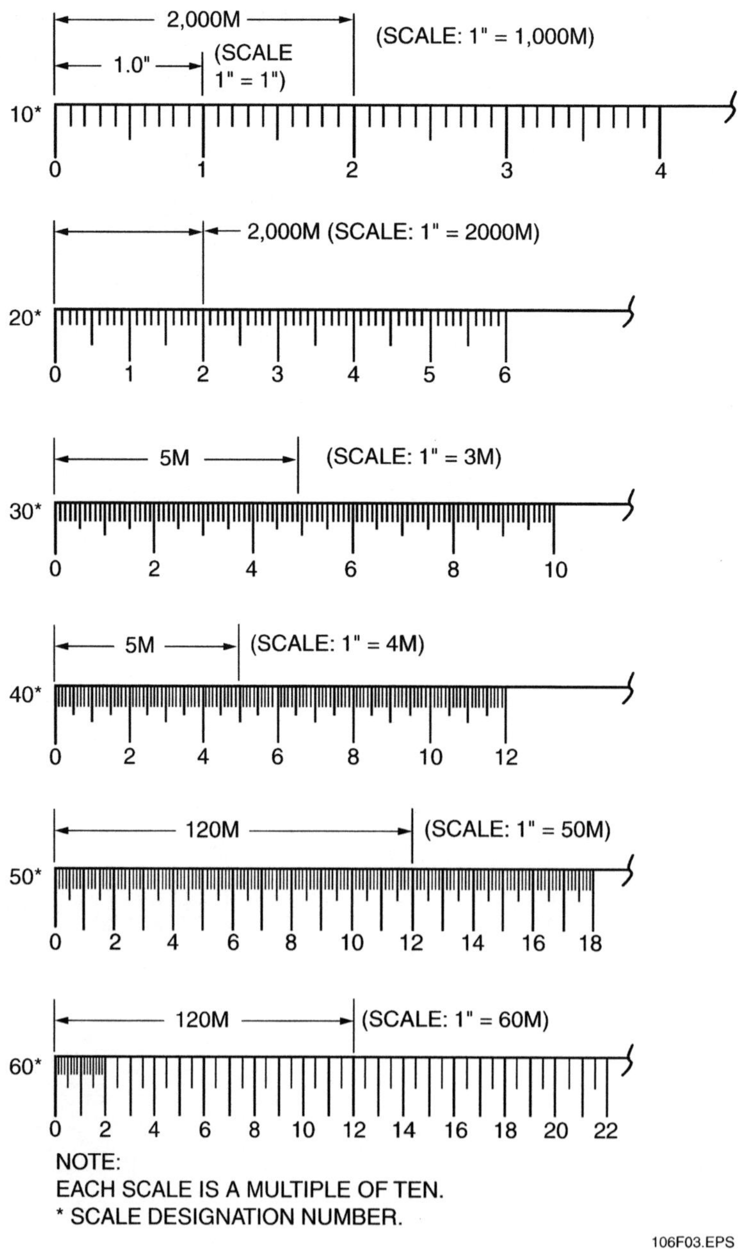

Figure 3 ◆ Scales on engineer's scale.

3.2.0 Mathematical Tables

Mathematical tables can simplify or eliminate the long calculations that are often necessary in mathematical problems. If mathematical tables are used, solutions for larger units can easily be calculated. For example, the following conversion table for English and metric **cubic** measurements converts basic units into other units of measurement *(Table 4)*.

Example:

1 cubic centimeter = 0.06102 cubic inches

1 gallon = 0.1337 cubic feet

Follow these steps to use mathematical tables. As an example, assume that 7 cubic feet are being converted to gallons.

Step 1 Locate the necessary table.

Step 2 Find the unit quantity in the first column.

Example: 7 units

Step 3 Find the heading over the column where the correct conversion is listed.

Example: Cubic Feet to Gallons

Step 4 Find the number in this column that is in the same row with the units located in Step 2.

Answer: 52.36

If the number of units to be converted is not on this type of table, it can be calculated with simple addition or multiplication. For example, if 40 cubic feet must be converted to cubic meters, the solution for 4 cubic feet can be found. If the steps given in this section are followed, 4 cubic feet can be converted to 0.1133 cubic meters. Since 4 × 10 = 40, 0.1133 can be multiplied by 10 to get 1.133, which is the solution to converting 40 cubic feet to meters.

These comparative values are handy in several ways. If a tool is digital, such as a programmable lathe or cut-off saw, it may not be possible to instruct it to cut a length of pipe to 36¼ inches. However, the comparative value table will tell you that ¼ inch is the same as 0.250 inches, which the tool will be able to recognize.

Suppose you need to look up ⅛ inch in the comparative value table:

Step 1 Look down through the fraction column until you come to ⅛.

Step 2 Look across that row to the English (inches) value, and you will see that the decimal equivalent of ⅛ is 0.125.

Step 3 If you needed the metric equivalent, you would look to the Metric (millimeters) column, and you would find the value to be 3.1750 millimeters.

Suppose you had to thread a pipe nipple on a lathe, so that there were 2⅛ inches of thread on each end:

Step 1 Put the nipple in the chuck.

Step 2 Look up the decimal value of ⅛ inch in the comparative value table. The decimal equivalent is 0.125.

Step 3 Key in that value, 2.125, or set a vernier to that value, for the length of the thread.

Step 4 Turn on the lathe, and make the thread. If this was a programmable lathe, it could be programmed to thread both ends. However, let's assume it is a manual automatic lathe.

Step 5 Turn off the lathe, take the nipple out, turn it around, tighten it back up in the chuck, and thread that end.

4.0.0 ◆ USING RATIOS AND PROPORTIONS

Ratios and proportions are methods of solving problems using comparisons. They can be used for practical calculations to determine the answers to problems involving scales on architectural drawings, grades and slopes, and pipe capacities.

4.1.0 Ratios

A ratio is a comparison of two quantities expressed in the same unit. For example, if a 3-inch square is compared to a 6-inch square, the ratio of their lengths is 3 inches to 6 inches. If a 3-inch square is compared to a 2-foot square, one of the measurements must first be converted to the same type of unit as the other. The ratio of 3 inches to 2 feet must be converted to either 3 inches to 24 inches or 0.25 feet to 2 feet. Since both numbers in the ratio are the same kind of unit, the units can be left out when the ratio is stated. Thus, 3 inches to 24 inches becomes 3 to 24.

Table 4 Conversion of English and Metric Cubic Measurements

Unit	Cubic Inches to Cubic Centimeters	Cubic Centimeters to Cubic Inches	Cubic Feet to Cubic Meters	Cubic Meters to Cubic Feet	Cubic Yards to Cubic Meters	Cubic Meters to Cubic Yards	Gallons to Cubic Feet	Cubic Feet to Gallons
1	16.39	0.06102	0.0283	35.31	0.7646	1.308	0.1337	7.481
2	32.77	0.1220	0.0566	70.63	1.529	2.616	0.2674	14.960
3	49.16	0.1831	0.0849	105.90	2.294	3.924	0.4010	22.440
4	65.55	0.2441	0.1133	141.30	3.058	5.232	0.5347	29.920
5	81.94	0.3051	0.1416	176.60	3.823	6.540	0.6684	37.400
6	98.32	0.3661	0.1699	211.90	4.587	7.848	0.8021	44.880
7	114.70	0.4272	0.1982	247.20	5.352	9.156	0.9358	52.360
8	131.10	0.4882	0.2265	282.50	6.116	10.460	1.069	59.840
9	147.50	0.5492	0.2549	371.80	6.881	11.770	1.203	67.320

106T04.EPS

Ratios are always read as X to Y, but they can be written in either of the following two ways:

- With a colon between the numbers: X:Y
- With a division sign between the numbers: X ÷ Y or X/Y

Ratios are usually stated as fractions; and, like fractions, the numbers or terms in them are compared by division. When creating a ratio, the terms should be compared to each other in the order in which they are given. The first term becomes the numerator of the fraction, and the second term becomes the denominator. For example, the number of teeth in two gears in mesh can be compared. If the driven gear has 64 teeth and the driving gear has 20 teeth, the ratio of the driven gear to the driving gear is 64/20.

Since a ratio is also a fraction, it can be reduced to its lowest possible terms without changing the value of the ratio. This is done by dividing the numerator and the denominator by a common number. For example, a ratio of 64/20 can be reduced to 16/5 by dividing both the numerator and denominator by 4 without changing the relationship of the two terms.

4.2.0 Proportions

A proportion is a mathematical statement that two ratios are equal. Since 6/8 and 36/48 can both be reduced to 3/4, they are equal, or in proportion. Ratios that cannot be reduced to the same lowest terms do not form a proportion. Proportions can be written in either of the following two ways:

- A:B::X:Y
- A/B = X/Y

The first is read: A is to B as X is to Y. The second is called the equation form and is generally used for craft applications.

A proportion has four terms. The first and last terms are called the extremes. The second and third terms are called the means. *Figure 4* shows means and extremes.

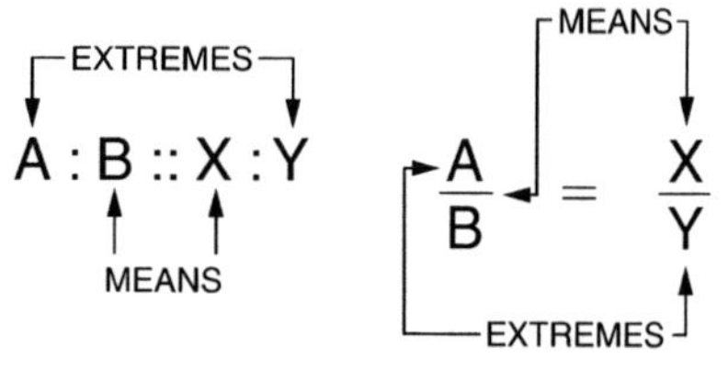

106F04.EPS

Figure 4 ◆ Means and extremes.

If two ratios form a proportion, the product of the means equals the product of the extremes.

Therefore, if A/B = X/Y, then AY = BX. This method of determining if two ratios form a proportion is called cross-multiplying, and the answers are called the cross products. Cross-multiplication and division can be used to find an unknown term when three of the terms in a proportion are known.

Follow these steps to find the missing term in a proportion:

> **NOTE**
> The total drop in any length of a graded pipeline is directly proportional to its slope. For the following steps, assume that 100 feet of pipe will drop 1 foot. This is equivalent to a 0.01-foot drop to every 1 foot of horizontal run. The problem is to determine how much a pipeline drops over 250 feet.

Step 1 Read the problem to determine the two ratios.

Example: 100 feet/250 feet and 1 foot/X feet

Step 2 Write the ratios as a proportion.

Example: 100/250 = 1/X

Step 3 Cross-multiply the means.

Example: 250 × 1 = 250
100 × X = 100X

> **NOTE**
> The products of the means must now equal each other.

Example: 100X = 250

Step 4 Divide the cross product of the means by the known extreme.

Example: 250 ÷ 100 = 2.5
100X ÷ 100 = X

X = 2.5

> **NOTE**
> If both extremes are known, divide the product of the extremes by the known mean.

Step 5 Check the answer by cross-multiplying. If the cross products are not the same, go back to Step 1.

Example: $100 \times 2.5 = 250$
$250 \times 1 = 250$

Many problems can be solved by setting up proportions of related quantities. These quantities must be analyzed to determine where to place the terms in the ratios. This is done by checking the problem to determine if the unknown term in the ratio will be larger or smaller than the known term. This ratio is stated to correspond with the ratio with two known terms. *Figure 5* shows this correspondence of ratios.

$$\frac{\text{SMALLER}}{\text{LARGER}} = \frac{\text{SMALLER}}{\text{LARGER}}$$

OR

$$\frac{\text{LARGER}}{\text{SMALLER}} = \frac{\text{LARGER}}{\text{SMALLER}}$$

106F05.EPS

Figure 5 ◆ Correspondence of ratios.

The two types of proportions used for solving different kinds of problems are direct proportions and inverse proportions.

Review Questions

Find the missing term in the proportion.

1. A sloped roof is given a 1 in 5 grade. If the roof is covering a 32-foot wide area, with 18-inch eaves on all sides, how much will the drop be from the peak to the edge of the roof?
   a. 1 foot
   b. 5 feet
   c. 7 feet
   d. 10 feet

2. A gas pipeline has a grade of 1 inch to one foot. If the line runs from the back wall of the shop 20 yards to the tank, what is the drop?
   a. 1 inch
   b. 10 inches
   c. 20 inches
   d. 60 inches

3. A pair of pulleys has a ¼ size ratio. If the smaller is 5 inches in diameter, how big is the larger?
   a. 5 inches
   b. 6 inches
   c. 20 inches
   d. 30 inches

4. Specifications for a drill press call for a drive pulley with a 1 to 5 ratio to the drill it drives. If the drill pulley is ten inches in diameter, what size is the drive pulley?
   a. 1 inch
   b. 2 inches
   c. 5 inches
   d. 50 inches

5. A 1-inch mild steel plate that is 1 foot by 1 foot weighs 40.8 pounds. If a steel plate 1 inch thick and 1 foot wide weighs 81.6 pounds, what is its length?
   a. 1 foot
   b. 2 feet
   c. 2.5 feet
   d. 3 feet

4.2.1 Direct Proportions

Two quantities are directly proportional when an increase in one causes a proportional increase in the other or a decrease in one causes a proportional decrease in the other. When a direct proportion is set up in equation form, the numerator of the first ratio must correspond with the numerator of the second ratio. The denominator of the first ratio must correspond with the denominator of the second ratio. For example, a 10-foot length of 8-inch carbon steel pipe weighs 286 pounds. The problem is to find the weight of a 12-foot length of the same pipe. The proportion would be stated as follows: $\frac{10}{12} = 286/X$.

The two numerators correspond since the 10-foot length of 8-inch carbon steel pipe weighs 286 pounds, and the two denominators correspond since the weight of the 12-foot length of pipe is unknown.

This example can be solved as follows:

$12 \times 286 = 3{,}432$

$3{,}432 \div 10 = 343.2$

$X = 343.2$; therefore, $10/12 = 286/343.2$

4.2.2 Inverse Proportions

Two quantities are indirectly, or inversely, proportional when an increase in one causes a decrease in the other or a decrease in one causes an increase in the other. For example, if one quantity increases to four times its original value, the other quantity decreases four times, or becomes one-fourth of its original value.

When an inverse proportion is set up in fractional form, the second ratio is inverted, or the terms are reversed. The numerator of the first ratio then corresponds with the denominator of the second ratio, and the denominator of the first ratio corresponds with the numerator of the second ratio. For example, pulley A is a 24-inch pulley that revolves at 200 rpm and is belted to pulley B, a 10-inch pulley. The problem is to determine the speed of pulley B. Since the smaller pulley, B, makes more rpms to work with pulley A, the ratio of their speeds is inversely proportional to the ratio of their diameters. The proportion is stated as follows:

$$\frac{\text{diameter of pulley B}}{\text{diameter of pulley A}} = \frac{\text{speed of pulley A}}{\text{speed of pulley B}}$$

or $10/24 = 200X$

The numerator of the first ratio corresponds with the denominator of the second since the speed of the 10-inch pulley is unknown. The denominator of the first pulley corresponds with the numerator of the second since the 24-inch pulley has a speed of 200 rpm.

This proportion can be solved as follows:

$24 \times 200 = 4{,}800$

$4{,}800 \div 10 = 480$

$X = 480$; therefore, $10/24 = 200/480$

Review Questions

1. A ten-foot length of 8-inch carbon steel pipe weighs 286 pounds. What is the weight of a 15-foot length of the same pipe?
   a. 190 pounds
   b. 429 pounds
   c. 668 pounds
   d. 2,860 pounds

2. A 20-horsepower motor weighs 430 pounds. If the relationship of weight to power is directly proportionate, a 100-horsepower motor would weigh _____.
   a. 860 pounds
   b. 1,500 pounds
   c. 2,000 pounds
   d. 2,150 pounds

3. A 24-inch pulley, turning at 60 rpms, is belted to a 10-inch pulley. How fast does the 10-inch pulley turn?
   a. 60 rpms
   b. 72 rpms
   c. 144 rpms
   d. 1,440 rpms

4. A 6-inch diameter gear has 18 teeth. It is driving a 20-inch gear with a proportionate number of teeth. By the time the 20-inch gear turns once the 6-inch gear will have turned _____ time(s).
   a. 1
   b. 3.33
   c. 20
   d. 60

5. A geared valve actuator turns a full turn to open the valve fully. If the actuator turns a gear with 10 teeth, and the gear directly turning the valve stem only turns a quarter turn, the stem gear has _____ teeth.
   a. 10
   b. 20
   c. 30
   d. 40

5.0.0 ◆ USING FORMULAS

Mathematics in industrial maintenance is frequently a matter of applying rules about the relationships between measurements. The rules are usually stated as formulas, statements that use letters to represent quantities. By using letters, we can state the relationships between numbers, such as the area of a **circle** and its width. This section explains the following principles of using mathematical formulas in industrial maintenance:

- Symbolism
- Expressing rules as formulas
- Factors
- Powers
- Square roots
- Evaluating formulas

5.1.0 Symbolism

Symbols are the language of mathematics. In blueprints, the symbols represent parts of a pipe assembly, such as valves and fittings. In mathematical formulas, symbols represent numbers. In blueprints, we use pictures; in mathematics, we use letters and numbers. Letters used to represent numbers in formulas are called **literal numbers,** as opposed to the **arithmetic numbers** we are used to. A literal number can represent a single arithmetic number, a wide range of numerical values, or all numerical values, depending on its function in a particular formula.

The multiplication sign × is commonly used in simple formulas but may not be used in formulas containing letters since it can be mistaken for the letter X. No sign of operation is required when a literal number is multiplied by an arithmetic number or when two or more literal numbers are multiplied. For instance, 5 × h is written 5h; b × h is written bh; 4 × h × w is written 4hw.

Parentheses, raised dots, or asterisks are often used instead of the multiplication sign when numerical numbers are multiplied. For instance, 3 × 6 may be written 3(6), 3·6, or 3*6, and ½ × 6.3 × 8 may be written ½(6.3)(8), ½·6.3·8, or ½*6.3*8.

5.2.0 Expressing Rules as Formulas

A formula is a short method of writing a mathematical rule. In formulas, the values from rules are represented by letters or symbols. A letter used to replace a value is often the first letter of the word representing that value. If that letter has already been used to represent a value in the formula, it cannot be used again to represent a different value. When a rule is changed into a formula, the values and mathematical operations are replaced with letters or symbols and mathematical signs. *Figure 6* shows changing a rule into a formula.

Since the letters or symbols in a formula represent the numbers in specific problems, these numbers can be written into the formula to replace the letters or symbols. This is called substitution. Usually, the values of all but one of the letters are known. These values can be used in the formula to find the unknown value. Follow these steps to solve a formula by substitution:

Step 1 Determine the formula needed to solve the problem. For this example, assume that the **radius** of a circle is 6 inches and the diameter is unknown. The formula shown in *Figure 6* can be used to solve the problem.

Example: $D = 2r$

Step 2 Substitute the known quantities for the letters or symbols in the formula.

Example: $D = 2 \times 6$

Step 3 Perform the mathematical operation needed to solve the problem.

Example: $12 = 2 \times 6$

Step 4 Check the answer.

Example: $12 \div 2 = 6$

Step 5 Label the answer with the correct unit of measure.

Example: D = 12 inches

NOTE

All measurements used in a formula must be the same type of unit.

If the diameter in the example was known and the radius was the unknown value, the formula would read 12 = 2r. It could be solved by dividing the diameter by 2, so that 12 ÷ 2 = 6.

Review Questions

1. The formula for gas mileage is mpg = miles ÷ gallons. If I drove 63 miles on 3 gallons of gas, my mpg is ______.
   a. 11
   b. 21
   c. 31
   d. 189

2. A industrial maintenance craftsperson is earning $22.50 per hour on the job. The formula for his paychecks is P = 22.5 × hours. If he worked 40 hours, he would earn ______.
   a. $180
   b. $900
   c. $1,800
   d. $2,250

3. The outside diameter of a 4-inch Schedule 40 pipe is 4½ inches. What is the radius of the pipe?
   a. 2 inches
   b. 2¼ inches
   c. 4½ inches
   d. 9 inches

4. A coil being made in a shop is bent to a 4-foot radius. What is the diameter of the coil?
   a. 2 feet
   b. 8 feet
   c. 12 feet
   d. 16 feet

5. A class 150 flange on 6-inch steel pipe has a 9½-inch diameter bolt circle. What is the radius of the bolt circle?
   a. 3 inches
   b. 3½ inches
   c. ½ inches
   d. 4¾ inches

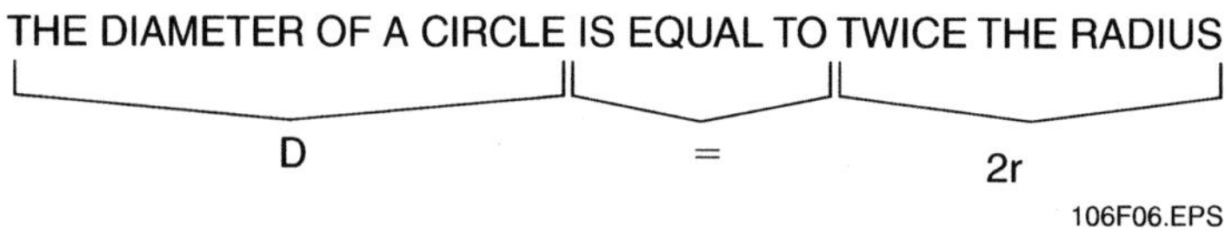

Figure 6 ◆ Changing a rule into a formula.

5.3.0 Factors

When multiplying two or more numbers to find a given number, the numbers being multiplied are **factors** of the given number. For instance, the factors of the number 12 are 1 and 12, 2 and 6, and 3 and 4, since $1 \times 12 = 12$, $2 \times 6 = 12$, and $3 \times 4 = 12$.

5.4.0 Powers

A power is the product of two or more equal factors. For example, 4×4 is the second power of 4; $6 \times 6 \times 6$ is the third power of 6; and $h \times h \times h \times h$ is the fourth power of h. For example, 5^4 is the same as $5 \times 5 \times 5 \times 5$. The **exponent** 4 tells you that 5 is taken as a factor four times and is read five to the fourth power; therefore, $5^4 = 625$.

5.5.0 Roots

The square root of a number, if multiplied by itself, equals the number of which it is the root. The radical sign ($\sqrt{}$) indicates a root of a number. The index number, which is written above and to the left of the radical sign, indicates the number of times that a root is to be taken as an equal factor to produce the given number. The index number for a square root is 2. The 2 is usually omitted in the square root sign. Therefore, the square root of 16 is written $\sqrt{16}$. It asks the question, "What number multiplied by itself equals 16?" Since $4 \times 4 = 16$, 4 is the square root of 16. So $\sqrt{16} = 4$.

When the square root of a fraction is to be computed, both the numerator and the denominator must be enclosed within the radical sign. For instance, $\sqrt{5/8}$ indicates that the square root of the complete fraction is to be taken. To compute the square root of a fraction, you must find the square root of both the numerator and the denominator.

Example: Compute the square root of 16/36:

$\sqrt{16} = 4$

$\sqrt{36} = 6$

$\sqrt{16/36} = 4/6 = 2/3$ reduced

Root expressions are sometimes written with two or more operations within the radical sign. For instance, the expression $\sqrt{1 + h}$ is read as "the square root of l+h." It is computed by first adding l+h and then computing the square root of the sum. Apply the following procedure to solve problems which involve operations within the radical sign:

Step 1 Perform the operations within the radical sign first.

Step 2 Compute the root.

Example: Compute $\sqrt{8+17}$

Add the numbers within the radical sign.

$8 + 17 = 25$

Compute: $\sqrt{25} : 5 \times 5 = 25$, so $\sqrt{8+17} = 5$

Any power of a number can be determined by multiplying the number by itself the number of times shown in the exponent. That is, the fourth power of a number, stated as n to the fourth power, is $n \times n \times n \times n$. By the same reasoning, the fourth root of a number is the number that when multiplied by itself four times equals the number you want the root of. The cube of a number is that number to the third power ($n \times n \times n$). The cube root of a number is the number that is multiplied by itself three times to obtain that number. For example, 2^3, read as two cubed, equals 8. The cube root of 27 is 3 because $3 \times 3 \times 3$ equals 27.

Review Questions

1. What is the value of 6 to the third power?
   a. 2
   b. 18
   c. 36
   d. 216

2. What is the square root of 9?
   a. 2
   b. 3
   c. 4½
   d. 81

3. The third power of 8 is $8 \times 8 \times 8$.
   a. True
   b. False

4. 2 to the fourth power is _____.
   a. 4
   b. 6
   c. 16
   d. 23

5. The exponent 3 means that the number is to be multiplied by itself 3 times.
   a. True
   b. False

5.6.0 Evaluating Formulas

Normally, formulas contain two or more arithmetic operations. It is essential that you perform the operations in the proper order to solve the formula. You should perform the operations in the following order:

Step 1 Perform the operations in parentheses first. If there are parentheses within parentheses, perform the operation within the innermost parentheses first.

Step 2 Perform the operations for powers and roots as they occur.

Step 3 Perform multiplication and division operations from left to right in the order in which they occur.

Step 4 Perform addition and subtraction operations from left to right in the order in which they occur.

One way to remember this order of operations is to remember PEMDAS (**P**lease **E**xcuse **M**y **D**ear **A**unt **S**ally): parentheses; exponent (powers and roots); multiply; divide; add; subtract.

To determine the numerical value of one letter in an expression when the numerical value of the other letter is known, use the following procedure:

Step 1 Write the expression.

Step 2 Replace each letter in the expression with its numerical value, and add a multiplication sign where multiplication is needed.

Step 3 Perform the operations in the proper order.

Example: Find the perimeter of a rectangle when h equals 6 inches and w equals 4 inches.

Step 1 Write the expression $p = 2h + 2w$

Step 2 Replace the letters in the expression with their numerical value, and add a multiplication sign where multiplication is needed.

$p = (2 \times 6) + (2 \times 4)$

ations in their proper

6 = 12 and 2 × 4 = 8

20

es

Another use for the formula for the perimeter of a rectangle would be to determine one of the sides if you know the perimeter of another side. Then the formula would state $p = 2l + 2w$. If the perimeter (p) is 18 and one of the sides is 4, then you could state the formula as $18 = 2l + 2(4)$ or $18 = 2l + 8$. You can do the same operation to both sides of the equation, so you subtract the 8 from both sides: $18 - 8 = 2l + 8 - 8$. Now you have $10 = 2l$. Divide both sides by two, and you have the length of the unknown side as $l = 5$. Remember, you can do the same operation on both sides of an equation and the two sides will still be equal.

6.0.0 ◆ SOLVING AREA PROBLEMS

Area is the amount of plane (flat) surface in a closed space. The area of any given surface can be calculated for practical applications, such as the following:

- Room layout
- Materials estimates
- Cost estimates
- Sizes of stock or parts

Area is measured by using standard areas of smaller units, such as the square inch and square foot. A square inch is a surface enclosed by a square that is 1 inch on each side. A square foot is enclosed by a square that is 1 foot on each side. There are 144 square inches in 1 square foot. These units can be used to solve problems in finding the area of rectangles, triangles, and circles.

6.1.0 Finding Area of Rectangles

A rectangle is a four-sided figure. Its opposite sides are equal, and all sides are joined at right angles. To find the area of a rectangle, it is necessary to find the number of surface units it contains. For example, if a rectangle contains 3 rows of square inches and there are 4 square inches in each row, the rectangle contains 12 square inches. *Figure 7* shows this rectangle.

The area of any rectangle can be found by multiplying the length of one side by the width of one side, or $A = lw$. Area is always expressed in square units. Remember that you can obtain the length or width if you know the area and either the length or width. In the case of a rectangle that is 2 feet long with an area of 10 square feet, the equation $A = lw$ becomes $10 = 2w$. Divide both sides by 2, and you get $5 = w$.

In an area formula, all measurements must be expressed in or converted to the same type of unit.

For example, if a room is 20 feet, 6 inches long and 15 feet, 9 inches wide, the area of floor is found as follows:

$A = lw$

$A = 20$ feet, 6 inches $\times$ 15 feet, 9 inches

$A = 20.5$ feet $\times$ 15.75 feet

$A = 322.875$ square feet

A square is a rectangle with four equal sides. A square that is 6 inches on each side is called a 6-inch square. The formula for finding the area of a rectangle can be used to find the area of a square. However, since each side is the same length, the length of any side can be multiplied by itself, or $A = S^2$. This is read as the area equals the sides squared. For example, the area of the 6-inch square is found as follows:

$A = S^2$

$A = 6 \times 6$

$A = 36$ square inches

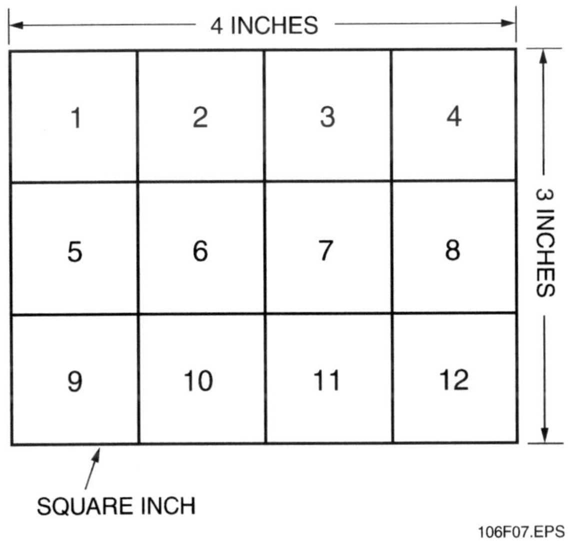

Figure 7 ◆ Rectangle.

Review Questions

1. What is the area of a rectangle that is 2½ feet by 4½ feet?
   a. 2 sq ft
   b. 8 sq ft
   c. 11¼ sq ft
   d. 12 sq ft

2. What is the area of a room that is 20 feet long by 15 feet wide?
   a. 250 sq ft
   b. 300 sq ft
   c. 600 sq ft
   d. 750 sq ft

3. What is the length of a rectangular platform that is 12 feet wide and has an area of 180 square feet?
   a. 15 ft
   b. 90 ft
   c. 190 ft
   d. 900 ft

4. One side of a square is 6 inches, so its area is _____.
   a. 6 sq in
   b. 12 sq in
   c. 36 sq in
   d. 216 sq in

5. The pad under a pump is 4½ feet wide and 5 feet long. What is its area?
   a. 22½ sq ft
   b. 25 sq ft
   b. 29 sq ft
   d. 40 sq ft

6.2.0 Finding Area of Triangles

A triangle is a three-sided figure with three angles. The area of any triangle can be found by multiplying the product of the base and the height by one half, or A = ½bh. The base of a triangle can be any one of its sides. The height is **perpendicular** to the base. *Figure 8* shows a triangle.

For example, if a triangle has a base of 15 feet and a height of 12 feet, the area is found as follows:

A = ½bh

A = 1⁄2 × 15 × 12

A = 90 square feet

If the base and the area are given, then you can plug them into the formula, and you would have 90 = 15h/2. First multiply both sides by 2, which gives you 180 = 15h. Now divide both sides by 15, and you will find 12 = h. The height is 12 feet.

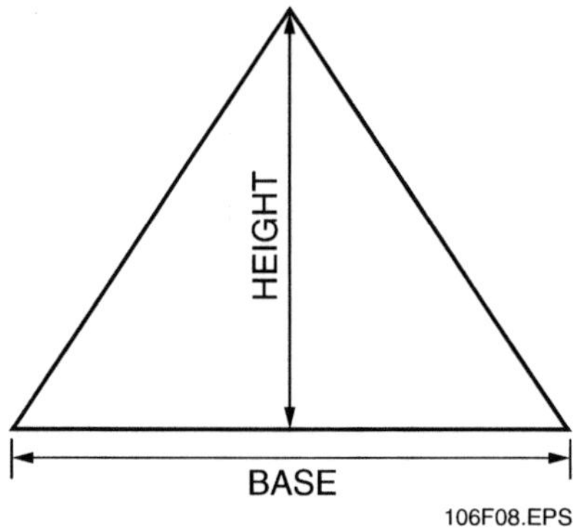

Figure 8 ◆ Triangle.

Review Questions

1. What is the area of a triangle that is formed when two of the legs, joined by a right angle, are each 4 feet long?
   a. 4 sq ft
   b. 8 sq ft
   c. 16 sq ft
   d. 32 sq ft

2. How much canvas will you need to cover the end of a triangular tent that is 6 feet high and 6 feet wide at the bottom?
   a. 6 sq ft
   b. 12 sq ft
   c. 18 sq ft
   d. 36 sq ft

3. The base of a triangle is 100 feet wide, and the area is 1250 square feet. What is its height?
   a. 10 feet
   b. 12.5 feet
   c. 25 feet
   d. 50 feet

4. How much area will be covered by a triangular pad, if all three sides are 10 feet, and the distance from the middle of one angle to the side opposite it is 8.8 feet (round to the closest foot)?
   a. 44 sq ft
   b. 50 sq ft
   c. 88 sq ft
   d. 100 sq ft

5. How many square feet of CDX sheathing will you need to cover the gable of your house? The triangle is 50 feet across and 10 feet high.
   a. 60
   b. 100
   c. 250
   d. 500

6.3.0 Finding Area of Circles

The area of a circle can be found by multiplying **pi** (pronounced "pie" and equal to 3.1416) by the product of the radius multiplied by itself, or $A = \pi r^2$. For example, if a pipe has an inside radius of 12 inches, the cross-sectional area of the pipe is found as follows:

$A = \pi r^2$

$A = 3.1416 \times 12 \times 12$

A = 452.39 or 452.4 square inches

This formula can also be used to find the pressure exerted on a piston. If the piston has a diameter of 6 inches and the pressure exerted on the piston head is 160 pounds per square inch, the total pressure is found as follows:

$A = \pi r^2$

$r = D \div 2 = 6 \div 2 = 3$

$A = 3.1416 \times 3 \times 3$

A = 28.2744 square inches

Pressure = 160 pounds per square inch

Pressure per square inch $\times$ area = total pressure

$160 \times 28.2744 = 4{,}523.904$

Review Questions

Use the value 3.1416 for pi.

1. A piece of 6-inch standard pipe has an outside diameter of $6\frac{5}{8}$ inches. The minimum area of a cover is _____.
   a. 16.33 sq in
   b. 28.27 sq in
   c. 34.47 sq in
   d. 36 sq in

2. What is the area of a manhole cover that is 24 inches in diameter?
   a. 6.1 sq in
   b. 73.9 sq ft
   c. 144 sq in
   d. 452.39 sq in

3. How many square inches of plywood will be needed to fabricate a circular cover for one end of a coil of pipe 24 inches outside diameter? Assume that you need to have a 1-inch overhang all around.
   a. 864
   b. 73.93
   c. 169
   d. 530.93

4. A circular tank is 11,310 sq ft. What is the radius?
   a. 6 ft
   b. 60 ft
   c. 120 ft
   d. 3,600 ft

5. If each person standing in a circus ring covers 2 square feet of ring, how many people can get into a 50-foot diameter ring?
   a. 25
   b. 100
   c. 205
   d. 981

7.0.0 ◆ SOLVING VOLUME PROBLEMS

Volume is the amount of space a solid figure occupies. Every solid figure has three dimensions: length, width, and height. The volume of any solid figure can be calculated for practical applications involving such objects as the following:

- Tanks
- Boxes
- Pipes
- Ducts
- Buildings

Volume, like area, is measured in standard units. A basic unit of measure for volume is the cubic inch. A cubic inch is the space occupied by a solid figure that is 1 inch long, 1 inch wide, and 1 inch high. The following sections explain using cubic measures to find the volume of the following:

- **Rectangular** solids
- **Cylinders**
- **Spheres**
- **Pyramids**
- Cones

7.1.0 Finding Volume of Rectangular Solids

A block with rectangular sides is a rectangular solid (*Figure 9*). To find the volume of a rectangular solid, it is necessary to find the number of cubic units it contains. For example, if a rectangular solid contains 4 cubic inches on each of three layers and the layers are 2 cubic inches deep, the rectangular solid contains 24 cubic inches.

The volume of any rectangular solid can be found by multiplying the product of the length and width by the height, or V = lwh. Volume is always expressed in cubic units.

For example, an excavation is 60¾ feet long, 28½ feet wide, and 15 feet deep; the problem is to find the amount of earth removed. The volume is found as follows:

$V = lwh$

$V = 60\frac{3}{4} \times 28\frac{1}{2} \times 15$

$V = 60.75 \times 28.5 \times 15$

V = 25,970.625 or 25,971 cubic feet

Volume problems can also be manipulated to find more than one of the dimensions. If you know the length, width, and volume, the height can be solved for. Another way of setting this equation up is to isolate the unknown on one side, with the known quantities on the other side. The equation for the volume of a rectangle (V = lwh) can be set up to determine the height by dividing both sides by lw, giving V/lw = h. You could also find the width by solving w = V/lh.

This formula can be used to find any of its variables if the other three variables are known. For example, a scrap box is to be built from steel plate to fit into a 10-foot by 12-foot space, and it must have a 480-cubic-foot volume. The height needed is found as follows:

$V = lwh$

$480 = 10 \times 12 \times h$

$480 = 120 \times h$

$480 \div 120 = 4$

Height = 4 feet

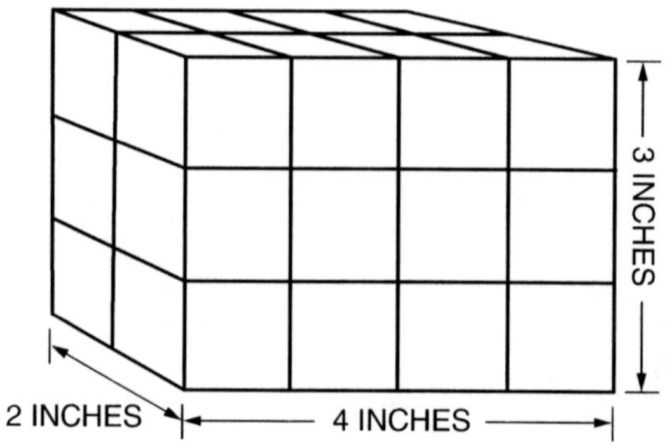

Figure 9 ◆ Rectangular solid.

Review Questions

1. You are pouring a rectangular pad of concrete 6 inches deep by 4 feet wide by 6 feet long. How many 1-cubic-foot bags of concrete mix will you need?
   a. 10
   b. 12
   c. 24
   d. 144

2. What is the volume of a solid concrete block that is 16 inches by 8 inches by 8 inches?
   a. 48 cu in
   b. 128 cu in
   c. 144 cu in
   d. 1,024 cu in

3. How many cubic feet are there in a cubic yard?
   a. 3
   b. 9
   c. 18
   d. 27

4. How many cubic feet of concrete will be required to pour a pump base pad 18 inches high by 10 feet by 20 feet?
   a. 180
   b. 200
   c. 300
   d. 3,600

5. You have to figure the capacity of the pump for a lift station. The rectangular sump is 10 feet wide by 12 feet long, and is to be 25 feet deep. What is the volume?
   a. 2,400 cu ft
   b. 3,000 cu ft
   c. 12,000 cu ft
   d. 30,000 cu ft

7.2.0 Finding Volume of Cylinders

A cylinder (*Figure 10*) is a solid figure with two identical circular bases. The height of a cylinder is the perpendicular distance between the two bases.

To find the volume of a cylinder, the area of one of the circular bases must first be calculated using the formula $A = \pi r^2$. The area is then multiplied by the height of the cylinder, or $V = \pi r^2 h$. For example, the volume of a cylinder with a radius of 4 inches and a height of 8 inches is calculated as follows:

$V = \pi r^2 h$

$V = 3.1416 \times 4 \times 4 \times 8$

$V = 402.1248$ or 402 cubic inches

Since π is a constant (a number that does not change), you can find the radius of a cylinder from the volume and the height. If you measured and found out that it took 402 cubic inches of liquid to fill a tank and the height of the tank is 8 inches, the equation would become:

$402 = 3.1416\, r^2(8)$

$402 \div (3.1416 \times 8) = r^2$

$r = \sqrt{16} = 4$

If a fuel oil tank with a diameter of 16 feet and a height of 18 feet is filled with oil, the volume of oil is found as follows:

$D \div 2 = r$

$16 \div 2 = 8$

$V = \pi r^2 h$

$V = 3.1416 \times 8 \times 8 \times 18$

$V = 3{,}619.1232$ or 3,619 cubic feet

If a cubic foot contains 7½ gallons and oil costs $0.98 per gallon, the cost to fill the tank is found as follows:

$3{,}619 \times 7.5 = 27{,}142.5$ gallons

$27{,}142.5 \times 0.98 = \$26{,}599.65$

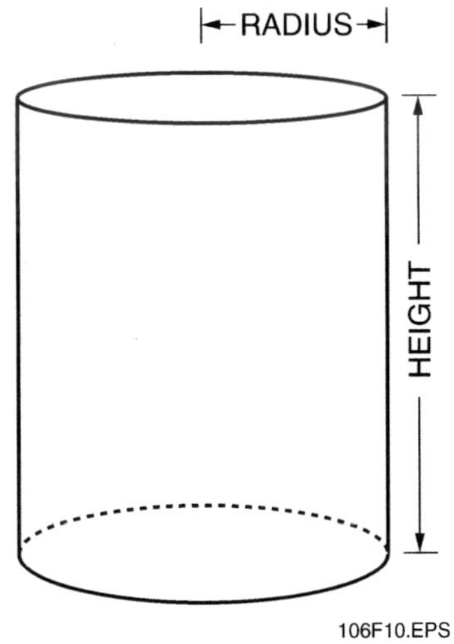

Figure 10 ◆ Cylinder.

Review Questions

Use the value 3.1416 for pi.

1. You need to choose a container to drain a section of pipe into. The pipe's inside diameter is 6 inches, and the section is 100 feet long. How much liquid, in cubic feet, do you need to be prepared for?
   a. 19.635
   b. 34
   c. 135.717
   d. 200

2. How much oil will be lost if the 4-foot-diameter pipe running the 700 miles from Mosul, in Northern Iraq, to the ports in Kuwait is drained by terrorists all at once? Hint: 1 mile is equal to 5,280 feet.
   a. 4,645 cu ft
   b. 46,445 cu ft
   c. 46,445,414 cu ft
   d. 93,000,000 cu ft

3. A natural gas pipeline runs from Russia across the Ukraine. If the pipe is 3 feet in diameter, and 500 miles long, how many cubic feet of gas can it hold at one time?
   a. 1,866
   b. 186,611
   c. 18,661,104
   d. 37,000,000

4. What is the diameter of a barrel that is 3½ feet tall and holds approximately 11 cubic feet of water?
   a. ½ ft
   b. 1 ft
   c. 2 ft
   d. 4 ft

5. In a chemical plant, you have to fill a cylindrical tank with liquid. If the tank is 20 feet in diameter and 10 feet high, what is the volume of fluid it will hold?
   a. 314.16 cu ft
   b. 1,256.6 cu ft
   c. 1,257 cu ft
   d. 3,141.6 cu ft

7.3.0 Finding Volume of Spheres

A sphere is a three-dimensional figure with a curved surface on which every point is an equal distance from the center. A line from the center to any point on the surface is called a radius. A diameter is a straight line from one point on the edge of a sphere, through its center, to another point on the edge of the sphere. *Figure 11* shows a sphere.

The volume of a sphere is found by multiplying the product of 4 times pi by the radius cubed and then dividing that product by 3, or $V = 4\pi r^3 \div 3$. The cube of the radius, written as r^3, means $r \times r \times r$. For example, the volume of a sphere with a radius of 3 inches is found as follows:

$$V = \frac{4 \times 3.1416 \times 3 \times 3 \times 3}{3}$$

$$V = 339.2928 \div 3$$

$$V = 113.0976 \text{ or } 113 \text{ square inches}$$

If a spherical tank used to store gas is 50 feet in diameter, the volume of the tank is found as follows:

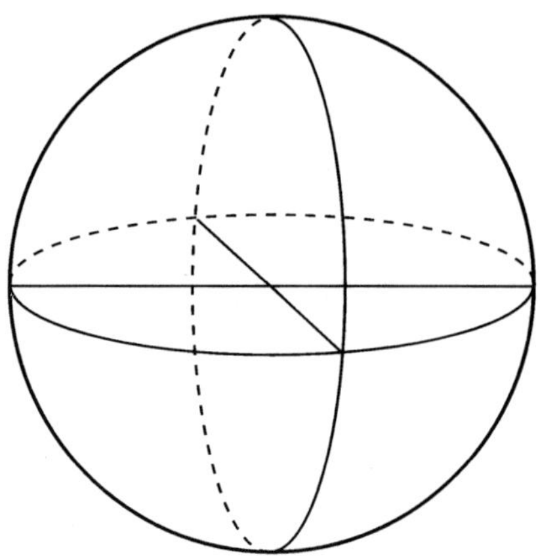

Figure 11 ◆ Sphere.

Diameter $\div$ 2 = radius

50 $\div$ 2 = 25

$V = 4\pi r^3 \div 3$

$$V = \frac{4 \times 3.1416 \times 25 \times 25 \times 25}{3}$$

V = 196,350 $\div$ 3

V = 65,450 cubic feet

Review Questions

Use the value 3.1416 for pi.

1. A ball 12 inches in diameter contains _____ of air.
   a. 37.6992 cu in
   b. 150 cu in
   c. 904.78 cu in
   d. 1,809.54 cu in

2. A spherical gasoline tank has been drained. If the result was 65,450 cubic feet of gasoline, what was the diameter?
   a. 25 ft
   b. 45 ft
   c. 50 ft
   d. 100 ft

3. A spherical reactor containment vessel, 200 feet in diameter, must be filled with coolant very quickly. How many cubic feet of coolant will be required?
   a. 3,141,600
   b. 4,188,800
   c. 12,566,400
   d. 418,880,000

4. A spherical bladder for water for firefighting is 20 feet in diameter. What is its volume in cubic feet?
   a. 314.16
   b. 418.88
   c. 3,141.6
   d. 4,188.8

5. A spherical chamber in a valve, 8 inches in diameter, must be included in your drainage calculations. What is its volume?
   a. 67 cu in
   b. 268.0832 cu in
   c. 536.32 cu in
   d. 804 cu in

7.4.0 Finding Volume of Pyramids

A pyramid is a solid figure with a base and three or more triangular faces that taper to one point opposite the base. This point is called the **apex**. The height is a straight line from the apex to the base. *Figure 12* shows a pyramid.

The volume of a pyramid can be found by multiplying the area of its base by the height and then dividing the product by 3, or V = Ah $\div$ 3. For example, if the area of the base is 24 square inches and the height is 15 inches, the volume is found as follows:

V = Ah $\div$ 3

V = 24 $\times$ 15 $\div$ 3

V = 360 $\div$ 3

V = 120 cubic inches

If the rectangular base of a pyramid has a length of 22 millimeters and a width of 17 millimeters and the height of the pyramid is 42 millimeters, the volume is found as follows:

Area = lw

Area = 22 $\times$ 17

Area = 374

V = Ah $\div$ 3

$$V = \frac{374 \times 42}{3}$$

$$V = \frac{15{,}708}{3}$$

V = 5,236 cubic millimeters

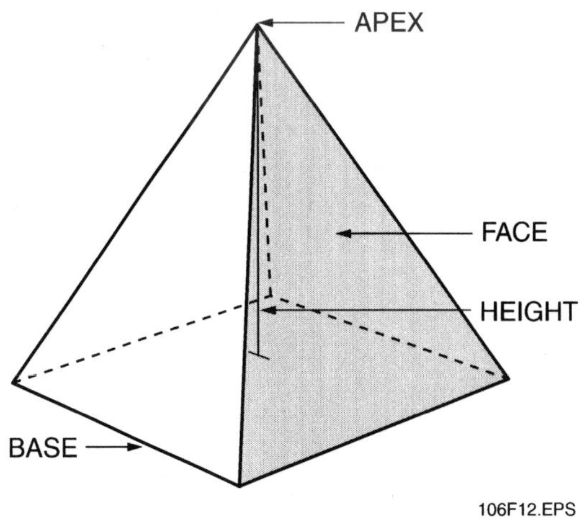

Figure 12 ◆ Pyramid.

Review Questions

1. While doing a takeoff for a water main, you find you are to figure the concrete for a pyramidal thrust block with a 2-foot by 2-foot base and a height of 18 inches. How many cubic feet of concrete do you need for each thrust block?
   a. 1
   b. 2
   c. 6
   d. 18

2. You need to figure the concrete to be pumped into a pyramidal form to support a pipe coil. The form is 6 feet by 6 feet at the base, and is 6 feet high. How much concrete will you need?
   a. 36 cu ft
   b. 72 cu ft
   c. 144 cu ft
   d. 216 cu ft

3. You have to calculate the piping system to pump concrete into a dam. The dam is basically an upside down pyramid 300 feet long, 100 feet wide, and 200 feet deep. What is its volume?
   a. 600,000 cu ft
   b. 1,200,000 cu ft
   c. 2,000,000 cu ft
   d. 6,000,000 cu ft

4. You need a thrust block for the riser for a fire hydrant. The pyramidal form is 1 foot high and 18 inches by 16 inches at the base. How many cubic inches of concrete will you need?
   a. 96 cu in
   b. 288 cu in
   c. 1,152 cu in
   d. 3,456 cu in

5. The Egyptian Pyramid of Khufu has a 756-foot by 756-foot base. Its volume is 91,636,272 cubic feet. What is its height?
   a. 200 ft
   b. 340 ft
   c. 481 ft
   d. 565 ft

7.5.0 Finding Volume of Cones

A cone (*Figure 13*) is a **solid** figure with a curved surface. One end of the surface is the apex, and the other is a circular base. Like the pyramid, the height of a cone is a straight line from the apex to the base.

The volume of a cone is found by multiplying pi times the radius of the base squared, times the height, and then dividing the product by 3, or $V = \pi r^2 h \div 3$.

For example, the volume of a conical tank with a base radius of 12 feet and a height of 30 feet is found as follows:

$$V = \pi r^2 h \div 3$$

$$V = \frac{3.1416 \times 12 \times 12 \times 30}{3}$$

$$V = \frac{13{,}571.712}{3}$$

$$V = 4{,}523.904 \text{ cubic feet}$$

A No. 3 taper standard lathe center has a tungsten carbide tip with a base diameter of 0.460 inch. If 0.75 inch of the tip is exposed, the cubic inches of carbide in the exposed tip is found as follows:

$$\text{Diameter} \div 2 = \text{radius}$$

$$0.460 \div 2 = 0.23$$

$$\text{radius} = 0.23$$

$$V = \pi r^2 h \div 3$$

$$V = \frac{3.1416 \times 0.23 \times 0.23 \times 0.75}{3}$$

$$V = \frac{0.1246429}{3}$$

$$V = 0.0415476 \text{ cubic inches}$$

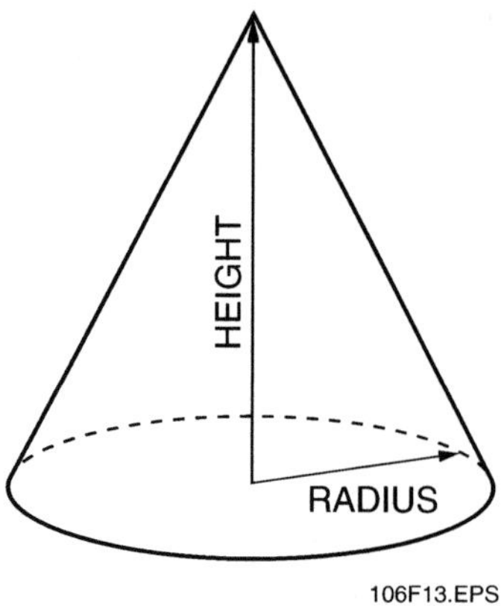

Figure 13 ◆ Cone.

Review Questions

Use the value 3.1416 for pi.

1. An ice cream cone is 2 inches wide at the opening and 4 inches long. After the scoop on top has been eaten down to level with the top, how much ice cream is left?
   a. 4.1888 cu in
   b. 8.8 cu in
   c. 12.5664 cu in
   d. 16.264 cu in

2. A conical inside support is to be machined for a coil. The inside radius for the bottom coil wrap is given as 18 inches. The height of the cone is to be 12 inches. What is the cone's volume?
   a. 3,141.6 cu in
   b. 3,888 cu in
   c. 4,071.5136 cu in
   d. 8,143.272 cu in

3. You are to set up a pump to fill a conical tank with liquid. The diameter of the cone's base is 20 feet, and the height is 10 feet. How much liquid will you need to pump?
   a. 1,047.2 cu ft
   b. 2,104.2 cu ft
   c. 3,141.6 cu ft
   d. 4,188.8 cu ft

4. The cone of a custom concentric reducer is to go from a pinhole on one end, to 6 inches inside diameter on the large end. It is to hold a little over 113 cubic inches of fluid. How long is it?
   a. 12 in
   b. 14 in
   c. 16 in
   d. 20 in

5. The nose cone for a missile is 2 feet in diameter at the base, and 3 feet high. What is its volume in cubic inches?
   a. 458.88
   b. 1,356.06
   c. 1,809.3146
   d. 5,428.6848

8.0.0 ◆ SOLVING CIRCUMFERENCE PROBLEMS

A circle is a closed, curved line on which every point is the same distance from the center. The distance around a circle is called its circumference. The diameter is the length of a straight line drawn from one point on a circle, through its center, to another point on the circle. A straight line drawn from the center of a circle to any point on the circle is the radius *(Figure 14)*.

The circumference of all circles is approximately 3.1416 times the diameter. The Greek letter π or pi, is used to represent this value. The circumference of any circle can be found by multiplying its diameter by pi, and the diameter can be found by dividing the circumference by pi. For example, if a pulley is 20 inches in diameter, $20 \times 3.1416 = 62.832$. Its circumference is 62.832 inches.

If the circumference of a pipe is 37.70 mm and the problem is to find its radius, it is found as follows:

$C \div \pi = D$

$37.70 \div 3.1416 = 12$

$12 \div 2 = 6$

Radius $= 6$

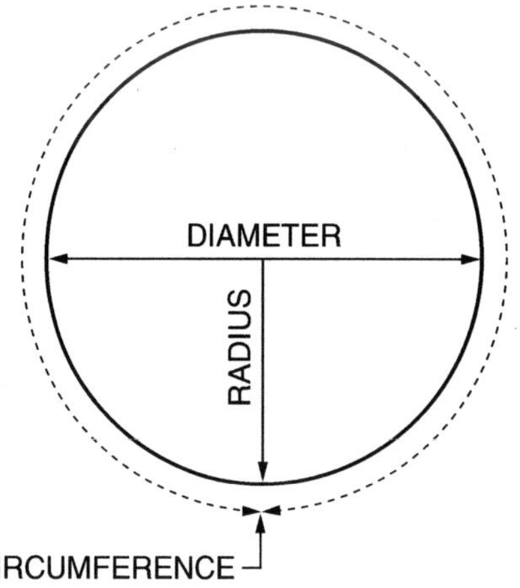

106F14.EPS

Figure 14 ◆ Parts of a circle.

Review Questions

Use the value 3.1416 for pi.

1. What is the circumference of a circle 12 inches in diameter?
   a. 18.8496 inches
   b. 37.6992 inches
   c. 113.36 inches
   d. 216 inches

2. You have to draw a cut line on a piece of 4½-inch outside diameter pipe. You do not have a wrap handy, so you pick up a piece of flexible banding. How long is the minimum amount of banding you need?
   a. 7.0686 inches
   b. 9 inches
   c. 14.1372 inches
   d. 28.2744 inches

3. How much pipe will be required for one full turn of a coil, if the radius of the coil is 15 feet?
   a. 30 feet
   b. 47.124 feet
   c. 94.248 feet
   d. 1,413.0788 feet

4. You have a pipe that is 56½ inches in circumference. What is its diameter?
   a. 6 inches
   b. 9 inches
   c. 12 inches
   d. 18 inches

5. How long a strap will you need to go all the way around a piece of 4½-inch-diameter pipe?
   a. 7.0686 inches
   b. 9 inches
   c. 14.1372 inches
   d. 28.2744 inches

9.0.0 ◆ PYTHAGOREAN THEOREM

The simplest triangles are right triangles. A right triangle has one 90-degree, or right, angle. This angle is usually indicated with a small box drawn in the angle. The right triangle is very important to industrial maintenance craftspersons because it is used to determine the components of a piping offset.

The sides of a right triangle have been named for reference. The side opposite the right angle is always called the **hypotenuse**, and the two sides adjacent to, or connected to, the right angle are called the legs. If one of the other angles is labeled angle A, the leg of the triangle that is not connected to angle A is called its **opposite side**. The remaining leg that is connected to angle A is called the **adjacent side**.

The sides of the piping offset have also been named for reference. These sides are called the **set**, **run**, and **travel**. The set is the distance, measured center to center, that the pipeline is to be offset. The run is the total lineal distance required for the offset. The travel is the center-to-center measurement of the offset piping. The angle of the fittings is the number of degrees the piping changes direction. *Figure 15* shows a right triangle and a piping offset.

The Pythagorean theorem states that the square of the hypotenuse is equal to the sums of the squares of the other two sides. For example, in triangle abc, with c being the hypotenuse and a a and b being the two legs, the Pythagorean theorem states that $a^2 + b^2 = c^2$. The following steps refer to triangle abc, in which one leg is 3 inches long and the other leg is 4 inches long. *Figure 16* shows triangle abc. Follow these steps to find the length of the hypotenuse, using the Pythagorean theorem:

Step 1 Insert the known values into the Pythagorean theorem.

Example: $a^2 + b^2 = c^2$

$3^2 + 4^2 = c^2$

Step 2 Square the known values in the formula.

Example: $9 + 16 = c^2$

Step 3 Add the squared values in the formula.

Example: $25 = c^2$

Step 4 Take the square root of both sides of the equation to determine the value for the unknown side.

Example: $\sqrt{25} = \sqrt{c^2}$

$5 = c$

The length of the hypotenuse is 5 inches.

The Pythagorean theorem can also be used to find the length of one of the legs if the other leg and the hypotenuse are known. To do this, start by isolating the unknown value on one side of the equation. As with any equation, isolate the unknown value to one side by doing the same operation to both sides of the equation. The following steps refer to triangle abc, in which one leg is 6 inches long and the hypotenuse is 10 inches long (*Figure 17*). Follow these steps to find the length of one of the legs of a triangle, using the Pythagorean theorem:

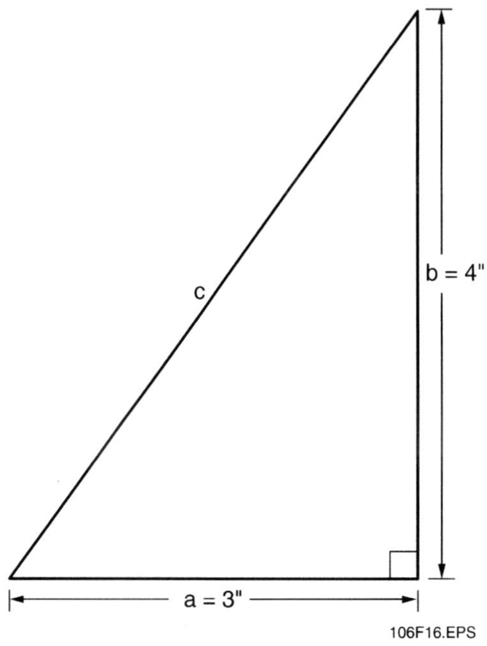

Figure 16 ◆ Triangle abc.

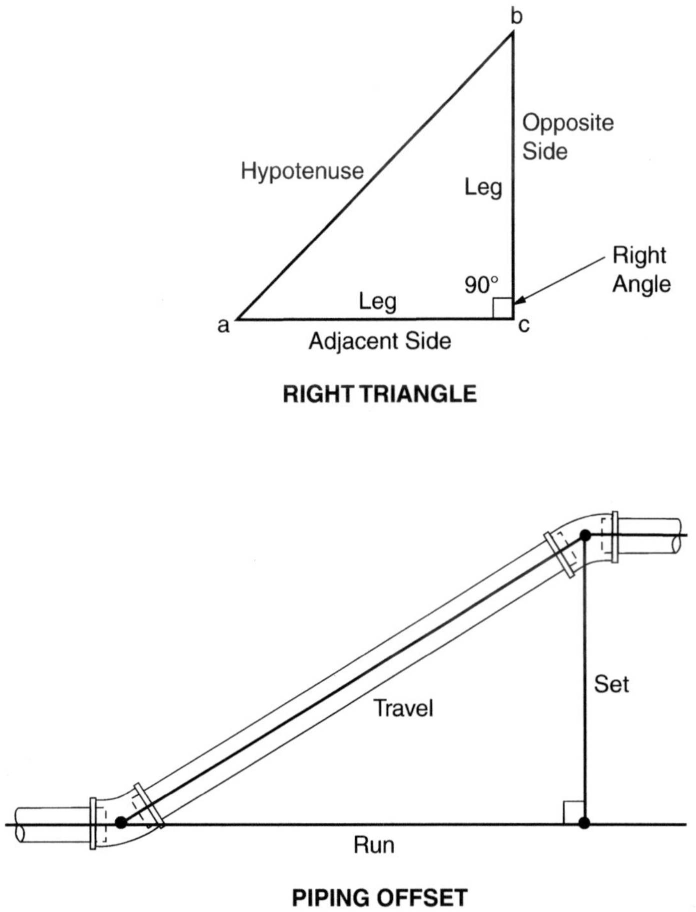

Figure 15 ◆ Right triangle and piping offset.

Step 1 Write the Pythagorean theorem, and isolate the unknown side.

Example: $a^2 + b^2 = c^2$

$a^2 = c^2 - b^2$

Step 2 Place the known values into the equation.

Example: $a^2 = 10^2 - 6^2$

Step 3 Square the known values in the equation.

Example: $a^2 = 100 - 36$

Step 4 Perform the subtraction to the right side of the equation.

Example: $a^2 = 64$

Step 5 Take the square root of both sides of the equation to determine the value of the unknown side.

Example: $\sqrt{a^2} = \sqrt{64}$

$a = 8$

The length of the unknown leg is 8 inches.

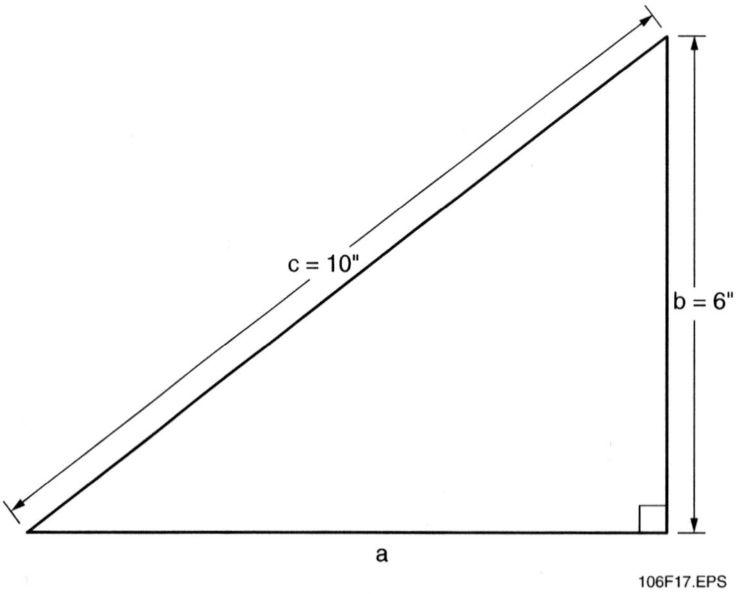

Figure 17 ◆ Triangle abc.

Review Questions

1. A triangle has one side adjacent to the right angle that is 3 feet long. The other adjacent side to the right angle is 4 feet long. How long is the third side?
   a. 5 feet
   b. 7 feet
   c. 12 feet
   d. 25 feet

2. A pipe runs between two 45-degree ells. The travel is slightly over 7 feet ¾ inches and the set is 5 feet. What is the run? Round off to even feet.
   a. 2 feet
   b. 5 feet
   c. 7 feet
   d. 12 feet

3. You are fitting up a length of welded pipe. You come to a point where you are to connect to another system. The set is 10 feet and the run is 6 feet. What is the travel?
   a. 8 feet
   b. 11.6619 feet
   c. 16 feet
   d. 36.3376 feet

4. A line of water pipe comes to the base of a tank. The connection for the tank will run up a leg and to the top of the tank, 100 feet above. The angle of the leg is such that the run is 20 feet. What length of pipe will you need?
   a. 99.876 feet
   b. 100 feet
   c. 101.98 feet
   d. 105.1416 feet

5. For a pipeline ditch, you need to know how long the horizontal distance is to an object. The set is 25 feet. The travel is 75 feet. What is the run?
   a. 50 feet
   b. 62.1416 feet
   c. 70.711 feet
   d. 81.99 feet

Summary

An industrial maintenance craftsperson uses mathematics to install and route piping systems, determine gear ratios, determine the volume and capacity of pipes and tanks, and calculate other requirements such as support pads. We use formulas to express the relationships that exist between quantities that we can measure, so that our measurements let us know what we need to do. Mathematics is an essential tool that you will use in many aspects of your job.

Notes

Trade Terms Introduced in This Module

Adjacent side: The side of a right triangle that is next to the reference angle.

Apex: The point at which the lines of a figure converge.

Arithmetic numbers: Numbers that have definite numerical values, such as 4, 6.3, and ⅝.

Circle: A continuous curved line that encloses a space, with every point on the line the same distance from the center of the circle.

Circumference: The distance around a circle.

Cubic: The designation of a given unit representing volume.

Cylinder: A shape created by a circle moving in a straight line through space perpendicular to the surface of the circle.

Exponent: A number or symbol placed to the right and above another number, symbol, or expression, denoting the power to which the latter is to be raised.

Factors: The numbers that can be multiplied together to produce a given product.

Formula: An equation that states a rule.

Hypotenuse: The longest side of a right triangle. It is always located opposite the right angle.

Literal numbers: Letters that represent arithmetic numbers, such as x, y, and h. Also known as algebraic numbers.

Opposite side: The side of a right triangle that is directly across from the reference angle.

Perpendicular: At a right angle to the plane of a line or surface.

Pi: A number that represents the ratio of the circumference to the diameter of a circle. Pi is approximately 3.1416 and is represented by the Greek letter π.

Pyramid: A shape with a multi-sided base, and sides that converge at a point.

Radius: A straight line from the center of a circle to a point on the edge of the circle.

Rectangular: Description of a shape having parallel sides and four right angles.

Run: The horizontal distance from one pipe to another.

Set: The vertical distance from the line of flow of a pipe and the line of flow of the pipe to which it is attached.

Solid: A figure enclosing a volume.

Sphere: A shape whose surface is everywhere the same distance from a central point.

Travel: The diagonal distance from one pipe to another.

Volume: The amount of space occupied by an object.

Resources & Acknowledgments

Additional Resources

This module is intended to be a thorough resource for task training. The following reference works are suggested for further study. These are optional materials for continued education rather than for task training.

Pipe Fitter's Math Guide, 1989. Johnny Hamilton. Clinton, NC: Construction Trade Press.

Applied Construction Math, Latest Edition. Upper Saddle River, NJ: Prentice Hall Publishing.

CONTREN® LEARNING SERIES — USER UPDATE

NCCER makes every effort to keep these textbooks up-to-date and free of technical errors. We appreciate your help in this process. If you have an idea for improving this textbook, or if you find an error, a typographical mistake, or an inaccuracy in NCCER's Contren® textbooks, please write us, using this form or a photocopy. Be sure to include the exact module number, page number, a detailed description, and the correction, if applicable. Your input will be brought to the attention of the Technical Review Committee. Thank you for your assistance.

Instructors – If you found that additional materials were necessary in order to teach this module effectively, please let us know so that we may include them in the Equipment/Materials list in the Annotated Instructor's Guide.

Write: Product Development and Revision
National Center for Construction Education and Research
3600 NW 43rd St. Bldg G, Gainesville, FL 32606

Fax: 352-334-0932

E-mail: curriculum@nccer.org

Craft ______ Module Name ______

Copyright Date ______ Module Number ______ Page Number(s) ______

Description

______

______

______

(Optional) Correction

______

______

(Optional) Your Name and Address

______

______

Industrial Maintenance E & I Technician Level One

40107-07

Construction Drawings

40107-07
Construction Drawings

Topics to be presented in this module include:

Overview

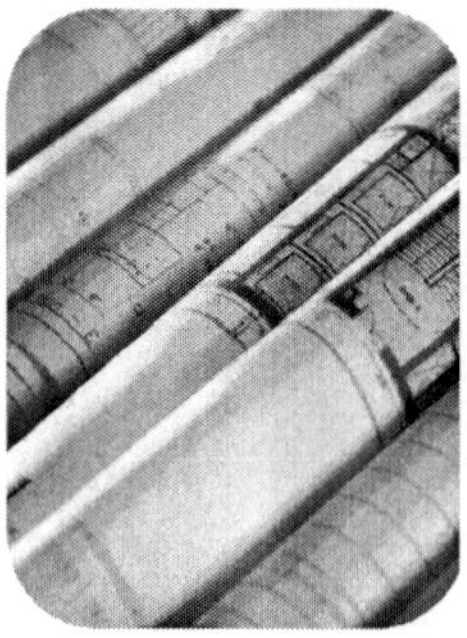

This module will introduce you to the basics of construction drawings. Drawings are both the way we communicate what is to be made, and the way we communicate how a machine was made. The different lines and symbols you will learn will all be necessary information in your career. You will learn about architect's and engineer's scales, and how to use them to understand drawings. You will also learn about reading circuit diagrams, so that you will understand the circuits of machinery you work on.

Objectives

When you have completed this module, you will be able to do the following:

1. Explain the basic layout of a blueprint.
2. Describe the information included in the title block of a blueprint.
3. Identify the types of lines used on blueprints.
4. Identify common symbols used on blueprints.
5. Understand the use of architect's and engineer's scales.
6. Demonstrate the use of an architect's scale.

Trade Terms

Architectural drawings
Block diagram
Blueprint
Construction drawings
Detail drawing
Dimensions
Electrical drawing
Elevation drawing
Exploded diagram
Floor plan
One-line diagram
Plan view
Scale
Schedule
Schematic diagram
Sectional view
Site plan
Written specifications

Required Trainee Materials

1. Pencil and paper
2. Appropriate personal protective equipment
3. Scientific calculator
4. Sample blueprints

Prerequisites

Before you begin this module, it is recommended that you successfully complete *Core Curriculum;* and *Industrial Maintenance E & I Technician Level One,* Modules 40101-07 through 40106-07.

This course map shows all of the modules in the first level of the *Industrial Maintenance E & I Technician* curriculum. The suggested training order begins at the bottom and proceeds up. Skill levels increase as you advance on the course map. The local Training Program Sponsor may adjust the training order.

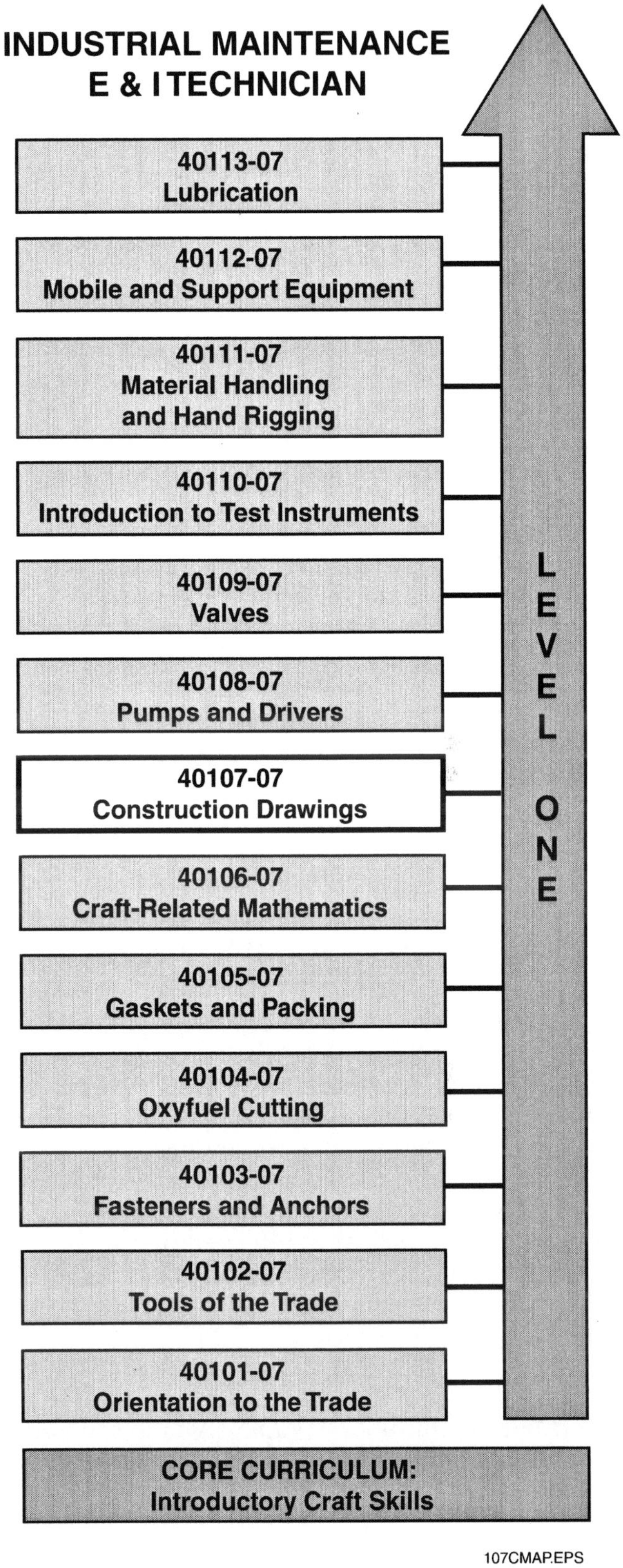

107CMAP.EPS

1.0.0 ◆ INTRODUCTION

In all large construction projects and in many of the smaller ones, a drawing set that includes site, architectural, mechanical, electrical, and structural drawings is normally prepared as required. An architect and, if necessary, special engineers, subcontractors, or consultants are employed to prepare the drawing set and specification along with any additionally required detail drawings. The drawing set, sometimes called architectural, mechanical, electrical (AME) or architectural, mechanical, electrical, structural (AMES) drawings, is used for bidding and during construction. When construction is complete and all in-process changes have been incorporated, the drawing set is referred to as the as-built drawings for the project. **Architectural drawings**, identified as A drawings in the sheet number of the title blocks for the drawing set, include the following elements:

- A **site plan** indicating the location of the building on the property
- **Floor plans** showing the walls and partitions for each floor or level
- Elevations of all exterior faces of the building
- Several vertical cross sections to indicate clearly the various floor levels and details of the footings, foundation, walls, floors, ceilings, and roof construction
- Large-scale **detail drawings** showing such construction details as required

If not covered in the A drawings, structural drawings, identified as S drawings, may be included. These drawings provide the necessary detail for structural support components such as footings, foundations, floors, ceilings, and other loadbearing elements, including concrete or structural steel supports. **Electrical drawings**, identified as E drawings, show the requirements for power, lighting, and other special electrical systems. While traditionally included in the E drawings, drawings for special electrical systems, such as alarm, sound, video, or security systems, also may be prepared as separate drawings. This is because of the technical complexity of the circuits, especially for large projects. Mechanical drawings, identified as M or ME drawings, include the heating, ventilating, air conditioning (HVAC), sprinkler, instrumentation, and plumbing system drawings.

> **NOTE**
>
> Years ago, blueprints were created by placing a hand drawing against light-sensitive paper and then exposing it to ultraviolet light. The light would turn the paper blue except where lines were drawn on the original. The light-sensitive paper was then developed, and the resulting print had white lines against a blue background. Modern blueprints are usually blue or black lines against a white background and are generated using computer-aided design programs. Newer programs offer three-dimensional modeling and other enhanced features.

1.1.0 Site Plans

Site plans provide a bird's-eye view of the building site. Site plans feature the property boundaries, the existing contour lines, the new contour lines (after grading), the location of the building on the property, new and existing roadways, all utility lines, and other pertinent details. The drawing **scale** is also shown. Descriptive notes may also be found on the site (plot) plan listing names of adjacent property owners, the land surveyor, and the date of the survey. A legend or symbol list is also included so that anyone who must work with the site plan can readily read the information. See *Figure 1.*

1.2.0 Floor Plans

The **plan view** of any object is a drawing showing the outline and all details as seen when looking directly down on the object. It shows only two **dimensions**, length and width. The floor plan of a building is drawn as if a horizontal cut were made through the building at about window height, and then the top portion removed to reveal the bottom part. See *Figure 2.*

If a plan view of a home's basement is needed, the part of the house above the middle of the basement windows is imagined to be cut away. By looking down on the uncovered portion, every detail and partition can be seen. Likewise, imagine the part above the middle of the first floor windows being cut away. A drawing that looks straight down at the remaining part would be called the first floor plan or lower level. A cut through the second floor windows would be called the second floor plan or upper level. See *Figure 3.*

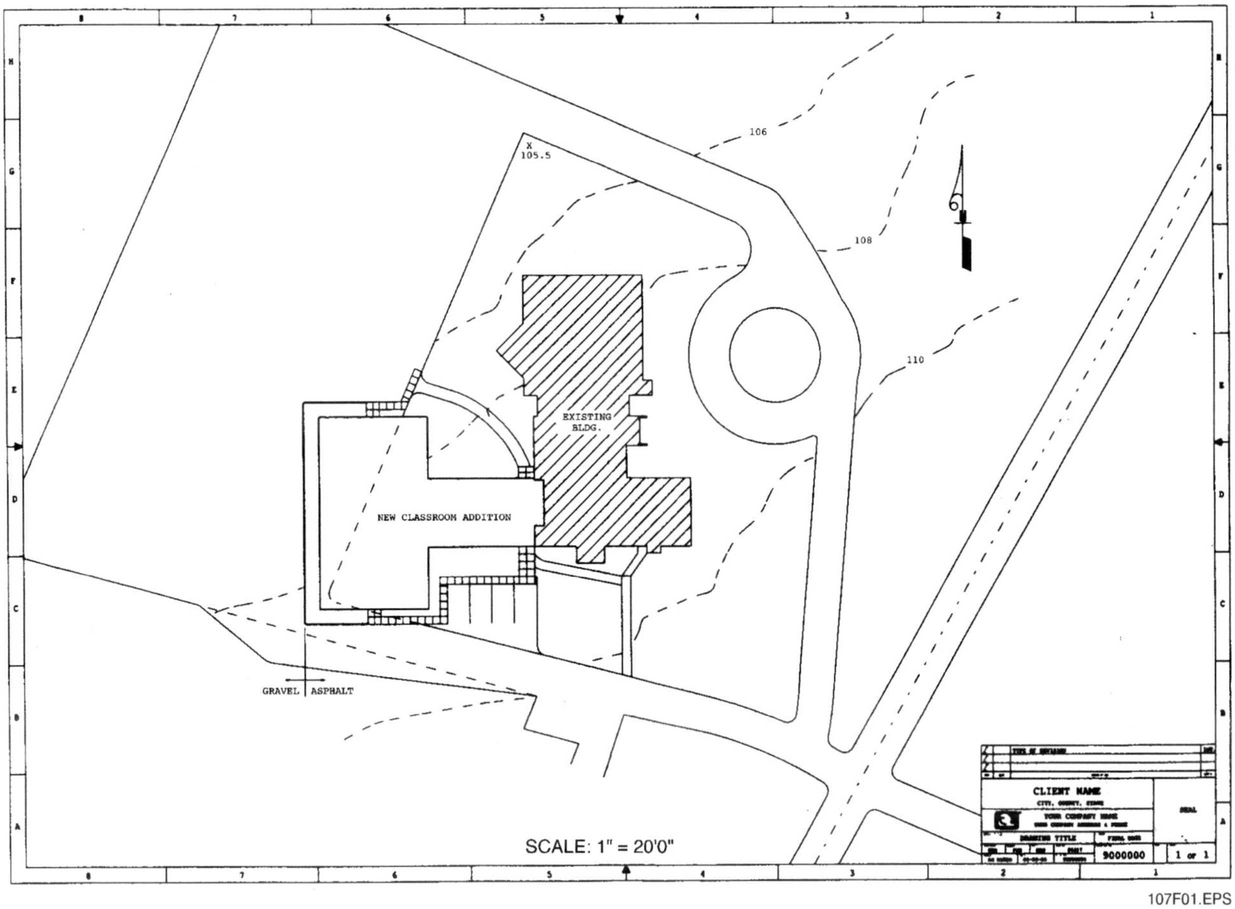

Figure 1 ◆ Typical site plan.

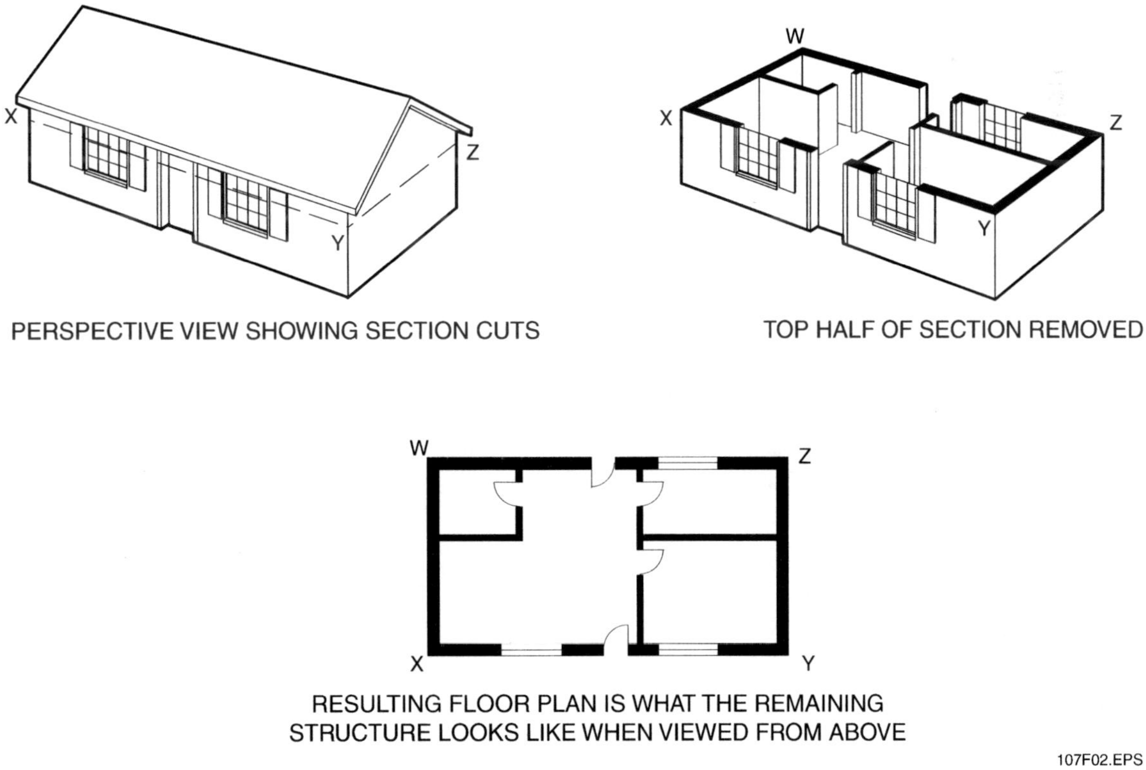

Figure 2 ◆ Principles of floor plan layout.

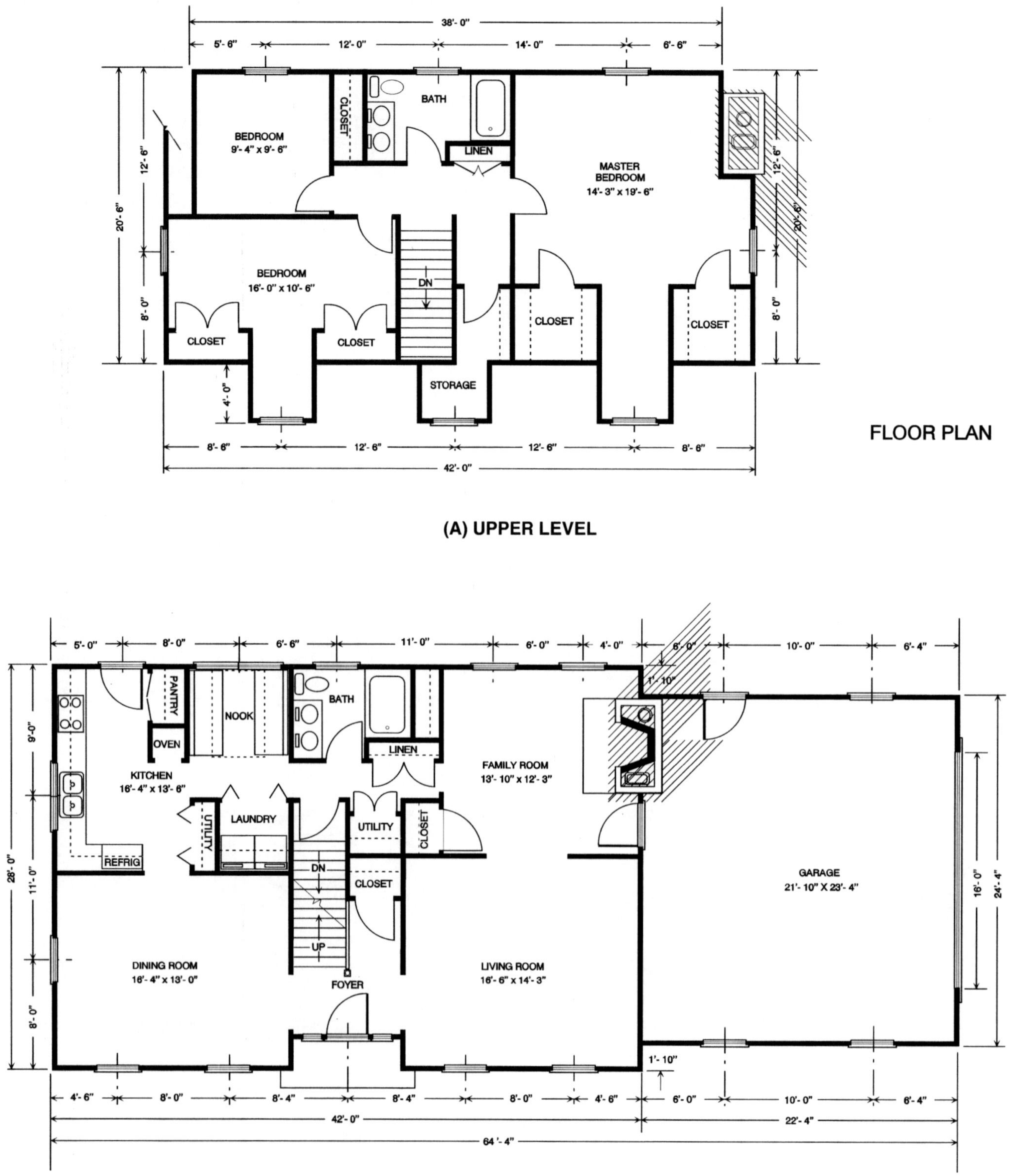

(A) UPPER LEVEL

(B) LOWER LEVEL

107F03.EPS

Figure 3 ◆ Floor plans of a building.

1.3.0 Elevations

The elevation is an outline of an object that shows heights and may show the length or width of a particular side, but not depth. *Figures 4* and *5* show **elevation drawings** for a building.

> **NOTE**
> These elevation drawings show the heights of windows, doors, and porches, the pitch of roofs, and so on because all of these measurements cannot be shown conveniently on floor plans.

1.4.0 Sections

A section or **sectional view** (*Figure 6*) is a cutaway view that allows the viewer to see the inside of a structure. The point on the plan or elevation showing where the imaginary cut has been made is indicated by the section line, which is usually a dashed line. The section line shows the location of the section on the plan or elevation. It is necessary to know which of the cutaway parts is represented in the sectional drawing. To show this, arrow points are placed at the ends of the section lines.

Section views provide important information not shown on other types of drawings. For example, section views show the structural members and materials used inside the walls and on the outside wall surfaces. The height, thickness, and shape of the walls are shown, all of which are important in the selection and installation of outlets and fixtures.

In architectural drawings, it is often necessary to show more than one section on the same drawing. The different section lines must be distinguished by letters, numbers, or other designations placed at the ends of the lines. These section letters are generally large so as to stand out on the drawings. To further avoid confusion, the same letter is usually placed at

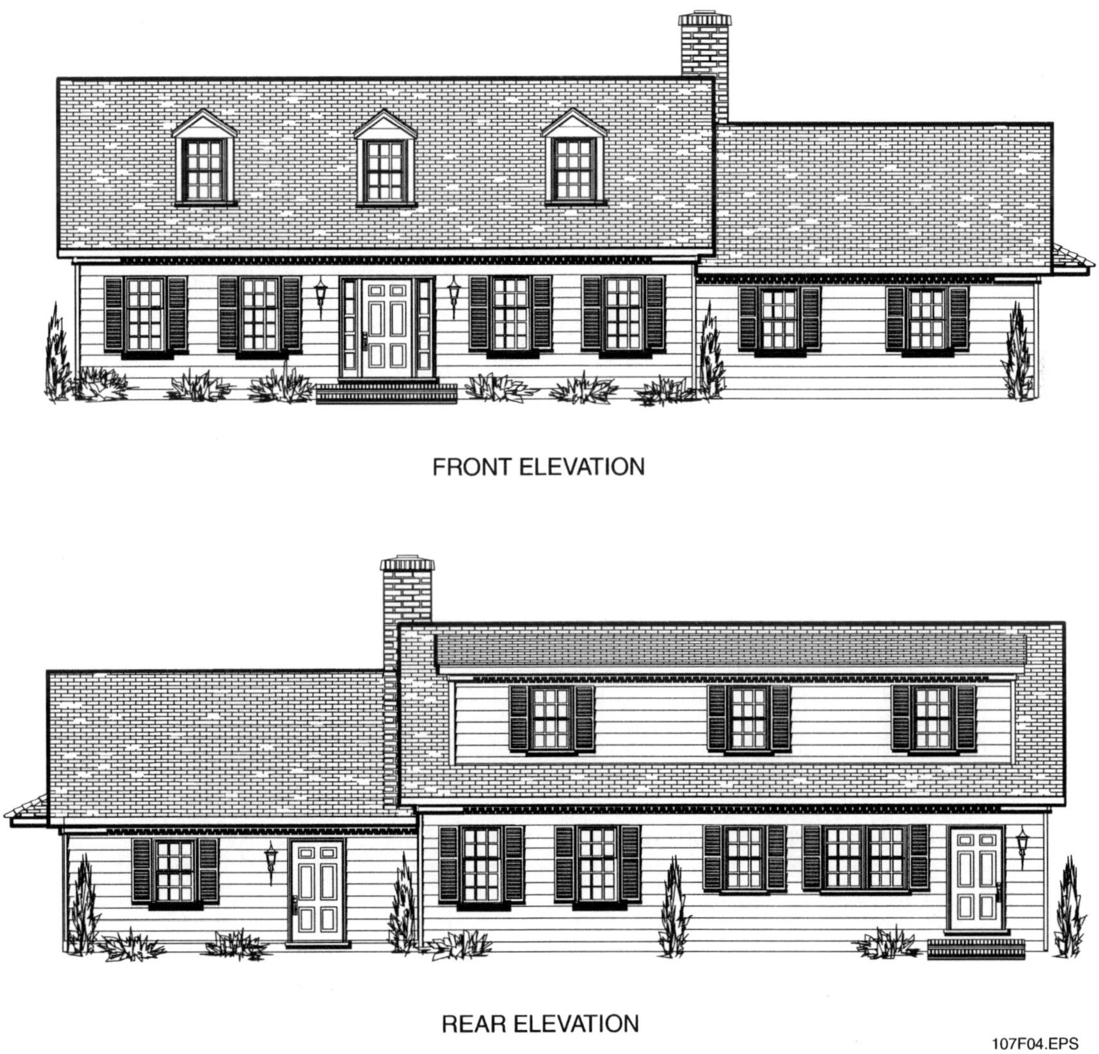

Figure 4 ◆ Front and rear elevations.

each end of the section line. The section is named according to these letters; for example, Section A-A, Section B-B, and so forth.

A longitudinal section is taken lengthwise while a cross section is usually taken straight across the width of an object. Sometimes, however, a section is not taken along one straight line. It is often taken along a zigzag line to show important parts of the object.

A sectional view, as applied to architectural drawings, is a drawing showing the building, or portion of a building, as though it were cut through on some imaginary line. This line may be either vertical (straight up and down) or horizontal. Wall sections are nearly always made vertically so that the cut edge is exposed from top to bottom. In some ways, the wall section is one of the most important of all the drawings to construction

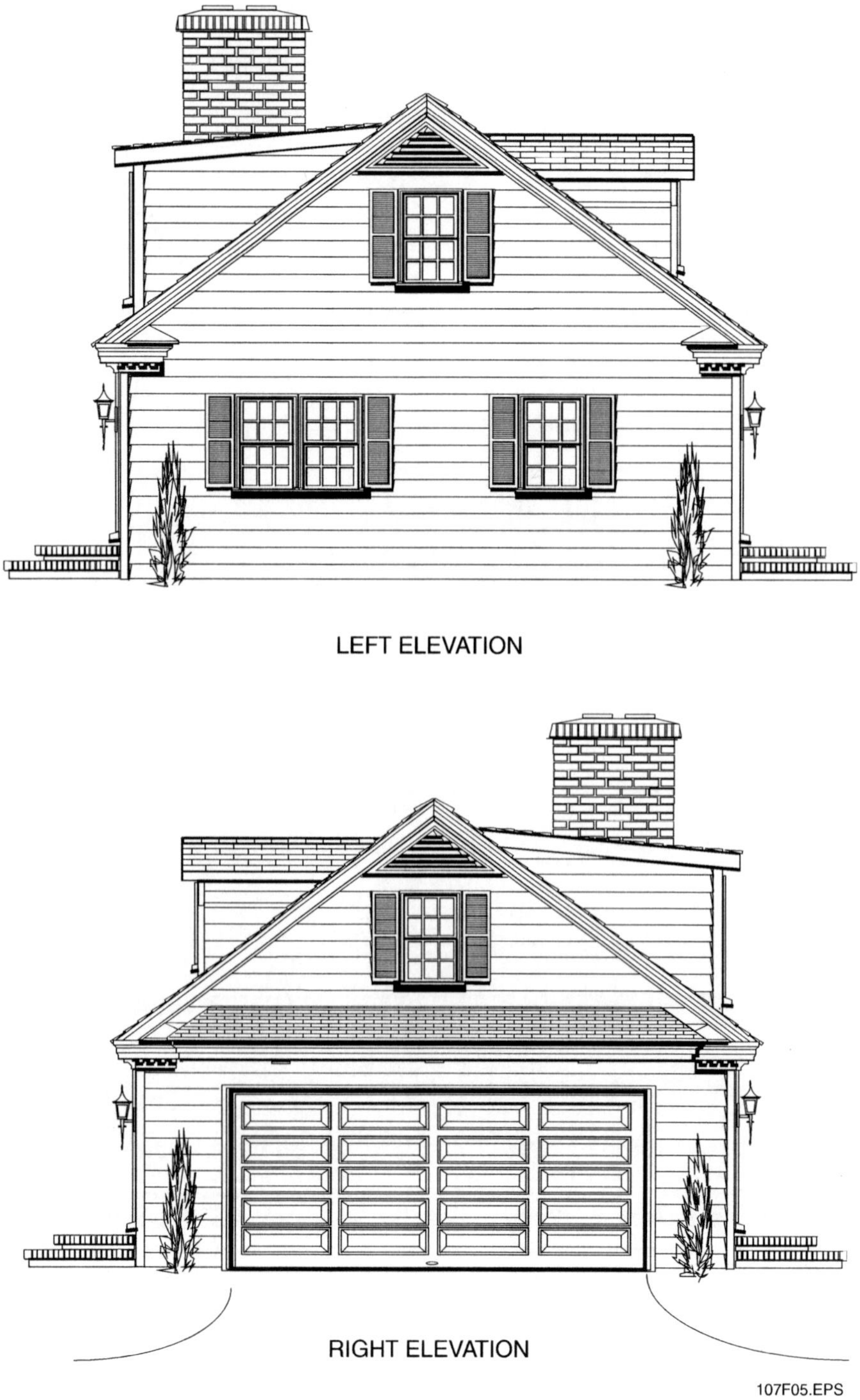

Figure 5 ◆ Left and right elevations.

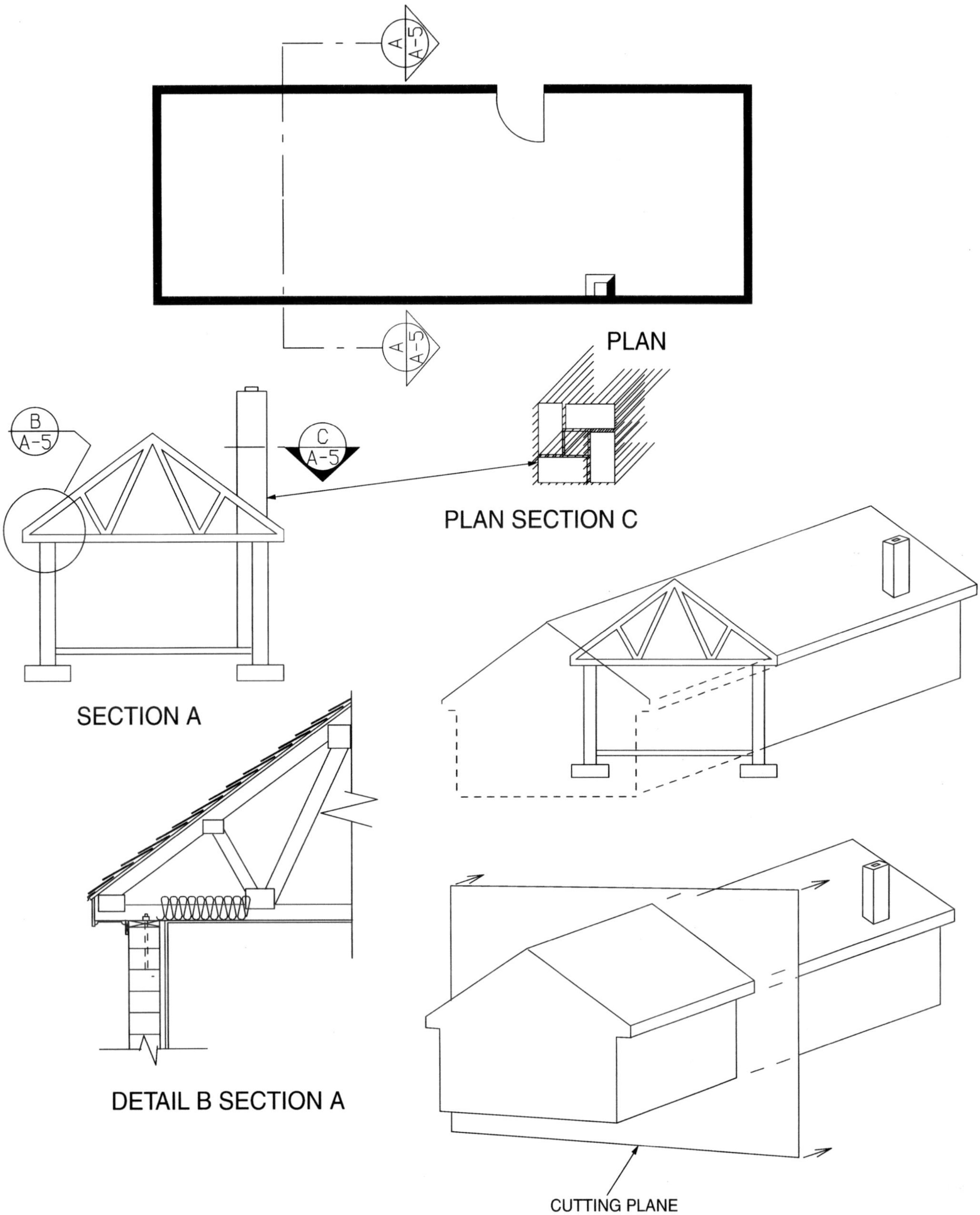

Figure 6 ◆ Sectional drawing.

workers, because it answers the questions as to how a structure should be built. The floor plans of a building show how each floor is arranged, but the wall sections tell how each part is constructed. Also, wall sections usually specify the building material.

On the drawings you will find callouts for details, usually enclosed in a circle, as shown in *Figure 6*. Where the callout does not have a directional arrow and is attached as a callout to a closed space on the drawing, it is a reference to a detail, a large-scale drawing on the area inside the closed line. On Section A, the Detail B is labeled. The line in the middle of the callout separates the detail name from the page on which it is shown; in this case, page A-5.

Where the callout has a directional arrow, that is, the direction you would be looking to see the section shown. An example is Detail C, which shows a vertical cross-section of the chimney on Section A. Note that the detail is to be found on drawing A-5.

> **NOTE**
> When reading a drawing, find the north arrow to orient yourself to the structure. Knowing where north is enables you to accurately describe locations of walls and other parts of the building.

1.5.0 Construction Drawings

Construction drawings show in a clear, concise manner exactly what is required of the mechanic. The amount of data shown on such drawings should be sufficient, but not overdone. A shop drawing, for example, may contain details of only one piece of equipment, while a set of working drawings for an industrial installation may contain dozens of drawing sheets detailing the electrical system for lighting and power, along with equipment, motor controls, wiring diagrams, **schematic diagrams**, equipment **schedules**, and other pertinent data.

In general, the working drawings for a given project serve three distinct functions:

- They provide contractors with an exact description of the project so that materials and labor may be estimated to project a total cost of the project for bidding purposes.
- They give workers installation instructions for the mechanical and electrical systems.
- They provide a map of the mechanical and electrical systems once the job is completed to aid in maintenance and troubleshooting for years.

Drawings from consulting engineering firms will vary in quality from sketchy, incomplete drawings to neat, precise drawings that are easy to understand. Few, however, will cover every detail of the mechanical and electrical systems. Therefore, a good knowledge of installation practices must go hand-in-hand with interpreting working drawings.

Some system contractors have drafters prepare special supplemental drawings for use by the contractors' employees. On certain projects, these supplemental drawings can save supervision time in the field once the project has begun.

2.0.0 ◆ BLUEPRINT LAYOUT

Although a strong effort has been made to standardize drawing practices in the building construction industry, **blueprints** prepared by different architectural or engineering firms will rarely be identical. Similarities, however, exist between most sets of blueprints, and with a little experience, you should have no trouble interpreting any set of drawings that might be encountered.

Most drawings used for building construction projects are drawn on sheets ranging from 11" × 17" to 24" × 36" in size. Each drawing sheet has border lines framing the overall drawing and one or more title blocks, as shown in *Figure 7*. The type and size of title blocks varies with each firm preparing the drawings. In addition, some drawing sheets will also contain a revision block near the title block, and perhaps an approval block. This information is normally found on each drawing sheet, regardless of the type of project or the information contained on the sheet.

You may hear drawings referred to as A-size, B-size, and so forth. These sizes refer to standard sheet dimensions used for construction drawings. *Table 1* shows the dimensions of the various sheets. When drawings are prepared on CAD systems, they are usually printed by plotters that are fed by long rolls of special paper. The standard widths for plotters are 24, 36, and 42 inches.

Table 1 Standard Drawing Dimensions

Size	Dimensions (inches)
A	8½ × 11
B	11 × 17
C	17 × 22
D	24 × 36
E	30 × 42
F	34 × 44

107T01.EPS

> **NOTE**
> Always treat a drawing set with care. It is best to keep two sets, one for the office and one for field use. After you use a sheet from a set of drawings, be sure to refold the sheet with the title block facing up.

2.1.0 Title Block

The architect's title block for a blueprint is usually boxed in the lower right-hand corner of the drawing sheet. The size of the block varies with the size of the drawing and with the information required. See *Figure 8*.

In general, the title block of a construction drawing should contain the following information:

- Name of the project
- Address of the project
- Name of the owner or client
- Name of the architectural firm
- Date of completion
- Scale(s)
- Initials of the drafter, checker, and designer, with dates under each
- Job number
- Sheet number
- General description of the drawing

Every architectural firm has its own standard for drawing titles, and they are often preprinted directly on the tracing paper or else printed on a sticker, which is placed on the drawing.

Often, the consulting engineering firm will also be listed, which means that an additional title block will be applied to the drawing, usually next to the architect's title block. *Figure 9* shows completed architectural and engineering title blocks as they appear on an actual drawing.

Always refer to the title block for scale, and be aware that it often changes from drawing to drawing within the same set. Also check for the current revision, and when you replace a drawing with a new revision, be sure to remove and file the older version.

Once you learn how to interpret blueprints, you can apply that knowledge to any type of construction, from simple residential applications to large industrial complexes.

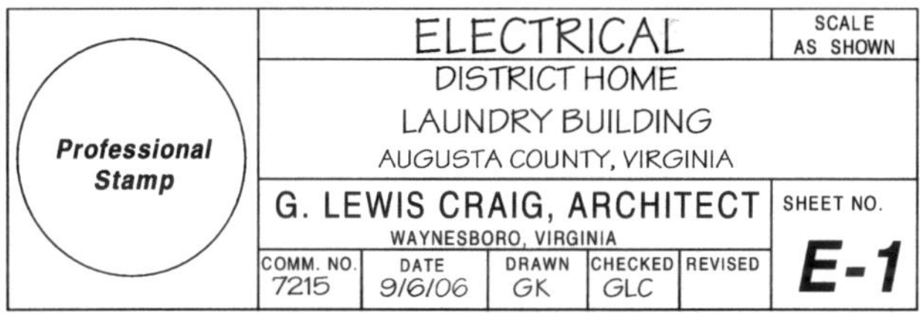

107F08.EPS

Figure 8 ◆ Typical architect's title block.

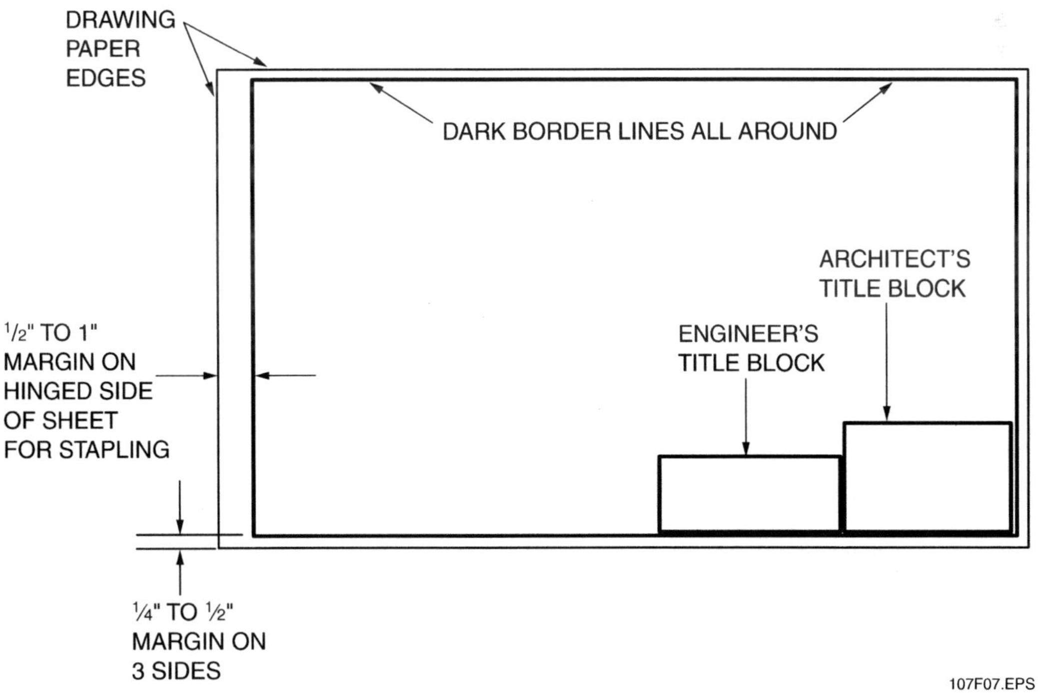

107F07.EPS

Figure 7 ◆ Typical blueprint layout.

2.2.0 Approval Block

The approval block, in most cases, will appear on the drawing sheet as shown in *Figure 10*. The various types of approval blocks (drawn, checked, etc.) will be initialed by the appropriate personnel. This type of approval block is usually part of the title block and appears on each drawing sheet.

On some projects, authorized signatures are required before certain systems may be installed, or even before the project begins. An approval block such as the one shown in *Figure 11* indicates that all required personnel have checked the drawings for accuracy and that the set meets with everyone's approval. Such an approval block usually appears on the front sheet of the blueprint set and may include the following information:

- *Professional stamp* – Registered seal of approval by the architect or consulting engineer
- *Design supervisor* – Signature of the person who is overseeing the design
- *Drawn (by)* – Signature or initials of the person who drafted the drawing and the date it was completed
- *Checked (by)* – Signature or initials of the person who reviewed the drawing and the date of approval
- *Approved* – Signature or initials of the architect/engineer and the date of the approval
- *Owner's approval* – Signature of the project owner or the owner's representative along with the date signed

2.3.0 Revision Block

Sometimes construction drawings have to be partially redrawn or modified during the construction of a project. It is extremely important that such modifications are noted and dated on the drawings to ensure that the workers have an up-to-date set of drawings from which to work. In some situations, sufficient space is left near the title block for dates and descriptions of revisions, as shown in *Figure 12*. In other cases, a revision block is provided near the title block, as shown in *Figure 13*.

COMM. NO.	DATE	DRAWN	CHECKED	REVISED
7215	9/6/06	GK	GLC	

107F10.EPS

Figure 10 ◆ Typical approval block.

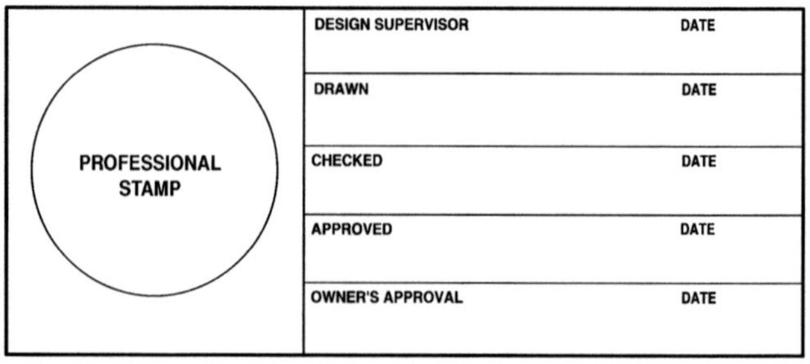

107F11.EPS

Figure 11 ◆ Alternate approval block.

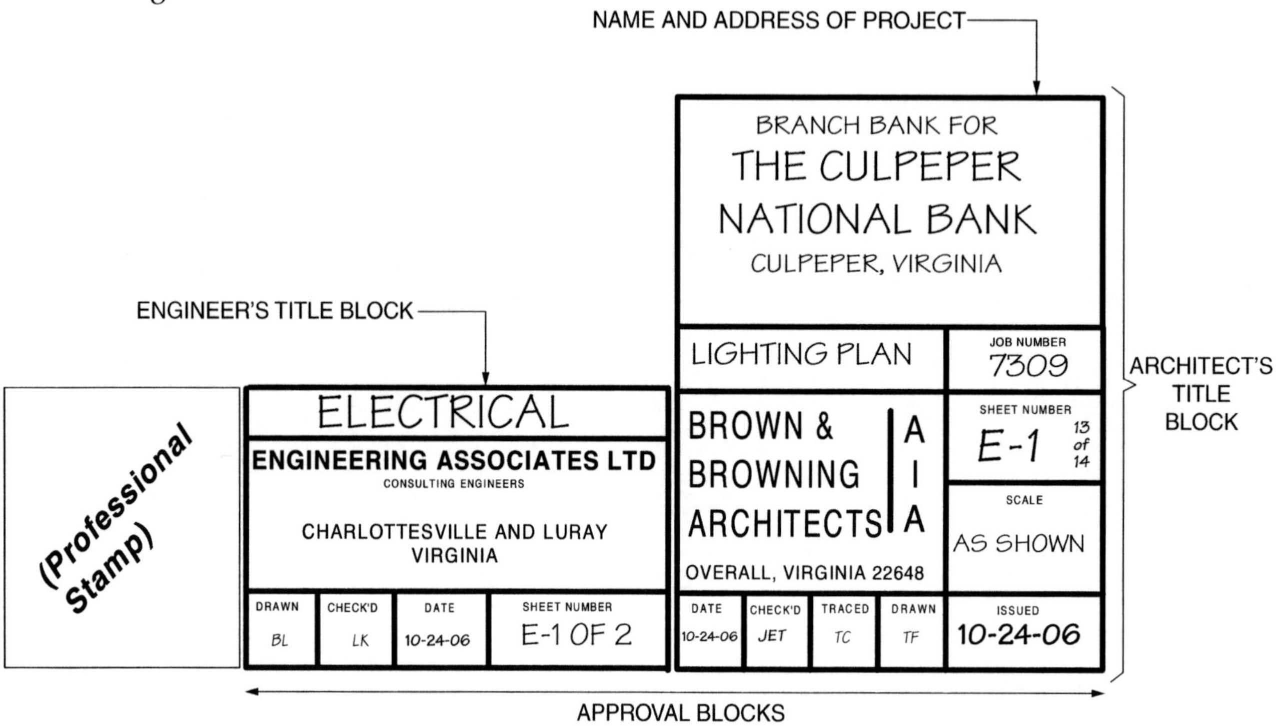

107F09.EPS

Figure 9 ◆ Title blocks.

> **NOTE**
> Architects, engineers, designers, and drafters have their own methods of showing revisions, so expect to find deviations from those shown here.

> **CAUTION**
> When a set of working drawings has been revised, always make certain that the most up-to-date set is used for all future layout work. Either destroy the old, obsolete set of drawings or else clearly mark on the affected sheets, *Obsolete Drawing—Do Not Use.* Also, when working with a set of working drawings and written specifications for the first time, thoroughly check each page to see if any revisions or modifications have been made to the originals. Doing so can save much time and expense to all concerned with the project.

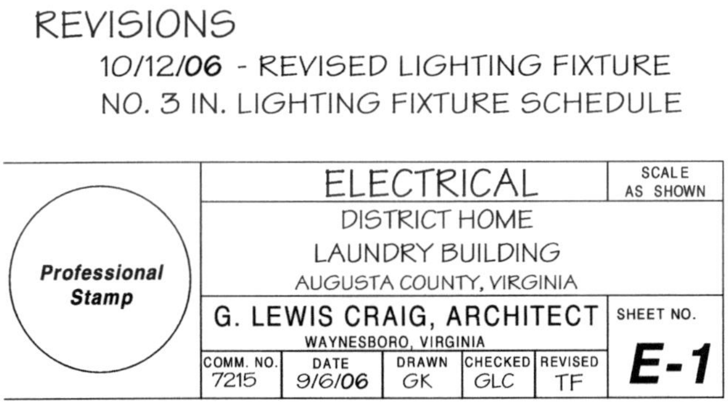

107F12.EPS

Figure 12 ◆ One method of showing revisions on working drawings.

3.0.0 ◆ DRAFTING LINES

Drawings contain many types of drafting lines. To specify the meaning of each type of line, contrasting lines can be made by varying the width of the lines or breaking the lines in a uniform way.

Figure 14 shows common lines used on architectural drawings. However, these lines can vary. Architects and engineers have strived for a common standard for the past century, but unfortunately, their goal has yet to be reached. Therefore, you will find variations in lines and symbols from drawing to drawing, so always consult the legend or symbol list when referring to any drawing. Also, carefully inspect each drawing to ensure that line types are used consistently.

The drafting lines shown in *Figure 14* are used as follows:

- *Light full line* – This line is used for section lines, building background (outlines), and similar uses where the object to be drawn is secondary to the system being shown, such as HVAC or electrical.
- *Medium full line* – This type of line is frequently used for hand lettering on drawings. It is also used for some drawing symbols, circuit lines, etc.
- *Heavy full line* – This line is used for borders around title blocks and schedules, and for hand lettering drawing titles. Some types of symbols are frequently drawn with a heavy full line.

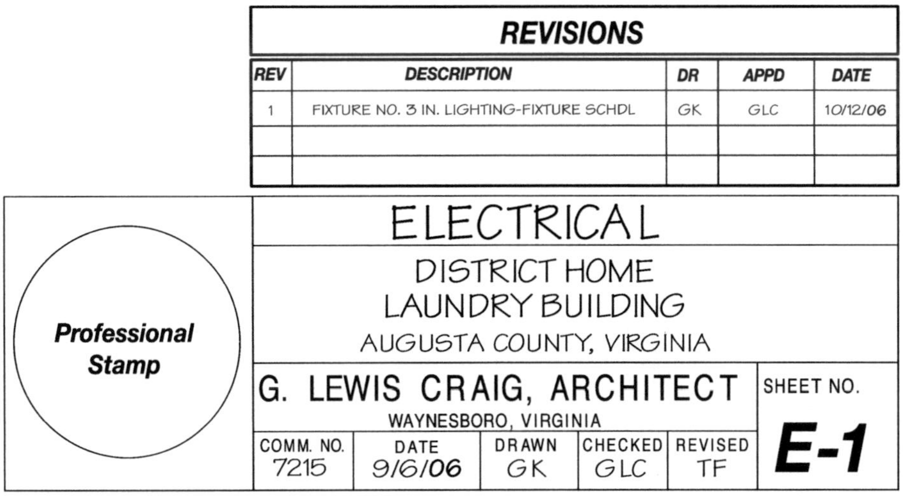

107F13.EPS

Figure 13 ◆ Alternate method of showing revisions on working drawings.

- *Extra heavy full line* – This line is used for border lines on architectural/engineering drawings.
- *Centerline* – A centerline is a broken line made up of alternately spaced long and short dashes. It indicates the centers of objects such as holes, pillars, or fixtures. Sometimes, the centerline indicates the dimensions of a finished floor.
- *Hidden line* – A hidden line consists of a series of short dashes that are closely and evenly spaced. It shows the edges of objects that are not visible in a particular view. The object outlined by hidden lines in one drawing is often fully pictured in another drawing.
- *Dimension line* – These are thin lines used to show the extent and direction of dimensions. The dimension is usually placed in a break inside the dimension lines. Normal practice is to place the dimension lines outside the object's outline. However, it may sometimes be necessary to draw the dimensions inside the outline.
- *Short-break line* – This line is usually drawn freehand and is used for short breaks.
- *Long-break line* – This line, which is drawn partly with a straightedge and partly with freehand zigzags, is used for long breaks.
- *Match line* – This line is used to show the position of the cutting plane. Therefore, it is also called the cutting-plane line. A match or cutting-plane line is a heavy line with long dashes alternating with two short dashes. It is used on drawings of large structures to show where one drawing stops and the next drawing starts.
- *Secondary line* – This line is frequently used to outline pieces of equipment or to indicate reference points of a drawing that are secondary to the drawing's purpose.

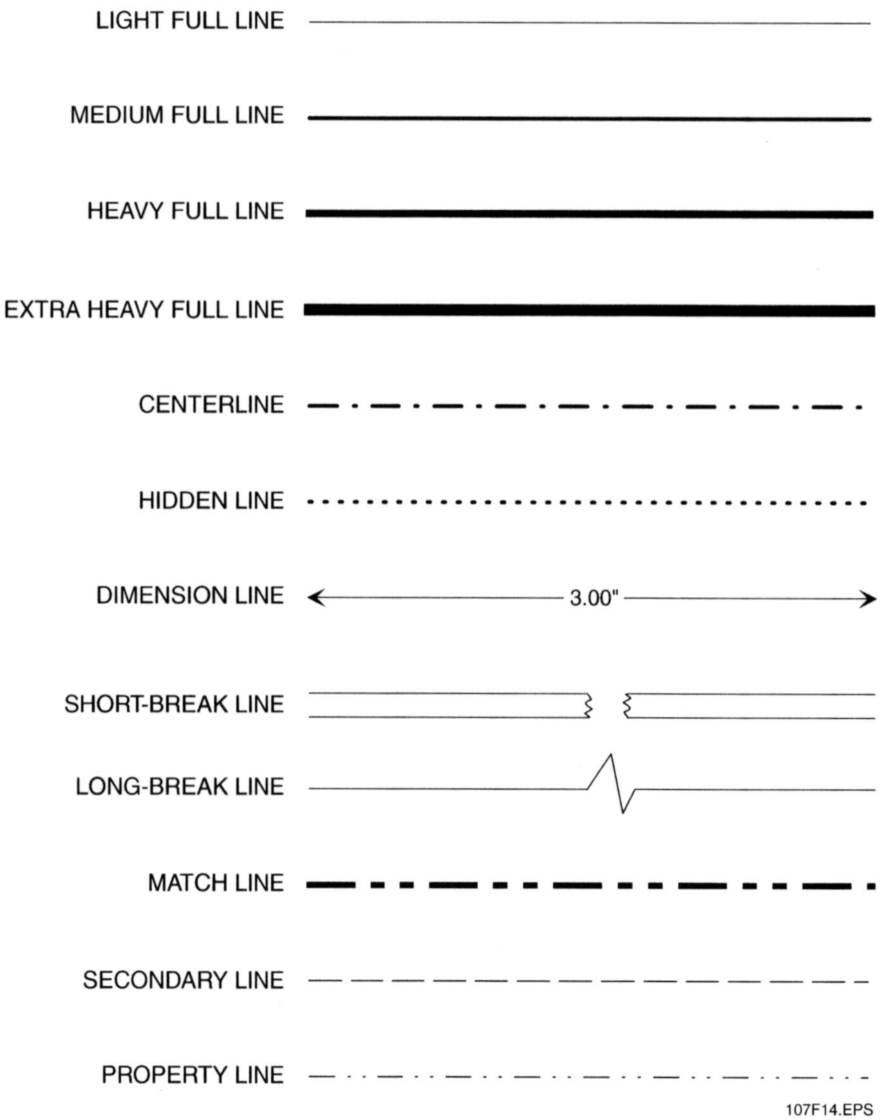

Figure 14 ◆ Typical drafting lines.

- *Property line* – This is a light line made up of one long and two short dashes that are alternately spaced. It indicates land boundaries on the site plan.

Other uses of the lines just mentioned include the following:

- *Extension lines* – Extension lines are lightweight lines that start about 1⁄16 inch away from the edge of an object and extend out. A common use of extension lines is to create a boundary for dimension lines. Dimension lines meet extension lines with arrowheads, slashes, or dots. Extension lines that point from a note or other reference to a particular feature on a drawing are called leaders. They usually end in either an arrowhead or a dot and may include an explanatory note at the end.
- *Section lines* – These are often referred to as cross-hatch lines. Drawn at a 45-degree angle, these lines show where an object has been cut away to reveal the inside.
- *Phantom lines* – Phantom lines are solid, light lines that show where an object will be installed. A future door opening or a future piece of equipment can be shown with phantom lines.

3.1.0 Electrical Drafting Lines

Besides the architectural lines shown in *Figure 14*, consulting electrical engineers, designers, and drafters use additional lines to represent circuits and their related components. Again, these lines may vary from drawing to drawing, so check the symbol list or legend for the exact meaning of lines on the drawing with which you are working. *Figure 15* shows lines used on some electrical drawings.

4.0.0 ◆ CIRCUIT DIAGRAMS

In order to repair and maintain machinery, you will need to be able to read circuit diagrams. A circuit diagram is much like a pipe schematic in that it uses lines to represent connections and symbols to show switches, transistors, and other parts of the electrical controls and power circuits of the machine. *Figure 16* shows some of the symbols used in circuit diagrams. *Figure 17* shows piping symbols. In later modules, you may study the functions of all of these parts, but for now, you can learn to recognize the symbols.

Another source of information on drawings is the abbreviations used to identify and describe components of mechanical systems. The *Appendix* contains a list of some common industrial abbreviations.

NOTE

In an effort to standardize drawings and symbols used in design and construction projects, the Construction Specifications Institute (CSI) released a new Computer Aided Design (CAD) software program in 2005. The software program is entitled *U.S. National CAD Standard (NCS) Version 3.1 – Uniform Drawing System™ (UDS).* This CAD standard allows project drawing sets and drawings prepared by subcontractors to be produced in a uniform format. The software also includes an extensive uniform symbol library. The CAD standard was developed based on understandings between CSI and the following organizations:

- Construction Specifications Canada (CSC)
- American Institute of Architects (AIA)
- National Institute of Building Sciences (NIBS)
- Sheet Metal and Air Conditioning Contractors National Association (SMACNA)
- U.S. General Services Administration (GSA)

Symbols for communication and signal systems, as well as symbols for light and power, are drawn to an appropriate scale and accurately located with respect to the building. This reduces the number of references made to the architectural drawings. Where extreme accuracy is required in

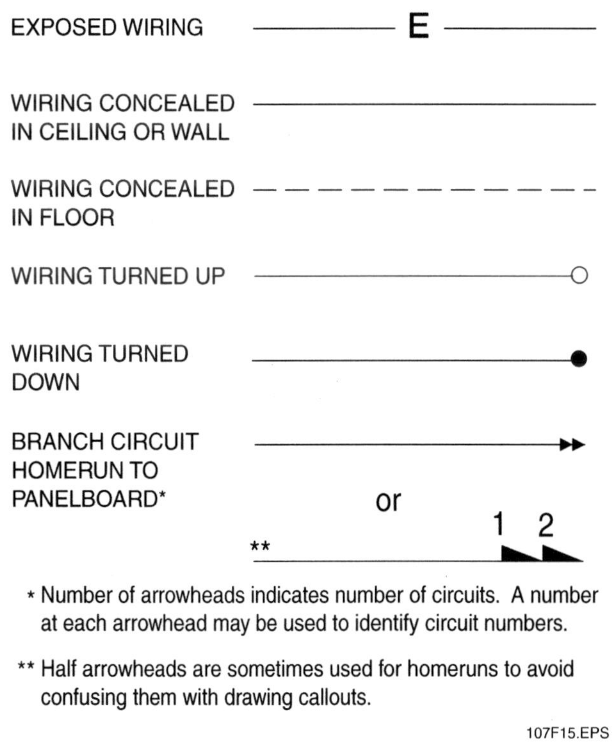

Figure 15 ◆ Electrical drafting lines.

locating outlets and equipment, exact dimensions are given on larger-scale drawings and shown on the plans.

Each different category in a system is usually represented by a basic distinguishing symbol. To further identify items of equipment or outlets in the category, a numeral or other identifying mark is placed within the open basic symbol. In addition, all such individual symbols used on the drawings should be included in the symbol list or legend. The electrical symbols shown in *Figure 18* were modified by a consulting engineering firm for use on a small industrial electrical installation.

5.0.0 ◆ SCALE DRAWINGS

In most drawings, the components are so large that it would be impossible to draw them actual size. Consequently, drawings are made to some reduced scale; that is, all the distances are drawn smaller than the actual dimensions of the object itself, with all dimensions being reduced in the same proportion. For example, if a floor plan of a building is to be drawn to a scale of ¼" = 1'–0", each ¼" on the drawing would equal 1 foot on the building itself; if the scale is ⅛" = 1'–0", each ⅛" on the drawing equals 1 foot on the building, and so forth.

When architectural and engineering drawings are produced, the selected scale is very important. Where dimensions must be held to extreme accuracy, the scale drawings should be made as large as practical with dimension lines added. Where dimensions require only reasonable accuracy, the object may be drawn to a smaller scale (with dimension lines possibly omitted).

In dimensioning drawings, the dimensions written on the drawing are the actual dimensions of the building, not the distances that are measured on the drawing. To further illustrate this point, look at the floor plan in *Figure 19*; it is drawn to a scale of ½" = 1'–0". One of the walls is drawn to an actual length of 3½" on the drawing paper, but since the scale is ½" = 1'–0" and since 3½" contains 7 halves of an inch (7 × ½ = 3½"), the dimension shown on the drawing will therefore be 7'–0" on the actual building.

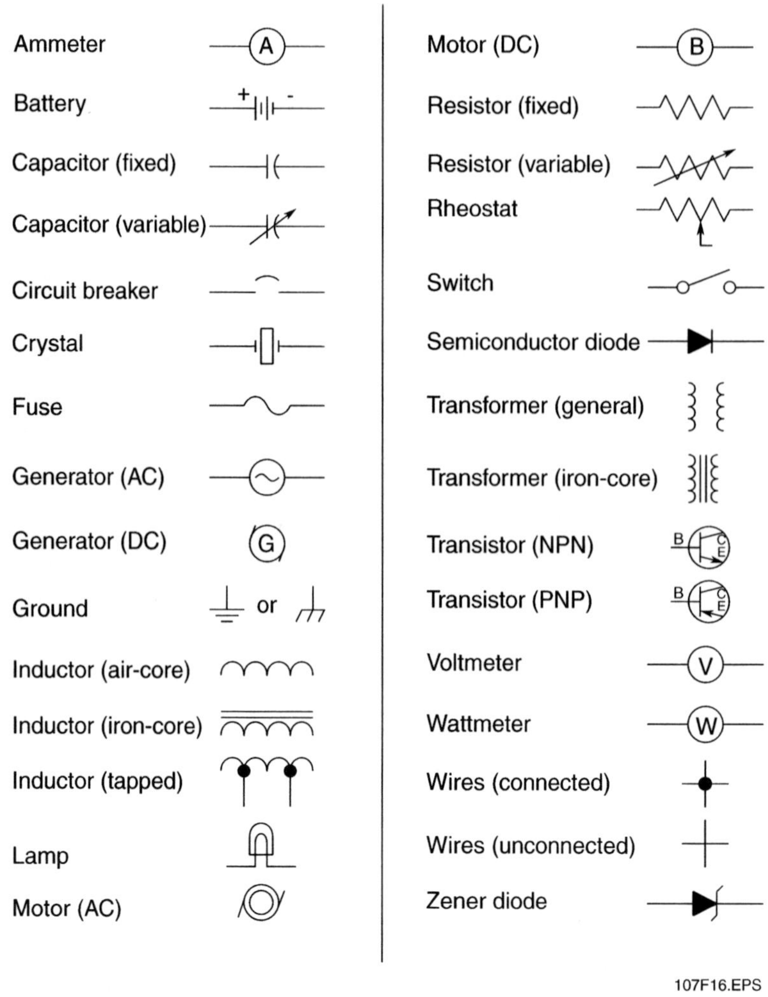

Figure 16 ◆ Standard schematic symbols.

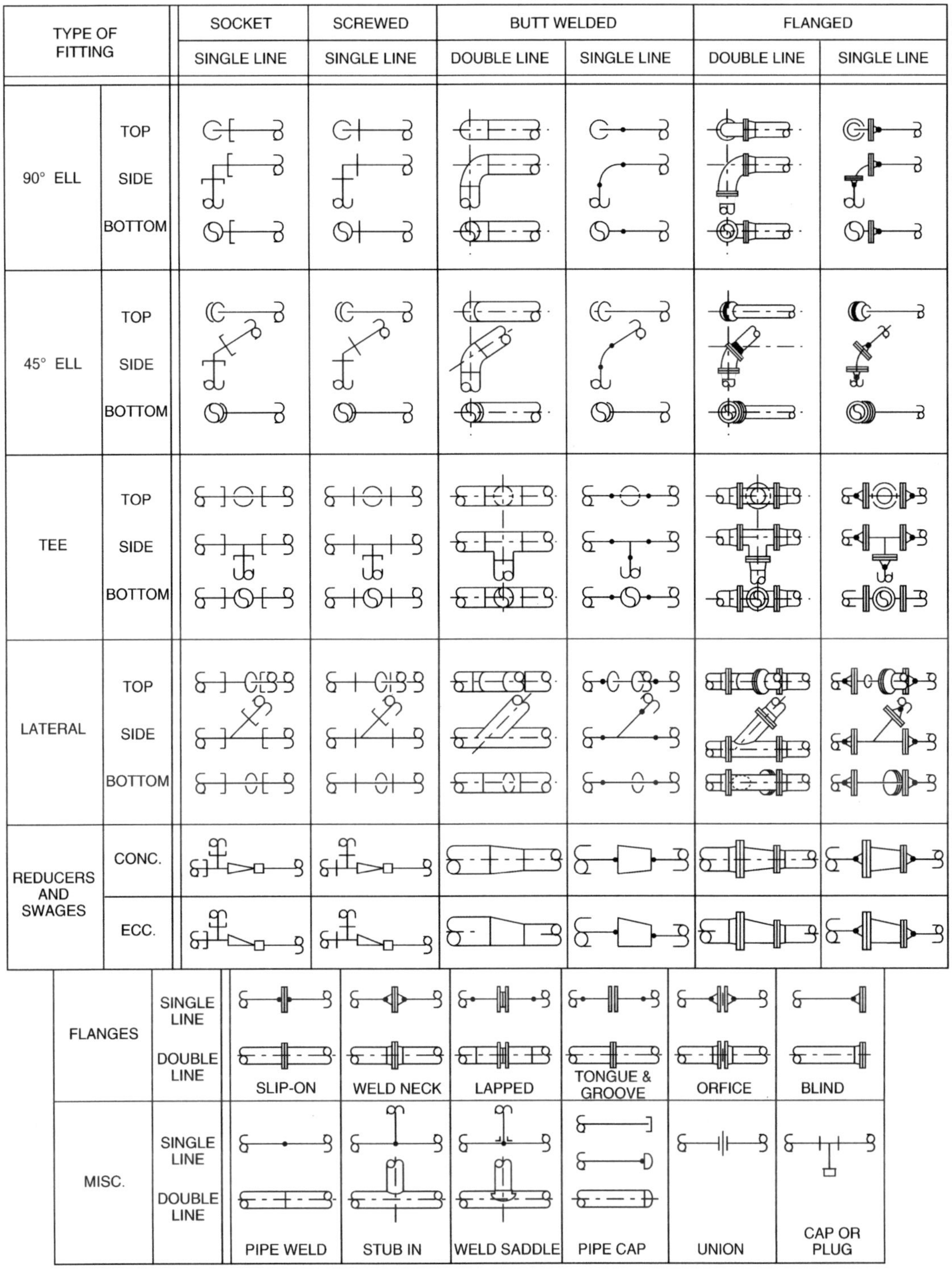

Figure 17 ◆ Piping symbols.

As shown in the previous example, the most common method of reducing all the dimensions (in feet and inches) in the same proportion is to choose a certain distance and let that distance represent one foot. This distance can then be divided into twelve parts, each of which represents an inch. If half inches are required, these twelfths are further subdivided into halves, etc. Now the scale represents the common foot rule with its subdivisions into inches and fractions, except that the scaled foot is smaller than the distance known as a foot and, likewise, its subdivisions are proportionately smaller.

When a measurement is made on the drawing, it is made with the reduced foot rule or scale; when a measurement is made on the building, it is made with the standard foot rule. The most common reduced foot rules or scales used in drawings are the architect's scale and the engineer's scale. Drawings may sometimes be encountered that use a metric scale, but using this scale is similar to using the architect's or engineer's scales.

5.1.0 Architect's Scale

Figure 20 shows two configurations of architect's scales. The one on the top is designed so that 1" = 1'–0", and the one on the bottom has graduations spaced to represent ⅛" = 1'–0". The ⅛" = 1'-0" scale in *Figure 20* is dependent on whether the numbers are read from left to right or from right to left. The ⅛" = 1'-0" scale, read from left to right, becomes the ¼" = 1'-0" when read from right to left.

Note that on the one-inch scale in *Figure 21*, the longer marks to the right of the zero (with a numeral beneath) represent feet. Therefore, the distance between the zero and the numeral 1 equals one foot. The shorter mark between the zero and 1 represents ½ of a foot, or six inches.

Referring again to *Figure 21*, look at the marks to the left of the zero. The numbered marks are spaced three scaled inches apart and have the numerals 0, 3, 6, and 9 for use as reference points. The other lines of the same length also represent scaled inches, but are not marked with numerals. In use, you can count the number of long marks to the left of the zero to find the number of inches, but after some practice, you will be able to tell the exact measurement at a glance. For example, the measurement A represents five inches because it is the fifth inch mark to the left of the zero; it is also one inch mark short of the six-inch line on the scale.

The lines that are shorter than the inch line are the half-inch lines. On smaller scales, the basic unit is not divided into as many divisions. For example, the smallest subdivision on some scales represents two inches.

5.1.1 Types of Architect's Scales

Architect's scales are available in several types, but the most common are the triangular scale (*Figure 22*) and the flat scale. The quality of architect's scales varies from cheap plastic scales costing a dollar or two to high-quality, wooden-laminated tools that are calibrated to precise standards.

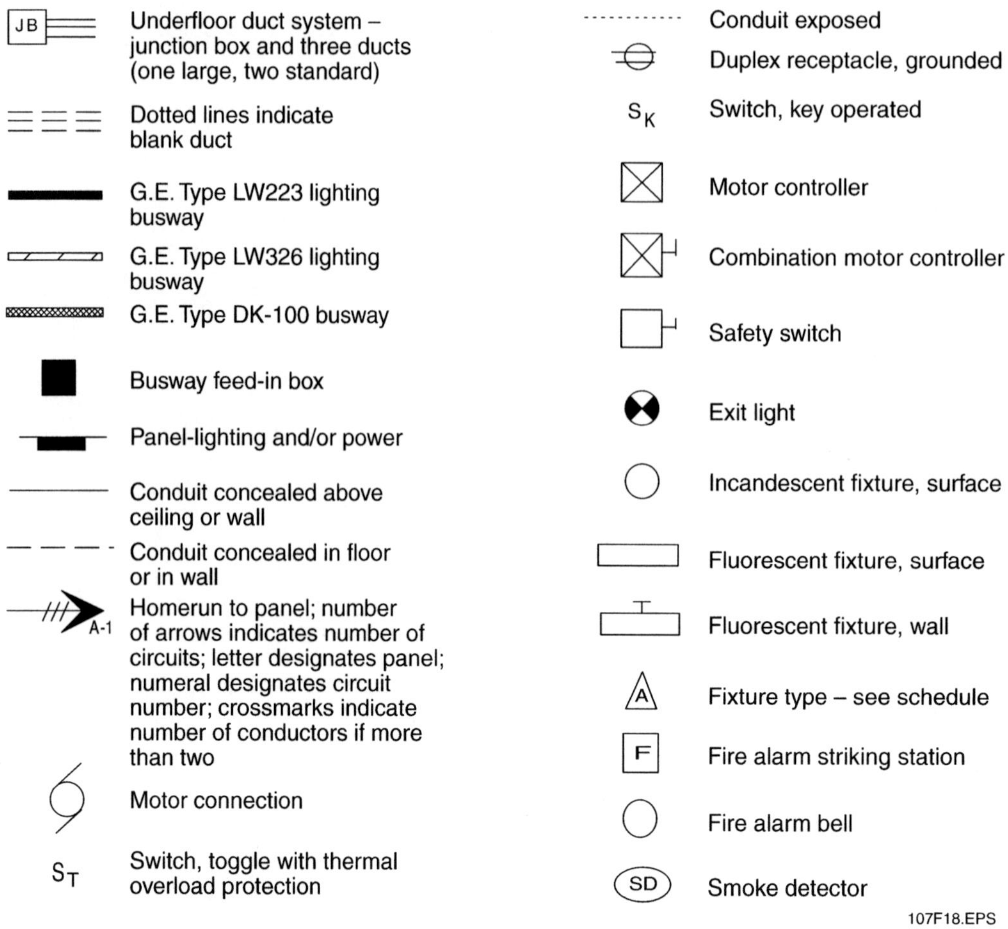

Figure 18 ◆ Electrical symbols used by one consulting engineering firm.

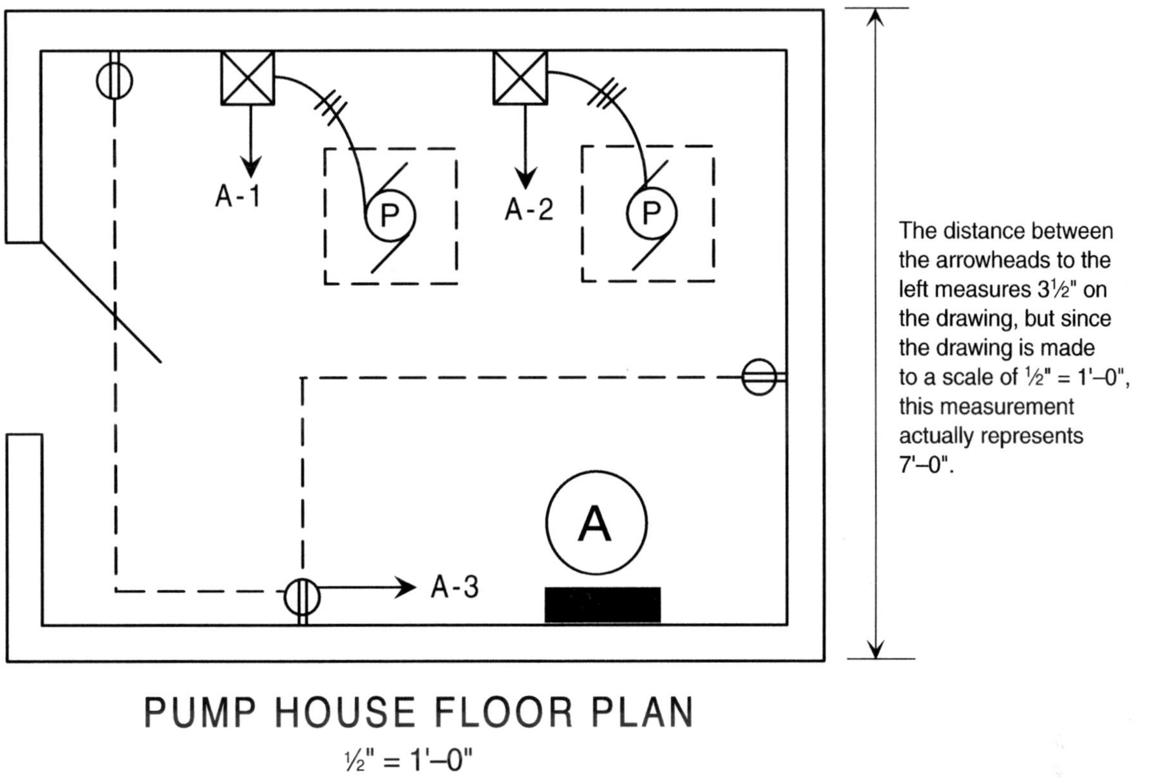

Figure 19 ◆ Typical floor plan showing drawing scale.

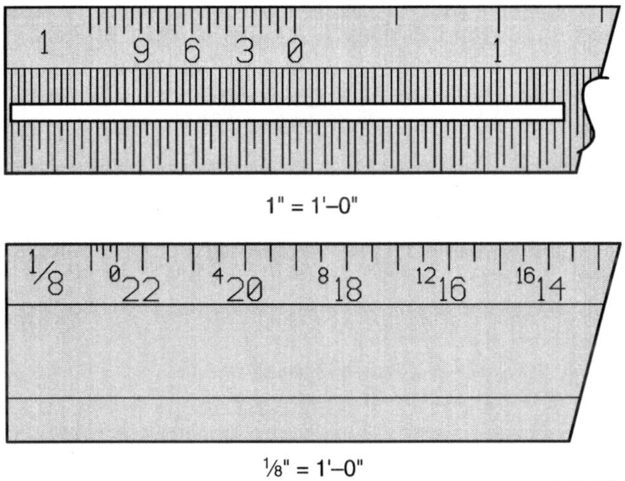

Figure 20 ◆ Two different configurations of architect's scales.

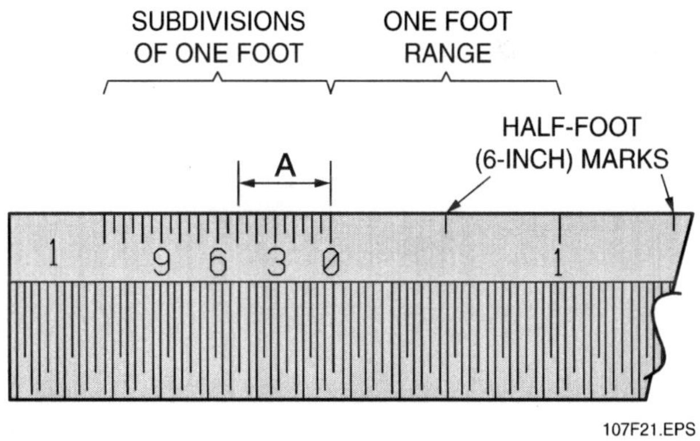

Figure 21 ◆ One-inch architect's scale.

The triangular scale (*Figure 23*) is frequently found in drafting and estimating departments or engineering and electrical contracting firms, while the flat scales are more convenient to carry on the job site.

Triangular architect's scales have 11 different scales as follows:

- Common foot rule (12 inches)
- 3/32" = 1'–0"
- 3/16" = 1'–0"
- 1/8" = 1'–0"
- 1/4" = 1'–0"
- 3/8" = 1'–0"
- 3/4" = 1'–0"
- 1" = 1'–0"
- 1/2" = 1'–0"
- 1 1/2" = 1'–0"
- 3" = 1'–0"

Two separate scales on one face may seem confusing at first, but after some experience, reading these scales becomes second nature.

In all but one of the scales on the triangular architect's scale, each face has one of the scales placed opposite to the other. For example, on the one-inch face, the one-inch scale is read from left to right, starting from the zero mark. The half-inch scale is read from right to left, again starting from the zero mark.

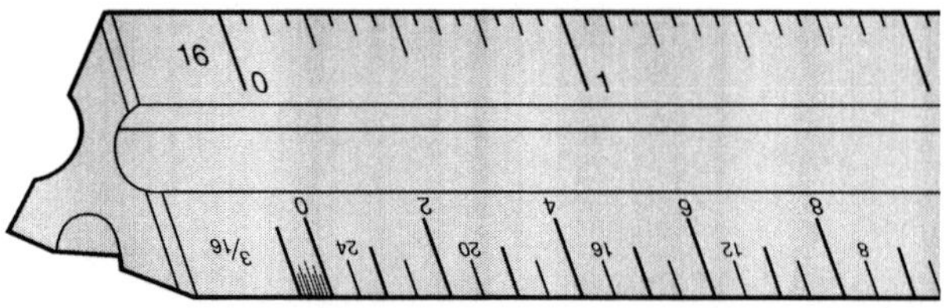

107F22.EPS

Figure 22 ◆ Typical triangular architect's scale.

Figure 23 shows all the scales found on the triangular architect's scale.

The flat architect's scale shown in *Figure 24* is ideal for workers on most projects. It is easily and conveniently carried in the shirt pocket, and the four scales (1/8", 1/4", 1/2", and 1") are adequate for the majority of projects that will be encountered.

The partial floor plan shown in *Figure 24* is drawn to a scale of 1/8" = 1'–0". The dimension in question is found by placing the 1/8" architect's scale on the drawing and reading the figures. It can be seen that the dimension is 24'–6".

Every drawing should have the scale to which it is drawn plainly marked on it as part of the drawing title. However, it is not uncommon to have several different drawings on one blueprint sheet where each drawing has a different scale. Therefore, always check the scale of each different view found on a drawing sheet.

5.2.0 Engineer's Scale

The civil engineer's scale is used in basically the same manner as the architect's scale, with the principal difference being that the graduations on the engineer's scale are decimal units, rather than feet as on the architect's scale.

The engineer's scale is used by placing it on the drawing with the working edge away from the user. The scale is then aligned in the direction of

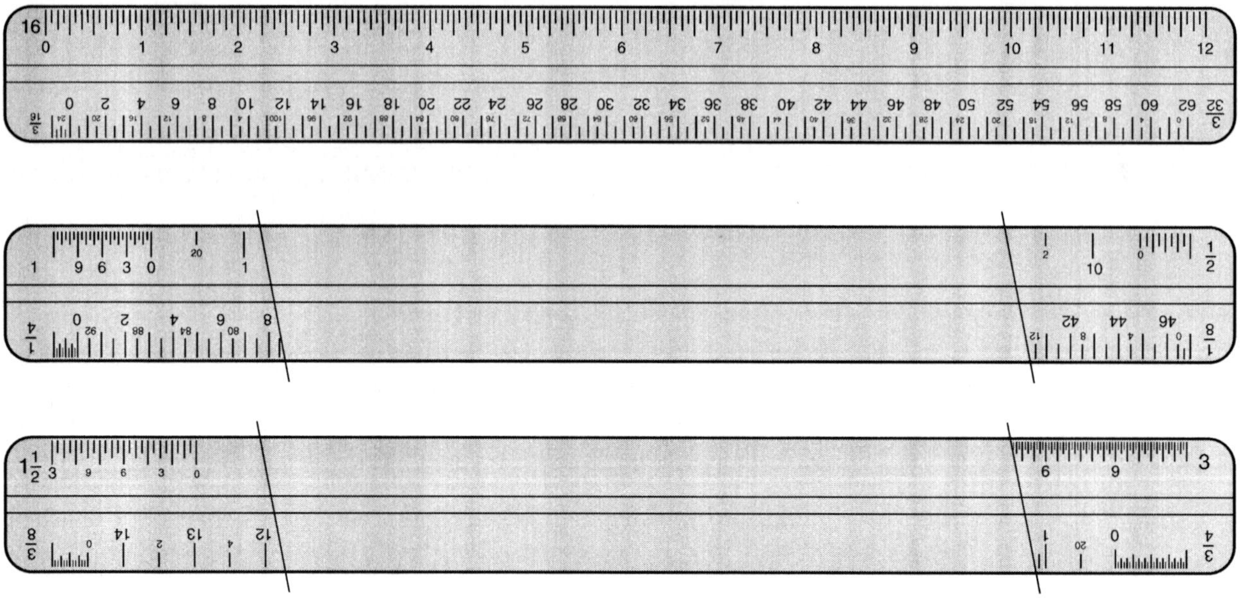

107F23.EPS

Figure 23 ◆ Various scales on a triangular architect's scale.

the required measurement. Then, by looking down at the scale, the dimension is read.

Civil engineer's scales commonly show the following graduations:

- 1" = 10 units
- 1" = 20 units
- 1" = 30 units
- 1" = 40 units
- 1" = 60 units
- 1" = 80 units
- 1" = 100 units

The purpose of this scale is to transfer the relative dimensions of an object to the drawing or vice versa. It is used mainly on site plans to determine distances between property lines, manholes, duct runs, direct-burial cable runs, and the like.

Site plans are drawn to scale using the engineer's scale rather than the architect's scale. On small lots, a scale of 1 inch = 10 feet or 1 inch = 20 feet is used. For a 1:10 scale, this means that one inch (the actual measurement on the drawing) is equal to 10 feet on the land itself.

On larger drawings, where a large area must be covered, the scale could be 1 inch = 100 feet or 1 inch = 1,000 feet, or any other integral power of 10. On drawings with the scale in multiples of 10, the engineering scale marked 10 is used. If the scale is 1 inch = 200 feet, the engineer's scale marked 20 is used, and so on.

Although site plans appear reduced in scale, depending on the size of the object and the size of the drawing sheet to be used, the actual dimensions must be shown on the drawings at all times. When you are reading the drawing plans to scale, think of each dimension in its full size and not in the reduced scale it happens to be on the drawing (*Figure 25*).

5.3.0 Metric Scale

The metric scale (*Figure 26*) is divided into centimeters (cm), with the centimeters divided into 10 divisions for millimeters (mm) or in 20 divisions for half millimeters. Scales are available with metric divisions on one edge while inch divisions are inscribed on the opposite edge. Many contracting firms that deal in international trade have adopted a dual-dimensioning system expressed in both metric and English symbols. Drawings prepared for government projects may also require metric dimensions.

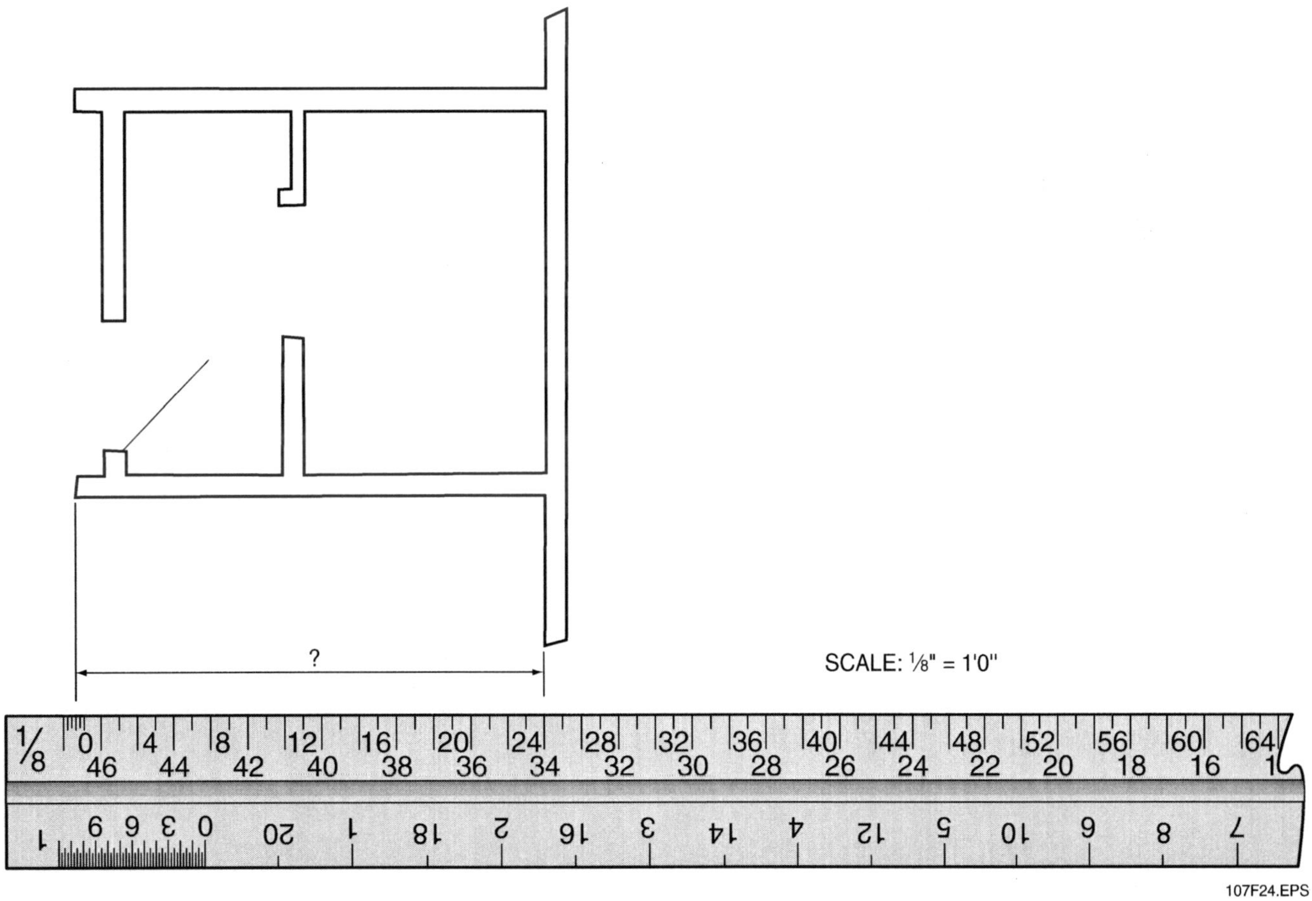

Figure 24 ◆ Using the 1/8" architect's scale to determine the dimensions on a drawing.

6.0.0 ◆ ANALYZING DRAWINGS

The most practical way to learn how to read construction documents is to analyze an existing set of drawings prepared by consulting or industrial engineers. Engineers are responsible for the complete layout of systems for most projects. Drafters then transform the engineer's designs into working drawings, using either manual drafting instruments or computer-aided design (CAD) systems. The following is a brief outline of what usually takes place in the preparation of electrical design and working drawings:

- The engineer meets with the architect and owner to discuss the electrical needs of the building or project and to discuss various recommendations made by all parties.
- After that, an outline of the architect's floor plan is laid out.
- The engineer then calculates the required power and lighting outlets for the project; these are later transferred to the working drawings.
- Schedules are then placed on the drawings to identify various pieces of equipment.
- Wiring diagrams and pipe schematics are made to show the workers how various components are to be connected.
- A symbol list or legend is drafted and shown on the drawings to identify all symbols used to indicate equipment.

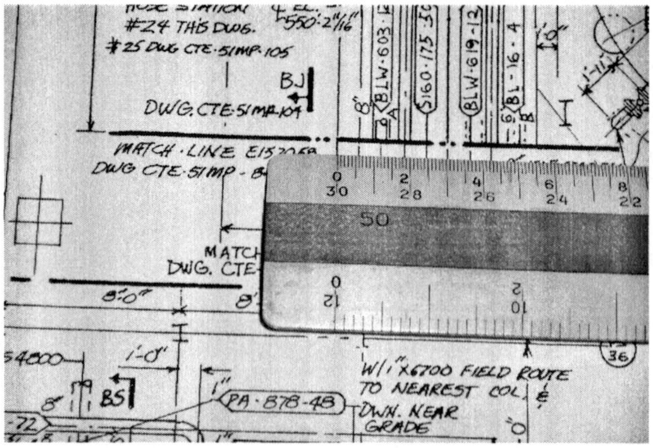

107F25.EPS

Figure 25 ◆ Practical use of the engineer's scale.

- Various large-scale details are included, if necessary, to show exactly what is required to complete the installation.
- Written specifications are then made to give a description of the materials and installation methods.

6.1.0 Development of Site Plans

In general practice, it is usually the owner's responsibility to furnish the architect/engineer with property and topographic surveys, which are made by a certified land surveyor or civil engineer. These surveys show the following:

- All property lines
- Existing public utilities and their location on or near the property, such as electrical lines, sanitary sewer lines, gas lines, water-supply lines, storm sewers, manholes, or telephone lines

A land surveyor does the property survey from information obtained from a deed description of the property. A property survey shows only the property lines and their lengths, as if the property were perfectly flat.

The topographic survey shows both the property lines and the physical characteristics of the land by using contour lines, notes, and symbols. The physical characteristics may include:

- The direction of the land slope
- Whether the land is flat, hilly, wooded, swampy, high, or low, and other features of its physical nature

All of this information is necessary so that the architect can properly design a building to fit the property. The electrical engineer also needs this information to locate existing electrical utilities and to route the new service to the building, provide outdoor lighting and circuits, etc.

Figure 27 shows a utilities plan for a site. Shown on it are the gas main, the existing and new electrical lines, new telephone service connections, and details of manholes and electrical ducts where communication and power lines would be brought to the building.

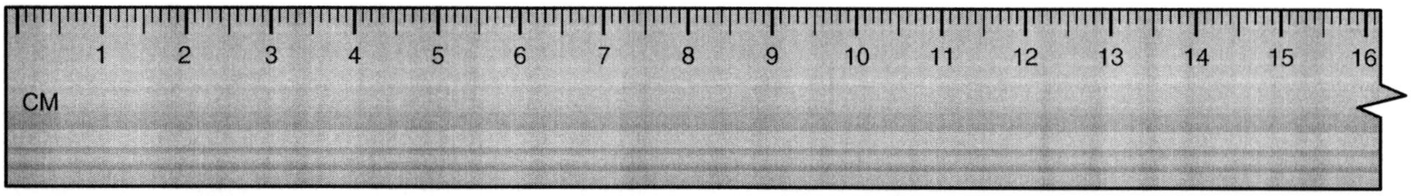

107F26.EPS

Figure 26 ◆ Typical metric scale.

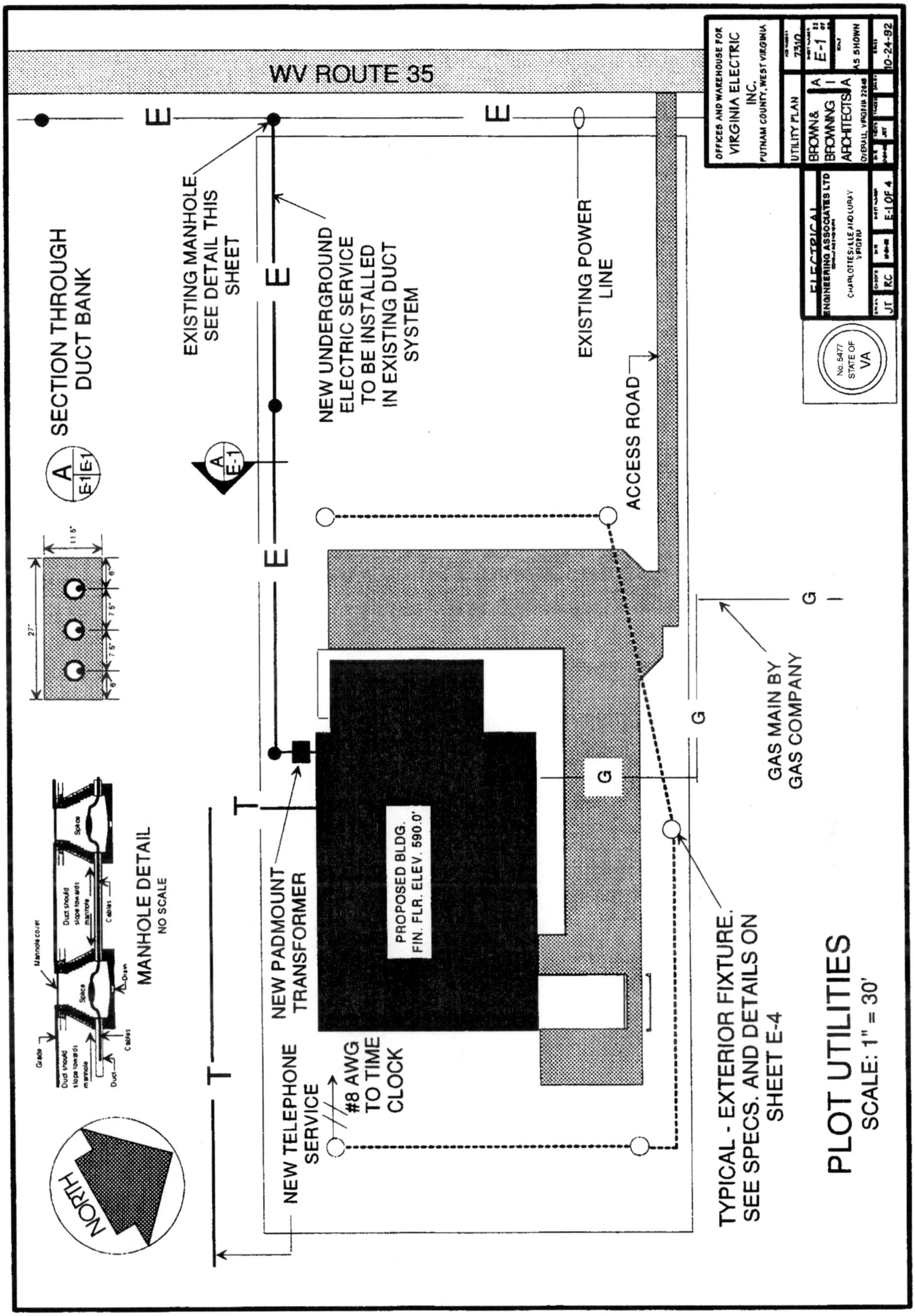

107F27.EPS

Figure 27 ◆ Typical site utilities plan.

7.0.0 ◆ FLOOR PLAN

Figure 28 shows a plan for electrical power in part of a building. Such a drawing would allow the maintenance mechanic to locate specific power lines within a building, either to gain access or to avoid damaging power lines. However, due to the size of the drawing in comparison with the size of the pages in this module, it is still difficult to see very much detail. This illustration is meant to show the partial layout of the power plan and how symbols and notes are arranged.

In general, this plan shows the service equipment (in plan view), receptacles, underfloor duct system, motor connections, motor controllers, electric heat, busways, and similar details. The electric panels and other service equipment are drawn close to scale. The locations of electrical outlets and other components are only approximated on the drawings because they have to be exaggerated to show up on the prints. For example, a common duplex receptacle is only about three inches wide. If such a receptacle were to be

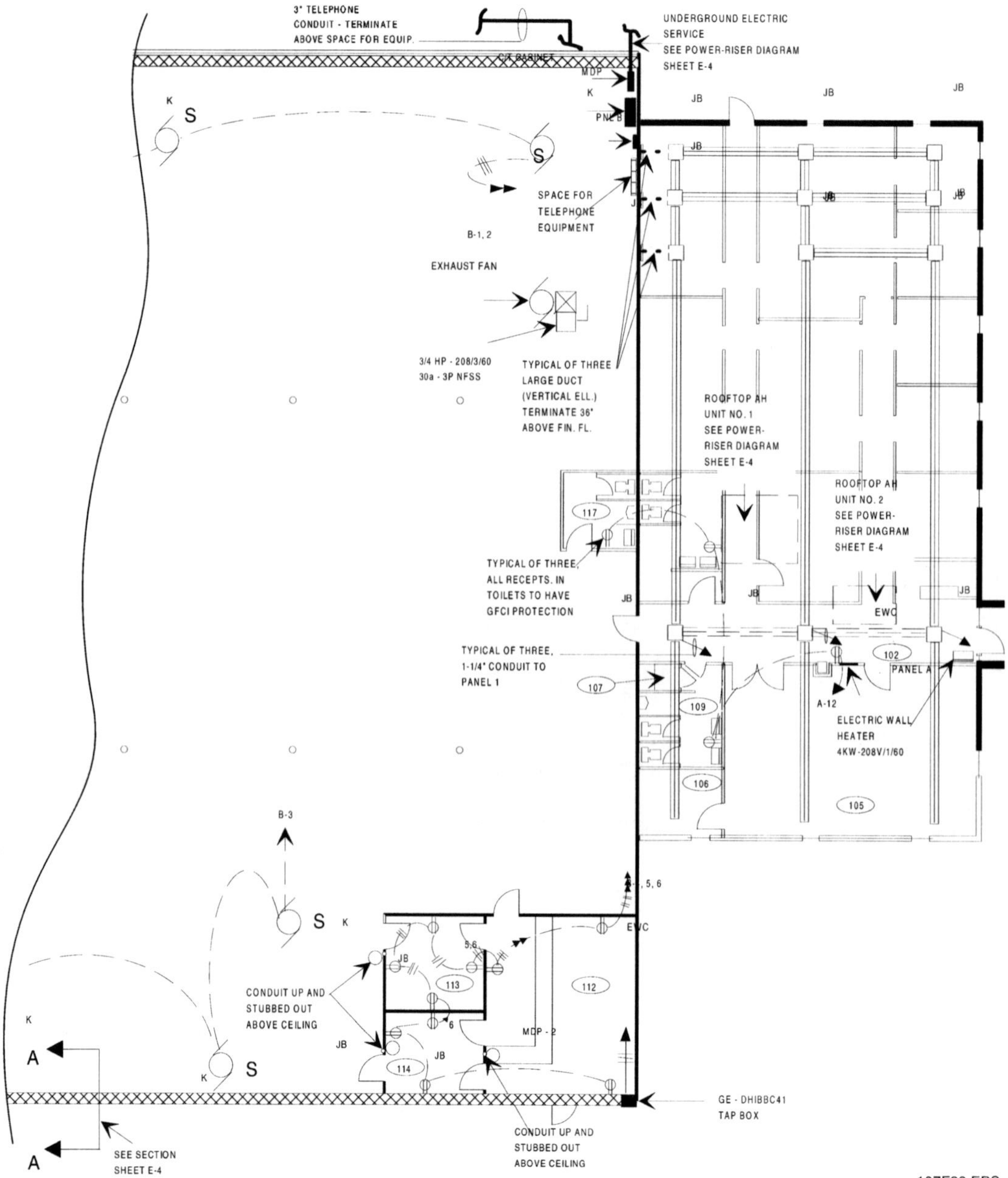

107F28.EPS

Figure 28 ◆ Partial view of electric power plan.

located on the floor plan of this building (drawn to a scale of ⅛" = 1'–0"), even a small dot on the drawing would be too large to draw the receptacle exactly to scale. Therefore, the receptacle symbol is exaggerated. When such receptacles are scaled on the drawings to determine the proper location, a measurement is usually taken to the center of the symbol to determine the distance between outlets. Junction boxes, switches, and other electrical or electronic devices shown on the plan will be exaggerated in a similar manner.

> **NOTE**
> When devices are to be located at heights specified above the finished floor (AFF), be sure to find out the actual height of the flooring to be installed. Some materials, such as ceramic tile, can add significantly to the height of the finished floor.

7.1.0 Notes and Building Symbols

Referring again to *Figure 28*, you will notice numbers placed inside an oval symbol in each room. These numbered ovals represent the room name or type and correspond to a room schedule in the architectural drawings. For example, room number 112 is designated as the lobby in the room schedule (not shown), room number 113 is designated as office No. 1, etc. On some drawings, these room symbols are omitted and the room names are written out on the drawings.

There are also several notes appearing at various places on the floor plan. These notes offer additional information to clarify certain aspects of the drawing. For example, only one electric heater is to be installed by the electrical contractor; this heater is located in the building's vestibule. Rather than have a symbol in the symbol list for this one heater, a note is used to identify it on the drawing. Other notes on this drawing describe how certain parts of the system are to be installed. For example, in the office area (rooms 112, 113, and 114), you will see the following note: CONDUIT UP AND STUBBED OUT ABOVE CEILING. This empty conduit is for telephone/communications cables that will be installed later by the telephone company.

> **NOTE**
> The notes are crucial elements of the drawing set. Receptacles, for example, are hard to position precisely based on a scaled drawing alone, and yet the designer may call for exact locations. For example, the designer may want receptacles exactly 6" above the kitchen counter backsplash and centered on the sink.

8.0.0 ◆ SCHEMATICS

The circuit diagram you will see is the schematic diagram (*Figure 29*), which represents circuit components using symbols. It shows how components

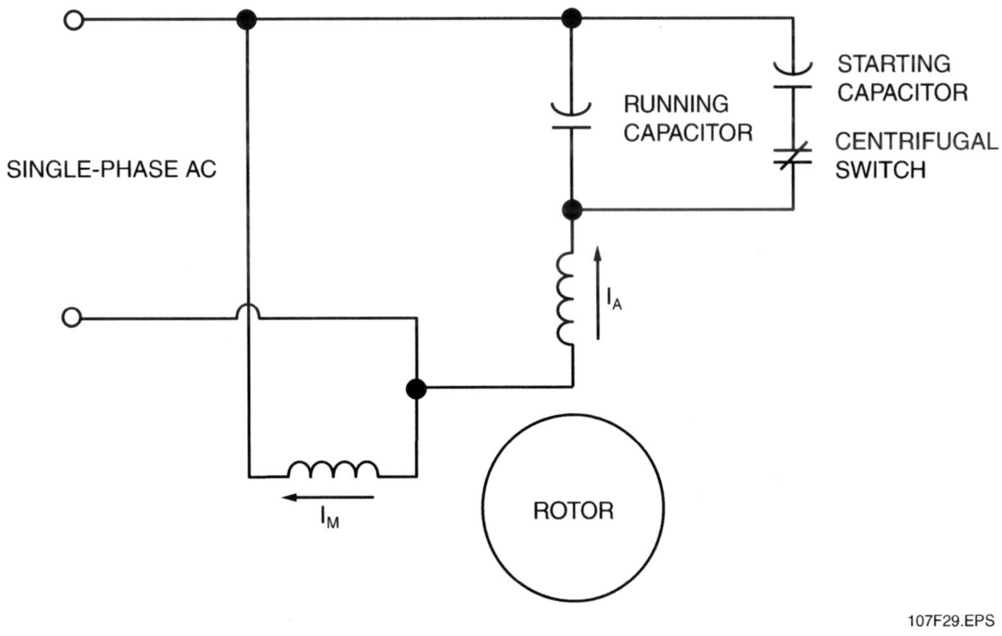

Figure 29 ◆ Capacitor-start, capacitor-run motor schematic.

are connected together electrically, but typically does not show complete point-to-point wiring. As an industrial maintenance mechanic, your schematic diagram reading will most likely be done in the course of troubleshooting. The schematic shown in *Figure 29* is a capacitor-start, capacitor-run motor. As shown, it runs on single-phase alternating current. The coils that produce the magnetic field that turns the rotor are labeled as I_m and I_a. If you had an electrical problem with this motor, you could follow the circuit from this diagram and test the components until you found what was wrong.

In this example, if a motor does not start when it should, the problem could be a burned-out motor or a malfunction in the control circuit for the motor. The obvious and easiest thing to do is to check the fuses. If that does not correct the problem, the next step is to trace the circuit from the motor back to the voltage source looking for a defective device or wire connection. In order to do this effectively, you must understand the logic and interdependencies designed into the circuit. *Figure 30* shows a schematic diagram of a simple lighting circuit. This type of diagram is called a ladder diagram because the voltage source is represented by uprights, and the load lines are like the rungs of a ladder. This method of diagramming clearly shows controls that affect each load, as well as the interdependencies of the loads. Let's assume this ladder represents the control circuit for parking lot lights and that there are three conditions for operating the lights:

1. The lights should be enabled whenever it is dark.
2. In order to save utility costs, the lights will only come on when the building is occupied.
3. A means of bypassing the automatic controls is required.

The lights themselves are activated by the closure of lighting contactor (LC1) in a 240-volt circuit. However, the contactor is controlled by a 24-volt circuit that obtains its supply voltage from step-down transformer T1. This is a very common

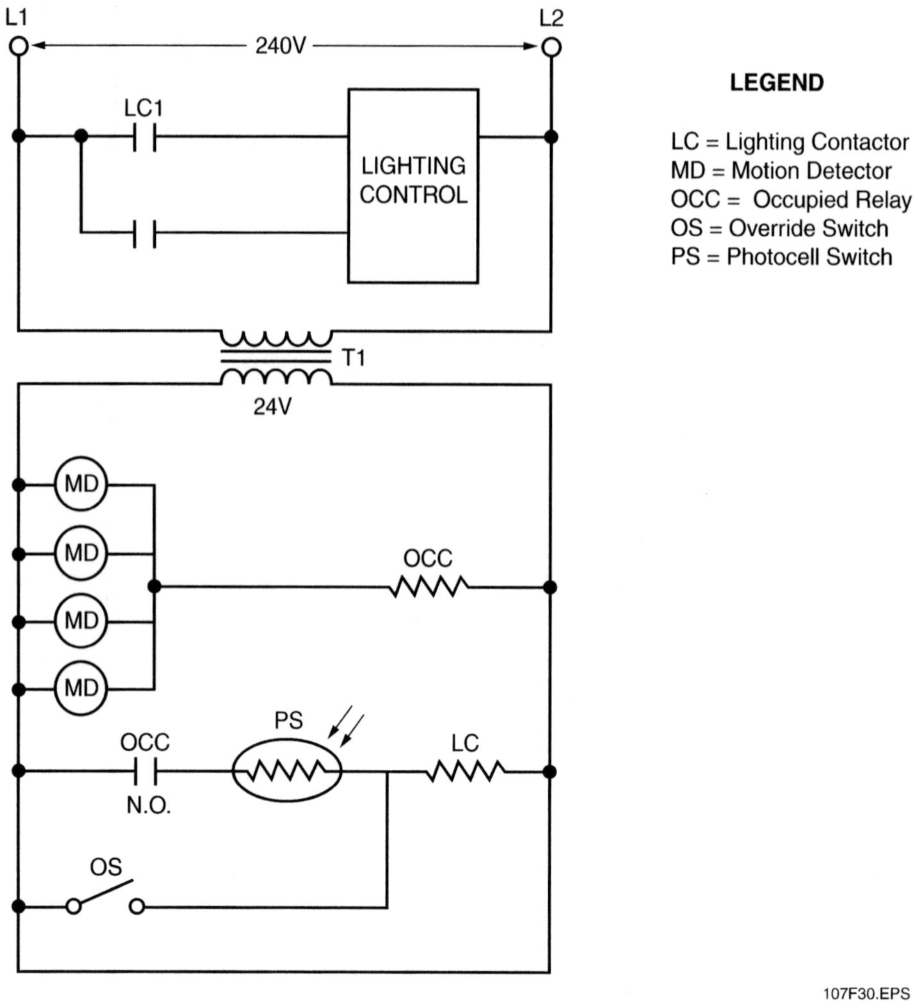

Figure 30 ◆ Ladder diagram.

design approach. The lower voltage is safer and the 24-volt circuit can use less expensive wiring and components than a 240-volt circuit.

A photoelectric relay is in series with the contactor coil. Motion sensors are used to prevent the light from being turned on when the building is unoccupied. Working back from the contactor coil, you can see that a photo switch must be activated in order to energize the lighting contactor. This satisfies condition 1.

The OCC relay is energized by motion detectors scattered around the building. Because the normally open contacts of the OCC relay are in series with the photo switch, the circuit is not complete until a motion detector energizes the OCC relay while the photo switch is activated. That is, it must be after dark and someone must be in the building. This satisfies condition 2.

A single-pole switch is wired into the contactor coil circuit so the lights can be manually turned on. This satisfies condition 3.

The diagram in *Figure 31* is a Piping and Instrumentation Drawing (P&ID) with the parts labeled. The valves are shown as standard valve symbols, with the connections shown to all the other pieces. This drawing will allow you to follow the connections of each component, and to know what each component does in the assembly. The circles represent instruments such as a thermostat, relays, or pressure sensors.

A key would be supplied to identify the various parts of the system by their specific designations such as LR6 (level recorder 6), which records the tank liquid level for the control room, or TI3A, which is a temperature indicator for mixed water to the tank.

8.1.0 Exploded Diagrams

The mechanic has to understand how a piece of equipment is put together and what all the parts are in order to repair and maintain the equipment. The **exploded diagram** is usually supplied with the paperwork that accompanies a unit. *Figure 32* shows an example of an exploded diagram of a compressor. As you can see, the diagram shows all the parts of the machine and where they connect together, with lines showing where each piece connects to the assembly.

8.2.0 Drawing Details

A detail drawing is a drawing of a separate item or portion of a system, giving a complete and exact description of its use and all the details needed to show exactly what is required for its installation.

A set of electrical drawings will sometimes require large-scale drawings of certain areas that are not indicated with sufficient clarity on the small-scale drawings. For example, the site plan may show exterior pole-mounted lighting fixtures that are to be installed by the contractor.

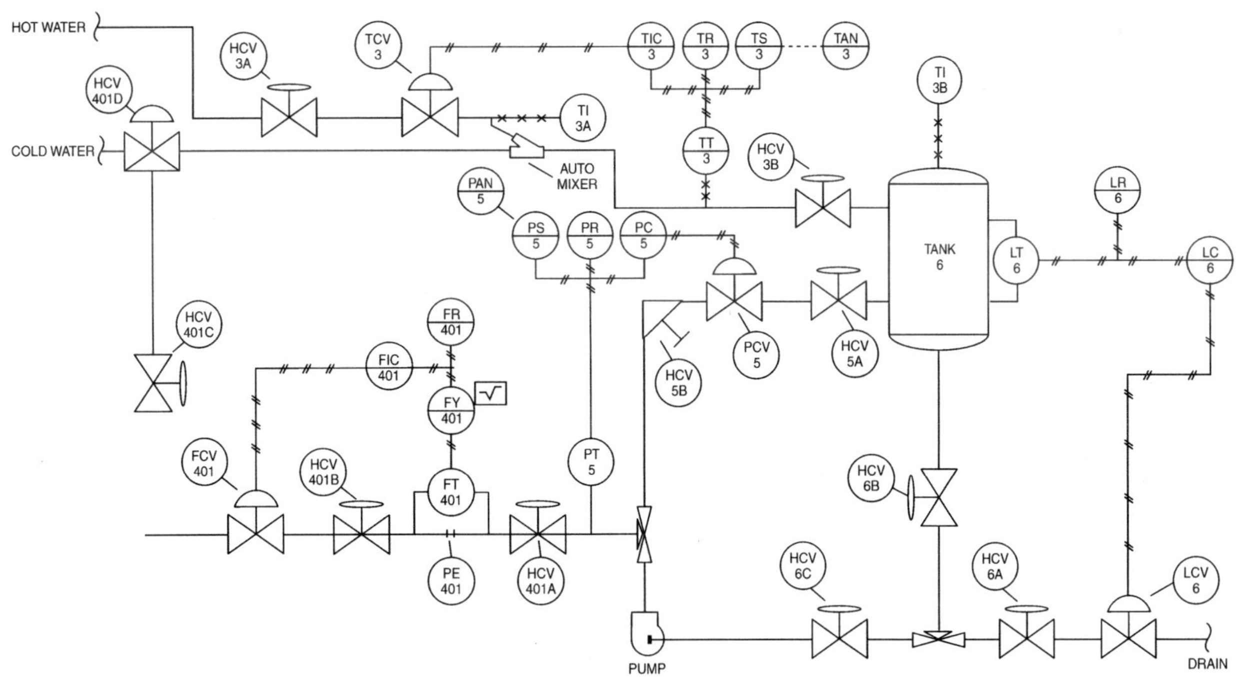

Figure 31 ◆ Example of a P&ID.

Although most of the work of an industrial maintenance mechanic involves the use of electrical drawings and schedules, other drawings in the set are important as well:

- The site plan often shows the locations of underground utilities and wall penetrations.
- Elevation drawings show exterior finish and trim.
- Section drawings show the construction of walls, floors, and ceilings. This is extremely important to installers who are creating pathways for conduit and cables because it provides guidance on where and how to make penetrations.
- Mechanical drawings show the location of heating and air conditioning equipment and ductwork, which can serve as obstacles to cabling pathways.
- Floor plans show the interior layout of the structure and are a valuable tool in planning an installation.
- Schedules, along with the other drawings in the set, are important when doing equipment and material takeoffs.
- The electrical plan usually includes details of the electronic systems and cabling.

9.0.0 ◆ WRITTEN SPECIFICATIONS

The **written specifications** for a building or project are the written descriptions of work and duties required of the owner, architect, and consulting engineer. Together with the working drawings, these specifications form the basis of the contract requirements for the construction of the building or project. Those who use the construction drawings and specifications must always be alert to discrepancies between the working drawings and the written specifications. These are some situations where discrepancies may occur:

- Architects or engineers use standard or prototype specifications and attempt to apply them without any modification to specific working drawings.
- Previously prepared standard drawings are changed or amended by reference in the specifications only and the drawings themselves are not changed.
- Items are duplicated in both the drawings and specifications, but an item is subsequently amended in one and overlooked in the other contract document.

In such instances, the person in charge of the project has the responsibility to ascertain whether the drawings or the specifications take precedence. Such questions must be resolved, preferably before the work begins, to avoid added costs to the owner, architect/engineer, or contractor.

9.1.0 How Specifications are Written

Writing accurate and complete specifications for building construction is a serious responsibility for those who design the buildings because the specifications, combined with the working drawings, govern practically all important decisions that are made

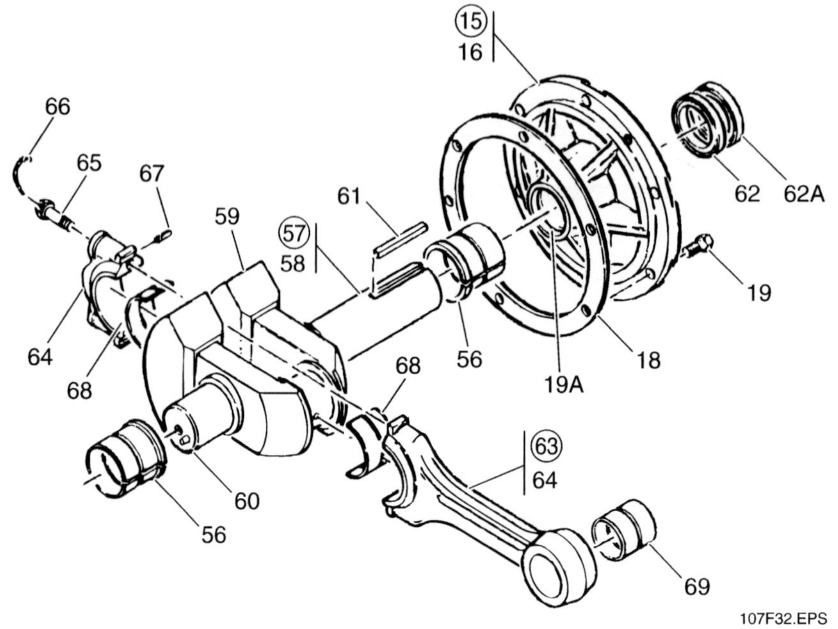

Figure 32 ◆ Exploded diagram of a compressor.

during the construction span of every project. Compiling and writing these specifications is not a simple task, even for those who have had considerable experience in preparing such documents. A set of written specifications for a single project will usually contain thousands of products, parts, and components, and the methods of installing them, all of which must be covered in either the drawings and/or specifications. No one can memorize all of the necessary items required to describe accurately the various areas of construction. One must rely upon reference materials such as manufacturer's data, catalogs, checklists, and, most of all, a high-quality master specification.

9.2.0 Format of Specifications

For convenience in writing, speed in estimating, and ease of reference, the most suitable organization of the specifications is a series of sections dealing with the construction requirements, products, and activities. It must be easily understandable by the different trades. Those people who use the specifications must be able to find all the information they need without spending too much time looking for it.

The most commonly used specification-writing format used in North America is the MasterFormat™. This standard was developed jointly by the Construction Specifications Institute (CSI) and Construction Specifications Canada (CSC). For many years prior to 2004, the organization of construction specifications and suppliers' catalogs has been based on a standard with 16 sections, otherwise known as divisions, where the divisions and their subsections were individually identified by a five-digit numbering system. The first two digits represented the division number, and the next three individual numbers represented successively lower levels of breakdown. For example, the number 13213 represents division 13, subsection 2, sub-subsection 1 and sub-sub-subsection 3. In this older version of the standard, electrical systems, including any electronic or special electrical systems, were lumped together under Division 16 – Electrical. Today, specifications conforming to the 16 division format may still be in use.

In 2004, the MasterFormat™ standard underwent a major change. What had been 16 divisions was expanded to four major groupings and 49 divisions with some divisions reserved for future expansion (*Figure 33*). The first 14 divisions are essentially the same as the old format. Subjects under the old Division 15 – Mechanical have been relocated to new divisions 22 and 23. The basic subjects under old Division 16 – Electrical have been relocated to new divisions 26 and 27. In addition, the numbering system was changed to six digits to allow for more subsections in each division, which allowed for finer definition. In the new numbering system, the first two digits represent the division number. The next two digits represent subsections of the division, and the two remaining digits represent the third level sub-subsection numbers. The fourth level, if required, is a decimal and number added to the end of the last two digits. For example, the number 132013.04 represents division 13, subsection 20, sub-subsection 13 and sub-sub-subsection 04.

NOTE

Written specifications supplement the related working drawings in that they contain details not shown on the drawings. Specifications define and clarify the scope of the job. They describe the specific types and characteristics of the components that are to be used on the job and the methods for installing some of them. Many components are identified specifically by the manufacturer's model and part numbers. This type of information is used to purchase the various items of hardware needed to accomplish the installation in accordance with the contractual requirements.

MasterFormat™ 2004 Edition – Numbers & Titles — Division Numbers & Titles

6/8/04

Division Numbers and Titles

PROCUREMENT AND CONTRACTING REQUIREMENTS GROUP

Division 00 Procurement and Contracting Requirements

SPECIFICATIONS GROUP

GENERAL REQUIREMENTS SUBGROUP

Division 01 General Requirements

FACILITY CONSTRUCTION SUBGROUP

Division 02 Existing Conditions
Division 03 Concrete
Division 04 Masonry
Division 05 Metals
Division 06 Wood, Plastics, and Composites
Division 07 Thermal and Moisture Protection
Division 08 Openings
Division 09 Finishes
Division 10 Specialties
Division 11 Equipment
Division 12 Furnishings
Division 13 Special Construction
Division 14 Conveying Equipment
Division 15 Reserved
Division 16 Reserved
Division 17 Reserved
Division 18 Reserved
Division 19 Reserved

FACILITY SERVICES SUBGROUP

Division 20 Reserved
Division 21 Fire Suppression
Division 22 Plumbing
Division 23 Heating, Ventilating, and Air Conditioning
Division 24 Reserved
Division 25 Integrated Automation
Division 26 Electrical
Division 27 Communications
Division 28 Electronic Safety and Security
Division 29 Reserved

SITE AND INFRASTRUCTURE SUBGROUP

Division 30 Reserved
Division 31 Earthwork
Division 32 Exterior Improvements
Division 33 Utilities
Division 34 Transportation
Division 35 Waterway and Marine Construction
Division 36 Reserved
Division 37 Reserved
Division 38 Reserved
Division 39 Reserved

PROCESS EQUIPMENT SUBGROUP

Division 40 Process Integration
Division 41 Material Processing and Handling Equipment
Division 42 Process Heating, Cooling, and Drying Equipment
Division 43 Process Gas and Liquid Handling, Purification, and Storage Equipment
Division 44 Pollution Control Equipment
Division 45 Industry-Specific Manufacturing Equipment
Division 46 Reserved
Division 47 Reserved
Division 48 Electrical Power Generation
Division 49 Reserved

Div Numbers - 1

107F33.EPS

Figure 33 ◆ 2004 MasterFormat™.

Review Questions

1. The location of a building on the property is shown on the _____.
   a. elevation drawings
   b. floor plan
   c. site plan
   d. section drawing

2. The drawing shown in *Figure 1* is a(n) _____.
   a. sectional view
   b. floor plan
   c. elevation drawing
   d. site plan

3. Government regulations require that electrical drawings from engineering firms be complete, precise, and easy to read.
   a. True
   b. False

4. A 24 × 36 drawing equates to letter size _____.
   a. A
   b. B
   c. C
   d. D

5. A dashed line on an electrical diagram represents _____.
   a. a busway
   b. wiring concealed in the floor
   c. wiring concealed in a wall
   d. a branch circuit homerun

6. An architect's scale shows relation of _____.
   a. temperatures
   b. decimal ratios
   c. inches to feet
   d. hexadecimal numbers

7. A symbol list on a drawing shows the _____ for various components.
   a. plan drawing
   b. engineer's seal
   c. symbols used
   d. panelboard schedule

8. One reason the working drawings must be checked against written specifications is that _____.
   a. the drawings are always right
   b. written specifications never change
   c. something may have changed in one, but not in the other
   d. it makes the engineers happy

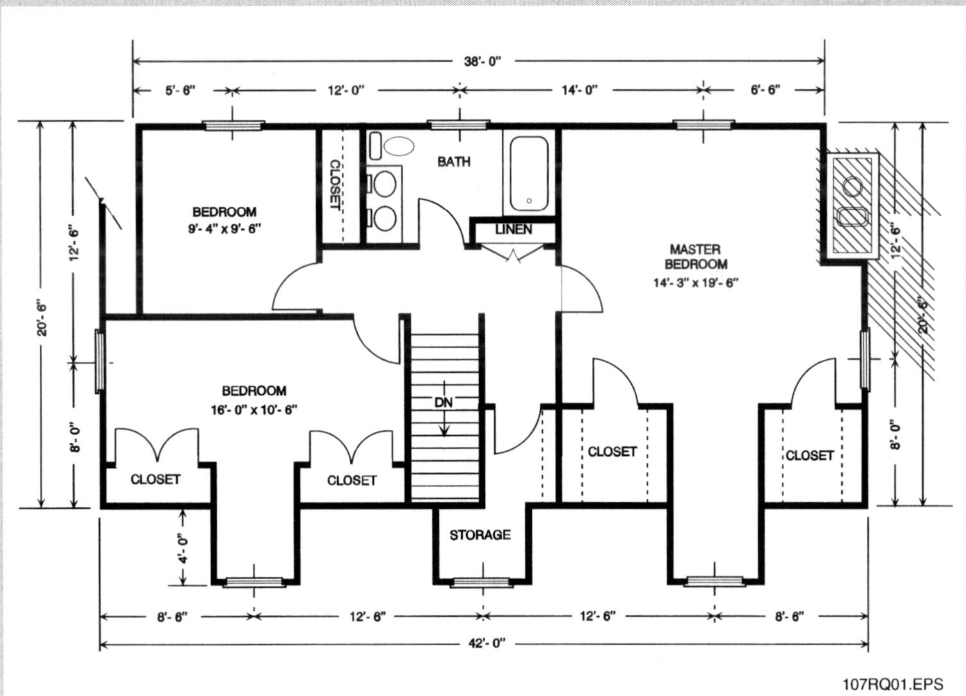

Figure 1

Review Questions

9. The plan that often shows underground utilities is the _____.
   a. site plan
   b. elevation drawing
   c. floor plan
   d. electrical plan

10. The interior layout of a structure appears in the _____.
    a. elevation drawing
    b. site plan
    c. floor plan
    d. schedules

Summary

In this module, the symbols and conventions used on architectural and engineering drawings are discussed. As an industrial maintenance mechanic, you need to know how to recognize the basic symbols used on electrical drawings and other drawings used in the building construction industry. You should also know where to find the meaning of symbols that you do not immediately recognize. Schedules, diagrams, and specifications often provide detailed information that is not included on the working drawings.

Reading architectural and engineering drawings takes practice and study. Now that you have the basic skills, take the time to master them.

Notes

Trade Terms Introduced in This Module

Architectural drawings: Working drawings consisting of plans, elevations, details, and other information necessary for the construction of a building. Architectural drawings usually include:

- A site (plot) plan indicating the location of the building on the property
- Floor plans showing the walls and partitions for each floor or level
- Elevations of all exterior faces of the building
- Several vertical cross sections to indicate clearly the various floor levels and details of the footings, foundations, walls, floors, ceilings, and roof construction
- Large-scale detail drawings showing such construction details as may be required

Blueprint: An exact copy or reproduction of an original drawing.

Construction drawings: Drawings that show in a clear, concise manner exactly what is required of the mechanic or builder.

Detail drawings: An enlarged, detailed view taken from an area of a drawing and shown in a separate view.

Dimensions: Sizes or measurements printed on a drawing.

Electrical drawings: A means of conveying a large amount of exact, detailed information in an abbreviated language. Consists of lines, symbols, dimensions, and notations to accurately convey an engineer's designs to electricians and electronic systems technicians who install the electrical system on a job.

Elevation drawings: An architectural drawing showing height, but not depth; usually the front, rear, and sides of a building or object.

Exploded diagram: A diagram that shows all the separated parts of the machine and where they connect together, with lines showing where each piece connects to the assembly.

Floor plan: A drawing of a building as if a horizontal cut were made through a building at about window level, and the top portion removed. The floor plan is what would appear if the remaining structure were viewed from above.

Plan view: A drawing made as though the viewer were looking straight down (from above) on an object.

Scale: On a drawing, the size relationship between an object's actual size and the size it is drawn. Scale also refers to the measuring tool used to determine this relationship.

Schedule: A systematic method of presenting equipment lists on a drawing in tabular form.

Schematic diagram: A detailed diagram showing complicated circuits, such as control circuits.

Sectional view: A cutaway drawing that shows the inside of an object or building.

Site plan: A drawing showing the location of a building or buildings on the building site. Such drawings frequently show topographical lines, electrical and communication lines, water and sewer lines, sidewalks, driveways, and similar information.

Written specifications: A written description of what is required by the owner, architect, and engineer in the way of materials and workmanship. Together with working drawings, the specifications form the basis of the contract requirements for construction.

Appendix

Abbreviations

Adapter	ADPT
Air preheating	A
American Iron and Steel Institute	AISI
American National Standards Institute	ANSI
American Petroleum Institute	API
American Society for Testing and Materials International	ASTM
American Society of Mechanical Engineers	ASME
Ash removal water, sluice or jet	AW
Aspirating air	AA
Auxiliary steam	AS
Bench mark	BM
Beveled	B
Beveled end	BE
Beveled, both ends	BBE
Beveled, large end	BLE
Beveled, one end	BOE
Beveled, small end	BSE
Bill of materials	BOM
Blind flange	BF
Blowoff	BO
Bottom of pipe	BOP
Butt weld	BW
Carbon steel or cold spring	CS
Cast iron	CI
Ceiling	CLG
Chain operated	CO
Chemical feed	CF
Circulating water	CW
Cold reheat steam	CR
Compressed air	CA
Concentric or concrete	CONC
Condensate	C
Condenser air removal	AR
Continue, continuation	CONT
Coupling	CPLG
Detail	DET
Diameter	DIA
Dimension	DIM
Discharge	DISCH
Double extra-strong	XX STRG
Drain	DR
Drain funnel	DF
Drawing	DWG
Drip leg or dummy leg	DL
Ductile iron	DI
Dust collector	DC
Each	EA
Eccentric	ECC
Elbolet	EOL
Elbow	ELB
Electric resistance weld	ERW
Elevation	EL
Equipment	EQUIP
Evaporator vapor	EV
Exhaust steam	E
Expansion	EXP
Expansion joint	EXP JT
Extraction steam	ES
Fabrication (dimension)	FAB
Face of flange	FOF
Faced and drilled	F&D
Factory Mutual	FM
Far side	FS
Feed pump balancing line	FB
Feed pump discharge	FD
Feed pump recirculating	FR
Feed pump suction	FS
Female	F
Female	FM
Female pipe thread	FPT
Field support or forged steel	FS

107A01A.EPS

Field weld	FW
Figure	FIG
Fillet weld	W
Finish floor	F/F
Finish floor	FIN FL
Finish grade	FIN GR
Fitting	FTG
Fitting makeup	FMU
Fitting to fitting	FTF
Flange	FLG
Flat face	FF
Flat on bottom	FOB
Flat on top	FOT
Floor drain	FD
Foundation	FDN
Fuel gas	FG
Fuel oil	FO
Gage	GA
Galvanized	GALV
Gasket	GSKT
Grating	GRTG
Hanger	HGR
Hanger rod	HR
Hardware	HDW
Header	HDR
Heat traced, heat tracing	HT
Heater drains	HD
Heating system	HS
Heating, ventilating, and air conditioning	HVAC
Hexagon	HEX
Hexagon head	HEX HD
High point	HPT
High pressure	HP
Horizontal	HORIZ
Hot reheat steam	HR
Hydraulic	HYDR
Increaser	INCR
Input/output	I/O
Inside diameter	ID
Insulation	INS
Invert elevation	IE
Iron pipe size	IPS
Isometric	ISO
Issued for construction	IFC
Lap weld	LW
Large male	LM
Length	LG
Long radius	LR
Long tangent	LT
Long weld neck	LWN
Low pressure	LP
Low-pressure drains	DR
Low-pressure steam	LPS
Lubricating oil	LO
Main steam	MS
Main system blowouts	BL
Makeup water	MU
Male	M
Malleable iron	MI
Manufacturer	MFR
Manufacturer's Standard Society	MSS
National Pipe Thread	NPT
Nipolet	NOL
Nipple	NIP
Nominal	NOM
Not to scale	NTS
Nozzle	NOZ
Outside battery limits	OSBL
Outside diameter	OD
Outside screw and yoke	OS&Y
Overflow	OF
Pipe support	PS
Pipe tap	PT
Piping and instrumentation diagram	P&ID
Plain	P
Plain end	PE
Plain, both ends	PBE
Plain, one end	POE
Point of intersection	PI
Point of tangent	PT
Pounds per square inch	PSI
Process flow diagram	PFD
Purchase order	PO
Radius	RAD
Raised face	RF
Raised face slip-on	RFSO

107A01B.EPS

Raised face smooth finish .... RFSF
Raised face weld neck .... RFWN
Raw water .... RW
Reducer, reducing .... RED
Relief valve .... PRV-PSV
Ring-type joint .... RTJ
Rod hanger or right hand .... RH

Safety valve vents .... SV
Sanitary .... SAN
Saturated steam .... SS
Schedule .... SCH
Screwed .... SCRD
Seamless .... SMLS
Section .... SECT
Service and cooling water .... SW
Sheet .... SH
Short radius .... SR
Slip-on .... SO
Socket weld .... SW
Sockolet .... SOL
Stainless steel .... SS
Standard weight .... STD WT
Steel .... STL
Suction .... SUCT
Superheater drains .... SD
Swage .... SWG
Swaged nipple .... SN

Temperature or temporary .... TEMP
That is .... I.E.
Thick .... THK
Thousand .... M
Thread, threaded .... THRD
Threaded .... T
Threaded end .... TE
Threaded, both ends .... TBE
Threaded, large end .... TLE
Threaded, one end .... TOE
Threaded, small end .... TSE
Threadolet .... TOL
Top of pipe or top of platform .... TOP
Top of steel or top of support .... TOS
Treated water .... TW
Turbine .... TURB
Typical .... TYP

Underwriters Laboratories .... UL

Vacuum .... VAC
Vacuum cleaning VC
Vents V
Vitrified tile .... VT

Wall thickness or weight .... WT
Weld neck .... WN
Weldolet .... WOL
Well water .... WW
Wide flange .... WF

107A01C.EPS

Resources & Acknowledgments

Additional Resources

This module is intended to be a thorough resource for task training. The following reference works are suggested for further study. These are optional materials for continued education rather than for task training.

American Electrician's Handbook, 2002. Terrell Croft, Wilfred Summers. New York, NY: McGraw-Hill.

National Electrical Code Handbook, Latest Edition. Quincy, MA: National Fire Protection Association.

The Pipefitter's Bluebook, Latest Edition. W.V. Graves: Clinton, NC: Construction Trades Press.

Figure Credits

Topaz Publications, Inc., 107F25

John Traister, 107F27, 107F28

Dresser-Rand Company, 107F29, 107F32

The Construction Specifications Institute, 107F32

The Numbers and Titles used in this textbook are from *MasterFormat*™ *2004* published by The Construction Specifications Institute (CSI) and Construction Specifications Canada (CSC), and are used with permission from CSI. For those interested in a more in-depth explanation of *MasterFormat*™ *2004* and its use in the construction industry visit www.csinet.org/masterformat or contact:

The Construction Specifications Institute (CSI)
99 Canal Center Plaza, Suite 300, Alexandria, VA 22314
800-689-2900; 703-684-0300; http://www.csinet.org

CONTREN® LEARNING SERIES — USER UPDATE

NCCER makes every effort to keep these textbooks up-to-date and free of technical errors. We appreciate your help in this process. If you have an idea for improving this textbook, or if you find an error, a typographical mistake, or an inaccuracy in NCCER's Contren® textbooks, please write us, using this form or a photocopy. Be sure to include the exact module number, page number, a detailed description, and the correction, if applicable. Your input will be brought to the attention of the Technical Review Committee. Thank you for your assistance.

Instructors – If you found that additional materials were necessary in order to teach this module effectively, please let us know so that we may include them in the Equipment/Materials list in the Annotated Instructor's Guide.

Write: Product Development and Revision
National Center for Construction Education and Research
3600 NW 43rd St. Bldg G, Gainesville, FL 32606

Fax: 352-334-0932

E-mail: curriculum@nccer.org

Craft ______________________ Module Name ______________________

Copyright Date __________ Module Number __________ Page Number(s) __________

Description

____

____

____

____

(Optional) Correction

____

____

____

(Optional) Your Name and Address

____

____

____

Industrial Maintenance E & I Technician Level One

40108-07

Pumps and Drivers

40108-07
Pumps and Drivers

Topics to be presented in this module include:

Overview

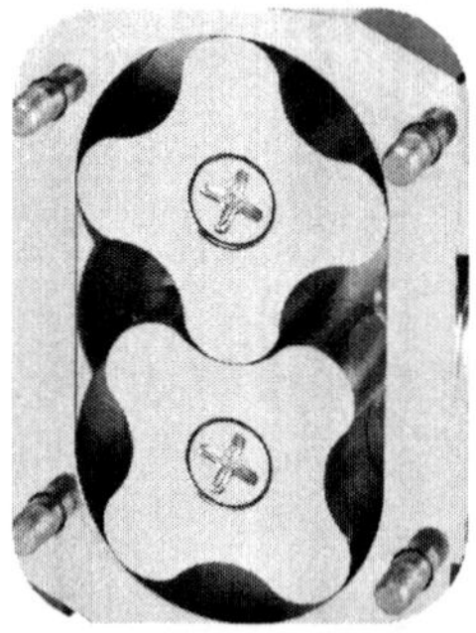

Pumps add energy to fluids to move them from place to place. This module will explain the basics of the many different types of pumps. You will learn what they are used for, some of their strengths and weaknesses, and what can be wrong with them. You will also learn about the kinds of drivers used in industry.

Objectives

When you have completed this module, you will be able to do the following:

1. Identify and explain centrifugal pumps.
2. Identify and explain rotary pumps.
3. Identify and explain reciprocating pumps.
4. Identify and explain metering pumps.
5. Identify and explain vacuum pumps.
6. Explain net positive suction head and cavitation.
7. Identify types of drivers.

Trade Terms

Capacity
Discharge
Head
Net positive suction head (NPSH)
Programmable logic controller
Static
Suction
Velocity

Required Trainee Materials

1. Pencil and paper
2. Appropriate personal protective equipment

Prerequisites

Before you begin this module, it is recommended that you successfully complete *Core Curriculum;* and *Industrial Maintenance E & I Technician Level One*, Modules 40101-07 through 40107-07.

This course map shows all of the modules in the first level of the *Industrial Maintenance E & I Technician* curriculum. The suggested training order begins at the bottom and proceeds up. Skill levels increase as you advance on the course map. The local Training Program Sponsor may adjust the training order.

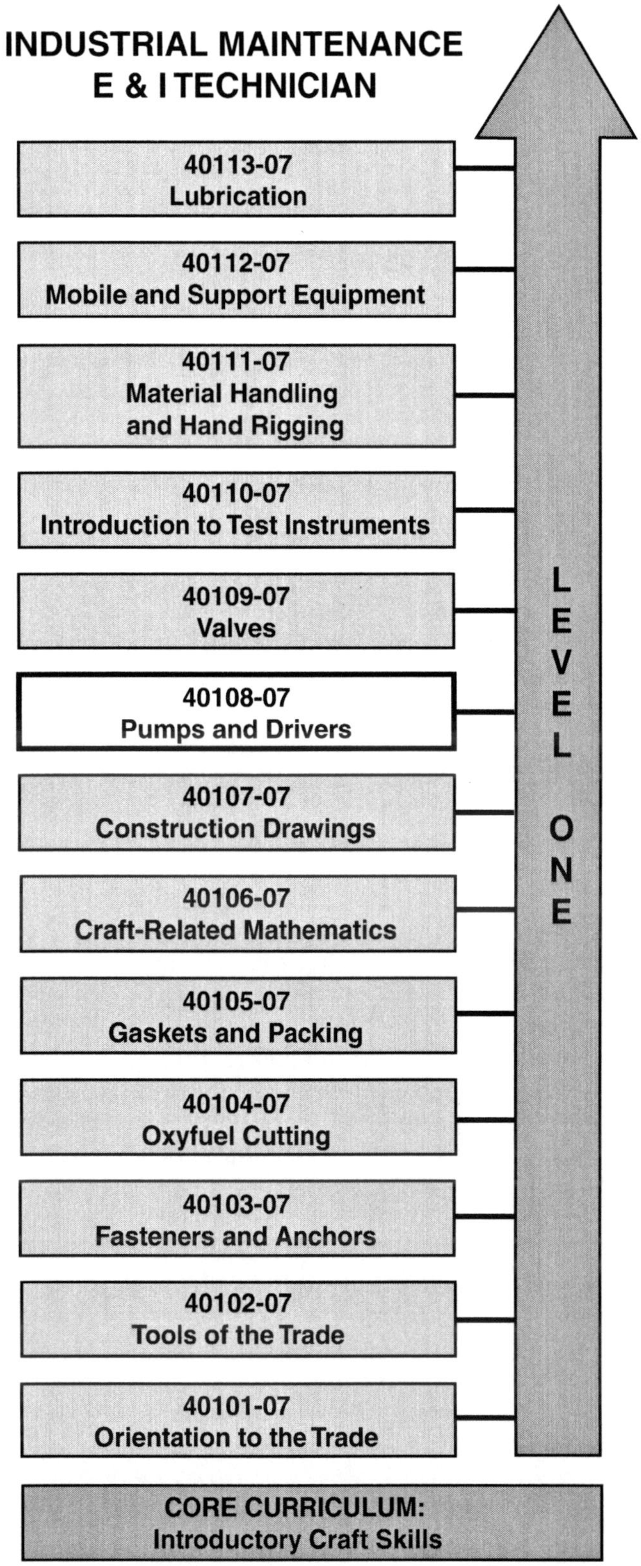

108CMAP.EPS

1.0.0 ◆ INTRODUCTION

Pumps are used in almost every industry to move liquids and semisolids from one location to another. Pumps come in many types and sizes for various applications. They range from small metering pumps that pump only ounces per day to very large, high-**capacity** pumps that pump thousands of gallons per hour.

Pumps are a major part of the maintenance workload in most plants and industries. Pump failure is a major cause of downtime and lost production. Premature pump failure can usually be attributed to improper installation or maintenance procedures.

The installation, repair, and maintenance of pumps represent a significant part of a maintenance worker's job. The care and precision with which these jobs are done greatly impacts the reliability and efficiency of the pumps.

This module presents general information about pumps and general procedures for installing pumps. For detailed information and repair procedures, the manufacturer's maintenance and repair manual is required.

2.0.0 ◆ PUMP TYPES

A pump is a mechanical device designed to increase the energy of a fluid so that a quantity of the fluid can be transported from one location to another. Pumps are generally driven by electric motors and are used to pump water, stock, coating, additives, oil, and hydraulic fluid.

All pumps operate under the general principle that fluid is drawn in at one end, the **suction** side, and forced out at the other end, the **discharge** side, at an increased **velocity**.

Pumps are classified as either positive-displacement or centrifugal. Positive-displacement pumps are generally used for pumping liquids that have very high viscosities; that is, they are thick and flow slowly. Positive-displacement pumps are generally high-pressure pumps. Centrifugal pumps are used for pumping liquids that have low viscosities, such as water or thin stock. Centrifugal pumps are basically low-pressure pumps.

A positive-displacement pump takes a discrete amount of liquid and moves it physically from one place to another. A centrifugal pump adds kinetic energy to the liquid, and provides an enclosure that forces the liquid to go in a particular direction. Positive-displacement pumps produce a pattern of intermittent flow; the liquid flow is interspersed with intervals of no flow, although this can be so nearly continuous as to be undetectable. Centrifugal pumps are continuous-flow pumps.

Because positive-displacement pumps create very high pressures very quickly in a closed discharge pipe, there is usually a relief valve between the pump and the first closing point, such as a valve or possible pipe obstruction.

In automated pumping applications, **programmable logic controllers** connected to flow and pressure sensors are used to automate pump control. If flow characteristics or pressure become unacceptable, a message is sent to the PLC controlling the pump, and pump speed or discharge flow rate is adjusted, preventing possible pump damage from cavitation.

The most common types of pumps are the following:

- Centrifugal
- Rotary
- Reciprocating
- Metering
- Vacuum

2.1.0 Centrifugal Pumps

A centrifugal pump operates by increasing the velocity of a liquid. Fluid entering the pump is rotated by an impeller. This rotation creates centrifugal force within a stationary casing. The force raises the pressure and causes the fluid to be discharged at high speed. *Figure 1* shows a centrifugal pump.

The parts of centrifugal pumps may vary in size and shape, depending on the manufacturer, but will have the same functions. Centrifugal pump parts include the following:

- *Pump casing* – The part surrounding the shaft, bearings, packing gland, and impeller. Pump casings can be of split or solid design.
- *Suction port* – The place where fluid enters the pump.
- *Discharge port* – The place where fluid is discharged into the piping system.
- *Pump shaft* – A bearing-supported part that holds and turns the impeller when the shaft is coupled to a motor.
- *Bearings* – The parts that support the shaft and impeller in the casing.
- *Impeller* – A rotating part that increases the speed of the fluid. There are many different types of impellers used for different purposes.
- *Impeller vanes or blades* – The parts of the impeller that direct the flow of fluid within the pump.
- *Impeller shrouds* – The parts that enclose the blades and keep the flow of fluid in the impeller area.

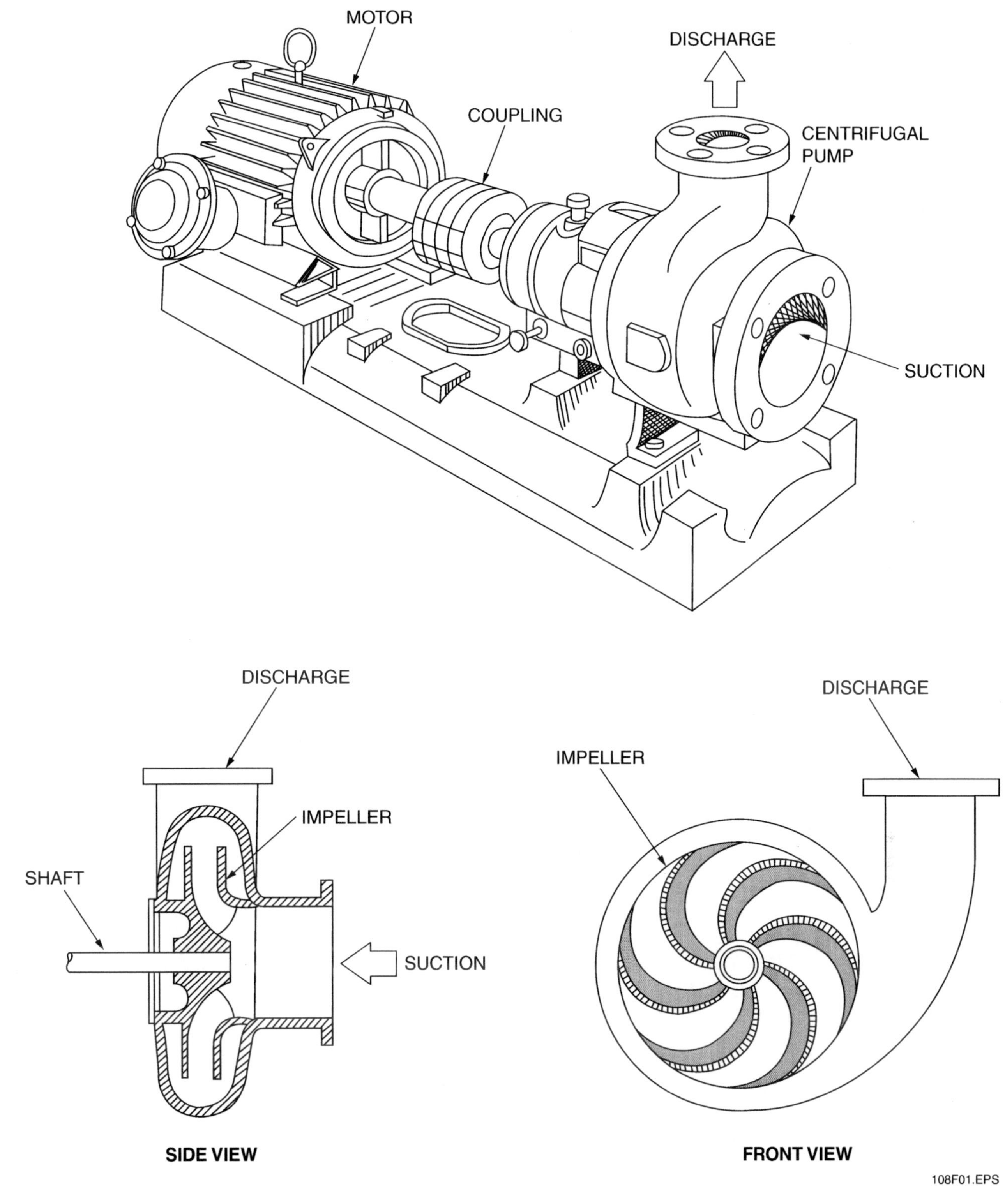

Figure 1 ◆ Centrifugal pump.

- *Wear rings* – Replaceable rings used in some pumps to allow some fluid leakage between the impeller and the casing in the suction area. The leakage makes a hydraulic seal and helps the pump operate more efficiently.
- *Packing gland* – Contains an adjustable follower that exerts force upon the packing to control fluid leakage around the shaft.
- *Mechanical seal* – Seals the fluid flow in the pump. Used instead of packing in some pumps.

Figure 2 shows a centrifugal pump with a vertically split case.

Figure 3 shows types of impellers.

The type of centrifugal pump used depends upon the requirements of the system. Some systems require that large amounts of fluids be pumped; others require high pressures. Two types of centrifugal pumps are double-suction and multistage.

2.1.1 Double-Suction Centrifugal Pumps

A double-suction centrifugal pump is used to pump large volumes of fluid. Fluid is drawn in through suction openings on both sides of the impeller and passes out through a single discharge opening. *Figure 4* shows a double-suction centrifugal pump with a horizontally split case.

2.1.2 Multistage Centrifugal Pumps

Multistage centrifugal pumps contain two or more stages and can be either single or double suction. Fluid is discharged from one stage to the next through passages in the pump casing. Each stage increases the speed of the fluid until the desired pressure is reached. *Figure 5* shows a multistage pump.

2.1.3 Slurry Pumping

Slurries are mixtures of fluids, most commonly water, carrying large quantities of particles of solids. Wastewater pumping is an obvious case in point, but mining and paper pulp are other examples. In such cases, the slurry moving through the pump can be very abrasive to the impellers and the volute of the pump. There are common solutions to the wear problem:

- Rubber lining or some other elastomeric coating, usually used with finer particles. The rubber lined pump is characteristically used for slurries with particles less than ¼ inch (6 mm) in diameter, because of the impact on the coating.
- Specially hardened metal pumps. The speed of these pumps is usually higher than that of rubber-coated pumps.

For applications such as wastewater or pulp processing, the solids in the flow can clog normal pumps. Grinding and chopping pumps have specially designed impellers that cut or grind solids while pumping. *Figure 6* shows a chopping pump, in which the sharpened impeller blades move against a straight cross bar.

The progressive capacity pump, known by the name of one of its pioneering manufacturers as the Moyno®, is especially effective for moving sludge and high-solids slurries. The principle is that of the Archimedes screw. A spiral rotor turns within a soft nitrile stator producing a moving cavity full of liquid. The result is a steady, non-pulsating flow, and the rounded edges of the rotor help keep the solids in the fluid from getting caught on the rotor and jamming the cavity (*Figure 7*). These pumps are often used in sewer applications with a grinder to break up solids.

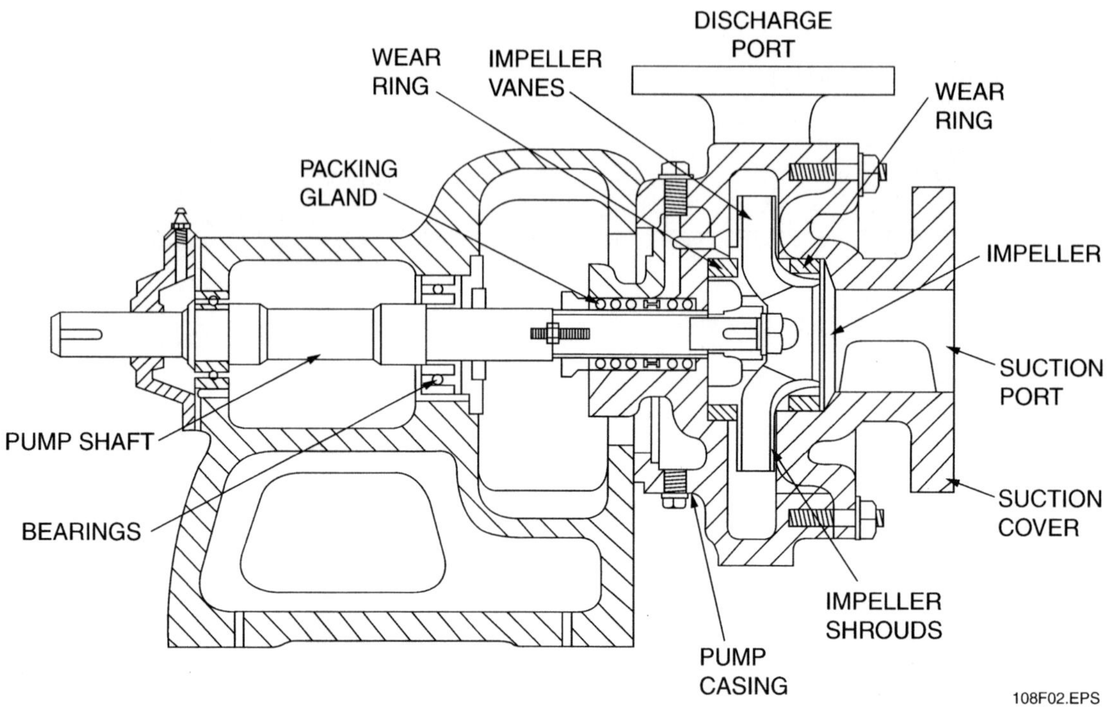

Figure 2 ◆ Centrifugal pump with a vertically split case.

2.2.0 Rotary Pumps

A rotary pump is a positive-displacement pump that operates by a turning motion. Rotary pumps may be gear, screw, vane, or flexible impeller types. Rotary pumps are used in hydraulic systems. They are also used to transfer fluids such as oils, solvents, chemicals, and paints.

2.2.1 Gear Pumps

The gear pump is the most common type of rotary pump. Gear pumps are used mainly in hydraulic systems. They are also used for bearing lubrication in industrial machines and automotive engines. Gear pumps may be either external or internal design.

In gear pumps, the fluid being pumped is drawn into the intake port by the gear teeth. The fluid is then forced through the space between the pump casing and the impellers and out the discharge port. The meshing of the gear teeth prevents the fluid from flowing back out the intake port. *Figure 8* shows a gear pump.

Spur gear impellers are the most common type used in gear pumps. They are economical to manufacture and maintain. Where a smooth fluid flow and transfer of power is needed, helical and herringbone gears are used. Helical and herringbone gears are often used in pumps that handle larger capacities and higher speeds than spur gear pumps.

Figure 9 shows the three types of gear impellers.

108F04.EPS

Figure 4 ◆ Double-suction centrifugal pump.

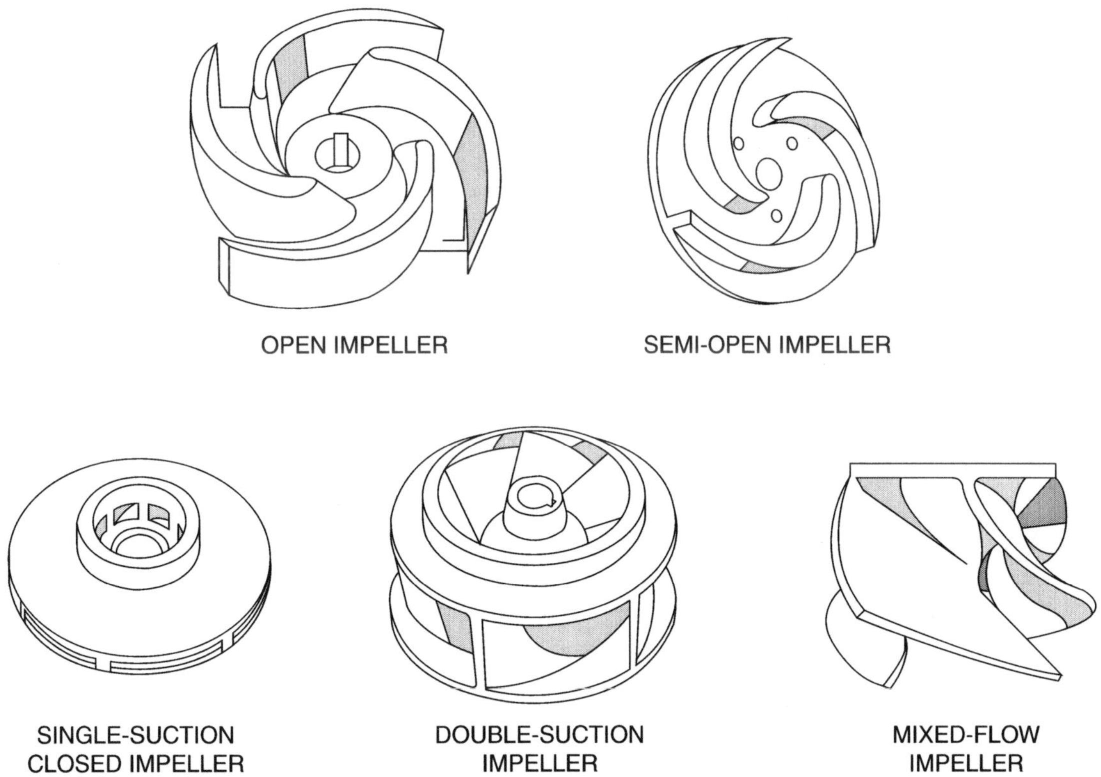

108F03.EPS

Figure 3 ◆ Impeller types.

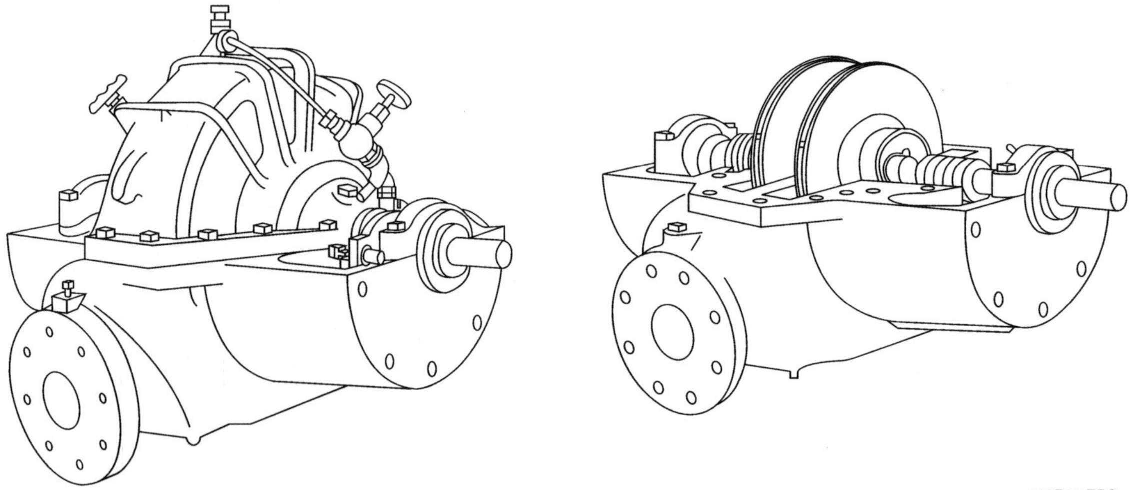

108F05.EPS

Figure 5 ◆ Multistage pump.

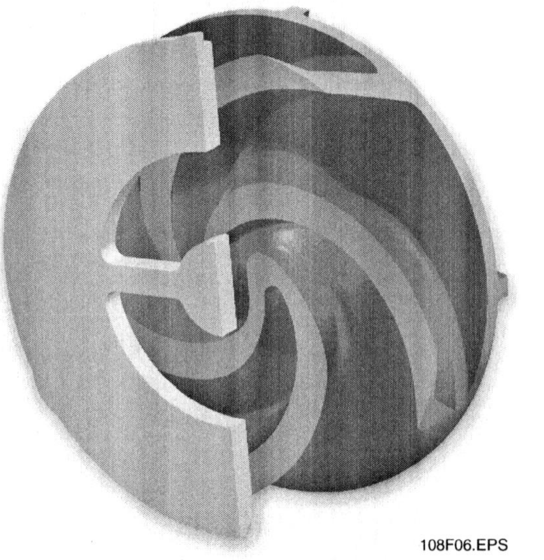

108F06.EPS

Figure 6 ◆ Chopping pump.

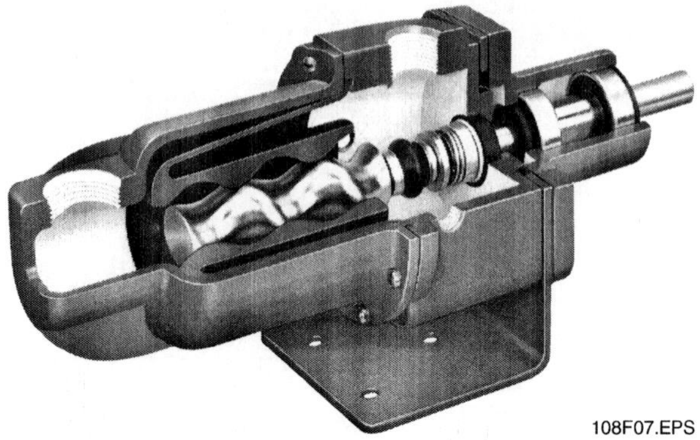

108F07.EPS

Figure 7 ◆ Moyno® pump.

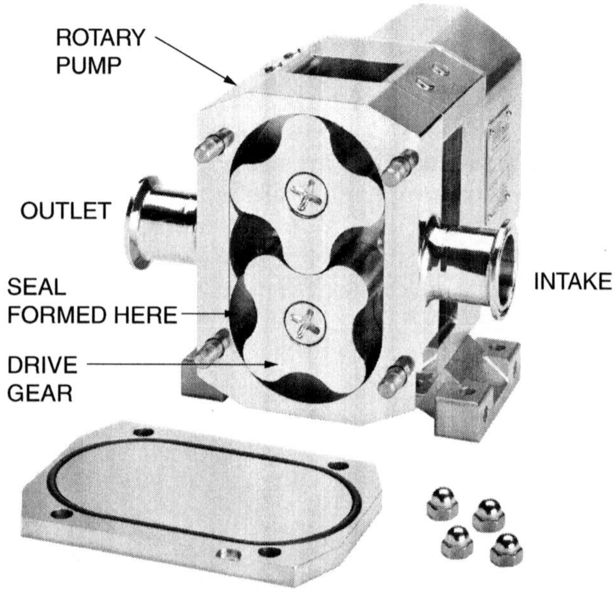

108F08.EPS

Figure 8 ◆ Gear pump.

2.2.2 External Gear Pumps

In external gear pumps, the casing forms the external housing of the pump and the impellers are fitted inside (*Figure 10*). The driving shaft extends beyond the casing and the driven shaft is contained within. External rotary gear pumps are designed to operate in either direction without major changes. Drive shaft rotation determines which is the suction port.

2.2.3 Internal Gear Pumps

Unlike the external gear pump, the internal gear pump (*Figure 11*) has two gears that mesh. The outer gear is the driving gear, and the inner gear is the driven gear. A crescent-shaped part keeps the gears separated to reduce eddy currents and increase pump efficiency. The crescent can be moved to allow the pump to operate in either direction.

2.2.4 Helical Screw Gear Pumps

Helical screw gear pumps (*Figure 12*) are used in industry for pumping oil and other heavy fluids. The pump is driven by the center rotor, or power rotor. The center rotor drives one or more idler rotors that mesh with it. In these pumps, the rotors and casing have close tolerances. Abrasive particles entering the pump can jam the rotors and can cause faster wearing of the rotors.

2.2.5 Vane Pumps

Vane pumps are rotary pumps used for transferring hydraulic oil, lubricating oil, solvents, and chemicals. Vane pumps are also used for viscous materials and other heavy fluids that may contain abrasive particles.

In these pumps, the vanes are made of soft materials such as molded neoprene or phenolic resin materials. The vanes are softer than the casing and are replaceable. Some vanes are spring-loaded (*Figure 13*) to help maintain contact with the casing during operation.

2.2.6 Flexible Impeller Pumps

Flexible impeller pumps are rotary pumps that are used to pump chemicals (*Figure 14*). Each pump has a flexible impeller made from rubber or plastic compounds.

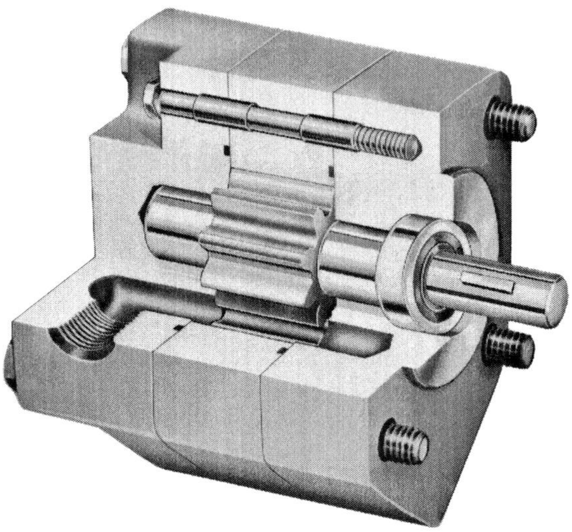

108F10.EPS

Figure 10 ◆ External gear pump.

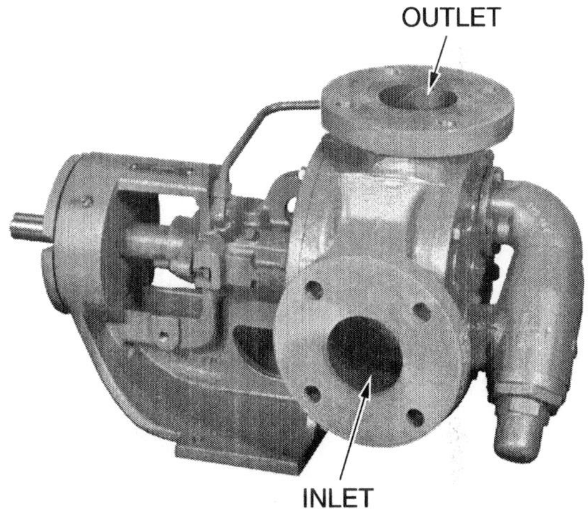

108F11.EPS

Figure 11 ◆ Internal gear pump.

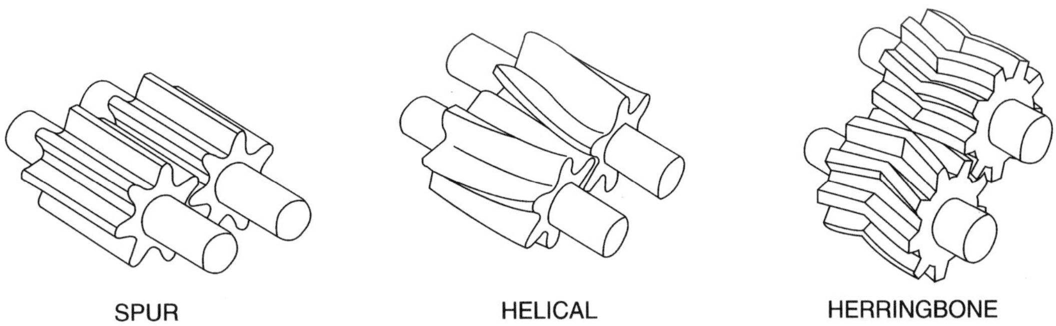

108F09.EPS

Figure 9 ◆ Gear impellers.

2.3.0 Reciprocating Pumps

Reciprocating pumps operate by back-and-forth or up-and-down, straight-line motion. Reciprocating pumps require a suction, or intake, stroke and a discharge stroke to move the fluid.

During the suction stroke, a check valve in the suction line opens and a check valve in the discharge line closes. During the discharge stroke, the suction check valve closes and the discharge check valve opens. The action of the check valves prevents the liquid from flowing back out the suction line on the discharge stroke. *Figure 15* shows check valves in the suction and discharge lines.

The pistons of reciprocating pumps create a back-and-forth movement to deliver a pulsating flow of fluid. This movement causes the flow to go from zero to maximum pressure and volume and then return to zero.

To compensate for this flow pulsation, an air chamber (*Figure 16*) is used in the discharge side of the reciprocating pumps. Air trapped in this chamber compresses as the pressure increases during the discharge stroke. The air expands as pressure decreases during the suction stroke. This reduces the extreme changes in pressure and results in a smoother fluid flow.

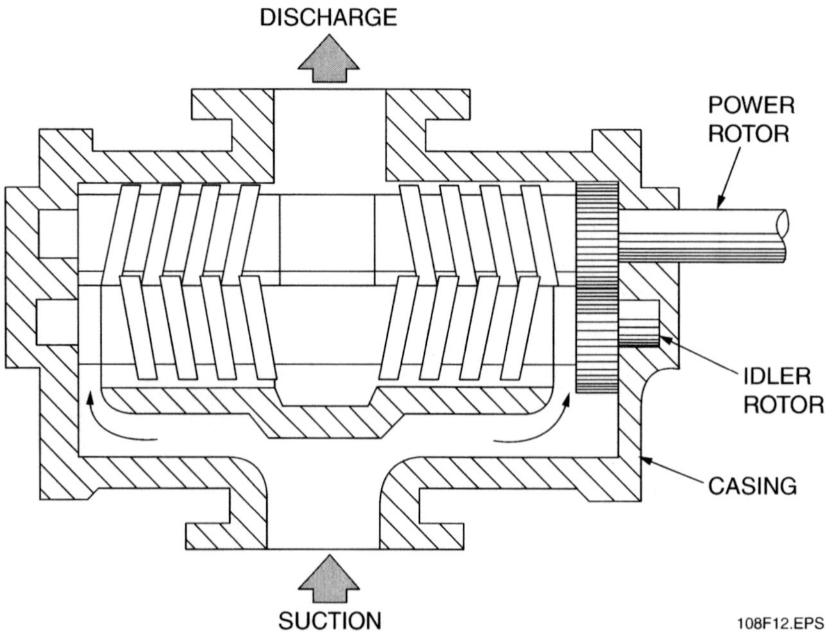

Figure 12 ◆ Helical screw gear pump.

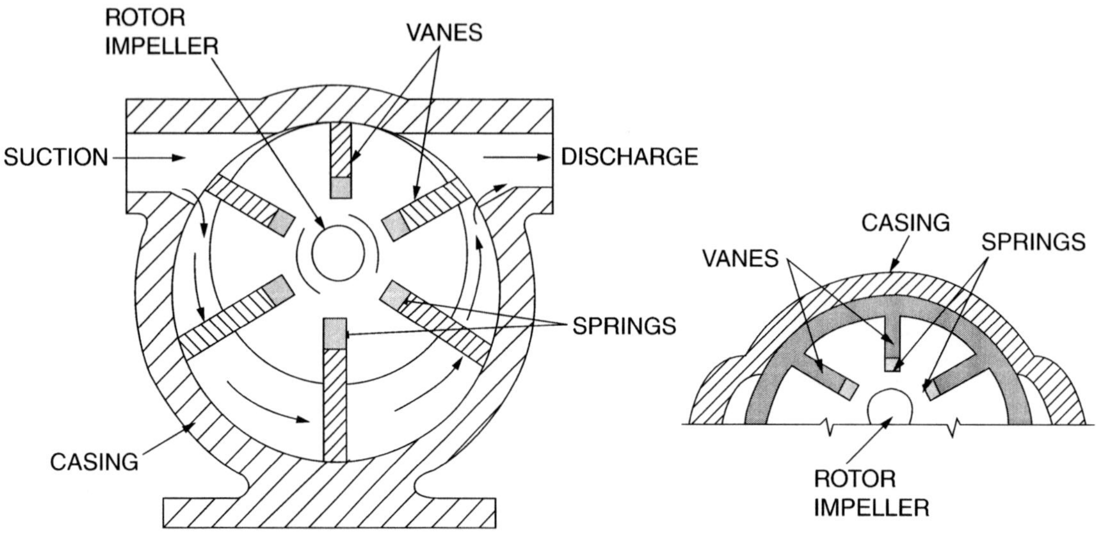

Figure 13 ◆ Spring-loaded impeller vanes.

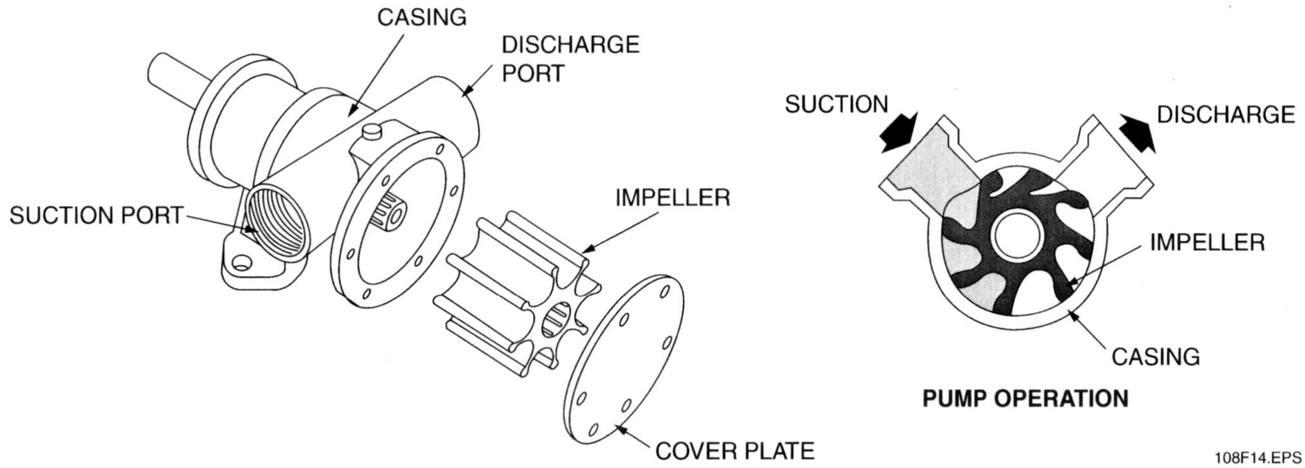

Figure 14 ◆ Flexible impeller pump parts.

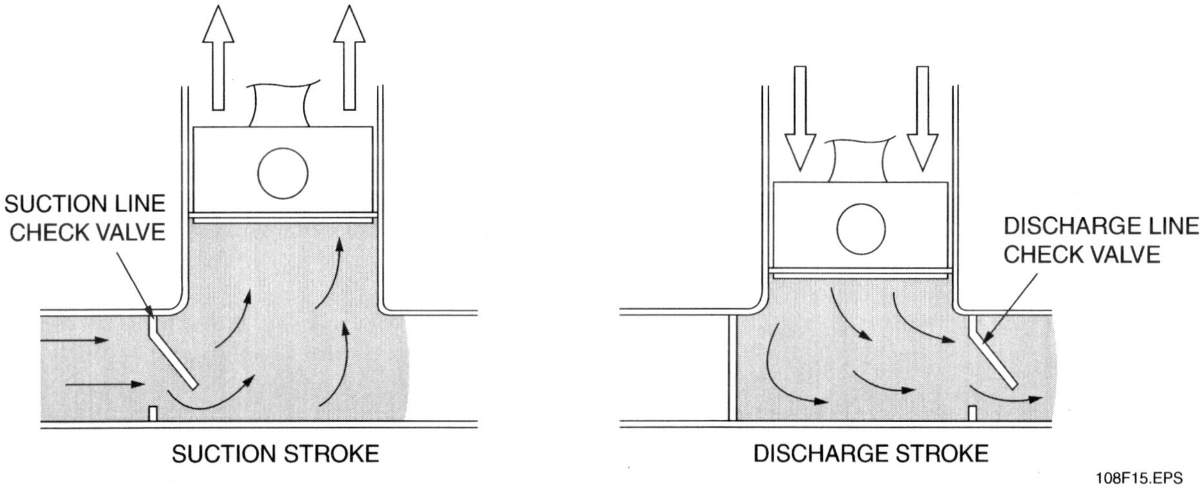

Figure 15 ◆ Check valves.

The reciprocating pump moves liquid by displacing the liquid with a piston. This operating principle is called positive displacement. *Figure 17* shows an example of a positive displacement plunger pump.

The three types of reciprocating pumps are piston, plunger, and diaphragm.

2.3.1 Piston Pumps

In piston pumps (*Figure 18*), the pumping element is a piston, which is a short, cylindrical part that moves back and forth in the pump chamber, or cylinder. The stroke of the piston is usually greater than the piston length. Leakage of fluid around the outside of the piston is controlled by piston rings or packing.

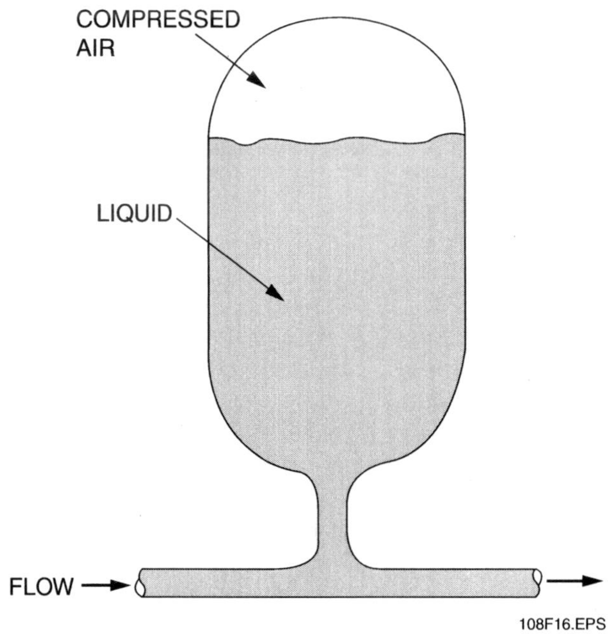

Figure 16 ◆ Air chamber.

2.3.2 Plunger Pumps

The plunger pump is often confused with the piston pump. The biggest difference is that a piston moves back and forth within a cylinder, while the plunger moves into and withdraws from a cylinder. *Figure 19* shows the difference between plungers and pistons in reciprocating pumps.

Unlike a piston, the length of a plunger is greater than its stroke. In the plunger pump, the packings in the cylinder control leakage around the outside of the plunger. *Figure 20* shows a plunger.

2.3.3 Diaphragm Pumps

A diaphragm pump has a flexible diaphragm used to displace fluids (*Figure 21*). The diaphragm covers the widest part of the pump cavity, sealing half of the cavity from the other half. The seal formed by the diaphragm makes the diaphragm pump ideal for pumping caustic chemicals and also for use as a metering pump. Care must be taken to ensure that the suction line never becomes blocked. If the suction valve is closed or some other blockage of the suction occurs, the diaphragm usually bursts.

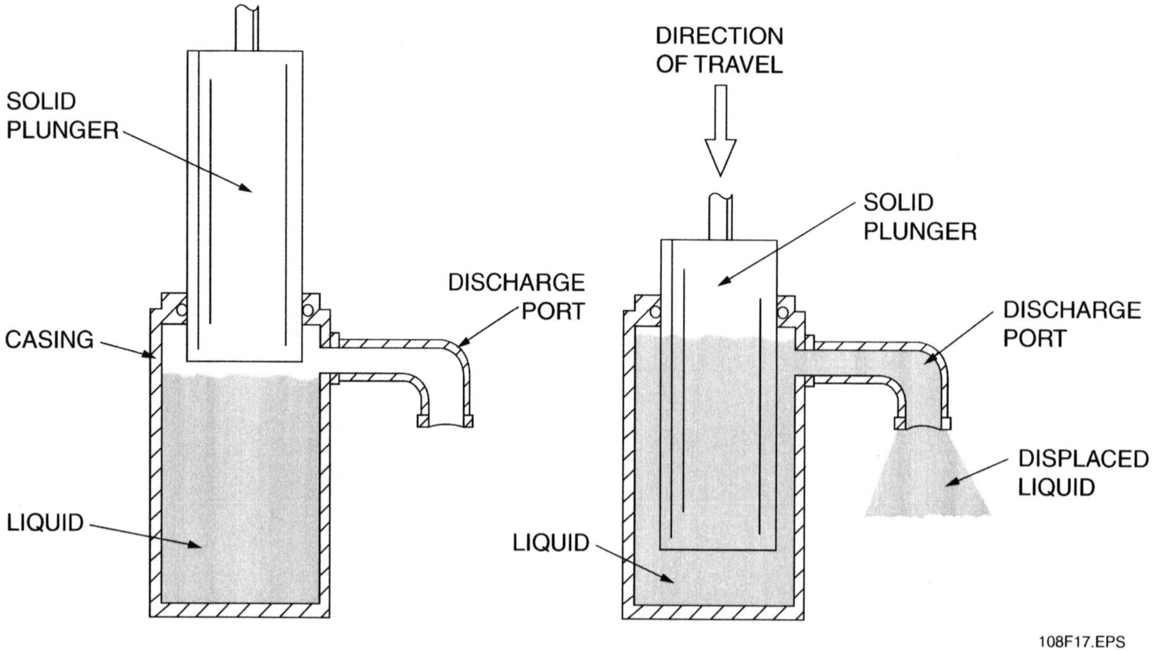

Figure 17 ◆ Positive displacement.

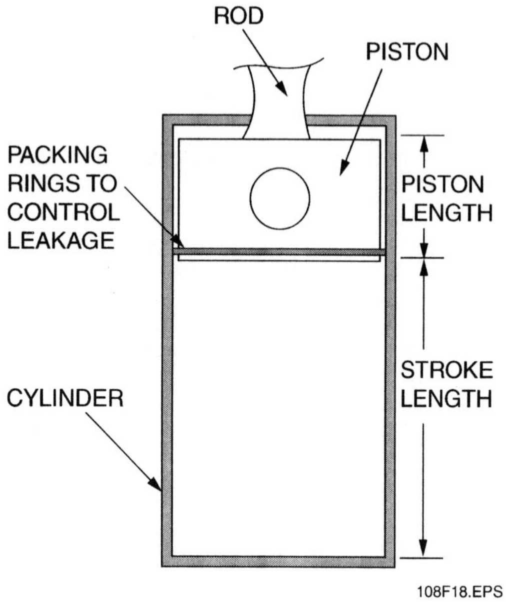

Figure 18 ◆ Piston.

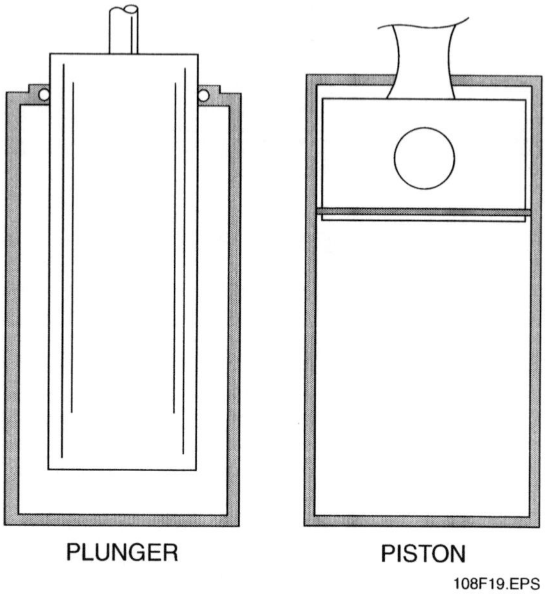

Figure 19 ◆ Plunger and piston.

A connecting rod connects the diaphragm with a motor-driven eccentric. As the eccentric rotates, the connecting rod pushes and pulls on the diaphragm. This action causes the diaphragm to bulge toward and then away from the suction and discharge valves on each stroke. The diaphragm is made of a flexible material, which is usually covered with a thin metal disc where the connecting rod is attached. By connecting to the metal disc, this spreads the force from the connecting rod over the diaphragm so that the connecting rod does not punch holes in the diaphragm during operation.

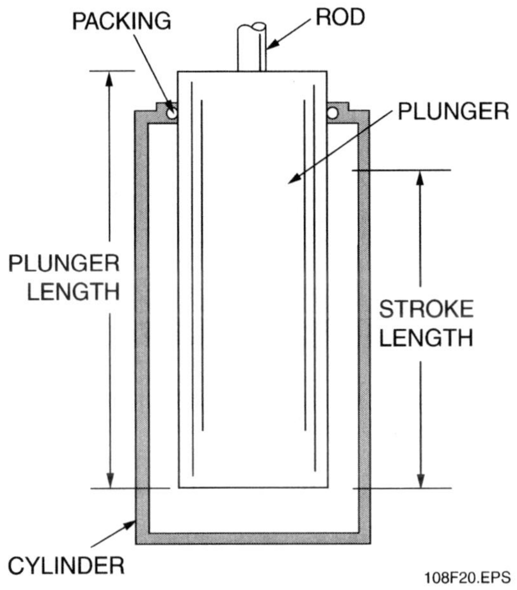

Figure 20 ◆ Plunger.

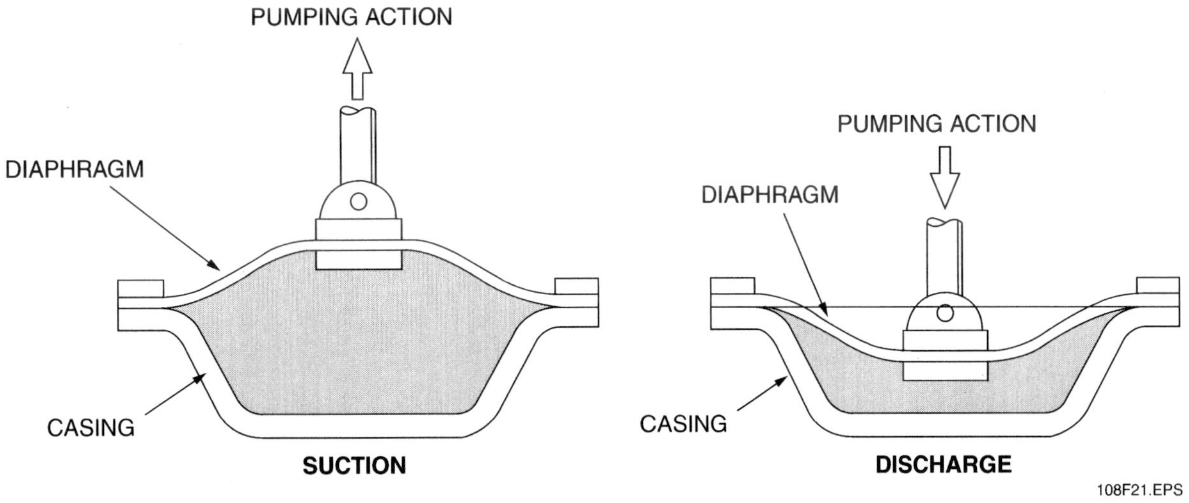

Figure 21 ◆ Diaphragm pump.

2.4.0 Metering Pumps

Metering pumps have close tolerances to pump exact amounts of fluids. A metering pump can discharge fluids at a rate ranging from zero to over 20 gallons per minute (gpm). Systems that require a large-capacity range may use more than one metering pump. Metering pumps are very sensitive to temperature changes and abuse. The two main types of metering pumps are the plunger and diaphragm. Other types, such as rotary and air cylinder piston pumps, are used for special applications. *Figure 22* shows two types of metering pumps.

2.4.1 Plunger Metering Pumps

Plunger metering pumps are the most common type used in industry. They can be adjusted mechanically to vary the amount of fluid being pumped. The plunger and the pump body are ground or polished as a set to form a close fit. This fit helps to meter the fluid discharge accurately and to reduce leakage within the pump. Suction and discharge check valves open and close as the fluid is pumped. The check valves may be ball or disc type. *Figure 23* shows a plunger pump.

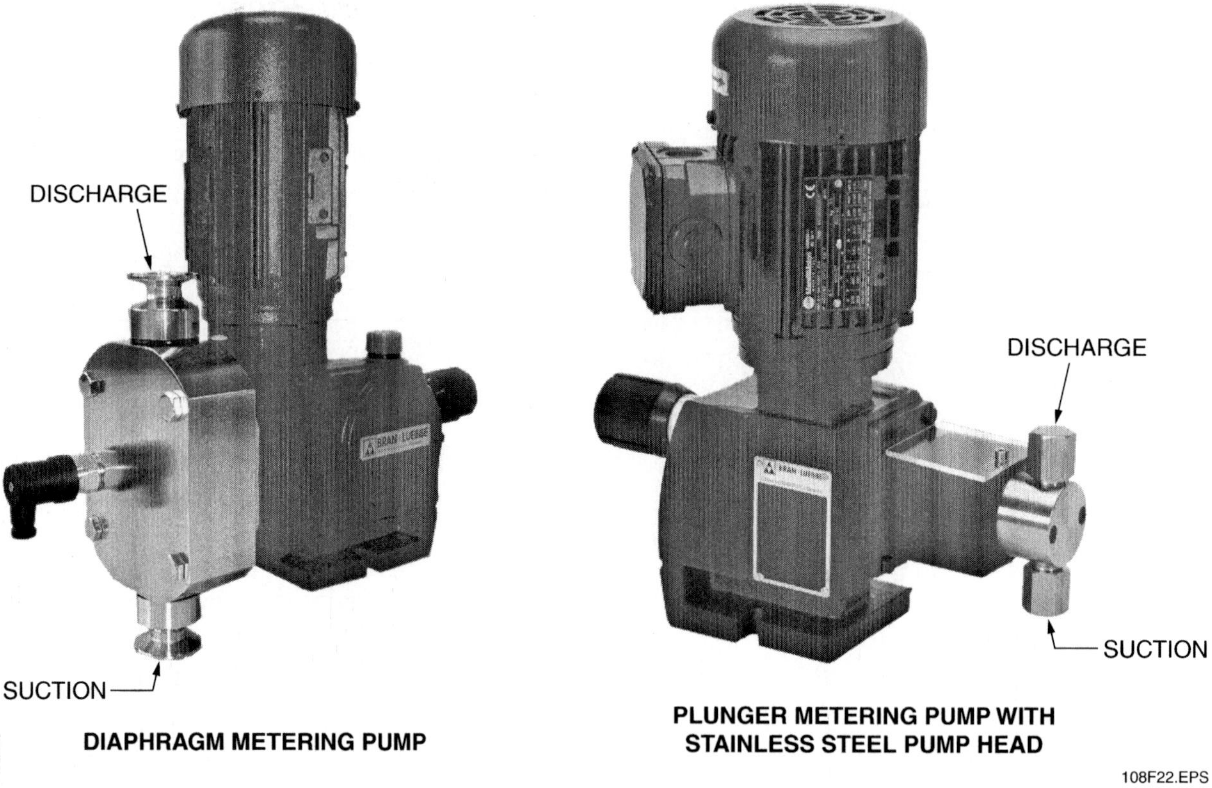

Figure 22 ◆ Metering pumps.

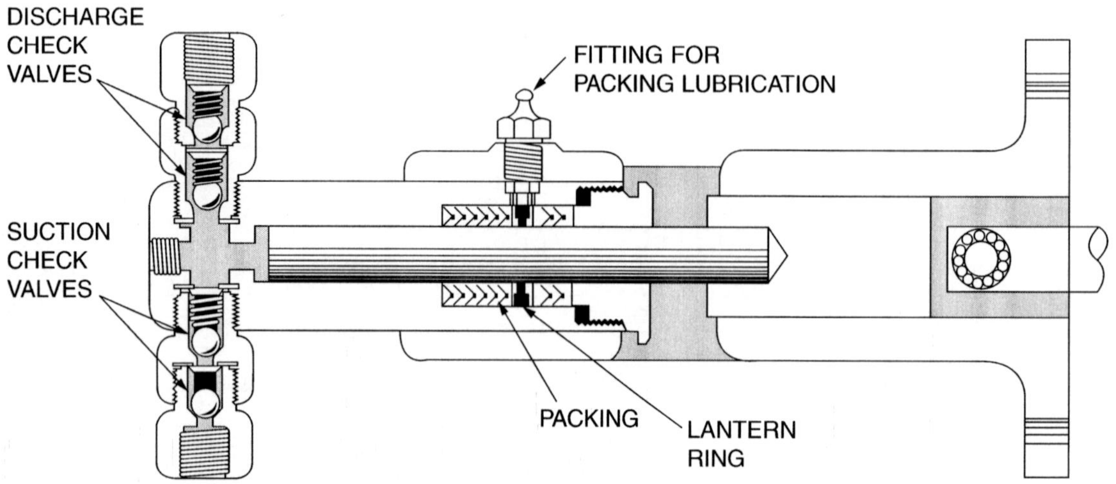

Figure 23 ◆ Plunger pump.

2.4.2 Diaphragm Metering Pumps

Most diaphragm metering pumps (*Figure 24*) use a piston action to move the diaphragm back and forth and move the fluid. Flow is controlled either by changing the speed of the drive or by making mechanical adjustments

Some metering pumps can be adjusted mechanically by cams or varying speed units. These can be controlled by electric or pneumatic actuating devices. These actuators are connected to the adjustable metering control devices and may be remotely controlled. Sensing devices installed in the lines pick up any variations in fluid flow. When a variation occurs, the sensing device sends a signal to the controller, which increases or decreases the amount of fluid.

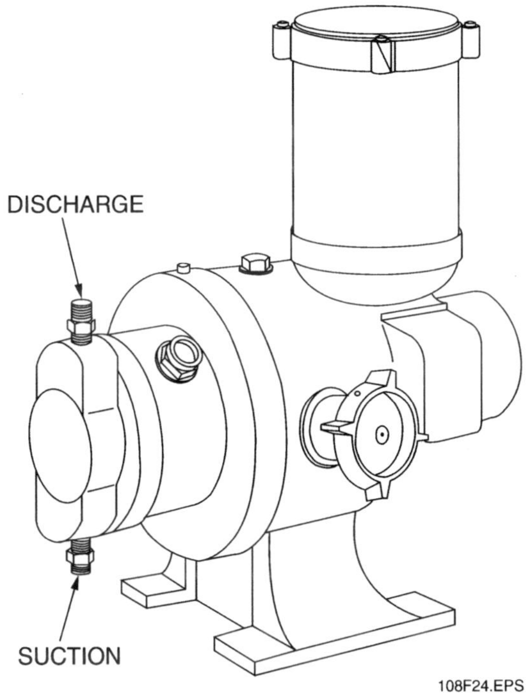

Figure 24 ◆ Diaphragm metering pump.

2.4.3 Peristaltic Pumps

Among specialty pumps for chemical applications is the peristaltic pump *(Figure 25)*, which uses two shoes on a rotor to squeeze fluids along a hose, thus reducing wear on the housing of the pump. The hose and shoes are usually easy to replace. Small peristaltic pumps are used as metering pumps in chemical and medical laboratories. The fact that the pump mechanism does not contact the fluid being pumped makes the peristaltic pump a good choice for chemical applications where high purity is required.

2.5.0 Vacuum Pumps

Vacuum pumps vary in size and weight. The functions of most vacuum pump parts are similar, although the parts vary in size.

A double-suction vacuum pump, also known as a liquid ring compressor, receives air in the inlet ports. The air flows down the bearing end heads and enters the rotor chamber. A revolving rotor inside the pump turns the air and a liquid, usually water, called liquid compressant or seal water. The liquid recedes to the outer wall of the pump, emptying the rotor chamber, and is replaced with air.

Figure 25 ◆ Peristaltic pump.

As the water follows the casing wall, the wall curves back toward the rotor, forcing the water toward the center of the pump shaft. The air that has collected between the water and the bottom of the rotor chamber is forced out the internal discharge port and out the external discharge. *Figure 26* shows airflow through a vacuum pump.

The liquid in a vacuum pump is necessary to compress the air and to seal clearances between the rotor and the cone. Some liquid goes out the discharge during pump operation. An external chamber receives the air and water from the pump.

The air is carried off the top while the liquid settles to the bottom and is circulated to the sewer system. Additional liquid is added to the pump at the air inlet. *Figure 27* shows the sealant/compressant liquid system.

The external parts of a vacuum pump and a brief description of each are as follows:

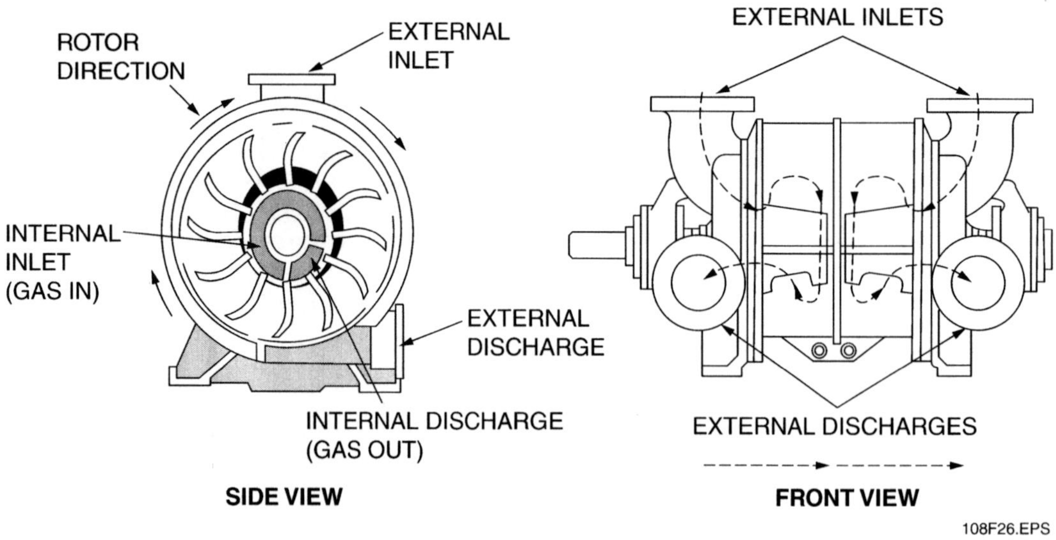

Figure 26 ◆ Airflow through vacuum pump.

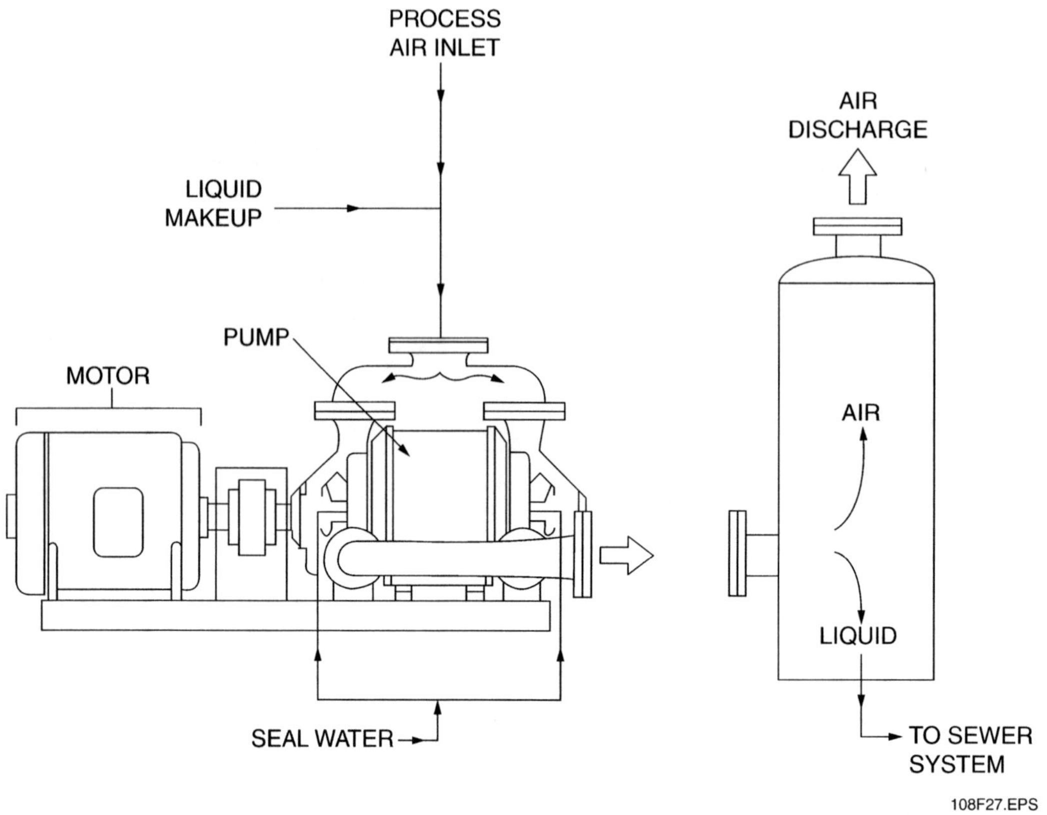

Figure 27 ◆ Sealant/compressant liquid system.

- *Circular casing* – The central metal housing that surrounds the rotor.
- *Floating bearing end head* – A cast metal end cap containing an inlet and discharge port; usually mounted on the end without the drive shaft.
- *Fixed bearing end head* – Cast metal end cap containing an inlet and discharge port; usually mounted on the end without the drive shaft; bearing fixed to eliminate side shift. A double-suction vacuum pump has two pump inlet ports.
- *Drive shaft* – The central shaft that is connected to the power source and turns the rotor.
- *Hold-down bolts* – Four of these are used to secure the vacuum pump assembly to a firm surface to minimize vibration.

Figure 28 shows the external parts of a double-suction vacuum pump.

The major internal parts of a double-suction vacuum pump move the air through the pump. The internal parts of a vacuum pump include the following:

- *Fixed bearing end cone* – A cast metal cone bolted to the end head. Directs incoming air to the rotor and outgoing air to the discharge through internal inlet and discharge ports.
- *Floating bearing end cone* – Performs the same function as the fixed bearing end cone. It is bolted to the opposite end head.
- *Rotor* – A series of blades that project from a hub. The blades are shrouded at the sides and form a series of chambers.
- *Shaft* – The central drive rod hooked to the power source and rotor. It is mounted in bearings at both ends.

Figure 29 shows the internal parts of a double-suction vacuum pump.

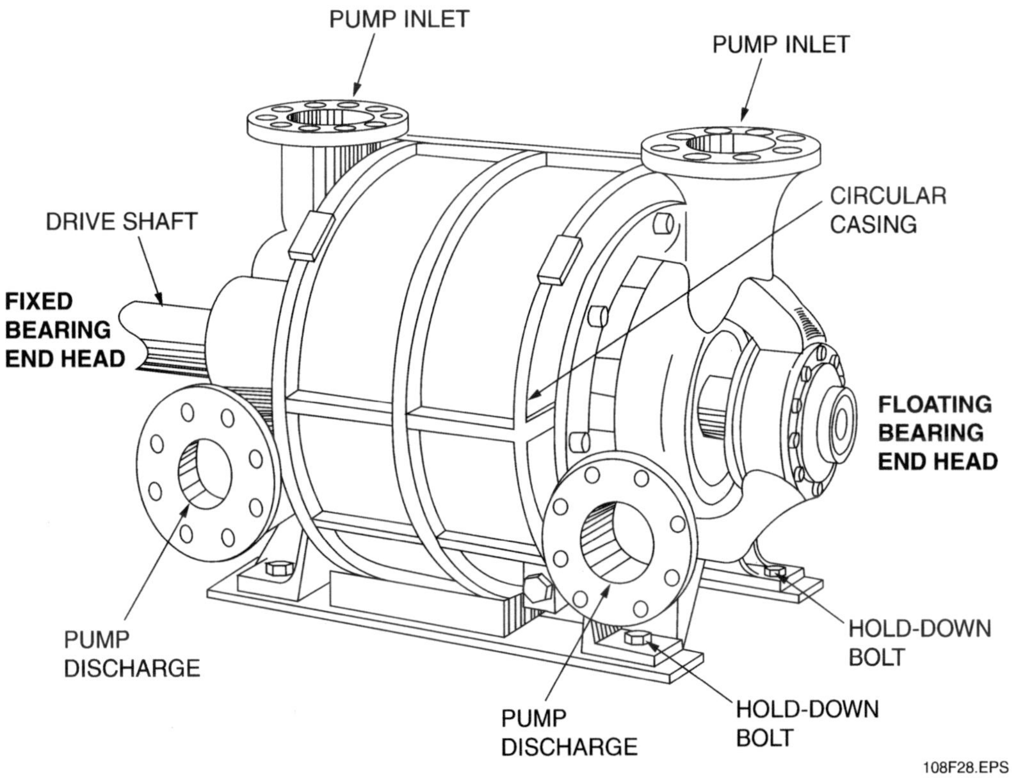

Figure 28 ◆ Double-suction vacuum pump external parts.

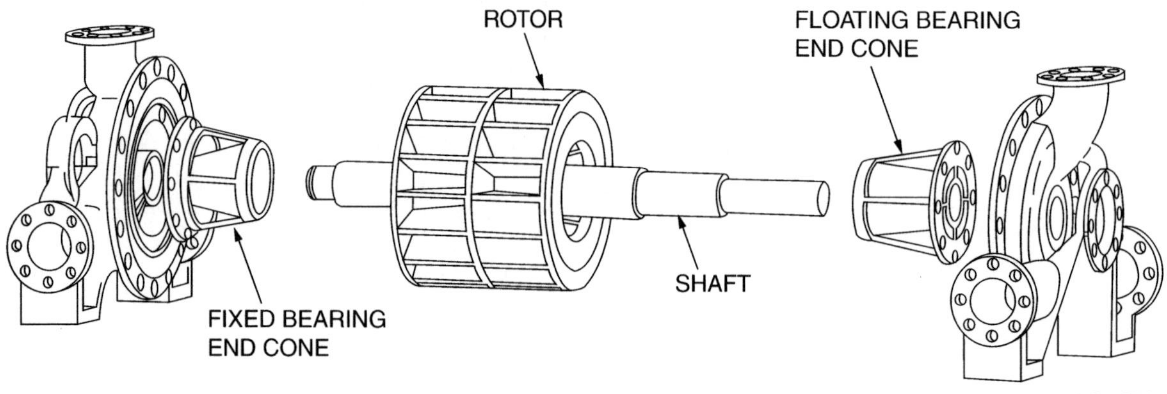

Figure 29 ◆ Double-suction vacuum pump internal parts.

3.0.0 ◆ NET POSITIVE SUCTION HEAD AND CAVITATION

Net positive suction head (NPSH) is the amount of suction **head** required to prevent vaporization of the pumped liquid. The suction head must be high enough to overcome friction losses at the intake and pressure drops through the impeller as the fluid increases in speed. The suction side must allow a smooth flow of liquid to enter the pump at a pressure high enough to prevent cavitation, which occurs when vapor bubbles form. It is important that the suction pressure at the pump inlet be well above the pressure at which cavitation occurs.

When the pressure conditions surrounding a liquid reach the vapor pressure of that liquid, the liquid begins to vaporize, turning into gas. Vapor bubbles form, and instead of a steady liquid flow, the pump draws a mixture of liquid and vapor. This decreases the amount of liquid being pumped. The volume change for a certain weight of liquid going to a gas is tremendous. For example, 1 cubic foot of water at 72°F expands to 1,700 cubic feet of vapor at the same temperature. It is clear, therefore, that any pumping system must be engineered to ensure that the liquid being pumped remains fluid.

3.1.0 Cures for Cavitation

Cavitation results when the NPSH available is less than the NPSH required. The solution for cavitation is to lower the required head by throttling the discharge. As you lower the flow through the pump, the NPSH requirement drops. This has the disadvantage of reducing efficiency, affecting the system operation, and possibly heating the fluid and raising its vapor pressure. If the pump is at the right operating point and it is changed, radial loads are generated that affect the bearings and seals on the pumps.

Cavitation can also be corrected by increasing the NPSH available.

- Increase the **static** head:
  - Lower the pump
  - Raise the level in the tank
  - Raise the tank
- Increase the pressure head:
  - Pressurize the tank
  - Increase the pressure in a pressurized tank by external means

In high-vapor pressure fluids, such as refrigerants and liquefied gases, the pressure in the tank comes from the vapor pressure of the fluid. Heating the tank increases the pressure gauge readings, but this does not help because the vapor pressure has increased and this figure is subtracted from the pressure head. The following items may help to decrease the vapor pressure:

- Reducing the piping and fitting loss
- Using larger pipe
- Using bell mouth inlets
- Using larger strainers
- Remove restrictions

When troubleshooting cavitation that started in a system where there previously was none, the above bulleted items are the usual areas to suspect. Common causes include blocked pumps, clogged strainers, buildup in the lines, and clogged foot valves. To eliminate cavitation, reduce the vapor pressure of the fluid. Vapor pressure of a fluid depends on the fluid and its temperature. Lowering the temperature lowers the vapor pressure. Pumps that are allowed to recirculate from discharge back to suction often have this type of problem. Recirculation builds up heat, which increases the vapor pressure.

3.2.0 Air or Vapor Blockage

Cavitation that is not caused by any of the four NPSH elements may be caused by air being sucked or dropped into the system or by vapors being trapped in some section of the system and restricting the flow.

- *Air in system* – Air in the suction side of the pump causes a noise that can be annoying but does not cause the same type of destruction caused by cavitation. When an air bubble enters the pump, it expands based on the law of gases (one-half the pressure, twice the volume). The air occupies space and therefore reduces the output of the pump and its efficiency. Sometimes, air accompanies classic cavitation, which causes a cracking noise, severe pump vibration, and internal damage. This happens when the air or vapor bubble gets trapped in the eye of the pump. Because the air or vapor bubble is lighter than the fluid, it has a tendency to stay at the eye while the fluid is thrown to the outside. This air restricts the input to the pump, causing a severe lowering of NPSH. If the pump is taking suction from a tank near its own level or below it, the air could cause the pump to lose its prime and stop pumping altogether.
- *Sources of air entering the suction* – The packing of the pump is the most likely spot for air to enter. This problem can be checked by placing grease, water, or shaving soap around the valve stem. Seals are preferable on a lift pump. If the lift pump has packing, the packing may require a flushing fluid to block the air. If there are any suction valves, check the packing in these. A soft packing ring between harder rings solves this problem.

4.0.0 ◆ INSTALLING PUMPS

All procedures involved in pump installation must be performed precisely to ensure a successful installation. When the pump is properly set, it should be precisely aligned on the equipment center lines, firmly bolted to its foundation, and properly aligned with its motor. The suction and discharge connections must be level, plumb, and properly aligned with the connecting pipelines. The foundation and baseplate should be rigid enough to withstand the vibration and strains that the operating pump produces.

The procedures involved in installing a pump vary from one type of pump to another. Even pumps of the same type made by different manufacturers may vary in the way they are installed. It is important to follow the manufacturer's installation procedures when installing any pump. The following sections provide some common guidelines for installing pumps that should be followed in conjunction with or in addition to the manufacturer's installation instructions.

4.1.0 Preinstallation Guidelines

There are certain checks that should be performed prior to pump installation to verify that all preinstallation conditions have been met. These checks are as follows:

- Ensure that the baseplate is aligned with the equipment center lines.
- Ensure that the baseplate is level in all directions.
- Ensure that the baseplate is at the proper elevation.
- Ensure that the baseplate is properly grouted.
- Ensure that the anchor bolts are tightened to the proper torque.
- Ensure that the baseplate is clean and free of foreign matter.
- Ensure that the baseplate is in good condition, with no damage, cracks, or extensive corrosion.
- Ensure that the pump mounting holes are properly laid out to the equipment center lines and pump base. Holes that are not precisely laid out may cause the bolts to become bolt-bound during the alignment procedure.
- Ensure that the pump mounting holes are in good condition, with no stripped or damaged threads.

4.2.0 Installation Guidelines

The following guidelines should be used in conjunction with the pump manufacturer's installation procedures.

- Lay out and drill the baseplate mounting holes if they do not already exist.

> **NOTE**
> The mounting holes must be laid out and drilled precisely according to the equipment center lines and the pump base holes to prevent them from becoming bolt-bound during alignment later.

- Set the pump on the base, and align it with the equipment center lines.
- Shim the pump to the proper elevation. The pump should be shimmed up to an elevation that is higher than the motor so that the motor can be aligned to the pump.

- Plumb the pump suction connection and level the pump discharge connection, using a precision level.
- Tighten the pump hold-down bolts (lugs) to the proper torque.
- Align the motor to the pump.
- Connect the suction and discharge piping to the pump. If piping is under stress, this can affect the function of the pump.

> **NOTE**
> Set up dial indicators on the pump and motor couplings and watch for coupling stress when connecting pipe to the pump.

- Perform other installation functions as directed by the manufacturer's installation procedures.

5.0.0 ◆ DRIVERS

A driver is a motor or engine that activates a pump or other equipment such as a generator. In general, most drivers are electric motors of the proper horsepower and speed rating, possibly incorporating a variable frequency drive. Other drivers such as internal combustion engines are fueled by gasoline, natural gas, or diesel fuel. Some companies employ turbine engines as drivers where high horsepower ratings, high speeds, or extremely large volumes are required.

5.1.0 Identifying Types of Drivers

Generally, drivers fall into one of three categories. These are electric drivers, including variable speed drives; gas and diesel reciprocating drivers; and turbine drivers.

5.2.0 Electric Drivers

When a pump installation is operating under normal conditions, the movement of product is determined by the rotations per minute (rpm) of the main-line pumping units by the drivers. It may be applied in several ways at several points, but the total horsepower determines pressure and flow through the pipeline (see *Figure 30*). The pump driver supplies a steady supply of power to the pump, but it may be more than required. In fixed-speed pumps, since the pump speed cannot be adjusted, throttling with valves must be used to control the pipeline flow rate. Some pressure is always lost due to throttling.

108F30.EPS

Figure 30 ◆ Electric driver on an industrial compressor.

Pumping orders specify the rate of flow required for the operation. Control center personnel calculate the amount of horsepower and the amount of throttling necessary to push the volume of product through the pipe at the desired rate of flow. Elevation and other obstacles must be included in the calculations. After the calculations, control center personnel can determine the number of pumps to bring online and settings for horsepower or throttling.

5.2.1 Variable Frequency Drives

Some main-line drivers, called variable frequency drives (VFD), have an adjustable speed control with which they may be set to pump at a specific rate and pressure. The horsepower at that rate is calculated and adjusted for efficiency and economic operation. VFDs control pump speed in graduated steps. The VFD is designed to change speed for control of pressure and flow by changing the frequency (the number of cycles per second) of the power supply.

The VFD makes it easier to improve efficiency and control pressure and flow. VFDs offer more efficiency because no throttling is required to control the pipeline. Secondly, VFDs offer significant savings in energy required because only partial power is needed and used in many operations. Variable frequency drives will work with most three-phase motors, so they can be used to control existing motor and pump combinations.

5.2.2 Motor Winding

Motor winding, properly called field winding, is made up of a number of turns of wire to create field coils that produce a constant-strength magnetic field in an electric motor. The windings produce the magnetic flux that reacts with the armature to turn the rotor (see *Figure 31*).

Motor speed is a factor that must be considered when setting the horsepower ratings for pressure and flow on main-line units. Motor speed affects the operating temperature. When motors are operated at lower speeds, the ability to dissipate heat is reduced due to a slower cooling fan speed. If a motor is set too slow or slows down, operations personnel must adjust the speed quickly to keep the motor from overheating and possibly burning the motor's winding. Operations personnel may need to switch to another main-line unit to prevent a complete motor failure.

5.2.3 Power Optimization

Power optimization is used to economically coordinate throughput and horsepower requirements. Electricity is an expensive necessity on the pipeline. Almost every pipeline activity consumes some type of electrical power. Operations personnel make efforts to economize operations by optimizing electrical efficiency. With training, experience, and good judgment, shipments are set up to make the most economical use of equipment and power. For example, variable frequency drives allow the operator to set up pumping operations that develop the most throughput for the amount of power used.

Switching main-line units might provide a more economical operation in some locations. If a shipment is running on more than one unit and requiring throttle, it might be more economical to shut down a unit and use less throttle. Switching from a constant speed pump to a pump with VFD could also result in significant power savings. Some units might be switched off while others downstream might be brought online to achieve line balance. The entire pipeline would then operate more efficiently.

5.3.0 Diesel Drivers

Diesel drivers are used in locations without access to commercial electricity and as auxiliary equipment for emergency use. Diesel drivers are often used to operate generators to supply power to electric units and installations during power failures. In some locations, diesel engines are connected directly to pumps or are used as portable units for backup systems. Diesel units are economical and dependable and offer a wide range of horsepower and speed ratings. Diesel drivers are suitable for many types of installations that can be mounted on skids as well as for portable truck- or trailer-mounted units. Diesel units are often used with gearboxes designed to change shaft speed.

Diesel engines are commonly used for fire and flood pumps and permanently installed generator sets for air-conditioning and refrigeration, as well as compressors, blowers for aeration, and pumps

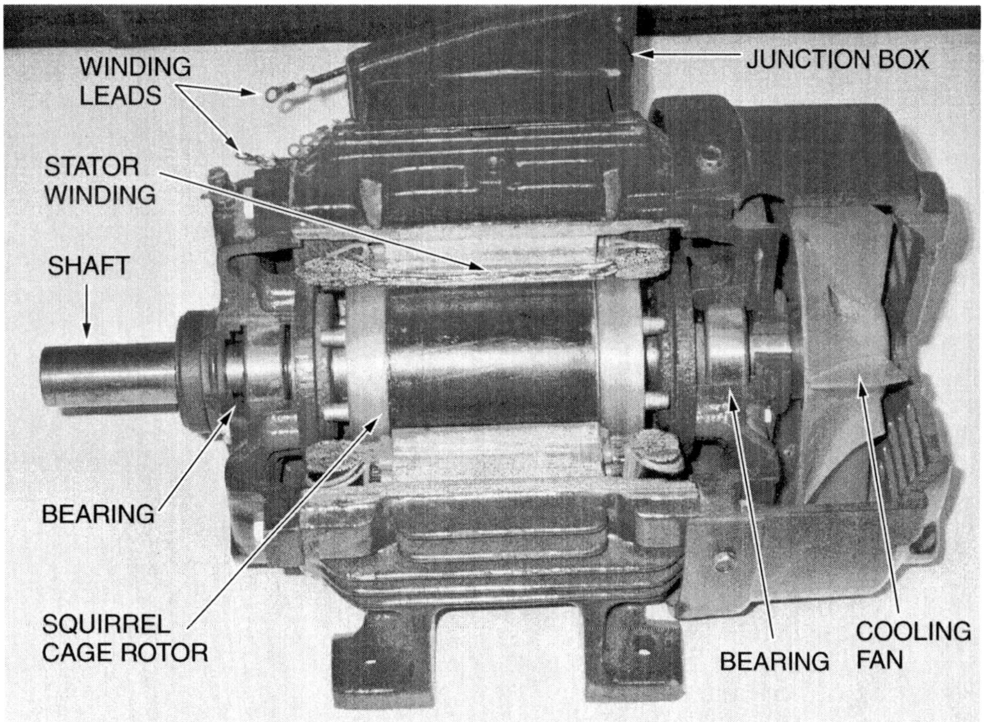

108F31.EPS

Figure 31 ◆ Main components of an induction motor.

for treatment plants. In some cases, engine systems have been developed that will reclaim and burn gas from landfills and sewage treatment facilities.

5.3.1 Diesel Characteristics

Diesel engine drivers have some characteristics that must be considered when selecting a unit. As with other equipment, the size, rating, and possible limitations of the unit being considered should be carefully examined and compared with the needs of the application. Diesel engines are affected by temperature, barometric pressure, and humidity; therefore, climate is an important factor in selecting a diesel driver.

Some factors that must be considered when selecting a diesel driver include:

- Size of the unit
- Weight of the unit
- Noise level, in some locations
- Convenient fuel supply

Diesel engines require less maintenance than other internal combustion engines, partly because they require no ignition system. However, you must provide clean fuel. Diesels rely on the air that fills the cylinders being compressed until the heat of compression ignites the injected fuel. Fuel ignited by the hot air expands, pushing the piston down into the cylinder on the power stroke. Compression ratios on some diesel engines may be as high as 23:1 to ensure ignition. Diesel engines are able to extract more power per gallon of fuel under full load than gas or gasoline engines. At the same time, the diesel is able to maintain the high compression under light loads, which is not possible with gas engines, and therefore run much more economically.

Although diesels can be designed to burn many different types of fuels, the common diesel grades 1 or 2 are preferred unless the manufacturer's recommendations specify otherwise. Fuel consumption rates can be a factor in the selection of units, so the company specifications should be compared. At the same time, fuel-handling plans must be considered because dirty or contaminated fuels are responsible for many diesel engine problems, largely due to poor fuel-handling procedures.

Many authorities agree that an internal combustion engine should be kept at a steady load of no more than 70 percent of its maximum power. If this load range is exceeded, the overload will result in premature failure of the bearings and other diesel or gas engine components. In intermittent loading, an extra 10 percent of the maximum can be tolerated. This keeps the engine within its most economical operating range of about 75 percent of its maximum. In keeping with diesel operating characteristics, a diesel of equivalent horsepower will normally be much heavier than a gas or gasoline engine due to the necessity of being able to withstand the greater pressure on bearings and cylinder components.

5.3.2 Diesel Ratings

Caution should also be used when considering an engine manufactured in a foreign country, since there are international differences in the way horsepower and other ratings are calculated. For example, one metric horsepower is slightly less than one horsepower as calculated in the United States. Therefore, when considering engine power ratings, specifications as to the conditions under which the rating was calculated must be examined closely. A rating by groups such as the Diesel Engine Manufacturer's Association (DEMA), the Society of Automotive Engineers (SAE), or International Standards Organization (ISO) will give exact information as to how the rating of an engine was derived. Horsepower ratings may be greatly increased by the addition of a turbocharger, a small, exhaust-driven compressor that forces extra air into the engine to increase efficiency. However, it must be determined whether the engine components can withstand the added stress of increased power over the long haul.

5.4.0 Gas Drivers

Gas drivers are set up to burn liquid petroleum gas, natural gas, or other gases as fuel. Gas drivers are considered cleaner, environmentally speaking, than other internal combustion engines. Gas drivers are often selected for service in locations where personnel will be exposed to the exhaust emissions. They can often be used in the same situations as diesel drivers and do the same work. However, gas-fueled drivers must be selected based on different criteria because their horsepower ratings are achieved somewhat differently.

Gas engines, as well as gasoline engines, are rated at lower temperatures than diesel, while the diesel engine gains much of its power from the higher compression and heat value of diesel fuel. Gas engines are not as productive when under full load because they cannot maintain total compression like diesels.

A diesel engine draws clean air directly into the cylinder in a volume near or equal to the total displacement of the piston. The fuel is mixed with this air at or during compression. The heat of the compressed air ignites the fuel for a complete burn. A gas engine utilizes a carburetor or valve body-type system that mixes fuel with the air before it is drawn into the cylinder. An explosive mixture already exists before it enters the cylinder, and therefore the compression must be kept low enough to allow combustion. A spark is then introduced to cause the cylinder to fire or explode. If the engine were allowed to self-ignite as in a diesel, there would not be proper control over ignition timing to operate the engine with any real efficiency. Therefore, compression is restricted below the self-ignition point, and a spark is used for ignition at the proper time. Such engines utilize only as much fuel as required by the engine load, and therefore the intake must be throttled to keep a consistent mixture of air and fuel supplied to the engine.

Some other considerations when selecting a gas driver include:

- Size of the unit
- Weight of the unit
- Noise level, in some locations
- Convenient fuel supply

5.5.0 Turbine Drivers

Turbine drivers are among the most modern advancements in providing power for pumps, compressors, and generators. Turbine drivers are specialized power plants. Although they are the ideal solution to many power needs, they are probably not suitable for most locations. Turbine drivers are actually modified versions of aircraft engines, fastened down and harnessed to provide rotation for shafts (see *Figure 32*).

Turbine drivers are constructed as simple internal combustion engines with two shafts. The jet engine is used to convert an air-fuel mixture into powerful exhaust gas and is called an exhaust gas generator. A reaction turbine with a separate shaft is installed directly behind the engine to convert the exhaust gas/heat energy into shaft rotation that can be harnessed to drive pumps, generators, compressors, or any other equipment needed. Normally, they are installed in isolated locations.

For industrial applications, most of the turbine driver's external systems and equipment are removed. Necessary systems are installed separately from the engine and connected from a location off the unit. Such systems handle lubricating oil, hydraulic supply, and fuel control. The engines are controlled by governor systems to protect them from over-speed and serious damage. The engine converts the energy of the fuel to

108F32.EPS

Figure 32 ◆ Turbine driver.

extremely high-energy exhaust gas. Some turbine units can supply power in the range of 20,000 exhaust gas horsepower.

Exhaust gas is directed through a transition duct into a diffuser chamber, where it is spread evenly against the blades of a reaction turbine. There is no solid drive shaft connection between the jet engine and the reaction turbine. The exhaust gas forces the reaction turbine to rotate in a comparable manner to the engine. The reactor turbine rotates a jackshaft, and some models can develop over 15,000 shaft horsepower. A governor controls the speed of the units, and with appropriate couplings and gearboxes, the power can be used to drive pumps, generators, and any other equipment needed. With the proper setup, one turbine driver can power more than one unit at a time. In most installations, however, a unit only powers one main-line pump and, possibly, a generator.

Some of the characteristics of turbine drivers require special consideration when deciding if a turbine driver is suitable. The initial costs of purchasing and installing turbine drivers are considerably higher than those associated with conventional drivers. Turbines require much more space than conventional drivers. Turbines generate tremendous noise, and the engine exhaust must be dealt with effectively. These are all an integral part of operating a turbine power system. If these factors are acceptable and such a powerful drive system is needed, turbine drivers are considered well worth the effort from engineering points of view, but they are expensive to operate.

Review Questions

1. All pumps take in fluids from the _____.
   a. top side
   b. bottom side
   c. suction side
   d. discharge side

2. Centrifugal pumps are basically _____ pumps.
   a. low-temperature
   b. cavitating
   c. vacuum
   d. low-pressure

3. Positive-displacement pumps are used to pump _____ liquids.
   a. high-viscosity
   b. high-temperature
   c. explosive
   d. water-solution

4. The shell surrounding the shaft, bearings, packing gland, and impeller of a centrifugal pump is the _____.
   a. packing gland
   b. impeller shrouds
   c. pump casing
   d. mechanical seal

5. The part of a centrifugal pump that increases the speed of the fluid is the _____.
   a. wearing ring
   b. impeller
   c. mechanical seal
   d. discharge port

6. A multistage centrifugal pump increases the speed of the fluid until the desired _____ is reached.
   a. velocity
   b. pressure
   c. vapor
   d. viscosity

7. A mixture of solid particles suspended in fluid is a _____.
   a. slurry
   b. glob
   c. loppy
   d. vortex

8. The problem of wear in slurry pumps is solved by _____.
   a. faster pumps
   b. rubber lining or hardened metal
   c. coal-tar epoxy
   d. slower pumps

9. The pump in a hydraulic system would be a _____ pump.
   a. rotary
   b. centrifugal
   c. double-suction
   d. recessed impeller vortex

10. The gear pump is used mainly in _____ systems.
    a. slurry
    b. hydraulic
    c. multi-stage
    d. vortex

11. The most common type of impeller used in gear pumps is the _____.
    a. shrouded
    b. helical
    c. herringbone
    d. spur gear

12. In external gear pumps, the direction of the drive shaft rotation determines the _____.
    a. suction port
    b. centrifugal force
    c. push rods
    d. piston travel

13. In helical screw gear pumps, the center rotor drives _____.
    a. a spur gear
    b. the Archimedes screw
    c. one or more idler rotors
    d. a fan

14. In reciprocating valves, the fluid is kept moving in one direction by the _____.
    a. impeller
    b. spur gear
    c. check valves
    d. vanes

Review Questions

15. The three types of reciprocating pumps are _____.
    a. centrifugal, gear, and diaphragm
    b. piston, plunger, and diaphragm
    c. helical, spur, and impeller
    d. multi-stage, double suction, and flexible vane

16. In plunger pumps, the plunger is _____ the stroke.
    a. shorter than
    b. outside
    c. under
    d. longer than

17. In the diaphragm pump, the connecting rod runs between the diaphragm and a(n) _____.
    a. vane
    b. impeller
    c. eccentric
    d. concentric

18. Metering pumps pump exact amounts of fluids because of _____.
    a. short pistons
    b. close tolerances
    c. vanes
    d. small impellers

19. Which of the following will *not* correct cavitation?
    a. Lowering the pump
    b. Lowering the flow through the pump
    c. Increasing the flow rate
    d. Throttling down discharge

20. Which of the following is *not* a common cause of cavitation?
    a. A blocked discharge pipe
    b. Blocked pumps
    c. Clogged strainers
    d. Buildup in the intake lines

Summary

Pumps are a major part of the maintenance workload in most plants and industries. Pump failure is a major cause of downtime and lost production. Premature pump failure can usually be attributed to improper installation or maintenance procedures.

The installation, repair, and maintenance of pumps represent a considerable part of a maintenance worker's job. The care and precision with which these jobs are done have a great impact on the reliability and efficiency of the pumps.

Notes

Trade Terms Introduced in This Module

Capacity: The volume of liquid handled by a pump per unit of time, expressed as gallons per minute (gpm) or cubic feet per minute (cfm).

Discharge: The exit side of the pump from which the fluid leaves the pump.

Head: The pressure at the base of a column of fluid of a specific height. For water, 100 feet of head equals 43 psi. The pressure at any point in a liquid can be thought of as being caused by a vertical column of the liquid which, due to its weight, exerts a pressure equal to the pressure at the point in question. The height of this column is called the static head and is expressed in terms of feet of liquid.

Net positive suction head (NPSH): The amount of suction head required to prevent vaporization of the pumped liquid.

Programmable logic controller: A computerized control device that operates a particular machine, based on input from sensors and instructions from the operator.

Static: Having no motion; at rest.

Suction: The inlet side of a pump where the fluid is supplied to the pump.

Velocity: Speed, expressed in distance over time, that is, ft/sec or ft/min.

Resources & Acknowledgments

Additional Resources

This module is intended to be a thorough resource for task training. The following reference work is suggested for further study. This is optional material for continued education rather than for task training.

Mechanical and Electrical Systems in Building, Third Edition. Upper Saddle River, NJ: Prentice Hall Publishing.

Figure Credits

Photo courtesy of Paragon Pump Co., 108F04

Wemco Model CF Chopper pump from Weir Specialty Pumps, 108F06

Photo was provided by Moyno, Inc., 108F07
Moyno® is a registered trademark of Moyno, Inc., A unit of Robbins & Myers, Inc.

Images reprinted with permission from Viking Pump Inc., 108F08, 10F8F10, 108F11
© 2006, Viking Pump, Inc., a Unit of IDEX Corporation

Bran+Luebbe, an SPX Process Equipment Operation, 108F22

Abaque Peristaltic Hose Pump – Blackmer, 108F25
www.blackmer.com

Dresser-Rand Company, 108F30, 108F32

CONTREN® LEARNING SERIES — USER UPDATE

NCCER makes every effort to keep these textbooks up-to-date and free of technical errors. We appreciate your help in this process. If you have an idea for improving this textbook, or if you find an error, a typographical mistake, or an inaccuracy in NCCER's Contren® textbooks, please write us, using this form or a photocopy. Be sure to include the exact module number, page number, a detailed description, and the correction, if applicable. Your input will be brought to the attention of the Technical Review Committee. Thank you for your assistance.

Instructors – If you found that additional materials were necessary in order to teach this module effectively, please let us know so that we may include them in the Equipment/Materials list in the Annotated Instructor's Guide.

Write: Product Development and Revision
National Center for Construction Education and Research
3600 NW 43rd St. Bldg G, Gainesville, FL 32606

Fax: 352-334-0932

E-mail: curriculum@nccer.org

Craft ______________ Module Name ______________

Copyright Date ______________ Module Number ______________ Page Number(s) ______________

Description ______________

____

____

____

(Optional) Correction ______________

____

____

(Optional) Your Name and Address ______________

____

____

HOW POWER IS PRODUCED

HYDRO PLANT

he mighty Colorado River vas dammed near Las Vegas, IV in the 1930s. Its bank of enerators produces electricity or cities in Nevada, Arizona, nd California.

NUCLEAR POWER

Nuclear power plants like the Enrico Fermi facility in Ohio use the heat generated by radioactive uranium to turn water into steam that drives giant steam turbine-generators.

COAL-FIRED POWER PLANT

Most power-generating plants in the U.S. burn coal to produce the steam that drives their turbine-generators.

ELECTRICIAN
POWER GENERATION MAINTENANCE

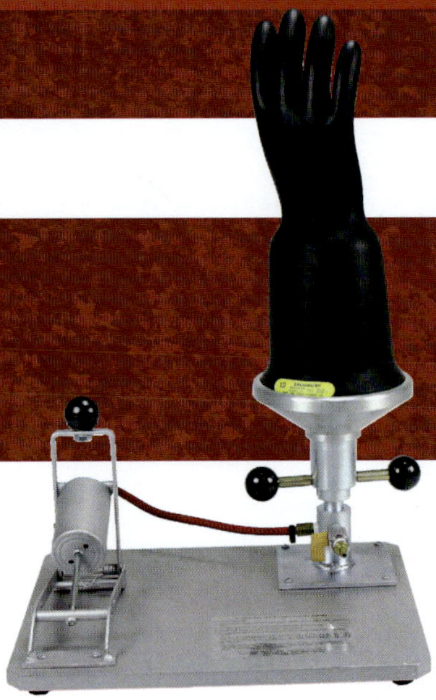

Protective equipment such as rubber gloves must be periodically tested.

VOLTAGE SENSOR

A proximity voltage sensor can alert you to the presence of high voltage.

PROTECTIVE EQUIPMENT

Wearing the proper PPE is critical when working with high voltage.

Temporary safety grounds are used to protect workers from electrical shock.

HIGH-VOLTAGE WORK REQUIRES SPECIAL SAFETY PRECAUTIONS

POWER PLANT ELECTRICIANS ARE RESPONSIBLE FOR A VARIETY OF EQUIPMENT

SWITCHYARD BREAKER

The size of a circuit breaker depends on its use.

MOTOR CONTROL CENTER BREAKER BUCKET

TRANSFORMER

A power plant transformer requires a specially designed enclosure and protective systems.

VOLTAGE REGULATOR

A voltage regulator like this one can maintain a voltage within +/– 1%.

BAGHOUSE

ELECTROSTATIC PRECIPITATOR

To protect the environment, coal-fired plants use pollution control devices to remove emissions from flue gases.

OPERATOR STATION

Plant operators are able to monitor the status of every plant system and respond to changing demands.

COOLING TOWER

Power plants rely on cooling towers like this one at a coal-fired plant to convert steam from the turbines back to water.

Page 1: *Hydro Plant (Hoover Dam and Generator Bank)* – U.S. Department of the Interior, Bureau of Reclamation; *Nuclear Power* – U.S. Nuclear Regulatory Commission, file photo; *Coal-fired Power Plant* – Topaz Publications, Inc.
Page 2: *Glove Tester, Voltage Sensor, Wearing Proper PPE* – Salisbury Electrical Safety; *Temporary Safety Grounds* – A.B. Chance Co., Hubbell Power Systems
Page 3: *Switchyard Breaker, Motor Control Center Breaker Bucket* – Topaz Publications, Inc.; *Voltage Regulator* – GE Consumer & Industrial; *Transformer* – Jim Mitchem
Page 4: *Baghouse* – Asphalt Drum Mixers, Inc.; *Electrostatic Precipitator* – Photo courtesy of McGill AirClean LLC; *Operator Station, Cooling Tower* – Topaz Publications, Inc.

Industrial Maintenance E & I Technician Level One

40109-07

Valves

40109-07
Valves

Topics to be presented in this module include:

Overview

Valves are the switches and brakes and steering wheels of industrial process. Valves control the direction of flow, regulate the rate of flow, and start or stop flow. You will learn in this module how the different kinds of valves work, and what each is used for. You will learn how to handle valves and what kinds of actuators they may have.

Objectives

When you have completed this module, you will be able to do the following:

1. Identify types of valves that start and stop flow.
2. Identify types of valves that regulate flow.
3. Identify valves that relieve pressure.
4. Identify valves that regulate the direction of flow.
5. Explain how to properly store and handle valves.
6. Explain valve locations and positions.

Trade Terms

Actuator
Angle valve
American Society for Testing Materials International (ASTM)
Ball valve
Bonnet
Butterfly valve
Check valve
Control valve
Deformation
Disc
Elastomeric
Galling
Gate valve
Globe valve
Kinetic energy
Lift
Linear flow
Packing
Phonographic
Plug
Plug valve
Positioner
Relief valve
Seat
Thermal transients
Throttling
Torque
Valve body
Valve stem
Valve trim
Wedge
Wire drawing
Yoke bushing

Required Trainee Materials

1. Pencil and paper
2. Appropriate personal protective equipment

Prerequisites

Before you begin this module, it is recommended that you successfully complete *Core Curriculum;* and *Industrial Maintenance E & I Technician Level One,* Modules 40101-07 through 40108-07.

This course map shows all of the modules in the first level of the *Industrial Maintenance E & I Technician* curriculum. The suggested training order begins at the bottom and proceeds up. Skill levels increase as you advance on the course map. The local Training Program Sponsor may adjust the training order.

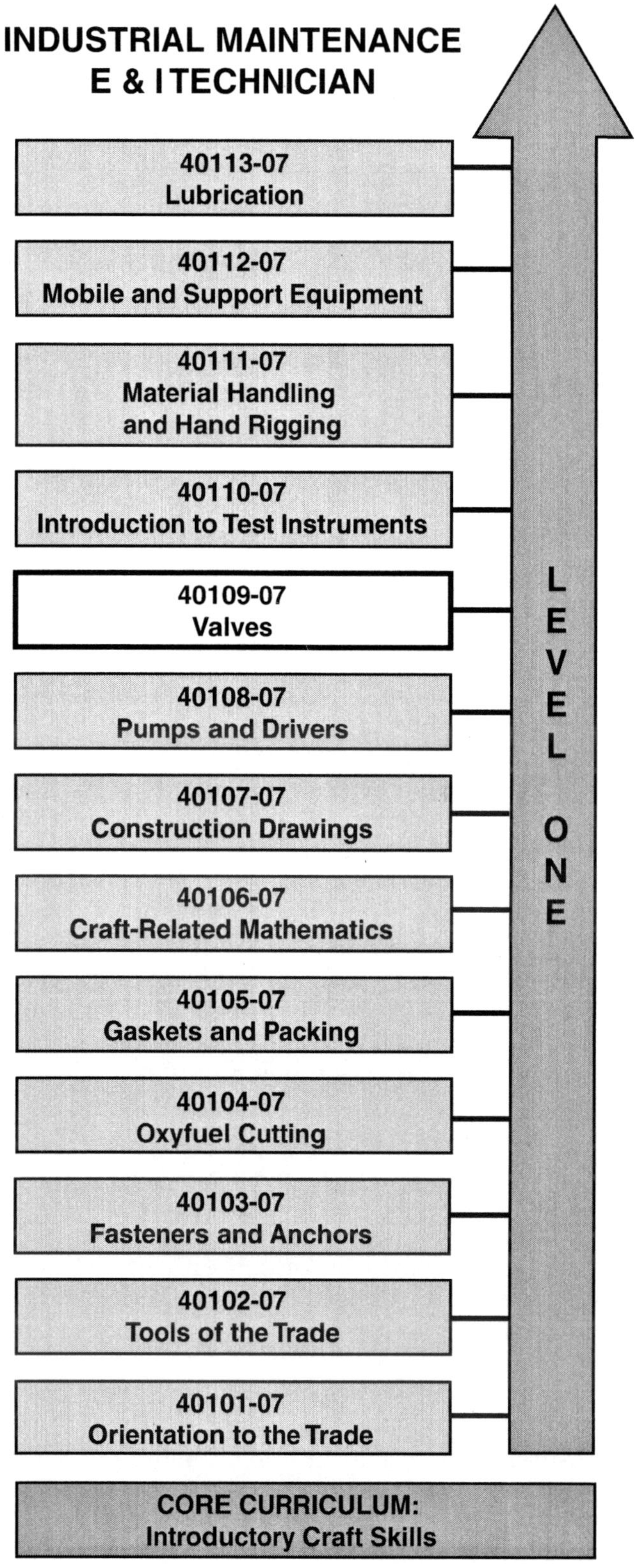

109CMAP.EPS

1.0.0 ◆ INTRODUCTION

Valves are devices that control the flow of fluids or gases through a piping system. While the designs of valves vary, all valves have two common features: a passageway through which fluid or gas flows and a moveable part that opens and closes the passageway. A valve can be used to provide on-off service only, can act as a throttling device to allow different flow rates, can relieve excess pressure in a pipeline, or can prevent reversal of flow through a line. There are basically four ways to control the flow through a piping system, and each type of valve uses one or more of these methods:

- Moving a **disc** or **plug** into or against a passageway
- Sliding a flat, round surface across a passageway
- Rotating an opening inside a shaft across the passageway
- Moving a flexible material into the passageway

Valves are made of a variety of materials, including bronze, iron, carbon steel, aluminum, alloy steels, and PVC. The application of the valve usually dictates what type of material is used, and many times, there is more than one material that meets the requirements of an application. Bronze or iron with bronze trim are used for valves in air or water services. Iron is also used for valves in low-pressure steam services. Steel is used for valves that regulate noncorrosive products, and ductile iron, which is less expensive than steel, offers better resistance to mechanical or thermal shock than other metals provide.

Stainless steel is often used to regulate the flow of corrosive materials and in services that require sterile conditions. Most systems that operate at extremely low temperatures also use stainless steel valves because they are less brittle at low temperatures than iron or carbon steel. Highly corrosive chemicals require specialized valve bodies that are made of or lined with plastic, rubber, ceramics, or special alloys.

This module introduces the types of valves used to start and stop flow, regulate flow, relieve pressure, and regulate the direction of flow. Various types of valve **actuators** are also explained.

2.0.0 ◆ VALVES THAT START AND STOP FLOW

Many valves are designed to operate either completely open or completely closed. These valves cannot practically be used to throttle or control the flow of fluid through a piping system. The valves used to start and stop flow through a system include the following types:

- Gate
- Knife
- Ball
- Plug
- Three-way

2.1.0 Gate Valves

Gate valves are used to start or stop fluid flow, but not to regulate or throttle flow. The term gate is derived from the appearance of the disc in the flow stream, which is similar to a gate. *Figure 1* shows a gate valve. The disc is completely removed from the flow stream when a gate valve is fully open. The disc offers virtually no resistance to flow when the valve is open, so there is little pressure drop across an open gate valve. When the valve is fully closed, a disc-to-seal-ring contact surface exists for 360 degrees and good sealing is provided. With proper mating of disc to seal ring, very little or no leakage occurs across the disc when the gate valve is closed.

Gate valves are not used to regulate flow because the relationship between **valve stem** movement and flow rate is nonlinear. Operating with the valve in a partially open position can cause disc and **seat** wear, which will eventually lead to valve leakage.

Gate valves are available with a variety of fluid-control elements. Classification of gate valves is usually made by the type of fluid-control element used. The fluid-control elements are available as:

- Solid **wedge**
- Flexible wedge
- Split wedge
- Double disc (parallel disc)

Solid, flexible, and split wedges (*Figure 2*), are employed in valves with inclined seats, while the double discs are used in valves with parallel seats. Regardless of the style of wedge or disc used, they are all replaceable. In services where solids or high velocity may cause rapid erosion of the seat or disc, these components should have a high surface hardness and replaceable seats as well as discs. If the seats are not replaceable, any damage would require refacing of the seat. Valves being used in corrosive service should always be specified with renewable seats.

The solid or single wedge, shown in *Figure 3*, is the most commonly used disc because of its simplicity and strength. A valve with this type of wedge may be installed in any position. It is suitable for almost all fluids. It is most practical for turbulent flow. The majority of solid wedge gate valves have resilient seats, seats covered with rubber, providing a better sealing action.

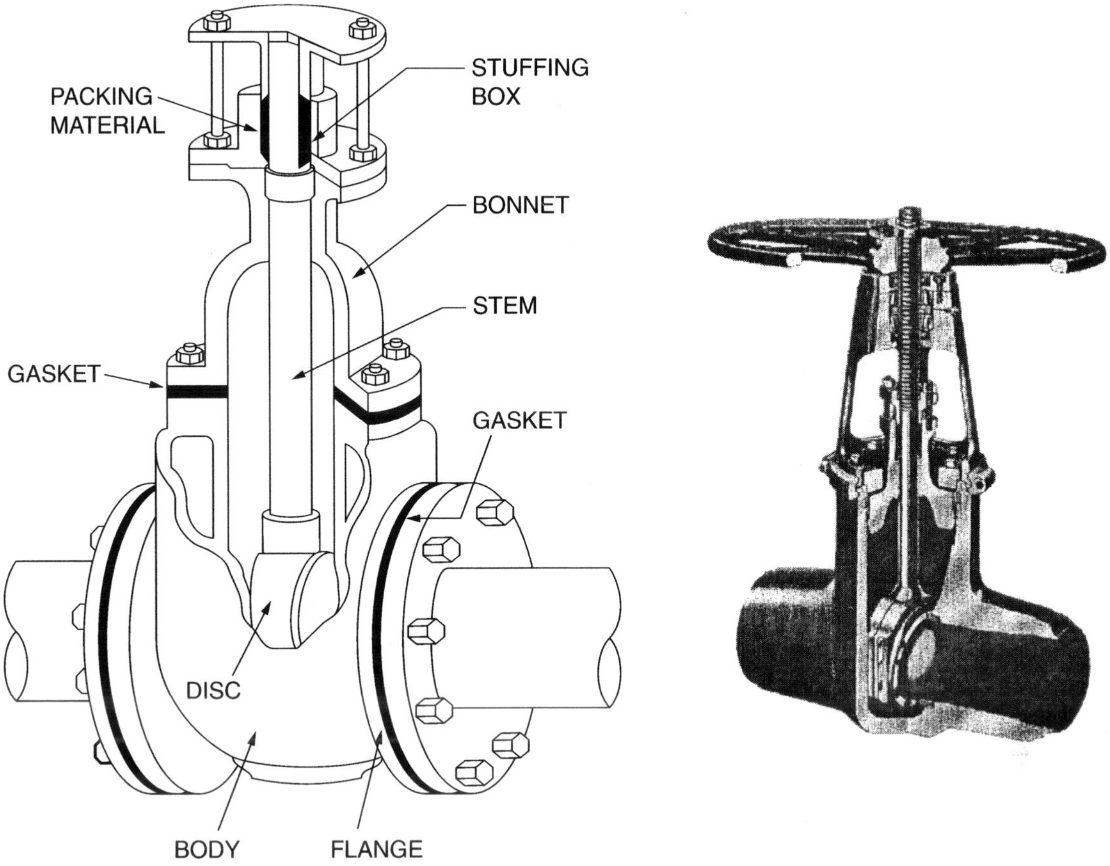

Figure 1 ◆ Gate valve.

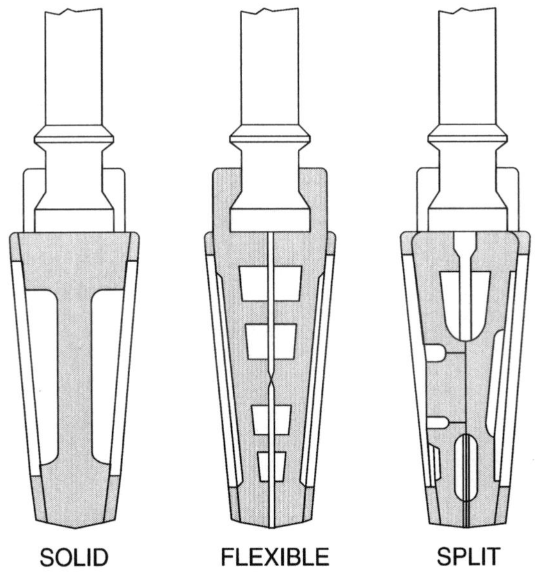

Figure 2 ◆ Wedge types.

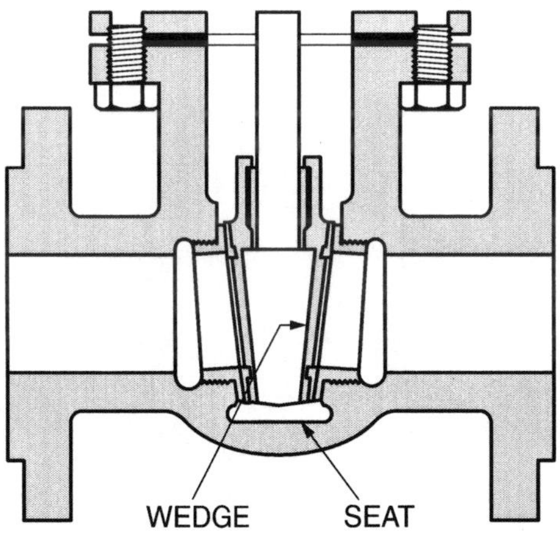

Figure 3 ◆ Solid or single wedge.

The flexible wedge (*Figure 4*) is a one-piece disc with a cut or groove around the edge to improve ability to match error or change in angle between the seats. The cut varies in size, shape, and depth. A shallow, narrow cut gives little flexibility but retains strength. A deeper and wider cut, or a cast-in recess, leaves little material at the center, allowing more flexibility but potentially compromising strength and risking permanent bending. A correct profile of the disc half in the flexible wedge design can allow uniform deflection at the disc edge, so that the wedging force applied in seating will force the disc seating surface uniformly and tightly against the seat.

Gate valves used in steam systems have flexible gates. This prevents binding of the gate within the valve when the valve is in the closed position. When steam lines are heated, they will expand, causing some distortion of valve bodies. If a solid gate fits snugly between the seats of a valve in a cold steam system, when the system is heated and pipes elongate, the seats will compress against the gate, wedging the gate between them and clamping the valve shut. A flexible gate, by contrast, flexes as the valve seat compresses it. This prevents clamping.

The major problem associated with flexible gates is that water tends to collect in the **body** neck. Under certain conditions, the admission of steam may cause the valve body neck to rupture, the **bonnet** to lift off, or the seat ring to collapse. It is essential that correct warming-up procedures be followed to prevent this. Some very large gate valves also have a three-position vent and bypass valve installed. This valve allows venting of the bonnet either upstream or downstream of the valve and has a position for bypassing the valve.

Split wedges, as shown in *Figure 5*, are of the ball and socket design, and are self-adjusting and self-aligning to both seating surfaces. The disc is free to adjust itself to the seating surface if one half of the disc is slightly out of alignment because of foreign matter lodged between the disc half and the seat ring. This type of wedge (disc) is suitable for handling noncondensing gases and liquids, particularly corrosive liquids, at normal temperatures. Freedom of movement of the discs in the carrier prevents binding, even though the valve may be closed when hot (later contracting as it cools). This type of valve should be installed with the stem in the vertical position.

The parallel disc (*Figure 6*) was designed to prevent valve binding due to **thermal transients**. Both low-pressure iron valves and high-pressure steel types have this disc. The principle of operation is that wedge surfaces between the parallel-faced disc halves press together under stem thrust and spread the discs to seal against the seats. The tapered wedges may be part of the disc halves or may be separate elements. The lower wedge must bottom out on a rib at the valve bottom so that the stem can develop seating force. In one version (*Figure 7*), the wedge contact surfaces are curved to keep the point of contact close to optimal.

In other parallel disc gates (*Figures 8* and *9*), the two halves do not move apart under wedge action. Instead, the upstream pressure holds the downstream disc against the seat. A carrier ring lifts the discs. A spring or springs hold the discs apart and seated when there is no upstream pressure.

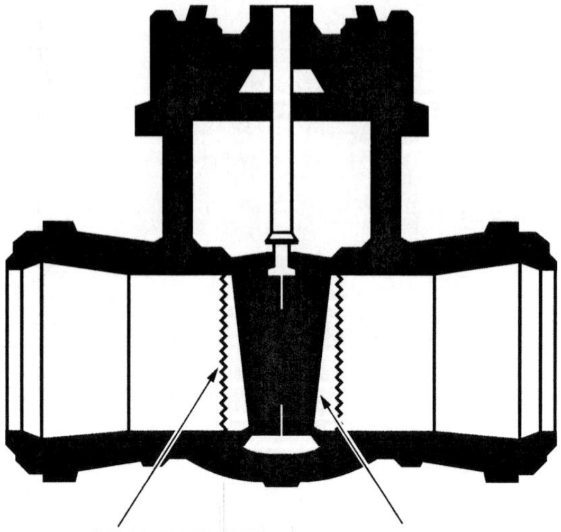

Figure 4 ◆ Flexible wedge.

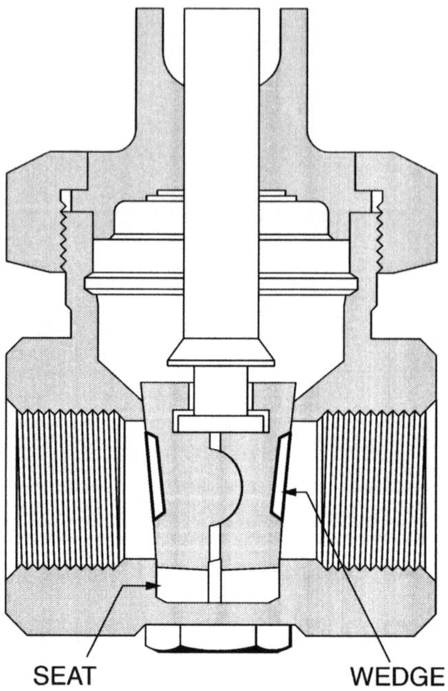

Figure 5 ◆ Split wedge.

Another design found on parallel gate discs provides for sealing only one port. In these designs, the high-pressure side pushes the disc open (relieving disc) on the high-pressure side, but forces the disc closed on the low-pressure side (*Figure 10*). With such designs, the amount of seat leakage tends to decrease as differential pressure across the seat increases.

These valves usually have a flow direction marking to show which side is the high-pressure (relieving) side. Make sure these valves are not installed backwards in the system.

Some parallel-disc gate valves used in high-pressure systems are equipped with an integral bonnet vent/bypass line. A three-way valve is used to position the line to bypass in order to equalize pressure across the discs prior to opening. When the gate valve is closed, the three-way valve is positioned to vent the bonnet to one side or the other. This prevents moisture from accumulating in the bonnet. The three-way valve is positioned to the high-pressure side of the gate

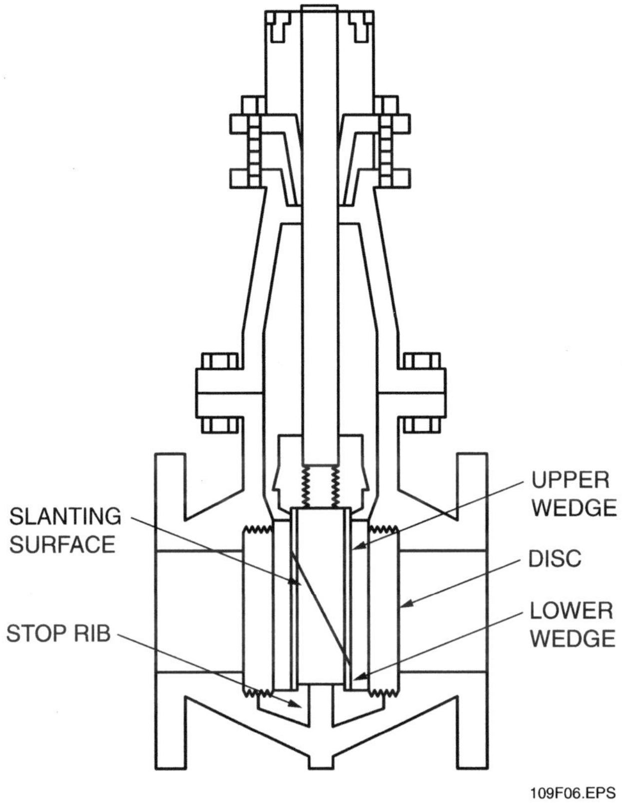

Figure 6 ◆ Parallel-disc gate valve.

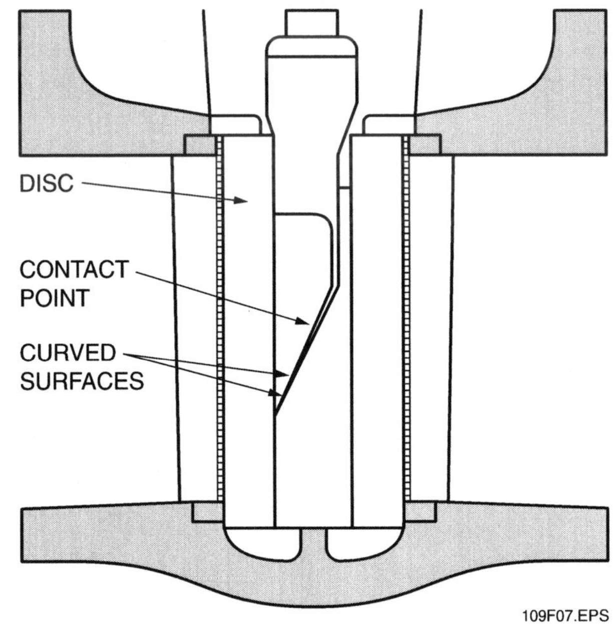

Figure 7 ◆ Parallel disc.

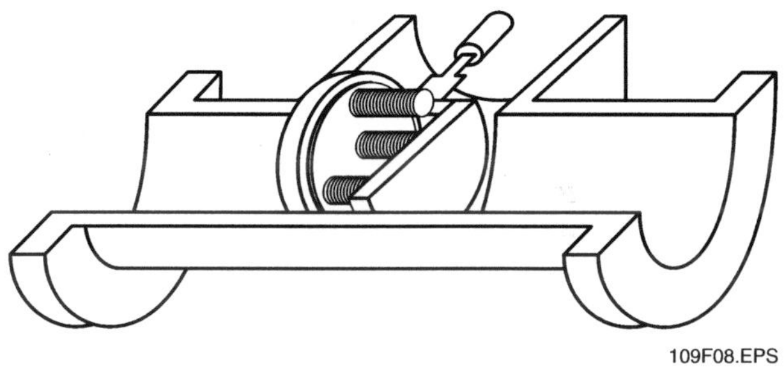

Figure 8 ◆ Spring-loaded parallel disc gate valve cutaway.

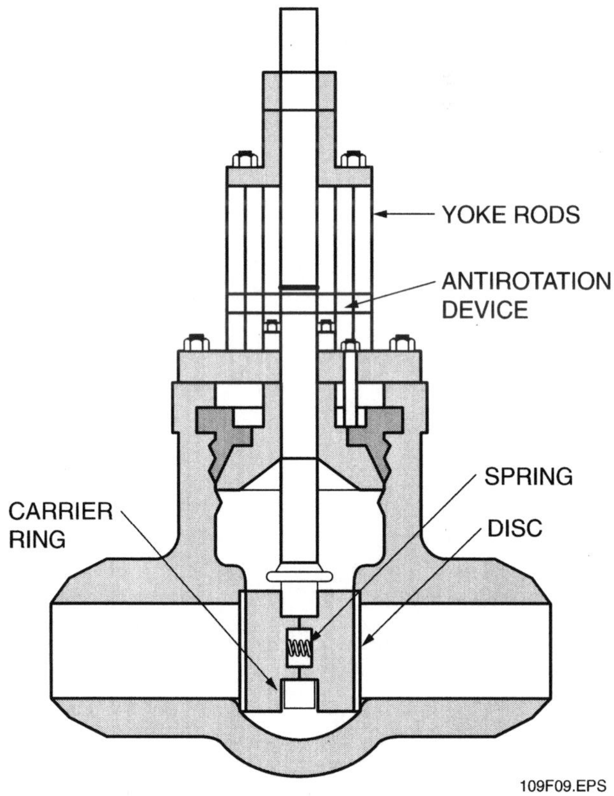

Figure 9 ◆ Parallel-disc gate valve with spring.

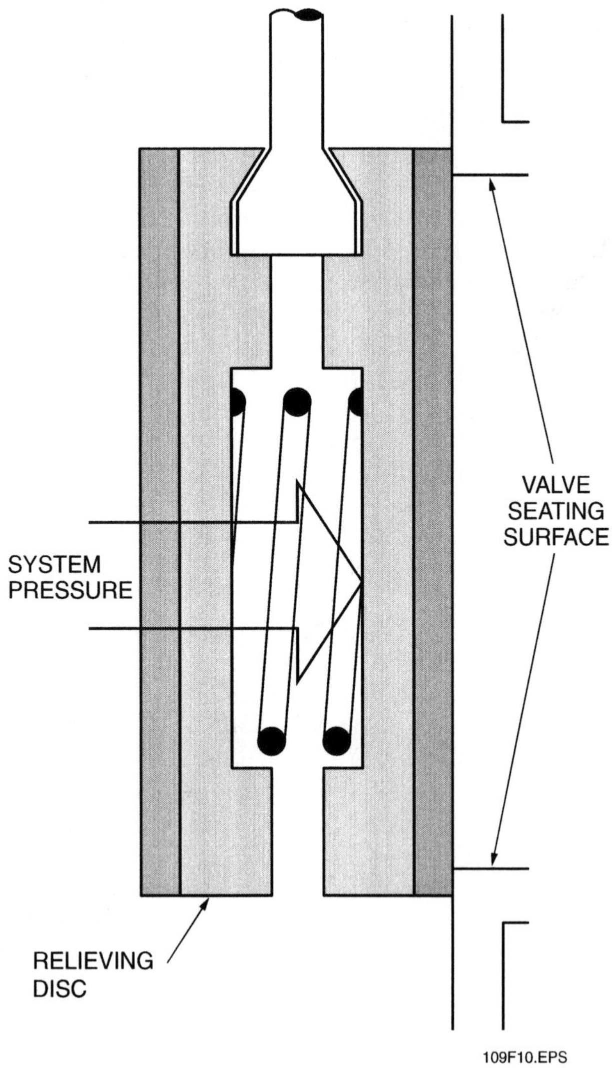

Figure 10 ◆ Relieving disc gate valve.

valve when the gate valve is closed to ensure that flow does not bypass the isolation valve. The high pressure acts against spring compression and forces one gate off its seat. The three-way valve vents this flow back to the pressure source.

2.1.1 Valve Stem

The valve stem provides a link between the handwheel and the gate. Many times, the industrial maintenance technician has to attach the handwheel to the stem after the valve is installed. Some handwheels bolt onto the stem; others slip over a keyway cut into the stem and are then secured to the stem by a nut or machine screw. Handwheels vary according to the type of stem they fit. Three general types of stems include the rising stem, the nonrising stem, and the outside screw-and-yoke (OS&Y) stem. *Figure 11* shows types of stems.

When a rising stem valve is opened, both the handwheel and the stem rise. The height of the stem gives an approximation of how far the valve is open. These stems can only be installed where there is sufficient clearance for the handwheel to rise. Rising stems come in contact with the fluid in the valve and are only used with fluids that will not harm the threads, such as hydrocarbons, water, and steam.

Nonrising stems are suitable in spaces where space is limited. As the name implies, neither the stem nor the handwheel rises when the valve is opened. A spindle inside the **valve body** turns when the handwheel is turned and raises the stem; therefore, stem wear is held to a minimum. This type of stem also comes in contact with the fluid in the valve.

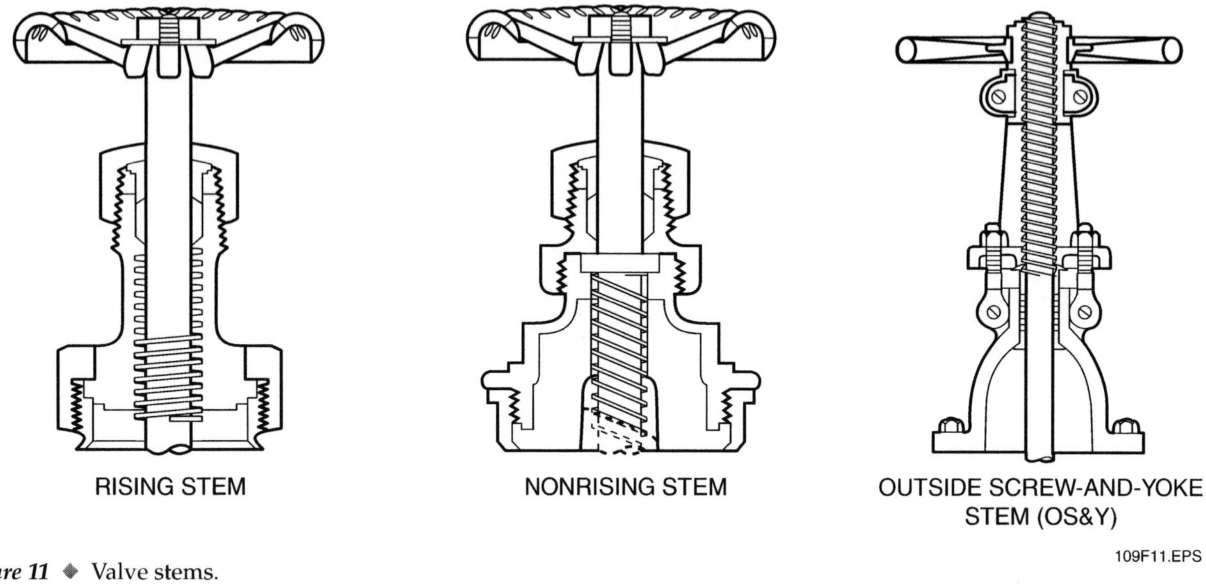

Figure 11 ◆ Valve stems.

The OS&Y stem is suitable for corrosive fluids because the stem does not come in contact with the fluid in the line. As the handwheel is turned, the stem moves up through the handwheel. The height of the stem indicates how far the valve is open.

2.2.0 Knife Gate Valves

A special type of gate valve is the knife gate valve (*Figure 12*). This valve serves in slurry and waste lines, in other low-pressure applications, and in paper stock. The sharp edge of the disc bottom is easily forced closed in contact with a metal or **elastomeric** seat. When open, the disc is in the air after passing through a full-width packing box.

Knife valves have a disc that can cut through deposits and flow-stream solids such as resin slurry or pulp. These valves, like most gate valves, can be positioned by manual, electrical, pneumatic, and hydraulic actuation.

The positioning of a knife gate valve is critical. If the valve is used to isolate equipment, the seat side of the valve must face in the direction of flow. If the valve is used to isolate flow, the seat side must be opposite the direction of flow.

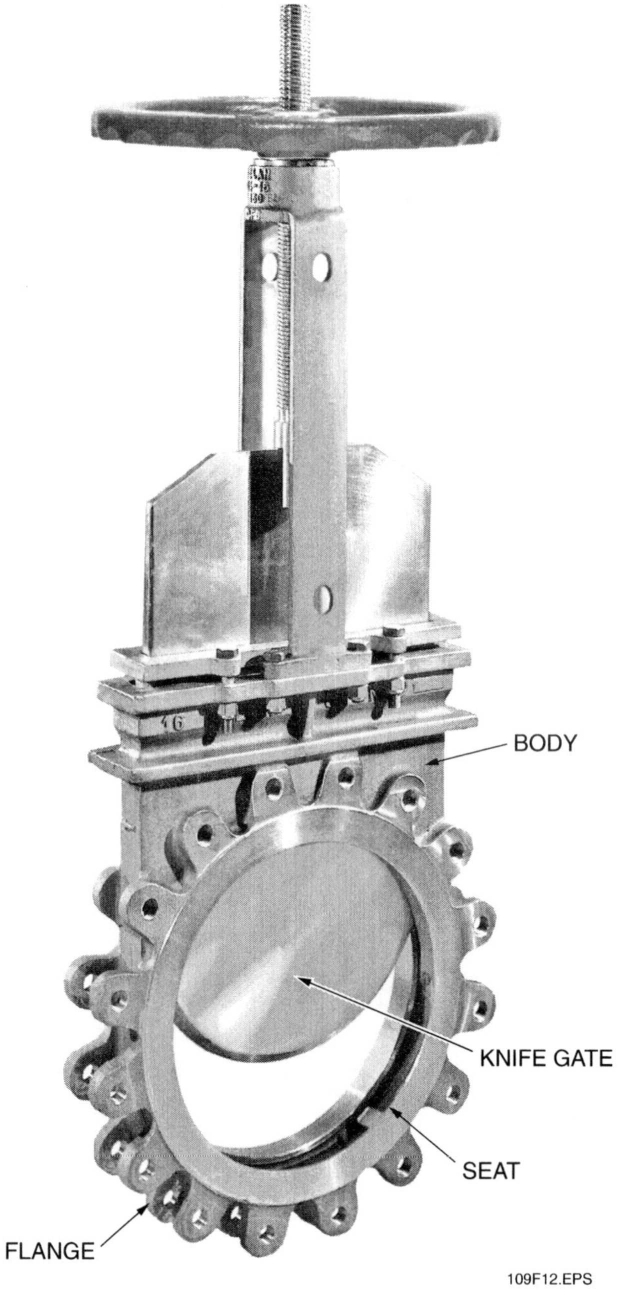

Figure 12 ◆ Knife gate valve.

2.3.0 Ball Valves

Ball valves, as the name implies, are stop valves that use a ball to stop or start the flow of fluid. They are rotary action valves. The ball performs the same function as the disc in the **globe valve**. However, it turns instead of traveling up and down. When the actuator is operated to open the valve, the ball rotates to a point where the hole through the ball is in line with the valve body inlet and outlet. When the valve is shut, which requires only a 90-degree rotation for most valves, the ball is rotated so that the hole is perpendicular to the flow openings of the valve body, and flow is stopped.

Most ball valves are quick-acting, requiring only a 90-degree turn to operate the valve either completely open or closed. *Figure 13* shows the major components of a ball valve. The ball valve, in general, is the least expensive of any valve configuration. In early designs having metal-to-metal seating, the valves could not provide bubble-tight

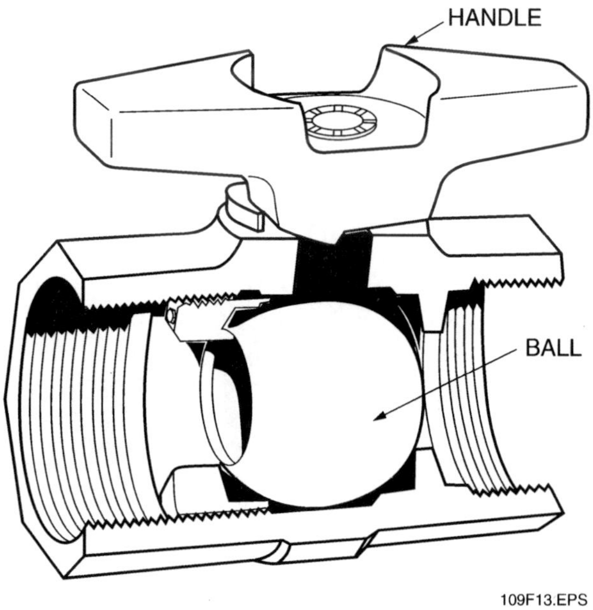

Figure 13 ◆ Ball valve.

sealing and were not fire-safe. With the development of elastomeric materials and with advances in plastics, the original metallic seats have been replaced with materials such as fluorinated polymers, nylon, neoprene, and Buna-N. Ball valves also have low maintenance costs.

In addition to quick on-off operation, ball valves are compact, require no lubrication, and provide tight sealing with low **torque**. With a soft seat on both sides of the ball, most ball valves provide equally effective sealing of flow in either direction. Many designs permit adjustment for wear.

Because conventional ball valves have relatively poor **throttling** characteristics, they are not generally satisfactory for this service. In a throttling position, the partially exposed seat rapidly erodes because of the high-velocity flow. However, a ball valve has been developed with a spherical surface-coated plug off to one side in the open position. This rotates into the flow passage until it blocks the flow path completely. Seating is accomplished by the eccentric movement of the plug. The valve requires no lubrication and can be used for throttling service. Ball valves are designed on the simple principle of floating a polished ball between two plastic seating surfaces, permitting free turning of the ball. Because the plastic is subject to **deformation** under load, some means must be provided to hold the ball against at least one seat. This is normally accomplished through spring pressure, differential line pressure, or a combination of both. *Figure 14* shows a combination of line pressure and a spring on the ball.

Ball valves are available in the venturi (*Figure 15*) and full-port patterns. The latter has a ball with a bore equal to the inside diameter of the pipe. A venturi tube is a tube that is narrowed in the middle, increasing the speed of the fluid and decreasing the pressure. A venturi pattern ball valve is narrower at the middle, where the ball is. Balls are usually metallic in metallic bodies with trim (seats) produced from elastomeric materials. All-plastic designs are also available. Seats and balls are replaceable.

Ball valves are available in top-entry (*Figure 16*) and split-body (end-entry) types (*Figure 17*). In the former, the ball and seats are inserted through the top, while in the latter, the ball and seats are inserted from the ends.

The resilient seats for ball valves are made from various elastomeric materials. The most common seat materials are TFE (virgin), filled TFE, nylon, Buna-N, neoprene, and combinations of these materials. Because of the presence of the elastomeric materials, these valves cannot be used at elevated temperatures. To overcome this disadvantage, a graphite seat has been developed that

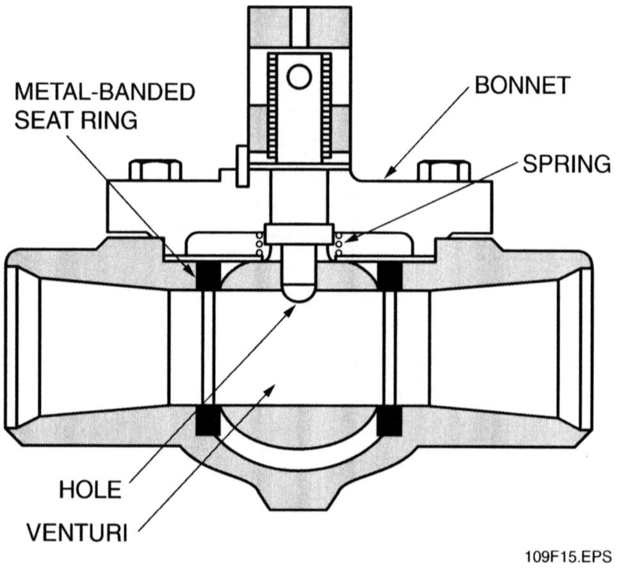

Figure 15 ◆ Venturi-type ball valve.

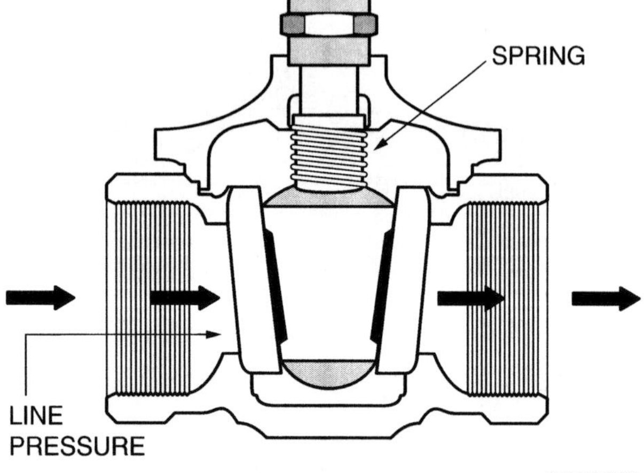

Figure 14 ◆ Ball valve with slanted seals.

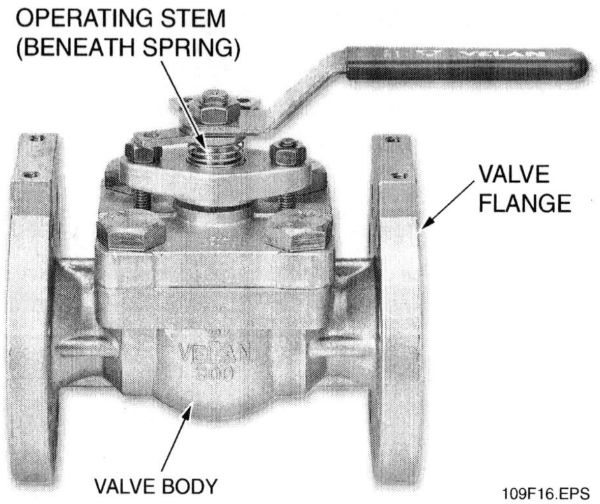

Figure 16 ◆ Top-entry ball valve.

will permit operation up to 1,000°F. Typical maximum operating temperatures of the valves at their full pressure ratings are shown in *Table 1* for various seat materials.

CAUTION

Use care when selecting the seat material to make sure it is compatible with the materials being handled by the valve and suitable according to its temperature limitations.

Table 1 Maximum Operating Temperature for Seat Materials

Seat Material	Operating Temperature
TFE (Virgin)	450°F
Filled TFE	400°F
Buna-N	180°F
Neoprene	180°F

109T01.EPS

2.4.0 Plug Valves

Plug valves are used to stop or start fluid flow. They are rotary-action valves. The name is derived from the shape of the disc, which resembles a plug. A plug valve is shown in *Figure 18*.

The body of a plug valve is machined to receive the tapered or cylindrical plug. The disc is a solid plug with a bored passage at a right angle to the longitudinal axis of the plug. In the open position, the passage in the plug lines up with the inlet and outlet ports of the valve body. When the plug is turned 90 degrees from the open position, the solid part of the plug blocks the ports and stops fluid flow.

Plug valves are available in either a lubricated or nonlubricated design and with a variety of styles of port openings through the plug. There are numerous plug designs.

An important characteristic of the plug valve is its easy adaptation to multiport construction. Multiport valves are widely used. Their installation simplifies piping. They provide a much more convenient operation than multiple gate valves do. They also eliminate pipe fittings. The use of a multiport valve eliminates the need for two, three, or even four conventional shutoff valves, depending on the number of ports in the plug valve.

Plug valves are normally used only for on-off operations, particularly where frequent operation of the valve is necessary. These valves are not normally recommended for throttling service because, as with the gate valve, a great percentage of flow change occurs near shutoff at high velocity. However, a diamond-shaped port has been developed for throttling service. Another valve operating on the same principle as the plug is the cone valve, in which a conical plug, with a port in the middle, can be turned to obstruct or permit flow. The difference is that the cone valve is eccentric; that is, it is not centered in the valve chamber. This is a very fast-operating valve, requiring only a half-turn to work.

Multiport valves are particularly good on transfer lines and for diverting services. A single multiport valve may be installed in lieu of three or four gate valves or other types of shutoff valves. Many of the multiport configurations do not permit complete shutoff of flow, however. In most cases one flow path is always open. These valves are intended to divert the flow to one line while shutting off flow from the other lines. If complete shutoff of flow is required, a suitable multiport valve should

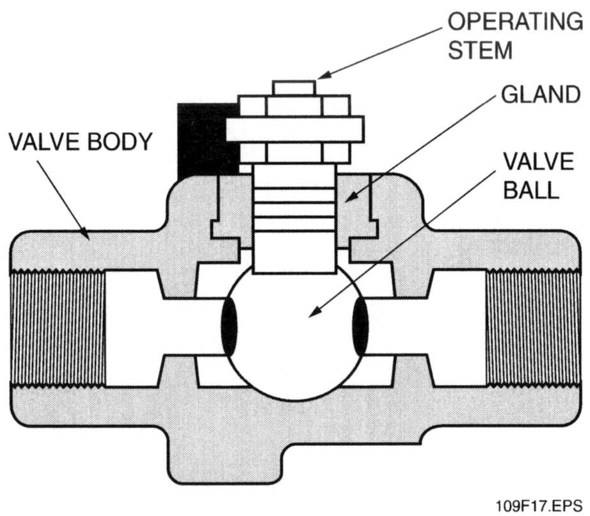

Figure 17 ◆ Split-body ball valve.

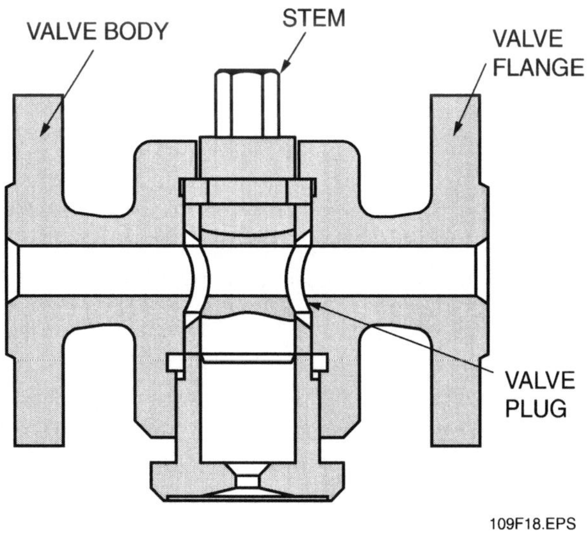

Figure 18 ◆ Plug valve.

be used, or a secondary valve should be installed on the main feed line ahead of the multiport valve to permit complete flow shutoff.

It should also be noted that in some multiport configurations, flow to more than one port simultaneously is also possible. Great care should be taken in specifying the port arrangement to guarantee proper operation.

Plugs are either round or cylindrical with a taper. They may have various types of port openings, each with a varying degree of free area relative to the corresponding pipe size (*Figure 19*).

- Rectangular port is the standard-shaped port with a minimum of 70 percent of the area of the corresponding size of standard pipe.
- Round port means that the valve has a full round opening through the plug and body, of the same shape as standard pipe.
  - Full port means that the area through the valve is equal to or greater than the area of standard pipe.
  - Standard opening means that the area through the valve is less than the area of standard pipe. These valves should be used only where restriction of flow is unimportant.
- Diamond port means that the opening through the plug is diamond-shaped. This has been designed for throttling service. All diamond port valves are venturi, restricted-flow type.

Clearances and leakage prevention are the chief considerations in plug valves. Many plug valves are of all-metal construction. In these versions, the narrow gap around the plug can permit leakage. If the gap is reduced by sinking the taper plug deeper into the body, actuation torque will climb rapidly, and **galling** can occur.

Lubrication remedies this. A series of grooves around the port openings in the plug or body is supplied with grease prior to actuation, not only to lubricate the plug motion but also to seal the gap (*Figure 20*). Grease injected into a fitting at the stem top travels down through a **check valve** in the passageway and then past the plug top to the grooves on the plug and down to a well below the plug.

The lubricant must be compatible with the temperature and nature of the fluid. The most common substances controlled by plug valves are gases and liquid hydrocarbons. Some water lines

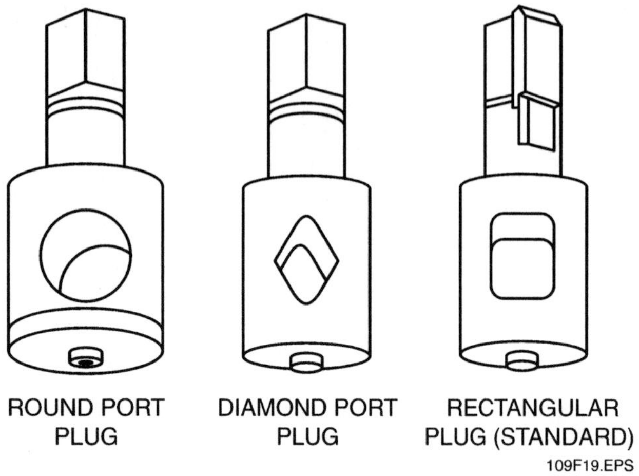

Figure 19 ◆ Plug valve – port types.

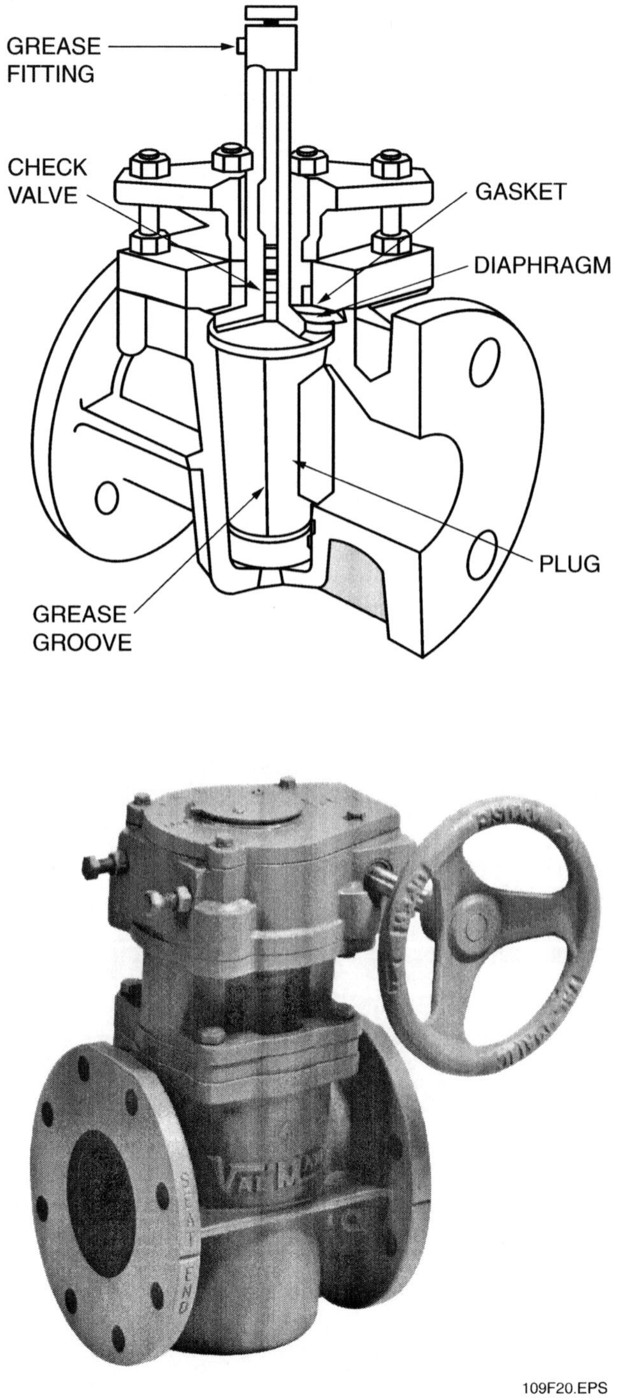

Figure 20 ◆ Lubricated taper plug valve.

have these valves too, if lubricant contamination is not a serious danger. This type can go to 24-inch size, with pressure capability of 6,000 psig. Steel and iron bodies are available. The plug can be cylindrical or tapered.

The correct choice of lubricant is extremely important for successful lubricated plug valve performance. In addition to providing adequate lubrication to the valve, the lubricant must not react chemically with the material passing through the valve. The lubricant must not contaminate the material passing through the valve, either. All manufacturers of lubricated plug valves have developed a series of lubricants that are compatible with a wide range of media. Their recommendations should be followed regarding which lubricant is best suited for the service.

To overcome the disadvantages of lubricated plug valves, two basic types of nonlubricated plug valves were developed. A nonlubricated valve may be a lift-type, or it may have an elastomer sleeve or plug coating that eliminates the need to lubricate the space between the plug and seat.

Lift-type valves provide a means of mechanically lifting the tapered plug slightly from its seating surface to permit easy rotation. The mechanical lifting can be accomplished through either a cam (*Figure 21*) or an external lever.

A typical nonlubricated plug valve with an elastomer sleeve is shown in *Figure 22*. In this particular valve, a sleeve of TFE completely surrounds the plug. It is retained and locked in place by the metal body. This results in a continuous primary seal between the sleeve and the plug, both while the plug is rotated and when the valve is in either the open or closed position. The TFE sleeve is durable and essentially inert to all but a few rarely encountered chemicals. It also has a low coefficient of friction and therefore is self-lubricating.

Lubricants are available in stick form and in bulk. Stick lubrication is usually employed when a small number of valves are in service or when they are widely scattered throughout the plant. However, for a large number of valves, gun lubrication is the most convenient and economical solution.

Valves are usually shipped with an assembly lubricant. This assembly lubricant should be removed and the valve completely relubricated with the proper lubricant before the valve is put into service.

Regular lubrication is critical for best results. Extreme care should be taken to prevent any foreign matter from entering the plug when inserting new lubricant.

2.5.0 Three-Way Valves

Three-way valves are multiport plug valves that are installed at the intersection of three lines. They are used to direct the flow between two of the lines only and to block off the third line. Situations requiring three-way valves include alternating the connections of two supply lines to a common delivery line or diverting a line into either of two

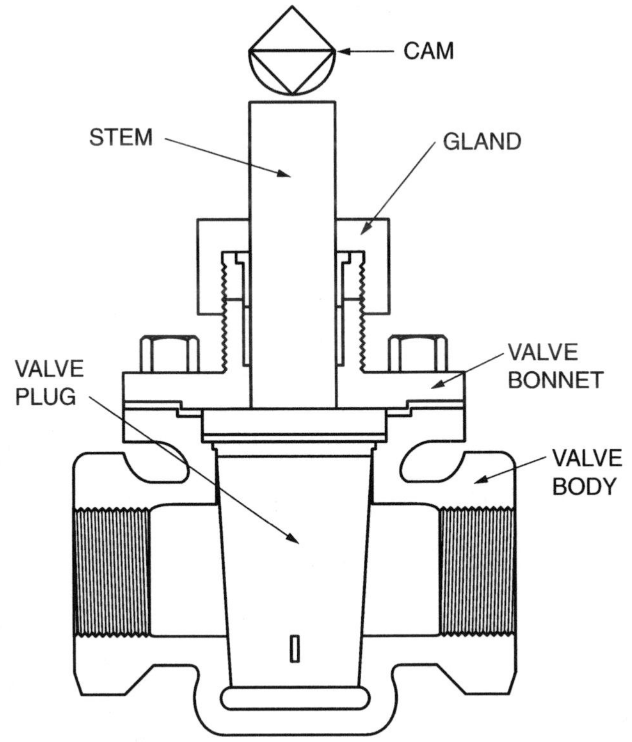

Figure 21 ◆ Cam-operated and nonlubricated plug valve.

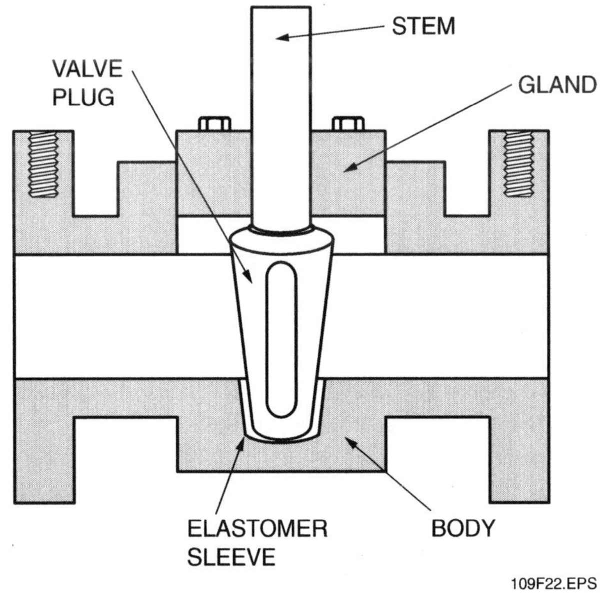

Figure 22 ◆ Typical nonlubricated plug valve with elastomer sleeve.

possible directions. Three-way valves are designed so that when the plug is turned from one position to another, the channels previously connected are completely closed off before the new channels begin to open, preventing mixture of fluids or loss of pressure. Three-way valves come in several different arrangements. They can be two- or three-port valves with stops that limit the turning of the plug to two, three, or four positions. *Figure 23* shows three-way valve applications.

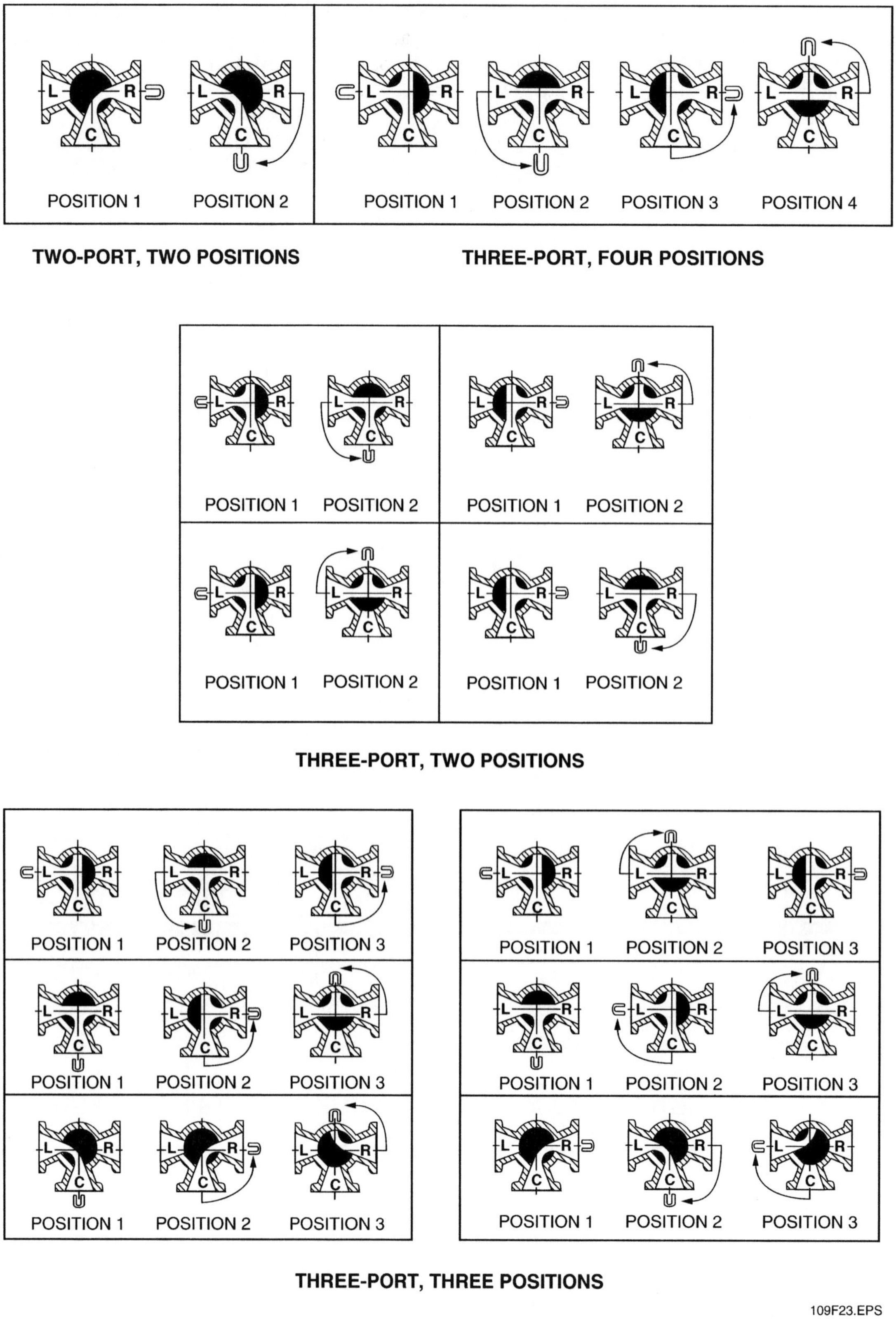

Figure 23 ◆ Three-way valve applications.

3.0.0 ◆ VALVES THAT REGULATE FLOW

Many valves are designed to provide accurate flow control through a system. The valves used to regulate flow through a system include the following:

- Globe (angle, Y-type, or needle)
- Butterfly
- Diaphragm

3.1.0 Globe Valves

Globe valves are used to stop, start, and regulate fluid flow. They are important elements in power plant systems and are commonly used as the standard against which other valve types are judged.

As shown in *Figure 24*, the globe valve disc can be totally removed from the flow path or it can completely close the flow path. The essential principle of globe valve operation is the perpendicular movement of the disc away from the seat. This causes the space between disc and seat ring to gradually close as the valve is closed. This characteristic gives the globe valve good throttling ability, which permits its use in regulating flow. Therefore, the globe valve is used not only for start-stop functions, but also for regulating flow.

The primary consideration in the application of a gate valve with relation to a globe valve is that the gate valve represents much less flow restriction than the globe valve. This reduced flow restriction is the result of straight-through body construction and the design of the disc. A gate valve can be used for a wide variety of fluids and provides a tight seal when closed. The major disadvantages of using a gate valve as opposed to a globe valve are as follows:

- It is not good for throttling.
- It is prone to vibration when partially open.
- It is more subject to seat and disc wear than a globe valve.
- Repairs, such as lapping and grinding, are generally more difficult.

It is generally easier to obtain very low seat leakage with a globe valve as compared to a gate valve. This is because the disc to seat ring contact is more at right angles, which permits the force of closing to tightly seat the disc.

Globe valves can be arranged so that the disc closes against the direction of fluid flow (flow-to-open) or so that the disc closes in the same direction as fluid flow (flow-to-close). When the disc closes against the direction of flow, the **kinetic energy** of the fluid impedes closing but aids opening of the

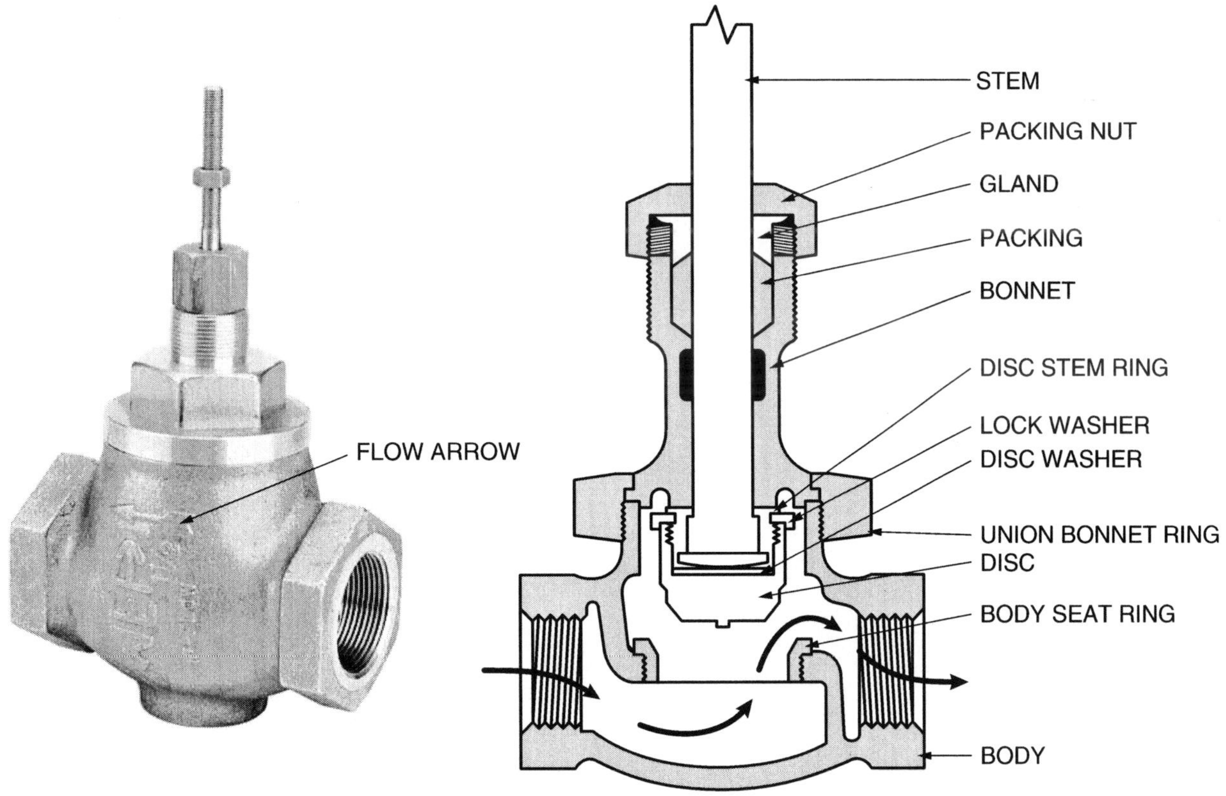

Figure 24 ◆ Globe valve.

valve. When the disc closes in the same direction as flow, the kinetic energy of the fluid aids closing but impedes opening of the valve. This characteristic makes the globe valve preferable to the gate valve when quick-acting stop valves are necessary.

Along with its advantages, the globe valve has a few drawbacks. Although valve designers can eliminate any or all of the drawbacks for specific services, the corrective measures are expensive and often narrow the valve's scope of service. The most evident shortcoming of the simple globe valve is the high head loss from the two or more right-angle turns of flowing fluid. Obstructions and breaks in the flow path add to the loss.

High-pressure losses in the globe valve can cost thousands of dollars a year for large high-pressure lines. The fluid-dynamic effects from the pulsation, impacts, and pressure drops in traditional globe valves can damage **valve trim**, stem **packing**, and actuators. Troublesome noise can also result. In addition, large sizes require considerable power to operate, which may make gearing or levers necessary.

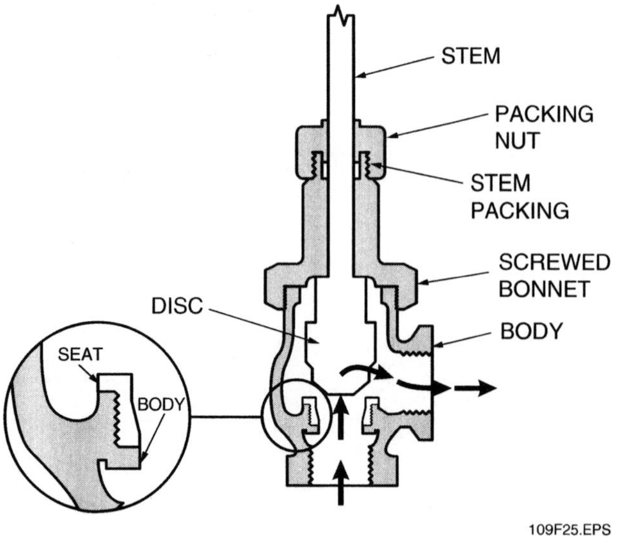

Figure 25 ◆ Angle globe valve.

Another drawback is the large opening needed for assembly of the disc. Globe valves are often heavier than other valves of the same flow rating. The cantilever mounting of the disc on its stem is also a potential trouble source. Each of these shortcomings can be overcome, but only at costs in dollars, space, and weight.

The **angle valve** (*Figure 25*) is a simpler modification of the basic globe form. With the ports at right angles, the diaphragm can be a simple flat plate. Fluid can flow through with only a single 90-degree turn, discharging more symmetrically than the discharge from an ordinary globe. Installation advantages also may suggest the angle valve. It can replace an elbow, for example.

For moderate conditions of pressure, temperature, and flow, the angle valve closely resembles the ordinary globe. Many manufacturers have interchangeable trim and bonnets for the two body styles, with the body differing only in the outlet end. The angle valve's discharge conditions are so favorable that many high-technology **control valves** use this configuration. They are self-draining and tend to prevent solid buildup inside the valve body, like straight flow-through globe valves.

Like valve bodies, there are also many variations of disc and seat arrangements for globe valves. The three basic types are shown in *Figure 26*.

The ball-shaped disc shown in *Figure 26(A)*, fits on a tapered, flat-surfaced seat and is generally used on relatively low-pressure, low-temperature systems. It is generally used in a fully open or shut position, but it may be employed for moderate throttling of a flow.

Figure 26(B) shows one of the proven modifications of seat/disc design, a hard nonmetallic insert ring on the disc to make closure tighter on steam and hot water. The composition disc is resistant to erosion and is sufficiently resilient and cut-resistant

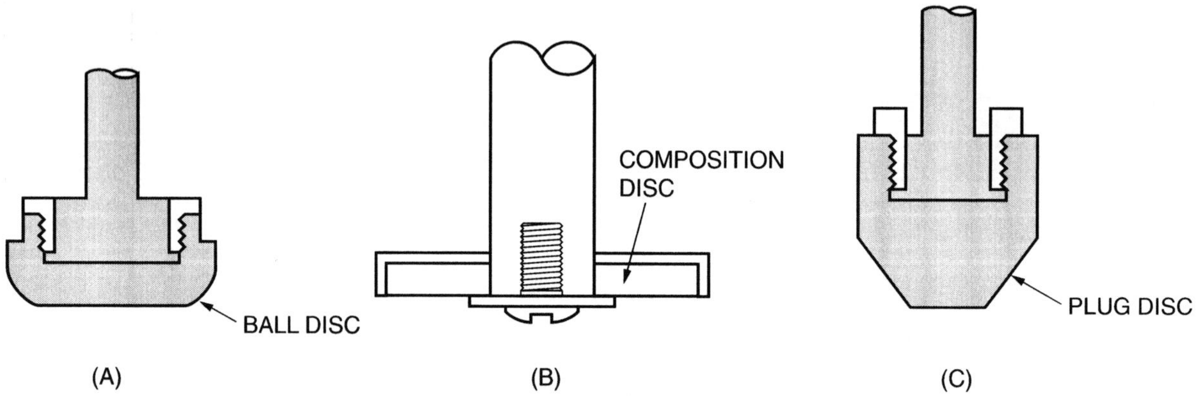

Figure 26 ◆ Seat/disc arrangements.

to close on solid particles without serious permanent damage.

The composition disc is renewable. It is available in a variety of materials that are designed for different types of service, such as high- and low-temperature water, air, or steam. The seating surface is often formed by a rubber O-ring or washer.

The plug-type disc, *Figure 26(C)*, provides the best throttling service because of its configuration. It also offers maximum resistance to galling, **wire drawing**, and erosion. Plug-type discs are available in a variety of specific configurations, but in general they all have a relatively long tapered configuration. Each of the variations has specific types of applications and certain fundamental characteristics. *Figure 27* shows the various types.

The equal percentage plug, as its name indicates, is used for equal percentage flow characteristics for predetermined valve performance. Equal increments of valve **lift** give equal percentage increases in flow.

Linear flow plugs are used for linear flow characteristics with high-pressure drops.

V-port plugs provide linear flow characteristics with medium- and low-pressure drops. This design also prevents wire drawing during low flow periods by restricting the flow through the orifices in the V-port plug when the valve is only partially open.

Needle plugs are used primarily for instrumentation applications and are seldom available in valves over 1 inch in size. These plugs provide high-pressure drops and low flows. The threads on the stem are usually very fine. Consequently, the opening between the disc and seat does not change rapidly with stem rise. This permits closer regulation of flow.

All of the plug configurations are available in either a conventional globe valve design or the angle valve design. When the needle plug is used, the valve name changes to needle valve. In all other cases the valves are still referred to as globe valves with a specific type of disc.

A special version of the globe valve uses a metallic bellows in place of packing to seal the stem and prevent contact between the medium being controlled and the atmosphere. The bellows valve is especially used for volatile organic compounds such as insecticides or for various hydrocarbon compounds, because the compounds may be poisonous or highly flammable.

Globe and angle valves should be installed so that the pressure is under the disc (flow-to-open). This promotes easy operation. It also helps to protect the packing and eliminates a certain amount of erosive action on the seat and disc faces. However, when high-temperature steam is the medium being controlled, and the valve is closed with the pressure under the disc, the valve stem, which is now out of the fluid, contracts on cooling. This action tends to lift the disc off the seat, causing leaks that eventually result in wire drawing on

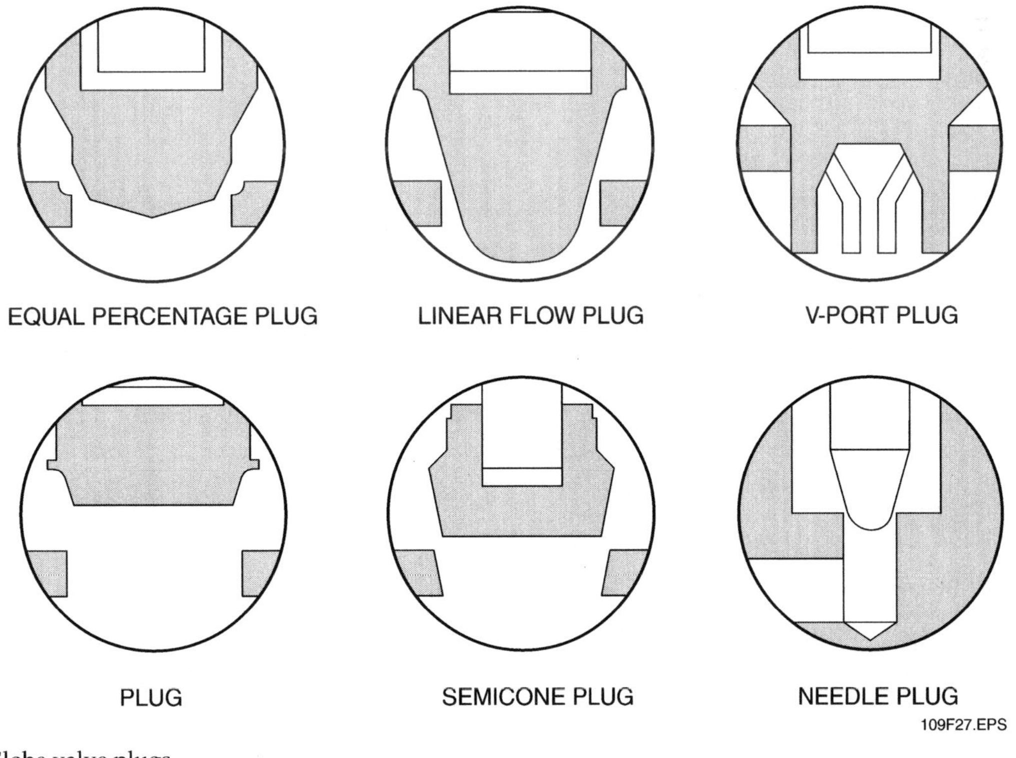

Figure 27 ◆ Globe valve plugs.

seat and disc faces. Therefore, in high-temperature steam service, globe valves may be installed so that the pressure is above the disc (flow-to-close).

Y-type valves are a cross between a gate valve and a globe valve. They provide the straight-through flow with minimum resistance of the gate valve and the throttling ability and flow control of a globe valve. The Y-type valve produces lower pressure drop and turbulence than a standard globe and is preferred over a globe valve for use in corrosive services. Y-type valves are also used in many high-pressure applications, such as boiler systems. All of the disc and seat designs available in globe valves are also available in Y-type valves. *Figure 28* shows a Y-type valve.

3.2.0 Butterfly Valves

A **butterfly valve** has a round disc that fits tightly in its mating seat and rotates 90 degrees in one direction to open and allow fluid to pass through the valve (*Figure 29*). The butterfly valve can be operated quickly by turning the handwheel or hand lever one-quarter of a turn, or 90 degrees, to open or close the valve. These valves can be used completely open, completely closed, or partially open for noncritical throttling applications.

Butterfly valves are typically used in low to medium pressure and low to medium temperature applications and generally weigh less than other types of valves because of their narrow body design. When the butterfly valve is equipped with a hand lever, the position of the hand lever indicates whether the valve is open, closed, or partially open. When the lever is parallel with the flow line through the valve, the valve is open. When the lever is perpendicular to the flow line through the valve, the valve is closed. Butterfly valves that are 12 inches in diameter and larger are usually equipped with a handwheel or gear-operated actuator because of the large amount of fluid flowing through the valve and the great amount of pressure pushing against the seat when the valve is being closed.

Butterfly valves have an arrow stamped on the side, indicating the direction of flow through the valve. They must be installed in the proper flow direction, or the seat will not seal and the valve will leak. Three types of butterfly valves include the wafer, wafer lug, and two-flange.

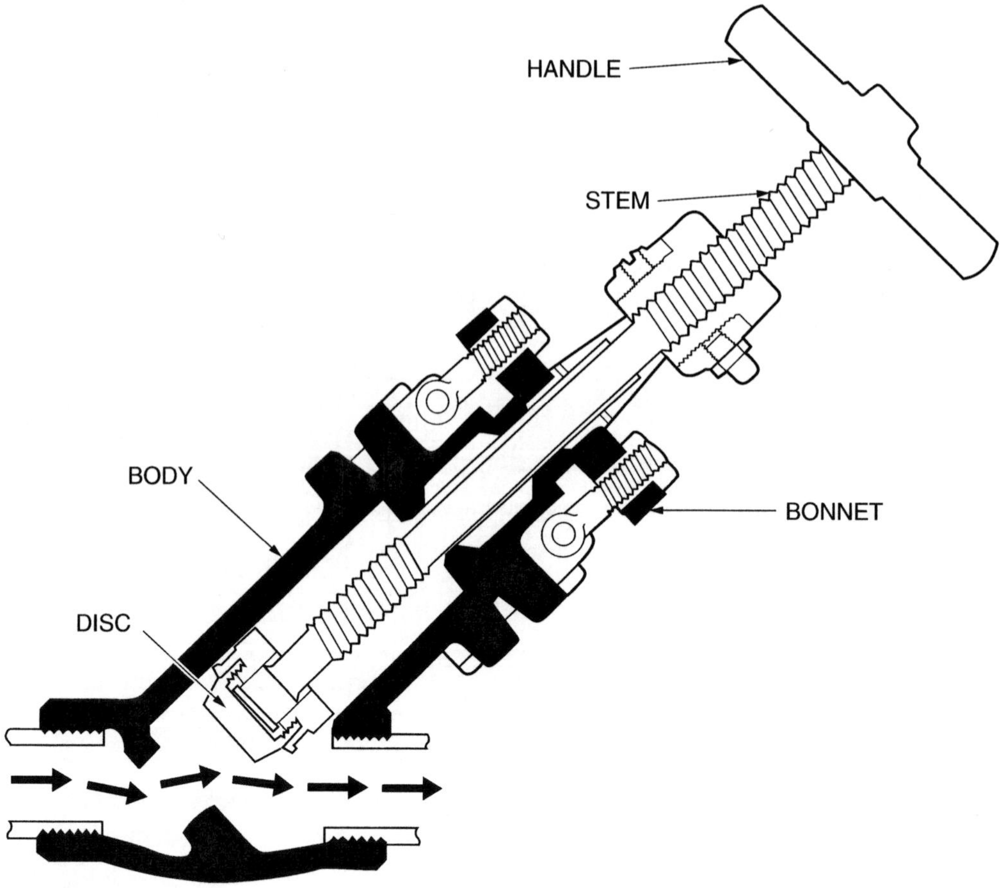

Figure 28 ◆ Y-type valve.

3.2.1 Wafer Valves

The wafer-type butterfly valve is designed for quick installation between two flanges. No gasket is needed between the valve and the flanges because the valve seat is lapped over the edges of the body to make contact with the valve faces. Bolt holes are provided in larger wafer valves to help line up the valve with the flanges. During installation of the valves, the valve must be in the open position prior to torquing to prevent damage to the valve seat. *Figure 30* shows a wafer butterfly valve.

3.2.2 Wafer Lug Valves

Wafer lug valves are the same as the wafer valves except that they have bolt lugs completely around the valve body. Like the wafer valve, no gasket is needed between the valve and the flanges because the valve seat is lapped over the edges of the body to make contact with the valve faces. The lugs are normally drilled to match ANSI 150-pound steel drilling templates. The lugs on some wafer lug valves are drilled and tapped so that when the valve is closed, downstream piping can be dismantled for cleaning or maintenance while the

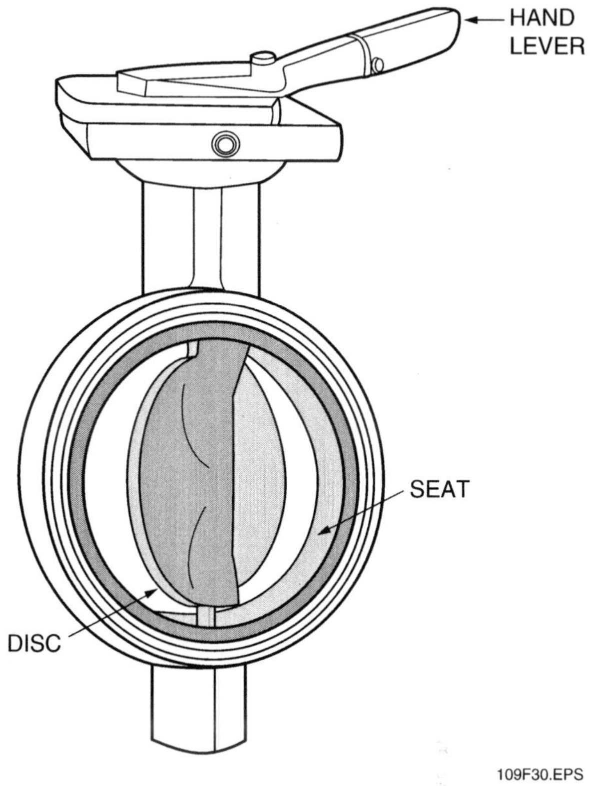

Figure 30 ◆ Wafer butterfly valve.

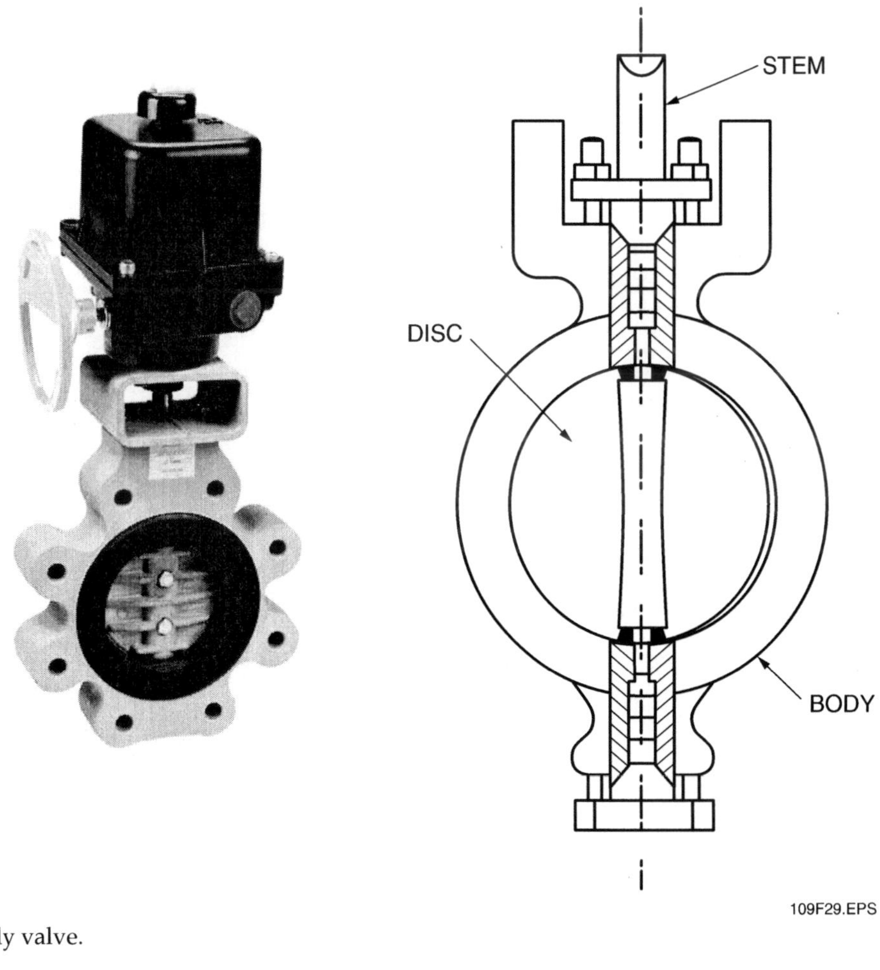

Figure 29 ◆ Butterfly valve.

upstream piping is left intact. When tapped wafer lug valves are used for pipe end applications, only one pipe flange is necessary. *Figure 31* shows a wafer lug butterfly valve.

3.2.3 Two-Flange Valves

The body of a two-flange butterfly valve is made with a flange on each end. The valve seat is not lapped over the flange ends of the valve, so gaskets are required between the flanged body and the mating flanges. The two-flange valve is made with either flat-faced flanges or raised-face flanges, and the mating flanges must match the valve flanges. *Figure 32* shows a two-flange butterfly valve.

3.3.0 Diaphragm Valves

Flow through a diaphragm valve is controlled by a flexible disc that is connected to a compressor by a stud molded into the disc. The valve stem moves the compressor up and down, regulating the flow. Diaphragm valves have no seats because the body of the valve acts as the seat for the flexible disc. The operating mechanism of the valve does not come in contact with the material within the pipeline.

These valves can be used fully opened, fully closed, or for throttling service. They are good for handling slurries, highly corrosive materials, or materials that must be protected from contamination. Many types of fluids that would clog other types of valves will flow through a diaphragm valve. *Figure 33* shows a weir-type diaphragm valve. In addition to the types shown, there is a straight-through type. A special case of the diaphragm valve is the pinch valve, in which the flexible lining of the opening is pinched closed by moving bars at the top and bottom of the valve. The linings are replaceable, and prevent exposure of the metal parts of the valve to the substance passing through.

3.4.0 Needle Valves

Needle valves, as shown in *Figure 34,* are used to make relatively fine adjustments in the amount of fluid allowed to pass through an opening. The needle valve has a long, tapering, needle-like point on the end of the valve stem. This needle acts as a disc. The longer part of the needle is smaller than the orifice in the valve seat. It therefore passes through it before the needle seats. This arrangement permits a very gradual increase or decrease in the size of the opening and thus allows a more precise control of flow than could be obtained with an ordinary globe valve. Needle valves are often used as component parts of other, more complicated valves. For example, they are used in some types of reducing valves. Most constant-pressure pump governors have needle valves to minimize the effects of fluctuations in pump discharge pressure.

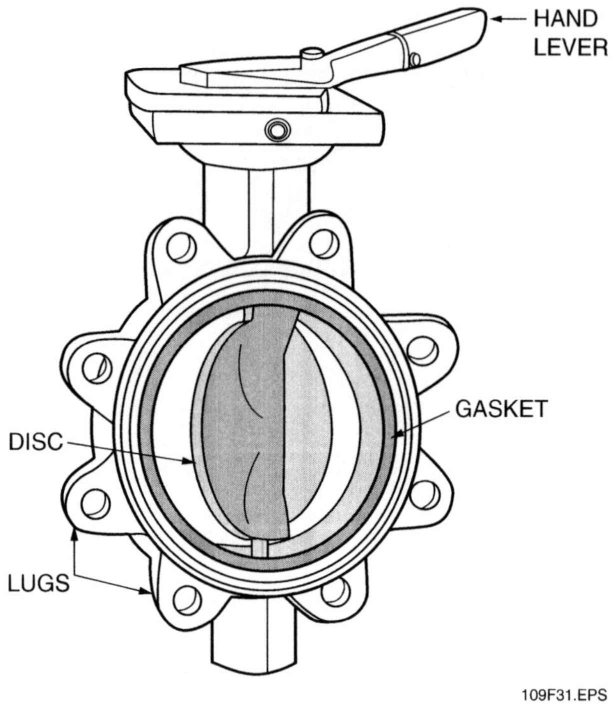

Figure 31 ◆ Wafer lug butterfly valve.

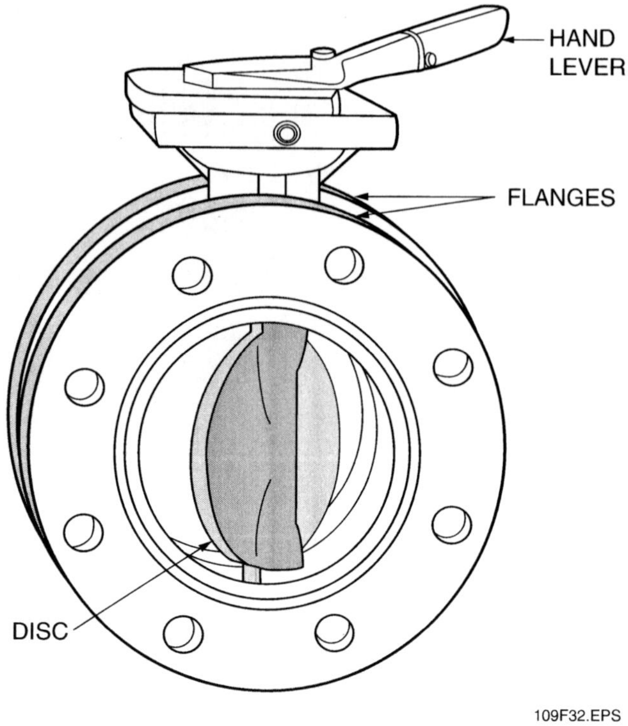

Figure 32 ◆ Two-flange butterfly valve.

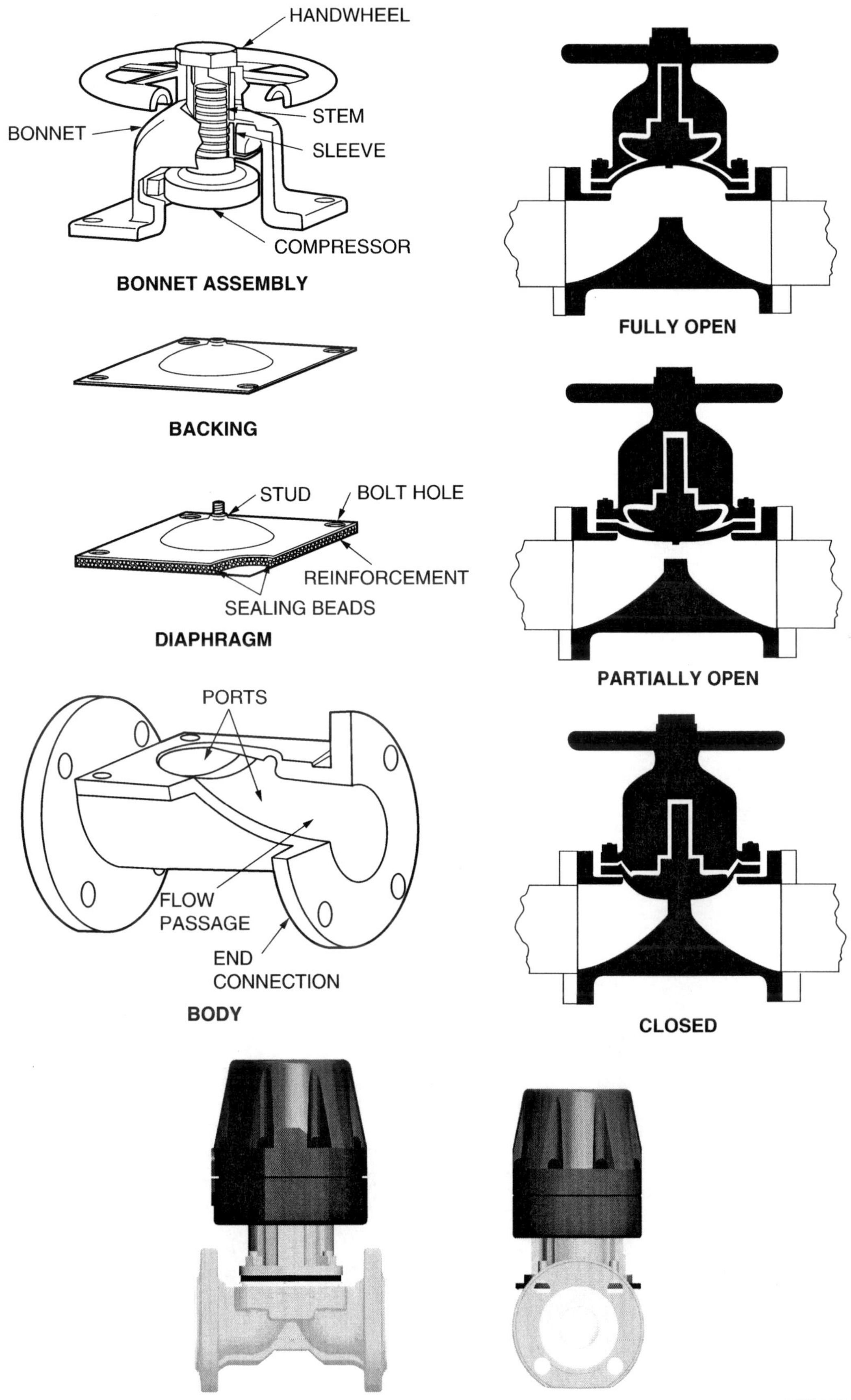

Figure 33 ◆ Diaphragm valve.

Needle valves are also used in some components of automatic combustion control systems where very precise flow regulation is necessary.

4.0.0 ◆ VALVES THAT RELIEVE PRESSURE

Valves used to relieve pressure in a pipeline, tank, or vessel are known as safety valves and pressure-**relief valves** (*Figure 35*). These are installed in pipelines to prevent excess pressure from rupturing the line and causing an accident. Both types are adjustable and operate automatically.

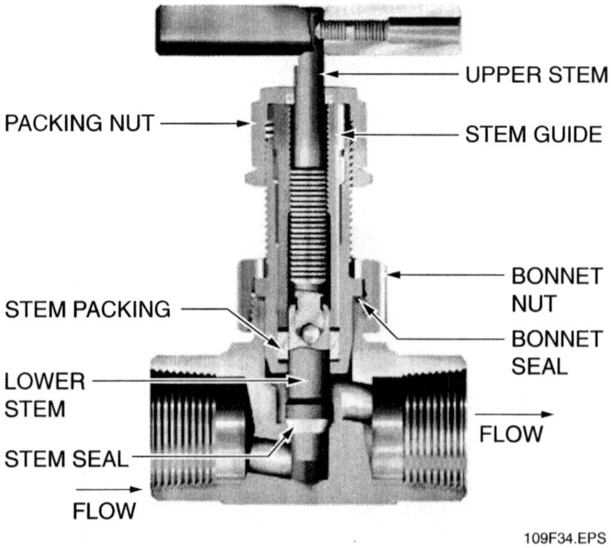

Figure 34 ◆ Needle valve.

4.1.0 Safety Valves

Safety valves are normally used in steam, air, or other gas service pipelines. They operate in the closed position until the pressure in the line rises above the preset pressure limit of the valve. At this point, the valve opens fully to relieve the pressure and remains open until the pressure drops, at which point the valve snaps shut. Because of this fully open and tight-closing feature, safety valves are commonly referred to as pop-off valves. *Figure 36* shows a cutaway view of a safety valve.

Safety valves are most commonly installed in the vertical position. For air or gas service, these valves can be installed upside down to allow moisture to collect, which seals the surfaces. Safety valves should be mounted directly to the tank or vessel being protected using connecting piping and block valves. Before the safety valve can close, its discharge pressure must drop several pounds below the opening pressure. A safety valve has an adjustment to control the closing of the valve, depending on the amount of pressure drop after the valve has opened. This adjustment is calibrated by the manufacturer and should not be adjusted in the field.

An item which should be included with safety valves is the vacuum breaker. This is a small version of a backflow preventer. The vacuum breaker is screwed onto a hose bib to prevent backflow from the hose attached to the bib. When sprayers containing chemicals are attached to a hose, or when the end of a hose is immersed in liquid, any loss of pressure upstream of the hose can cause

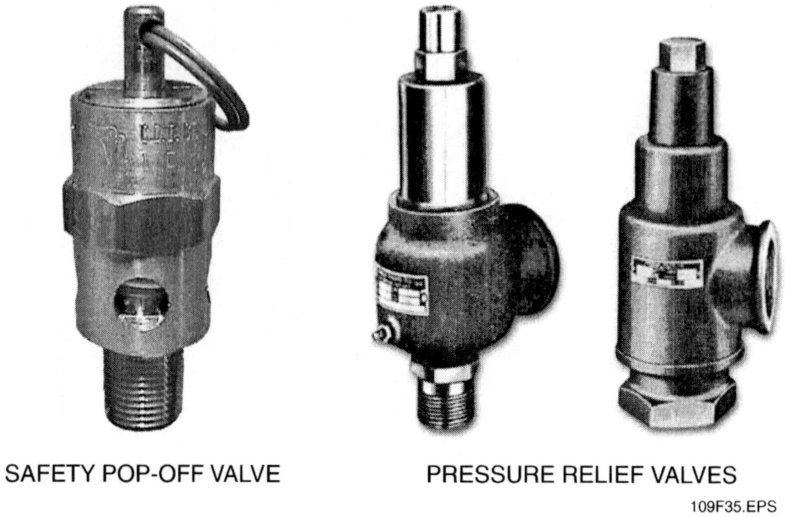

Figure 35 ◆ Examples of pressure-relieving valves.

backflow into the system. A vacuum breaker does not allow the backflow to take place.

4.2.0 Pressure-Relief Valves

Relief valves are normally used to relieve pressure in liquid services. They are operated by pressure acting directly on the bottom of a disc that is held on its seat by a spring. The amount of spring pressure on the disc is determined at the factory when the valve is manufactured. When liquid pressure becomes great enough to overcome the force of the spring on the disc, the disc rises, allowing the liquid to escape. After enough liquid has escaped to allow the pressure to drop, the system pressure becomes too low to overcome the force of the spring on the disc. The spring then pushes the disc back onto its seat, stopping the flow of liquid through the valve. The force of the spring is called the compression of the valve. *Figure 37* shows a pressure-relief valve.

Spring-loaded relief valves are also most commonly installed with the stem in the vertical position. The piping connected to these valves must be as large or larger than the valve inlet and outlet openings. You can use a reducer to reduce down to the inlet of a relief valve, but you cannot increase the line size to the inlet of the relief valve. The piping must also be well-supported so that the line strains will not cause the valve to leak at the seat.

One variety of pressure-relief valve is called a rupture disk. The rupture disk is used in high-pressure applications, such as the discharge end of a high-pressure pump. The component consists of a small metallic disk, which is designed to be blown out at a certain pressure level. The disk is a one-time valve. If it is blown, you must obtain and install another disk. Do not use any disk other than one designed for the purpose. Another disk, such as a coin, may not blow in time to prevent damage or personal injury.

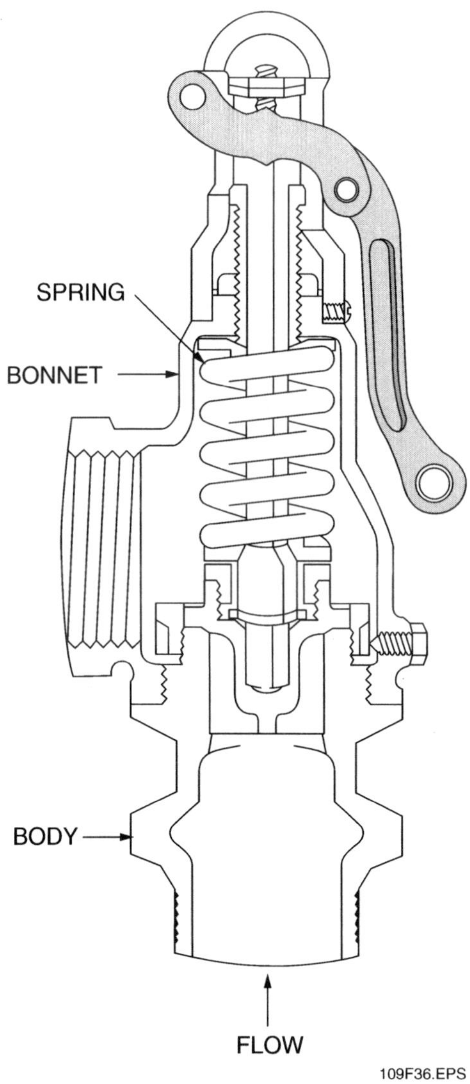

Figure 36 ◆ Safety valve.

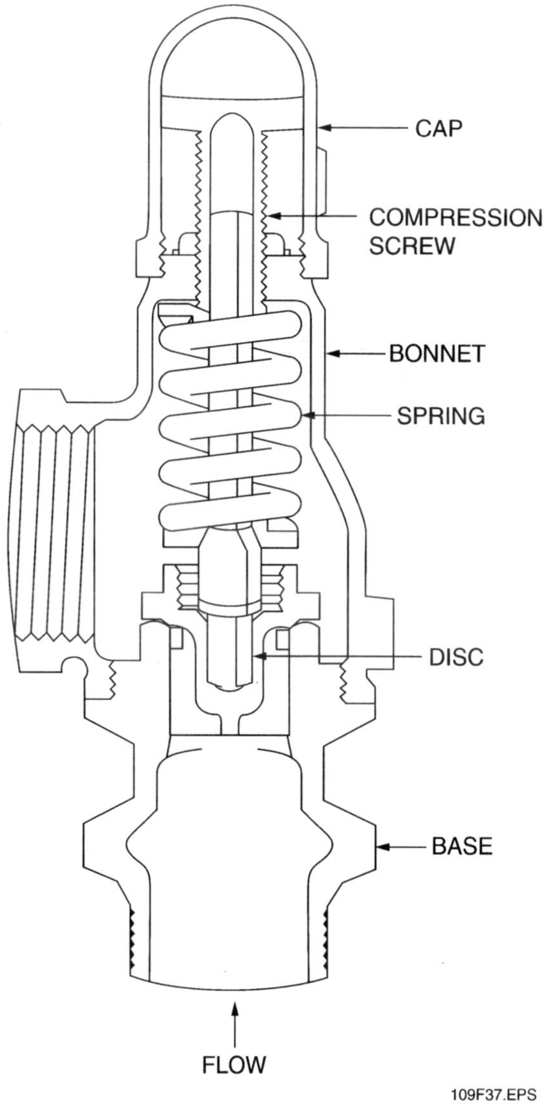

Figure 37 ◆ Pressure-relief valve.

5.0.0 ◆ VALVES THAT REGULATE DIRECTION OF FLOW

Valves that prevent backflow are called check valves. Check valves regulate flow by preventing the backflow of liquids or gases in a pipeline. Check valves are intended to permit the flow in only one direction, which is indicated by an arrow stamped on the side of the valve. The force of flow and the action of gravity cause the valve to open and close. Check valves are manufactured in several different designs, including the following:

- Swing
- Lift
- Ball
- Butterfly
- Foot

5.1.0 Swing Check Valves

The most commonly used type of check valve is the swing check valve. This valve has a disc that swings, or pivots, to open and close the valve. Since swing check valves provide straight line flow through the valve body, they offer less resistance to flow than other types of check valves. Swing check valves can be mounted horizontally or vertically with the flow upward. *Figure 38* shows a swing check valve. If this valve is used in a vertical flow, a counterweight or spring assist is needed to make sure it does not stick open.

The disc, which is hinged at the top, seats against a machined seat in the tilted wall bridge opening. The disc swings freely from the fully closed position to the fully open position, which provides unobstructed flow through the valve. The flow through the valve causes the disc to stay open, and the amount that the disc is open depends on the volume of liquid moving through the valve. Gravity or reversal of flow cause the disc to close, preventing backflow through the valve. A special case of this valve is the flap valve, used in some slurry and waste pipe-to-tank applications to prevent backflow. The flap valve does not have an enclosing housing, only a flap and a seat. It is essentially the self-closing end of the pipe line.

Swing check valves can also be equipped with an outside lever and counterweight to assist the valve in closing. These valves are often referred to as weighted check valves.

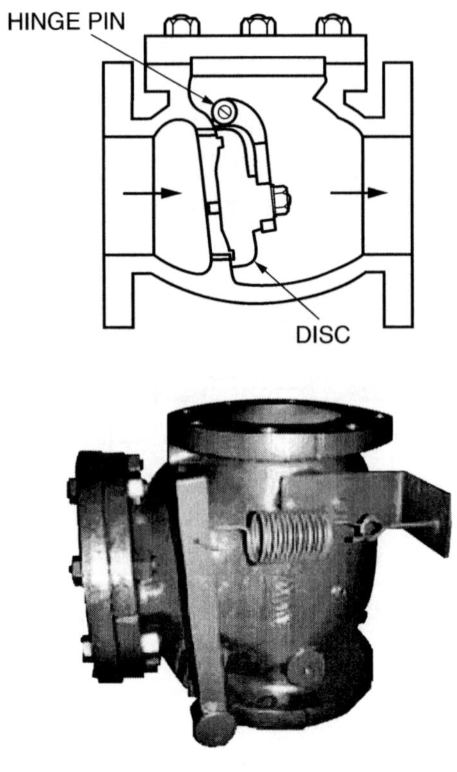

ANGLE CHECK VALVE

106F38.EPS

Figure 38 ◆ Swing check valve.

5.2.0 Lift Check Valves

The lift check valve is designed like a globe valve, with an indirect line of flow through the valve. The lift check valve also works automatically by line pressure and prevents the reversal of flow through the valve. A lift check valve consists of a disc that is held against the disc seat by gravity, by a spring or counterweight, or by the pressure of the fluid flowing in the opposite direction through the valve. When fluid flows through the valve in the correct direction, the force of the fluid lifts the disc away from the seat to allow the fluid to pass through the valve. Lift check valves can be designed for horizontal or vertical applications and must only be used for the application they are designed for. *Figure 39* shows lift check valves.

Another variety, the angle lift check, has the disk and seat at an angle to the direction of flow.

5.3.0 Ball Check Valves

Ball check valves operate in the same manner as lift check valves except they use a ball instead of a disc to allow the fluid to pass through the valve. Ball check valves that are not spring-loaded should only be installed in the vertical position. Gravity and the reversal of flow cause the ball to rest against the valve seat, restricting flow through the valve.

Ball check valves are recommended for use in lines in which the fluid pressure changes rapidly since the action of the ball is practically noiseless.

The ball rotates during operation and therefore tends to wear at a uniform rate over its entire surface. The ball valve also stops flow reversal more rapidly than other types of check valves. *Figure 40* shows a ball check valve.

5.4.0 Butterfly Check Valves

A butterfly check valve has two vanes that resemble the wings of a butterfly. These vanes fold back from a central hinge to open and allow flow but close to prevent backflow. *Figure 41* shows a butterfly check valve.

> **NOTE**
> If the butterfly check valve is used in the horizontal position, it must have a spring to assist in closing.

5.5.0 Foot Valves

Foot valves are used at the bottom of the suction line of a pump to maintain the prime of the pump. This type of valve operates similarly to the lift check valve. On the inlet side of the foot valve is a strainer that keeps foreign material out of the line. The weight of the liquid in the pipeline between the pump and the foot valve keeps the suction pipeline full when the pump is shut down. The weight pushes the seats closed in the valve, preventing the liquid from flowing out through the valve. When the pump starts, the force of the suction causes the liquid coming into the valve to push against the outside of the seats to open the valve. *Figure 42* shows a foot valve.

> **NOTE**
> If the foot valve is used in the horizontal position, it must have a spring to assist in closing.

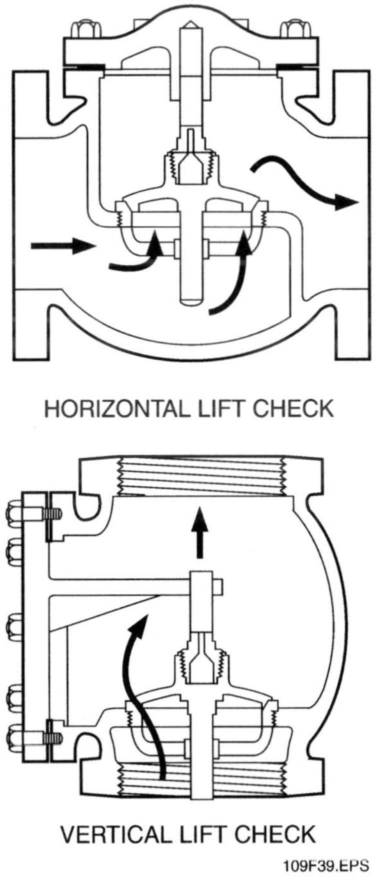

Figure 39 ◆ Lift check valves.

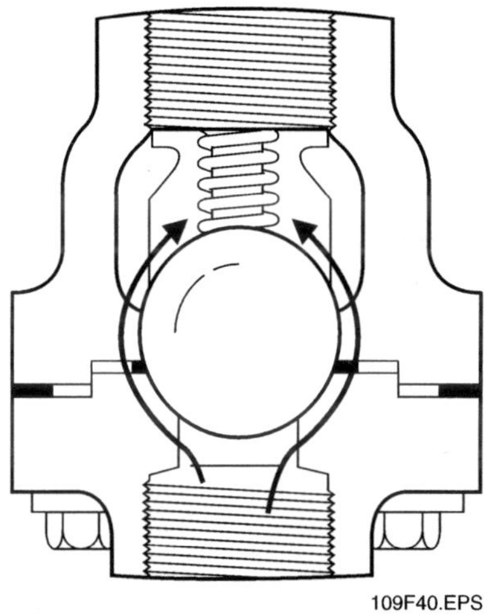

Figure 40 ◆ Ball check valve.

Figure 41 ◆ Butterfly check valve.

6.0.0 ◆ VALVE ACTUATORS

The primary purpose of a valve actuator, also known as an operator, is to provide automatic control of a valve or to reduce the effort required to manually operate a valve. A standard handwheel or hand lever attached to a valve stem is one type of actuator commonly used on smaller valves that do not require great effort to operate. Larger valves must be equipped with some other type of actuator to reduce the effort to operate the valve.

In today's automated industries and central control stations, many valves are mechanically operated and powered by electric, pneumatic, or hydraulic actuators or operators. Electric, pneumatic, and hydraulic valve actuators are used when a valve is remote from the main working area, when the frequency of operation of the valve would require unreasonable human effort, or if rapid opening or closing of the valve is required. There are many types of valve actuators in use today, including the following:

- Gear operators
- Chain operators
- Pneumatic and hydraulic actuators
- Electric or air motor-driven actuators

6.1.0 Gear Operators

Gear operators minimize the effort required to operate large valves that work at unusually high pressures. Gear operators are also used to connect to valves located in inaccessible areas. Three basic types of gear operators are spur-gear operators, bevel-gear operators, and worm-gear operators.

6.1.1 Spur-Gear Operators

Spur-gears are used to connect parallel shafts. They have straight teeth cut parallel to the shaft axis and use gears to transfer motion and power from one shaft to another parallel shaft. Spur-gear operators can be as simple as one gear transmitting power to another gear attached to the valve stem, or they can consist of many gears, depending on how much power needs to be transmitted. *Figure 43* shows a spur-gear operator.

Although you may see valves with open gear trains in older installations, this design is no longer permitted by OSHA. Gear-driven valves in current use have enclosed gear boxes like the one shown in *Figure 44*.

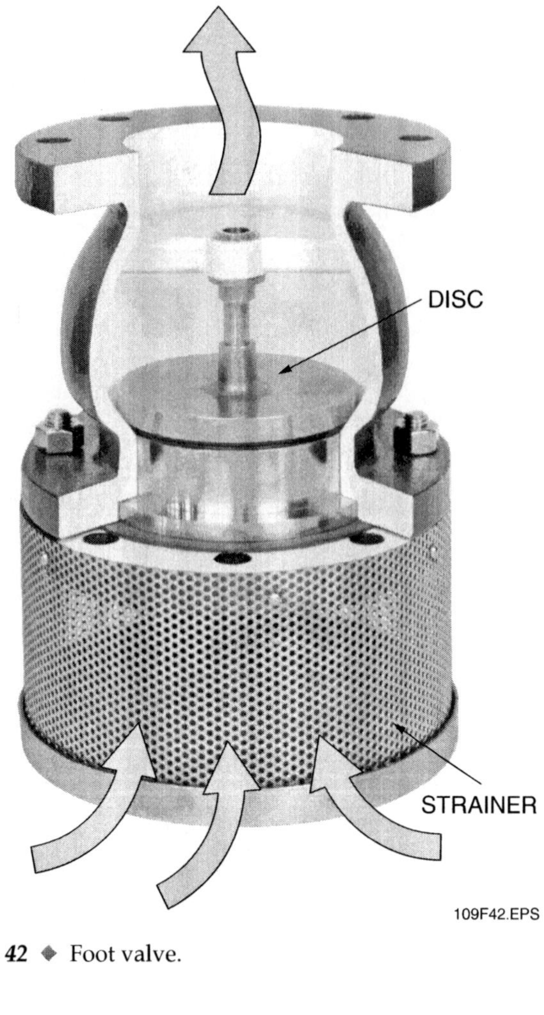

Figure 42 ◆ Foot valve.

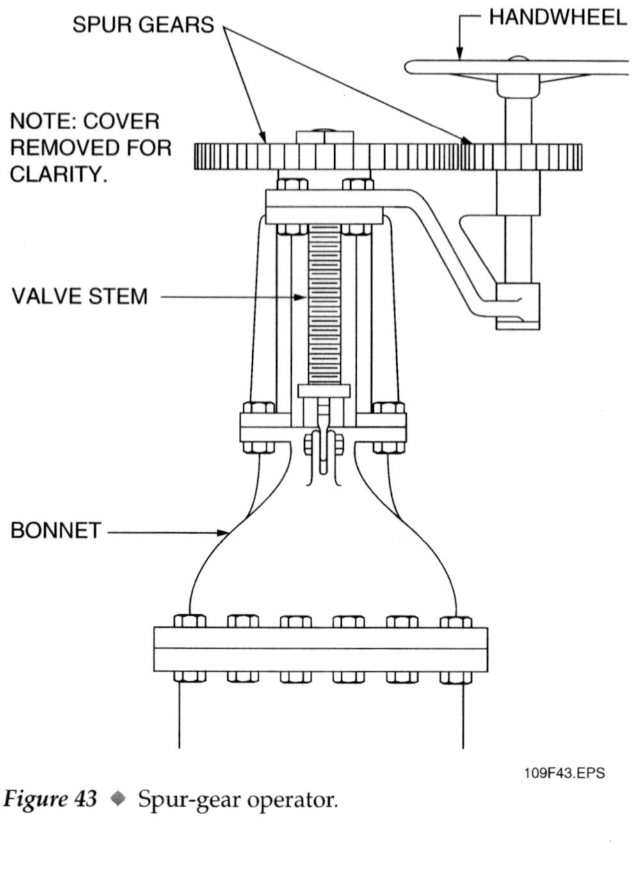

Figure 43 ◆ Spur-gear operator.

6.1.2 Bevel-Gear Operators

Bevel-gears are used to transmit power between two shafts that intersect at a 90-degree angle. The bevel gear resembles a cone in that the teeth on each gear are angled at 45 degrees to mesh with the gears on the mating gear. Bevel-gear operators transmit power from a handwheel and a shaft that is perpendicular to the valve stem. *Figure 45* shows a bevel-gear operator.

6.1.3 Worm-Gear Operators

Worm gears are used in butterfly valves to transmit power at right angles. This means that the shafts of the connecting gears are at a 90-degree angle to each other. The worm is a cylindrical-shaped gear similar to a screw and is the driver. The worm gear is the larger, circular gear in the assembly. Worm-gear operators transmit power from the handwheel, which is attached to the worm, to the worm gear, which is attached to the valve stem. Worm gears are set up like bevel gears, with a side-mounted handwheel. A large speed ratio is obtainable with a worm-gear operator because the operator allows rapid opening and closing of large valves with minimum effort.

Figure 44 ◆ Enclosed gear box valve.

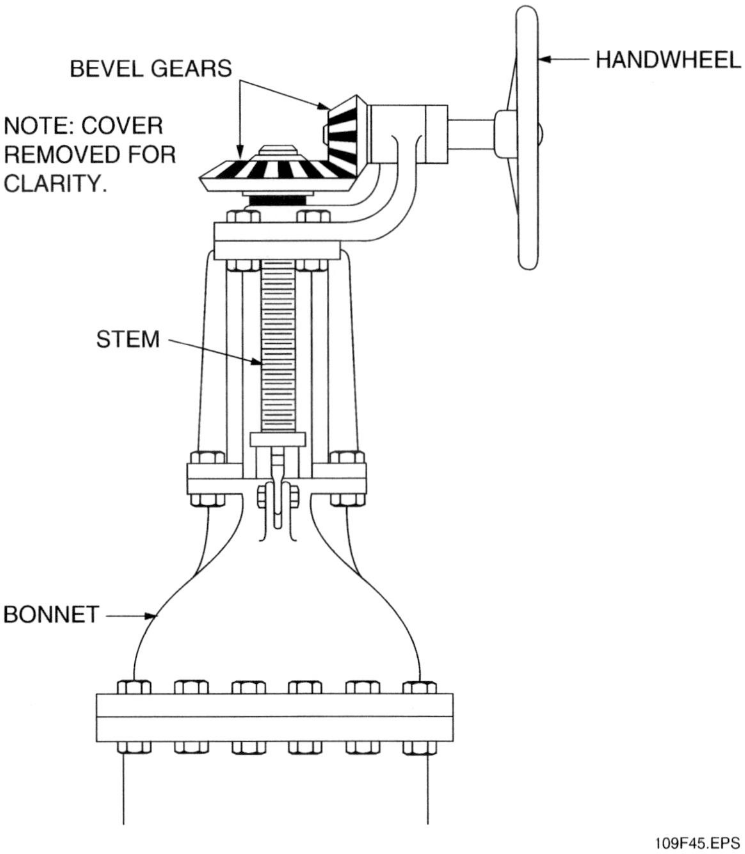

Figure 45 ◆ Bevel-gear operator.

It takes only one-quarter turn of the handwheel to open or close the valve. *Figure 46* shows a worm-gear operator.

6.2.0 Chain Operators

A chain operator is installed in some situations when the valve is mounted too high to reach the handwheel. The stem of the valve is mounted with a chain wheel, and the chain is brought to within 3 feet of the working floor level.

Chain operators are used only when they are absolutely necessary. Universal-type chain wheels that attach to the regular valve handwheel have been blamed for many industrial accidents. In corrosive atmospheres where an infrequently operated valve is located, the attaching bolts of this type of chain wheel have been known to fail. Chain wheels attach to the stem and replace the regular valve handwheel. *Figure 47* shows a chain operator.

WARNING!

OSHA now requires that the wheels for chain operators be secured with a safety cable (lanyard). This is done to prevent them from falling if their fasteners have been weakened by rust or corrosion.

6.3.0 Pneumatic and Hydraulic Actuators

Electric, hydraulic, or pneumatic actuators are frequently used in contemporary highly-automated systems. If you are required to work on such a system, make sure that the actuator is locked or tagged out while you are working, so that the valve will not be opened while you are downstream.

Pneumatic and hydraulic actuators operate in basically the same manner, except the pneumatic actuator operates off air pressure and the hydraulic actuator operates off fluid pressure. A cylinder assembly that contains a piston is attached to the valve stem. Air or fluid pressure above and below the piston moves the piston and the valve stem up and down, opening and closing the valve. Many pneumatic and hydraulic actuators also have spring-loaded pistons in either the naturally open or naturally closed position, depending on which position is considered the fail-safe position. These springs allow the valve to fail open or fail closed. If the control medium is lost, the spring will force the valve to the selected fail-safe position, which is determined as part of the system design process. This safety feature is required so that the valve will return to its safe position in case of pressure failure. *Figure 48* shows a spring-loaded pneumatic valve actuator.

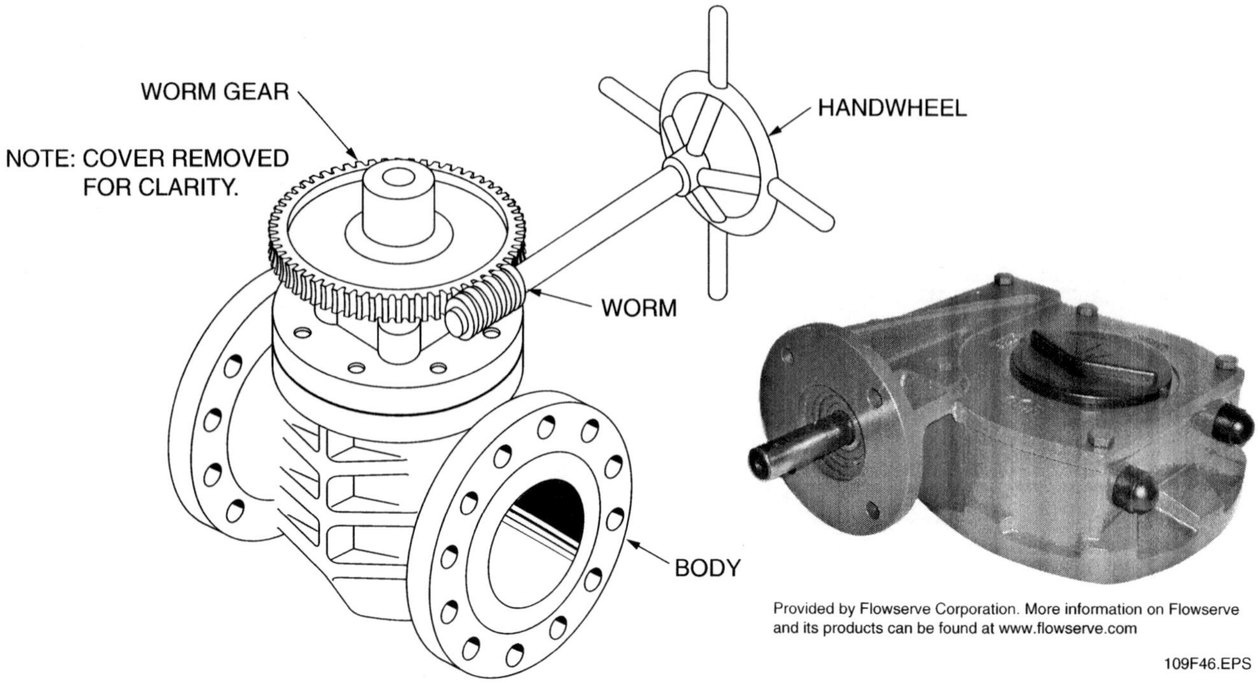

Figure 46 ◆ Worm-gear operator.

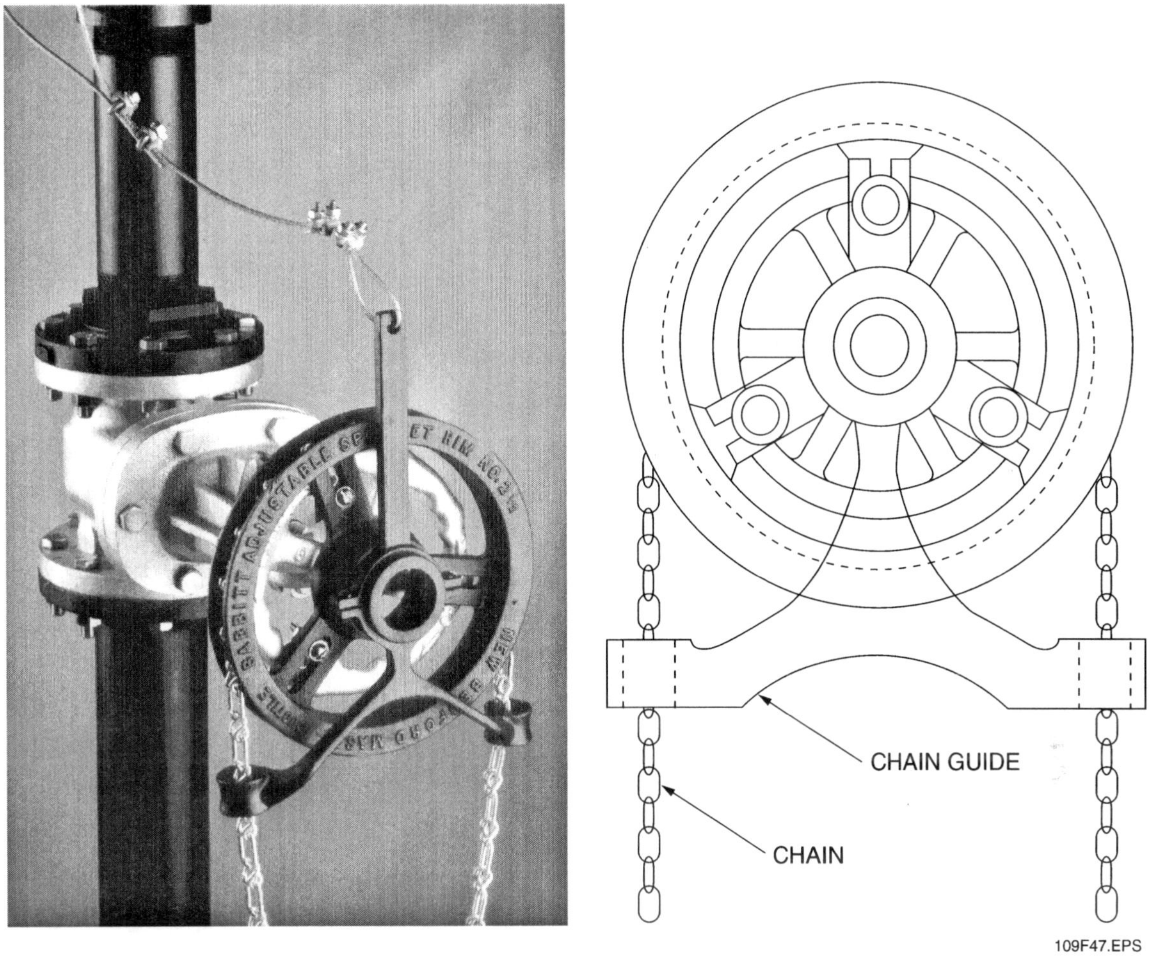

Figure 47 ◆ Chain operator.

> **WARNING!**
> The springs in pneumatic valve actuators are under strong tension and can cause serious injury if let loose. Do not disassemble a valve unless you have been trained for that work.

6.4.0 Electric or Air Motor-Driven Actuators

Electric or air motor-driven valve actuators contain a motor that is linked through reduction gears to the valve stem. The electric motor is equipped with electrical limit switches that shut off the motor when the motor has turned the valve stem as far as it can go in either direction. This prevents unnecessary wear and tear on the motor. Electric valve actuators are usually equipped with a handwheel for controlling the valve manually if the power fails. Actuators should only be removed, turned, and adjusted by qualified technicians.

6.5.0 Control Valves

Control valves are variations of angle, globe, or ball valves that are controlled by pneumatic, electronic, or hydraulic actuators. They operate in a partially open position and are most commonly used for pressure limiting or flow control. Control valves are precision-built for increased accuracy in flow control. The control valve is usually smaller than the pipeline to avoid throttling and constant wear of the valve seat. The actuators for control valves are explained in more detail later in the module. *Figure 49* shows a control valve.

Control valves are often used with pressure-sensing elements that measure the pressure drop at a given point in the pipeline. The flow rate is directly related to the pressure drop. The sensing element sends a pressure signal to a controller that compares the pressure drop with the pressure drop for the desired flow rate. If the actual flow is different, the controller adjusts the valve through the valve actuator to increase or decrease the flow. *Figure 50* shows a schematic for a control valve in a pipeline.

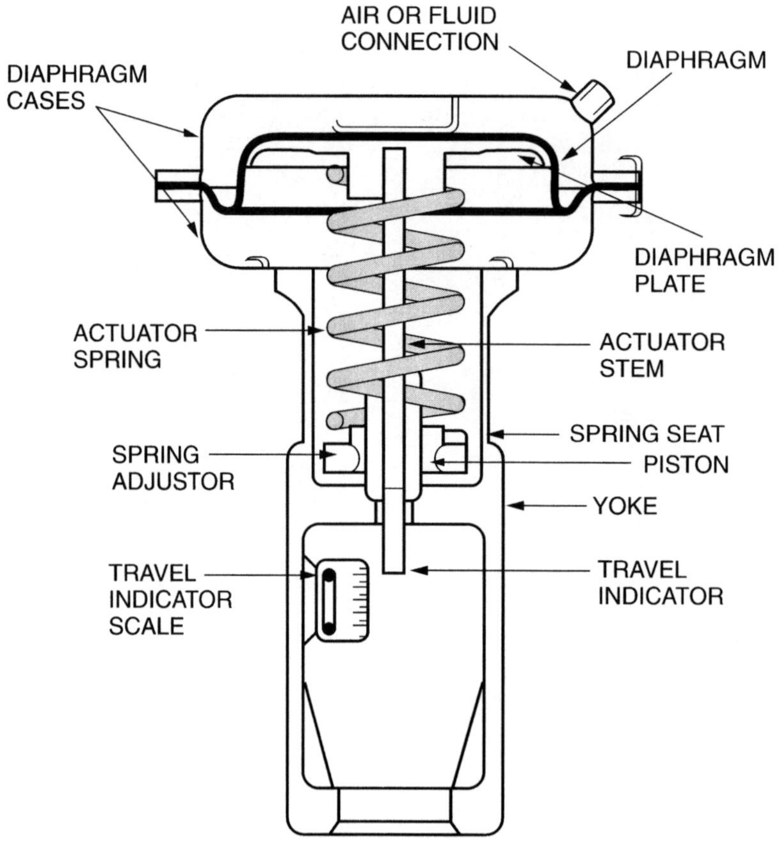

AIR- OR FLUID-TO-LOWER
ACTUATOR

AIR- OR FLUID-TO-RAISE
ACTUATOR

109F48.EPS

Figure 48 ◆ Spring-loaded pneumatic valve actuators.

7.0.0 ◆ STORING AND HANDLING VALVES

Once a valve has been received at the job site, it must be properly stored until it is installed in the pipeline. If the valve has been abused in storage, it may fail after installation. Caution must be taken to ensure that the valves are handled properly and safely.

7.1.0 Safety Considerations

When working with and around valves, you must be alert and work cautiously to ensure your own safety and the safety of your co-workers. Follow these guidelines when handling or working around valves:

- Be aware of all pinch points.
- Do not stand under a load.

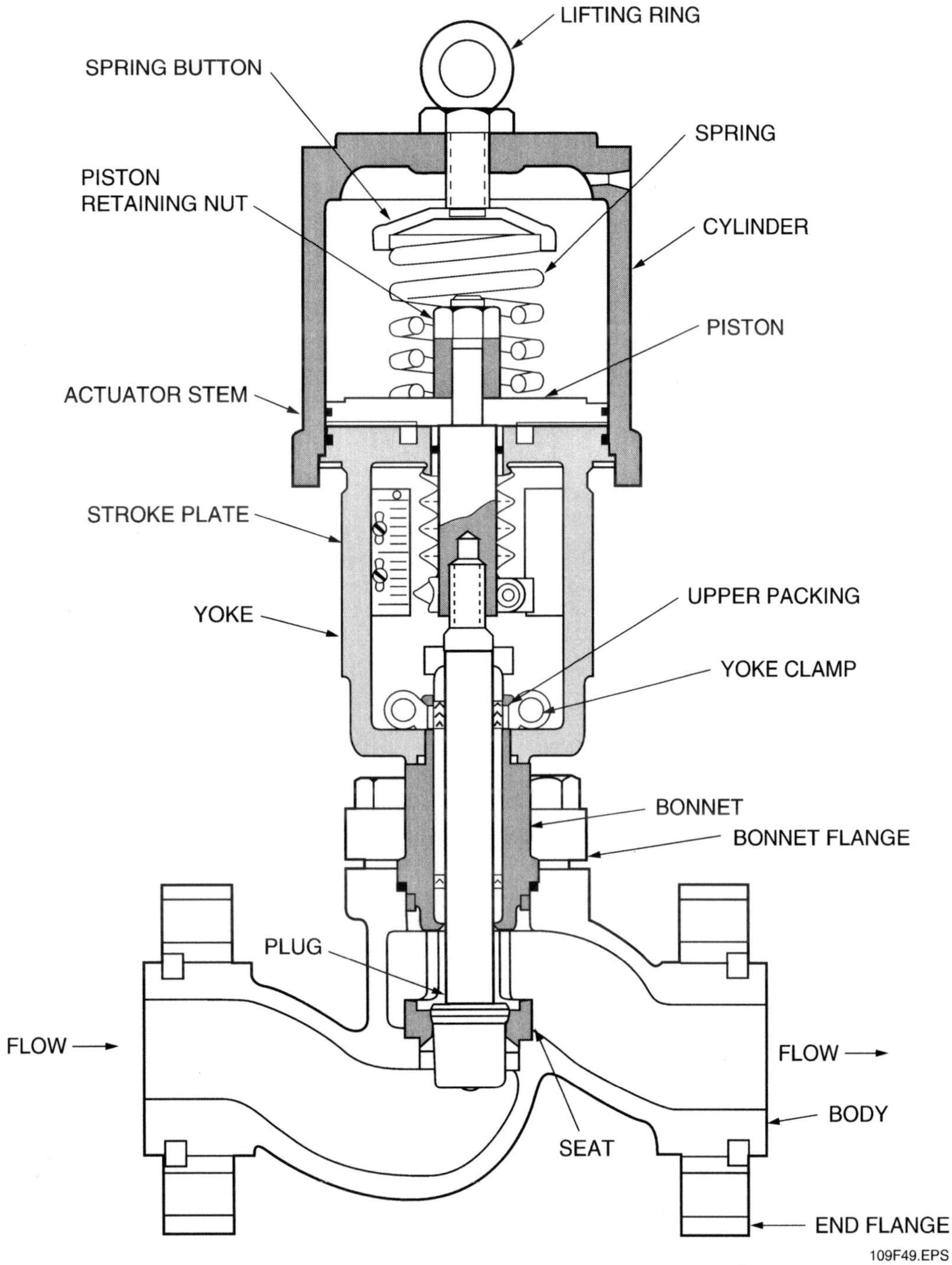

Figure 49 ◆ Control valve.

- Watch for overhead power lines and other equipment.
- Do not overload temporary work platforms.
- Ensure that final supports and hangers are in place before installing a large valve.
- Never operate any valve in a live system without proper authorization.
- Always use a spud wrench or drift pin when aligning bolt holes in a flanged valve. Never align bolt holes with your fingers.
- Never stand in front of a safety valve relief discharge.

7.2.0 Storing Valves

Most valve manufacturers wrap the valves in protective wrapping for shipment. This wrapping should remain on the valve until installation. Follow these guidelines when storing and handling valves:

- Clearly label all valves with identification tags or stenciled placards.
- Store all valves on appropriate hardwood dunnage to facilitate transportation of the materials and to keep the valves off the ground. Never store valves on the ground.
- Store all valves on the basis of their compatibility with other valves. For example, do not store stainless steel valves with carbon steel valves because the rust from carbon steel can cause the stainless steel to corrode.
- Pressure relief valves and safety valves must be stored upright and under cover.
- Store valves in an area where corrosive fumes, freezing weather, and excessive water can be kept to a minimum.

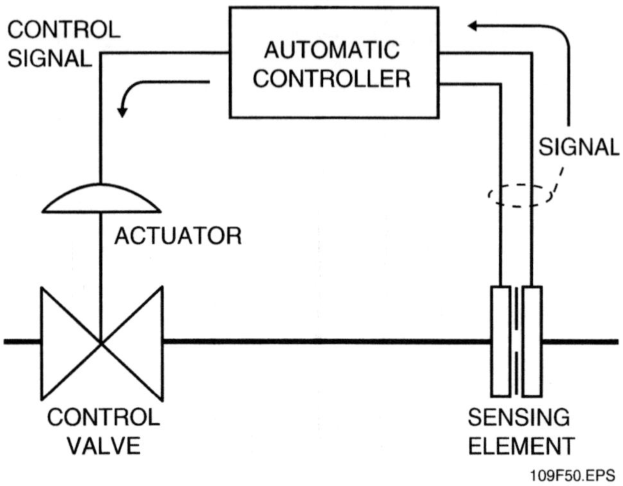

Figure 50 ◆ Control valve schematic.

- Store the valves in an area where no objects can fall and damage them.
- Do not remove any tags from valves.
- Ensure that all open ends are protected with end protectors.
- Store valve handles with their mating valves.
- When storing valves outside, always lay them on their side so that water cannot get trapped inside the valve. If water inside the valve freezes, it can shatter the body material.

7.3.0 Rigging Valves

Special precautions must also be taken when rigging valves. Follow these guidelines when rigging valves:

- Clean and protect all threads and weld ends before lifting.
- Select your rigging equipment based on the weight of the valve.
- Do not rig a valve by the stem, handle, actuator, or through the body opening. Rig only to the valve body.
- Rig stainless steel valves using nylon straps only.
- Install a tag line to control the lift.
- Rig for the proper position of the valve in the final installation.
- Rig to allow for installation of bolts and nuts.
- Remove all shipping materials before installation.

8.0.0 ◆ INSTALLING VALVES

Proper installation of valves is very important to the efficient operation of the piping system. Valves can be installed in piping systems with welded, threaded, or flanged joints. The procedure for installing a valve is the same for installing a fitting. However, there are added factors to consider when installing valves.

The location of a valve in a piping system is very important. Most piping drawings indicate exactly where to install the valve. The direction in which the stem and handwheel are to be located when working with small-bore piping is often the industrial maintenance technician's responsibility. When working with large-bore piping, the stem direction and orientation is normally shown on the drawings. To allow regular maintenance procedures, the valve must be easily accessible when possible. The height of the valve must be accessible without causing hazards. The best installation height for valve handles is between 2 feet and 4 feet 6 inches off the floor or working level. Valves installed below and above this area (up to 6 feet 6 inches) create either a tripping hazard or a face hazard; therefore, the valve handwheel must have some type of guard around it. *Figure 51* shows the order of preference for valve location.

Valves that are installed with the stem in the upright position tend to work best. The stem can be rotated down to the horizontal position, but should not point down. If the valve is installed with the stem in the downward position, the bonnet acts like a trap for sediment, which may cut and damage the stem. Also, if water is trapped in the bonnet in cold weather, it may freeze and crack the body of the valve.

Another factor that must be considered when installing valves is the direction of flow through the valve. Butterfly, safety, pressure-relief, and some other valves either have arrows indicating direction of flow stamped on the side, or the ports are labeled as the inlet or the outlet. When the valve is not marked, you must determine which side of the disc you want the pressure against. Gate valves can have the pressure on either side. Globe valves should be installed so that the pressure is below the disc unless pressure above the disc is required in the job specifications (*Figure 52*).

9.0.0 ◆ VALVE SELECTION, TYPES, AND APPLICATIONS

Because of the diverse nature of valves, with valve types overlapping each other in both design and application, the valve selection process must be examined. This section discusses valve selection, valve types, and valve applications.

9.1.0 Valve Selection

With valve selection, there are many factors that must be taken into consideration. Cost is often an overriding factor, although experience has shown that sparing expense now may result in additional expense later. When selecting a valve during system design, overall system performance must be taken into consideration. Questions that must be asked include these:

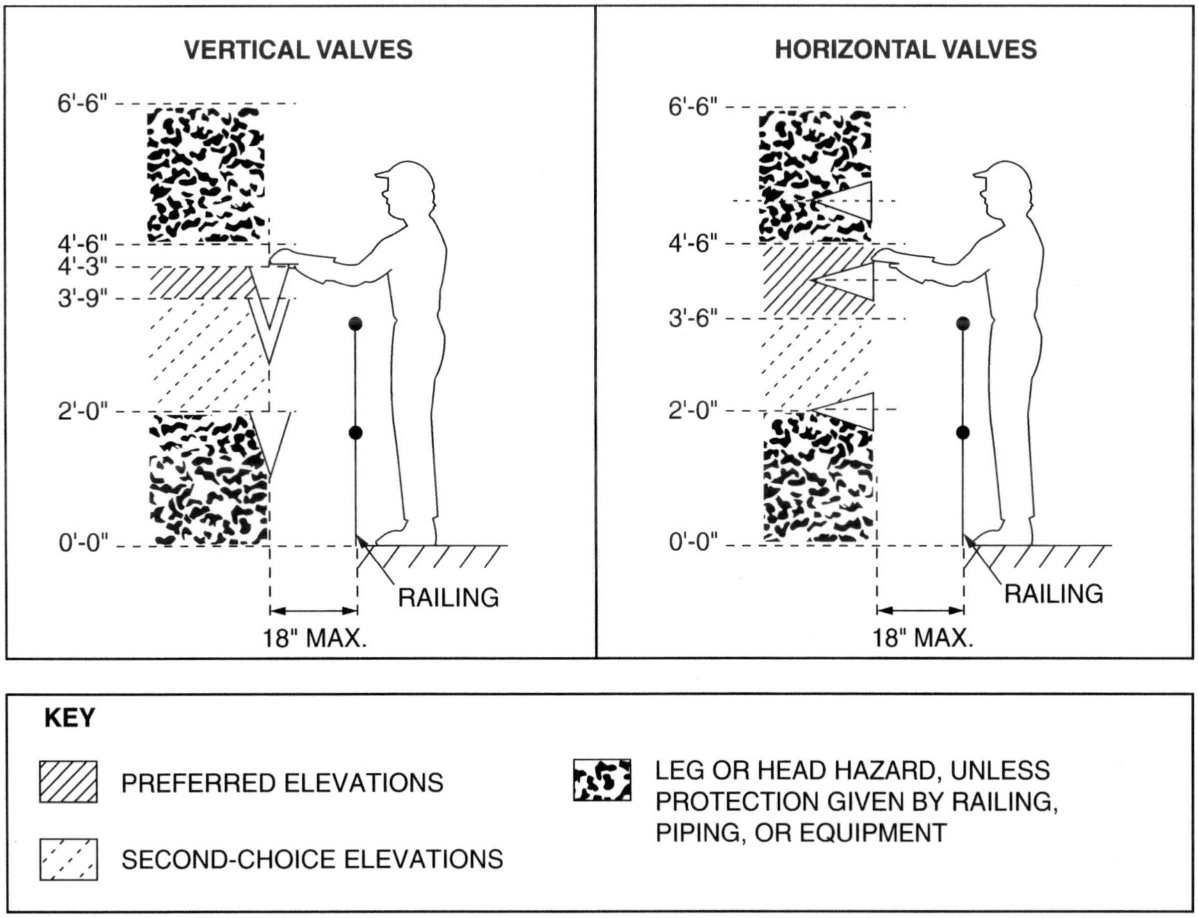

109F51.EPS

Figure 51 ◆ Order of preference for valve location.

- At what temperature will the system be operating? Are there any internal parts that would be adversely affected by the temperature? Valves designed for high temperature steam systems are not necessarily suited for the extreme low temperatures that may be found in a liquid nitrogen system.
- At what pressure (or vacuum) will this valve be operating? How does the temperature affect the valve's pressure rating? System integrity is a major concern on any system. The valve must be rated at or above the maximum system pressure anticipated. Due to factors such as valve design, packing construction, and end attachments, the valve is often considered a weak point in the system.
- Are there any sizing constraints? It seems obvious that a 2-inch valve would not be installed in a 10-inch pipe, but what may not be obvious is how the yoke size, actuator, or **positioner** figures in the scenario. Valve manufacturers provide dimensional tables to aid in valve selections.
- Will this valve be used for on-off or throttle application? Throttle valves are generally globe valves, although in some applications a ball valve or butterfly valve may be used.

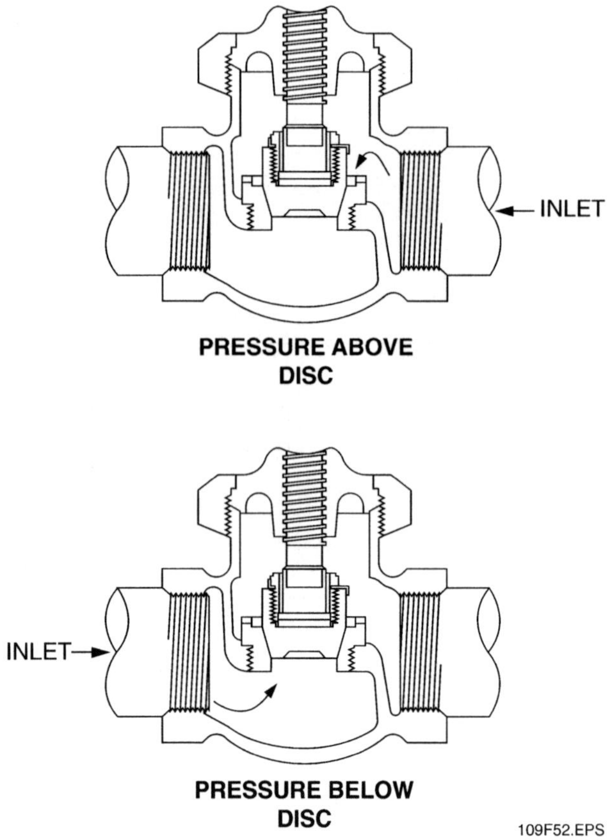

Figure 52 ◆ Pressure above and below disc.

- To what type of erosion will the valve be exposed? Will it require hardened seats and discs? Will it be throttled close to its seat and need a special pressure drop valve?
- What kind of pressure drop is allowed? Globe valves exhibit the largest pressure drop or head loss, whereas ball valves exhibit the least.
- What kind of differential pressure will this valve be operated against? Will this differential pressure be used to seat or unseat the valve? Will the high differential pressure deform the body or disc and bind the valve? Will this also require a bypass valve?
- How will this valve be connected to the system? Will it be welded, screwed or flanged? Should it be butt-welded or socket welded? Will it be union threaded or pipe threaded? Should the flanges be raised, flat, **phonographic**, male/female, or tongue-and-groove?
- In what type of environment will the valve be installed? Is it a dirty environment where an exposed stem would score the **yoke bushing** and cause premature failure? Is it a clean environment where different stem lubricant should be used?
- What kind of fluid is being handled? Is it hazardous in such a way that packing leakage may be detrimental? Is it corrosive to the packing or to the valve itself?
- What is the life expectancy required? Will it require frequent maintenance? If so, is it easily repairable, or does the cost of labor justify replacement instead of repair?

Of course, there are many other questions that may be required. After these are answered, a suitable valve may be selected.

If an installed valve is to be replaced, a valve identical to the one removed should be installed. If that valve is no longer manufactured, valve selection should be made in the same fashion as for a new application, except that the valve dimensions are the limiting factor unless piping alterations can be made. Several questions should be asked:

- Are the system parameters the same as when the system was designed, or has the system intent changed?
- Have any problems been noted since system fabrication that could be remedied by installing a different valve design at this time?
- With what type of operator should the new valve be fitted? Is the new valve compatible with the installed operator?

10.0.0 ◆ VALVE MARKINGS AND NAMEPLATE INFORMATION

Before the present system of valve and flange coding, manufacturers had their own systems. With the development of components rated at higher temperatures and pressures, in conjunction with more stringent regulations, a standard was needed. The Manufacturers Standardization Society (MSS) first developed SP-25 in 1934. In 1978, SP-25 was revised to incorporate all the changes that had developed since 1934. To preclude errors in cross-referencing, the American National Standards Institute (ANSI) and the **American Society for Testing Materials International (ASTM)** have adopted the MSS marking system.

Two markings that are frequently used on valves are the flow direction arrow, indicating which way the flow is going, and the bridgewall marking, shown in *Figure 53*. The bridgewall marking is usually found on globe valves and is an indication of how the seat walls are angled in relation to the inlet and outlet ports of the valve. Specifically, it shows whether the wall of the seat on the inlet side angles up or down. The wall of the seat on the outlet side of the globe valve will always be angled opposite to the angle on the inlet side, as indicated in *Figure 54*.

Not all globe valves are designed with angled bridgewalls. However, some applications may specifically require the process to enter either on the top side or the bottom side of the disc in a globe valve.

If the process enters on the top side (bridgewall angled up on the inlet), the force of the process will assist in the closing of the valve. However, if the process enters on the bottom side (bridgewall angled down on the inlet), the force of the process will assist in the opening of the valve.

Markings for flow are not normally used on gate (except knife gate), plug, or ball valves. If a gate valve has a flow arrow, it is because the gate valve has a double gate, or is a knife gate. Double-gate valves are capable of relieving fluid pressure in the event that a high pressure difference exists across the shut gate. Standard practice is for the outlet-side gate to relieve to the inlet side. This type of valve is used for specific applications. Therefore, system plans should be consulted for correct valve orientation.

There are normally two identification sets: one permanently embossed, welded, or cast into the valve body, and the other a valve identification plate (*Figure 55*). Typically, as a minimum, the following information will be included within the two sets:

- Rating designation markings
- Material designation markings
- Material ASTM number
- Trim identification markings (if applicable)
- Size markings
- Thread identification markings (if applicable)

10.1.0 Rating Designation

The rating designation of a valve gives the pressure and temperature rating as well as the type of service. *Table 2* shows commonly used service designations.

The product rating may be designated by the class numbers alone, as with a steam pressure rating or pressure class designation. The ratings for products that conform to recognized standards but are not suitable for the full range of pressures or temperatures of those standards may be marked as appropriate. The numbers and letters representing the pressure rating at the limiting conditions may also be shown.

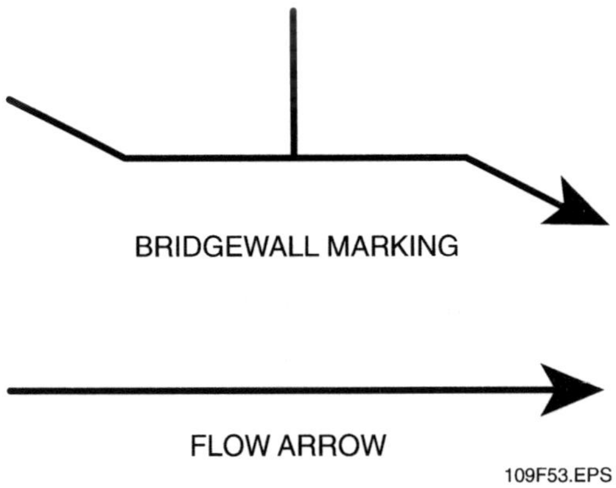

Figure 53 ◆ Valve markings.

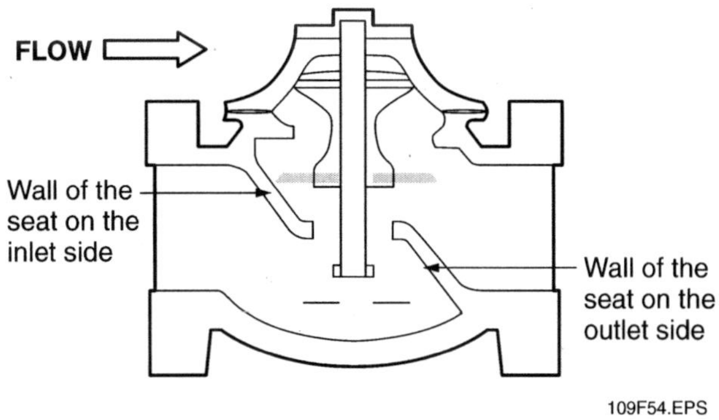

Figure 54 ◆ The meaning of bridgewall markings on valves.

The rating designation for products that do not conform to recognized national product standards may be shown by numbers and letters representing the pressure ratings at maximum and minimum temperatures. If desired, the rating designation may be shown as the maximum pressure followed by cold working pressure (CWP) and the allowed pressure at the maximum temperature (for example, 2,000 CWP 725/925°F).

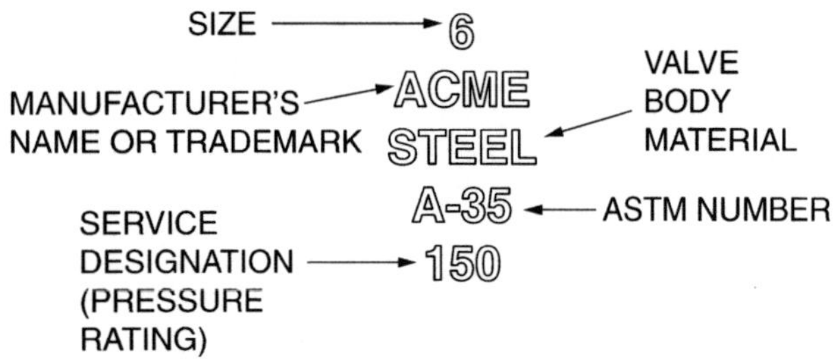

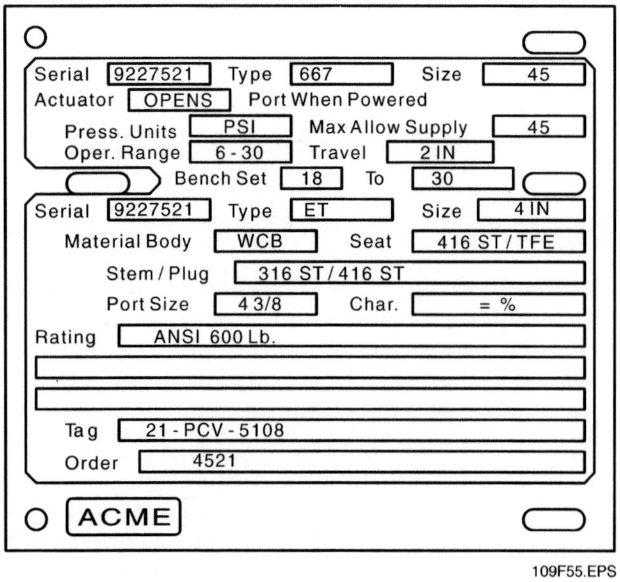

Figure 55 ◆ Actuator nameplate.

Other typical designations are given as the first letter of the system for which they are designated:

- *A* – Air service
- *G* – Gas service
- *L* – Liquid service
- *O* – Oil service
- *W* – Water service
- *DWV* – Drainage waste and vent service

10.2.0 Trim Identification

Trim identification marking is required on the identification plate for all flanged-end and butt welding end steel or flanged-end ductile iron body valves with trim material that is different than the body material. Symbols for materials are the same. The identification plate may be marked with the word trim followed by the appropriate material symbol.

Trim identification marking for gate, globe, angle, and cross valves, or valves with similar design characteristics, consists of three symbols. The first indicates the material of the stem. The second indicates the material of the disc or wedge face. The third indicates the material of the seat face. The symbol may be preceded by the words stem, disc, or seat, or it may be used alone. If used alone, the symbols must appear in the order given.

Plug valves, ball valves, butterfly valves, and other quarter-turn valves require no trim identification marking unless the plug, disc, closure member, or stem is of different material than the body. In such cases, trim identification symbols on the nameplate first indicate the material of the stem and then the material of the plug, ball, disc, or closure member.

Table 2 Valve Rating Designations

Correspond to Steam Working Pressure (SWP)	Correspond to Cold Working Pressure (CWP)
Steam pressure (SP)	Water, oil pressure (WO)
Working steam pressure (WSP)	Oil, water, gas pressure (OWG)
Steam (S)	Water, oil, gas pressure (WOG)
	Gas, liquid pressure (GLP)
	Working water pressure (WWP)
	Water pressure (WP)

109T02.EPS

Those valves with seating or sealing material different from the body material must have a third symbol to indicate the material of the seat. In these cases, symbol identification must be preceded by the words stem, disc (or plug, ball, or gate, as appropriate) and the word seat. If used alone, the symbols must appear in the order given.

10.3.0 Size Designation

Size markings are in accordance with the product-referenced marking requirements. For size designation for products with a single nominal pipe size of the connecting ends, the word nominal indicates the numerical identification associated with the pipe size. It does not necessarily correspond to the valve, pipe, or fitting inside diameter.

Products with internal elements that are the equivalent of one pipe size or are different than the end size may have dual markings unless otherwise specified in a product standard. Unless these exceptions exist, the first number indicates the connecting end pipe size. The second indicates the minimum bore diameter, or the pipe size corresponding to the closure size (for example, 6 × 4, 4 × 2½, 30 × 24).

At the manufacturer's option, triple marking size designation may be used for valves. If triple size designation is used, the first number must indicate the connecting end size at the other end. For example, 24 × 20 × 30 on a valve designates a size 24 connection, a size 20 nominal center section, and a size 30 connection.

Fittings with multiple outlets may be designated at the manufacturer's option in a run × run × outlet size method. For example, 30 × 30 × 24 on a fitting designates a product with size 30 end connections and a nominal size 24 connection between.

10.4.0 Thread Markings

Fittings, flanges, and valve bodies with threaded connecting ends other than American National Standard Pipe Thread or American National Standard Hose Thread will be marked to indicate the type. The style or marking may be the manufacturer's own symbol provided confusion with standard symbols is avoided. Fittings with left-hand threads must be marked with the letters LH on the outside wall of the appropriate opening.

Marking of products with ends threaded for API casing, tubing, or drill pipe must include the nominal size, the letters API, and the thread type symbol as listed in *Table 3*.

Marking of products using other pipe threads must include the following:

- Nominal pipe, tubing, drill pipe, or casing size
- Outside diameter of pipe, tubing, drill pipe, or casing
- Name of thread
- Number of threads per inch

10.5.0 Valve Schematic Symbols

The last and most important aspect of valve identification is the ability to identify different types of valves from blueprints and schematics. In general, the symbols that denote various control valves, actuators, and positioners are standard symbols as shown in *Figure 56*. However, in certain cases these symbols vary, depending on site-specific prints. The legend of a typical system print or schematic will show the symbols that represent all components on the drawing.

Table 3 Examples of Threaded-Type Symbols

Namr/Description	Symbol
Casing – Short round thread	CSG
Casing – Long round thread	LCSG
Casing – Buttress thread	BCSG
Casing – Extreme-line	XCSG
Line pipe	LP
Tubing – Non-upset	TBG
Tubing – External-upset CSG	UP TBG
Britiah Standard Thread	BST
Britisht Standard Pipe Thread	BSPT

109T03.EPS

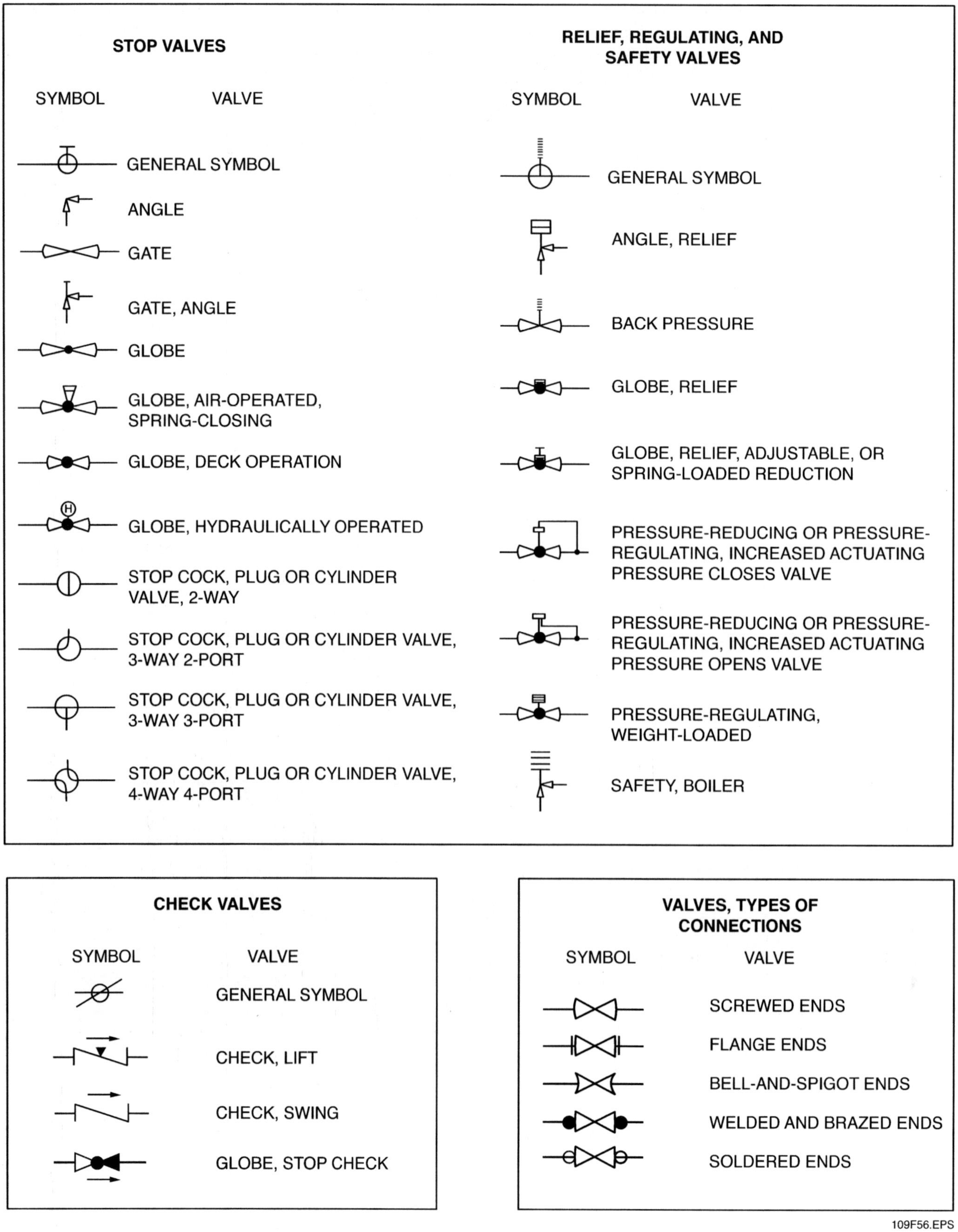

Figure 56 ◆ Typical piping system schematic symbols.

Review Questions

1. A gate valve offers very little resistance to flow when open, so it produces very little _____.
   a. cost
   b. pressure increase
   c. pressure drop
   d. laminar flow

2. Gate valves should *not* be used to _____.
   a. start flow
   b. stop flow
   c. throttle flow
   d. control an on/off process

3. A knife gate valve is likely to be used in which of the following applications?
   a. High pressure
   b. Fine liquids only
   c. Slurry or waste
   d. Throttling steam

4. Which of the following valves is a rotary-action valve?
   a. Gate
   b. Needle
   c. Globe
   d. Ball

5. Which of the following valves is easily adapted to multiport construction?
   a. Gate
   b. Plug
   c. Globe
   d. Knife

6. One disadvantage of the globe valve is _____.
   a. high head loss
   b. inability to regulate flow
   c. high seat leakage
   d. inability to serve in quick-acting start-stop applications

7. When looking at an illustration of valve plug and seat installation, if the taper of the plug goes from wide to narrow from top to bottom, the plug must travel _____.
   a. up to close, down to open
   b. down to close, up to open
   c. down to close, down to open
   d. up to close, up to open

8. A needle valve may be used when an application requires _____.
   a. fine adjustment of flow
   b. coarse flow
   c. on-off control
   d. slurry flow

9. Relief valves are normally used in liquid services.
   a. True
   b. False

10. A swing check valve requires a counterweight if used in a _____ application.
    a. horizontal
    b. vertical
    c. steam
    d. liquid

11. Butterfly valves use _____ operators to transmit power at right angles.
    a. spur-gear
    b. chain
    c. worm-gear
    d. bevel

12. The fail-safe position for any automatically actuated valve is the closed position.
    a. True
    b. False

13. When choosing among valves, the _____ valve exhibits the largest pressure drop.
    a. globe
    b. gate
    c. ball
    d. butterfly

14. If a gate valve has a flow arrow indicated on it, it is because the gate valve has a _____.
    a. single gate
    b. double gate
    c. bridgewall marking
    d. vent port

15. Fittings that are marked LH on the outside wall indicate _____.
    a. low heat
    b. liquid hydrogen only
    c. left-hand threads
    d. low hardening material

Summary

Piping systems use valves to start and stop flow, regulate flow, relieve pressure, and prevent reversal of flow. The type of valve used depends on the type of piping system, the nature of the fluid in the system, the temperature and pressure of the fluid in the system, and the desired operation of the system. Valves are manufactured from the same materials as the piping systems. Proper storage, handling, and installation of valves ensure safe and efficient operation of the system.

Valves are an integral part of the operation and control of fluid flow systems. Each device has characteristics that make it specially suited for certain applications. The applications for which they are used vary widely and include fine control of flow, temperature regulation, pressure regulation, and flow isolation.

Proper selection of the control valve for a specific system is determined by several factors such as system design pressure and temperature, piping size, and system flow conditions.

Actuators and positioners serve as energy transmission devices that cause valve stem movement. They are designed to use several different energy sources to perform their functions. These devices provide a means through which valves may be operated remotely or against extremely high differential pressures.

The information presented in this module provides detailed descriptions of the construction and operation of various control valves, actuators, and positioners. This information will prove invaluable in the proper selection and correct installation of any of these valves, actuators, and positioners in a fluid system.

Notes

Trade Terms Introduced in This Module

Actuator: The part of a regulating valve that converts electrical or fluid energy to mechanical energy to position the valve.

Angle valve: A type of globe valve in which the piping connections are at right angles.

American Society for Testing Materials International (ASTM): Founded in 1898, a scientific and technical organization, formed for the development of standards on the characteristics and performance of materials, products, systems, and services.

Ball valve: A type of plug valve with a spherical disc.

Body: The main part of the valve. It contains the disc, seat, and valve ports. The body of the valve is directly connected to the piping by threaded, welded, or flanged ends.

Bonnet: The part of a valve containing the valve stem and packing.

Butterfly valve: A quarter-turn valve with a plate-like disc that stops flow when the outside area of the disc seals against the inside of the valve body.

Check valve: A valve that allows flow in one direction only.

Control valve: A globe valve automatically controlled to regulate flow through the valve.

Deformation: A change in the shape of a material or component due to an applied force or temperature.

Disc: Part of a valve used to control the flow of system fluid.

Elastomeric: Elastic or rubber-like; flexible, pliable.

Galling: An uneven wear pattern between trim and seat that causes friction between the moving parts.

Gate valve: A valve with a straight-through flow design that exhibits very little resistance to flow. It is normally used for open/shut applications.

Globe valve: A valve in which flow is always parallel to the stem as it goes past the seat.

Kinetic energy: Energy of motion.

Lift: The actual travel of the disc away from the closed position when a valve is relieving.

Linear flow: Flow in which the output is directly proportional to the input.

Packing: Material used to make a dynamic seal, preventing system fluid leakage around a valve stem.

Phonographic: When referring to the facing of a pipe flange, serrated grooves cut into the facing, resembling those on a phonograph record.

Plug: The moving part of a valve trim (plug and seat) that either opens or restricts the flow through a valve in accordance with its position relative to the valve seat, which is the stationary part of a valve trim.

Plug valve: A quarter-turn valve with a ported disc.

Positioner: A field-based device that takes a signal from a control system and ensures that the control device is at the setting required by the control system.

Relief valve: A valve that automatically opens when a preset amount of pressure is exerted on the valve disc.

Seat: The part of a valve against which the disc presses to stop flow through the valve.

Thermal transients: Short-lived temperature spikes.

Throttling: The regulation of flow through a valve.

Torque: A twisting force used to apply a clamping force to a mechanical joint.

Valve body: The part of a valve containing the passages for fluid flow, valve seat, and inlet and outlet connections.

Valve stem: The part of a valve that raises, lowers, or turns the valve disc.

Valve trim: The combination of the valve plug and the valve seat.

Wedge: The disc in a gate valve.

Wire drawing: The erosion of a valve seat under high velocity flow through which thin, wire-like gullies are eroded away.

Yoke bushing: The bearing between the valve stem and the valve yoke.

Resources & Acknowledgments

Additional Resources

This module is intended to be a thorough resource for task training. The following reference works are suggested for further study. These are optional materials for continued education rather than for task training.

Choosing the Right Valve, Crane Company; 300 Park Avenue, New York, NY.

Piping Pointers; Application and Maintenance of Valves and Piping Equipment, Crane Company; 300 Park Avenue, New York, NY.

Figure Credits

Flowserve Corporation, 109F01 (photo), 109F46 (photo)

Provided by Flowserve Corporation, a global leader in the fluid motion and control business. More information on Flowserve and its products can be found at www.flowserve.com.

Velan Valve Corp., 109F12, 109F16, 109F44

Val-Matic Valve and Manufacturing Corp., 109F20 (photo)

Dwyer Instruments, Inc., 109F24 (photo), 109F29 (photo), 109F35 (safety pop-off valve), 109F48 (photo)

ITT Engineered Valves, 109F33 (photo)

Parker Hannifin Corp., 109F34

©2002 Swagelok Company, 109F35 (pressure-relief valve)

Crispin Valve Co., 109F38 (photo), 109F42

Crane Valves North America, 109F41

Babbitt Steam Specialty Company, 109F47 (photo)

CONTREN® LEARNING SERIES — USER UPDATE

NCCER makes every effort to keep these textbooks up-to-date and free of technical errors. We appreciate your help in this process. If you have an idea for improving this textbook, or if you find an error, a typographical mistake, or an inaccuracy in NCCER's Contren® textbooks, please write us, using this form or a photocopy. Be sure to include the exact module number, page number, a detailed description, and the correction, if applicable. Your input will be brought to the attention of the Technical Review Committee. Thank you for your assistance.

Instructors – If you found that additional materials were necessary in order to teach this module effectively, please let us know so that we may include them in the Equipment/Materials list in the Annotated Instructor's Guide.

Write: Product Development and Revision
National Center for Construction Education and Research
3600 NW 43rd St. Bldg G, Gainesville, FL 32606

Fax: 352-334-0932

E-mail: curriculum@nccer.org

Craft ____________ Module Name ____________

Copyright Date ____________ Module Number ____________ Page Number(s) ____________

Description

____

____

____

(Optional) Correction

____

____

(Optional) Your Name and Address

____

____

40110-07

Introduction to Test Instruments

40110-07
Introduction to Test Instruments

Topics to be presented in this module include:

Overview

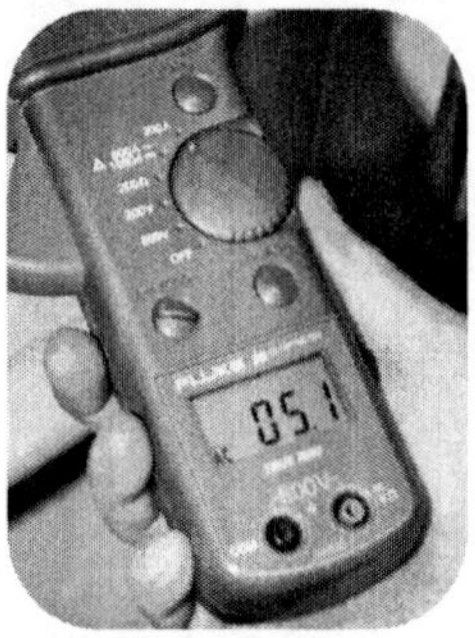

In order to repair and maintain equipment, you need to know what kind of test instruments are available, and how they work, ranging from multimeters to strobe lights to tachometers. You will learn how to test various characteristics of electrical systems with meters, how to use a stroboscope to analyze the motion of a spinning motor shaft, and how to diagnose a bearing assembly with a stethoscope and a non-contact infrared device.

Objectives

When you have completed this module, you will be able to do the following:

1. Explain the operation of and describe the following pieces of test equipment:
   - Tachometer
   - Pyrometers
   - Multimeters
   - Automated diagnostics tools
   - Wiggy® voltage tester
   - Stroboscope
2. Explain how to read and convert from one scale to another using the above test equipment.
3. Define frequency and explain the use of a frequency meter.

Trade Terms

Amperes	Ohms
Circuit	Parallel
Continuity	Polarity
Frequency	Series
Hertz	

Required Trainee Materials

1. Pencil and paper
2. Appropriate personal protective equipment

Prerequisites

Before you begin this module, it is recommended that you successfully complete *Core Curriculum; Industrial Maintenance E & I Technician Level One,* Modules 40101-07 through 40109-07.

This course map shows all of the modules in the first level of the *Industrial Maintenance E & I Technician* curriculum. The suggested training order begins at the bottom and proceeds up. Skill levels increase as you advance on the course map. The local Training Program Sponsor may adjust the training order.

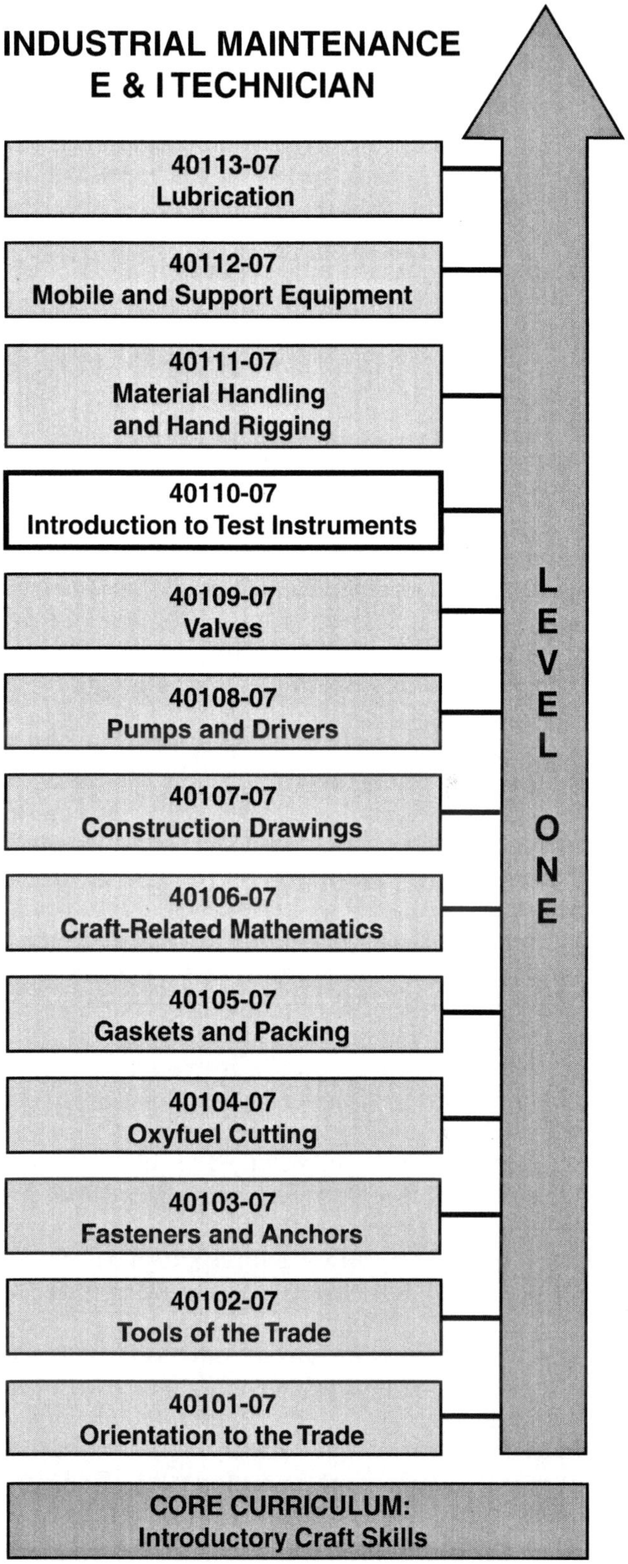

110CMAP.EPS

1.0.0 ◆ INTRODUCTION

Industrial maintenance craftworkers must be able to determine how well the machinery they maintain is working. A major part of good mechanical maintenance is knowing about problems before they happen. This allows the craftsperson to prevent expensive downtime. Electrical systems must also be tested for proper operation. If a problem develops, it must be located and repaired. Industrial maintenance workers use tools of different types to test for and diagnose different problems, both mechanical and electrical.

This module will focus on some of the test equipment that you will be required to use in your job as a craftsperson. The intent is to familiarize you with the use and operation of such equipment and to provide you with practical experience involving that equipment. Upon completion of this module, you should be able to select the appropriate test equipment and effectively use that equipment to perform an assigned task.

1.1.0 Testing and Electricity Basics

Many industrial maintenance testing requirements consider electrical characteristics, such as the following: voltage, which is thought of as electrical pressure; current, which is the amount of electricity flowing; and resistance, which is resistance to flow. Voltage, also called electromotive force (emf) is measured in volts, and tells how much resistance can be overcome by a particular **circuit**. Current is measured in **amperes**, and tells how much work can be done by the source in the circuit. Resistance is measured in **ohms**, and it tells how much voltage will be required to travel through the circuit.

The two types of power source you will encounter are alternating current (AC) and direct current (DC). A DC power source supplies current traveling in only one direction. An example of DC power is battery-operated equipment, such as flashlights or automobile sound systems. AC equipment has a source that switches directions continually, usually at the standard **frequency** of 60 cycles per second or 60 **hertz**.

Electrical equipment carries electrical current within a circuit. If the circuit is broken (open), it will not carry electricity. A switch stops or redirects the flow of electricity. A resistor, whether fixed or variable, changes the resistance to the flow of electricity within the circuit. A transformer changes the voltage of the electric current.

Wiring is color–coded to tell you what part of the circuit it represents. This is especially critical with DC circuits, since the current is always traveling in the same direction. The convention is that a black wire is used for the negative wires, while red or white are used for the positive wires. A minus sign (–) on a terminal is an indication that the wire is negative, while a plus sign (+) tells you that the terminal is positive. This **polarity** is especially critical in working with DC voltage testing, as a reversed polarity test could bend the pointer on the meter.

The two ways of hooking up to a circuit are **series** and **parallel**. In a series circuit, the current has to pass through each segment, so it is not possible for any portion to be removed and the circuit to continue to function. In a parallel circuit, the connections all provide an alternative path, so that the other parts will work if one fails. An example of connecting a meter in series would be disconnecting a wire from a terminal and connecting the meter leads to the loose wire and to the terminal. In a parallel connection, the wires would remain connected to the terminal, and the meter leads would be connected to the attachment points of a component while the component was connected to the circuit. Voltage measurements should always be made with the leads in parallel, so that only part of the voltage will go through the meter.

You may be permitted to work with electrical equipment as a craftsperson, so you must know the basics of electrical circuits to protect yourself and your equipment. The most basic precaution is to be sure that any circuit with which you may come in contact is de-energized. Do not work with a hot circuit unless you are a qualified electrician. Be sure to follow lockout/tagout procedures, both to be certain that the power is turned off, and to be certain that someone else does not turn it back on while you are exposed to risk.

If you wish to measure any quantity in a system, you need an appropriate measuring tool. To measure voltage, you will use a voltmeter or voltage tester. To measure current, you will use an ammeter, and to measure resistance, you will use an ohmmeter. Spinning motor shafts are inspected with a tachometer or a strobe light. Heat is measured with a thermometer or a pyrometer, while vibration is measured with an accelerometer. Each will provide you, as a craftsperson, with information on how a machine or system is working.

2.0.0 ◆ VOLT-OHM-MILLIAMMETER

The functioning of conventional electrical measuring instruments is based upon electromechanical principles. The instruments' mechanical components usually work on direct current (DC). Mechanical frequency meters are an exception. A meter that measures alternating current (AC) has a built-in rectifier to change the AC to DC and resistors to correct for the various ranges.

Today, many meters are solid-state digital systems; they are superior because they have no moving parts. These meters will work in any position, unlike mechanical meters, which must remain in one position to be read accurately.

The volt-ohm-milliammeter (VOM), also known as the multimeter, is a multipurpose instrument, capable of measuring ohms, amperes, and voltage. One common analog multimeter is shown in *Figure 1*. There are many different models of this basic multimeter. To prevent having to discuss each and every meter, one version will be explained here. Any controls or functions on your meter that are not covered here should be reviewed in the applicable owner's manual.

The typical volt-ohm-milliammeter is a rugged, accurate, compact, easy-to-use instrument. The instrument can be used to make accurate measurements of DC and AC voltages, current, and resistance.

This meter has the following features: a 0–1 volt DC range, 0–500 volt DC and AC ranges, a transit position on the range switch, rubber plug bumpers on the bottom of the case to reduce sliding, and an externally accessible battery and fuse compartment.

2.1.0 Specifications

The specifications of the example multimeter are:

- DC voltage:

  Sensitivity: 20KW per volt
  Accuracy: 1¾ percent of full scale

- AC voltage:

  Sensitivity: 5KW per volt
  Accuracy: 3 percent of full scale

- DC current:

  250mV to 400mV drop
  Accuracy: 1¾ percent of full scale

- Resistance:

  Accuracy: 1.75 degree of arc
  Nominal open circuit voltage 1.5V
  (9V on the 10KW ohm range)

- Nominal short circuit current:

  1W range: 1.25mA
  100W range: 1.25mA
  10KW range: 75mA

- Meter frequency response:

  Up to 100kHz

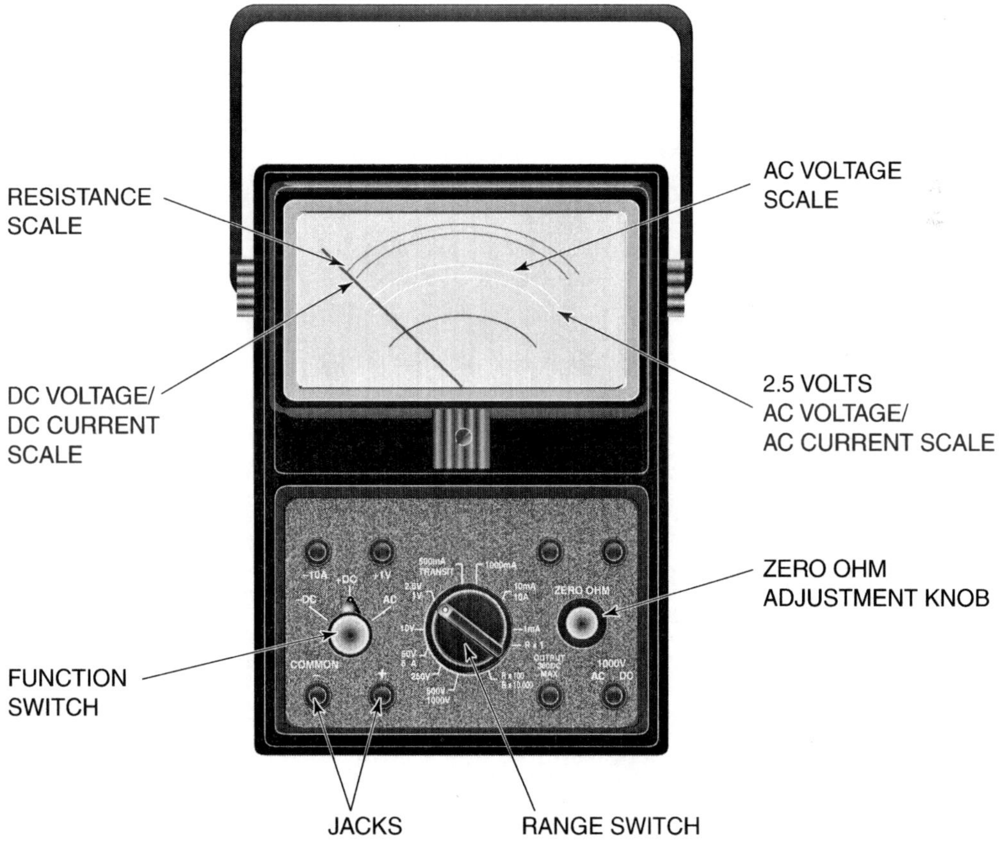

110F01.EPS

Figure 1 ◆ Multimeter.

2.2.0 Overload Protection

In the example multimeter, a 1A, 250V fuse is provided to protect the circuits on the ohmmeter ranges. It also protects the milliampere ranges from excessive overloads. If the instrument fails to indicate, the fuse may be burned out. The fuse is mounted in a holder in the battery and fuse compartment. A spare fuse is located in a well between the 1 terminal of the D cell and the side of the case. Access to the compartment is obtained by loosening the single screw on the compartment cover. To replace a burned-out fuse, remove it from the holder and replace it with a fuse of the exact same type. When removing the fuse from its holder, first remove the battery.

In addition to the fuse, a varistor (variable resistor) protects the indicating instrument circuit. The varistor limits the current through the moving coil in case of overload.

The fuse and varistor will prevent serious damage to the meter in most cases of accidental overload. However, no overload protection system is completely foolproof, and misapplication on high-voltage circuits can damage the instrument. Care and caution should always be exercised to protect both you and the VOM.

Beside the actual steps used in making a measurement, some other points to consider while using a multimeter are as follows:

- Keep the instrument in a horizontal position when storing and away from the edge of a workbench, shelf, or other area where it may be knocked off and damaged.
- Avoid rapid or extreme temperature changes. For example, do not leave the meter in your vehicle during hot or cold weather. Rapid and extreme temperature changes will advance the aging of the meter components and adversely affect meter life and accuracy.
- Avoid overloading the measuring circuits of the instrument. Develop a habit of checking the range position before connecting the test leads to a circuit. Even slight overloads can damage the meter. Even though it may not be noticeable in blown fuses or a bent needle, damage has been done. Slight overloads will advance the aging of components, again causing changes in meter life and accuracy.
- Place the range switch in the transit position when the instrument is not in use or when it is being moved. This reduces the swinging of the pointer when the meter is carried. Every meter does not have a transit position, but if the meter does, it should be used. Random, uncontrolled swings of the meter movement may damage the movement, bend the needle, or reduce its accuracy.
- If the meter has not been used for a long period of time, rotate the function and range switches in both directions to wipe the switch contacts. Most switch contacts are plated with copper or silver. Over a period of time, these materials will oxidize (tarnish). This will create a high resistance through the switch, causing a large inaccuracy. Rotating through the switch positions will clean the tarnish off and provide good electrical contact.

When using a VOM, make sure your leads match the function. The meter can be in the voltage function and even display volts, but if the leads are in the amp jack, the meter will short circuit if connected to a voltage source.

2.3.0 Making Measurements

Take care when using the meter as a millivoltmeter to prevent damage to the indicating instrument from excessive voltage. Before using the 250mV range, first use the 1.0V range to determine that the voltage measured is no greater than 250mV (or 0.25VDC).

2.3.1 Measuring DC Voltage, 0–250 Millivolts

Step 1 Set the function switch to +DC.

Step 2 Plug the black test lead into the – (COMMON) jack and the red test lead into the +50μA/250mV jack.

Step 3 Set the range switch to 50μA (dual position with 50V).

Step 4 Connect the black test lead to the negative side of the circuit being measured and the red test lead to the positive side of the circuit.

Step 5 Read the voltage on the black scale marked DC and use the figures marked 0–250. Read directly in millivolts.

2.3.2 Measuring DC Voltage, 0–1 Volt

Step 1 Set the function switch to –DC. Plug the black test lead into the + (COMMON) jack and the red test lead into the +1V jack.

Step 2 Set the range switch to 1V (dual position with 2.5V).

Step 3 Connect the black test lead to the negative side of the circuit being measured and the red test lead to the positive side of the circuit.

Step 4 Read the voltage on the black scale marked DC. Use the figures marked 0–10 and divide the reading by 10.

2.3.3 Measuring DC Voltage, 0–2.5 Through 0–500 Volts

WARNING!
Be extremely careful when working with higher voltages. Do not touch the instrument test leads while power is on in the circuit being measured.

Step 1 Set the function switch to +DC.

Step 2 Set the range switch to one of the five voltage range positions marked 2.5V, 10V, 50V, 250V, or 500V. If the unit does not have autoranging capability, when in doubt as to the voltage present, always use the highest voltage range as a protection for the instrument. If the voltage is within a lower range, the switch may be set for the lower range to obtain a more accurate reading.

Step 3 Plug the black test lead into the – (COMMON) jack and the red test lead into the + jack.

Step 4 Connect the black test lead to the negative side of the circuit being measured and the red test lead to the positive side of the circuit.

Step 5 Read the voltage on the black scale marked DC. For the 2.5V range, use the 0–250 figures and divide by 100. For the 10V, 50V, and 250V ranges, read the figures directly. For the 500V range, use the 0–50 figures and multiply by 10.

2.3.4 Measuring DC Voltage, 0–1,000 Volts

WARNING!
Be extremely careful when working with higher voltages. Do not touch the instrument test leads while power is on in the circuit being measured.

Step 1 Set the function switch to +DC.

Step 2 Set the range switch to 1,000V (dual position with 500V).

Step 3 Plug the black test lead into the – (COMMON) jack and the red test lead into the 1,000V jack.

Step 4 Be sure power is off in the circuit being measured and all capacitors have been discharged. Connect the black test lead to the negative side of the circuit being measured and the red test lead to the positive side of the circuit.

Step 5 Turn on the power in the circuit being measured.

Step 6 Read the voltage using the 0–10 figures on the black scale marked DC. Multiply the reading by 100.

2.3.5 Measuring AC Voltage, 0–2.5 Through 0–500 Volts

WARNING!
Be extremely careful when working with higher voltages. Do not touch the instrument test leads while power is on in the circuit being measured.

CAUTION
When measuring line voltage such as from a 120V, 240V, or 480V source, be sure that the range switch is set to the proper voltage position.

Step 1 Set the function switch to AC.

Step 2 Set the range switch to one of the five voltage range positions marked 2.5V, 10V, 50V, 250V, or 500V. When in doubt as to the actual voltage present, always use the highest voltage range as a protection to the instrument. If the voltage is within a lower range, the switch may be set for the lower range to obtain a more accurate reading.

Step 3 Plug the black test lead into the – (COMMON) jack and the red test lead into the + jack.

Step 4 Connect the test leads across the voltage source (in parallel with the circuit).

Step 5 Turn on the power in the circuit being measured.

Step 6 For the 2.5V range, read the value directly on the AC scale marked 2.5V. For the 10V, 50V, and 250V ranges, read the red scale marked AC and use the black figures immediately above the scale. For the 500V range, read the red scale marked AC and use the 0–50 figures. Multiply the reading by 10.

2.3.6 *Measuring AC Voltage, 0–1,000 Volts*

> **WARNING!**
> Be extremely careful when working with higher voltages. Do not touch the instrument test leads while power is on in the circuit being measured.

Step 1 Set the function switch to AC.

Step 2 Set the range switch to 1,000V (dual position with 500V).

Step 3 Plug the black test lead into the – (COMMON) jack and the red test lead into the 1,000V jack.

Step 4 Be sure the power is off in the circuit being measured and that all capacitors have been discharged. Connect the test leads to the circuit.

Step 5 Turn on the power in the circuit being measured.

Step 6 Read the voltage on the red scale marked AC. Use the 0–10 figures and multiply by 100.

2.4.0 Direct Current Measurements

Direct current measurements are performed differently from AC measurements because the current runs in one direction only. This polarity demands that the leads be placed in correct polarity. This also prevents damage to the meter and incorrect readings.

2.4.1 *Voltage Drop*

The voltage drop across the meter on all milliampere current ranges is approximately 250mV measured at the jacks. An exception is the 0–500mA range with a drop of approximately 400mV. This voltage drop will not affect current measurements. In some transistor circuits, however, it may be necessary to compensate for the added voltage drop when making measurements.

2.4.2 *Measuring Direct Current, 0–50 Microamperes*

> **CAUTION**
> Never connect the test leads directly across voltage when the meter is used as a current-indicating instrument. Always connect the instrument in series with the load across the voltage source.

Step 1 Set the function switch to +DC.

Step 2 Plug the black test lead into the – (COMMON) jack and the red test lead into the +50μA/250mV jack.

Step 3 Set the range switch to 50μA (dual position with 50V).

Step 4 Open the circuit in which the current is being measured. Connect the instrument in series with the circuit. Connect the red test lead to the positive side and the black test lead to the negative side.

Step 5 Read the current on the black DC scale. Use the 0–50 figures to read directly in microamperes. In all direct current measurements, be certain the power to the circuit being tested has been turned off before disconnecting test leads and restoring circuit **continuity**.

2.4.3 *Measuring Direct Current, 0–1 Through 0–500 Milliamperes*

Step 1 Set the function switch to +DC.

Step 2 Plug the black test lead into the – (COMMON) jack and the red test lead into the + jack.

Step 3 Set the range switch to one of the four range positions (1mA, 10mA, 100mA, or 500mA).

Step 4 Open the circuit in which the current is being measured. Connect the VOM in series with the circuit. Connect the red test lead to the positive side and the black test lead to the negative side of the part of the circuit you are measuring.

Step 5 Read the current in milliamperes on the black DC scale. For the 1mA range, use the 0–10 figures and divide by 10. For the 10mA range, use the 0–10 figures and multiply by 10. For the 500mA range, use the 0–50 figures and multiply by 10.

2.4.4 *Measuring Direct Current, 0–10 Amperes*

Step 1 Plug the black test lead into the –10A jack and the red test lead into the +10A jack.

Step 2 Set the range switch to 10A (dual position with 10mA).

Step 3 Open the circuit in which the current is being measured. Connect the instrument in series with the circuit. Connect the red test lead to the positive side and the black test lead to the negative side.

NOTE

The function switch has no effect on polarity for the 10A range.

Step 4 Read the current on the black DC scale. Use the 0–10 figures to read directly in amperes.

CAUTION

When using the 10A range, never remove a test lead from its panel jack while current is flowing through the circuit. Otherwise, damage may occur to the plug and jack.

2.4.5 *Zero Ohm Adjustment*

When resistance is measured, the VOM batteries furnish power for the circuit. Since batteries are subject to variation in voltage and internal resistance, the instrument must be adjusted to zero prior to measuring a resistance.

Step 1 Set the range switch to the desired ohms range.

Step 2 Plug the black test lead into the – (COMMON) jack and the red test lead into the + jack.

Step 3 Connect the ends of the test leads to short the VOM resistance circuit.

Step 4 Rotate the ZERO OHM control until the pointer indicates zero ohms. If the pointer cannot be adjusted to zero, one or both of the batteries must be replaced.

Step 5 Disconnect the ends of the test leads and connect them to the component being measured.

2.4.6 *Measuring Resistance*

The ohmmeter needle rests on the infinite resistance mark on the left-hand side of the meter until the leads are connected to something that will carry a current. The circuit is created when some kind of conductor is touching the leads. If the needle is deflected all the way to the right, the resistance is zero. If the leads are touched together, for example, the meter will show full-scale deflection, because there is zero resistance.

CAUTION

Before measuring resistance, be sure power is off to the circuit being tested. Disconnect the component from the circuit before measuring its resistance.

Step 1 Set the range switch to one of the resistance range positions:

- Use R x 1 for resistance readings from 0 to 200 ohms.
- Use R x 100 for resistance readings from 200 to 20,000 ohms.
- Use R x 10,000 for resistance readings above 20,000 ohms.

Step 2 Set the function switch to either the –DC or +DC position. The operation is the same in either position.

Step 3 Adjust the ZERO OHM control for each resistance range.

- Observe the reading on the OHMS scale at the top of the dial. Note that the OHMS scale reads from right to left for increasing values of resistance.
- To determine the actual resistance value, multiply the reading by the factor at the switch position (K on the OHMS scale equals one thousand).

3.0.0 ◆ DIGITAL METERS

Digital meters have revolutionized the test equipment world. Improved accuracy is very easily attainable, more functions can be incorporated into one meter, and both autoranging and automatic polarity indication can be used. Technically, digital multimeters are classified as electronic multimeters; however, digital multimeters do not use a meter movement. Instead, a digital meter's input circuit converts a current into a digital signal, which is then processed by electronic circuits and displayed numerically on the meter face.

A major limitation with many meters that use meter movements is that the scale reading must be estimated if the meter pointer falls between scale divisions. This is called interpolation. Digital multimeters eliminate the need to estimate these readings by displaying the reading as a numerical display.

With digital meters, technicians must revise the way the indications are viewed. For example, if a technician were reading the AC voltage on a normal wall outlet with an analog voltmeter, any indication within the range of 120VAC would be considered acceptable. But, when reading with a digital meter, the technician might think something was wrong if the meter showed a reading of 114.53VAC. Bear in mind that the digital meter is very precise in its reading, sometimes more precise than is called for, or is usable. Also, be aware that the indicated parameter may change with the range used. This is primarily due to the change in accuracy and where the meter is rounding off.

There are many types of digital multimeters. Some are bench-type multimeters, while others are designed to be handheld. Most types of digital multimeters have an input impedance of 10 megohms and above. They are very sensitive to small changes in current and are therefore very accurate.

An example of a digital meter is shown in *Figure 2*. The internal operation of this meter is basically the same as other digital meters. The following sections discuss the operation and use of this particular meter. For specific instructions, always refer to the owner's manual supplied with your meter.

3.1.0 Features

The example meter offers the following features:

- *Autorange/manual range modes* – The meter features autoranging for all measurement ranges. Press the range button to enter manual range mode. A flashing symbol may be used to show that you are in the manual range mode. Press the range button as required to select the desired range. To switch back to autorange, press the ON/CLEAR button once (clear mode) or select another function.
- *Automatic off* – The example meter turns itself off after one hour of non-use. The current draw while the meter is turned off does not affect battery life. If the meter turns itself off while a parameter is being monitored, press the ON/CLEAR button to turn it on again. To protect against electrical damage, the meter also turns itself off if a test lead is inserted into the 10A jack while the meter is in any mode other than A⎓ or A ~.
- *Dangerous voltage indication* – The meter shows the symbol for any range over 20V. In the autoranging mode, the meter also beeps when it changes to any range over 20V.
- *Out of limits (OL)* – The meter displays OL and a rapidly flashing decimal point (position determined by range) when the measured value is greater than the limit of the instrument or selected range.
- *Audible acknowledgment* – The meter acknowledges each press of a button or actuation of the selector switch with a beep.

NOTE

Never assume that a blank meter represents a reading of zero volts. Some digital meters automatically cut off the power to preserve the battery. Similarly, if you press the HOLD button on some meters when you are reading 0V, it will lock on that reading and will continue to read 0V, regardless of the actual voltage present. Always check the meter for proper operation before using it.

3.2.0 Operation

This section will discuss the use of various controls and explain how measurements should be taken.

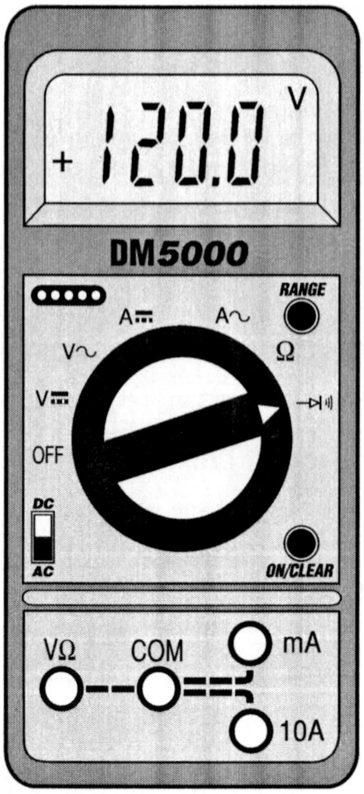

110F02.EPS

Figure 2 ◆ Digital meter.

3.2.1 Dual Function ON/CLEAR Button

Press the ON/CLEAR button to turn the meter on. Operation begins in the autorange mode, and the range for maximum resolution is selected automatically. Press the ON/CLEAR button again to turn the meter off.

3.2.2 Measuring Voltage

Step 1 Select V⎓ or V~.

Step 2 Connect the test leads as shown in *Figure 3*.

Step 3 Observe the voltage reading on the display. Depending on the range, the meter displays units in mV or V.

To avoid shock hazard or meter damage, do not apply more than 1,500VDC or 1,000VAC to the meter input or between any input jack and earth ground.

3.2.3 Measuring Current

Step 1 Select A⎓ or A~.

Step 2 Insert the meter in series with the circuit with the red lead connected to either the mA jack (for input up to 200 milliamps) or the 10A jack (for input up to 20 amps).

Step 3 Make hookups as shown in *Figures* 4 and 5.

Step 4 Observe the current reading.

> **NOTE**
>
> The meter shuts itself off if a test lead is inserted into the 10A jack when the meter is in any function other than A⎓ or A ~.

3.2.4 Measuring Resistance

When measuring resistance, any voltage present will cause an incorrect reading. For this reason, the capacitors in a circuit in which resistance measurements are about to be taken should first be discharged.

Step 1 Select Ω (ohms).

Step 2 Connect the test leads as shown in *Figure 6*.

Step 3 Observe the resistance reading.

3.2.5 Display Test

To test the LCD display, hold the ON/CLEAR button down when turning on the meter. Verify that the display shows all segments (see *Figure 7*).

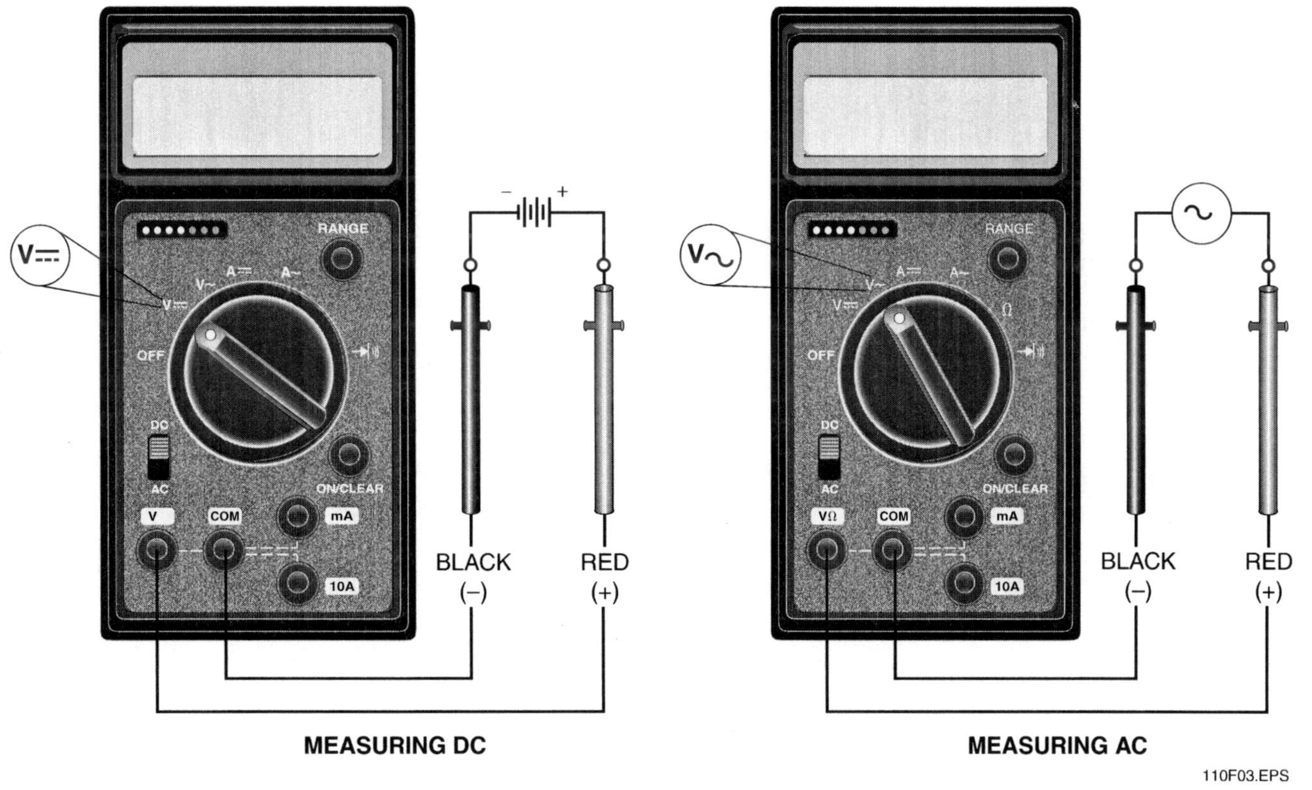

Figure 3 ◆ Measuring voltage.

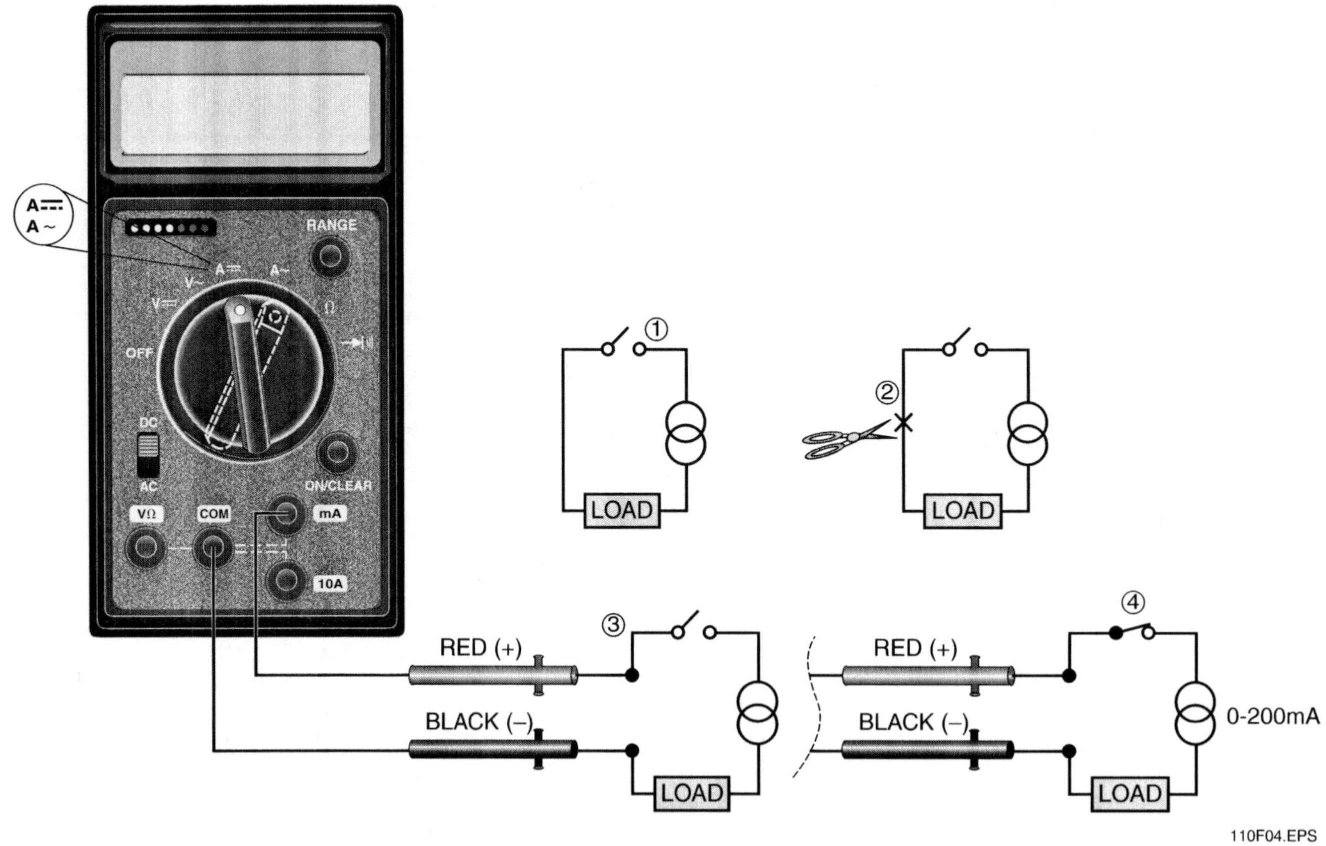

Figure 4 ◆ Measuring current (mA).

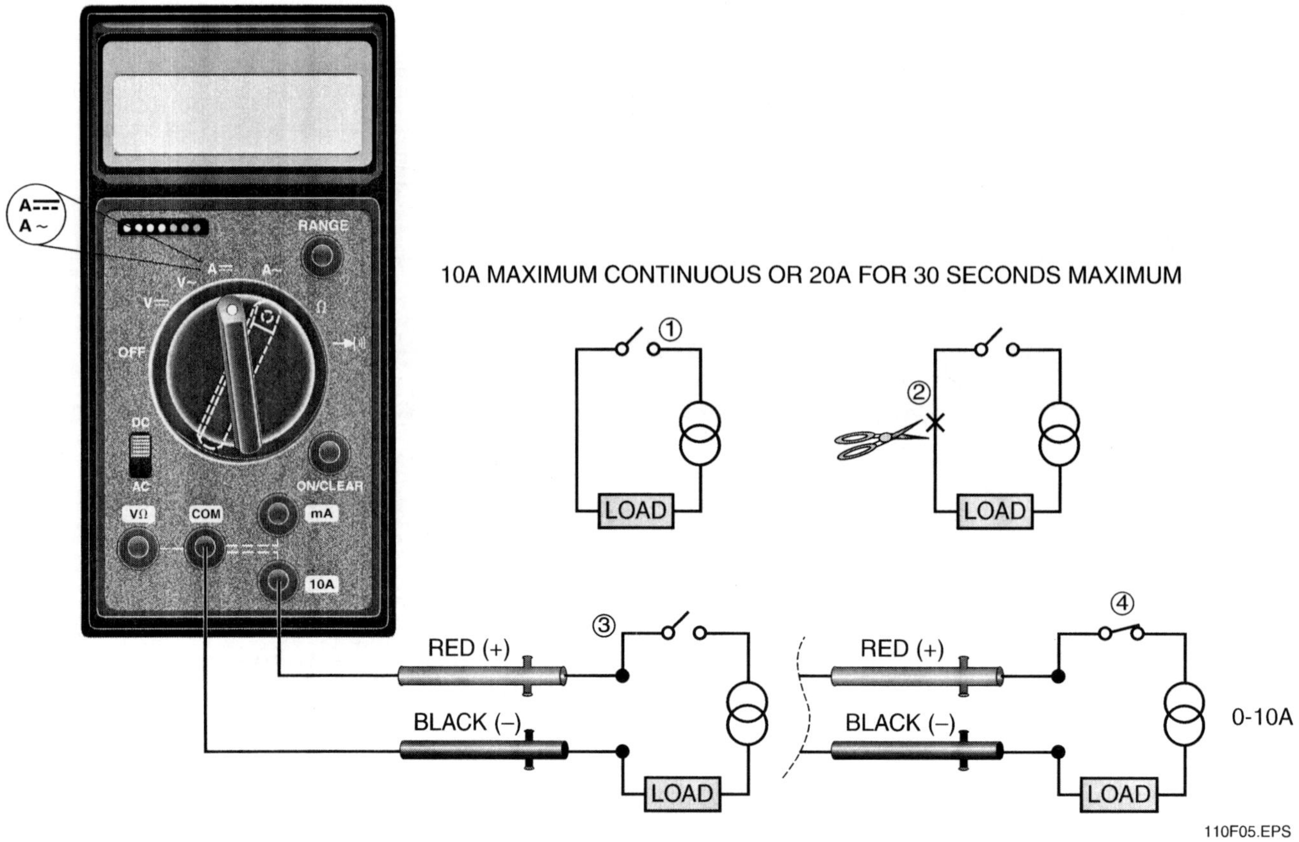

Figure 5 ◆ Measuring current (amps).

3.2.6 Clamp Meters

Modern clamp-type multimeters can measure current and voltage without having to make contact with uninsulated wires (*Figure 8*). These meters are used for currents in excess of 1 amp, and operate by sensing the strength of the electromagnetic field around the wire. The meter must be in contact with only one conductor at a time.

3.3.0 Maintenance

The following sections discuss the necessary maintenance for a multimeter.

3.3.1 Battery Replacement

Replace the battery as soon as the meter's decimal point starts blinking during normal use; this indicates that less than 100 hours of battery life remain. Remove the case back and replace the battery with the same or equivalent 9V alkaline battery.

3.3.2 Fuse Replacement

The meter uses two input protection fuses for the mA and 10A inputs. Remove the case back to gain access to the fuses. Replace with the same type only. The large fuse should be readily available. A spare for the smaller fuse is included in the case. If necessary, this fuse must be reordered from the factory.

3.3.3 Calibration

Have a qualified technician calibrate the meter once a year. This should ensure correct readings.

4.0.0 ◆ FREQUENCY METER

Frequency is the number of cycles completed each second by a given AC voltage, and it is usually expressed in hertz (one hertz = one cycle per second). The frequency meter is used in AC power-producing devices such as generators to ensure that the correct frequency is being produced. Failure to produce the correct frequency will result in excess heat and component damage.

There are two common types of frequency meters. One operates with a set of reeds having natural vibration frequencies that respond in the range being tested. The reed with a natural frequency closest to that of the current being tested

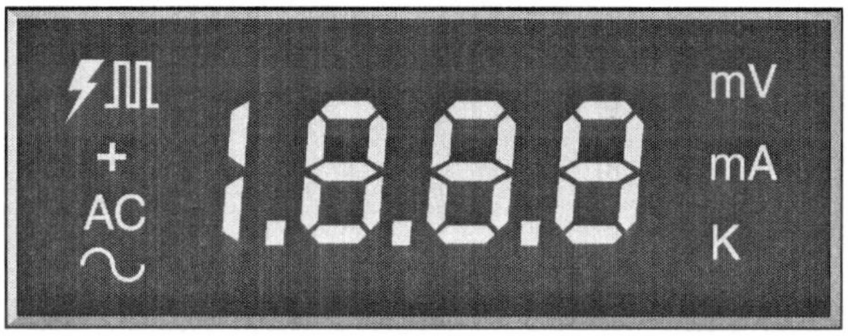

110F07.EPS

Figure 7 ◆ Display test.

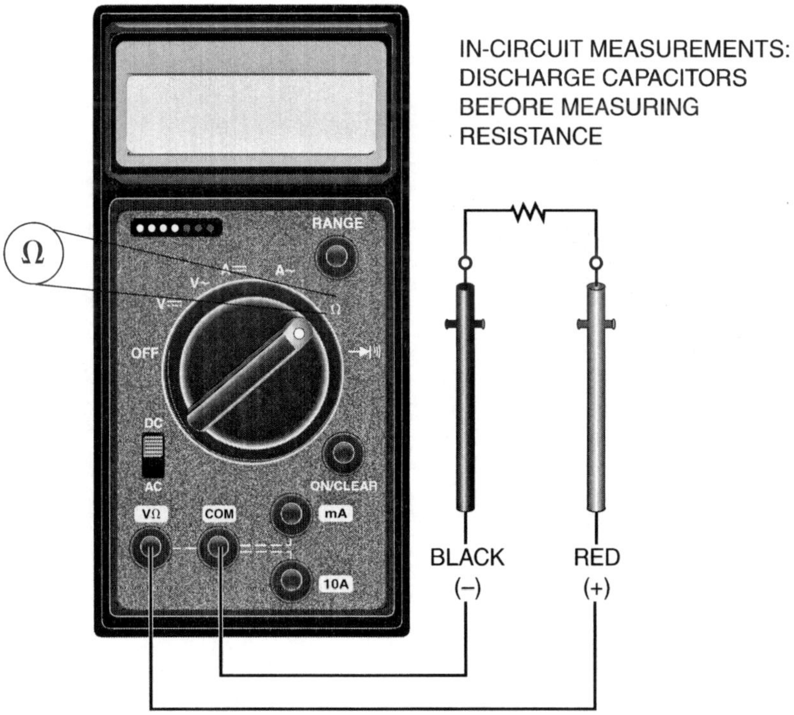

110F06.EPS

Figure 6 ◆ Measuring resistance.

will vibrate most strongly when the meter operates. The frequency is read from a calibrated scale.

A moving-disk frequency meter works with two coils, one of which is a magnetizing coil whose current varies inversely with the frequency. A disk with a pointer mounted between the coils turns in the direction determined by the stronger coil. Solid-state frequency meters are also available.

4.1.0 Tachometers

Tachometers (*Figure 9*) are used to measure motor or fan rpm. As shown here, there are two types: contact and non-contact. Use of the no-contact type is safer and it is more convenient when the motor is located in a hard-to-reach place. Some manufacturers make a combination contact/non-contact model tachometer that can be used to make rpm measurements either by the contact or no-contact method.

Use the following procedure to measure motor rpm with the contact-type tachometer.

Step 1 Turn the motor on.

Step 2 Contact the end of the motor shaft with the tachometer sensor tip.

Step 3 Allow the reading to stabilize, then read the rpm.

To measure motor rpm with the non-contact tachometer:

Step 1 Turn the motor off.

Step 2 Place a reflecting mark with a piece of metal tape on the motor shaft or object to be measured.

Step 3 Turn the motor on.

Step 4 Point the tachometer light beam at the shaft or object, then read the rpm.

5.0.0 ◆ CONTINUITY TESTER

An ohmmeter can be used to test continuity, but this means carrying an expensive, often bulky test instrument with you. Pocket-type continuity testers are just as reliable and much more compact and portable.

There are two types of continuity testers: audio and visual. These are specialized devices for identifying conductors in a conduit by checking continuity.

5.1.0 Audio Continuity Tester

This type of tester is used to ring out wires in conduit runs. At one end of the conduit run, select one wire, strip off a little insulation, and connect that wire to the conduit. At the other end of the conduit run, clip one lead of the tester to the conduit.

110F08.EPS

Figure 8 ◆ Clamp-type multimeter.

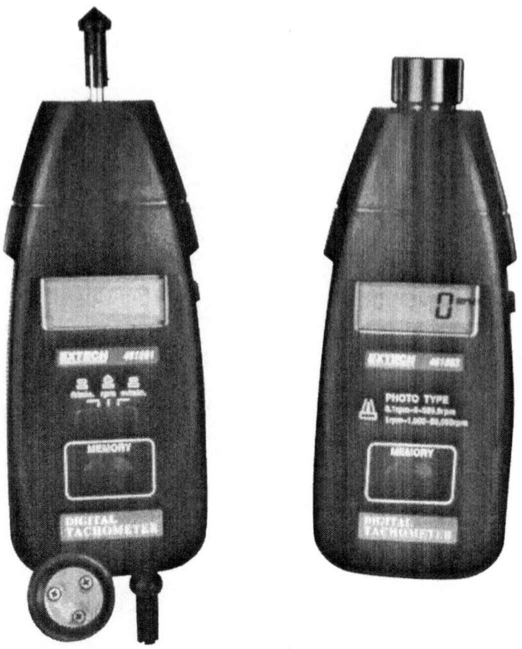

110F09.EPS

Figure 9 ◆ Tachometers.

Touch the other lead to one wire at a time until the audible alarm sounds, which indicates continuity (a closed circuit). Then, put matching tags on the wire. Continue this procedure, one wire at a time, to identify the other wires.

5.2.0 Visual Continuity Tester

This type of tester is useful if you are working in an area where background noise might make it hard to hear the audio tester. The procedure is the same. When the proper wire is tested, the light will come on.

6.0.0 ◆ VOLTAGE TESTER

A voltage tester is a simple aid that determines whether there is a potential difference between two points. It does not calculate the value of that difference. If the actual value of the potential difference is needed, use a voltmeter.

6.1.0 Wiggy®

Pocket-type voltage testers are inexpensive and portable. They can easily be carried in your tool pouch, eliminating the need to carry a delicate voltmeter on the job. Simple neon testers are becoming quite popular. Another type is known as a Wiggy® (see *Figure 10*).

6.1.1 Principle of Operation

The operation of a Wiggy® is fairly simple. The basic component is a solenoid. When a current flows through the coil, it will produce a magnetic field, which will pull the plunger down against a spring. The spring will limit how far the plunger can be drawn into the cylinder. As the current increases, the plunger will move farther. The amount of current depends upon the potential difference applied across the coil. A pointer on the plunger indicates the potential difference on the scale.

CAUTION

This is an approximation only. If you need to know the actual value of the voltage, use a voltmeter.

The scale on the tester has voltage indications for AC on one side of the pointer and DC on the other side. The AC scale indicates 120, 240, 480, and 600 volts. The DC scale indicates 125, 240, and 600 volts.

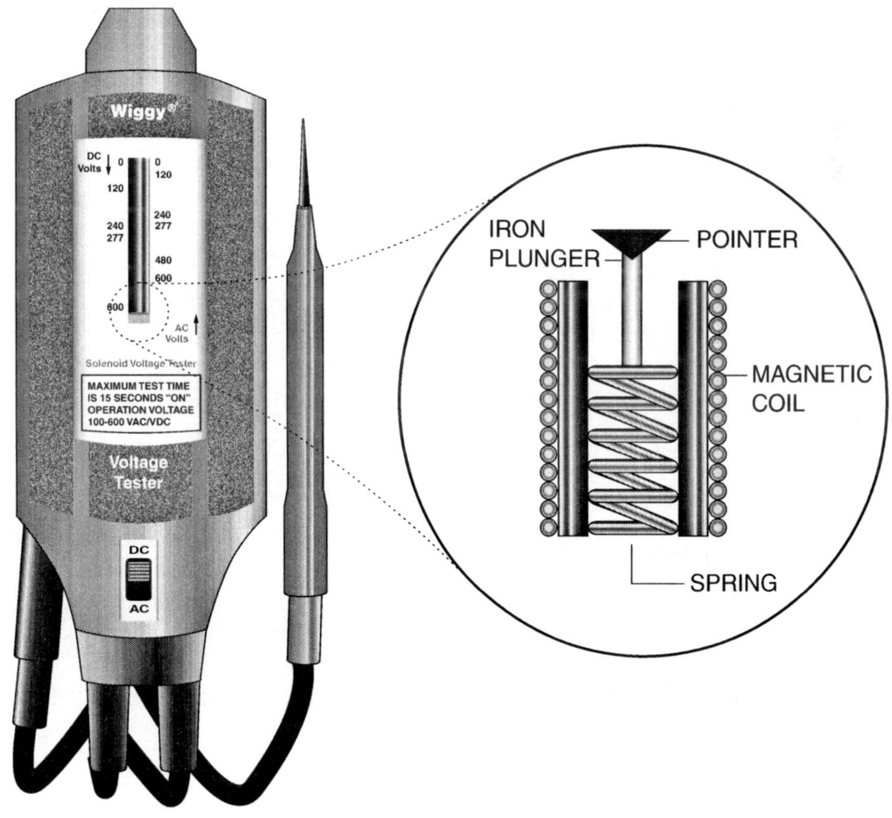

110F10.EPS

Figure 10 ◆ Wiggy® voltage tester.

> **CAUTION**
> To avoid damage to the tester, do not use a Wiggy® above its stated range.

To use a voltage tester, place the probes across a possible source of voltage. If there is voltage present, current flows through the coil inside the tester, creating a magnetic field. The magnetized coil pulls the indicator along the scale until it reaches a point corresponding to the approximate voltage. If there is no voltage, there is no current flow, so the indicator on the scale does not move, and no voltage reading is displayed.

The range of voltage and the type of current (AC and/or DC) that a voltage tester is capable of measuring are usually indicated on the scales that display the reading. The scales for each type of current are marked accordingly.

The methods used to show voltage readings vary from one type of voltage tester to another. For example, there are voltage testers that have lights to indicate the approximate amount of voltage registered. Both lights and scales have relatively broad readout ranges because these voltage testers can only indicate approximate values, not precise voltage measurements.

A voltage tester should always be checked before each use to make sure that it is in good condition and is operating correctly. The external check of the tester should include a careful inspection of the insulation on the leads for cracks or frayed areas. Faulty leads constitute a safety hazard, so they must be replaced. As a check to make sure that the voltage tester is operating correctly, the probes of the tester are connected to a power source that is known to be energized. The voltage indicated on the tester should match the voltage of the power source. If there is no indication, the voltage tester is not operating correctly, and it must be repaired or replaced. It must also be repaired or replaced if it indicates a voltage different from the known voltage of the source.

It is essential to check a voltage tester before use. A faulty voltage tester can be dangerous to the craftsperson using it and to other personnel. For example, damaged insulation or a cracked casing could expose the craftsperson to electrical shock. Also, a faulty voltage tester might indicate that power is off when it is really on. This would create a serious safety hazard for personnel involved in equipment repair. The face plate or the scale on the front of the tester should be checked before a voltage tester is used to be sure that the tester can handle the amount of voltage that the power source may contain. Care should be taken when placing the probes of the tester across the power source. A voltage tester is designed to take a quick reading. If the probes are left in the circuit too long, the tester will burn out. A voltage tester should never be connected for more than a few seconds at a time.

> **CAUTION**
> Do not use a voltage tester in hazardous locations.

Voltage testers are used to make sure that power is available when it is needed and to make sure that power has been cut off when it should have been. In a troubleshooting situation, it might be necessary to verify that power is available in order to be sure that lack of power is not the problem. For example, if there were a problem with a power tool, such as a drill, a voltage tester might be used to make sure that power is available to run the drill. A voltage tester might also be used to verify that there is power available to a three-phase motor that will not start.

For safety reasons, it is always necessary to make sure that the power is turned off before working on any electrical equipment. A voltage tester can be used for such a test.

Keep the following in mind when using a voltage tester:

- Check the tester before each use.
- Handle the tester as if it were a calibrated instrument.
- Use good safety practices when operating the tester.
- Do not use circuits expected to be above the scale on the tester.
- Do not use if damage is indicated to the tester.
- Do not use in classified, hazardous areas, or in high-frequency circuits.

7.0.0 ◆ SAFETY

In the interest of safety, all test equipment must be inspected and tested before use. A thorough visual inspection, checking for broken meters or knobs, damaged plugs, or frayed cords is important.

Perform an operational check. For example:

- On an ohmmeter, short the probes and ensure that you can zero the meter.
- A voltmeter can be checked against an AC wall receptacle or a battery.

If a meter has a calibration sticker, check to see if it has been calibrated recently. For precise measurements, a recently calibrated meter is a more reliable instrument.

Every person who works with electronic equipment should be constantly alert to the hazards to which personnel may be exposed, and should also be capable of rendering first aid. The hazards considered in this section are: electric shock, burns, and related hazards.

Safety must be the primary responsibility of all personnel. The installation, maintenance, and operation of electrical equipment enforces a stern safety code. Carelessness on the part of the craftsperson or operator can result in serious injury or death due to electrical shock, falls, burns, flying objects, etc. After an accident has occurred, investigation almost invariably shows that it could have been prevented by the exercise of simple safety precautions and procedures. Each person concerned with industrial equipment is responsible for reading and becoming thoroughly familiar with the safety practices and procedures contained in all safety codes and equipment technical manuals before performing work on equipment. It is your personal responsibility to identify and eliminate unsafe conditions and unsafe acts that cause accidents.

You must bear in mind that de-energizing main supply circuits by opening supply switches will not necessarily de-energize all circuits in a given piece of equipment. A source of danger that has often been neglected or ignored, sometimes with tragic results, is the input to electrical equipment from other sources, such as backfeeds. Moreover, the rescue of a victim shocked by the power input from a backfeed is often hampered because of the time required to determine the source of power and isolate it. Therefore, turn off all power inputs before working on equipment and tag and lock out, then check with an operating tester to be sure that the equipment is safe to work on. Take the time to be safe when working on electrical circuits and equipment. Carefully study the schematics and wiring diagrams of the entire system, noting what circuits must be de-energized in addition to the main power supply. Remember, electrical equipment commonly has more than one source of power. Be certain that all power sources are de-energized before servicing the equipment. Do not service any equipment with the power on unless absolutely necessary. Remember that the 115V power supply voltage is not a low, relatively harmless voltage but is the voltage that has caused more deaths than any other medium.

Safety can never be stressed enough. There are times when your life literally depends on it. The following is a listing of commonsense safety precautions that must be observed at all times:

- Use only one hand when turning power switches on or off. Keep the doors to switch and fuse boxes closed except when working inside or replacing fuses. Use a fuse puller to remove cartridge fuses, after first making certain that the circuit is dead.
- Your company will make the determination as to whether or not you are qualified to work on an electrical circuit.
- Do not work with energized equipment by yourself; have another person (safety observer), qualified in first aid for electrical shock, present at all times. The person stationed nearby should also know which circuits and switches control the equipment, and that person should be given instructions to pull the switch immediately if anything unforeseen happens.
- Always be aware of the nearness of high-voltage lines or circuits. Use rubber gloves where applicable and stand on approved rubber matting. Not all rubber mats are good insulators.
- Inform those in charge of operations as to the circuit on which work is being performed.
- Keep clothing, hands, and feet dry. When it is necessary to work in wet or damp locations, use a dry wooden platform and place a rubber mat or other nonconductive material on top of the wood. Use insulated tools and insulated flashlights of the molded type when required to work on exposed parts.
- Do not work on energized circuits unless absolutely necessary.
- All power supply switches or cutout switches from which power could possibly be fed must be secured in the OPEN (safety) position and locked and tagged.
- Never short out, tamper with, or block open an interlock switch.
- Keep clear of exposed equipment; when it is absolutely necessary to work on it, use only one hand as much as possible.
- Avoid reaching into enclosures except when absolutely necessary. When reaching into an enclosure, use rubber blankets to prevent accidental contact with the enclosure.
- Do not use bare hands to remove hot vacuum tubes from their sockets. Wear protective gloves or use a tube puller.
- Have a qualified electrician check that the equipment is properly grounded.
- Turn off the power before connecting alligator clips to any circuit.

7.1.0 Use of High-Voltage Protection Equipment

Anyone working on or near energized circuitry must use special equipment to provide protection from electrical shock. Protective equipment includes gloves, rubber sleeves, rubber blankets, and rubber mats. This electrical protective equipment is in addition to the regular protective equipment normally required for maintenance work. Regular protective equipment typically includes hard hats that are rated for electrical resistance, eye protection, safety shoes, and long sleeves.

> **NOTE**
> Any exposed conductive articles, such as key or watch chains, rings, watches, or other jewelry should be removed.

Gloves that are approved for protection from electrical shock are made of rubber. A separate leather cover protects the rubber from punctures or other damage. They protect the worker by insulating the hands from electrical shock. Gloves are rated as providing protection from certain amounts of voltage. Whenever an individual is going to be working around exposed conductors, the gloves chosen should be rated for at least as much voltage as the conductors are carrying.

Rubber sleeves are used along with gloves to provide additional protection. The combination of sleeves and gloves protects the hands and arms from electrical shock. Both the high-voltage rubber gloves and sleeves should be tested and certified annually.

Rubber blankets and floor mats have many uses. Blankets are used to cover energized conductors while work is going on around them. They might be used to cover the energized main buses in a breaker panel before working on a de-energized breaker. Rubber floor mats are used to insulate workers from the ground. If a worker is standing on a rubber mat and then contacts an energized conductor, the current cannot flow through the body to the ground, so the worker will not get shocked.

Finally, when working on anything that might produce flames or electric arcs, the material of the worker's clothing should also be considered. Clothing should meet standards for flame resistance or be made from a fabric that would not increase the severity of an injury, should an accident occur.

7.2.0 General Testing of Electrical Equipment

As stated earlier, personnel must be familiar with the proper use of available test equipment. Also, personnel must be familiar with and use the local instructions governing the testing of electrical circuits to protect both the person and the equipment.

The next section includes an example of instructions from a local utility and covers work permits, test permits, authorized employees for testing, circuit isolation, and relay, instrument, and meter testing. Each topic is discussed individually.

7.2.1 Example Policy—Work Permits or Test Permits

A written work permit must be obtained by an authorized technician or supervisor from the operator in charge before testing equipment under the operator's control. A test permit is required to make a dielectric proof test.

The following must be on the work or test permit:

- Designation of the equipment to be tested
- Scope and limits of tests to be made
- Method of isolation and protection provided

If it is necessary to alter the nature or to extend the scope of tests for which a work permit or test permit has been granted, return the permit to the operator and request a new permit before proceeding with the revised tests. When the test is completed, return the work permit or test permit to the operator, and report whether the results of the test were satisfactory or unsatisfactory.

Whenever a work permit or test permit remains in effect beyond the working hours of the person to whom it was issued and a different person will be in charge of the tests, consult the operator for the proper turnover procedure. The person assuming the responsibility for continuance of the work must verify that the items on the work or test permit are correct.

If a work permit or test permit is issued for testing that is to continue over a period of several days, the permit must be returned to the operator at the end of each working day and picked up at the beginning of the next working day.

7.2.2 Example Policy—Authorized Employees for Testing

Only employees authorized to perform operations at a given station shall:

- Open, remove, close, or replace doors or covers on electrical compartments.
- Remove or replace potential transformer fuses.
- Open or close switches.
- Open or close fuse cutouts or disconnecting potheads.
- Make tests for the presence of potential.
- Place or remove blocks and protective tags.
- Apply or remove grounds or short circuits.
- Attach or remove leads to high- or intermediate-voltage conductors.
- Make operating tests.
- Test links, fuses, switches associated with generators, buses, feeders, transformers, etc. normally shall be removed, replaced, or operated only under the direct supervision of an authorized employee when the associated equipment is live or is available for normal operation.

In some cases, when it is necessary for the equipment to remain in service, or in any case when the associated equipment has been removed from service, the operator-in-charge may authorize the person to whom the work permit was issued to remove and replace test links and fuses and to operate the heel and toe switches and potential switches in low-voltage circuits.

7.2.3 Example Policy—Clearances

Adequate clearances are to be maintained between energized and exposed conductors and personnel (refer to *NEC Section 110.26*). Where DC voltages are involved, clearances specified shall be used with specified voltages considered as DC line-to-ground values.

If adequate clearances cannot be maintained from exposed live parts of apparatus in the normal course of free movement within the area during test, then access to that area shall be restricted by fences and barricades. Signs clearly indicating the hazard shall be posted in conspicuous locations. This requirement applies to equipment in service as well as to equipment to which test voltages are applied.

Whenever there is any question of the adequacy of clearance between the specific area in which work is to be done and exposed live parts of adjacent equipment, a field inspection shall be made by management representatives of the group involved before starting the job. The result of this inspection should be to outline the protection necessary to complete the work safely, including watchers where needed.

7.2.4 Electrical Circuit Tests

Testing of electrical circuits can be required at new installations or existing installations where wiring has been modified or new wiring has been interfaced with existing circuits. It is broken down into two sections:

- Electrical circuit tests performed on AC relay protection and metering circuits include the following:
  - Point-to-point wire checks to verify that the wiring has been installed according to the diagram of connections
  - Current transformer polarity, impedance, continuity, and where applicable, tap progression and intercore coupling tests
  - Voltage transformer polarity, ratio, and self-induced high-potential (hi-pot) tests
  - Insulation resistance tests
- Electrical circuit tests performed on control, power, alarm, and annunciator circuits include the following:
  - Insulation resistance tests
  - Operation tests

An important piece of information is the minimum distance allowed when working near energized electrical circuits because large voltages can arc across an air gap. Personnel must maintain a distance that is greater than that arc distance. This is especially true when using a hot stick to open a disconnect. These distances are listed in *Table 1*.

Table 1 Minimum Safe Working Distances.

Voltage Range (Phase-to-Phase) Kilovolts	Minimum Working and Clear Hot Stick Distance
2.1 – 15	2 ft 0 in
15.2 – 35	2 ft 4 in
35.1 – 46	2 ft 6 in
46.1 – 72.5	3 ft 0 in
72.6 – 121	3 ft 4 in
138 – 145	3 ft 6 in
161 – 169	3 ft 8 in
230 – 242	5 ft 0 in
345 – 362	*7 ft 0 in
500 – 552	*11 ft 0 in
700 – 765	*15 ft 0 in

* For voltages above 345kV, the minimum working and clear hot stick distances may be reduced provided that such distances are not less than the shortest distance between the energized part and a grounded surface.

110T01.EPS

8.0.0 ◆ TROUBLESHOOTING MOTORS

All types of machinery that contain moving parts produce friction. The most common form of high-speed motion in industrial machinery is rotary motion. Rotary motion includes a shaft of some sort turning in some sort of bearing. There is not, as yet, a completely frictionless bearing system. It is the job of the maintenance craftsperson to find out where wear has begun to degrade the capability of the bearings to operate, before the actual failure.

The key to prevention is a combination of regular maintenance, such as lubrication, and physical repairs such as tightening loose nuts, and testing for problems. Tools have been developed to make the testing possible, often without shutting down the machines. First, however, you need to determine what parts of the machine you are testing, and what characteristics you are testing for.

Symptoms of excessive wear, bearing problems, misalignment of shafts, and loading of drive motors include noisy operation, visible vibration, and heat generation. To detect noisy operation, craftspersons can simply listen to the sound of the bearings. A grinding or growling noise would indicate that the bearings were losing the ability to turn freely.

Both forms of energy loss, vibration and friction wear, also generate heat and infrared light. Infrared light can be observed with a non-contact thermometer, frequently known as a pyrometer (after one of the brand-named instruments). This allows the craftsperson to look with useful precision for hot areas in a piece of equipment from several feet away *(Figure 11)*. These devices use a laser light to indicate the center of the area whose temperature is being measured. A digital readout shows the temperature at that point. This highlights the importance of recording such information, to provide a baseline temperature from which deviation would indicate possible problems.

Vibration in shafts and other cyclic moving parts can be detected by observing the part with a strobe light, also known as a stroboscope *(Figure 12)*. The strobe light can be set to flash at a rate exactly the same as the cyclic rate of a motor shaft, for example. The observer will then see what appears to be a motionless shaft. If there is excess play or wobble in the shaft, the shaft will visibly move when the strobe light shines on it, so that the operator can determine the fact and rate of wobble. If the strobe light is set to flash slightly faster than the rate of the shaft's revolution, the shaft will appear to turn forward in slow motion. This allows each part of the shaft to be inspected although the shaft is turning at high speed. If the strobe flashes a little slower than the rate of the cycle, the shaft will appear to turn backward.

If the strobe light is set to the frequency of an industrial motor, the light will always show the same point on the motor's shaft. If that point appears to move from side to side, the shaft is vibrating. If a belt on a shaft moves when seen with a strobe light, the belt is traveling. If the strobe is set to slightly less or more frequent flashing than the rpm of the motor, the process of the motor turning can be seen in slow motion because a slightly different point will be seen on each flash.

A transistorized stethoscope is equipped with a transistor-type amplifier and is used to determine the condition of motor bearings. A little practice in interpreting its usage may be required, but in gen-

HAND-HELD THERMOMETER

110F11.EPS

Figure 11 ◆ Infrared thermometer.

110F12.EPS

Figure 12 ◆ Strobe light.

eral, it is relatively simple to use. When the stethoscope is applied to a motor bearing housing, a smooth purring should be heard if the bearing is operating normally. However, if a thumping, grinding, or growling sound is detected, it could be an indication of a failing bearing.

Modern plants frequently attach accelerometers to rotating machinery to detect vibration. The small devices operate by what is called the piezoelectric effect, using crystals that respond to physical pressure by producing electrical currents. By maintaining records of the direction and volume of vibrations, computers with analytic software can determine any changes that take place, alerting the maintenance personnel to future problems. Such devices can be an excellent tool, supplementing the activity of the maintenance craftsperson in inspecting plant machinery.

8.1.0 Infrared Devices

Any object that is not at a temperature of absolute zero (–273°C, or –460°F) will emit invisible electromagnetic energy in the form of IR. A detector can pick up this transmitted energy and produce an electrical signal proportional to the amount of IR detected. This signal can be used to control other devices.

IR detectors are commonly used for motion/people detectors, TV remote control detectors, and IR scanners that detect hot spots in electrical equipment. The most common application is the motion/people detector.

IR sensors are used in the instrumentation of temperature-sensitive industrial processes, including paper production and stack temperature monitoring in cement and petrochemical facilities. They are particularly useful when the use of temperature probes is impossible or impractical. *Figure 13* shows a typical fixed IR monitoring system that includes controller functions for industrial processes. Also shown is a calibrated IR handheld thermometer used to perform random temperature checks. A visible laser pointing beam is used to identify the area measured by the instrument's IR sensor.

IR TEMPERATURE MONITORING SYSTEM

HAND-HELD THERMOMETER

110F13.EPS

Figure 13 ◆ Typical IR instrumentation equipment.

Review Questions

1. An ammeter is used to measure _____.
   a. current
   b. voltage
   c. resistance
   d. insulation value

2. Electromotive force (emf) is measured using a(n) _____.
   a. ammeter
   b. wattmeter
   c. voltmeter
   d. ohmmeter

3. A voltmeter is connected _____ with the circuit being tested.
   a. in parallel
   b. in series
   c. in a sequential configuration
   d. in a looped configuration

4. All of the following values can be directly measured using a VOM *except* _____.
   a. current
   b. voltage
   c. resistance
   d. wattage

5. When in doubt as to the voltage present, always check the _____ position to protect the VOM.
   a. volume
   b. range
   c. fuse
   d. power

6. Before testing for low voltages, for safety, test for _____.
   a. resistance
   b. infrared light
   c. vibration
   d. higher ranges

7. The red lead on a multimeter should be connected to the _____.
   a. positive terminal
   b. negative terminal
   c. black wire
   d. live wire

8. Before connecting a meter to a circuit, be sure the circuit is _____.
   a. positive
   b. energized
   c. de-energized
   d. connected to the power source

9. Digital meters do *not* have a(n) _____.
   a. red lead wire
   b. battery
   c. meter movement
   d. autoranging setting

10. Frequency is expressed as cycles per second or _____.
    a. hertz
    b. volts
    c. ohms
    d. amperes

11. The tachometer is used to determine _____.
    a. voltage
    b. rpms
    c. current
    d. alignment

12. An ohmmeter can be used to test for continuity.
    a. True
    b. False

13. A Wiggy® is used to _____.
    a. measure wattage
    b. measure impedance
    c. test for potential difference
    d. provide an exact voltage reading

14. One way to check for bearing degradation is with a(n) _____.
    a. ohmmeter
    b. tachometer
    c. pyrometer
    d. ammeter

15. Stroboscopes can be used to measure _____.
    a. rpms
    b. speed
    c. horsepower
    d. misalignment

Summary

The use of test equipment is an important part of your job as an industrial maintenance craftsperson. Selecting the proper instrument will help you to fully perform your task. Use the information provided by these instruments to help evaluate the work.

Notes

Trade Terms Introduced in This Module

Amperes: The unit of measure for current.

Circuit: A complete path of electrical components and conductors.

Continuity: An uninterrupted electrical path for current flow.

Frequency: The number of cycles completed each second by a given AC voltage; usually expressed in hertz; one hertz = one cycle per second.

Hertz: One cycle per second.

Ohms: The unit of electrical resistance.

Parallel: A circuit wired in parallel has multiple paths for current to flow.

Polarity: In DC circuits, the direction of current flow is from negative to positive; it is critical that the connections be identically arranged.

Series: A circuit wired in series has only one path for current to flow.

Resources & Acknowledgments

Additional Resources

This module is intended to present thorough resources for task training. The following reference works are suggested for further study. These are optional materials for continued education rather than for task training.

Electronics Fundamentals: Circuits, Devices, and Applications, Thomas L. Floyd. New York: Prentice Hall.

Principles of Electric Circuits, Thomas L. Floyd. New York: Prentice Hall.

Figure Credits

Topaz Publications, Inc., 110F08

Raytek, 110F11, 110F13 (bottom)

Photo courtesy of Monarch Instrument, 110F12

Dwyer Instruments, Inc., 110F13 (top)

CONTREN® LEARNING SERIES — USER UPDATE

NCCER makes every effort to keep these textbooks up-to-date and free of technical errors. We appreciate your help in this process. If you have an idea for improving this textbook, or if you find an error, a typographical mistake, or an inaccuracy in NCCER's Contren® textbooks, please write us, using this form or a photocopy. Be sure to include the exact module number, page number, a detailed description, and the correction, if applicable. Your input will be brought to the attention of the Technical Review Committee. Thank you for your assistance.

Instructors – If you found that additional materials were necessary in order to teach this module effectively, please let us know so that we may include them in the Equipment/Materials list in the Annotated Instructor's Guide.

Write: Product Development and Revision
National Center for Construction Education and Research
3600 NW 43rd St. Bldg G, Gainesville, FL 32606

Fax: 352-334-0932

E-mail: curriculum@nccer.org

Craft ______________________ Module Name ______________________

Copyright Date ______________ Module Number ______________ Page Number(s) ______________

Description

____

____

____

____

(Optional) Correction

____

____

____

(Optional) Your Name and Address

____

____

____

Industrial Maintenance E & I Technician Level One

40111-07

Material Handling and Hand Rigging

40111-07
Material Handling and Hand Rigging

Topics to be presented in this module include:

Overview

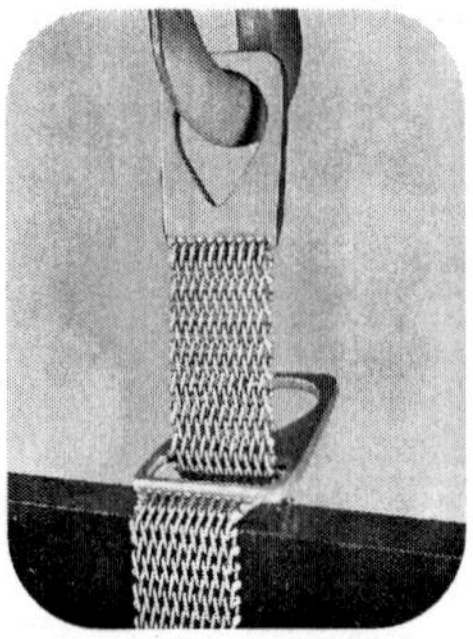

Part of your job as an industrial maintenance craftworker will be to move machinery and parts of machinery. Rigging is the safe and proper handling of material, so as to eliminate damage in the transportation process. You will learn to work with cranes, chains, wire rope, and slings. Always be careful when heavy objects are off the ground, because they can always fall.

Objectives

When you have completed this module, you will be able to do the following:

1. Identify and describe the uses of common rigging hardware and equipment.
2. Inspect common rigging equipment.
3. Select, use, and maintain special rigging equipment, including:
   - Jacks
   - Block and tackle
   - Chain hoists
   - Come-alongs
4. Tie knots used in rigging.
5. Use and understand the correct hand signals to guide a crane operator.
6. Identify basic rigging and crane safety procedures.

Trade Terms

Anneal
Anti-two-blocking devices
Bird caging
Equalizer beam
Fixed block
Hauling line
Hydraulic
Kinking
Parts of line
Sling angle

Required Trainee Materials

1. Pencil and paper
2. Appropriate personal protective equipment

Prerequisites

Before you begin this module, it is recommended that you successfully complete *Core Curriculum;* and *Industrial Maintenance E & I Technician Level One,* Modules 40101-07 through 40110-07.

This course map shows all of the modules in the first level of the *Industrial Maintenance E & I Technician* curriculum. The suggested training order begins at the bottom and proceeds up. Skill levels increase as you advance on the course map. The local Training Program Sponsor may adjust the training order.

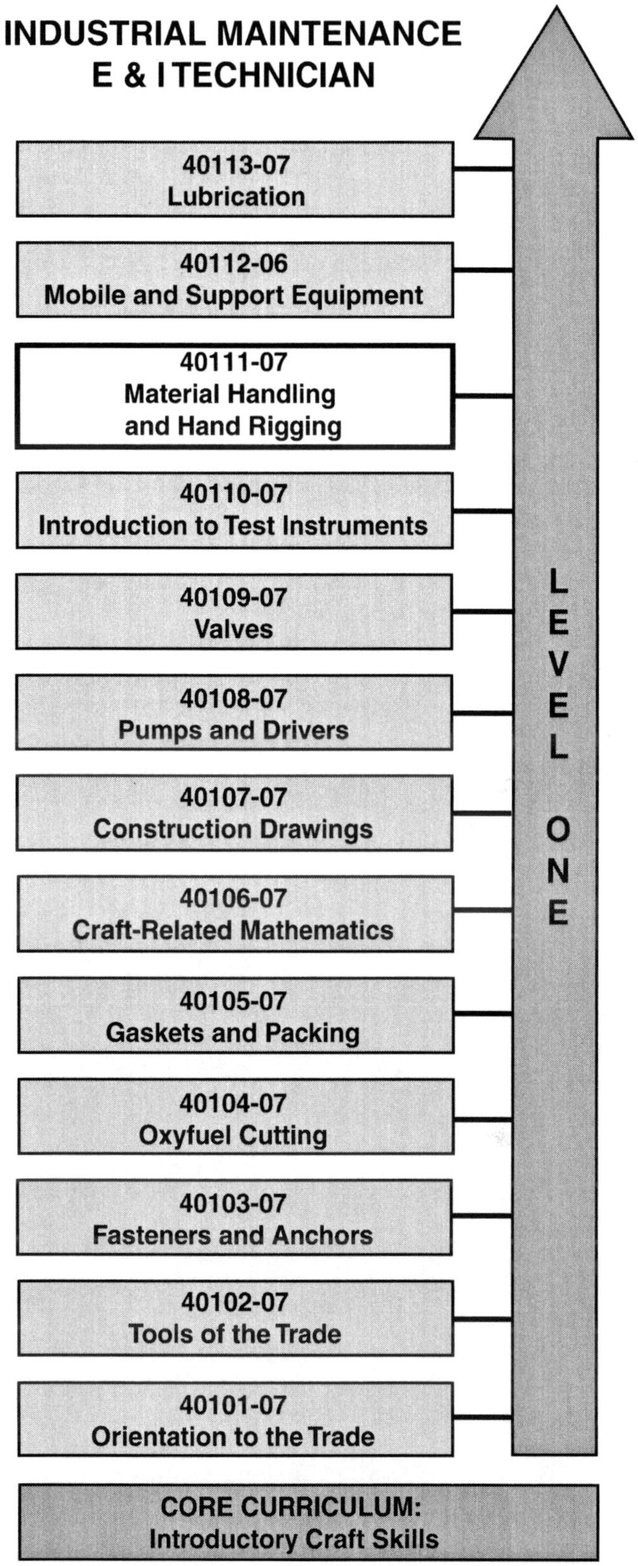

111CMAP.EPS

1.0.0 ◆ INTRODUCTION

As a maintenance craftsperson, you will frequently be required to move very heavy objects into position. This will typically include compressors, generators, motors, valves, and heavy machine tools. The skills you must have to do so are designated as rigging techniques. The tools you will use for rigging include jacks, come-alongs, cranes, turnbuckles, shackles, and simple things like eyebolts and lifting lugs. If you know how to use all these things safely and correctly, your job will be safe and your work will go well.

Rigging involves the lifting and moving of heavy, and often bulky, objects (*Figure 1*). For that reason, safety must be the foremost consideration in any rigging job.

One important job that a rigger performs is guiding the crane operator. This is often done using a standard set of hand signals, which everyone involved in rigging must memorize.

Anyone doing rigging work must know how to select the right equipment, with the right capacity, for every rigging job. Failure to use the right equipment and correct rigging method can lead to serious consequences, including death or injury to workers and damage to equipment or materials.

111F01.EPS

Figure 1 ◆ A rigging application.

Every crane and lifting device has capacity limits and conditions in which they are designed to operate safely. If you are not aware of these limits, you could make a fatal error. If you make the effort to learn the right and safe way to do the job, you will be successful. Always err on the side of caution.

This module is intended to instruct you on the basic safety precautions and rigging practices that will enable you to assist in rigging equipment and materials related to your craft.

2.0.0 ◆ RIGGING HARDWARE

Hardware used in rigging includes hooks, shackles, eyebolts, spreader and **equalizer beams**, and blocks. These hardware items must be carefully matched to the lifting capability of the crane or hoist and the slings used in each application. Mixing items with different safe working loads makes it impossible to determine if the lift will be safe. Careful inspection and maintenance of all lifting hardware is essential for continuous operation. Hardware should always be inspected before each use.

> **NOTE**
>
> Rigging hooks must be properly marked or tagged with the manufacturer's ratings and application information. If a hook is not properly marked, don't use it. Also, rigging hardware must never be modified because modifications could affect its capacity.

2.1.0 Hooks

A rigging hook can be used to attach to a sling or a load. As you learned in the *Core Curriculum*, there are six basic types of rigging hooks, with the eye-type hook being the most commonly used. Rigging hooks are equipped with safety latches to prevent a sling attachment from coming off of the hook when a sling is slackened. The capacity of a rigging hook, in tons, is determined by its size and physical dimensions. Information about a specific hook's capacity is readily available from the hook manufacturer.

The safe working load of the hook is accurate only when the load is suspended from the saddle of the hook. If the working load is applied anywhere between the saddle and the hook tip, the safe working load is considerably reduced, as shown in *Figure 2*. Some manufacturers do not permit their hooks to be used in any other position.

Always inspect hooks before each use. Look for wear in the saddle of the hook. Also, look for cracks, severe corrosion, and twisting of the hook body. The safety latch should be in good working order. Measure the hook throat opening; if a hook has been overloaded or if it is beginning to weaken, the throat will open. OSHA regulations require that the hook be replaced if the throat has opened 15 percent from its original size or if the body has been twisted 10 degrees from its original plane. *Figure 3* shows the location of inspection points for a hook.

Never use a sling eye over a hook with a body diameter larger than the natural width of the eye. Never force the eye onto the hook. Always use an eye with at least the nominal diameter of the hook.

Whenever you are using a choker arrangement with hooks, such as when picking up a bundle of pipes, whether with chains or wire rope, always turn the hooks away from each other. This places the direction of the pull into the saddle of the hook, instead of toward the mouth.

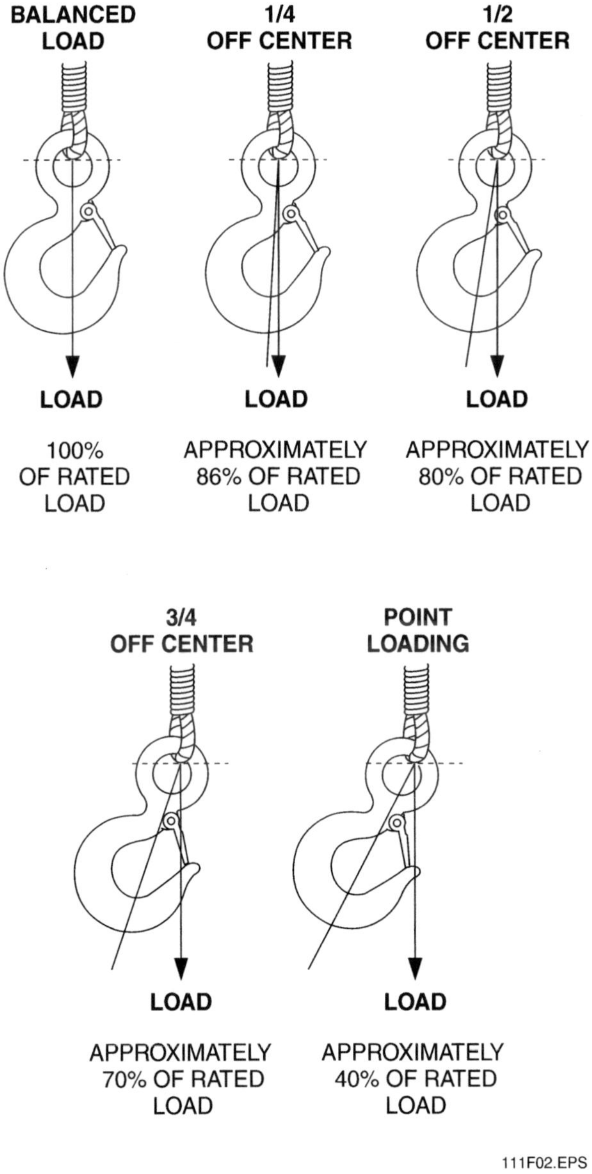

Figure 2 ◆ Rigging hook rated load versus load location.

2.2.0 Shackles

A shackle, sometimes called a clevis, is used to attach an item to a load or to attach slings together. It can be used to attach the end of a wire rope to an eye fitting, hook, or other type of connector. Shackles are made in several configurations (*Figure 4*).

Shackles used for overhead lifting are made of forged alloy steel. They are sized by the diameter of the steel in the bow section rather than the pin size. When using shackles, be sure that all pins are straight, all screw pins are completely seated, and cotter pins are used with all round pin shackles. A shackle must not be used in excess of its rated load.

Shackle pins should never be replaced with a bolt. Bolts cannot take the bending force normally applied to a pin. Shackles that are stretched, or that have crowns or pins worn more than 10 percent, should be destroyed and replaced. If a shackle is pulled out at an angle, the capacity is reduced. Use spacers to center the load being hoisted on the pin. Only shackles of suitable load ratings can be used for lifting. When using a shackle on a hook, the pin of the shackle should be hung on the hook. Spacers can be used on the pin to keep the shackle hanging evenly on the hook (*Figure 5*). Never use a screw-pin shackle in a situation where the pin can roll under load, as shown in *Figure 6*.

WARNING!

Shackles may not be loaded at an angle unless approved by the manufacturer. Contact the manufacturer for any angular loading. Crosby™ shackles have markings at the 45-degree angles.

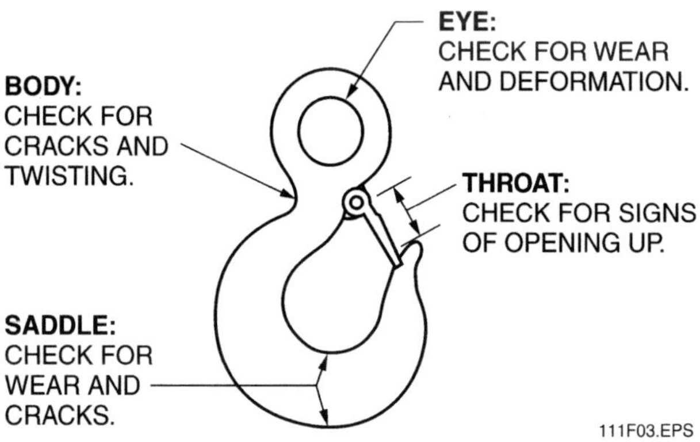

Figure 3 ◆ Rigging hook inspection points.

2.3.0 Eyebolts

Eyebolts (*Figure 7*) are often attached to heavy loads by a manufacturer in order to aid in hoisting the load. One type of eyebolt, called a ringbolt, is equipped with an additional movable lifting ring. Eyebolts and ringbolts can be of either the shoulder or shoulderless type. The shoulder type is recommended for use in hoisting applications because it can be used with angular lifting pulls, whereas the shoulderless type is designed only for lifting a load vertically. The safe working load of shoulder eyebolts and ringbolts is reduced with angular loading. Loads should always be applied to the plane of the eye to reduce bending. This procedure is particularly important when bridle slings are used.

When installed, the shoulder of the eyebolt must be at right angles to the axis of the hole and must be in full contact with the working surface

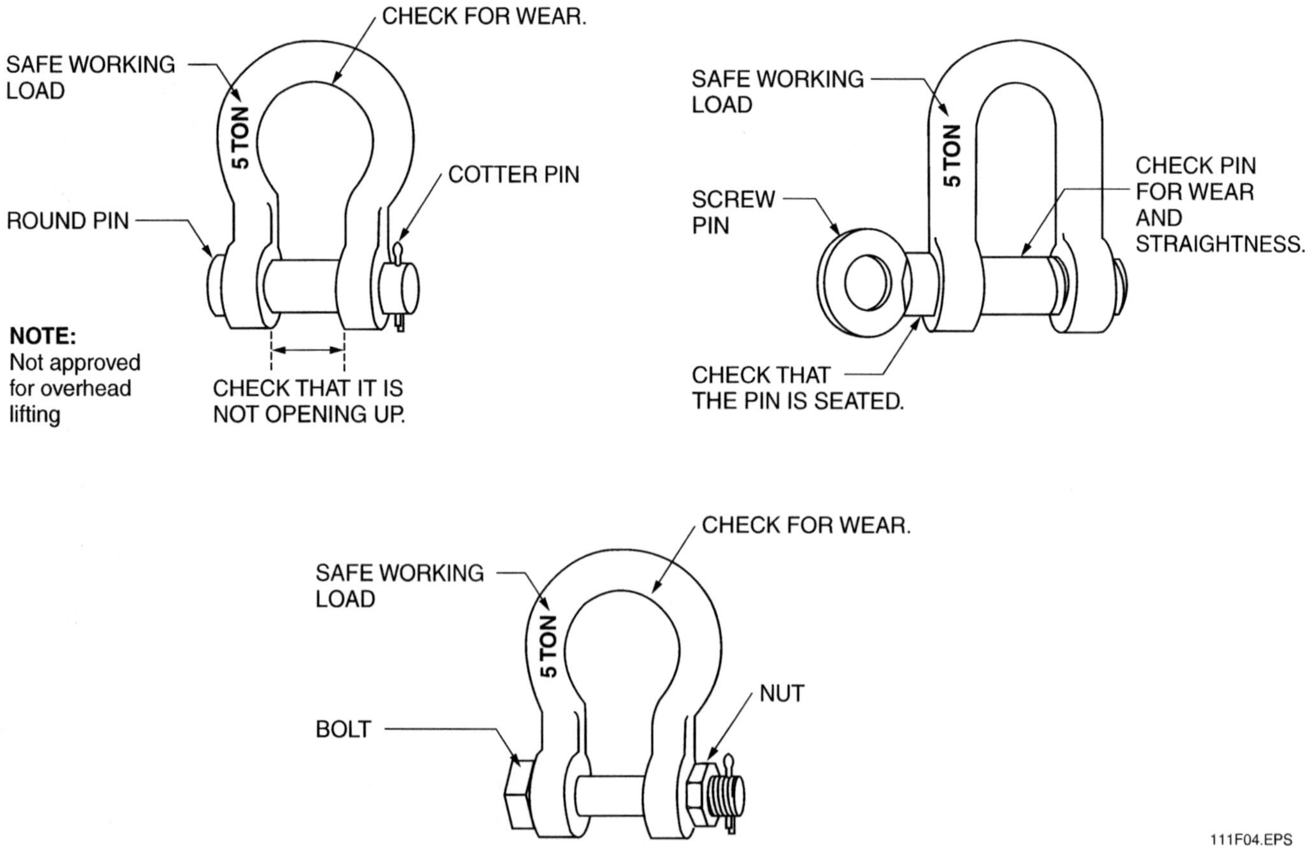

Figure 4 ◆ Round pin and screw pin shackle.

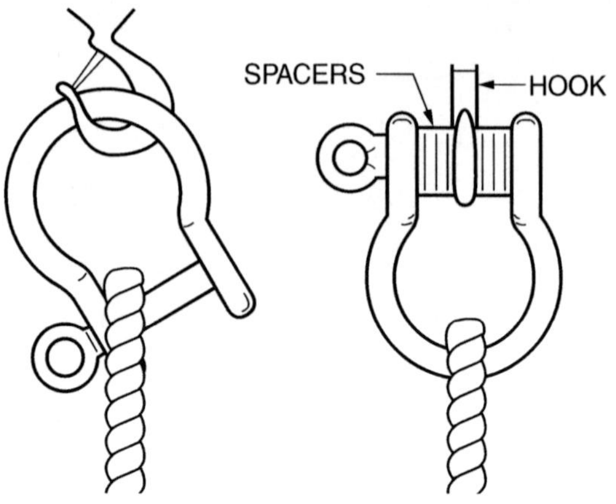

111F05.EPS

Figure 5 ◆ Spacers used with a shackle to keep it hanging evenly.

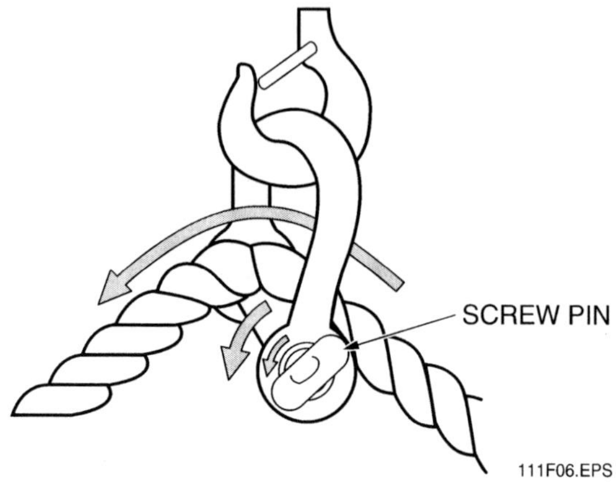

111F06.EPS

Figure 6 ◆ Example of how a rope can cause a shackle screw-pin to roll under load.

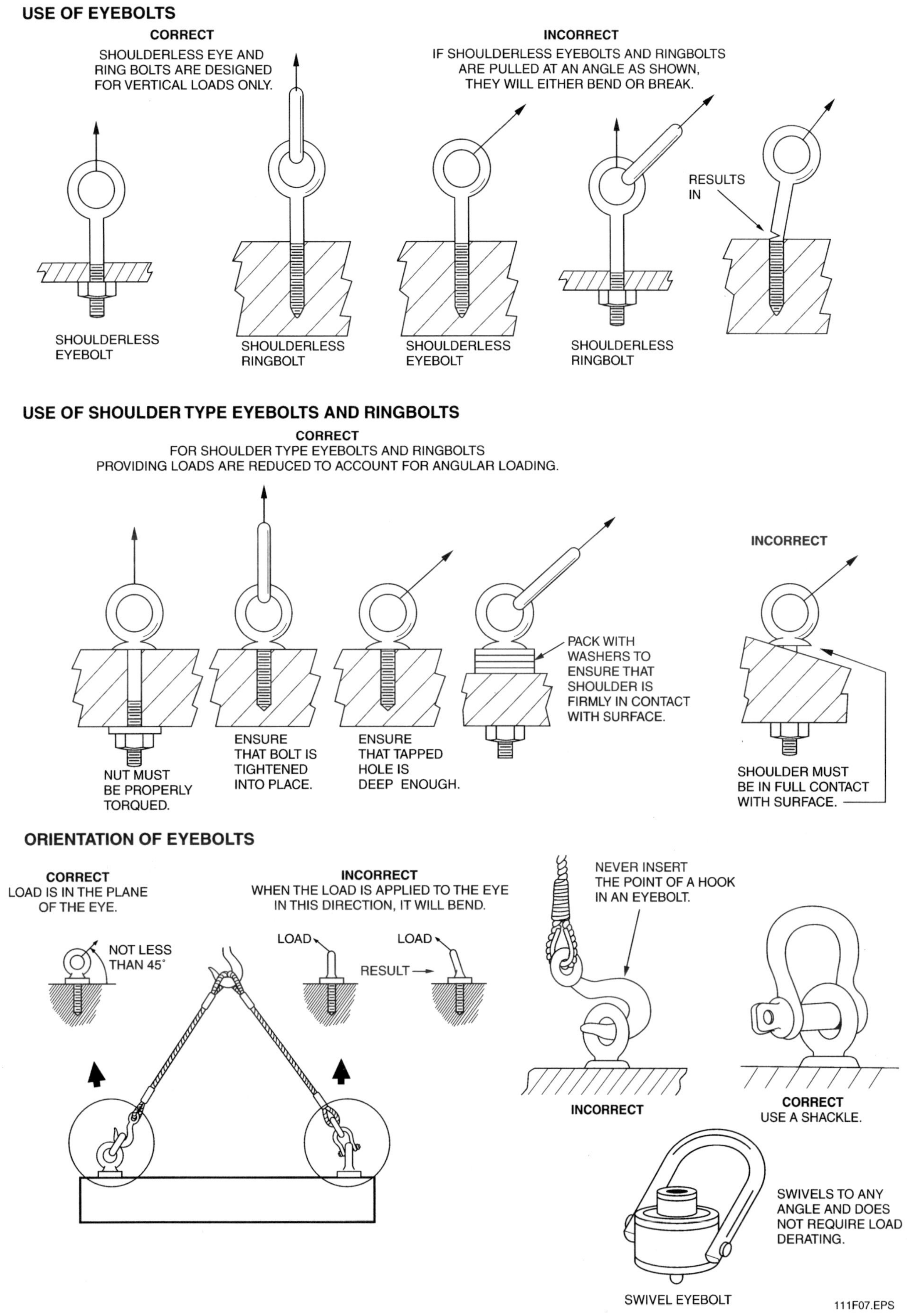

Figure 7 ◆ Eyebolt and ringbolt installation and lifting criteria.

when the nuts are properly fastened. Washers or other suitable spacers may be used to ensure that the shoulders are in firm contact with the working surface. Tapped holes used with screwed-in eyebolts should have a minimum depth of 1½ times the bolt diameter. Swivel eyebolts may be installed instead of fixed eyebolts. These devices swivel to the desired lift position and do not require any load rating reduction.

2.4.0 Lifting Lugs

Lifting lugs (*Figure 8*) are typically welded, bolted, or pinned by a manufacturer to the object to be lifted. They are designed and located to balance the load and support it safely. Lifting lugs should be used for straight, vertical lifting only, unless specifically designed for angular loads. This is because they can bend if loaded at an angle.

Lifting lugs should be inspected before each use for deformation, cracks, corrosion, and defective welds. A lifting lug must be removed from service if such conditions are found.

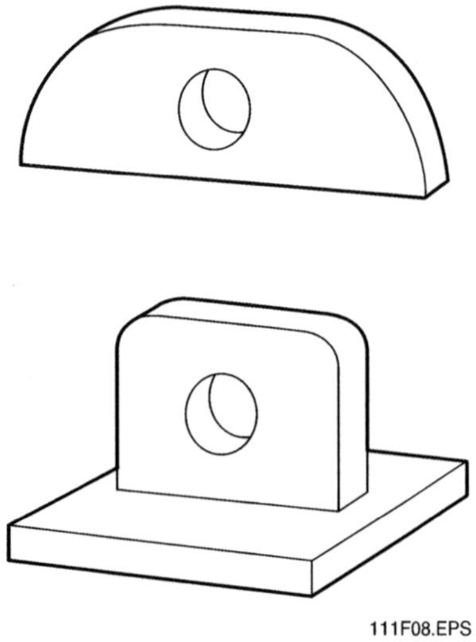

Figure 8 ◆ Two types of lifting lugs.

2.5.0 Turnbuckles

Turnbuckles are available in a variety of sizes. They are used to adjust the length of rigging connections. Three common types of turnbuckles are the eye, jaw, and hook ends (*Figure 9*). They can be used in any combination. The safe working load for turnbuckles is based on the diameter of the threaded rods. The safe working load can be found in the manufacturer's catalog. The safe working load of turnbuckles with hook ends is less than that of the same size turnbuckles with other types of ends.

Consider the following when selecting turnbuckles for rigging:

- Turnbuckles should be made of alloy steel and should not be welded.
- When using turnbuckles with multi-leg slings, do not use more than one turnbuckle per leg.
- Do not use jam nuts on turnbuckles that do not come equipped with them.
- Turnbuckles should not be overtightened. Do not use a cheater to tighten them. Tighten with a wrench of the proper size, using as much force as a person can achieve by hand.
- Turnbuckles must not be used for angular loading unless permitted by the manufacturer, or if the angular load has been calculated.

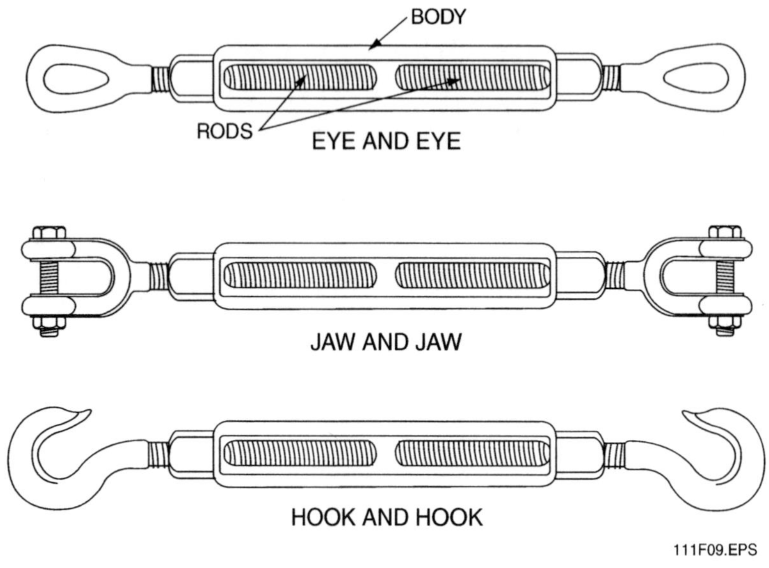

Figure 9 ◆ Turnbuckles.

Turnbuckles should be inspected as shown in *Figure 10*. If a turnbuckle is damaged, remove it from service.

2.6.0 Beam Clamps

Beam clamps (*Figure 11*) are used to connect hoisting devices to beams so the beams can be lifted and positioned in place. The following are guidelines for using beam clamps:

- Do not use homemade clamps unless they are designed, load tested, and stamped by an engineer.
- Ensure that the clamp fits the beam and is of the correct capacity.
- Make sure beam clamps are securely fastened to the beam.
- Be careful when using beam clamps where angle lifts are to be made. Most are designed for essentially straight vertical lifts only. However, some manufacturers allow two clamps to be used with a long bridle sling if the angle of lift does not exceed 25 degrees from the vertical.
- Be certain the capacity appears on the beam clamp.
- Do not place a hoist hook directly in the beam clamp lifting eye.
- Never overload a beam clamp beyond its rating capacity.

Beam clamps must be inspected before each use, and removed from service if you observe any of the following problems:

- The jaws of the beam clamp have been opened more than 15 percent of their normal opening.
- The lifting eye is bent or elongated.
- The lifting eye is worn.
- The capacity and beam size are unreadable.

2.7.0 Plate Clamps

Plate clamps or plate grabs (*Figure 12*) attach to structural steel plates to allow for easier rigging attachment and handling of the plate. There are two basic types: the camming plate clamp and the screw lock. The screw lock variety is basically a clamp with a lifting eye. The camming plate clamp or plate grab shown is the most common plate clamp, with a serrated jaw that holds the plate. The other clamp shown is a cross between the two types, with a screw adjustment that helps to make the clamp grip more tightly. The tightening screw makes this one hold more positively when raising or lowering a plate to or from a horizontal position.

111F11.EPS

Figure 11 ◆ Typical beam clamp application.

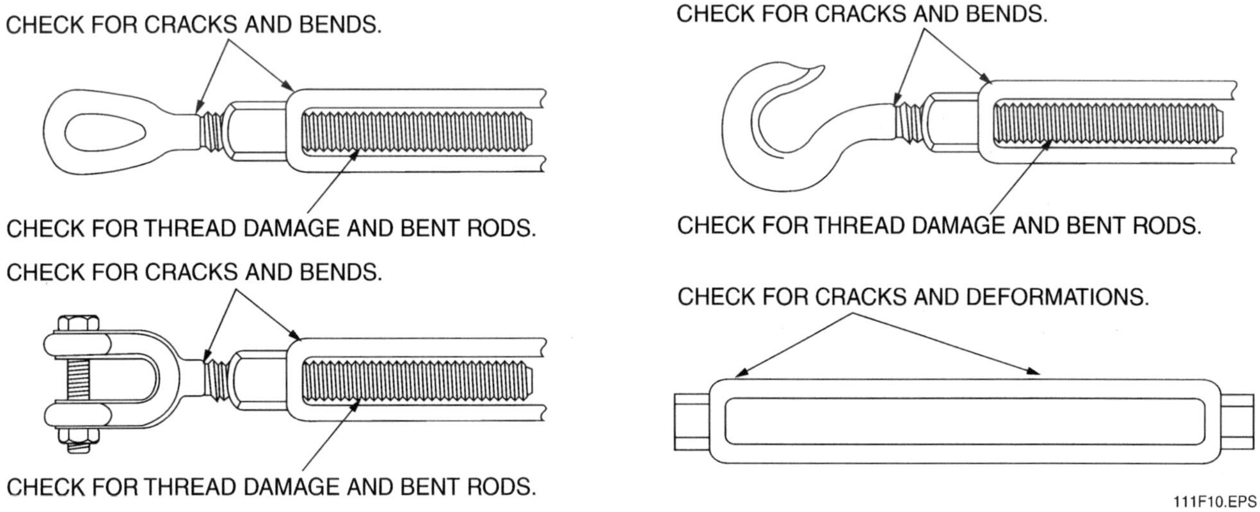

111F10.EPS

Figure 10 ◆ Turnbuckle inspection points.

Camming clamps are designed to grip a single plate for hoisting and are available with a locking device. Screw clamps are considered the safest. They rely on a clamping action of a screw against the plate to secure it. Camming clamps are used for vertical lifting, whereas screw clamps can be used from a horizontal position through 180 degrees. Plate clamps are designed to lift only one plate at a time. Always follow the manufacturer's recommendations for use and safe working load.

Never pick up more than one plate at a time with any plate clamp. Plate clamps are more likely to fail with lighter gauge material. Do not carry sheet metal or plate over anyone using a plate grab, as the clamp can slip and drop the sheet.

Remove plate clamps from service if any of the following are present during inspection:

- Changes in the opening at the jaw plate or wear of cam teeth
- Cracks in body
- Loose or damaged rivets
- Worn, bent, or elongated lifting eye
- Excessive rust or corrosion
- Unreadable capacity size

2.8.0 Rigging Plates and Links

Rigging plates and links (*Figure 13*) are made for specific uses. The holes in the plates or links may be different sizes and may be placed in different locations in the plates for different weights and types of lifts. Plates with two holes are called rigging links. Plates with three or more holes are

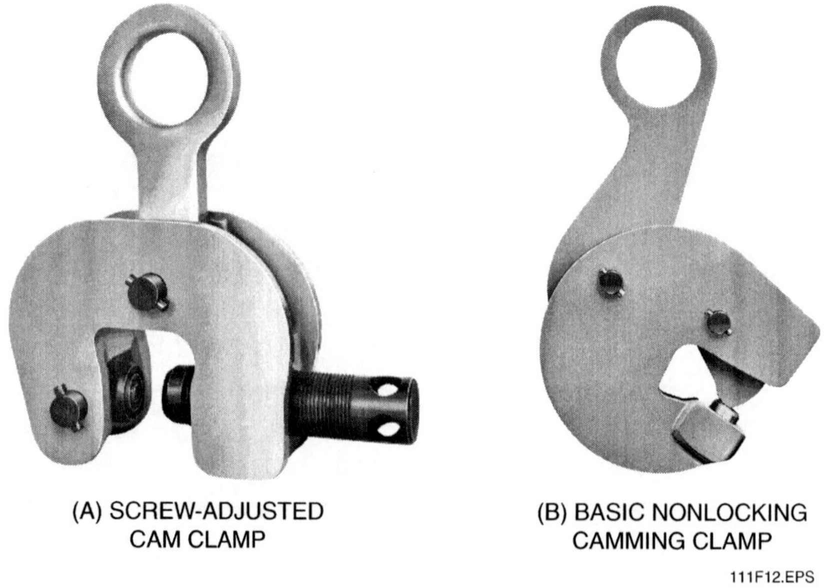

Figure 12 ◆ Standard lifting clamps.

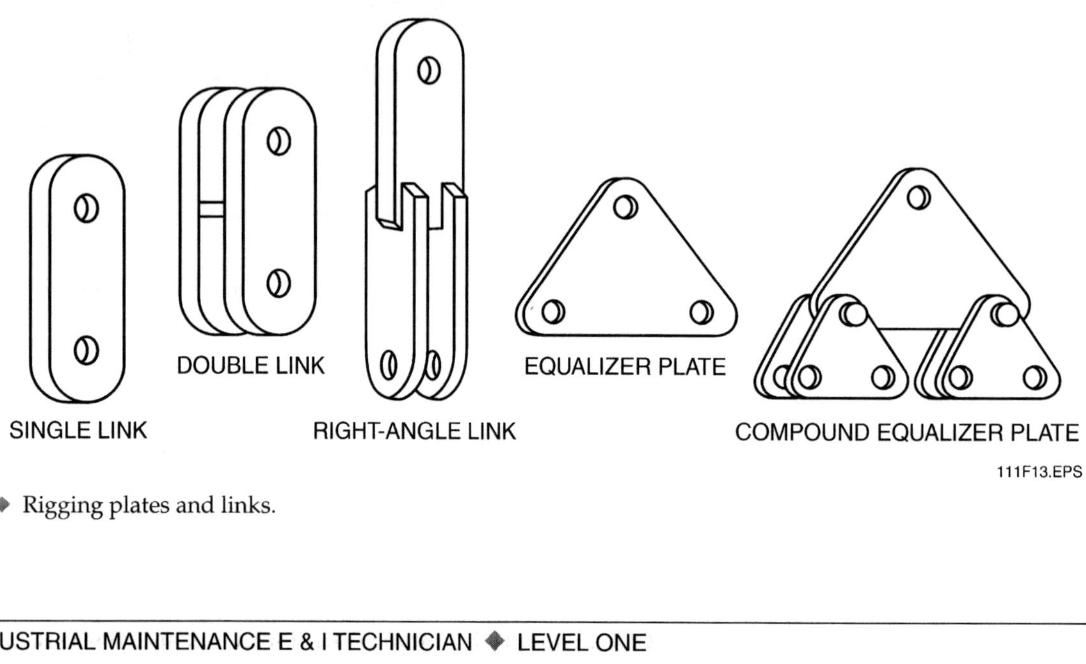

Figure 13 ◆ Rigging plates and links.

called equalizer plates. Equalizer plates can be used to level loads when the legs of a sling are unequal. Plates are attached to the rigging with high-strength pins or bolts. Rigging plates and links must be engineer approved.

Inspect rigging plates and links, and remove them from service if any of the following are present:

- Cracks in body
- Worn or elongated lifting eye
- Excessive rust or corrosion

2.9.0 Spreader and Equalizer Beams

Spreader beams (*Figure 14*) are used to support long loads during lifting operations. If used properly, they help eliminate the hazard of the load tipping, sliding, or bending. They reduce low **sling angles** and the tendency of the slings to crush loads. Equalizer beams are used to balance the load on sling legs and to maintain equal loads on dual hoist lines when making tandem lifts. Lifting beams, spreader beams, or spider frames (frames that allow the crane to lift circular or rectangular structures) all require engineering calculations to be made. They must be used as they have been designed to be used.

Both types of beams are usually fabricated to suit a specific application. They are often made of heavy pipe, I-beams, or other suitable material. Custom-fabricated spreader or equalizer beams must be designed by an engineer and have their capacity clearly stamped on the side. They should be tested at 125 percent of rated capacity. Information on the beams should be kept on file. The capacity of beams designed for use with multiple attachment points depends upon the distance between attachment points.

111F14.EPS

Figure 14 ◆ Typical use of a spreader beam.

Before use, a spreader or equalizer beam should be inspected for the following:

- Solid welds
- No cracks, nicks, gouges, or corrosion
- Condition of attachment points
- Capacity rating
- Sling angle tension at the points of attachment

3.0.0 ◆ SLINGS

The common types of slings include wire rope slings, synthetic slings, chain slings, and metal mesh slings. Wire rope and synthetic slings were covered in the *Core Curriculum* module, *Basic Rigging*. Metal mesh slings (*Figure 15*) are typically made of wire or chain mesh. They are similar in appearance to web slings and are suited for situations where the loads are abrasive, hot, or tend to cut other types of slings. Metal mesh slings resist abrasion and cutting, grip the load firmly without stretching, can withstand temperatures up to 550°F, are smooth, conform to irregular shapes, do not kink or tangle, and resist corrosion. These slings are available in several mesh sizes and can be coated with a variety of substances, such as rubber or plastic, to help protect the load they are handling.

3.1.0 Sling Capacity

Sling capacities depend on the sling material, construction, size of hitch configuration, quantity, and angle for the specific type of sling being used. The

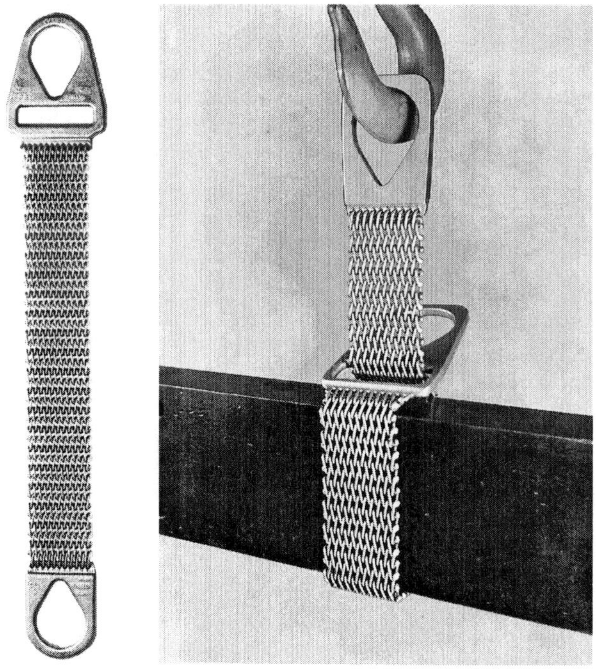

111F15.EPS

Figure 15 ◆ Metal mesh sling.

amount of tension on the sling is directly affected by the angle of the sling (*Figure 16*). For this reason, proper sling angles are crucial to safe rigging. This information, along with other pertinent rigging information, is available from rigging equipment manufacturers and rigging trade organizations in the form of easy-to-use pocket guides like the ones shown in *Figure 17*. Sling capacity information is also given in *OSHA Regulation 29 CFR, 1926.251, Rigging Equipment for Material Handling*. *Table 1* shows an example of a typical capacity table used with wire rope slings. The vertical columns on *Table 1* give (from left to right) the International Wire Rope Class; the size of the wire rope; the allowable load for a vertical lift over the eyebolt; the allowable load for a choker on the eye; allowable loads for the given angle of a basket hitch; and the eye dimensions to be used.

The D/d ratio applies to wire rope slings. It refers to the relationship between the diameter of the surface (D) the sling is around and the diameter of the sling (d). This relationship, specifically the severity of the bends in the sling, has an impact on the capacity of the sling. Efficiency charts have been developed to help determine the actual sling capacity. *Figure 18* shows load and hook diameter examples and a D/d ratio chart.

For example, consider a 1-inch-diameter rope sling in a basket hitch around a 2-inch shackle pin. The capacity of the sling in the basket hitch is listed as 17 tons. Look across the bottom of the chart for the 2-inch pin size and find the line that meets the ratio curve. It is opposite 65 percent efficiency. Multiply the rated load of 17 tons by 65 percent and you will get an actual capacity of 11 tons. Sling eyes with pin or body sizes that are at least equal to the rope diameter should be used with shackles and hooks. Always refer to the tag when selecting a wire rope sling.

3.2.0 Sling Care and Storage

You learned about inspection criteria for slings in the *Core Curriculum* module, *Basic Rigging*. The following are some important reminders for sling inspection, care, and storage:

- Store slings in a rack to keep them off the ground. The rack should be in an area free of moisture and away from acid or acid fumes and extreme heat.
- Never let slings lie on the ground in areas where heavy machinery may run over them.
- Slings should be inspected at each use for broken wires, kinks, rust, or damaged fittings. Any slings found to be defective should be destroyed.

111F17.EPS

Figure 17 ◆ Rigging pocket guides.

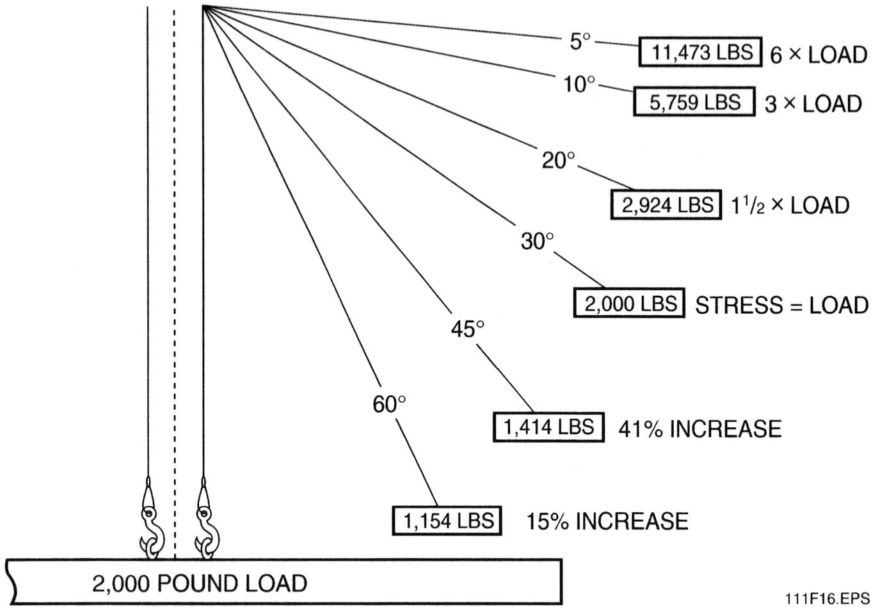

111F16.EPS

Figure 16 ◆ Sling angles.

Table 1 Example of a Wire Rope Sling Capacity Table

CLASS	SIZE (IN.)	RATED CAPACITY - LBS.*						EYE DIMENSIONS (APPROXIMATE)	
		VERTICAL	CHOKER**	BASKET HITCH				WIDTH (IN.)	LENGTH (IN.)
					30°	60°	90°		
6 x 19 IWRC	1/4	1,120	820	2,200	2,200	1,940	1,580	2	4
	5/16	1,740	1,280	3,400	3,400	3,000	2,400	2 1/2	5
	3/8	2,400	1,840	4,800	4,600	4,200	3,400	3	6
	7/16	3,400	2,400	6,800	6,600	5,800	4,800	3 1/2	7
	1/2	4,400	3,200	8,800	8,600	7,600	6,200	4	8
	9/16	5,600	4,000	11,200	10,800	9,600	8,000	4 1/2	9
	5/8	6,800	5,000	13,600	13,200	11,800	9,600	5	10
	3/4	9,800	7,200	19,600	19,000	17,000	13,800	6	12
	7/8	13,200	9,600	26,000	26,000	22,000	18,600	7	14
	1	17,000	12,600	34,000	32,000	30,000	24,000	8	16
	1 1/8	20,000	15,800	40,000	38,000	34,000	28,000	9	18
6 x 37 IWRC	1 1/4	26,000	19,400	52,000	50,000	46,000	36,000	10	20
	1 3/8	30,000	24,000	60,000	58,000	52,000	42,000	11	22
	1 1/2	36,000	28,000	72,000	70,000	62,000	50,000	12	24
	1 5/8	42,000	32,000	84,000	82,000	72,000	60,000	13	26
	1 3/4	50,000	38,000	100,000	96,000	86,000	70,000	14	28
	2	64,000	48,000	128,000	124,000	110,000	90,000	16	32
	2 1/4	78,000	60,000	156,000	150,000	136,000	110,000	18	36
	2 1/2	94,000	74,000	188,000	182,000	162,000	132,000	20	40

*RATED CAPACITIES FOR UNPROTECTED EYES APPLY ONLY WHEN ATTACHMENT IS MADE OVER AN OBJECT NARROWER THAN THE NATURAL WIDTH OF THE EYE AND APPLY FOR BASKET HITCHES ONLY WHEN THE D/d RATIO IS 25 OR GREATER, WHERE D=DIAMETER OF CURVATURE AROUND WHICH THE BODY OF THE SLING IS BENT, AND d=NOMINAL DIAMETER OF THE ROPE.
**SEE CHOKER HITCH RATED CAPACITY ADJUSTMENT CHART.

111T01.EPS

3.2.1 Wire Rope Slings

Wire rope slings should be destroyed if you discover any of the following conditions:

- *Broken wires* – Destroy a wire rope sling if you find ten randomly distributed broken wires in one rope lay, five broken wires in one strand in one rope lay, or any broken wire at any fitting.
- *Diameter reduction* – Some wear is normal; however, if the wear exceeds one-third of the original rope diameter, the sling should be destroyed. This type of wear is often caused by excessive abrasion of the outside wires, loss of core support, internal or external corrosion, inner wire failure, loosening of the rope lay, or stretching due to overloading.
- *Heat damage* – This is usually indicated by the discoloration and pitting associated with high-temperature sources.
- *Corrosion* – Any noticeable rusting, pitting, or discoloration indicates corrosion in the rope or at the end fittings. Corrosion around end fittings is typically associated with broken wires near the end fittings. Corrosion can be difficult to detect when it develops internally in areas not visible from outside inspection.

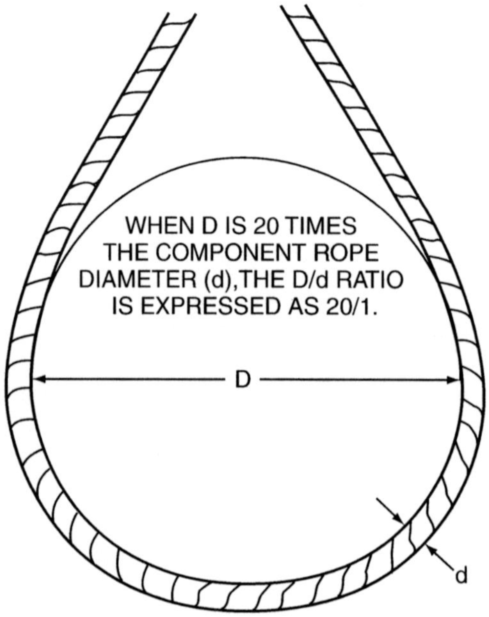

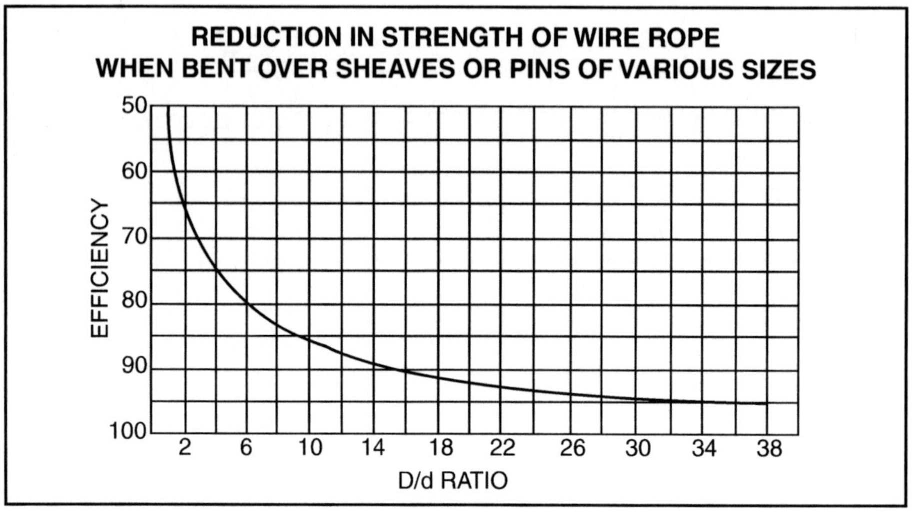

111F18.EPS

Figure 18 ◆ Load and hook diameter and D/d ratio chart for wire rope.

- *Damaged end fittings* – This type of damage is easily detected and includes cracks, gouges, nicks, severe corrosion, and evidence that the end fitting is creasing into the rope.
- *Rope distortion* – This includes **kinking**, crushing, and **bird caging**.
- *Core protrusion* – This damage involves the core sticking out from between the strands or the strands separating, exposing the core.
- *Unlaying of a splice* – This is the unraveling of a splice.

3.2.2 Synthetic Web and Round Slings

When inspecting synthetic and round slings, look for the following conditions. If any of these appear, discard the slings immediately:

- Acid or caustic burns
- Melting or charring of any part of the sling
- Holes, cuts, tears, or snags
- Broken or worn stitching in loadbearing splices
- Excessive abrasive wear (to the point where the colored yarns are showing)
- Knots in any part of the sling
- Excessive pitting or corrosion, or cracked, distorted, or broken fittings
- Other visible damage that causes doubt as to the strength of the sling
- Missing or illegible tag

3.2.3 Metal Mesh Slings

Although metal mesh slings withstand abrasive or hot loads, they are still vulnerable to damage. Metal mesh slings should be discarded if any of the following conditions appear during inspection:

- A broken welded or brazed joint along the sling edge
- A broken wire in any part of the mesh
- A reduction of wire diameter of 25 percent due to abrasion or 15 percent due to corrosion
- Lack of flexibility due to distortion of the mesh
- Distortion of the choker fitting so that the depth of the slot is increased by 10 percent
- Distortion of either end fitting so that the width of the eye opening is decreased by more than 10 percent
- A 15 percent reduction in the original cross-sectional area of metal at any point around

3.3.0 Chain Slings

Some jobs are better suited for chain slings than for wire rope or web slings. Use of chain slings is recommended when lifting rough castings that would quickly destroy wire slings by bending the wires over rough edges. They are also used in high-heat applications or where wire chokers are not suitable, and for dredging and other marine work because they withstand abrasion and corrosion better than cable. Information pertaining to sling angles and sling capacities described earlier for rope and web slings also apply to chain slings. *Figure 19* shows some common configurations of chain slings and hooks.

A chain link consists of two sides. The failure of either side would cause the link to open and drop the load. Wire rope is frequently composed of as many as 114 individual wires, all of which must fail before it breaks. Chains have less reserve strength and should be more carefully inspected. They must never be used for routine lifting operations.

Chains will stretch under excessive loading. This causes elongating and narrowing of the links until they bind on each other, giving visible warning. If overloading is severe, the chain will fail with less warning than a wire rope. If a weld should break, there is little, if any, warning.

WARNING!

Chains must be carefully inspected for weak or damaged links. The failure of one link can cause the entire chain to fail.

Typically, iron-hoisting chains should be **annealed** every two years to relieve work hardening. Chains used as slings should be annealed every year. After being annealed six times, the chain should be destroyed. Steel chains should not be heat-treated after leaving the factory.

Note that only Grade 8 or higher chain is permitted for overhead lifting.

3.3.1 Chain Sling Storage

To store chains properly, hang them inside a building or vehicle on racks to reduce deterioration due to rust or corrosion from the weather. Never let chain slings lie on the ground in areas where heavy machinery can run over them.

Be aware that some manufacturers suggest lubrication of alloy chains while in use; however, slippery chains increase handling hazards. Chains coated with oil or grease accumulate dirt and grit which may cause abrasive wear. For chains to be stored in exposed areas, coat them with a film of oil or grease for rust and corrosion protection.

3.3.2 Chain Sling Care and Inspection

Chain slings should be visually inspected before every lift. Annual inspections are also required and records of inspections must be maintained. Discard chain slings if any of the following conditions are found during inspection:

- *Wear* – Any portion of the chain worn by 15 percent or more should be removed from service immediately. Wear will usually occur at the points where chain links bear on each other or on the outside of the link barrels.

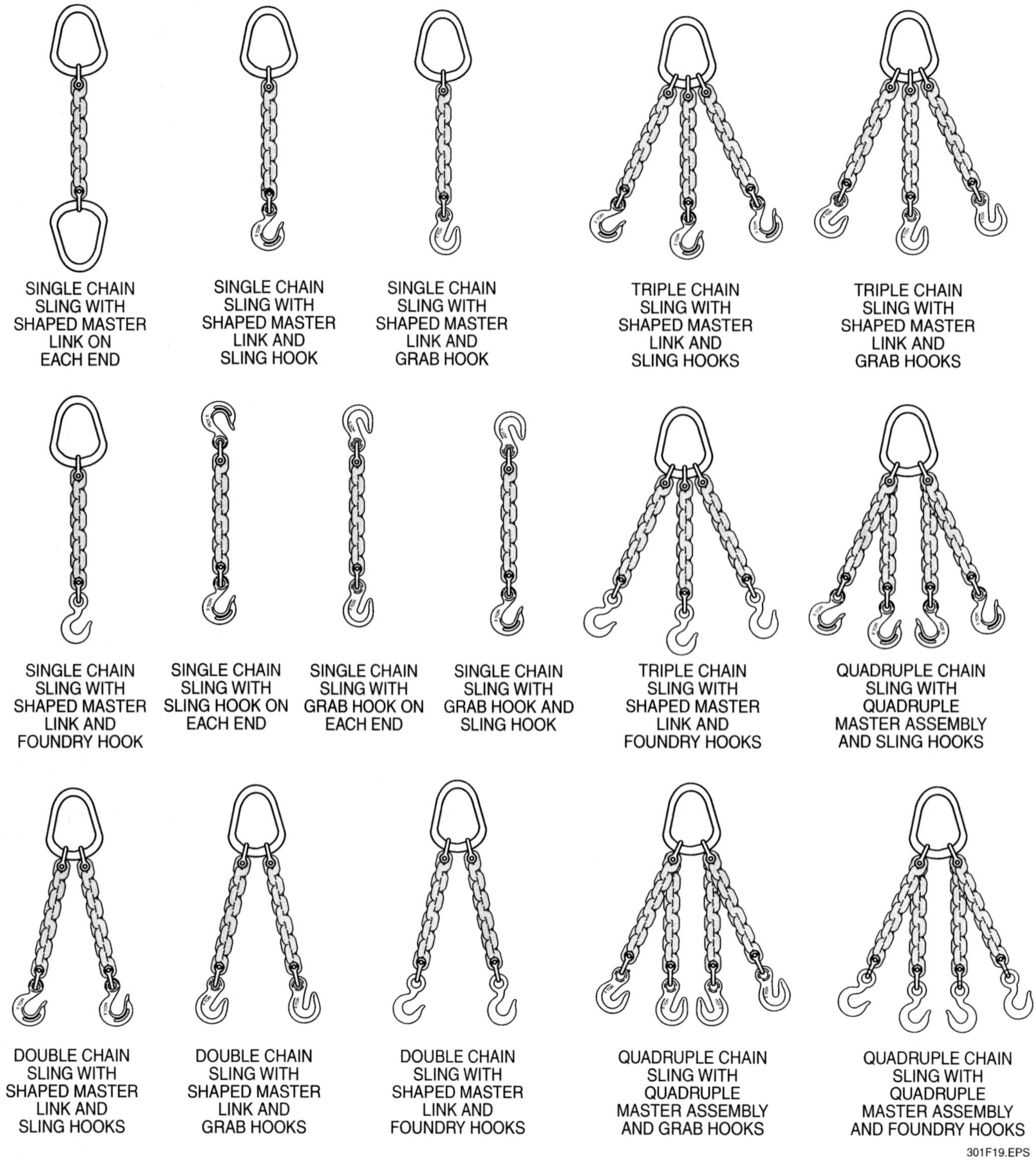

Figure 19 ◆ Common chain slings and hooks.

- *Stretch* – Compare the chain with its rated length or with a new length of the same type chain. Any length increase means wear or stretch has occurred. If the length has increased 3 percent, further careful inspection is needed. If it is stretched more than 5 percent, it should be removed from service immediately. Significantly stretched links have an hourglass shape, and links tend to bind on each other. Be sure to check for localized stretching since a chain link can be overlooked easily.
- *Link condition* – Look for twisted, bent, cut, gouged, or nicked links.
- *Cracks* – Discard the chain if any cracks are found in any part.
- *Link welds* – Lifted fins at the weld edges signify overloading.
- *End fittings* – Check for signs of stretching, wear, twisting, bending, opening up, and corrosion.
- *Capacity tag* – Check for missing or illegible tags.

4.0.0 ◆ TAG LINES

Tag lines typically are natural fiber or synthetic rope lines used to control the swinging of the load in hoisting activities (*Figure 20*). Improper use of tag lines can turn the simplest hoisting operation into a dangerous situation. Tag lines should be used to control swinging of the load when a crane is traveling. Tag lines should also be used when the crane is rotated if rotation of the load is hazardous. A non-conductive tag line is required during operation of a mobile crane within the vicinity of power lines.

111F20.EPS

Figure 20 ◆ Use of a tag line to control swinging of a suspended load.

When selecting a tag line, take certain factors into consideration. Natural fiber rope is inexpensive when compared to synthetic rope and is the most common type of rope used for tag lines. However, it is notably weaker than synthetic fiber (nylon, polyester, polypropylene, or polyethylene) ropes of the same size, and it is subject to ultraviolet deterioration and damage from heat and chemicals. Manila rope can also become a conductor of electricity when wet. Synthetic ropes are light and strong for their size, and most are resistant to chemicals. When dry, they are poor conductors of electricity, although some synthetic ropes will readily absorb water and conduct electricity. Dry polypropylene line is preferred for use as a tag line.

The diameter of a tag line should be large enough so that it can be gripped well even when wearing gloves. Rope with a diameter of ½ inch is common, but ¾- and 1-inch-diameter rope are sometimes used on heavy loads or where the tag line must be extremely long. Tag lines should be of sufficient length to allow control of the load from its original lift location until it is safely placed or control is taken over by co-workers. Special consideration should be given to situations where a long tag line would interfere with safe handling of loads, such as steel erection or catalyst pours.

Tag lines should be attached to loads at a location that gives the best mechanical advantage in controlling the load. Long loads should have tag lines attached as close to the ends as possible. The tag line should be located in a place that allows personnel to remove it easily after the load is placed. Knots used to attach tag lines should be tied properly to prevent slipping or accidental loosening, but they need to be easily untied after the load is placed. Some recommended knots are the clove hitch with overhand safety and bowline. Riggers use different knots for different purposes. *Figure 21* shows several knots commonly used in rigging. As much as possible, tag lines should be of one continuous length, free of knots and splices, and seized on both ends. If joining two tag lines together is necessary, it should be done by splicing the lines. Knots tied in the middle or with free ends of tag lines can create difficulties.

To properly handle a tag line, determine the mechanical advantage intended by the tag line and stand away from the load in order to have a clear area. When possible, stand where you can see the crane being used for lifting. Keep yourself and the tag line in view of the crane operator. Stay alert. Do not become complacent during the lift. Be aware of the location of any excess rope, and do not allow it to become fouled or entangled on anything.

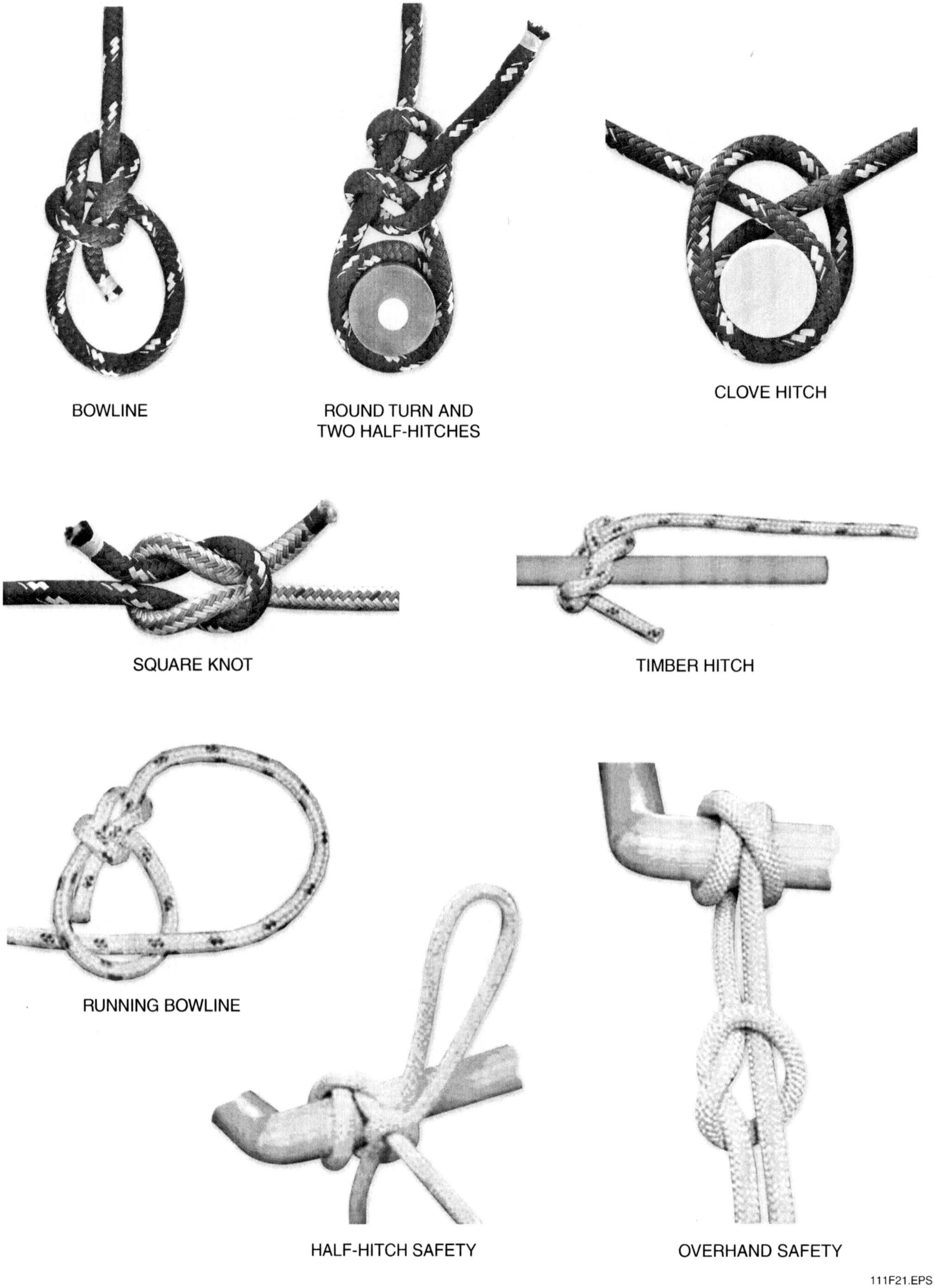

Figure 21 ◆ Common rigging knots.

Large loads often require the use of more than one tag line. In such cases, tag line personnel must work as a team and coordinate their actions.

5.0.0 ◆ BLOCK AND TACKLE

The block and tackle is the most basic lifting device. It is used to lift or pull light loads. While you should know what a block and tackle is, and it may be necessary for you to use one, it is very old technology and not often used now. A block consists of one or more sheaves or pulleys fitted into a wood or metal frame with a hook attached to the top. The tackle is the line that runs through the block and is used for lifting and pulling. Some block and tackle rigs have a brake that holds the load once it is lifted; others do not. The types that do not have a brake require continuous pull on the **hauling line**, or the hauling line must be tied off to hold the load. There are two types of block and tackle rigs: simple and compound.

A simple block and tackle consists of one sheave and a single line. It is used to lift or pull very light loads. The line hook is attached to the load, and the line is pulled by hand to lift the load. The load capacity of this type of block and tackle is equal to the capacity of the load line. The block must be attached to a building structure or other support by a method that provides adequate load capacity to support the load and the tackle. *Figure 22* shows a simple block and tackle.

A compound block and tackle uses more than one block. It has an upper, **fixed block** that is attached to the building structure or other support and a lower, movable block that is attached to the load. Each block may have one or more sheaves. The more sheaves the blocks have, the more **parts of line** the block and tackle has, and the higher the lifting capacity. The compound block and tackle is capable of handling heavier loads than the simple block and tackle and requires less effort to raise the load. *Figure 23* shows a compound block and tackle.

6.0.0 ◆ CHAIN HOISTS

A chain hoist, also called a chain fall, is another very useful and commonly used lifting device. It is used for much the same purposes as the block and tackle. Chain hoists should be used for straight, vertical lifts only. They are designed for straight lifts and may be damaged if used for angled lifts. Always use chokers with chain hoists. The load chain of a chain hoist should never be wrapped around an object and used as a choker.

The load capacity of any chain hoist should be marked on the chain hoist. The capacity of a chain hoist must never be exceeded. Chain hoists are standard equipment in most shops and rigging departments because they are dependable, portable, and easy to use. Some common types of chain hoists are spur-geared and electric.

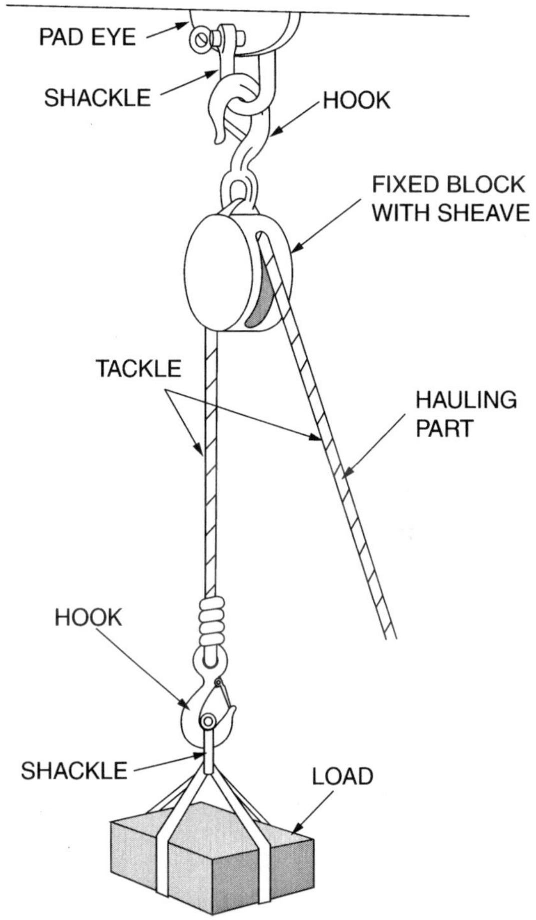

Figure 22 ◆ Simple block and tackle.

Figure 23 ◆ Compound block and tackle.

6.1.0 Spur-Geared Chain Hoists

The spur-geared chain hoist (*Figure 24*) has two chains. An endless chain, the hand chain, drives a single pocketed sheave, which in turn drives a gear-reduction unit. The load chain must have a mechanical brake. It is fitted to the gear-reduction unit and has a hook that attaches to the load. Roller chain is sometimes used as the load chain instead of link-type chain. The spur-geared chain hoist is the most efficient type of manual chain hoist and is most commonly used for heavier loads.

6.2.0 Electric Chain Hoists

Electric chain hoists (*Figure 25*) work much like manual chain hoists, except that they have an electric motor instead of a pull chain to raise and lower the load. Electric chain hoists are the fastest and most efficient type of chain hoist. Common electric chain hoists are available in capacities of ¼ to 5 tons, but special-application electric chain hoists are available in much higher capacities. They are controlled from a handheld push button control that is suspended on a wire from the chain hoist. Special care should be taken when raising a load with an electric chain hoist, since it is hard to tell how much force is being exerted by the electric motor. If the load gets caught on something, the chain hoist could be overloaded and damaged. The electrically driven pneumatic hoist is operated by a rotary air pump, and the motor of the hoist will stall before the chain or hoist mechanism can break.

6.3.0 Care of Chain Hoists

Like other lifting devices, chain hoists must be carefully inspected before each use. In particular, the chain must be checked to ensure that it has no defects. Follow these steps to select, inspect, use, and maintain a chain hoist:

Step 1 Select a chain hoist of adequate capacity to handle the load.

Step 2 Inspect the load chain and hook to ensure that they are not excessively worn, bent, or deformed in any way.

Step 3 Inspect the sheaves to ensure that they are not bent or excessively worn.

Step 4 Ensure that the chain hoist has proper lubrication.

Step 5 Hang the chain hoist on a suitable support using the proper rigging. If it is an electric chain hoist, plug the chain hoist motor into an electrical outlet.

Figure 24 ◆ Spur-geared chain hoists.

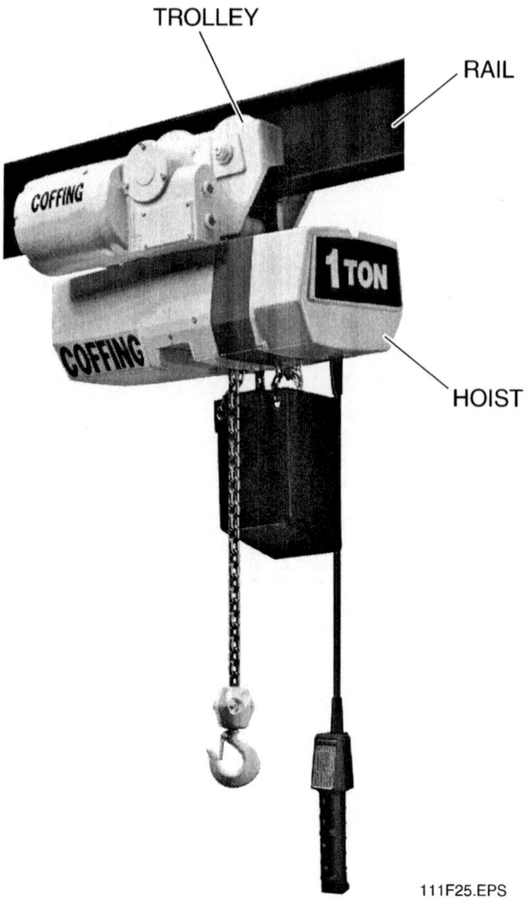

Figure 25 ◆ Electric chain hoist.

Step 6 Lower the load hook, and connect it to the load using the proper rigging.

Step 7 Raise the load just barely off the ground, and stop to check whether the brake will hold the load.

Step 8 Raise and place the load.

> **NOTE**
> To raise the load using a manual chain hoist, pull the hand chain. To raise the load using an electric chain hoist, press and hold the UP push button on the handheld control.

Step 9 Disconnect the load hook from the load.

Step 10 Remove the chain hoist from its support.

Step 11 Coil the chain so that it will not get tangled.

Step 12 Store the chain hoist in its proper place.

7.0.0 ◆ RATCHET-LEVER HOISTS AND COME-ALONGS

Ratchet-lever hoists and come-alongs (*Figure 26*) are used for short pulls on heavy loads. The term come-along is widely used to identify both types, but they are not the same thing. A come-along uses a cable, whereas a ratchet-lever hoist uses a chain. Ratchet-lever hoists can be used for vertical lifts, but cable-type come-alongs can only be used for horizontal pulls. Come-alongs and ratchet-lever hoists are portable, easy to store, and easy to use. They are available in capacities ranging from ½ to 6 tons.

A ratchet-lever hoist consists of the following parts:

- *Suspension hook* – A steel hook used to hang the hoist.
- *Load hook* – A steel hook that attaches to the load.
- *Load chain* – The chain that attaches to the load hook and lifts the load.
- *Ratchet handle* – A handle that operates the ratchet that takes up the chain and lifts or lowers the load.
- *Fast wind handle* – A handle that takes up or lets out the chain without using the ratchet handle. It cannot be used to raise or lower the load.
- *Ratchet release* – A device that releases the ratchet so the chain can be pulled out. It also switches the device to the up or down position.

Follow these steps to select, inspect, use, and maintain a ratchet-lever hoist:

Step 1 Select a device of adequate capacity to handle the load.

Step 2 Inspect the chain and hooks to ensure that they are not excessively worn, bent, or deformed in any way.

Step 3 Inspect the device to ensure that it is not damaged in any way.

Step 4 Hang the device on a suitable support.

Step 5 Turn the ratchet release to the middle position.

Step 6 Pull the chain out enough to attach it to the load.

Step 7 Attach the load hook to the load, using the proper rigging.

Step 8 Turn the fast wind handle to take the slack out of the chain.

Step 9 Turn the ratchet release to the up position.

Step 10 Pump the ratchet handle to raise and position the load. Stop and check the brake as soon as the load clears the ground, then continue the lift.

Step 11 Turn the ratchet release to the down position.

Step 12 Pump the ratchet handle to lower the load into position and until there is slack in the chain.

Step 13 Disconnect the load hook from the load.

Step 14 Store the device in its proper place.

Figure 26 ◆ Come-along and ratchet-lever hoist.

8.0.0 ◆ JACKS

A jack is a device used to raise or lower equipment. Jacks are also used to move heavy loads a short distance, with good control over the movement. The following are the three basic types of jacks:

- Ratchet
- Screw
- Hydraulic

111F27.EPS

Figure 27 ◆ Ratchet jack.

8.1.0 Ratchet Jacks

The ratchet jack (*Figure 27*), also called a railroad jack, is used only to raise loads under 25 tons. It uses the lever-and-fulcrum principle. The downward stroke of the lever raises the rack bar one notch at a time. A latching mechanism, called a pawl, automatically springs into position, holding the load and releasing the lever for the next lifting stroke. Ratchet jacks permit safe lifting, lowering, and leveling.

8.2.0 Screw Jacks

Screw jacks (*Figure 28*) are used to lift heavier loads. There are two general types of screw jacks: upright and inverted. The screw jack uses the screw-and-nut principle. For lighter loads, a simple lever will apply enough power to turn the screw. For heavier loads, gear-reduction units and ratchet devices are used to increase the operator's strength. In the heaviest jacks, the screw jack is operated by an air motor for faster lifting and lowering.

8.3.0 Hydraulic Jacks

Hydraulic jacks (*Figure 29*) are the most useful general purpose jacks. They are operated by the pressure of an enclosed liquid. There are two types of hydraulic jacks. One type has the pump built into the jack. The other uses a separate pump. Both jacks can lift heavy loads with little effort. The hydraulic jack with a separate pump has the following advantages:

- The craftworker can operate the pump at a safe distance from the load.

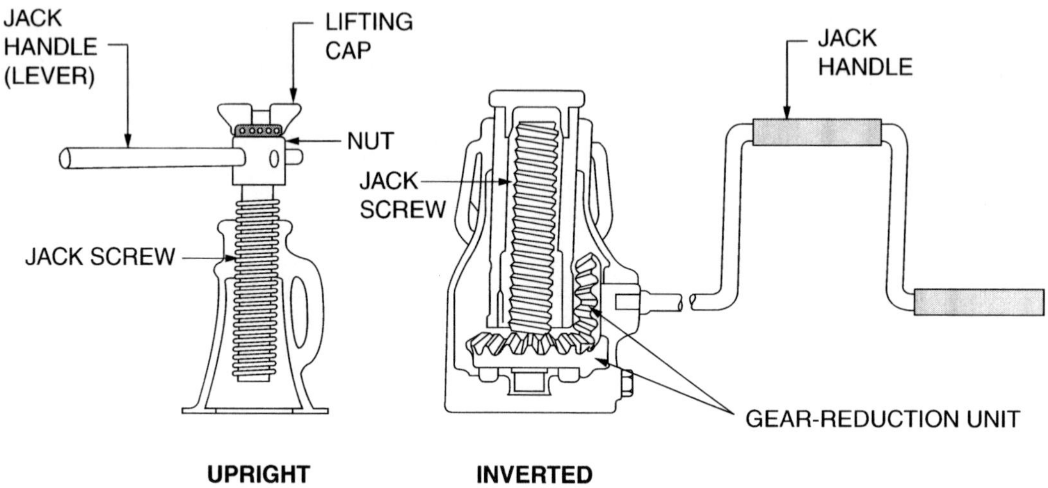

111F28.EPS

Figure 28 ◆ Upright and inverted screw jacks.

- The ram provides the greatest amount of travel for a given closed height.
- Pumps can be powered by hand, electric motor, air motor, or gasoline engine.

The hydraulic jack is not as fast as the ratchet jack, but it is excellent for lifting heavy loads.

8.4.0 Inspecting and Using Jacks

In time, a jack can become damaged or weakened and can fail under a load. To avoid such failures, all jacks should be carefully inspected before each use. Apply the following guidelines when using jacks:

- Inspect jacks before using them to ensure that they are not damaged in any way.
- Use wood softeners when jacking against metal.
- Use the proper jack handle, and remove it from the jack when it is not in use.
- Never step on a jack handle to create additional force.
- Use blocking or cribbing under the load when jacking. Never leave a jack under a load without having the load blocked up.
- Position jacks properly, and raise the load evenly to prevent the load from shifting or falling.
- Never place jacks directly on the ground when lifting. Use a solid footing.
- Position jacks so the direction of force is perpendicular to the base and the surface of the load.
- Never exceed the load capacity of the jack.
- Do not use extensions to the jack handles.
- Brace loads to prevent the jacks from tipping.
- When you first lift the load, check to see if the jack holds.
- Lash or block jacks when using them in a horizontal position to move an object.
- Never jack against rollers.
- Match ratchet jacks for uniform lifting.
- Keep fingers away from ratchet parts when raising or lowering an object.
- Thoroughly clean all hydraulic hose connectors before connecting them.

9.0.0 ◆ TUGGERS

Tuggers are either electric or pneumatic winches. They can be used for lifting and lowering or for pulling. The tugger must be securely anchored to the floor or a steel beam by a method that will support any load that may be put on the tugger. When securing a tugger to a steel beam, use a softener on the choker. Tuggers can be located in almost any convenient location in a building, and the load line can be run through a series of blocks to the area where the lift is to be made. The blocks and all hardware must be of adequate load capacity to handle the load being lifted. The controls may be located on the tugger, or the tugger may have remote controls. Many times, the tugger is in a location where the operator cannot see the load being lifted. In this case, a flagman is needed to direct the operator, using the same hand signals that are used when flagging a crane. The tugger has a band-type brake to hold the load. *Figure 30* shows a common tugger.

100 TON CAPACITY

2 TO 10 TON CAPACITY

111F29.EPS

Figure 29 ◆ Hydraulic jacks.

Follow these steps to select, inspect, use, and maintain a tugger:

Step 1 Ensure that the tugger is of adequate capacity to handle the load. Check the card that is attached to ensure preinspection has been completed.

Step 2 Ensure that an adequate air supply is connected to the tugger and is in service.

Step 3 Push the control handle toward the down position and reel off enough cable to attach to the load.

Step 4 Attach the cable to the load using the proper rigging.

> **NOTE**
> Many times, the load is far removed from the tugger. In this case, the tugger operator will need the help of a rigger to attach the cable to the load and direct the movement of the object.

Step 5 Ensure that the cable is running properly through all blocks.

Step 6 Push the control handle toward the up position to raise the load.

111F30.EPS

Figure 30 ◆ Tugger.

> **WARNING!**
> Always stand to the side of the tugger behind the safety shield while operating the tugger.

> **NOTE**
> Tuggers are variable-speed. Pushing the control handle slightly toward the up position raises the load slowly. Start slowly; then, push the handle further toward the up position to speed up. Slow down as the load approaches its destination.

Step 7 Push the control handle toward the down position to position the load at its destination.

Step 8 Push the control handle toward the down position to slacken the cable.

Step 9 Disconnect the load hook from the load.

Step 10 Operate the control handle to position the cable in its normal stored position.

10.0.0 ◆ CRANES

As a maintenance craftworker in plants, you will probably have available a variety of lifting devices already in place, such as overhead beam cranes, gantry cranes, and elevator lifts. The requirements of supply for production will have placed such equipment at your service. These cranes include freestanding jib and gantry cranes *(Figure 31)*, usually in the one- or two-ton range, overhead beam cranes *(Figure 32)*, which are most commonly rated in the five- to ten-ton range, and work station cranes *(Figure 33)*, which are generally in the one-ton range. Overhead cranes are operated with either wireless or corded control boxes *(Figure 34)* with buttons for up and down, side to side, and beam movement up and down the plant. Jib and gantry cranes frequently only lift up and down, but may have an in-and-out control to move objects closer to or away from the pivot; workstation cranes vary, much like the jibs. You may or may not ever have need for a mobile crane, but must be fully familiar with them. When you do need to work with a mobile crane operator, you will need to know how to properly communicate and rig for them.

111F31.EPS

Figure 31 ◆ Overhead cranes.

111F33.EPS

Figure 33 ◆ Work station cranes.

111F32.EPS

Figure 32 ◆ Freestanding cranes.

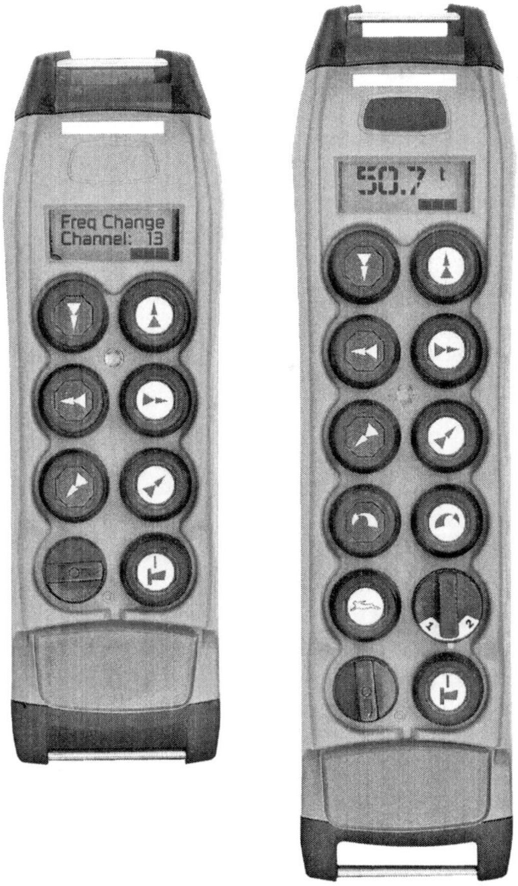

111F34.EPS

Figure 34 ◆ Control box.

10.1.0 Verbal Modes of Communication

Verbal modes of communication vary depending on the requirements of the situation. One of the most common modes of verbal communication used is a portable radio (walkie-talkie). Compact, low-power, inexpensive units enable the crane operator and signal person to communicate verbally. These units are rugged and dependable and are widely used on construction sites and in industrial plants.

There are some disadvantages, however, to using low-power and inexpensive equipment in an industrial setting. One disadvantage is interference. With low-power, inexpensive units, the frequency used to carry the signal may have many other users. This crowding of the frequency could disrupt the signal from a more powerful unit. Another disadvantage is high background noise. In attempting to send a signal in a high-noise area, the person sending the signal may transmit unintended noise, resulting in a garbled, unintelligible signal for the receiver. On the receiving end, the individual may not be able to hear the transmission due to a high level of background noise in the cab of the crane.

There are several solutions to the problems associated with radio use. To overcome the shortcomings associated with low-power units, more expensive units with the ability to program specific frequencies and transmit at a higher power level may be needed. Some of these more expensive units may require licensing. To overcome the background noise problem, the use of an ear-mounted, noise-canceling microphone/headphone combination may be required (*Figure 35*).

Another solution is the use of an optional throat microphone. This device feeds the transmitted sound directly to the ear and picks up the voice communication from the jawbone at the ear junction. This prevents noise from entering the microphone and blocks out any background noise when listening. To avoid missed communication, the signal person's radio is usually locked in transmit so that the crane operator can tell if the unit is not transmitting. In any event, a feedback method should be established between the signal person

VOICE-OPERATED (VOX) OR PUSH-TO-TALK (PTT) RADIO SYSTEM WITH THROAT MICROPHONE

DUPLEX HARDWIRED SYSTEM WITH IN-LINE AMPLIFIER AND EXTENSION CORDS FOR SIMULTANEOUS TWO-WAY COMMUNICATION

111F35.EPS

Figure 35 ◆ Electronic communications systems.

and the crane operator so that the signal person knows the crane operator has received the signal.

Another mode of verbal communication is a hardwired system (*Figure 35*). These units overcome some of the disadvantages of radio use. When using this type of system, interference from another unit is unlikely because this system does not use a radio frequency to transmit information. Like a telephone system, occasional interference may be encountered if the wiring is not properly shielded from very strong radio transmissions. A hardwired unit is not very portable or practical when the crane is moved often. These units can also use an ear-mounted noise-canceling microphone/headphone combination to minimize the effects of background noise.

10.2.0 Nonverbal Modes of Communication

There is a wide range of nonverbal modes of communication. This is the most common type of communication used when performing crane operations. Several modes are available for use under this method. One mode is the use of signal flags, which may mean different colored flags or a specific positioning of the flags to communicate the desired message. Another mode is the use of sirens, buzzers, and whistles in which the number of repetitions and duration of the sound convey the message. The disadvantage of these two modes is that there is no established meaning to any of the distinct signals unless they are prearranged between the sender and receiver. Also note that when sirens, buzzers, and whistles are used, background noise levels can be a problem.

The most common mode of communication reference used during crane operations is the *ASME B30.5 Consensus Standard of Hand Signals* (see *Figures 36* through *52*). In accordance with *ASME B30.5*, the operator must use standard hand signals unless voice communication equipment is utilized. It also requires that the hand signal chart be posted conspicuously at the job site. The advantage to using these hand signals is that they are well established and published in an industry-wide standard. This means that these hand signals are recognized by the industry as the standard hand signals to be used on all job sites. This helps ensure that there is a common core of knowledge and a universal meaning to the signals when lifting operations are being conducted. Agreed-upon signals eliminate significant barriers to effective communication.

HOIST: With forearm vertical, forefinger pointing up, move hand in small horizontal circle.

EXPECTED MACHINE MOVEMENT: The load attached to the block or ball rises vertically, accelerating and decelerating smoothly.

111F36.EPS

Figure 36 ◆ Hoist.

LOWER: With arm extended downward, forefinger pointing down, move hand in small horizontal circle.

EXPECTED MACHINE MOVEMENT: The load block or ball smoothly lowers vertically.

111F37.EPS

Figure 37 ◆ Lower.

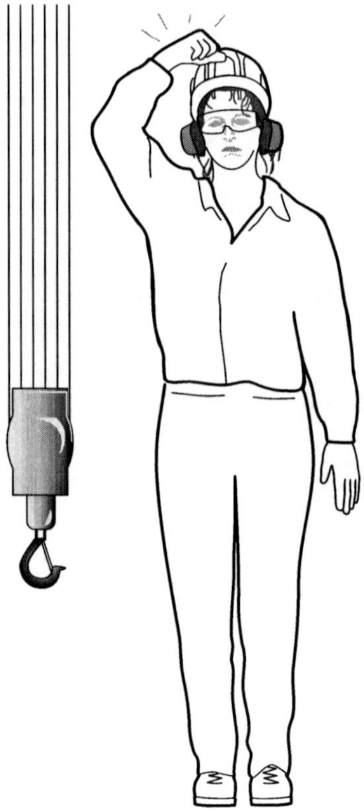

USE MAIN HOIST: Tap fist on head, then use regular hand signal.

EXPECTED MACHINE MOVEMENT: None. This signal is used only to inform the operator that the signal person has chosen the main hoist for the action to be performed as opposed to the auxiliary hoist.

111F38.EPS

Figure 38 ◆ Use main hoist.

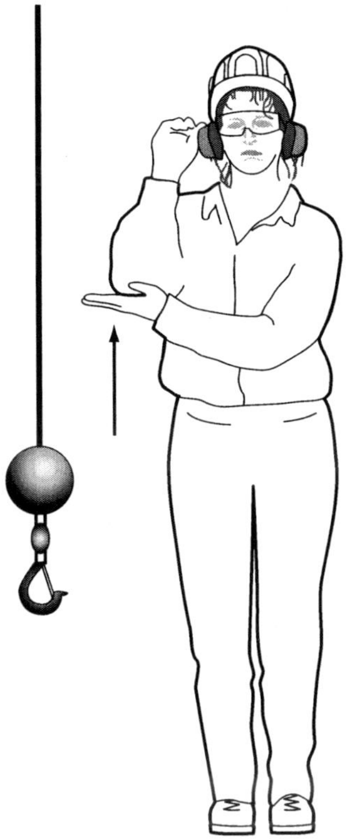

USE WHIP LINE (AUXILIARY HOIST): Tap elbow with open palm of one hand, then use regular hand signal.

EXPECTED MACHINE MOVEMENT: None. This signal is used only to inform the operator that the signal person has chosen the auxiliary hoist for the action to be performed as opposed to the main hoist.

111F39.EPS

Figure 39 ◆ Use whip line.

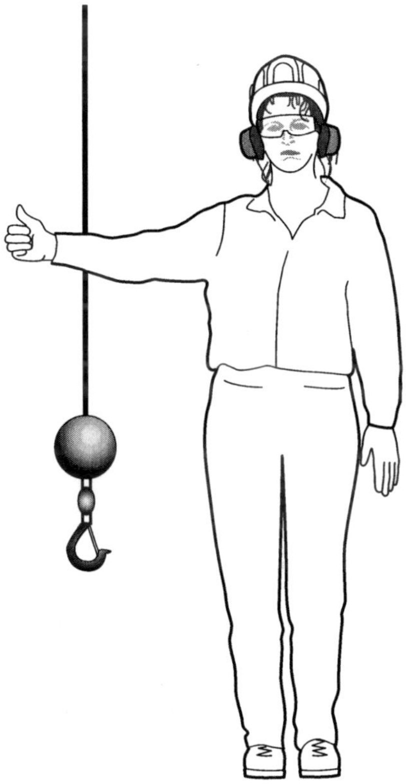

RAISE BOOM: Arm extended, fingers closed, thumb pointing upward.

EXPECTED MACHINE MOVEMENT: The boom rises, increasing the hook height and reducing the overall machine height clearance. The operating radius is slowly decreased, thus possibly increasing machine capacity and stability.

111F40.EPS

Figure 40 ◆ Raise the boom.

LOWER THE BOOM: Arm extended, fingers closed, thumb pointing downward.

EXPECTED MACHINE MOVEMENT: The boom will lower, decreasing the hook height and reducing the overall machine horizontal clearance. The operating radius is slowly increased, thus possibly decreasing machine capacity and stability.

111F41.EPS

Figure 41 ◆ Lower the boom.

MOVE SLOWLY: Use one hand to give any motion signal and place the other hand motionless over the hand giving the motion signal.

EXPECTED MACHINE MOVEMENT: Machine movement will vary depending on the signal being given.

111F42.EPS

Figure 42 ◆ Move slowly.

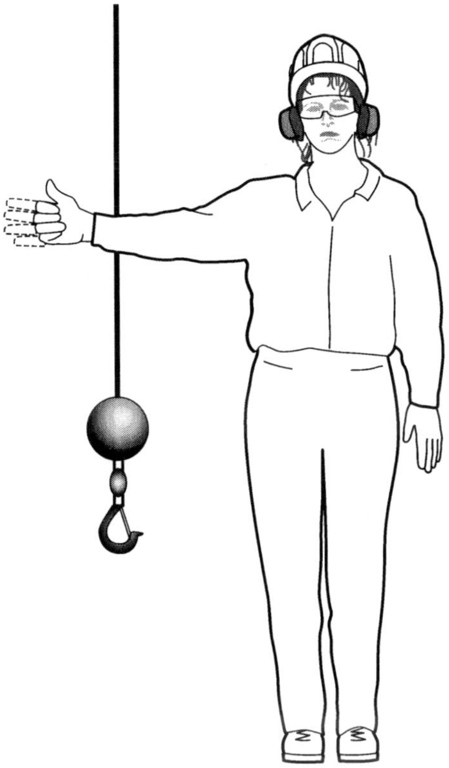

RAISE BOOM AND LOWER LOAD: Arm extended, fingers closed, thumb pointing up, flex fingers in and out as long as load movement is desired.

EXPECTED MACHINE MOVEMENT: The boom rises, reducing overall machine height clearance, as the load moves horizontally toward the crane. The operating radius is slowly decreased, thus possibly increasing machine capacity and stability.

111F43.EPS

Figure 43 ◆ Raise the boom and lower the load.

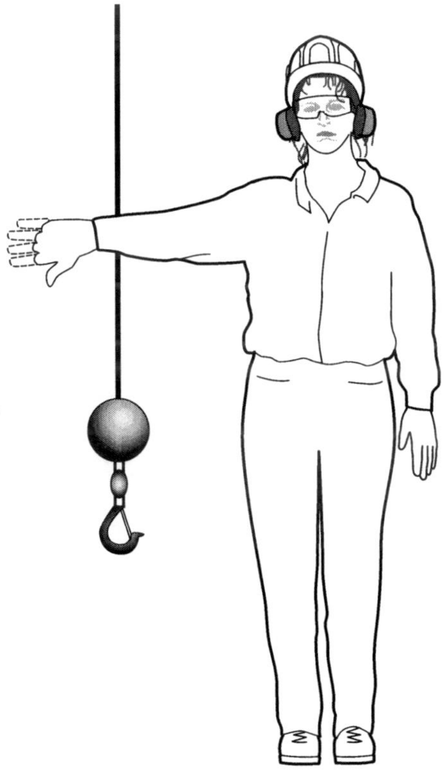

LOWER BOOM AND RAISE LOAD: Arm extended, fingers closed, thumb pointing down, flex fingers in and out as long as load movement is desired.

EXPECTED MACHINE MOVEMENT: The boom lowers, increasing overall machine height clearance, as the load moves horizontally away from the crane. The operating radius is slowly increased, thus possibly reducing both machine capacity and stability.

111F44.EPS

Figure 44 ◆ Lower the boom and raise the load.

SWING: Arm extended, point with fingers in direction of boom swing. (Swing left is shown as viewed by the operator.) Use appropriate arm for desired direction.

EXPECTED MACHINE MOVEMENT: The boom moves about the center of rotation with the load (block or ball) swinging in an arc, either toward the right or left, while remaining approximately equidistant to a level plane.

111F45.EPS

Figure 45 ◆ Swing.

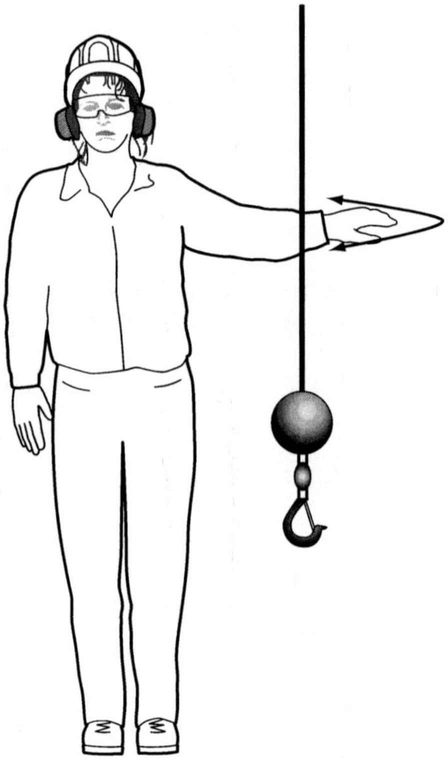

STOP: Arm extended, palm down, move arm back and forth horizontally.

EXPECTED MACHINE MOVEMENT: None. All movement of the machine ceases.

111F46.EPS

Figure 46 ◆ Stop.

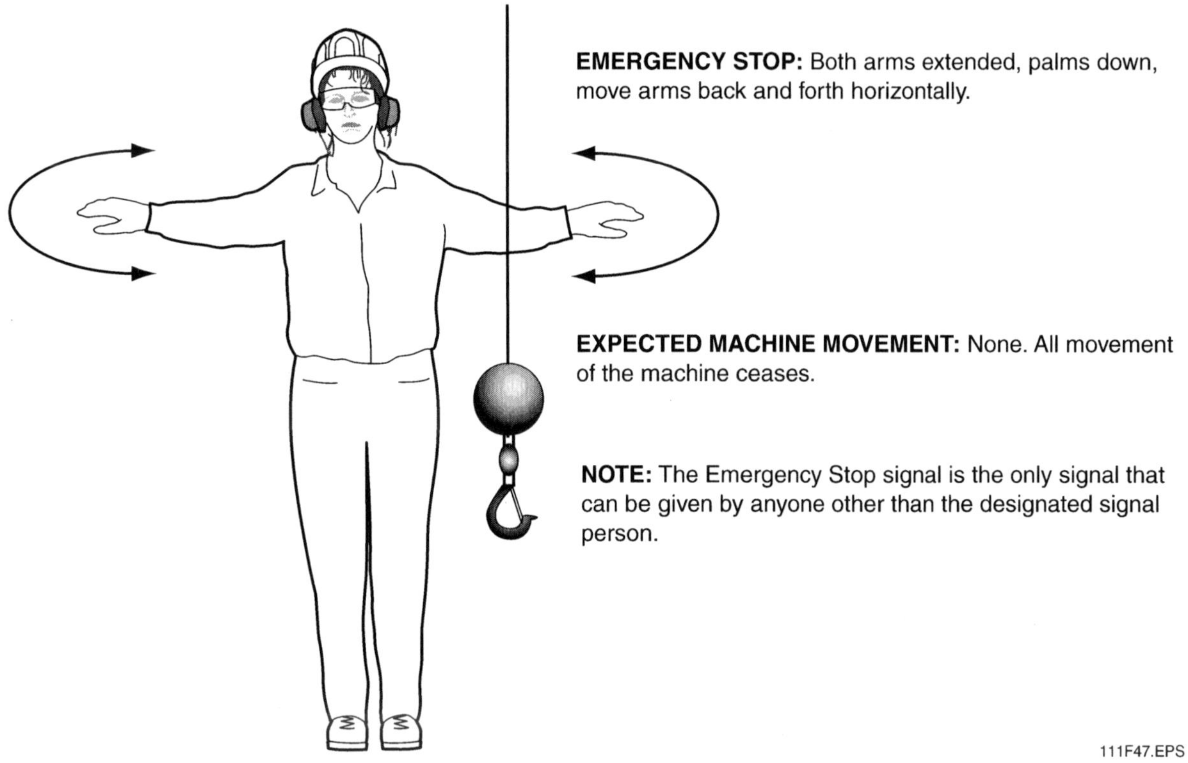

Figure 47 ◆ Emergency stop.

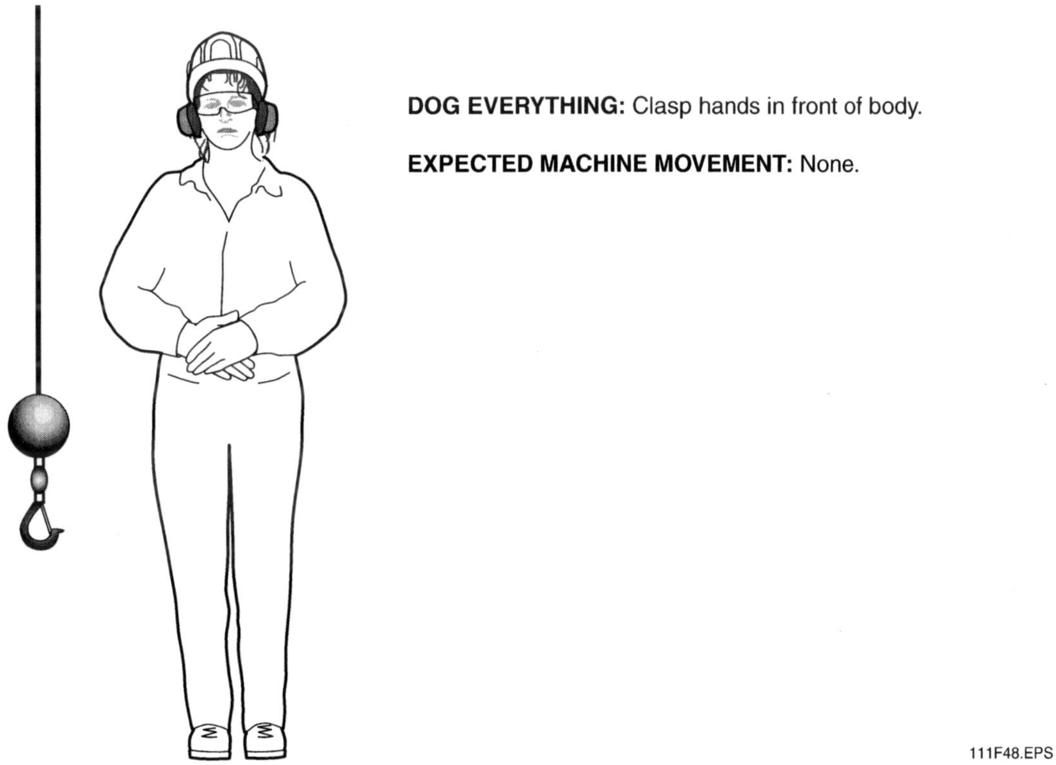

Figure 48 ◆ Dog everything.

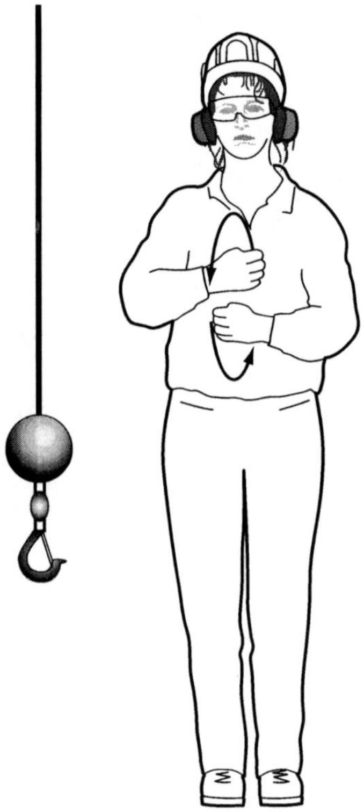

TRAVEL (BOTH TRACKS): Position both fists in front of body and move them in a circular motion, indicating the direction of travel (forward or backward).

EXPECTED MACHINE MOVEMENT: Machine travels in the direction chosen.

111F49.EPS

Figure 49 ◆ Travel (both tracks).

TRAVEL (ONE TRACK): Lock track on side of raised fist. Travel opposite track in direction indicated by circular motion of other fist, rotated vertically in front of body.

EXPECTED MACHINE MOVEMENT: Machine turns in the direction chosen.

111F50.EPS

Figure 50 ◆ Travel (one track).

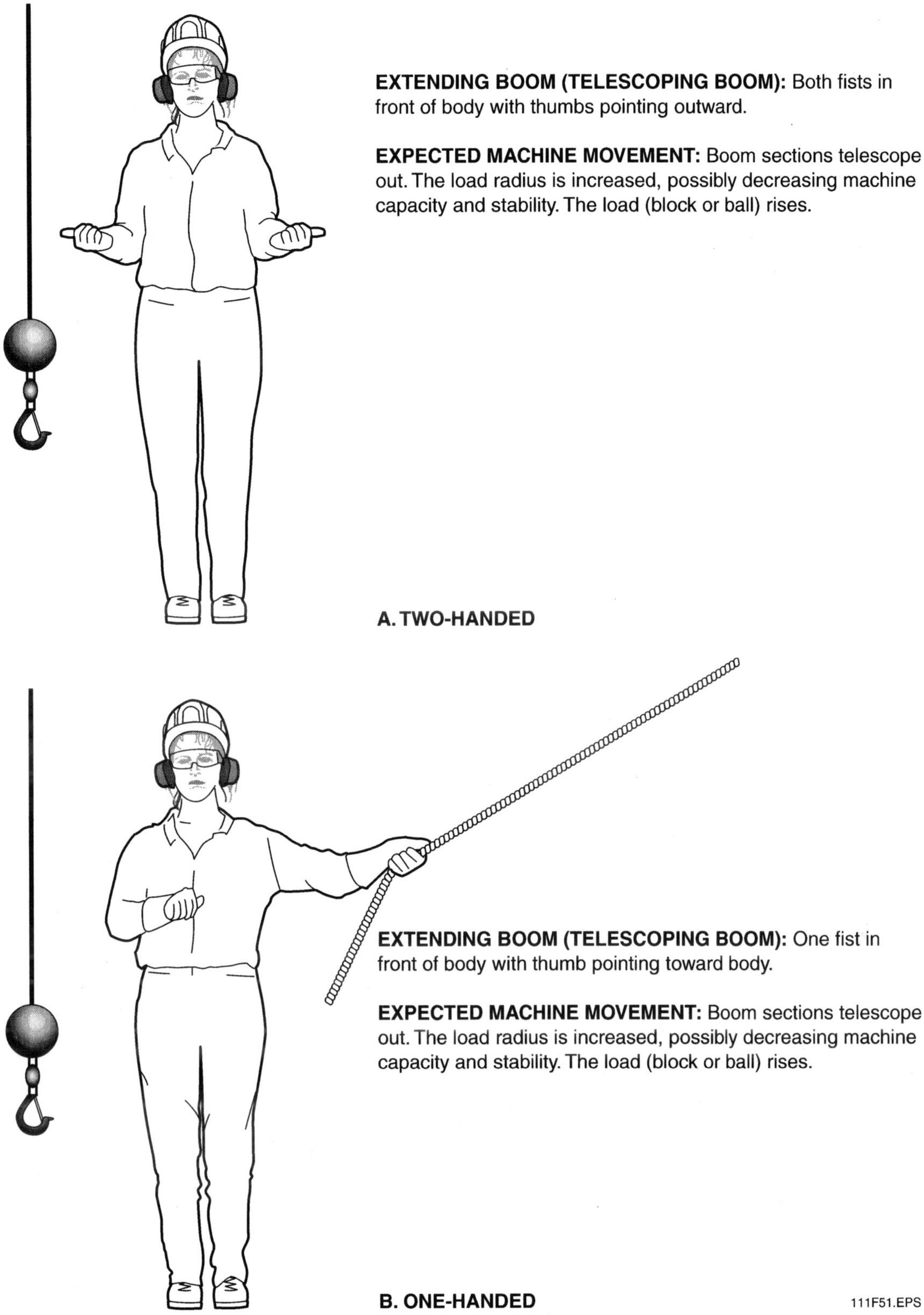

Figure 51 ◆ Extend boom.

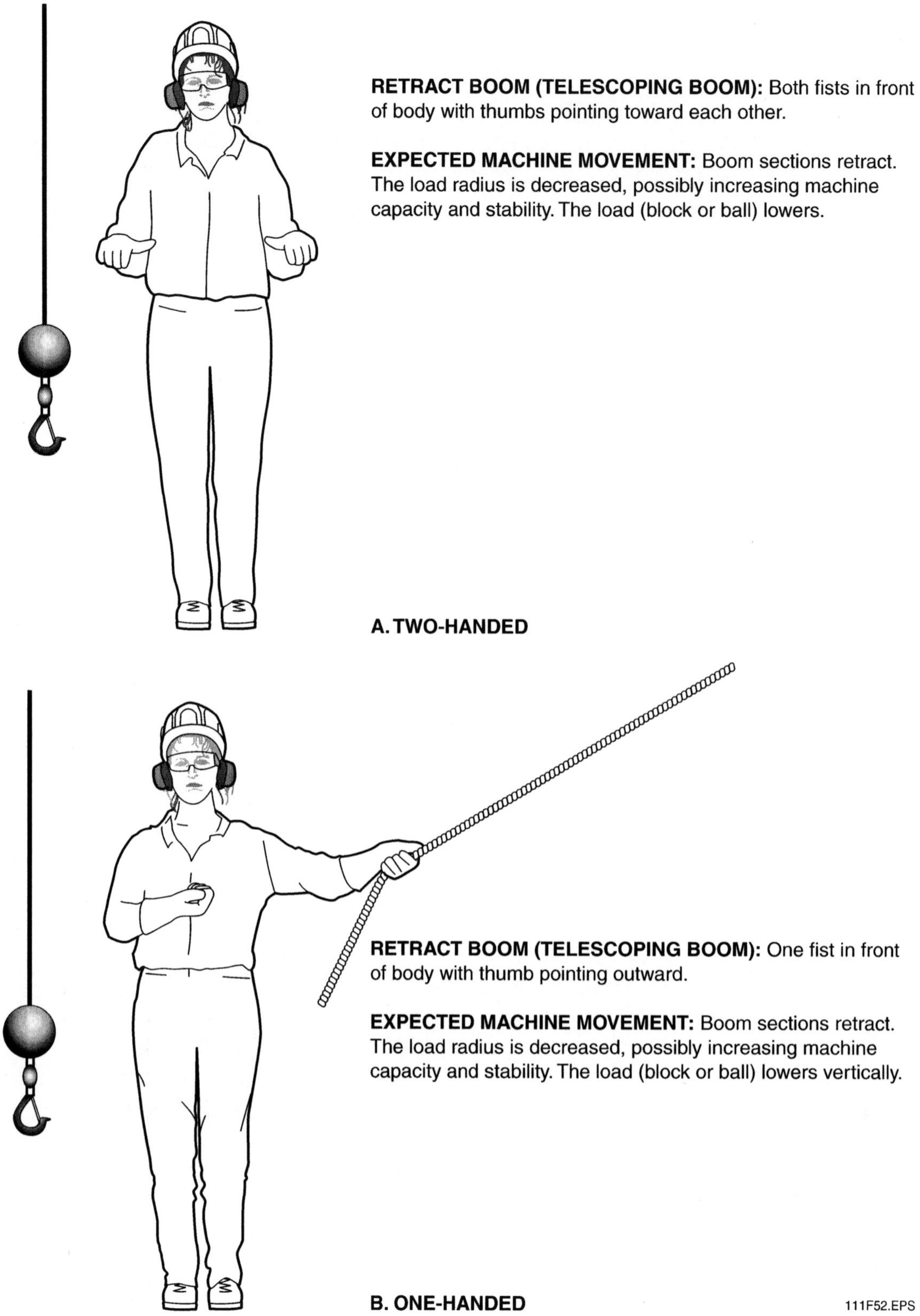

Figure 52 ◆ Retract boom.

Additions or modifications may be made for operations not covered by the illustrated hand signals, such as deployment of outriggers. The operator and signal person must agree upon these special signals before the crane is operated, and these signals should not be in conflict with any standard signal.

If it is apparent that the operator is not following the signal, immediately signal a stop. If you need to give instructions verbally to the operator instead of by hand signals, all crane motions must be stopped before doing so.

When a mobile crane is being moved without direction from the rigger, audible travel signals must be given using the crane's horn:

- *Stop* – One audible signal
- *Forward* – Two audible signals
- *Reverse* – Three audible signals

There are certain requirements that mandate the presence of a signal person. When the operator of the crane can't see the load, the landing area, or the path of motion, or can't judge distance, a signal person is required. A signal person is also required if the crane is operating near power lines or another crane is working in close proximity.

Not just anyone can be a signal person. Signal persons must be qualified by experience, be knowledgeable in all established communication methods, be stationed in full view of the operator, have a full view of the load path, and understand the load's intended path of travel in order to position themselves accordingly. In addition, they must wear high-visibility gloves and/or clothing, be responsible for keeping everyone out of the operating radius of the crane, and never direct the load over anyone.

NOTE

Do not give signals to the crane operator unless you have been designated as the signal person.

Although personnel involved in lifting operations are expected to understand these signals when they are given, it is acceptable for a signal person to give a verbal or nonverbal signal to an operator that is not part of the *ASME B30.5* standard. In cases where such non-standardized signals are given, it is important that both the operator and the signal person have a complete understanding of the message that is being sent.

11.0.0 ◆ GENERAL RIGGING SAFETY

As a rigging worker, you need to be aware of the unavoidable hazards associated with the trade. You will be directing the movement of loads above and around other workers, where falling equipment and material can present a grave safety hazard (*Figure 53*). You may also work during extreme weather conditions where winds, slippery surfaces, and unguarded work areas exist. When working near cranes, always look up and be mindful of the hazards above and around you.

As a result of tighter regulations established by the Occupational Safety and Health Administration (OSHA), and more lost work days experienced by construction employees, individual construction trade contractors have put more stringent safety policies into place.

Safety consciousness is extremely important for the employee. The earning ability of injured employees may be reduced or eliminated for the rest of their lives. The number of employees injured can be lowered if each employee is committed to

111F53.EPS

Figure 53 ◆ Overhead hazards.

safety awareness. Full participation in the employer's safety program is your personal responsibility.

Safety consciousness is the key to reducing accidents, injuries, and deaths on the job site. Accidents can be avoided because most result from human error. Mistakes can be reduced by developing safe work habits derived from the principles of on-the-job safety.

11.1.0 Personal Protection

Workers on the job have responsibilities for their own safety and the safety of their fellow workers. Management has a responsibility to each worker to ensure that the workers who prepare and use the equipment, and who work with or around it, are well trained in operating procedures and safety practices.

Always be aware of your environment when working with cranes. Stay alert and know the location of equipment at all times when moving about and working within the job area.

Standard personal protective equipment, including hard hats, safety shoes, and barricaded work areas (*Figure 54*) are among the important safety requirements at any job site.

11.2.0 Equipment and Supervision

Your employer is responsible for ensuring that all hoisting equipment is operated by experienced, trained operators. Rigging workers must also be capable of selecting suitable rigging and lifting equipment, and directing the movement of the crane and the load to ensure the safety of all personnel. Rigging operations must be planned and supervised by competent personnel who ensure the following:

- Proper rigging equipment is available.
- Correct load ratings are available for the material and rigging equipment.
- Rigging material and equipment are well maintained and in good working condition.

A rigging supervisor is responsible for the following functions:

- Proper load rigging
- Crew supervision
- Ensuring that the rigged material and equipment meet the required capacity and are in safe condition
- Ensuring that the lifting bolts and other rigging materials and equipment are installed correctly
- Guaranteeing the safety of the rigging crew and other personnel

111F54.EPS

Figure 54 ◆ Barricaded swing zone.

11.3.0 Basic Rigging Precautions

The most important rigging precaution is determining the weight of all loads before attempting to lift them. When the assessment of the weight load is difficult, safe-load indicators or weighing devices should be attached to the rigging equipment. It is equally important to rig the load so that it is stable and the center of gravity is below the hook.

The personal safety of riggers and hoisting operators depends on common sense. Always observe the following safety practices:

- Always read the manufacturer's literature for all equipment with which you work. This literature provides information on required startup checks and periodic inspections, as well as inspection guidelines. Also, this literature provides information on configurations and capacities in addition to many safety precautions and restrictions of use.
- Determine the weight of loads, including rigging and hardware, before rigging. Site management must provide this information if it is not known.
- Know the safe working load of the equipment and rigging, and never exceed the limit.
- Examine all equipment and rigging before use. Discard all defective components.
- Immediately report defective equipment or hazardous conditions to the supervisor. Someone in authority must issue orders to proceed after safe conditions have been ensured.

- Stop hoisting or rigging operations when weather conditions present hazards to property, workers, or bystanders, such as when winds exceed manufacturer's specifications; when the visibility of the rigger or hoist crew is impaired by darkness, dust, fog, rain, or snow; or when the temperature is cold enough that hoist or crane steel structures could fracture upon shock or impact.
- Recognize factors that can reduce equipment capacity. Safe working loads of all hoisting and rigging equipment are based on ideal conditions. These conditions are seldom achieved under working conditions.
- Remember that safe working loads of hoisting equipment are applicable only to freely suspended loads and plumb hoist lines. Side loads and hoist lines that are not plumb can stress equipment beyond design limits, and structural failure can occur without warning.
- Ensure that the safe working load of equipment is not exceeded if it is exposed to wind. Avoid sudden snatching, swinging, and stopping of suspended loads. Rapid acceleration and deceleration greatly increase the stress on equipment and rigging.
- Follow all manufacturer's guidelines, and consult applicable standards to stabilize mobile cranes properly (*Figure 55*).

11.4.0 Barricades

Barricades should always be used to isolate the area of an overhead lift to prevent the possibility of injuring personnel who may walk into the area. Always follow the individual site requirements for proper barricade erection. If in doubt as to the proper procedure, ask your supervisor for guidance before proceeding with any overhead lifting operation. It is important to remember that accessible areas within the swing radius of the rear of the crane's rotating structure must be barricaded in such as manner as to prevent an employee or others from being struck or crushed by the crane (*Figure 56*).

11.5.0 Load-Handling Safety

The safe and effective control of the load involves the rigger's strict observance of load-handling safety requirements. This includes making sure that the swing path or load path is clear of personnel and obstructions (*Figure 57*). Keep the front and rear swing paths (*Figure 58*) of the crane clear for the duration of the lift. Most people watch the load when it is in motion, which prevents them from seeing the back end of the crane coming around.

Make sure the landing zone is clear of personnel, with the exception of the tag line tenders. Also make sure that the necessary blocking and cribbing for the load are in place before you position the load for landing. The practice of lowering the

111F55.EPS

Figure 55 ◆ Mobile crane with outriggers.

111F56.EPS

Figure 56 ◆ Use of a barrier to isolate the swing circle area of a crane.

load just above the landing zone and then placing the cribbing and blocking can be dangerous. No one should work under the load. The layout of the cribbing can be completed in the landing zone before you set the load. Blocking of the load may have to be done after the load is set. In this case, do not take the load stress off the sling until the blocking is set and secured. Do not attempt to position the load onto the cribbing by manhandling it. In some circumstances, such as when the crane is moving a load some distance, it may be necessary to have flagmen serve as a moving barricade around and ahead of the load.

After the rigging has been set and whenever loads are to be handled, follow these procedures:

- Before lifting, make sure that all loads are securely slung and properly balanced to prevent shifting of any part.
- Use one or more tag lines to keep the load under control.
- Safely land and properly block all loads before removing the slings.
- Only use lifting beams for the purpose for which they were designed. Their weight and working load abilities must be visible on the beams.
- Never wrap hoist ropes around the load. Use only slings or other adequate lifting devices.
- Do not twist multiple-part lines around each other.
- Bring the load line over the center of gravity of the load before starting the lift.
- Make sure the rope is properly seated on the drum and in the sheaves if there has been a slack rope condition.
- Load and secure any materials and equipment being hoisted to prevent movement which, in turn, could create a hazard.
- Keep hands and feet away from pinch points as the slack is taken up.
- Wear gloves when handling wire rope.
- See that all personnel are standing clear while loads are being lifted and lowered or when slings are being drawn from beneath the load. The hooks may catch under the load and suddenly fly free. It is prohibited to pull a choker out from under a load that has been set on the choker.
- Never ride on a load that is being lifted.
- Never allow the load to be lifted above other personnel.
- Never work under a suspended load.
- Never leave a load suspended in the air when the hoisting equipment is unattended.
- Never make temporary repairs to a sling.
- Never lift loads with one or two legs of a multileg sling until the unused slings are secured.
- Ensure that all slings are made from the same material when using two or more slings on a load.
- Remove or secure all loose pieces from a load before it is moved.
- Lower loads onto adequate blocking to prevent damage to the slings.

111F57.EPS

Figure 57 ◆ Front swing path.

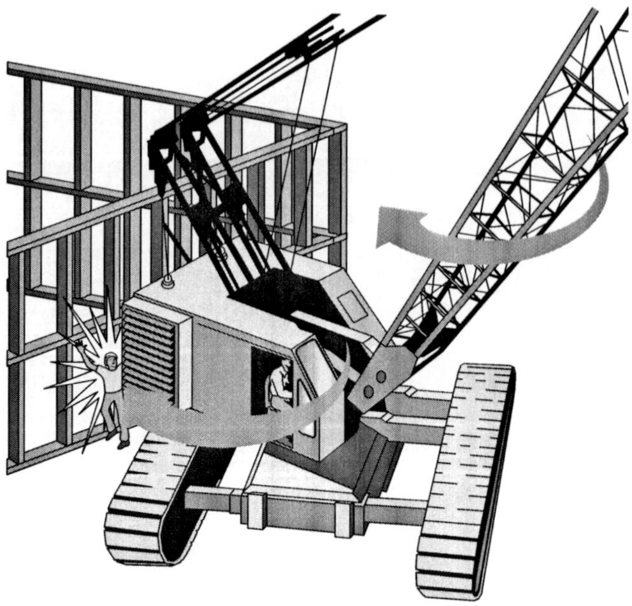

111F58.EPS

Figure 58 ◆ Rear swing path.

12.0.0 ◆ WORKING AROUND POWER LINES

A competent signal person must be stationed at all times to warn the operator when any part of the machine or load is approaching the minimum safe distance from the power line. The signal person must be in full view at all times. *Table 2* and *Figure 59* show the minimum safe distance from power lines.

> **WARNING!**
> The most frequent cause of death of riggers and material handlers is electrocution caused by contact of the crane's boom, load lines, or load with electric power lines. To prevent personal injury or death, stay clear of electric power lines. Even though the boom guards, insulating links, or proximity warning devices may be used or required, these devices do not alter the precautions given in this section.

The preferred working condition is to have the owner of the power lines de-energize and provide grounding of the lines that are visible to the crane

Table 2 High-Voltage Power Line Clearances

CRANE IN OPERATION (1)	
POWER LINE (kV)	BOOM OR MAST MINIMUM CLEARANCES (feet)
0 to 50	10
50 to 200	15
200 to 350	20
350 to 500	25
500 to 750	35
750 to 1000	45
CRANE IN TRANSIT (with no load and the boom or mast lowered) (2)	
POWER LINE (kV)	BOOM OR MAST MINIMUM CLEARANCES (feet)
0 to 0.75	4
0.75 to 50	6
50 to 345	10
345 to 7500	16
750 to 1000	20

Note 1: For voltages over 50kV, clearance increases 5 feet for every 150 kV.
Note 2: Environmental conditions such as fog, smoke, or precipitation may require increased clearances.

111T02.EPS

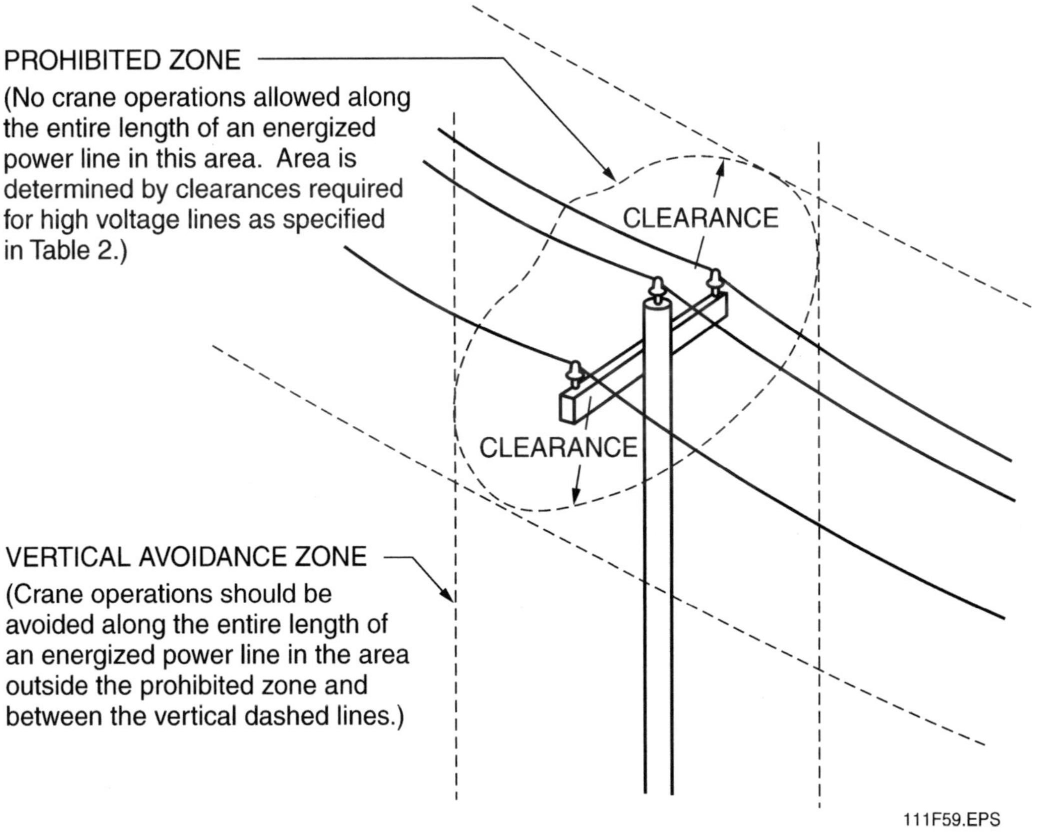

Figure 59 ◆ Prohibited zone and avoidance zone.

operator. When that is not possible, observe the following procedures and precautions if any part of your boom can reach the power line:

- Make sure a power line awareness permit or equivalent has been prepared.
- Erect non-conductive barricades to restrict access to the work area.
- Use tag lines of a non-conductive type if necessary for load control.
- The qualified signal person(s), whose sole responsibility is to verify that the proper clearances are established and maintained, shall be in constant contact with the crane operator.
- The person(s) responsible for the operation shall alert and warn the crane operator and all persons working around or near the crane about the hazards of electrocution or serious injury, and instruct them on how to avoid these hazards.
- All non-essential personnel shall be removed from the crane work area.
- No one shall be permitted to touch the crane or load unless the signal person indicates it is safe to do so.

If a crane or load comes in contact with or becomes entangled in power lines, assume that the power lines are energized unless the lines are visibly grounded. Any other assumption could be fatal.

The following guidelines should be followed if the crane comes in contact with an electrical power source:

- The operator should stay in the cab of the crane unless a fire occurs.
- Do not allow anyone to touch the crane or the load.
- If possible, the operator should reverse the movement of the crane to break contact with the energized power line.
- If the operator cannot stay in the cab due to fire or arcing, the operator should jump clear of the crane, landing with both feet together on the ground. Once out of the crane, the operator must take very short steps with feet together and on the ground until well clear of the crane.
- Call the local power authority or owner of the power line.
- Have the lines verified as secure and properly grounded within the operator's view before allowing anyone to approach the crane or the load.

13.0.0 ◆ SITE HAZARDS AND RESTRICTIONS

There are many site hazards and restrictions related to crane operations. These hazards include the following:

- Underground utilities such as gas, oil, electrical, and telephone lines; sewage and drainage piping; and underground tanks
- Electrical lines or high-frequency transmitters
- Structures such as buildings, excavations, bridges, and abutments

WARNING!
Power lines and environmental issues such as weather are common causes of injury and death during crane operations.

The operator and riggers should inspect the work area and identify hazards or restrictions that may affect the safe operation of the crane. This includes the following actions:

- Ensuring the ground can support the crane and the load
- Checking that there is a safe path to move the crane around on site
- Making sure that the crane can rotate in the required quadrants for the planned lift

The operator must follow the manufacturer's recommendations and any locally established restrictions placed on crane operations, such as traffic considerations or time restrictions for noise abatement.

14.0.0 ◆ EMERGENCY RESPONSE

Operators and riggers must react quickly and correctly to any crane malfunction or emergency situation that might arise. They must learn the proper responses to emergency situations. The first priority is to prevent injury and loss of life. The second priority is to prevent damage to equipment or surrounding structures.

14.1.0 Fire

Judgment is crucial in determining the correct response to fire. The first response is to cease crane operation and, if time permits, lower the load and secure the crane. In all cases of fire, evacuate the area even if the load cannot be lowered or the crane secured. After emergency services have been notified, a qualified individual may judge if the fire can be combated with a fire extinguisher. A fire extinguisher can be successful at fighting a small fire in its beginning stage, but a fire can get out of control very quickly. Do not be overconfident, and keep in mind that priority number one is preventing loss of life or injury to anyone. Even trained firefighters using the best equipment can be overwhelmed and injured by fires.

14.2.0 Malfunctions During Lifting Operations

Mechanical malfunctions during a lift can be very serious. If a failure causes the radius to increase unexpectedly, the crane can tip or the structure could collapse. Loads can also be dropped during a mechanical malfunction. A sudden loss of load on the crane can cause a whiplash effect that can tip the crane or cause the boom to fail.

The chance of these types of failures occurring in modern cranes is greatly reduced because of system redundancies and safety backups. However, failures do happen, so stay alert at all times.

If a mechanical problem occurs, the operator should lower the load immediately. Next, the operator should secure the crane, tag the controls out of service, and report the problem to a supervisor. The crane should not be operated until it is repaired by a qualified technician.

14.3.0 Hazardous Weather

Mobile crane operations generally take place outdoors. Under certain environmental conditions, such as extreme hot or cold weather or in high winds, work can become uncomfortable and possibly dangerous. For example, snow and rain can have a dramatic effect on the weight of the load and on ground compaction. During the winter, the tires, outriggers, and crawlers can freeze to the ground. This may lead the operator or rigger to the false conclusion that the crane is on stable ground. In fact, as weight is added during the lifting operation, it may cause an outrigger float, tire, or crawler to sink into the ground below the frozen surface. Severe rain can cause the ground under the crane to become unstable, resulting in instability of the crane due to erosion or loss of compaction of the soil. There are other specific things to be aware when of working under these adverse conditions.

14.3.1 High Winds and Lightning

High winds and lightning may cause severe problems on the job site (*Figure 60*). They are major weather hazards and must be taken seriously. Crane operators and riggers must be prepared to handle extreme weather in order to avoid accidents, injuries, and damages. It is very rare for high winds or lightning to arrive without some warning. This gives operators and riggers time to react appropriately.

High winds typically start out as less dramatic gusts. Operators and riggers must be aware of changing weather conditions, such as worsening winds, to determine when the weather becomes a hazardous situation. With high winds, the operator must secure crane operations as soon as it is practical. This involves placing the boom in the lowest possible position and securing the crane. Once this is done, all personnel should seek indoor shelter away from the crane.

Because cranes' booms extend so high and are made of metal, they are easy targets for lightning. Operators and riggers must be constantly aware of this threat. Lightning can usually be detected when it is several miles away. As a general rule of thumb, when you hear thunder, the lightning associated with it is 6 to 8 miles away. Be aware, however, that successive lightning strikes can touch down up to 8 miles apart. That means once you hear thunder or see lightning, it is close enough to present a hazard.

In some high-risk areas, proximity sensors provide warnings when lightning strikes within a 20-mile radius. Once a warning is given or lightning is spotted, crane operations must be secured as soon as practical, following the crane manufacturer's recommendations for doing so.

> **WARNING!**
> Even with the boom in the lowest position, it may be taller than surrounding structures and could still be a target for lightning strikes.

Once crane operations have been shut down, all personnel should seek indoor shelter away from the crane. Always wait a minimum of 30 minutes from the last observed instance of lightning or thunder before resuming work.

15.0.0 ◆ USING CRANES TO LIFT PERSONNEL

Although using cranes to lift people was common in the past, OSHA regulations, as spelled out in *29 CFR 1926.550*, now discourage the practice. Using a crane to lift personnel is not specifically prohibited by OSHA, but the restrictions are such that it is only permitted in special situations where no other method is suitable. When it is allowed, certain controls must be in place, including the following:

- The rope design factor is doubled.
- No more than 50 percent of the crane's capacity, including rigging, may be used.
- **Anti-two-blocking devices** are required on the crane boom.
- The platform must be specifically designed for lifting personnel.
- Before the personnel basket is used, it must be tested with appropriate weight, and then inspected.
- Every intended use must undergo a trial run with weights rather than people.

Figures 61 and *62* illustrate the requirements for personnel lifts.

15.1.0 Personnel Platform Loading

The personnel platform must not be loaded in excess of its rated load capacity. When a personnel platform does not have a rated load capacity, the personnel platform must not be loaded in excess of its maximum intended load.

The number of employees, along with material, occupying the personnel platform must not exceed the limit established for the platform and the rated load capacity or the maximum intended load.

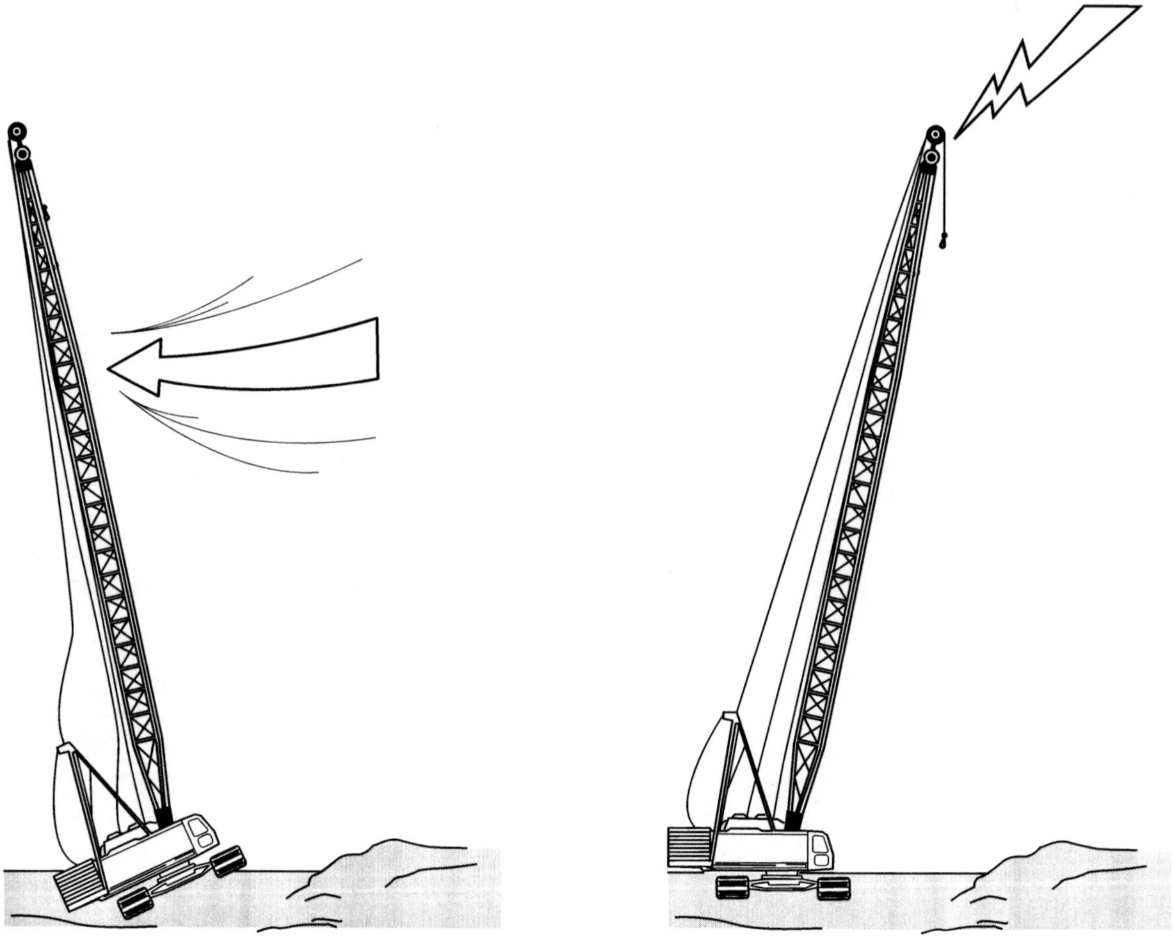

Figure 60 ◆ Wind and lightning hazards.

Personnel platforms must be used only for employees, their tools, and the materials necessary to do their work, and must not be used to hoist only materials or tools when not hoisting personnel.

Materials and tools for use during a personnel lift must be secured to prevent displacement. These items must be evenly distributed within the confines of the platform while the platform is suspended.

Operators may be required to shut down at a certain wind speed; some machines will automatically stop lifting.

15.2.0 Personnel Platform Rigging

When a wire-rope bridle sling is used to connect the personnel platform to the load line, each bridle leg shall be connected to a master link or shackle in such a manner as to ensure that the load is evenly divided among the bridle legs (*Figures 63* and *64*).

Hooks on headache ball assemblies, lower load blocks, or other attachment assemblies shall be of a type that can be closed and locked, eliminating the hook throat opening. Alternatively, an alloy anchor type shackle with a bolt, nut, and retaining pin may be used.

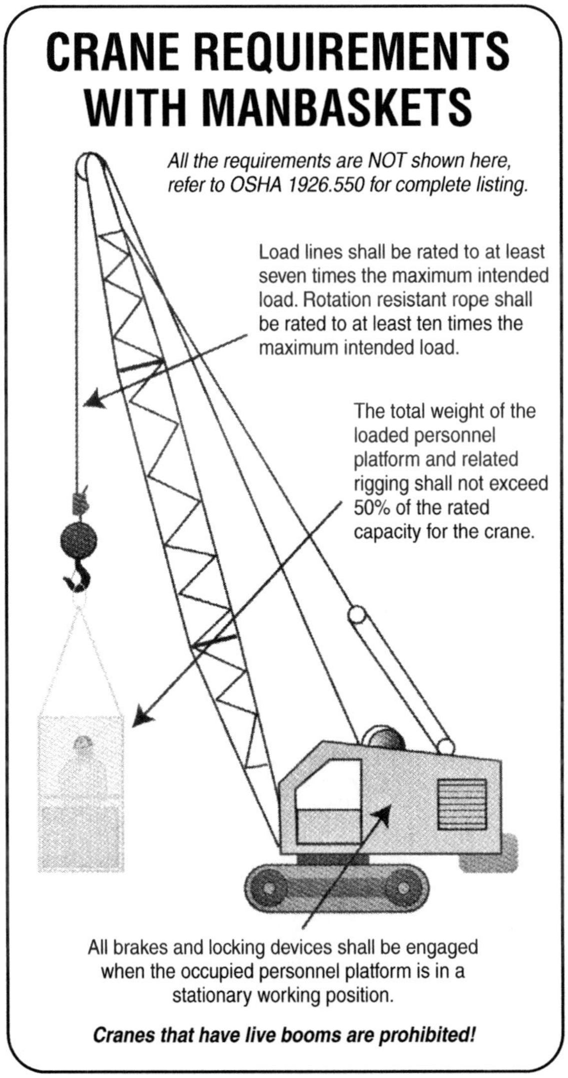

This information provides a generic, non-exhaustive overview of the OSHA standard on suspended personnel platforms. Standards and interpretations change over time, you should always check current OSHA compliance requirements for your specific requirements.

29 CFR 1926.550 addresses the use of personnel hoisting in the construction industry, and *29 CFR 1910.180* addresses the use of personnel hoisting in general industry.

111F61.EPS

Figure 61 ◆ Crane requirements with manbaskets.

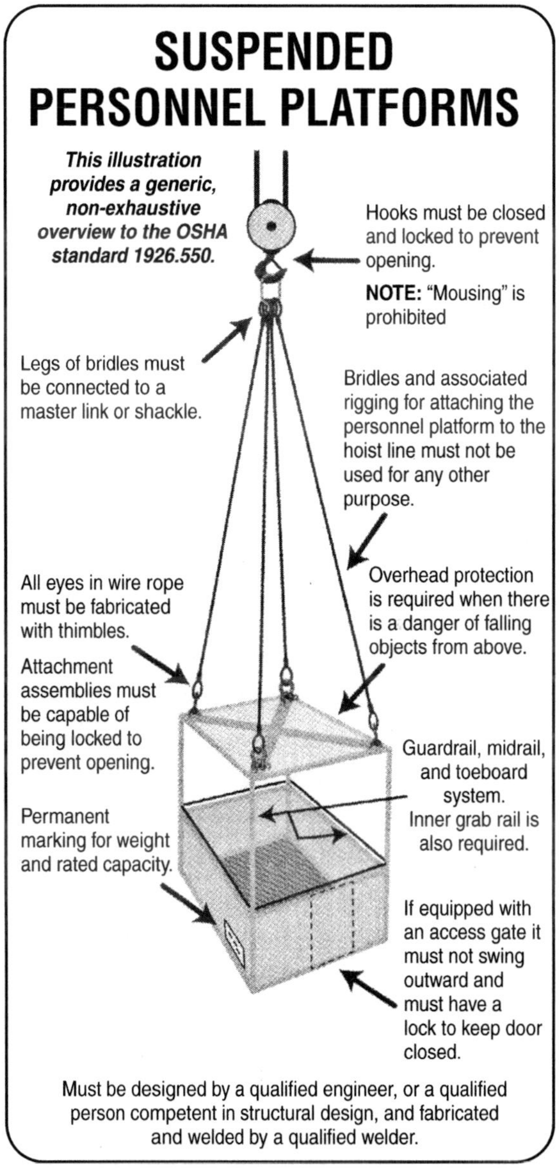

The OSHA rules on crane suspended personnel platforms contain many specifics that are not covered in this book. Refer to *29 CFR 1926.550* for the current OSHA compliance requirements.

111F62.EPS

Figure 62 ◆ Suspended personnel platforms.

Wire rope, shackles, rings, master links, and other rigging hardware must be capable of supporting, without failure, at least five times the maximum intended load applied or transmitted to that component. Where rotation-resistant rope is used, the slings shall be capable of supporting without failure at least ten times the maximum intended load. All eyes in wire rope slings shall be fabricated with thimbles.

Bridles and associated rigging for attaching the personnel platform to the hoist line shall be used only for the platform and the necessary employees, their tools, and the materials necessary to do their work, and they shall not be used for any other purpose when not hoisting personnel.

16.0.0 ◆ LIFT PLANNING

Before conducting a lift, most construction sites or companies require that a lift plan be completed and signed by competent personnel. Lift plans are mandatory for steel erection and multiple-crane lifts. A lift plan contains information relative to the crane, load, and rigging, and it lists any special instructions or restrictions for the lift. It is important to remember that any deviation from the original lift plan, no matter how small, requires a new lift plan.

16.1.0 Lift Plan Data

Critical lifts always require a lift plan (*Figure 65*). As a matter of policy, all personnel involved in a lift should require or prepare a plan for all lifts. A typical lift plan for either a critical or ordinary (minor) lift breaks down the following elements and information. The rigger is particularly concerned with crane placement, rope, and sizing of slings.

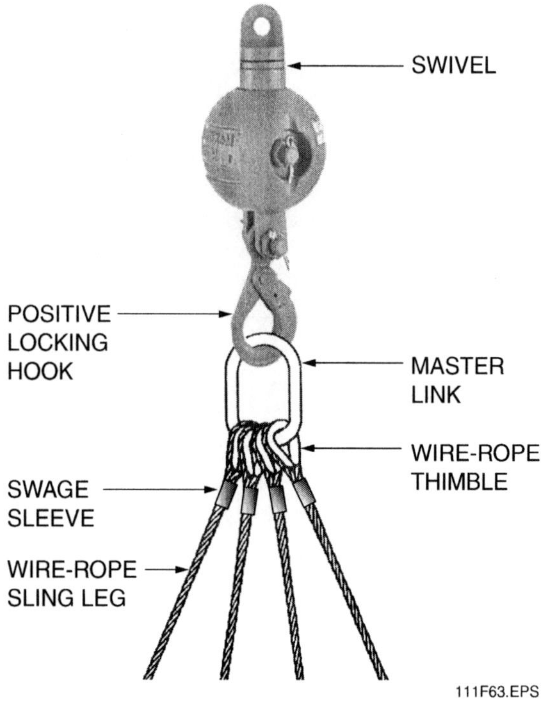

Figure 63 ◆ Bridle sling using a master link.

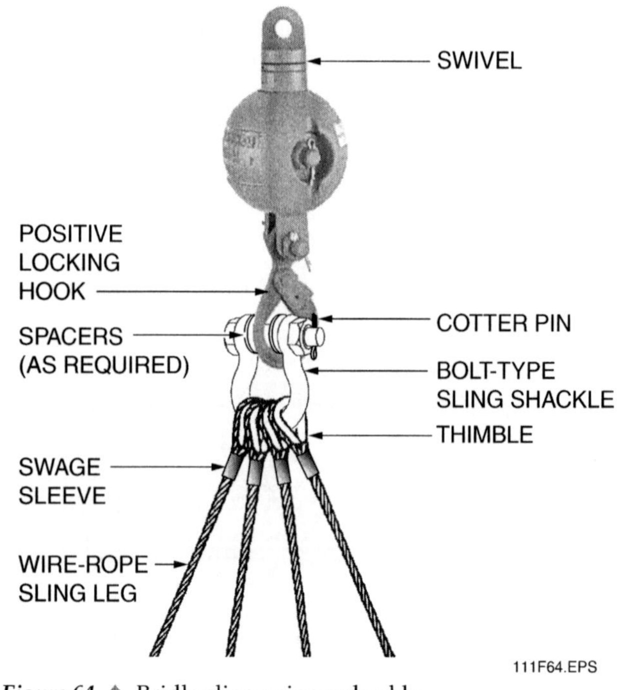

Figure 64 ◆ Bridle sling using a shackle.

PRE-LIFT CHECKLIST **<u>USE FOR LIFTS EXCEEDING 75% OF CRANE'S RATED CHART CAPACITY</u>**
LAND BASED CRANES ONLY. CRANE APPROPRIATELY SUPPORTED AND LEVEL WITHIN 1%.

1. Crane Type: ______________________
2. Load Description: ______________________
3. Operating Radius: __________ Boom Length: __________ Boom tip Elevation: __________
4. Loaded Boom Angle: __________ Minimum Boom Length: __________
5. Elevation of Lift: __________
6. Crane Capacity at Working Radius: (6) __________
   What is the Maximum Radius with total load? __________
7. Weight of Load: (7) – (__________)
   Subtotal __________
8. Crane Deductions:
   a. Block Capacity __________ Block WT. + __________
   b. Ball(s) Capacity __________ Ball WT. + __________
   c. Jib Effective WT + __________
   d. Wire Rope # of parts Required: __________
   e. Other+ __________ + __________
   WT/Ft. x # of ft. x parts being used ____ x ____ x ____ = + __________
9. Total Crane Deductions: (9) – (__________)
   Subtotal __________
10. Rigging Deductions:
   a. Slings: Qty.____ Type ____ Size ____ Cap. ____ WT. +____
   Qty.____ Type ____ Size ____ Cap. ____ WT. +____
   b. Shackles: Qty.____ Type ____ Size ____ Cap. ____ WT. +____
   Qty.____ Type ____ Size ____ Cap. ____ WT. +____
   c. Lifting Beam(s) Qty.____ Cap. ____ WT. +____
   d. Softeners: Qty.____ Type ____ Size ____ WT. +____
   e. Misc. Rigging: (Snatch Blocks, Turn Buckles, etc.) WT. +____
   (11) – (__________)
11. Total Rigging Deductions:
12. For Total Load (add 7 + 9 + 11) = (12) __________
   To double check safety margin subtract total load # 12 from line # 6 __________
   This number and line # 13 should match.
13. Safety Margin (13) __________

Lifting Information

14. # of Cranes in lift: __________ Use additional checklist for each crane.
15. Ground conditions: __________ Use appropriate blocking mats required. Y____ N____
16. Weather conditions: __________ Wind speed ______ Direction ______
17. Tag lines required: Y_____ N_____ Size_____ Length ______ No. of lines_____
18. Designated Signal Person: Name __________ Spotter (s) Y____ N____ No. of ______
19. Signal Method: Hand ____ Radio _____
20. Has matching Activity Plan been done and reviewed by crew? Y____ N____

Sign – off
Crane Operator: __________ Competent Rigger: __________
Crew: __________ __________ __________
__________ __________ __________
Superintendent: __________ Date: ________ Time: ________

111F65A.EPS

Figure 65 ◆ Sample lift plan. (1 of 2)

17.0.0 ◆ CRANE COMPONENT TERMINOLOGY

Figures 66, 67, and *68* show the components of most truck- or crawler-mounted lattice boom cranes.

Crane component terminology varies in different locations and sometimes between manufacturers. The following describes common crane components and identifies some of the alternate terms used for components. Some of the components are described in more detail later in this module.

- *Base mounting* – This structure consists of the carbody, swing circle, crawler frames, axles and track, and the propelling mechanism that forms the lowest element of a crane. It transmits loads to the ground. It is the crawler mounting for mobile cranes.

ADDED INFORMATION:

WIRE ROPE— IWRC –EIPS 6 X 19 & 6 X 37
MECHANICAL SPLICE SLINGS CAPACITY IN TONS

Diameter (inches)	Weight (lbs/ft)	Vertical	Choker	Vertical Basket
1 3/8	0.26	1.4	1.1	2.9
1/2	0.46	2.5	1.9	5.1
5/8	0.72	3.9	2.9	7.8
3/4	1.04	5.6	4.1	11
7/8	1.42	7.6	5.6	15
1	1.85	9.8	7.2	20
1 1/8	2.34	12	9.1	24
1 1/4	2.89	15	11	30
1 3/8	3.50	18	13	36
1 1/2	4.16	21	16	42
1 5/8	4.88	24	18	49
1 3/4	5.67	28	21	57
1 7/8	6.50	32	24	64
2	7.39	37	28	73
2 1/8	8.35	40	31	80
2 1/4	9.36	44	35	89
2 3/8	10.4	49	38	99
2 1/2	11.6	54	42	109
2 3/4	14.0	65	51	130

CROSBY SCREW PIN ANCHOR SHACKLES
(G-209, S-209)

Size	Capacity (t)	Weight (lbs)
3/4	4 3/4	3
7/8	6 1/2	4
1	8 1/2	5
1 1/8	9 1/2	8
1 1/4	12	10
1 3/8	13 1/2	14
1 1/2	17	18
1 3/4	25	28
2	35	45
2 1/2	55	86

111F65B.EPS

Figure 65 ◆ Sample lift plan. (2 of 2)

- *Base section* – The base section of a telescoping boom is attached to the base mounting of a crane. It does not telescope, but contains the boom-foot, pin-mountings, and the boom-hoist cylinder upper-end mountings.
- *Boom* – A boom is a crane member used to project the upper end of the hoisting tackle in reach or in combination of height and reach. The boom supports the weight of the load and gives the crane its lift and reach capabilities. Boom sections can be added to extend the boom length. The capacity of a boom decreases as it is lowered from a vertical position.
- *Boom base* – The boom base is the lowermost section of a sectional-latticed boom having the attachment (boom-foot pins) mounted at its lower end. It is also called the boom butt or butt section, or a crawler tip and butt.
- *Boom sheaves* – Boom sheaves are the pulleys that the load lines pass over.
- *Carbody* – A carbody is the part of a crawler crane base mounting that carries the upper structure. The crawler side frames are attached to it.
- *Counterweights* – Counterweights are weights added to a crane to counteract the weight of a load and to achieve lifting capacity.
- *Falls of line* – See parts of line.

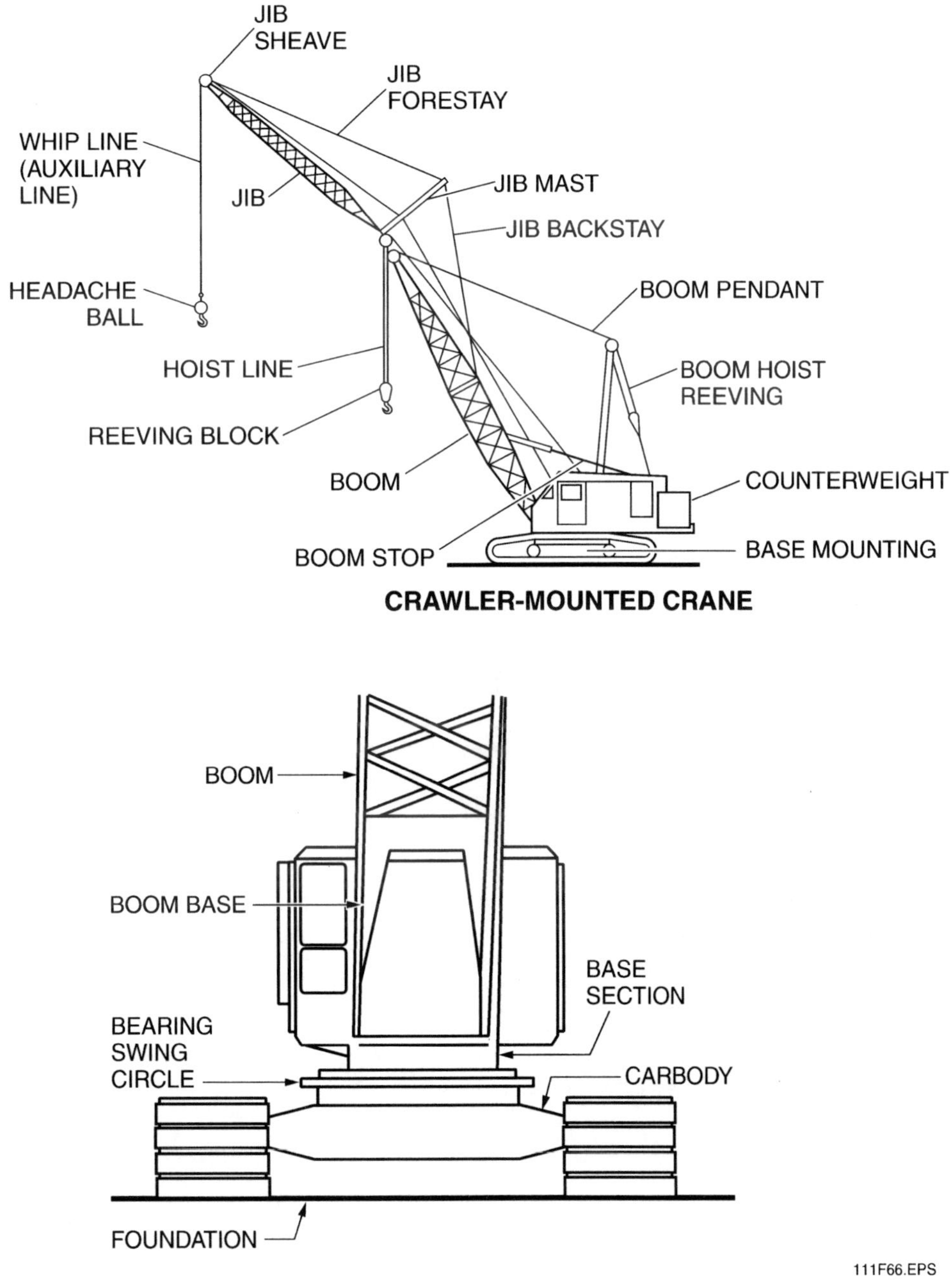

Figure 66 ◆ Crawler- and wheel-mounted crane components.

- *Headache ball* – A headache ball is a weight added to a load fall to overcome resistance and permit unspooling at the rope drum when no live load is being supported. A headache ball is usually equipped with a hook.
- *Hoist line reeving* – Hoist reeving is a method usually used to multiply the pulling or lifting capability of a rope (line) by using multiple pulleys in sheaves. Single-line reeving is used for a whip line, and multiple line reeving is used for main load, boom, and jib lines. Also see parts of line.
- *Jib* – A jib is an extension that is added to the boom and mounted at the boom tip. It may be in line with the boom long axis, or it may be offset from it. It is equipped with its own fixed wire rope suspensions (pendants) that are fastened to a mast at the boom tip. A fixed jib is supported by jib pendants, called a jib backstay and forestay.
- *Jib mast or jib strut* – The jib mast or strut is a short strut or frame mounted on the boom head to provide an attachment for the jib pendants. It may also be called a gantry, horse, or A-frame. With a luffing jib, the pendants and associated masts together may be called the forestay and backstay.

Figure 68 ◆ A heavy-lift crane with counterweights.

Figure 67 ◆ Swing cab, lattice boom truck crane.

- *Jib sheaves* – Jib sheaves serve the same function as boom sheaves.
- *Latticed boom or conventional boom* – This boom is constructed of four longitudinal corner members, called chords, assembled with transverse and/or diagonal members called lacing, to form a trusswork in two directions. The chords carry the axial (perpendicular) boom forces and bending moments, while the lacings resist shear forces.
- *Load block* – A load block is the lower block in a crane reeving system that moves with the load. The load block is sometimes called the traveling block.
- *Luffing jib* – This jib can have its angle to the boom changed by the crane operator via jib reeving.
- *Outriggers* – Outriggers are extendible arms attached to a crane base mounting that relieve the wheels of crane weight and increase stability.
- *Parts of line* – These are the number of wire ropes supporting a load block in a crane reeving system. They are also referred to as just parts or falls.
- *Pendant* – A pendant is a fixed-length rope or solid rod forming part of the boom or jib suspension system. It may be called the boom guy line, hog line, boom stay, standing line, forestay, or backstay.
- *Running line* – A running line is a wire rope that moves over sheaves or drums.
- *Standing line* – A standing line is a fixed line that supports loads without being spooled on or off a drum. It is a line on which both ends are dead. It is also referred to as a guy line, stay rope, or pendant.
- *Tag line* – A tag line is usually a fiber rope that is attached to the load and used for controlling load spin or alignment from the ground. When a tag line is used for clamshell operations, it consists of a wire rope needed to retard rotations and the pendulum action of the bucket.
- *Upperworks* – The upperworks is the entire rotating structure less the front end attachment. It may also be referred to as upper superstructure, revolving superstructure, or upper structure.
- *Whip line* – See hoist line reeving.

Review Questions

1. Mixing rigging hardware of different lifting capacities is okay, as long as the capacity of one of the hardware items exceeds the weight of the load.
   a. True
   b. False

2. The safe working load of a rigging hook is accurate only when the load is suspended from the _____ of the hook.
   a. body
   b. eye
   c. throat
   d. saddle

3. OSHA requires that a hook be replaced if the throat has opened _____ percent or more from its original size.
   a. 5
   b. 10
   c. 15
   d. 20

4. A hook that is point-loaded can carry about ____ percent of the rated load.
   a. 20
   b. 40
   c. 60
   d. 75

5. Slings are attached together using _____.
   a. spacers
   b. sling eyes
   c. shackles
   d. lifting lugs

6. All of the following are true *except* _____.
   a. shackles with pins worn more than 10 percent should be replaced
   b. the size of a shackle is determined by its pin size
   c. when using a shackle on a hook, the pin of the shackle should be hung on the hook
   d. if a shackle is pulled at an angle, its capacity is reduced

7. Tapped holes used with screwed-in eyebolts should have a minimum depth of _____ times the bolt diameter.
   a. 1¼
   b. 1½
   c. 1¾
   d. 2

8. Which of the following statements about lifting lugs is correct?
   a. They can only be used for vertical lifts.
   b. They can only be used for angled lifts.
   c. They are used primarily for vertical lifts, but some are also used for angled lifts if so designed.
   d. They are used primarily for angled lifts, but some are also designed for vertical lifts.

9. Screw-type plate clamps are used for vertical lifting of plates only.
   a. True
   b. False

10. To level loads when the legs of a sling are unequal, a(n) _____ can be used.
    a. rigging link
    b. equalizer plate
    c. plate clamp
    d. beam clamp

11. Custom-fabricated spreader and equalizer beams should be tested at _____ percent of their rated capacity.
    a. 100
    b. 125
    c. 150
    d. 200

12. Long loads are supported during lifting operations with _____.
    a. spreader beams
    b. equalizer beams
    c. compound equalizer plates
    d. serrated clamps

13. One of the slings best suited for situations where the load is abrasive, hot, or tends to cut is a _____ sling.
    a. wire rope
    b. synthetic web
    c. endless grommet
    d. metal mesh

Review Questions

14. The term D/d ratio refers to the relationship between the sling and the _____.
    a. distance from the hook to the load
    b. diameter of the surface the sling is wrapped around
    c. distance from the load to the ground
    d. diameter of the load

15. Dredging and other marine work requires _____ slings because they withstand corrosion better.
    a. chain
    b. synthetic web
    c. wire rope
    d. mesh

16. A chain sling should be discarded if any portion of the chain is worn by _____ percent or more.
    a. 5
    b. 10
    c. 15
    d. 20

17. Which of the following statements about a compound block and tackle is correct?
    a. It has greater capacity, but requires more effort.
    b. The fixed block is attached to the load.
    c. The more sheaves the blocks have, the greater the load capacity.
    d. It is always necessary to tie off the hauling line to keep the load from falling.

18. The capacity range of come-alongs is _____ tons.
    a. ½ to 6
    b. 2 to 5
    c. 3 to 7
    d. 8 to 10

19. The ratchet jack can lift heavier loads than the screw jack or hydraulic jack.
    a. True
    b. False

20. The most useful general-purpose jack is the _____ jack.
    a. inverted screw
    b. hydraulic
    c. upright screw
    d. ratchet

21. What action is signaled by tapping your fist on your head, then using regular hand signals?
    a. Use main hoist.
    b. Lower.
    c. Use whip line.
    d. Raise boom and lower load.

22. When the signal person gives the dog everything signal, the operator is expected to _____.
    a. turn the machine
    b. travel the machine
    c. retract the boom sections
    d. make no machine movement

23. If it is apparent that the crane operator has misunderstood a hand signal, the signal person should _____.
    a. climb up on the cab and knock on the door
    b. find a supervisor
    c. signal the operator to stop
    d. get away from the crane as quickly as possible

24. A one-handed signal that is given with the fist in front of the body and the thumb pointing toward the signaler's body means _____.
    a. extend the boom
    b. move the crane toward me
    c. swing the load toward me
    d. I'm taking a break

25. A(n) _____ is used to keep the load under control.
    a. tag line
    b. equalizer beam
    c. hoist line
    d. safe-load indicator

Summary

One of the important tasks a rigger performs is guiding the crane operator during a lift. In some instances, radio communications will be used. In others, it will be necessary to use the *ASME B30.5 Consensus Standard* hand signals. Both the rigger and operator must know these signals by heart.

The selecting and setting up of hoisting equipment, hooking cables to the load to be lifted or moved, and directing the load into position are all part of the process called rigging. Performing this process safely and efficiently requires selecting the proper equipment for each hoisting job, understanding safety hazards, and knowing how to prevent accidents. Riggers must have the greatest respect for the hardware and tools of their trade because their lives may depend on them working correctly.

The size of a crane and the huge loads it handles create a large potential for danger to anyone in the vicinity of the crane. The crane operator and rigging workers must be vigilant to ensure that the crane avoids power lines and that site personnel are not endangered by the movement of the crane or the load. Because of tipping hazards, it is important to make sure a crane is on level ground with outriggers extended, if applicable, before lifting a load. Although using cranes to lift personnel was once a common practice, it is discouraged by OSHA, and can only be done in special situations, and under special conditions.

Notes

Trade Terms Introduced in This Module

Anneal: To soften a metal by heat treatment.

Anti-two-blocking devices: Devices that provide warnings and prevent two-blocking from occurring. Two-blocking occurs when the lower load block or hook comes into contact with the upper load block, boom point, or boom point machinery. The likely result is failure of the rope or release of the load or hook block.

Bird caging: A deformation of wire rope that causes the strands or lays to separate and balloon outward like the vertical bars of a bird cage.

Equalizer beam: A beam used to distribute weight on multi-crane lifts.

Fixed block: The upper block of a block and tackle. The block that is attached to the support.

Hauling line: The line of a lifting device that is pulled by hand to raise the load.

Hydraulic: Operated by fluid pressure.

Kinking: Bending a rope so severely that the bend is permanent and individual wires or fibers are damaged.

Parts of line: The number of ropes between the upper and lower blocks of a block and tackle. These lines carry the load. Parts of line are also called falls.

Sling angle: The angle formed by the legs of a sling with respect to the horizontal when tension is put upon the load.

Resources & Acknowledgments

Additional Resources

This module is intended to be a thorough resource for task training. The following reference works are suggested for further study. These are optional materials for continued education rather than for task training.

Machinery's Handbook, Latest Edition. Erik Oberg, Franklin D. Jones, Holbrook L. Horton, and Henry H. Ryffel. New York, NY: Industrial Press Inc.

Occupational Safety and Health Standards for the Construction Industry, 29 CFR Part 1926. Washington, DC: OSHA Department of Labor, U.S. Government Printing Office.

Figure Credits

Topaz Publications, Inc., 111F11, 111F14, 111F17, 111F53, 111F54, 111F56

J.C. Renfroe & Sons, Inc., 111F12

Lift-All Company, Inc., 1115F15

Alan W. Grogono M.D., 111F21 (bowline, round turn, clove hitch, square knot)

David Root, 111F21 (running bowline, timber hitch)

Lincoln Fire & Rescue, Lincoln, NE, 111F21 (half-hitch safety, overhand safety)

Lehman's Non-Electric Catalog, 111F23

Chester Hoist Company, 111F24

Coffing Hoists, 111F25, 111F26

Duff-Norton, 111F27, 111F29

Courtesy of North American Industries, 111F31

Gorbel, Inc., 111F32, 111F33

Courtesy of Cervis, Inc., 111F34
www.cervis.net

Aearo Company, 111F35

Link-Belt Construction Equipment Co., 111F55, 111F67, 111F68

Rigging Handbook, 2003 by Jerry Klinke, ACRA Enterprises, Stevensville, MI, 111F61, 111F62

The Crosby Group, Inc., 111F63 (photo), 111F64 (photo)

Wire Rope Sling Users Manual, Second Edition © 1997, The Wire Rope Technical Board, 111F63 (line art), 111F64 (line art)

CONTREN® LEARNING SERIES — USER UPDATE

NCCER makes every effort to keep these textbooks up-to-date and free of technical errors. We appreciate your help in this process. If you have an idea for improving this textbook, or if you find an error, a typographical mistake, or an inaccuracy in NCCER's Contren® textbooks, please write us, using this form or a photocopy. Be sure to include the exact module number, page number, a detailed description, and the correction, if applicable. Your input will be brought to the attention of the Technical Review Committee. Thank you for your assistance.

Instructors – If you found that additional materials were necessary in order to teach this module effectively, please let us know so that we may include them in the Equipment/Materials list in the Annotated Instructor's Guide.

Write: Product Development and Revision
National Center for Construction Education and Research
3600 NW 43rd St. Bldg G, Gainesville, FL 32606

Fax: 352-334-0932

E-mail: curriculum@nccer.org

Craft ____________________ Module Name ____________________

Copyright Date __________ Module Number __________ Page Number(s) __________

Description

____

____

____

____

(Optional) Correction

____

____

____

(Optional) Your Name and Address

____

____

____

40112-07

Mobile and Support Equipment

40112-07
Mobile and Support Equipment

Topics to be presented in this module include:

Overview

In your work as an industrial maintenance craftworker, you will be required to work on machines in many different locations, sometimes in pits, and sometimes some distance above ground. You will use and work on many different machines. Light equipment includes personnel lifts, forklifts, stockpickers, compressors, and generators—all tools that you would use every day. You will learn about using and maintaining them in this module.

Objectives

When you have completed this module, you will be able to do the following:

1. State the safety precautions associated with the use of motor-driven equipment in industrial plants.
2. Explain the operation and applications of the following motor-driven equipment commonly used in industrial plants:
   - Portable generators
   - Air compressors
   - Aerial lifts
   - Forklifts
   - Mobile cranes
3. Operate and perform preventive maintenance on the following equipment:
   - Portable generators
   - Air compressors
   - Aerial lifts

Trade Terms

Aerial lift
Ampere
Cherry picker
Circuit breaker
Compressor
Forklift
Fuse holder
Generator
Governor
Ground fault interrupter (GFI)
Pneumatic
Powered industrial trucks
Proportional control
Straight blade duplex connector
Twist-lock connector
Watt

Required Trainee Materials

1. Pencil and paper
2. Appropriate personal protective equipment

Prerequisites

Before you begin this module, it is recommended that you successfully complete *Core Curriculum;* and *Industrial Maintenance E & I Technician Level One,* Modules 40101-07 through 40111-07.

This course map shows all of the modules in the first level of the *Industrial Maintenance E & I Technician* curriculum. The suggested training order begins at the bottom and proceeds up. Skill levels increase as you advance on the course map. The local Training Program Sponsor may adjust the training order.

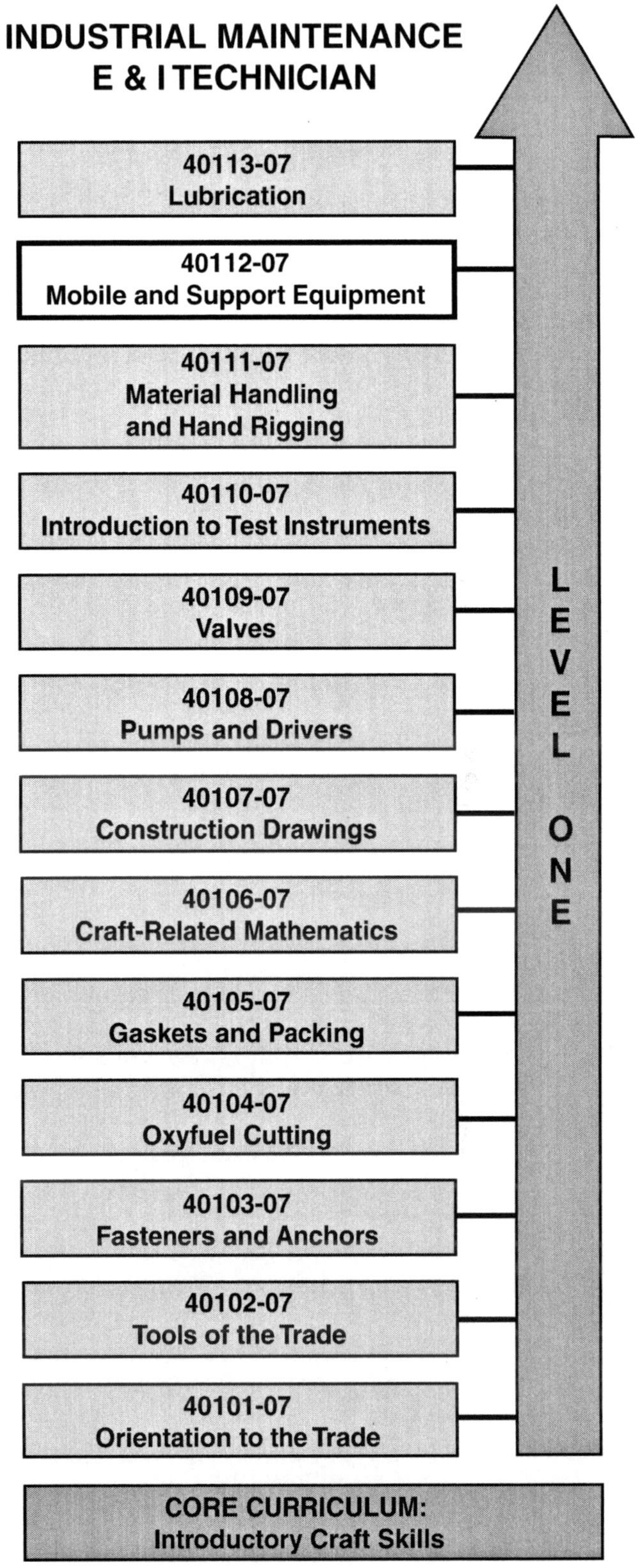

112CMAP.EPS

1.0.0 ◆ INTRODUCTION

Maintenance craftworkers work with and around various types of motorized equipment throughout their careers. Motorized equipment includes portable equipment such as **generators**, **compressors**, and pumps, as well as larger equipment, such as trucks, mobile cranes, and personnel lifts. A piece of motorized equipment is considered portable if it can be transported from job site to job site or to different areas of the same job site. Portable equipment can either be moved or rolled to a different location by hand or transported on a trailer behind a truck or tractor. A piece of equipment is also considered to be portable if it can provide its own power source in the field. This module explains the use, care, and preventive maintenance of portable motorized equipment. It also covers **forklifts** and mobile cranes that you will be working with and around throughout your career.

2.0.0 ◆ SAFETY PRECAUTIONS

Safety is the primary concern no matter what type of equipment you may be operating. Most of the equipment contained in this module shares common safety precautions. These safety precautions are discussed in this section. As each piece of equipment is discussed, remember this safety discussion and think about how these precautions apply to the equipment being covered. Additional safety precautions that are unique to particular equipment are discussed later. Additionally, appropriate personal protective equipment must be worn while operating or working near any equipment, and all use must be performed under the direct supervision of your instructor or other qualified personnel.

The manufacturer provides safety precautions for each piece of equipment. These safety precautions should be observed whenever you are using any equipment. All of the equipment discussed can cause serious injury or death if the safety precautions are not followed.

2.1.0 Interlocking Systems

Some equipment comes with safety interlocks to prevent unsafe operation. Keep the following in mind when using this type of equipment:

- Interlock systems must not be defeated or modified. Interlock systems help prevent incorrect control operation.
- If an interlock system fails, contact the manufacturer's repair representative immediately.
- Runaway equipment is always possible; learn how to use all the controls and emergency procedures.

2.2.0 Transporting Equipment

When transporting equipment between jobs, remember these safety guidelines:

- Park, unload, and load the equipment from a trailer on level ground.
- To prevent tipping, connect the trailer to the tow vehicle before loading or unloading.
- Follow the manufacturer's requirements for towing and transporting the equipment.
- Explosive separation of a tire and/or rim parts can cause injury or death.
- Always maintain the correct tire pressure.

WARNING!

Anyone operating power equipment must be properly trained in its use. Part of this training includes understanding the associated safety precautions. Some localities may require the operator to be trained and qualified in the use of certain equipment. The primary points to remember are:

- Read and fully understand the operator's manual and follow the instructions and safety precautions.
- Know the capacity and operating characteristics of the equipment being operated.
- Inspect the equipment before each use to make sure everything is in proper working order. Have any defects repaired before using the equipment. Never modify or remove any part of the equipment unless authorized to do so by the manufacturer.
- Check for hazards above, below, and all around the job site. Be sure to maintain a safe distance from electrical power lines and other electrical hazards.
- Learn as much about the work area as you can before beginning work.
- Fasten seat belt or operator restraints before starting.
- Set up warning barriers and keep others away from the equipment and job site.
- Know where to get assistance.
- Know how to use a first aid kit and fire extinguisher.
- In tight areas, use a spotter to cover blind spots.

- Inspect the tires and wheels before towing. Do not tow with high or low tire pressures, cuts, excessive wear, bubbles, damaged rims, or missing lug bolts or nuts.
- Do not inflate tires with flammable gases or from systems using an alcohol injector.

2.3.0 Hydraulic Systems

When operating equipment that uses hydraulics, always remember the following:

- Before disconnecting any hydraulic lines, relieve system pressure by cycling controls.
- Before pressurizing the system, be sure all connections are tight and the lines are undamaged.
- Do not perform any work on the equipment unless you are authorized and qualified to do so.
- Check for leaks using a piece of cardboard or wood; never use bare hands.

Hydraulic fluid escaping through pinholes in hoses and hose fittings can be almost invisible. Escaping hydraulic fluid under pressure can penetrate the skin or eyes, causing serious injury. When inspecting light construction equipment, hydraulic hoses, and hose couplings for fluid leaks, always wear protective clothing and eye protection. In the event that hydraulic fluid penetrates the skin, seek immediate medical attention from a doctor familiar with this type of injury.

2.4.0 Fueling Safety

Fueling safety precautions are applicable to all light equipment using gasoline. Make sure you follow all the manufacturer's instructions before fueling the equipment. The following are general safety precautions associated with fueling equipment:

- Never fill the fuel tank with the motor running, while smoking, or when near an open flame.
- Be careful not to overfill the tank or spill fuel. If fuel is spilled, clean it up immediately.
- Use clean fuel only.
- Do not operate the equipment if fuel has been spilled inside or near the unit.
- Ground the fuel funnel or nozzle against the filler neck to prevent sparks.
- Replace the fuel tank cap after refueling.

2.5.0 Battery Safety

Battery safety precautions are applicable to all light equipment equipped with a battery for starting or operating electric components.

- Battery chargers can ignite flammable materials and vapors. They should not be used near fuels, grain dust, solvents, or other flammables.
- To reduce the possibility of electric shock, a charger should only be connected to a properly grounded single-phase outlet. Do not use an extension cord longer than 25'.
- If the battery is frozen, do not charge or attempt to jump-start the equipment, as the battery may explode.
- Battery and fuel fumes could ignite and cause explosions and burns. Keep batteries away from flames or sparks. Do not smoke.
- If you are jumpstarting a manlift from another vehicle, be sure to hook the ground to the frame of the manlift. Do not charge batteries in confined spaces or around possibly flammable atmospheres.

3.0.0 ◆ GENERATORS

Generators are used to provide electrical power and/or lighting at the job site. Generators are available in many different sizes and configurations (*Figures 1* and 2). They range from small portable machines to large installed backup power systems for emergency power generation. Construction site requirements for power will vary depending on the scale of the work being performed. This section describes a typical tow-behind generator used to provide electrical power to a job site.

TOW-BEHIND GENERATOR

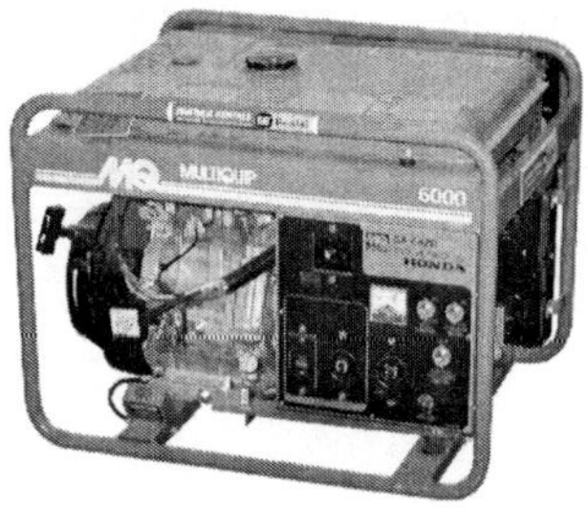

PORTABLE GENERATOR

112F01.EPS

Figure 1 ◆ Generators.

3.1.0 Generator Operator Qualifications

Only trained, authorized persons are permitted to operate a generator. Personnel who have not been trained in generator operation may only operate them for the purpose of training. The training must be conducted under the direct supervision of a qualified trainer.

3.2.0 Typical Generator Controls

Tow-behind generators have operator controls for the engine and an electrical control panel to control and monitor the power produced by the generator. *Figure 3* shows a typical engine control panel. This panel provides the operator with the controls required to start and stop the motor and monitor critical motor functions.

112F02.EPS

Figure 2 ◆ Tow-behind generator for lighting.

Controls and indicators include:

- *Engine over-speed indicator* – This indicator monitors engine speed. If speed exceeds the manufacturer's set limit, the engine will shut down automatically.
- *High engine temperature indicator* – This indicates when the engine coolant temperature has exceeded the manufacturer's set limit. The engine will shut down automatically.
- *Low engine oil pressure indicator* – This indicates when the engine oil pressure has fallen below the manufacturer's preset limit. The engine will shut down automatically.
- *Alternator not charging* – This indicates the engine alternator is not producing enough energy to charge the unit's battery.
- *Ignition switch* – Place this switch in the ON position to run; place it in the OFF position to stop.
- *Start switch* – Pressing this switch activates the engine starter motor.
- *Safety circuit bypass* – Pressing this switch bypasses automatic shutdowns when starting the engine.
- *Emergency stop* – Pressing this switch causes the engine to shut down, with no other operator action required.
- *Engine tachometer gauge* – This gauge indicates engine rpm.
- *Engine oil pressure gauge* – This gauge displays engine oil pressure.
- *Engine coolant temperature* – This gauge indicates the temperature of the engine coolant.

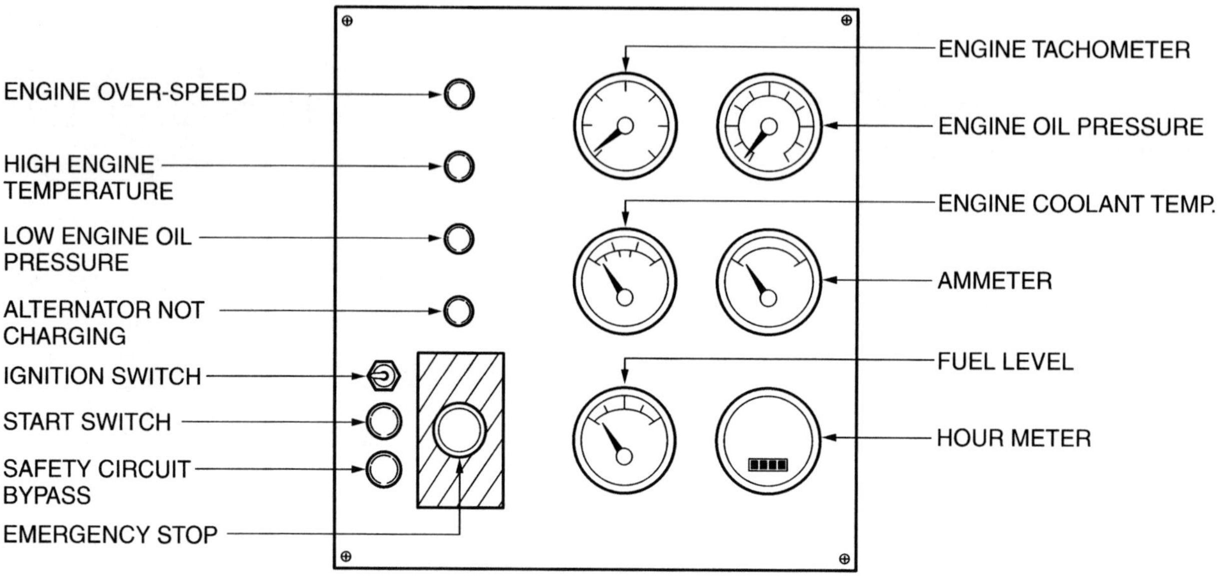

112F03.EPS

Figure 3 ◆ Generator engine control panel.

- *Ammeter gauge* – This gauge indicates the charge rate of the engine alternator.
- *Fuel level gauge* – This gauge indicates the level of fuel in the fuel tank(s).
- *Hour meter* – This meter records total engine operating hours. It is used in determining the maintenance schedule on the unit.

The second control panel in a tow-behind generator is the generator control panel. This panel contains meters, monitor switches, a voltage regulator, **circuit breakers**, and receptacles to control and monitor the output of the generator. *Figure 4* shows a typical generator control panel.

- *AC voltmeter* – This meter indicates the generator output voltage level.
- *AC amperes meter* – This meter indicates the generator output load in **amperes**.
- *Hertz meter* – This meter indicates the frequency of the generator output.
- *Voltage output monitor switch* – This switch selects the reference for the generator voltage displayed on the AC voltmeter.
- *Amperage output monitor switch* – This switch selects the line-to-line (phase) amperage displayed on the AC ammeter.
- *Voltage regulator on/off switch* – In the OFF position, this switch removes excitation to the generator field, stopping the generation of power.
- *Voltage regulator fuse holders* – The **fuse holders** house the fuses for the voltage regulator.
- *Voltage regulator voltage adjust shaft* – This shaft is turned to adjust generator output voltage.
- *Circuit breakers* – Various styles of circuit breakers are used, including push-to-reset, **ground fault interrupter (GFI)**, and flip-to-reset styles.
- *Receptacles* – Various types of receptacles are provided, depending on the size and style of generator used. They will include GFI **straight blade duplex connectors** and **twist-lock connectors**.

3.3.0 Generator Safety Precautions

The safe operation of a generator is the operator's responsibility. Remember the other safety precautions, already discussed in this module, also apply.

WARNING!

High voltages are present when the generator is operating. Exercise caution to avoid electric shock.

The following are general safety precautions that must be observed when operating a generator:

- Do not operate electrical equipment when standing in water or on wet ground, or with wet hands or shoes.

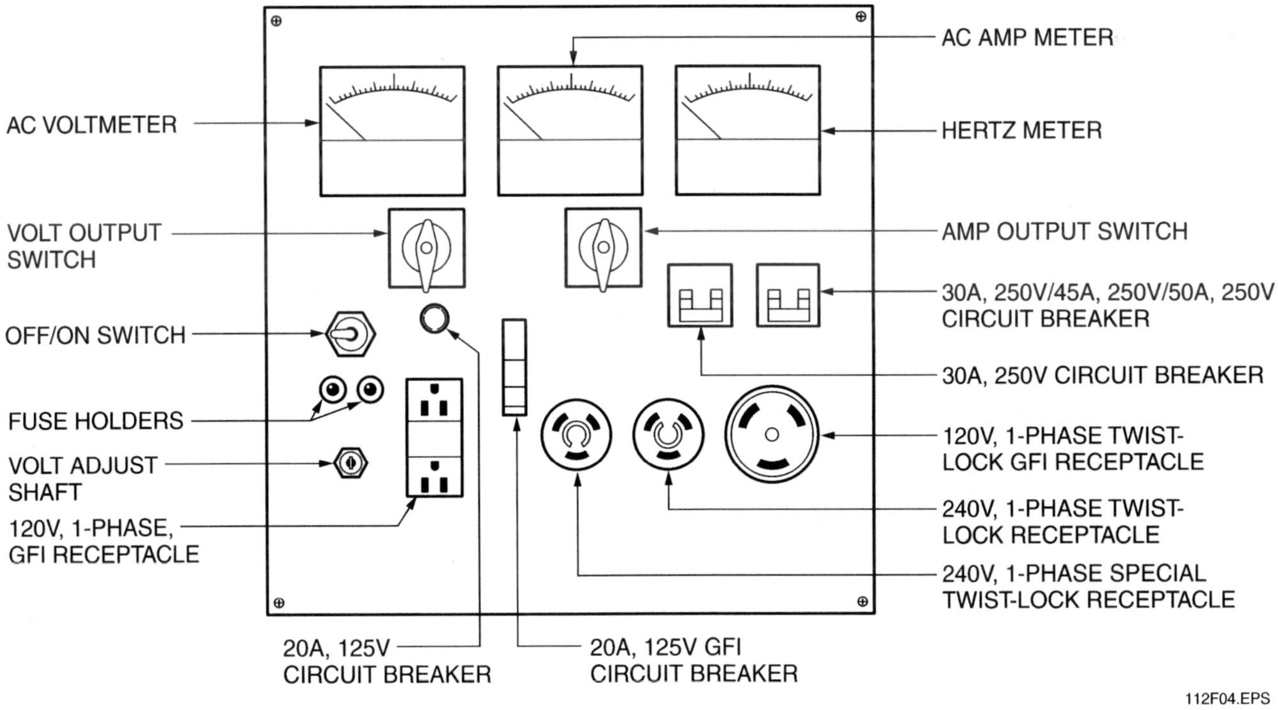

Figure 4 ◆ Generator control panel.

- Grounding should be performed in compliance with local electrical codes and in accordance with the manufacturer's manual.
- Use a generator as an alternate power supply only after the main feed at the service entrance panel has been opened and locked.
- Do not change voltage selection while the engine is running. Voltage selection, adjustment, and electrical connections may only be performed by qualified personnel.
- Do not exceed the generator power (**watt**) rating during operation.
- If welding is required on the unit, follow the manufacturer's instructions to prevent damage to the circuitry.
- Never make electrical connections with the unit running.
- Some localities require specific certification for an operator to run a generator. Be sure that you are aware of any local requirements.

3.4.0 Generator Operation

Before operating the generator, local requirements for grounding must be investigated and followed. The generator set can produce high voltages, which can cause severe injury or death to personnel and damage to equipment. The generator should have proper internal and external grounds when required by the *National Electrical Code®* (*NEC®*). A qualified, licensed electrical contractor, knowledgeable in local codes, should be consulted. Follow all manufacturer's requirements to ensure that the generator is connected properly before operation.

> **NOTE**
> The grounding requirements for portable and vehicle-mounted generators are listed in ***NEC Section 250.34. NEC Section 250.20*** governs the grounding of portable generators for applications that supply fixed wiring systems.

The following discussion covers generic operation of a generator. Each manufacturer provides detailed operating instructions with their equipment. Refer to the operator's manual provided with the generator for specific instructions. The general guidelines given here are intended to help you understand the operator controls. The operator should perform a complete inspection of the generator and correct any deficiencies before beginning operation.

3.4.1 Generator Setup and Preoperational Checks

Many sites have a prestart checklist which must be completed and signed prior to starting or operating an engine-driven generator. Check with your supervisor, and if your site has such a checklist, complete and sign it. A typical procedure for setting up a generator for use at the job site is as follows:

Step 1 Place the unit as level as possible. Follow the manufacturer's directions concerning equipment placement and any special considerations for equipment location.

Step 2 Disconnect the generator from the towing vehicle.

Step 3 Chock the wheel of the generator.

Step 4 Unlock the jack and lower it to the service position.

Step 5 Lock the jack in the service position.

Step 6 Disconnect the safety chains and crank the jack to raise the coupling off the hitch.

Before putting the generator in operation, check it for evidence of arcing on or around the control panel. If any arcing is noted, the problem must be located and repaired before beginning operation. Check for loose wiring or loose routing clamps within the housing.

If your site does not have a prestart checklist, perform the following checks before starting the engine:

- Check the oil, using the engine oil dipstick. If the oil is low, add the appropriate grade oil for the time of year.
- Check the coolant level in the radiator if the engine is liquid-cooled. If the coolant level is low, add coolant.

> **CAUTION**
> Do not add plain water to radiators that contain antifreeze. Antifreeze not only protects radiators from freezing in cold weather, it also has rust inhibitors and additives to aid in cooling. If the antifreeze is diluted, it will not function properly. If the weather turns cold, the system may freeze, causing damage to the radiator, engine block, and water pump.

- Check the fuel. The unit may have a fuel gauge or a dipstick. If the fuel is low, add the correct fuel (diesel or gasoline) to the fuel tank.

The type of fuel required should be marked on the fuel tank. If it is not marked, contact your supervisor to verify the fuel required and have the tank marked.

WARNING!
Never add fuel to an engine-driven generator while the motor is running. Always make sure that no one is using the generator, and shut down the generator to refuel it.

CAUTION
Adding gasoline to a diesel engine or diesel to a gasoline engine causes severe engine problems. It can also cause a fire hazard. Always be sure to add the correct fuel to the fuel tank.

- Check the battery water level unless the battery is sealed. Add room-temperature water if the battery water level is low. Never add cold water to a battery.
- If this is a welding machine, check the electrode holder to be sure it is not grounded. If the electrode holder is grounded, it will arc and overheat the welding system when the welding machine is started. This is a fire hazard and can cause damage to the equipment.
- Open the fuel shutoff valve if the equipment has one. The fuel shutoff valve is located in the fuel line between the fuel tank and the carburetor.
- Record the hours from the hour meter if the equipment has one. An hour meter records the total number of hours the engine runs. This information is used to determine when the engine needs to be serviced. The hours are displayed on a gauge similar to an odometer.
- Clean the unit. Use a compressed air hose to blow off the engine and generator or alternator. Use a rag to remove heavier deposits that cannot be removed with the compressed air.

WARNING!
Whenever using compressed air to clean equipment, you must wear safety glasses with side shields or mono-goggles to protect your eyes from flying debris. You should also warn all others in the immediate area to ensure their safety.

NOTE
Cleaning may not be required on a daily basis. Clean the unit as required.

3.4.2 Starting the Engine

Read and follow the manufacturer's starting procedures found in the operator's manual. Most engines have an on/off ignition switch and a starter. They may be combined into a key switch similar to the ignition on a car. To start the engine, turn on the ignition switch and press the starter. Release the starter when the engine starts. Larger diesel engines have glow plugs that must be warmed up before starting. The engine speed is controlled by the **governor**. If the governor switch is set for idle, the engine will slow to an idle after a few seconds. Small engine-driven generators may have an on/off switch and a pull cord to start the engine. Engine-driven generators should be started 5 to 10 minutes before they are needed to allow the engine to warm up before a load is placed on it.

NOTE
Small engines are equipped with a choke, which provides extra combustion air during startup. The choke is closed once the engine is running.

If no power is required for 30 minutes or more, stop the generator by turning off the ignition switch. Never shut down an engine-driven generator before making sure that all workers using the generator have stopped.

CAUTION
It is important that the extension cords used to supply power from a portable generator to electric power tools or other devices have an adequate current-carrying capacity for the job. If undersized extension cords are used, excessive voltage drops will result. This causes excessive heating of the extension cords and the portable tools, as well as additional generator loading.

3.5.0 Generator Operator's Maintenance Responsibility

Figure 5 shows a typical preventive maintenance schedule for a generator. The schedule is laid out with the performance period across the top and the items to be checked listed down the side of the page. For example, under the Daily column, the following items need to be checked:

- Evidence of arcing around electrical terminals
- Loose wire routing clamps
- Engine oil and coolant levels
- Grounding circuit
- Instruments
- Fan belts, hoses, and wiring insulation
- Air vents
- Fuel/water separator
- Service air indicator

Proper maintenance will extend the life and performance of the equipment.

PREVENTIVE MAINTENANCE SCHEDULE

	DAILY	WEEKLY	MONTHLY/ 150 HRS.	3 MONTHS/ 250 HRS.	6 MONTHS/ 500 HRS.	YEARLY 1000 HRS.
Evidence of Arcing Around Electrical Terminals	✓					
Loose Wire Routing Clamps	✓					
Engine Oil and Coolant Levels	✓					
Proper Grounding Circuit	✓					
Instruments	✓					
Frayed/Loose Fan Belts, Hoses, and Wiring Insulation	✓					
Obstructions in Air Vents	✓					
Fuel/Water Separator (drain)	✓					
Service Air Indicator	✓					
Precleaner Dumps		✓				
Tires		✓				
Battery Connections		✓				
Engine Radiator (exterior)			✓			
Air Intake Hoses and Flexible Hoses			✓			
Fasteners (tighten)			✓			
Emergency Stop Switch Operation			✓			
Engine Protection Shutdown System			✓			
Diagnostic Lamps			✓			
Voltage Selector and Direct Hook-up Interlock Switches				✓		
Air Cleaner Housing				✓		
Control Compartment (interior)					✓	
Fuel Tank (fill at end of each day)					Drain	
Fuel/Water Separator Element					Replace	
Wheel Bearings and Grease Seals					Repack	
Engine Shutdown System Switches (settings)						✓
Exterior Finish				(as needed)		
Engine				Refer to Engine Operator		
Decals				Replace decals if removed, damaged, or missing.		

✓ = Check or Clean (and Adjust or Replace, if necessary)

112F05.EPS

Figure 5 ◆ Example of a generator preventive maintenance schedule.

4.0.0 ◆ AIR COMPRESSORS

Compressors are used to provide compressed air for **pneumatic** tools at the plant. Compressors are available in many different sizes and configurations (see *Figure 6*). They range from portable home-use machines to large installed units for industrial applications. Requirements for compressed air vary depending on the scale of the work being performed. This section describes a typical tow-behind compressor. The primary rating of air compressor capacity is expressed in pounds per square inch (psi) and in cubic feet per minute (cfm). The pressure in psi will depend on the pressure ratings of the tools or equipment to be operated by the compressor. Construction compressors typically range from 50 to 125 psi, but can go up to more than 390 psi. Most hand-held pneumatic tools run on 90 psi. The term cfm refers to the amount of air delivered at the required pressure. The more tools and equipment, the more air delivery is required. A knowledgeable person must determine the size and type of compressor based on the tools and equipment to be used in the plant.

Many people size compressors based on a horsepower rating. This is incorrect. Selection of a larger horsepower compressor motor/engine allows the compressor to run at faster speeds or larger displacement to produce greater airflow at a rated pressure. The size of a compressor should be based on the total air requirements of the various tools or equipment to be used with the compressor. These are the total airflow in cubic feet per minute (cfm) and the required air pressure in pounds per square inch (psi). Pneumatic tool manufacturers provide this information for each of their tools. For example, a typical framing nailer gun operates at a pressure between 70 and 100 psi and consumes an average airflow of 9.6 cfm.

A rule of thumb for sizing a compressor is to determine the total cfm required, then multiply it by 1.5. For example, if the total airflow required is 20 cfm, the compressor should be sized to produce 30 cfm ($20 \times 1.5 = 30$). This means that the compressor must produce a minimum of 30 cfm at maximum rated pressure to operate this tool properly.

The compressor rated pressure in psi is determined by the pressure requirements of the tools and/or equipment that the compressor is supplying. It is important that the compressor maintain pressures that are higher than what is required. For example, if the tools/equipment being used require a pressure of 120 psi, then a compressor capable of producing and maintaining a pressure higher than 120 psi is required.

The operating duty cycle of the tools being used with the compressor is another factor to be considered. Tools such as nail guns tend to be operated intermittently so that they only consume small amounts of air over a long time interval. However, tools like sanders and grinders tend to be operated continuously, causing them to consume large amounts of air over a long time interval. If the application involves tools or equipment that will be used continuously, select a compressor capable of producing a higher cfm output.

4.1.0 Compressor Assemblies

Figure 7 shows a tow-behind compressor. Its major components include:

- Towing assembly and frame
- Protective cover and doors
- Engine and compressor assembly
- Operator control panel located behind the access door

Figure 6 ◆ Types of compressors.

4.2.0 Compressor Operator Qualifications

Only trained, authorized persons are permitted to operate a compressor. Personnel who have not been trained in the operation of compressors may only operate them for the purpose of training. The training must be conducted under the direct supervision of a qualified trainer.

4.3.0 Typical Compressor Controls

Compressors normally have operator controls for the engine and compressor. This section discusses controls associated with a tow-behind compressor. *Figure 8* shows a typical compressor control panel. This panel provides the operator with the controls required to start and stop the motor and monitor critical motor and compressor functions.

Controls and indicators include the following:

- *Engine over-speed indicator* – This indicator monitors engine speed. If the speed exceeds the manufacturer's preset limit, the engine will shut down automatically.
- *High engine temperature indicator* – This indicates that the engine coolant temperature has exceeded the manufacturer's preset limit. The engine will shut down automatically.
- *Low engine oil pressure indicator* – This indicates that the engine oil pressure has fallen below the manufacturer's preset limit. The engine will shut down automatically.
- *Alternator not charging* – This indicates that the engine alternator is not outputting enough energy to charge the unit's battery.
- *Ignition switch* – Place this switch in the ON position to run; place to the OFF position to stop.
- *Start switch* – Pressing the start switch activates the engine starting motor.

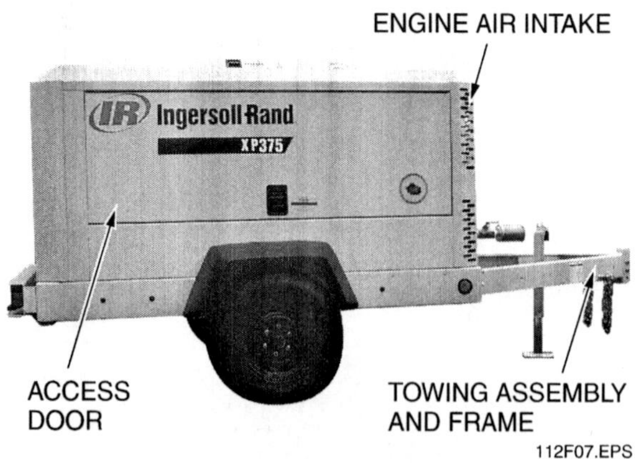

Figure 7 ◆ Large tow-behind compressor.

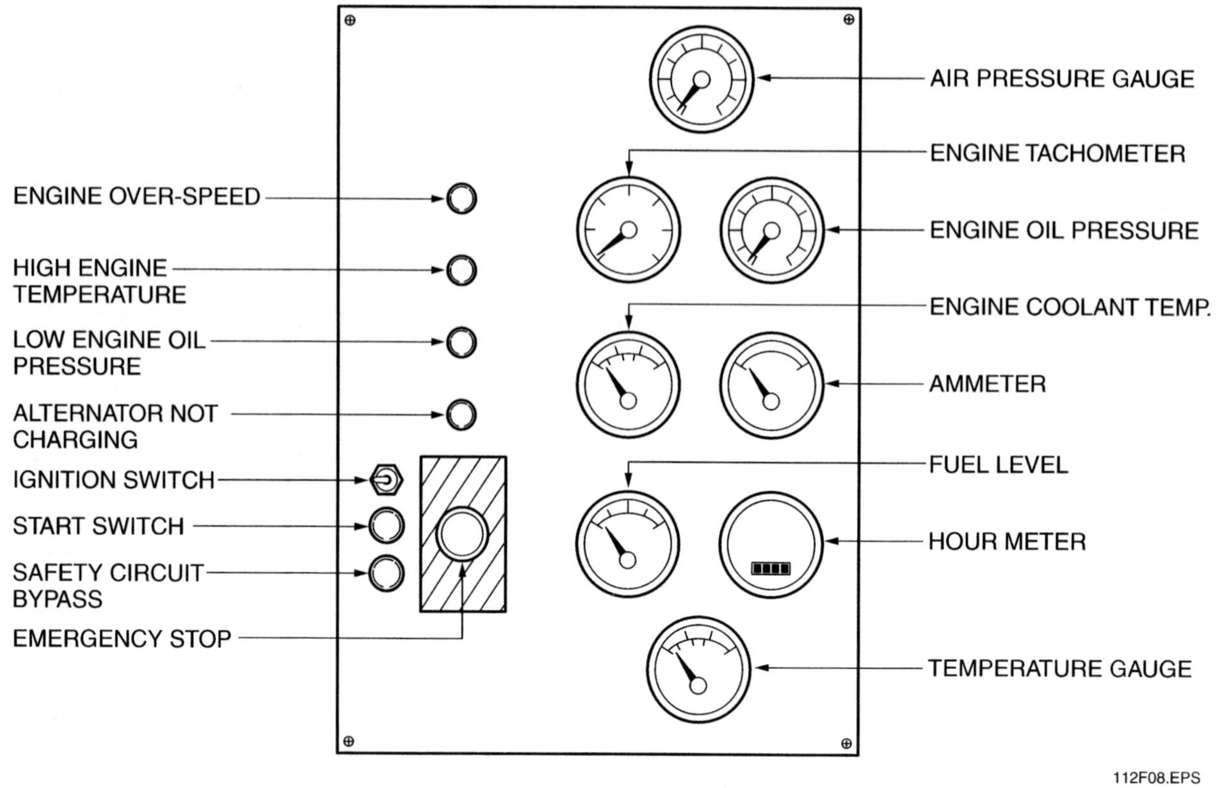

Figure 8 ◆ Compressor control panel.

- *Safety circuit bypass* – Pressing this switch bypasses automatic shutdowns when starting the engine.
- *Emergency stop* – Pressing this switch causes the engine to shut down with no other operator action required.
- *Air pressure gauge* – This gauge indicates the output air pressure level.
- *Engine tachometer gauge* – This gauge indicates engine rpm.
- *Engine oil pressure gauge* – This gauge displays engine oil pressure.
- *Engine coolant temperature* – This gauge indicates the temperature of the engine coolant.
- *Ammeter gauge* – This gauge indicates the charge rate of the engine alternator.
- *Fuel level gauge* – This gauge indicates the level of fuel in the fuel tank(s).
- *Hour meter* – This meter records total engine operating hours. It is used for scheduling maintenance on the unit.
- *Temperature gauge* – This gauge indicates the temperature of the compressed air.

4.4.0 Compressor Safety Precautions

The safe operation of the air compressor is the operator's responsibility. The following are general safety precautions that are applicable to air compressors. Remember the safety precautions already discussed in this module also apply. Refer to the operator's manual for the particular machine being used for specific safety precautions.

- Compressed air venting to the atmosphere can cause hearing damage. Make sure proper hearing protection is worn by everyone in the area.
- Ether is highly flammable. Do not inject ether into a hot engine or an engine equipped with a glow-plug type of preheater.
- Do not inject ether into the compressor air filter or a common air filter for the engine and compressor.
- Do not attempt to move the compressor or lift the drawbar without adequate personnel or equipment to handle the weight.
- Do not exceed the machine's air pressure rating.
- Do not use compressed air for breathing.
- Never direct compressed air at anyone.
- Be sure no pressure is in the system before removing the compressor filler cap.
- Do not use tools that are rated for a pressure lower than that provided by the compressor.
- Be sure all connections are securely made and hoses under pressure are secured to prevent whipping. Use appropriate safety devices to secure hoses.
- Do not weld or perform any modifications on the air compressor receiver tank.
- Use a safe, nonflammable solvent when cleaning parts.

4.5.0 Compressor Operating Procedure

This section covers the generic operation of a compressor.

Each manufacturer provides detailed operating instructions with their equipment. Refer to the operator's manual provided with the compressor for specific instructions. The general outline provided here is intended to help you understand the controls. Perform a complete inspection of the compressor and correct any deficiencies before beginning operation.

4.5.1 Compressor Setup and Preoperational Checks

Before beginning operation, the operator must perform a setup and preoperational checklist. Each manufacturer will provide a checklist with the equipment. This section covers the setup and preoperational checks for compressors.

- Place the unit in as level a position as possible.
- Follow the manufacturer's directions concerning equipment placement and any special considerations for equipment location.
- Disconnect the compressor from the towing vehicle.
- Chock the wheels of the compressor.
- Unlock the jack and lower it to the service position. Lock the jack in the service position.
- Disconnect the safety chains and crank the jack to raise the coupling off the hitch.

Vent all internal pressure before opening any air line, fitting, hose, valve, drain plug, connector, oil filler, or filler caps, and before refilling anti-ice systems. Be sure all hoses, shutoff valves, flow-limiting valves, and other attachments are connected according to the manufacturer's instructions. If the air compressor is to be used along with other sources of air, be sure there is a check valve installed at the service valve.

The following checks should be performed before starting the engine:

- Check the oil, using the engine oil dipstick. If the oil is low, add the appropriate grade oil for the time of year.
- Check the coolant level in the radiator if the engine is liquid-cooled. If the coolant level is low, add coolant.

CAUTION

Do not add plain water to radiators that contain antifreeze. Antifreeze not only protects radiators from freezing in cold weather, it also has rust inhibitors and additives to aid in cooling. If the antifreeze is diluted, it will not function properly. If the weather turns cold, the system may freeze, causing damage to the radiator, engine block, and water pump.

- Check the fuel. The unit may have a fuel gauge or a dipstick. If the fuel is low, add the correct fuel (diesel or gasoline) to the fuel tank. The type of fuel required should be marked on the fuel tank. If it is not marked, contact your supervisor to verify the fuel required and have the tank marked.

CAUTION

Adding gasoline to a diesel engine or diesel to a gasoline engine causes severe engine problems. It can also cause a fire hazard. Always be sure to add the correct fuel to the fuel tank.

- Check the battery water level unless the battery is sealed. Add room-temperature water if the battery water level is low. Do not add cold water to a battery.
- Open the fuel shutoff valve if the equipment has one. The fuel shutoff valve is located in the fuel line between the fuel tank and the carburetor.
- Record the hours from the hour meter if the equipment has one. An hour meter records the total number of hours the engine runs. This information is used to determine when the engine needs to be serviced. The hours are displayed on a gauge similar to an odometer.
- Check the tension and condition of the belts. If the belts are loose, tighten them. If they are worn, replace them.
- Check the safety valve setting to make sure it does not stick and is working properly.
- Make sure that the regulating valve between the compressor and the outlet valves is closed.

4.5.2 Starting the Air Compressor

Most engines have an on/off ignition switch and a starter, which may be combined into a key switch similar to the ignition on a car. To start the engine, turn on the ignition switch and press the starter. Release the starter when the engine starts. On diesel-engine generators, hold the glow plug button to warm up the glow plugs starting the compressor engine. Small air compressors may have an on/off switch and a pull cord to start the engine. Engine-driven compressors should be allowed to build up to the operating pressure before they are used. While the compressor is building up pressure, connect all outlet lines and tools to the compressor. As soon as the operating pressure is reached, open the regulating valve to the outlet valves.

CAUTION

Before disconnecting any tools from the air lines, close the regulating valve to that particular line and bleed the air from the line. It is now safe to disconnect a tool from that line and connect another tool. Open the regulating valve before trying to use the tool.

When shutting down the air compressor, close the regulating valve at the outlet ports and then bleed the air from the air lines. Turn off the ignition switch and disconnect all tools and air lines from the compressor.

4.6.0 Air Compressor Operator's Maintenance Responsibility

The manufacturer's manual provides a schedule and procedures to be followed when performing maintenance on the air compressor. Before performing any maintenance, the operator must be trained, authorized, and have the proper tools to perform any procedures. The following are examples of the types of maintenance that could be performed on an air compressor:

- Changing the engine oil and filter
- Changing the engine and compressor air filters
- Changing the engine coolant
- Performing battery maintenance
- Lubricating parts
- Inspecting all guards and safety devices

5.0.0 ◆ AERIAL LIFTS

Aerial lifts are used to raise and lower workers to and from elevated job sites. There are two main types of lifts: boom lifts and scissor lifts. Both types are made in various models. Some are transported on a vehicle to a job site, where they are unloaded. Others are trailer-mounted and towed to the job site by a vehicle, and some are permanently mounted on a vehicle. Depending on their design, they can be used for indoor work, outdoor work, or both. *Figure 9* shows two commonly used types of aerial lifts.

Boom lifts are designed for both indoor and outdoor use. Boom lifts have a single arm that extends a work platform/enclosure capable of holding one or two workers. Some models have a jointed (articulated) arm that allows the work platform to be positioned both horizontally and vertically. Scissor lifts raise a work enclosure vertically by means of crisscrossed supports.

Most models of aerial lifts are self-propelled, allowing workers to move the platform as work is performed. The power to move these lifts is provided by several means, including electric motors, gasoline or diesel engines, and hydraulic motors.

5.1.0 Aerial Lift Assemblies

Aerial lifts normally consist of three major assemblies: the platform, a lifting mechanism, and the base. *Figure 10* shows these components for a scissor lift.

The platform of an aerial lift is constructed of a tubular steel frame with a skid-resistant deck surface, railings, toe board, and midrails. Entry to the platform is normally from the rear. The entry opening is closed either with a chain or a spring-returned gate with a latch. The work platform may also be equipped with a retractable extension platform. The lifting mechanism is raised and

BOOM SUPPORTED WORK PLATFORM (BOOM LIFT)

SELF-PROPELLED ELEVATING WORK PLATFORM (SCISSOR LIFT)

112F09.EPS

Figure 9 ◆ Aerial lifts.

lowered either by electric motors and gears or by one or more single-acting hydraulic lift cylinder(s). A pump, driven by either an AC or DC motor, provides hydraulic power to the cylinder(s). The base provides a housing for the electrical and hydraulic components of the lift. These components are normally mounted in swingout access trays. This allows easy access when performing maintenance or repairs to the unit. The base also contains the axles and wheels for moving the assembly. In the case of a self-propelled platform, electrical or hydraulic motors will drive two or more of the wheels to allow movement of the lift from one location to another. Brakes will be incorporated on one or more of the wheels to prevent inadvertent movement of the lift.

5.2.0 Aerial Lift Operator Qualifications

Only trained and authorized workers may use an aerial lift. Safe operation requires the operator to understand all limitations and warnings, operating procedures, and operator requirements for maintenance of the aerial lift. The following is a list of requirements the operator must meet:

- Understand and be familiar with the associated operator's manual for the lift being used.
- Understand all procedures and warnings within the operator's manual and posted on decals on the aerial lift.
- Be familiar with employer's work rules and all related government (OSHA) safety regulations.
- Demonstrate this understanding and operate the associated model of aerial lift during training in the presence of a qualified trainer.

5.3.0 Typical Aerial Lift Controls

Figure 11 shows an example of an electrical panel and its associated controls. This is only an example: other models of lifts may have different control configurations. This electrical panel contains the following switches and controls:

- *Up/down toggle switch* – By holding this switch in the up position, the platform can be raised to the desired level. The platform will stop moving when the switch is returned to the center position. By holding this switch in the down position, the platform can be lowered.
- *Buzzer alarm* – An audible alarm sounds when the platform is being lowered. On some models, this alarm may sound when any control function is being performed.
- *Hour meter* – This meter records the number of hours the platform has been operating. The meter will only register when the electric or hydraulic motor associated with operating the aerial lift is running.

One other control that may be available to the operator is an emergency battery disconnect switch. When this switch is placed in the OFF position, it will disconnect power to all control circuits.

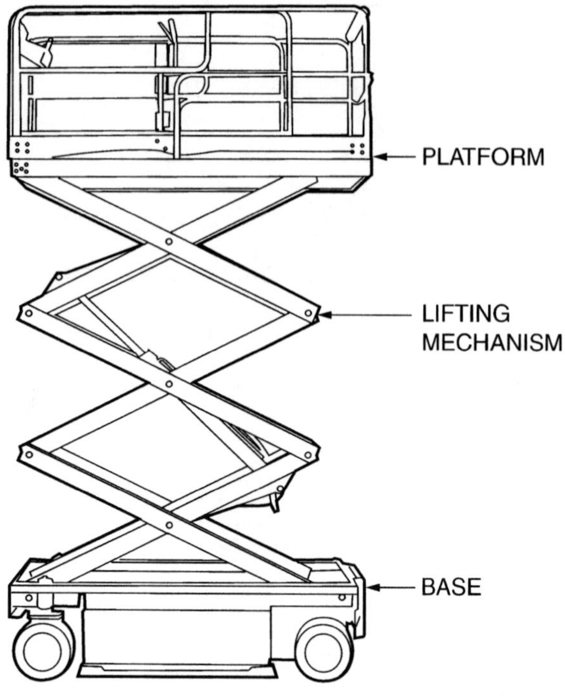

Figure 10 ◆ Aerial lift components.

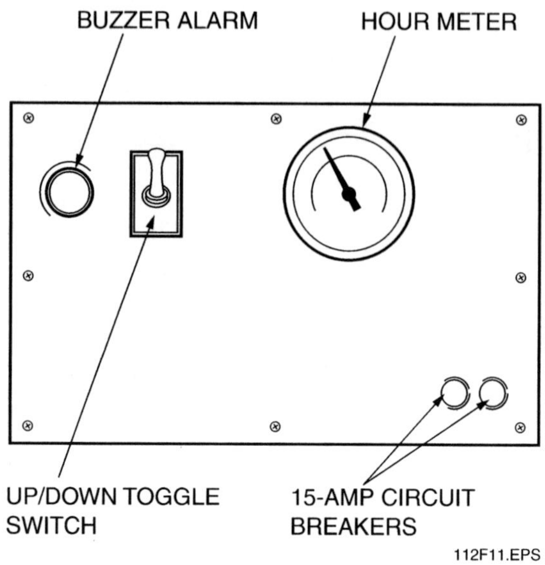

Figure 11 ◆ Aerial lift electrical panel.

The basic controls for an aerial lift with a hydraulic system would generally include:

- *Emergency lowering valve* – This valve allows for platform lowering in the event of an electrical/hydraulic system failure.
- *Free wheeling valve* – Opening this valve allows hydraulic fluid to flow through the wheel motors. This allows the aerial lift to be pushed by hand and prevents damage to the motors when the aerial lift is moved between job site locations. There are usually strict limits to how fast the aerial lift may be moved without causing damage to hydraulic system components.

Other types of controls that may be available on hydraulic aerial lifts include:

- *Parking brake manual release* – This control allows manual release of the parking brake. This control should only be used when the aerial lift is located on a level surface.
- *Safety bar* – The safety bar is used to support the platform lifting hardware in a raised position during maintenance or repair.
- *Emergency stop button* – This button is used to cut off power to both the platform and base control boxes.

Stabilizer hardware is required on some aerial lifts. At a minimum, this hardware includes a stabilizer leg, stabilizer lock pin and cotter key, and a stabilizer jack at each corner of the aerial lift.

Most aerial lifts are equipped with a base control box. The base control box includes a switch to select whether operation will be controlled from the platform or the base control box. The base control overrides the platform control.

The worker on the platform designated as the operator uses the platform controls. In some aerial lifts, operation from the platform is limited to raising and lowering the platform. On others, the aerial platform is designed to be driven by the operator using controls provided on the platform. In either case, the operator will have an operator control box located on the platform assembly. *Figure 12* shows a typical platform control box for a self-propelled aerial lift.

Controls available for the operator include (but are not limited to) the following:

- *Drive/steer controller* – This is a one-handed toggle-type (joystick) lever control for controlling speed and steering of the aerial platform. This is usually a deadman switch that returns to neutral and locks when released. The handle is moved forward to drive the aerial platform forward. The platform speed is determined by how far forward the handle is moved. The handle is moved backward to drive the aerial platform backward; speed is selected as before. Releasing the stick will stop the motion of the aerial lift. Steering is performed by depressing a rocker switch on the top of the stick in the desired direction of travel, either right or left.
- *Up/down selector switch* – Placing the switch in the up position raises the platform; placing the switch in the down position lowers the platform. When released, the switch returns to the middle position and stops the movement of the platform.
- *Lift/off/drive selector switch* – The lift position energizes the lift circuit; the off position removes power to the control box; the drive position energizes the drive and steering controls.
- *Emergency stop pushbutton* – When pushed, this button disconnects power to the platform control circuit. In an emergency, push the button in. To restore power, pull the button out.
- *Lift enable button* – When pushed and held down, this button enables the lift circuit. The button must be held down when raising or lowering the platform. Releasing the button stops the motion of the platform.

Different aerial lifts will have different controls. It is important that the operator fully understand all controls and their functions before operating the aerial lift. The operator must read and fully understand all control and safety information provided by the manufacturer before attempting to operate the equipment.

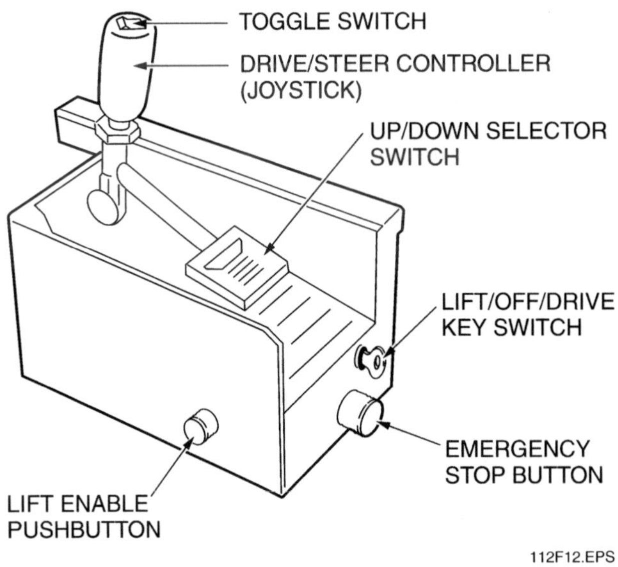

Figure 12 ◆ Aerial lift platform control box.

5.4.0 Aerial Lift Safety Precautions

Safety precautions unique to aerial lifts are listed here. Remember that the other safety precautions, already discussed in this module, also apply. Each manufacturer will provide specific safety precautions in the operator's manual provided with their equipment. Specifically, *OSHA Standard 1926.453* defines and governs the use of aerial lifts.

- Avoid using the lift outdoors in stormy weather or in strong winds.
- Prevent people from walking beneath the work area of the platform.
- Use personal fall arrest equipment (body harness and lanyard) as required for the type of lift being used. Use approved anchorage points.
- Lower the lift and lock it into place before moving the equipment. Also, lower the lift, shut off the engine, set the parking brake, and remove the key before leaving it unattended.
- Stand firmly on the floor of the basket or platform. Do not lean over the guardrails of the platform, and never stand on the guardrails. Do not sit or climb on the edge of the basket or use planks, ladders, or other devices to attain additional height.
- In some circumstances, a lift may be used on a barge or dock. If that is the case, the lift must be safely chained down.
- In any case where it is possible to keep the lift in one place while working, you must chock the wheels so that the machine cannot roll.
- When working on certain kinds of structures, you may be required to use two lanyards, one attaching you to points on the structure and the other attached to a point on the manlift.
- Understand and follow manufacturers' instructions at all times.

5.5.0 Aerial Lift Operation

This section describes operations for a scissor-type, self-propelled aerial lift. Remember that operating procedures will vary from model to model and the operator must become familiar with all operating procedures, controls, safety features, and associated safety precautions before operating any type of aerial lift.

A typical **proportional control** procedure involves the operation of a lever or foot pedal to cause the aerial lift to move. The further the control is moved, the more power is applied to the motor and the faster the aerial lift will move.

The following tasks must be performed before operating the aerial lift:

- Carefully read and fully understand the operating procedures in the operator's manual and all warnings and instruction decals on the work platform.
- Check for obstacles around the work platform and in the path of travel such as holes, dropoffs, debris, ditches, and soft fill. Operate the aerial lift only on firm surfaces.
- Check overhead clearances. Make sure to stay at least 10' away from overhead power lines.
- Make sure batteries are fully charged (if applicable).
- Make sure all guardrails are in place and locked in position.
- Perform an operator's checklist.
- Never make unauthorized modifications to the components of an aerial lift.

Figure 13 shows an example of an operator's checklist.

To drive the aerial lift forward, the operator selects the drive position with the lift drive select switch. The operator then lifts the handle lock ring and moves the drive/steer controller forward. The speed can be adjusted by continuing to move the controller forward until the desired speed is reached. By releasing the controller, the forward motion of the lift is stopped. To drive in reverse, the lever is moved in the opposite direction.

5.6.0 Aerial Lift Operator's Maintenance Responsibility

Death or injury to workers and damage to equipment can result if the aerial platform is not maintained in good working condition. Inspection and maintenance should be performed by personnel who are authorized for such procedures. The operator must be assured that the work platform has been properly maintained before using it.

As a minimum, if the operator is not responsible for the maintenance of the aerial lift, the operator should perform the daily checks. *Figure 14* shows an example of a maintenance and inspection schedule for a typical aerial lift. The daily checks should be performed at the beginning of each shift or at the beginning of the day if only one shift is worked. The aerial lift must never be used until these checks are completed satisfactorily. Any deficiencies found during the daily check must be corrected before using the aerial lift.

Figure 14 is broken into four major categories. Each category is further broken down into the components that should be checked daily, weekly, monthly, every three months, every six months, and yearly. Footnotes (the numbers in parentheses) following each component tell the operator what type of inspection is required. For example, under the Electrical category, battery fluid level should be checked daily or at the beginning of each shift. The (1) refers the operator to the Notes portion of the schedule. Note 1 tells the operator to perform a visual inspection of the battery fluid level.

Different models of aerial lifts will have different maintenance requirements. Always refer to the operator's manual to determine exactly what checks are required before operation. General maintenance rules applicable to any aerial lift include the following:

- Disconnect the battery ground negative (–) lead before performing any maintenance.
- Properly position safety devices before performing maintenance with the work platform in the raised position.

Preventive maintenance is easier and less expensive than corrective maintenance.

OPERATOR'S CHECKLIST

INSPECT AND/OR TEST THE FOLLOWING DAILY OR AT BEGINNING OF EACH SHIFT

___ 1. OPERATING AND EMERGENCY CONTROLS

___ 2. SAFETY DEVICES

___ 3. PERSONNEL PROTECTIVE DEVICES

___ 4. TIRES AND WHEELS

___ 5. OUTRIGGERS (IF EQUIPPED) AND OTHER STRUCTURES

___ 6. AIR, HYDRAULIC, AND FUEL SYSTEM(S) FOR LEAKS

___ 7. LOOSE OR MISSING PARTS

___ 8. CABLES AND WIRING HARNESSES

___ 9. DECALS, WARNINGS, CONTROL MARKINGS, AND OPERATING MANUALS

___ 10. GUARDRAIL SYSTEM

___ 11. ENGINE OIL LEVEL (IF SO EQUIPPED)

___ 12. BATTERY FLUID LEVEL

___ 13. HYDRAULIC RESERVOIR LEVEL

___ 14. COOLANT LEVEL (IF SO EQUIPPED)

112F13.EPS

Figure 13 ◆ Example of an aerial lift operator's checklist.

	Daily	Weekly	Monthly	3 Months	6 Months	*Annually
Mechanical						
Structural damage/welds (1)	✓					✓
Parking brakes (2)	✓					✓
Tires and wheels (1)(2)(3)	✓					✓
Guides/rollers/slides (1)	✓					✓
Railings/entry chain/gate (2)(3)	✓					✓
Bolts and fasteners (3)	✓					✓
Rust (1)			✓			✓
Wheel bearings (2) King pins (1)(8)	✓					✓
Steer cylinder ends (8)				✓		✓
Electrical						
Battery fluid level (1)	✓					✓
Control switches (1)(2)	✓					✓
Cords and wiring (1)	✓					✓
Battery terminals (1)(3)	✓					✓
Terminals and plugs (3)	✓					✓
Generator and receptacle (2)	✓					✓
Limit switches (2)	✓					✓
Hydraulic						
Hydraulic oil level (1)	✓					✓
Hydraulic leaks (1)	✓					✓
Lift/lowering time (10)				✓		✓
Hydraulic cylinders (1)(2)		✓				✓
Emergency lowering (2)	✓					
Lift capacity (7)			✓			✓
Hydraulic oil/filter (9)					✓	✓
MIscellaneous						
Labels (1)(11) Manual (12)	✓					✓

Notes:

(1) Visually inspect. (3) Check tightness. (5)(6) N/A. (8) Lubricate.

(2) Check operation. (4) Check oil level. (7) Check relief valve setting. Refer to serial number nameplate. (9) Replace.

(10) General specifications. (12) Proper Operating Manual *must* be in the manual tube.

(11) Replace if missing or illegible.

*Record Inspection Date

112F14.EPS

Figure 14 ◆ Example of an aerial lift maintenance and inspection schedule.

6.0.0 ◆ FORKLIFTS

Forklifts are used to move, unload, and place material (such as pallets of shingles or stacks of plywood) at required locations on the job site. Forklifts are also sometimes called **powered industrial trucks**. In this module, the term forklift will be used. This module introduces you to safe forklift operation. Forklifts are often misused. A firm foundation in forklift safety ensures greater productivity and a safer working environment.

6.1.0 Forklift Assemblies

Many different types of forklifts are available to meet different needs. Most forklifts used in construction are diesel-powered and fall into one or two broad categories: fixed mast and telescoping boom. Within each of these categories, forklifts are further differentiated by their drive train, steering, and capacity.

There are two types of fixed-mast forklifts: rough terrain and warehouse. Rough terrain forklifts are by far the most common type of fixed-mast forklift used in construction (see *Figure 15*). These forklifts are made for outside use. They have higher ground clearances than warehouse forklifts, larger tires, and leveling devices. Warehouse forklifts are designed for inside use. They have hard rubber tires that are usually of the same size.

When the term fixed-mast forklift is used in this module, assume it refers to the rough terrain forklift.

6.2.0 Forklift Operator Qualifications

Only trained, authorized persons are permitted to operate a powered forklift. Operators of powered forklifts must be qualified as to visual, auditory, physical, and mental ability to operate the equipment safely. Personnel who have not been trained and authorized in the operation of forklifts may only operate them for the purpose of training. The training must be conducted under the direct supervision of a qualified trainer.

6.3.0 Forklift Typical Controls

Forklifts have either two-wheel drive or four-wheel drive. The distinction is the same as that used with other vehicles.

A two-wheel-drive forklift has a drive train that transmits power to the front wheels. Many times, the front wheels are larger than the rear wheels and do not steer the forklift. These forklifts are good for general use, but may lack power in rough terrain.

A four-wheel-drive forklift has a drive train that transmits power to all four wheels. These forklifts are well suited to rough terrain and other conditions that might require additional traction.

The rear wheels of a forklift with two-wheel steering move when the steering wheel is turned. This feature is standard on most forklifts and provides increased maneuverability. The rear wheels are generally smaller than the front wheels.

When the four-wheel steering is engaged on a forklift with this feature, all the wheels move. However, some forklifts offer the following three options:

- The rear wheels may be locked to allow the front wheels to move.
- All wheels may move in the same direction. This is called crab steering or oblique steering.
- The front and rear wheels may move in opposite directions. This is called articulated steering or four-wheel steering.

112F15.EPS

Figure 15 ◆ Fixed-mast rough terrain forklifts.

6.3.1 Fixed-Mast Forklifts

The upright member along which the forks travel is called the mast. A typical fixed-mast forklift has a mast that may tilt as much as 19 degrees forward and 10 degrees rearward. These types of forklifts are suitable for placing loads vertically and traveling with loads, but their horizontal reach is limited to how close the machine can be driven to the pick-up or landing point. For example, a fixed-mast forklift cannot place a load of purlins beyond the edge of the roof because of its limited reach.

6.3.2 Telescoping-Boom Forklifts

Telescoping-boom forklifts (*Figure 16*) provide more versatility in horizontal and vertical placement than a fixed-mast forklift. A telescoping-boom forklift is really a combination of a telescoping-boom crane and a forklift.

The mast can be either two-stage or three-stage. This designation refers to the number of telescoping channels built in the mast. A two-stage mast has one telescoping channel and a three-stage mast has two telescoping channels. The purpose of the telescoping channels is to provide greater lift height.

Some models of telescoping-boom forklifts have a level-reach fork carriage, which allows the fork carriage to be moved horizontally while the boom remains in a stationary position.

6.4.0 Forklift Safety Precautions

Safe operation is the responsibility of the operator. Operators must develop safe working habits and recognize hazardous conditions to protect themselves and others from death or injury. Always be aware of unsafe conditions to protect the load and the forklift from damage. Be familiar with the operation and function of all controls and instruments before operating a forklift. Before operating any forklift, read and fully understand the operator's manual.

The following safety rules are specific to forklift operation. Remember the other safety precautions already discussed in this module.

- Never put any part of your body into the mast structure or between the mast and the forklift.
- Never put any part of your body within the reach mechanism.
- Understand the limitations of the forklift.
- Do not permit passengers to ride unless a safe place to ride has been provided by the manufacturer.
- Never leave the forklift running unattended or with a suspended load.

Safeguard pedestrians at all times by observing the following rules:

- Always look in the direction of travel.
- Do not drive the forklift up to anyone standing in front of an object or load.
- Make sure that personnel stand clear of the rear swing area before turning.
- Exercise particular care at blind spots, cross aisles, doorways, and other locations where pedestrians may step into the travel path. Unless otherwise instructed, blow your horn when approaching such points.
- The use of a spotter is recommended when landing an elevated load with a telescoping-boom forklift.
- Put the forks on the ground when you park.

6.5.0 Forklift Operation

The most important factor to consider when using any forklift is its capacity. Each forklift is designed with an intended capacity, and this capacity must never be exceeded. Exceeding the capacity jeopardizes not only the equipment but also the safety of

112F16.EPS

Figure 16 ◆ Telescoping-boom forklifts.

everyone on or near the equipment. Each manufacturer supplies a capacity chart for each forklift model. Be sure to read and follow the capacity chart.

6.5.1 Picking Up a Load

Some forklifts are equipped with a sideshift device that allows the operator to horizontally shift the load several inches in either direction with respect to the mast. A sideshift device enables more precise placing of loads, but it changes the center of gravity and must be used with caution. If the forklift being used is equipped with a sideshift device, be sure to return the fork carriage to the center position before attempting to pick up a load.

Step 1 Check the position of the forks with respect to each other. They should be centered on the carriage. If the forks have to be moved, check the operator's manual for the proper procedure. Usually, there is a pin at the top of each fork that, when lifted, allows each fork to be slid along the upper backing plate until the fork centers over the desired notch.

Step 2 Travel to the area at a safe rate of speed. Always keep the forks lowered when traveling with a forklift.

Step 3 Before picking up a load with a forklift, make sure the load is stable. If it looks like the load might shift when picked up, secure the load. Knowing the center of gravity is crucial, especially when picking up tapered sections. Make a trial lift, if necessary, to determine and adjust the center of gravity.

Step 4 Approach the load so that the forks straddle the load evenly. It is important that the weight of all loads be distributed evenly on the forks. Overloading one fork at the expense of the other can damage the forks. In some cases, it may be advisable to measure the load and mark its center of gravity.

Step 5 Drive up to the load with the forks straight and level. If the load being picked up is on a pallet, be sure the forks are low enough to clear the pallet boards.

Step 6 Move forward until the leading edge of the load rests squarely against the back of both forks. If you cannot see the forks engage the load, ask someone to signal for you. This prevents expensive damage and injury.

Step 7 Raise the carriage, then tilt the mast rearward until the forks contact the load. Raise the carriage until the load safely clears the ground. Then tilt the mast fully rearward to cradle the load. This minimizes the chance that the load may slip during travel.

6.5.2 Traveling with a Load

Always travel with a load at a safe rate of speed. Never travel with a raised load. Keep the load as low as possible and be sure the mast is tilted rearward to cradle the load.

As you travel, keep your eyes open and stay alert. Watch the load and the conditions ahead of you, and alert others of your presence. Avoid sudden stops and abrupt changes in direction.

Be careful when downshifting because sudden deceleration can cause the load to shift or topple. Be aware of front and rear swing when turning.

When carrying cylindrical objects, such as oil drums, keep the mast tilted rearward to cradle the load. If necessary, secure the load to keep it from rolling off the forks.

If you are traveling with a telescoping-boom forklift, be sure the boom is fully retracted.

If you have to drive on a slope, keep the load as low as possible. Do not drive across steep slopes. If you have to turn on an incline, make the turn wide and slow. In most cases, it is safer to back down an incline if your view is at all obstructed.

6.5.3 Placing a Load

Position the forklift at the landing point so that the load can be placed where you want it. Be sure everyone is clear of the load.

The area under the load must be clear of obstructions and must be able to support the weight of the load. If you cannot see the placement, use a signaler to guide you.

With the forklift in the unloading position, lower the load and tilt the forks to the horizontal position. When the load has been placed and the forks are clear from the underside of the load, back away carefully to disengage the forks or retract the boom on variable-reach units.

6.5.4 Placing Elevated Loads

Special care needs to be taken when placing elevated loads. Some forklifts are equipped with a leveling device that allows the operator to rotate the fork carriage to keep the load level during travel. When placing elevated loads, it is extremely important to level the machine before lifting the load.

One of the biggest potential safety hazards during elevated load placement is poor visibility. There may be workers in the immediate area who cannot be seen. The landing point itself may not be visible. Your depth perception decreases as the height of the lift increases. To be safe, use a signaler to help you spot the load.

Use tag lines to tie off long loads.

Drive the forklift as closely as possible to the landing point with the load kept low. Set the parking brake. Raise the load slowly and carefully while maintaining a slight rearward tilt to keep the load cradled. Under no circumstances should the load be tilted forward until the load is over the landing point and ready to be set down.

If the forklift's rear wheels start to lift off the ground, stop immediately, but not abruptly. Lower the load slowly and reposition it or break it down into smaller components, if necessary. If surface conditions are bad at the unloading site, it may be necessary to reinforce the surface conditions to provide more stability.

6.5.5 Traveling with Long Loads

Traveling with long loads presents special problems, particularly if the load is flexible and subject to damage. Traveling multiplies the effect of bumps over the length of the load. A stiffener may be added to the load to give it extra rigidity

To prevent slippage, secure long loads to the forks. This may be done in one of several ways.

A field-fabricated cradle may be used to support the load. While this is an effective method, it requires that the load be jacked up.

In some cases, long loads may be snaked through openings that are narrower than the load itself. This is done by approaching the opening at an angle and carefully maneuvering one end of the load through the opening first. Avoid making quick turns because abrupt maneuvers cause the load or its center of gravity to shift.

6.5.6 Using the Forklift to Rig Loads

The forklift can be a very useful piece of rigging equipment if it is properly and safely used. Loads can be suspended from the forks with slings, moved around the job site, and placed.

All the rules of careful and safe rigging apply when using a forklift to rig loads. Be sure not to drag the load or let it swing freely. Use tag lines to control the load.

Never attempt to rig an unstable load with a forklift. Be especially mindful of the load's center of gravity when rigging loads with a forklift.

6.6.0 Forklift Operator's Maintenance Responsibility

This section of the module is intended to familiarize you with some of the preventive maintenance procedures required of forklift operators:

- Check the operator's manual for lubrication points and suggested lubrication periods. These are given in terms of the number of hours of operation.
- Always check all fluid levels and use the recommended fluids when refilling.
- Because forklifts often operate in dusty and dirty environments, it is important to keep the moving parts of the machine well greased. This reduces wear, prolongs the life of the machine, and ensures safe operation. Follow the manufacturer's recommendations for lubricants.

Forklifts are also provided with a daily checklist from the manufacturer. The checklist covers all the items that must be checked on the forklift every day. *Figure 17* shows an example of a daily checklist and the items that are usually checked. These may differ from unit to unit. At the end of the day, park the forklift on level ground. Lower the forks to the ground and set the parking brake. If you have to park on a slope, position the forklift at right angles to the slope and block the wheels

6.7.0 Pallet Movers and Pallet Jacks

On concrete floors, such as those found in many industrial sites, heavy objects or supplies can often be moved with manual or powered pallet jacks or with electric pallet lifts. It is also frequently possible to gain access to machinery or materials with the powered pallet lift. As a mechanic, you will quite often have to move materials, machinery, or tools to and from a shop or work area. While the forklift would often be a good solution, from the point of view of minimizing effort, forklifts tend to be at a premium in factories in the production and shipping sectors, and you may find yourself better able to get the job done with a hand pallet jack or a lifter truck.

6.7.1 Pallet Jack

The simplest pallet mover is the manual pallet jack or pallet truck (*Figure 18*). The manual pallet jack can lift as much as 5,000 pounds by closing the hydraulic release valve and pumping the handle up and down. Once the material and pallet are clear of the ground, just pull on the handle to set the jack in motion.

OPERATOR'S DAILY CHECKLIST

Check Each Item Before Start Of Each Shift Date:___________

Check One: ☐ Gas/LGP/Diesel Truck ☐ Electric Sit-down ☐ Electric Stand-up ☐ Electric Pallet

Truck Serial Number:_________________ Operator:___________________ Supervisor's OK: ________________

Hour Meter Reading: _________________

Check each of the following items before the start of each shift. Let your supervisor and/or maintenance department know of any problem. DO NOT OPERATE A FAULTY TRUCK. Your safety is at risk.

After checking, mark each item accordingly. Explain below as necessary.

Check boxes as follows: ☐ OK ☐ NG, needs attention, or repair. Circle problem and explain below.

OK	NG	Visual Checks
		Tires/Wheels: wear, damage, nuts tight
		Head/Tail/Working Lights: damage, mounting, operation
		Gauges/Instruments: damage, operation
		Operator Restraint: damage, mounting, operation, oily, dirty
		Warning Decals/Operator's Manual: missing, not readable
		Data Plate: not readable, missing adjustment
		Overhead Guard: bent, cracked, loose, missing
		Load Back Rest: bent, cracked, loose, missing
		Forks: bent, worn, stops OK
		Engine Oil: level, dirty, leaks
		Hydraulic Oil: level, dirty, leaks
		Radiator: level, dirty, leaks
		Fuel: level, leaks
		Battery: connections loose, charge, electrolyte low
		Covers/Sheet Metal: damaged, missing
		Brakes: linkage, reservoir fluid level, leaks
		Engine: runs rough, noisy, leaks

OK	NG	Visual Checks
		Steering: loose/binding, leaks, operation
		Service Brake: linkage loose/binding, stops OK, grab
		Parking Brake: loose/binding, operational, adjustment
		Seat Brake (if equipped): loose/binding, operational, adjustment
		Horn: operation
		Backup Alarm (if equipped): mounting, operation
		Warning Lights (if equipped): mounting, operation
		Lift/Lower: loose/binding, excessive drift, leaks
		Tilt: loose/binding, excessive drift, "chatters," leaks
		Attachments: mounting, damaged, operation, leaks
		Battery Test (electric trucks only): indicator in green
		Battery: connections loose, charge, electrolyte low while holding full forward tilt
		Control Levers: loose/binding, freely return to neutral
		Directional Control: loose/binding, find neutral OK

Explanation of problems marked above: ______________________________

____

____

____

____

____

112F17.EPS

Figure 17 ◆ Example of a forklift operator's daily checklist.

When you are pulling the pallet jack, you must watch for irregularities in the floor that might be tall enough to catch the pallet or the jack itself, causing material to fall off the pallet. Remember that the path of the pallet and jack is frequently wider than yours. The pallet or the jack may catch on the machinery or structure around you, and it must be steered widely enough to keep from catching on corners. When the pallet is where you want it, simply release the hydraulic valve, and the jack will let the pallet down.

Some pallet jacks are provided with an electric motor and brakes, so that heavier pallets can safely be moved. The operator simply steers the truck and walks ahead (walkie), or rides on the truck. *Figure 19* shows an example of a walkie.

6.7.2 Pallet Lifter Truck

The pallet lifter truck is an electric pallet truck. Most models are capable of carrying pallets to the top of a 16-foot rack. The truck (*Figure 20)* steers by turning the rear wheels, and has a lift button and a forward-and-reverse throttle. *Figure 21* shows the controls of a pallet lifter. The wheel is spun to turn the steering wheels, while the throttle controls movement and speed. Driving the lifter truck, you must keep the pallet close to the ground. The lift can easily turn over if you make a sudden turn at any height and speed.

When you arrive at the intended location, raise the lift to the height where you are either dropping or lifting a pallet, and turn the machine to

112F18.EPS

Figure 18 ◆ Manual pallet jack.

112F19.EPS

Figure 19 ◆ Pallet jack with electric motor and brakes.

112F20.EPS

Figure 20 ◆ Pallet lifter truck.

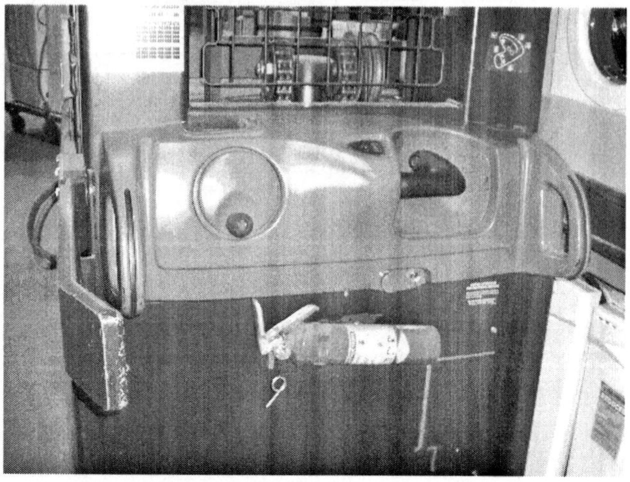

112F21.EPS

Figure 21 ◆ Pallet lifter controls.

align the forks to where the load is to go. If you are dropping a pallet, be careful that the forks clear the pallet when you withdraw them from the slots in the pallet. Be certain that the forks have completely come clear of contact with the top or bottom of the pallet slot, so that you do not drag the pallet off the racks. Be sure that the pallet is safely resting on the rack at front and back, especially if the racks do not have timbers across them.

Lifter trucks are also capable of being used to carry a cage for a worker, either for stocking or to work on a piece of equipment that is off the floor. Do not stand on the forks; use a cage or platform mounted safely on the forks (*Figure 22*). You must wear a protective harness while working from a lift above the ground.

7.0.0 ◆ CRANE TYPES AND USES

In most plants, there will be either an overhead crane on a beam, with or without a specialized operator, or a gantry crane. If this is not available to you, you may have to use a mobile crane to lift and move heavy machinery. The primary function of a mobile crane is to lift and swing loads. A mobile crane consists of a rotating superstructure, operating machinery, and a boom, mounted on a vehicle such as a truck, crawler, or railcar. Since their introduction in the 1800s, mobile cranes have gone through constant change, from steam power to air-cooled diesel engines. The manufacturers of these mobile cranes continuously strive to meet the needs of an ever-changing industry. Because of the wide variety of uses for cranes, many different types of cranes have been developed.

112F22.EPS

Figure 22 ◆ Cage mounted on lift forks.

7.1.0 Boom Trucks

Boom trucks are unlike any of the other mobile cranes in that they are mounted on carriers that are not solely designed for crane service. These cranes are mounted onto a commercial truck chassis that has been strengthened to carry the crane. The boom truck can also be used to haul loads and be driven on the highway system when the boom is in the transport position. Boom trucks typically range in capacity from 6 to 22 tons. *Figure 23* shows a typical boom truck.

7.2.0 Cherry Pickers

Cherry pickers (mobile hydraulic cranes) are rough-terrain mobile cranes. They have oversized tires that allow them to move across the rough terrain of construction sites and other broken ground. Two types of cherry pickers are the fixed cab and the rotating cab. *Figure 24* shows cherry pickers.

7.3.0 Carry Decks

A carry deck is a small hydraulic mobile crane. Carry decks are primarily used in industrial applications where working surfaces are significantly better than those found on most construction sites. The compact size of the carry deck, 15 feet long by 7 feet wide, allows it to be maneuvered easily about the job site. Most carry decks offer up to 12 tons of lifting capacity with the outriggers in place

112F23.EPS

Figure 23 ◆ Typical boom truck.

and up to 7½ tons for lifting and transporting work. Most carry decks come with water-cooled diesel engines, but they are also available with gasoline-powered and propane (LP), diesel, and scrubber engines. The carry deck has a transport area that is designed to lock down and carry up to 15,000 pounds of materials. *Figure 25* shows a carry deck. Carry decks come in two-wheel, four-wheel, and all-way steering.

7.4.0 Lattice Boom Cranes

Lattice boom cranes are capable of much greater boom lengths than hydraulic boom cranes. Site preparation and erection of lattice boom cranes may take anywhere from a day for a smaller crane to weeks for larger cranes. The length of setup time depends on the height of the boom and the lift attachments to be used. The newer lattice boom cranes are powered by computer-controlled hydraulic systems. They have redundant fail-safe devices to ensure operational safety. These devices consist of, but are not limited to: automatic braking systems, load moment measuring devices, mechanical system monitoring devices, and function lockout systems that stop operation when a system fault occurs. These controls provide for better control and accuracy than older friction-operated devices.

7.4.1 Lattice Boom Crawler Cranes

Lattice boom crawler cranes are specifically designed for the extreme duty associated with the use of a crawler crane. Lattice boom crawler cranes are used to pick up a heavy load and track with it to a nearby location. The reliability and versatility of the crawler-mounted lattice boom crane makes it the most widely applied crane design in use today. Besides lifting heavy loads to great heights, these cranes are also ideal for applications with a high duty cycle. *Figure 26* shows a lattice boom crawler crane.

7.4.2 Lattice Boom Truck Cranes

Lattice boom truck cranes provide the mobility of a truck crane with the extreme lifting capacity of a lattice boom crane. Depending on the size of the crane and its gross vehicle weight, some components such as counterweights and outriggers may have to be removed before highway travel to meet local, state, and federal weight restrictions. *Figure 27* shows a lattice boom truck crane.

112F24.EPS

Figure 24 ◆ Cherry pickers.

112F25.EPS

Figure 25 ◆ Carry deck.

7.5.0 Mobile Crane Safety

Safely lifting and moving machinery and equipment is not automatic. It involves making several decisions regarding the object being moved and the lifting and rigging equipment to be used. Safely lifting and moving equipment is the responsibility of everyone involved. Everyone must be alert, responsible, and aware of all the factors involved to use mobile cranes safely. The following guidelines should be followed when working with or around mobile cranes:

- Operators must be properly trained, competent, physically fit, alert, free from the influence of alcohol, drugs, or medications, and, if required, licensed to operate mobile cranes. Good vision, judgment, coordination, and mental ability are also required. Operators who do not possess all these qualities must not be allowed to operate the equipment.
- Signalmen must have good vision and sound judgment. They must know the standard crane signals and be able to give the signals clearly. They must also have enough experience to be able to recognize hazards and signal the operator to avoid them.
- Outriggers must always be used when lifting loads, and good judgment must be used for the placement of outriggers.
- If you do not understand the hand signals, do not lift the load with a crane.
- Riggers must be trained to determine weights and distances and to properly select and use lifting tackle.
- Crew members must understand their specific safety responsibilities and report any unsafe conditions or practices to the proper personnel immediately.
- All crew members working around mobile cranes must obey all warning signs and watch out for their own safety and the safety of others.
- Crew members setting up machines or handling loads must be properly trained and aware of proper machine erection and rigging procedures.
- Watch for hazards during operations, and alert the operator and signalman of dangers, such as power lines, the presence of people, other equipment, or unstable ground conditions.

112F26.EPS

Figure 26 ◆ Lattice boom crawler crane.

112F27.EPS

Figure 27 ◆ Lattice boom truck crane.

Review Questions

1. The operator should check for leaks in a pressurized hydraulic system by _____.
   a. checking the hydraulic oil level in the supply tank
   b. feeling all the hoses and connectors
   c. using a piece of wood or cardboard
   d. using a flashlight and an inspection mirror

2. Before disconnecting any hydraulic system hoses, the operator should shut down the engine and _____.
   a. check the hydraulic fluid level
   b. cycle the hydraulic controls
   c. set the parking brake
   d. chock the wheels

3. When refueling a piece of equipment, keep the nozzle or fuel funnel in contact with the filler neck to _____.
   a. prevent the possibility of sparks
   b. lower the possibility of a fuel spill
   c. protect you from breathing the fuel fumes
   d. make it easier to see when the fuel tank is full

4. When a battery is frozen, _____.
   a. disconnect the negative lead before attaching the charger
   b. do not charge it or attempt to jump-start it
   c. jump-start the equipment following the manufacturer's instructions
   d. replace the battery and charge the old battery

5. A _____ should be consulted about the grounding requirements of a generator.
   a. fellow worker
   b. qualified operator
   c. licensed electrical contractor
   d. representative from OSHA

6. If you find evidence of arcing on the control panel of a tow-behind generator, _____.
   a. continue with operation of the generator
   b. replace the control panel
   c. locate and repair the problem before operating
   d. call the manufacturer for additional information

7. When a portable air compressor is used along with another compressed air source, make sure _____.
   a. there is a check valve installed at the service valve
   b. all air vents are clear
   c. the engine and air compressor air filters are clean
   d. the compressor is on level ground

8. Before they are used, engine-driven compressors should be allowed to _____.
   a. warm up to 105°F
   b. run for 30 minutes
   c. build to operating pressure
   d. start running

9. The components of an aerial lift include a _____.
   a. base, auger, and bucket
   b. base, platform, and lifting mechanism
   c. platform, boom, and compactor
   d. compactor, control panel, and forks

10. The _____ allows hydraulic fluid to flow and allows the aerial lift to be pushed by hand.
    a. emergency lowering valve
    b. free wheeling valve
    c. stabilizer hardware
    d. steer controller

11. Aerial lifts should be kept at least _____ away from overhead lines.
    a. one foot
    b. five feet
    c. ten feet
    d. fifty feet

Review Questions

12. Before performing maintenance on an aerial lift, always disconnect the _____.
    a. hydraulic hoses
    b. battery negative lead
    c. control panel
    d. lift switch

13. The mast of a typical fixed-mast forklift can be tilted _____ degrees forward.
    a. 19
    b. 39
    c. 59
    d. 79

14. Always leave the forklift turned off and with the _____.
    a. mast tilted back
    b. forks tilted up
    c. forks down
    d. mast tilted forward

15. The machine shown in *Figure 1* is a _____.
    a. cherry picker
    b. carry deck
    c. boom truck
    d. lattice boom crane

112RQ01.EPS

Figure 1

Summary

Motorized equipment is another useful tool to make the industrial maintenance craftworker's job easier. This module has introduced you to some of the various types of motorized equipment you will use throughout your career. You should always remember to read and follow the manufacturer's operating procedures and preventive maintenance schedules before attempting to use any type of motorized equipment. If you follow the manufacturer's instructions, you can obtain a long life of safe service from the motorized equipment on the job site.

Keep in mind that anyone operating motorized equipment must be trained, and in many cases certified, to operate the specific equipment. The purpose of this module is to introduce you to the motorized equipment used in industrial maintenance work. Whether or not you are actually trained to use some or all of this equipment will depend on the policies of your school and/or employer.

Notes

Trade Terms Introduced in This Module

Aerial lift: A mobile work platform designed to transport and raise personnel, tools, and materials to overhead work areas.

Ampere: A unit of electrical current, also known as amp.

Cherry picker: A truck-mounted or trailer-mounted lift designed for both indoor and outdoor use.

Circuit breaker: A device designed to protect circuits from overloads and able to be reset after tripping.

Compressor: A motor-driven machine used to supply compressed air for pneumatic tools.

Forklift: A machine designed to move bulk items around the job site.

Fuse holder: A device used to hold a fuse that may be part of a circuit or may simply hold a spare fuse.

Generator: A machine designed to generate electricity.

Governor: A device used to provide automatic control of speed or power for an internal combustion engine.

Ground fault interrupter (GFI): A device that detects grounds on a power line and interrupts power flow to the attached outlet(s).

Pneumatic: Run by or using compressed air.

Powered industrial trucks: See forklift.

Proportional control: A control that increases speed in proportion to the movement of the control.

Straight blade duplex connector: An electrical connector or style of outlet.

Twist-lock connector: A type of electrical connector.

Watt: A unit of power that is equal to one joule per second.

Resources & Acknowledgments

Additional Resources

This module is intended to be a thorough resource for task training. The following reference works are suggested for further study. These are optional materials for continued education rather than for task training.

Construction Equipment Guide. New York, NY: John Wiley & Sons.

Machinery's Handbook, Latest Edition. Erik Oberg, Franklin D. Jones, Holbrook L. Horton, and Henry H. Ryffel. New York, NY: Industrial Press Inc.

Figure Credits

Ingersoll Rand, 112F01 (tow-behind generator), 112F07

Topaz Publications, Inc., 112F01 (portable generator), 112F02, 112F06, 112F21, 112F22

JLG Industries, Inc., 112F09

Sellick Equipment LTD, 112F15

Manitou North America, 112F16

Reprinted courtesy of Cat Lift Trucks, 112F18, 112F19, 112F20

Terex Cranes, 112F23, 112F24 (bottom)

Manitowoc Crane Group, 112F24 (top), 112F25

Link-Belt Construction Equipment Co., 112F26, 112F27

CONTREN® LEARNING SERIES — USER UPDATE

NCCER makes every effort to keep these textbooks up-to-date and free of technical errors. We appreciate your help in this process. If you have an idea for improving this textbook, or if you find an error, a typographical mistake, or an inaccuracy in NCCER's Contren® textbooks, please write us, using this form or a photocopy. Be sure to include the exact module number, page number, a detailed description, and the correction, if applicable. Your input will be brought to the attention of the Technical Review Committee. Thank you for your assistance.

Instructors – If you found that additional materials were necessary in order to teach this module effectively, please let us know so that we may include them in the Equipment/Materials list in the Annotated Instructor's Guide.

Write: Product Development and Revision
National Center for Construction Education and Research
3600 NW 43rd St. Bldg G, Gainesville, FL 32606

Fax: 352-334-0932

E-mail: curriculum@nccer.org

Craft ______ Module Name ______

Copyright Date ______ Module Number ______ Page Number(s) ______

Description ______

(Optional) Correction ______

(Optional) Your Name and Address ______

40113-07

Lubrication

40113-07
Lubrication

Topics to be presented in this module include:

Overview

Lubrication has become a scientific endeavor. There are specific best practices for lubricating different machines, and best lubricants for different environments and applications. In addition to different lubricants and equipment, there are computerized scheduling software and automatic delivery systems. Intelligent lubrication is one of the primary tools of machine maintenance, and a constantly evolving field of study.

Objectives

When you have completed this module, you will be able to do the following:

1. Explain OSHA hazard communication as pertaining to lubrication.
2. Read and interpret a material data safety sheet (MSDS).
3. Explain the EPA hazardous waste control program.
4. Explain lubricant storage.
5. Explain lubricant classification.
6. Explain lubricant film protection.
7. Explain properties of lubricants.
8. Explain properties of greases.
9. Explain how to select lubricants.
10. Identify and explain types of additives.
11. Identify and explain types of lubricating oils.
12. Identify and use lubrication equipment to apply lubricants.
13. Read and interpret a lubrication chart.

Trade Terms

Adsorption	Friction
Atomize	Grease
Bungs	Hydrocarbons
Carbon dioxide (CO_2)	Inhibitor
Centistoke (cSt)	NLGI consistency number
Corrosion	Oil
Decomposition	Oxidation
Dispersant	Paraffin
Disposal	Polymerization
Dropping point	Pour point
Dunnage	Saybolt universal seconds (SUS, SSU)
Emulsion	Viscosity
Environmental Protection Agency (EPA)	Viscosity index (VI)
Fire point	
Flash point	

Required Trainee Materials

1. Pencil and paper
2. Appropriate personal protective equipment

Prerequisites

Before you begin this module, it is recommended that you successfully complete *Core Curriculum;* and *Industrial Maintenance E & I Technician Level One,* Modules 40101-07 through 40112-07.

This course map shows all of the modules in the first level of the *Industrial Maintenance E & I Technician* curriculum. The suggested training order begins at the bottom and proceeds up. Skill levels increase as you advance on the course map. The local Training Program Sponsor may adjust the training order.

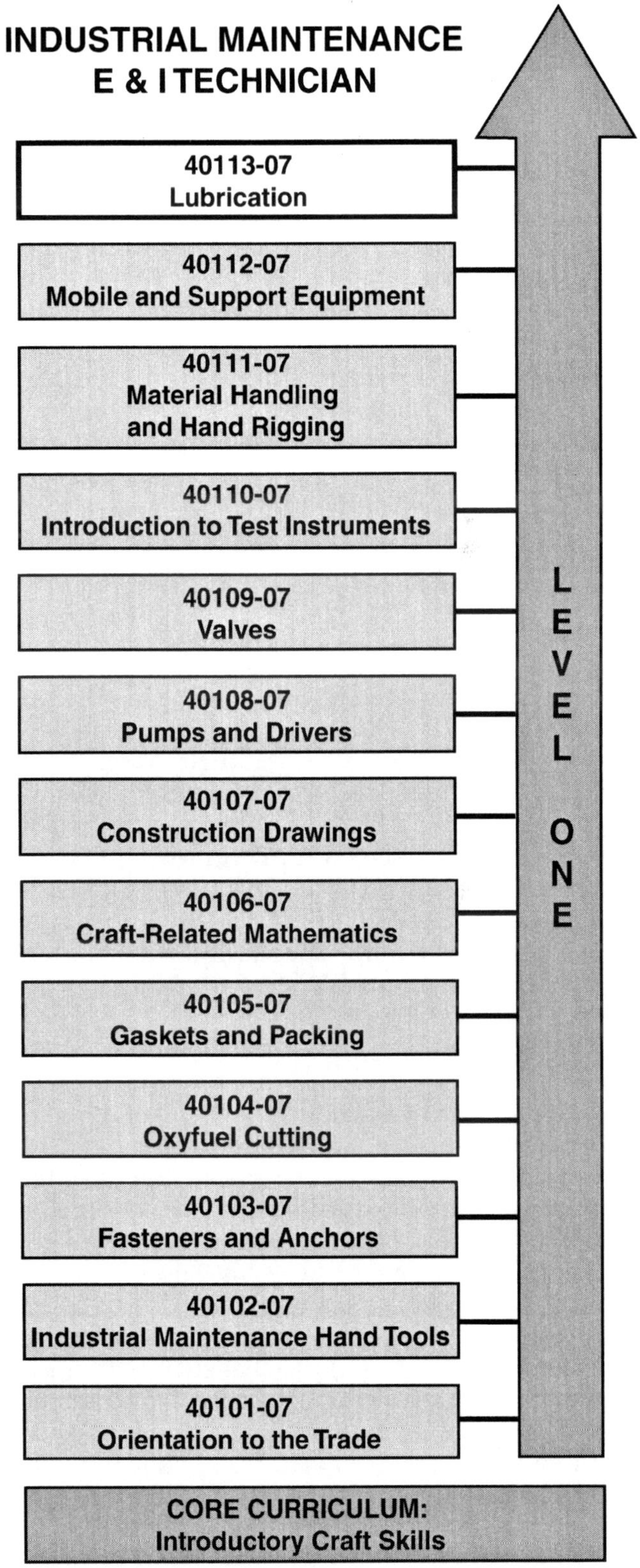

113CMAP.EPS

1.0.0 ◆ INTRODUCTION

Whenever movement takes place between surfaces in contact with each other, resistance to this movement occurs. This resistance is known as **friction**. Any piece of equipment, machine, tool, or other mechanism that has rotating or moving parts experiences friction. Friction is necessary in some cases, as with clutches or brakes that require friction to do their job. On the other hand, friction between moving parts, such as shafts and bearings and pistons and cylinders, causes excessive heat and wear. This type of friction must be reduced and controlled as much as possible to prevent premature wear and damage to the parts and equipment.

Lubrication is required wherever friction is a problem. Lubrication is the application of any substance that reduces friction by creating a slippery film between surfaces. Lubricants may be solid, semisolid, liquid, or gas. This module explains lubrication safety, lubricant types and properties, lubrication equipment, lubricating methods, and lubricating charts.

2.0.0 ◆ LUBRICATION SAFETY

Lubricants are classified as liquids (**oils**) and solids or semisolids (**grease**). Lubricants are made from many types of materials. They may be made from animal fats, such as cattle, hog, and sheep fat; vegetable oils, such as soybean and cottonseed oils; mineral oil; refined crude oil; or synthetics. Chemicals are sometimes blended with the lubricants to create special characteristics. Because of the many different types of materials used to make lubricants, they are considered hazardous materials. When using lubricants, special precautions must be taken to prevent bodily injury and illness.

2.1.0 OSHA Standards

In order to handle lubricants safely, you must be able to determine the composition, characteristics, and health hazards of the lubricant. You have the right to know this information and to work in safe surroundings and under safe conditions. The federal government passed the *Occupational Safety and Health Act* in 1970 to protect this right. This act established the Occupational Safety and Health Administration (OSHA) as an agency within the United States Department of Labor, which develops and enforces standards that minimize hazards in the work place. A hazardous chemical is any chemical that can cause health problems, explosions, fires, or other dangerous situations.

OSHA enacted the Hazard Communication standard (*29 CFR 1910.1200*) in November of 1983. The Hazard Communication standard defines employees' rights to know about the hazards of chemicals they work with. This standard sets national requirements for evaluating the hazards of chemicals used in the work place. It also requires that the hazard information be communicated to employees who will be handling the chemicals.

OSHA has defined seven actions that those who manufacture, import, or distribute chemicals or use chemicals in their operations must take. These actions are the following:

- Manufacturers and importers of chemicals must identify and evaluate the hazards of the chemicals they make or sell. Employers who use the chemicals may rely on the hazard determination of the supplier, or they may make their own hazard determination.
- Chemical manufacturers and importers must provide their customers with a material safety data sheet (MSDS) that documents the chemical hazards and the methods for controlling them. Hazardous chemical distributors must also ensure that all of their customers are provided with a copy of an MSDS for each chemical they purchase.

The following actions must be taken by employers whose employees use hazardous chemicals:

- Ensure that MSDSs are available to the employees.
- Label hazardous materials containers.
- Provide a list of all hazardous chemicals used in the work place.
- Provide employees with information and training on chemical hazards.
- Write a hazard communications program that describes specific plans for carrying out the six preceding actions.

2.2.0 Material Safety Data Sheets

Each material safety data sheet (MSDS) provides technical information about one chemical hazard. The MSDS may describe the hazards of a single chemical or a mixture of chemicals. The MSDS describes the composition, characteristics, and health hazards of the chemical(s). It also provides information about how to safely handle and store the chemical as well as any special procedures required. Under the rules of the Hazard Communications standard, your employer must ensure that the MSDSs are available for each hazardous chemical that is used or produced in the work place, and you must be permitted to use the MSDS at any time. The MSDS is your guide for working safely with a hazardous chemical.

Not all MSDSs look the same. They may be arranged differently or may have different titles, but all MSDSs contain the same information. There must be no blank spaces on an MSDS. When there is no information available, the party preparing the MSDS must indicate that no information was found. Always check the MSDS when you are unsure of the hazards of a material. The following sections explain the sections of an MSDS. The *Appendix* includes an example of an MSDS for a lubricant.

2.2.1 Chemical Product and Company Identification

The first section of an MSDS is the Chemical Product and Company Identification section. This section contains the name, address, and emergency telephone number of the party who prepared the MSDS. That party is usually the chemical manufacturer or importer. This information is provided in case there is an emergency and you require further information about the chemical. Section 1 also identifies the chemical as it is identified on the container label. The chemical identity must be given unless it is a trade secret. The chemical must be identified by both its chemical name and its common name.

2.2.2 Composition, Information on Ingredients

This section lists all of the ingredients in the product, specifically identifying the chemicals that are hazardous.

2.2.3 Hazards Identification

The hazards are identified for various forms of exposure to the substance, such as ingestion (eating or drinking), inhalation (breathing), or skin exposure.

2.2.4 First-Aid Measures

This section describes first-aid measures. These include recommended procedures for treating inhalation, eye exposure, skin exposure, and ingestion.

2.2.5 Fire-Fighting Measures

Fire-fighting measures give the flash point (the lowest temperature at which the material will ignite when a spark or flame is applied). The upper and lower flammability limits are given, and what can be used to put it out. Special fire-fighting procedures are described, as well as associated fire and explosion hazards and products of combustion that may increase the hazards, such as poisonous gases. Protective equipment for fire fighting is also detailed here.

2.2.6 Accidental Release Measures

This section explains what to do in case of accidental spills or releases, such as ways to prevent personal harm, and ways to keep the material from harming the environment. It also contains procedures for cleaning up a spill.

2.2.7 Handling and Storage

The Handling and Storage section explains how to keep the product safe, both in use and in storage.

2.2.8 Exposure Controls, Personal Protection

This section of the MSDS explains how to protect yourself when using dangerous substances. There are sections on ventilation, personal protective equipment, and engineering controls.

2.2.9 Physical and Chemical Properties

This section describes the physical properties of the material, including its appearance, color, and odor. It also lists chemical properties such as solubility in water and rate of evaporation.

2.2.10 Stability and Reactivity

The reactivity section of an MSDS includes information on the stability, incompatibility, **decomposition**, and **polymerization** of the chemical. The first part of the section tells whether the chemical is stable or unstable. A stable chemical does not change its chemical structure under common temperature, light, and pressure conditions and does not react easily with air. An unstable chemical may easily undergo chemical structure changes, causing a fire or an explosion, or may release another chemical hazard. The MSDS also identifies chemicals that are not compatible with the specified chemical. The chemical may undergo hazardous reactions when brought in contact with incompatible chemicals. If applicable, this section also explains what causes the chemical to decompose and the by-products of decomposition. The polymerization part of the section tells if polymerization can occur and what conditions to avoid. Polymerization can produce dangerous amounts of heat.

2.2.11 Toxicological Information

Toxicological information tells whether the product is poisonous (toxic), and to what extent.

2.2.12 Ecological Information

This section discusses possible environmental hazards associated with this material.

2.2.13 Disposal Considerations

This section provides the requirements for safe disposal of the material.

2.2.14 Transport Information

This section lists the DOT hazard classification. Specific sections deal with air transport, sea transport, and trucking or train requirements, if there are any such limits.

2.2.15 Regulatory Information

This section of the MSDS provides information on Federal regulation of the product. Under the *Superfund Amendments and Reauthorization Act* (SARA), your company is required to report amounts and other information on chemicals stored on the premises. *SARA 302* defines the emergency planning requirements, while *SARA 313* describes the reporting requirements. The other section of information required here is whether or not it is in the Comprehensive Environmental Response, Compensation, and Liability Information System (CERCLIS) database. It tells what hazards are known, and what action is being taken about it.

2.2.16 Other Information

This section provides National Fire Protection Association and Hazardous Materials Identification System information, along with the name of the preparer of the MSDS.

2.3.0 EPA Program

The substances used to make lubricants are not only a potential hazard to your health but may also be harmful to the environment. Every year our society produces billions of tons of waste that are a result of manufacturing, cleaning, lubricating, and production processes. These wastes are potentially hazardous to the delicate balance of the environment in the atmosphere, on the earth, and in our water supplies if they are not properly handled.

The handling, **disposal**, cleanup, and transportation of waste is regulated by state and federal laws. The United States government provides strict regulatory programs to control waste and protect the environment. The *Resource Conservation and Recovery Act (RCRA) of 1976, Public Laws 94 through 580*, were amended in 1984. The **Environmental Protection Agency (EPA)**, under the authority of the *RCRA*, developed a three-part program to deal with wastes currently being generated, with abandoned or uncontrolled waste dump sites, and with the cleanup of toxic materials at these dump sites and scenes of accidental spills.

The first part of the EPA program is intended to help states establish and implement regulations to ensure proper management of waste from its production to its final disposal. This part of the program identifies the waste and provides a tracking system with specific requirements for anyone involved with the waste. These specific requirements include the following:

- Identifying waste materials known to be hazardous to human health or to the environment
- A system for developing codes describing and characterizing these materials
- Standards for operating generators, transporters, and storage facilities
- Performance, design, and operating requirements for treatment and disposal facilities
- A system of identification numbers and permits for generators, transporters, and treatment, storage, and disposal (TSD) facilities
- Regulatory guidelines for retention of records
- Guidelines for states to follow in order to be authorized by the EPA to conduct their own hazardous waste management programs

Part two of the program consists of conducting investigations to identify dangerous, abandoned, or uncontrolled dump sites or illegal releases. The owners of these sites or the person(s) responsible for the release are held responsible for the cleanup.

Part three of the program provides funds to state and local governments to help them clean up hazardous dump sites and spills when the responsible party cannot be identified and when no other funds are available.

Controlling waste and hazardous materials in the work place is everyone's responsibility. You should know the EPA requirements for the proper handling and disposal of any waste or hazardous material. Your company should provide you with this information. Many companies develop their own requirements in addition to those of the EPA.

If you are not sure how to handle or dispose of a material, you should ask your supervisor. Any substance that is released into the environment has the potential of being a hazardous waste.

2.4.0 Lubricant Storage

Proper storage of lubricants is very important to prevent lubricant contamination, fires, spills, accidents, and unnecessary waste. The following sections list some guidelines for lubricant storage.

2.4.1 Guidelines for Inside Storage

The following guidelines should be followed when storing lubricants inside a building:

- Store lubricants in a fireproof room or building, because lubricants are combustible.
- Ensure that the storage area has a sprinkler system that produces a fine spray or a **carbon dioxide (CO_2)** extinguishing system that can be automatically or manually controlled.
- Ensure that there is a portable Class C CO_2 fire extinguisher in the room.
- Position storage racks or shelves so that all lubricants can be easily reached.
- Store small cans of lubricant on shelves.
- Store large drums on racks in the horizontal position, with a drip-proof valve to draw the lubricant from, or store them in the vertical position, with a pump to draw out the lubricant.
- Store grease that comes in small cans on shelves or on the floor in the upright position.
- Plainly mark all containers to identify the contents, and store them so that the markings can be clearly seen.
- Ensure that there are as few flammable objects in the room as possible.
- Repair leaks promptly, and clean up minor spills immediately. In the case of major spills, try to contain the material.
- Ensure that the storage room is well-ventilated.
- Keep the storage room clean and uncluttered at all times. There are different standards for different states.

2.4.2 Guidelines for Outside Storage

The major problems with outside storage are weather and dirt. The following guidelines should be followed when storing lubricants outside:

- Place containers on racks, pallets, or other **dunnage**, not on the ground.
- Do not store containers in an area where standing water tends to accumulate.
- Store lubricants in a location where there is a minimum of dust and dirt.
- Store barrels in the horizontal position if possible, with the **bungs** at the 3-o'clock and 9-o'clock positions.
- Cover containers to protect them from weather. Store containers in a small shed, or cover them with a canvas tarpaulin or plastic sheet.
- If a drum is stored in the upright position, place a block of wood under one side so that the drum will be tilted for water to run off.
- Do not store lubricants in an area where there is welding, cutting, or an open flame.
- Keep a Class C fire extinguisher in the storage area.

3.0.0 ◆ LUBRICANTS

To lubricate means to make a surface smooth and slippery. Lubrication is very important in any plant or industry. Lubricants reduce wear to moving parts by reducing friction and removing heat from parts and bearings. They also seal parts and bearings from dirt, grit, and moisture. Poor lubrication will lead to excessive equipment breakdowns and downtime, which result in higher maintenance costs and production losses. A good lubrication program increases the operating life of equipment and reduces maintenance efforts and costs.

Lubricants are classified as liquids, semisolids, and solids. Liquid lubricants are called oils and can be made of animal fats, vegetable oils, mineral oils, or synthetics. Liquid lubricants have all the properties of other liquids. They occupy a definite volume of space that can be changed only by adding or taking away lubricant. They can be poured from one container to another; they take the shape of the container that holds them, and they can be pumped from one place to another.

Semisolid lubricants, more commonly known as greases, are neither liquid nor solid, but are between a liquid and a solid. They become more like a liquid, or less viscous, when the temperature rises and more like a solid, or more viscous, when the temperature falls. When poured into a container, a semisolid will take awhile to take the shape of the container.

A solid is a substance that retains its shape under normal conditions. Solid lubricants often come in the form of granules or powders and very fine flakes. They are often used in applications where extremely low temperatures would freeze liquid lubricants or in applications where high temperatures would cause oils to burn.

Lubricants are also classified according to their sources, such as natural oils, mineral oils, and synthetics. Natural oils come from a wide variety of compounds and substances that are produced by animals and vegetables. Animal fats, such as stearin, lard, and tallow, come from common animals, such as cattle, hogs, and fish. Vegetable oils are refined from certain seed and plant parts. Castor oil, cottonseed oil, linseed oil, olive oil, and grape seed oil are good examples of vegetable oil lubricants. From a chemical standpoint, these oils are rather complex, unstable, and easily broken down. Before mineral oils became available, natural oils and fats were the most commonly used lubricants, and several are still widely used because they do have special properties that make them the best choice for some special applications.

For the last 150 years, the availability, cost, good performance, and variety of mineral oils have made them the first choice of lubricants for most applications. Mineral oils, which come from refined crude oil, represent over 90 percent of total lubricant use. They are chemically simpler than natural oils, are more stable than the natural oils, and can be used in extreme environments for extended periods of time. Mineral oils are preferred in most industrial applications, especially when high speeds, temperatures, and pressures are involved.

Synthetic lubricants, such as ester lubricants and silicones, are artificial lubricants. These lubricants are made up entirely of synthetic, nonpetroleum products, or they are made from combinations of such products mixed with natural oils. The greatest advantage of the synthetic lubricants is their wide temperature range. They are used extensively to provide lubrication in situations requiring fire resistance or superior cooling.

3.1.0 Lubricant Film Protection

Lubricant film protection is important when there is significant contact between sliding surfaces. To understand lubricant film protection, it is first necessary to understand what happens when metal surfaces that are not lubricated slide against each other. In extreme cases in which there is no lubricant and the metal surfaces are not contaminated by an oxide film or by any other foreign substance that could offer some lubrication, the metal surfaces tend to adhere to each other. This tendency is very strong for some types of metals and weaker for others. A few guidelines for common metals are as follows:

- Identical metals in contact have a stronger tendency to adhere to each other than dissimilar metals.
- Softer metals have a stronger tendency to adhere than harder metals.
- Nonmetallic alloying elements, such as a high content of carbon in cast iron, tend to reduce adhesion.
- Iron and its alloys have a low tendency to adhere to lead, silver, cadmium, and copper.
- Iron and its alloys have a strong tendency to adhere to aluminum, zinc, titanium, and nickel.

Friction and wear between moving surfaces is reduced when a film of lubricant is applied to the surfaces. The following are the four levels of lubricant film protection:

- Dry friction
- Mixed film lubrication
- Boundary lubrication
- Fluid film lubrication

3.1.1 Dry Friction

Dry friction is the condition in which each surface is unprotected from the abrasion of the other surface by any lubricant. The dirt or scale on the surfaces, as well as any other material that may be trapped between them, may serve to keep the friction from being exactly the same everywhere, but no intentional lubrication is present.

3.1.2 Mixed Film Lubrication

Mixed film lubrication is a condition that occurs when mating surfaces are partially lubricated. The lubricant film between the surfaces has decreased, and there is metal-to-metal contact between the high points of the surfaces moving across each other. Part of the load is relieved by the lubricant, but the high points take most of the load, resulting in friction and wear of the surfaces. Mixed film lubrication is by no means an ideal situation, but it is a common occurrence. Situations in which mixed film lubrication often occurs include the following:

- At low operating speeds
- During operations with frequent starting and stopping
- From inadequate lubricant-supplying methods
- When a low-viscosity lubricant is used
- With very heavy loads on the mating surfaces
- When a bearing is misaligned
- When a shaft is not straight
- When mating surfaces are not machined properly

3.1.3 Boundary Lubrication

Boundary lubrication is similar to mixed film lubrication, but the layer of film is thicker and there is less friction generated between the mating surfaces. Most of the low spots in the metal are filled with lubricant, and there is less friction between the higher spots because they are covered with a heavier film than exists in a mixed film situation. Boundary lubrication usually occurs when a machine first starts, and the film that first forms is made up of impurities instead of lubricating oil. This condition normally continues until operating speed is reached. Once operating speed is reached, fluid film lubrication is achieved. Some equipment, such as compressor cylinders, are designed to work continuously with boundary lubrication.

3.1.4 Fluid Film Lubrication

In fluid film lubrication, also known as full film lubrication, the load is supported entirely by the pressure within the separating lubricant film. Fluid film lubrication is the ideal condition in which all moving parts are completely separated from each other and friction is held at a minimum. Moving surfaces contact only the lubricant film, and the high points of the mating surfaces are kept apart. Fluid film lubrication is generated in two ways: hydrodynamically and hydrostatically.

Hydrodynamic fluid film lubrication occurs when the movement of the mating parts forces the lubricant between the surfaces. Pressure is created by lubricant resistance to this movement and the compression of the lubricant. This pressure causes the two surfaces to lift and separate. *Figure 1* shows fluid film lubrication between a plain journal bearing and a shaft.

As the shaft shown in *Figure 1* starts to rotate, a lubricant wedge forms between the shaft and the bottom of the bearing. As motion increases, the shaft slides up on the wedge of lubricant. The lubricant continually resists the effort of the shaft to squeeze the lubricant out of the way because it is in a confined space. This resistance to movement is the pressure that keeps the surfaces separated.

Hydrostatic fluid film lubrication also keeps the moving parts completely separated through pressure and resistance of the lubricant. In hydrostatic fluid film lubrication, the pressure is supplied by an outside source, such as a pump, instead of by the action of the rotating parts. The main advantage of hydrostatic lubrication is the ability to control the pressure of the lubricant. The pressure of the lubricant determines the amount of clearance between the moving parts.

3.2.0 Properties of Lubricants

Lubricants have certain unique properties that determine how well the lubricant reduces or controls friction in a given situation. The properties that need to be considered when selecting a lubricant include the following:

- **Viscosity**
- **Viscosity index (VI)**
- **Pour point**
- **Flash point**
- **Fire point**
- Oxidation resistance
- **Emulsion** resistance

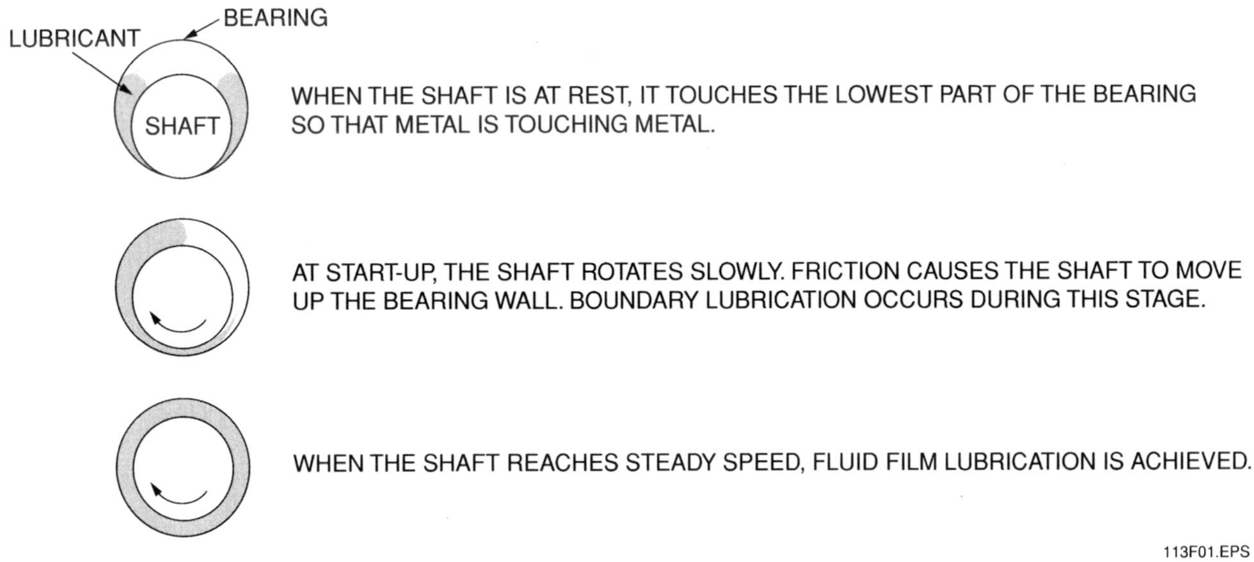

Figure 1 ◆ Fluid film lubrication.

3.2.1 Viscosity

The measure of the internal friction of a fluid is known as the viscosity. This friction is created when a layer of fluid is forced to move in relation to another layer. The greater the friction, the greater the force required to cause this movement, which is called shear. The shear rate is the measurement of the speed at which the intermediate layers move with respect to each other. The shear strength is the force measured to resist the shearing action. Shearing occurs whenever the fluid is physically moved, such as when it is poured, pumped, or flushed through a gearbox. More force is needed to move highly viscous lubricants than is needed to move less viscous lubricants.

Viscosity is the most important property of a lubricating oil. Viscosity is the thickness of a liquid or the ability of the liquid to flow at a specific temperature. Lubricants may have low or high viscosity. Low-viscosity lubricants are light oils that flow freely when poured. High-viscosity lubricants are heavy oils that flow slowly when poured. Medium-viscosity lubricants are oils that range between low and high. Lubricant viscosity changes with the temperature. When the temperature rises, lubricants flow more freely and the viscosity is lower. When the temperature drops, the lubricants flow more slowly and the viscosity is higher. Oils with different viscosities are used in different temperature conditions. This is why you should use a low-viscosity oil in your car in the winter and a high-viscosity oil in the summer.

Viscosity rating is expressed in one of two ways. Automotive oils and gear-lubricating oils are expressed in SAE ratings. SAE stands for Society of Automotive Engineers. This rating is stamped on motor oil cans, for example SAE 20W-50. This rating indicates that the oil viscosity has been measured by the SAE system. The other viscosity rating system, which applies to industrial lubricants, is the **Saybolt universal seconds (SSU, SUS)** system. This rating system was developed by the American Society for Testing and Materials. *Table 1* lists some typical operating viscosity ranges. The metric system has a unit for the relative resistance to flow of materials, called the kinematic viscosity of the material. The unit, called the **centistoke (cSt),** is the relationship between the simple viscosity and the density of the material.

Figure 2 shows viscosity equivalents.

3.2.2 Viscosity Index

The viscosity index (VI) refers to the viscosity-temperature relationship of a given lubricant because the viscosities of some lubricants change when they are subjected to changes in temperature. All lubricants do not change viscosity at the same rate when subjected to the same temperature changes. The VI measures the rate at which the viscosity changes as the temperature changes. If a lubricant has a low VI, the viscosity of the lubricant changes rapidly with changes in the temperature. A high VI represents a lower change of viscosity with temperature changes. The temperature changes of a system must always be considered before selecting a lubricant. When the system operates at a stable temperature, the VI is not a major consideration.

3.2.3 Pour Point

The pour point of a lubricant refers to the lowest temperature at which a lubricant will flow freely. This must be considered in cold weather applications and in refrigeration systems. A low pour point indicates that the lubricant flows freely at very low temperatures. A high pour point indicates that the lubricant stops flowing freely at low temperatures.

3.2.4 Flash Point

The flash point of a lubricant is the temperature at which the vapor or steam of the lubricant will flash into flame. This temperature is not high enough for the lubricant to support combustion. When the flash point is reached, the lubricant film between the mating surfaces is destroyed, and scoring of the metal can occur.

3.2.5 Fire Point

The fire point of a lubricant refers to the highest temperature that a lubricant can withstand before the vapors in the lubricant will support combustion and burn steadily. In most lubricant applications, neither the flash point nor the fire point are major concerns. These factors must be considered only in high-temperature applications.

Table 1 Typical Operating Viscosity Ranges

Lubricant	Viscosity Range (cSt)
Clock and instrument oils	5 to 20
Motor oils	10 to 50
Roller bearing oils	10 to 300
Plain bearing oils	20 to 1,500
Medium-speed gear oils	50 to 150
Hypoid gear oils	50 to 600
Worm gear oils	200 to 1,000

113T01.EPS

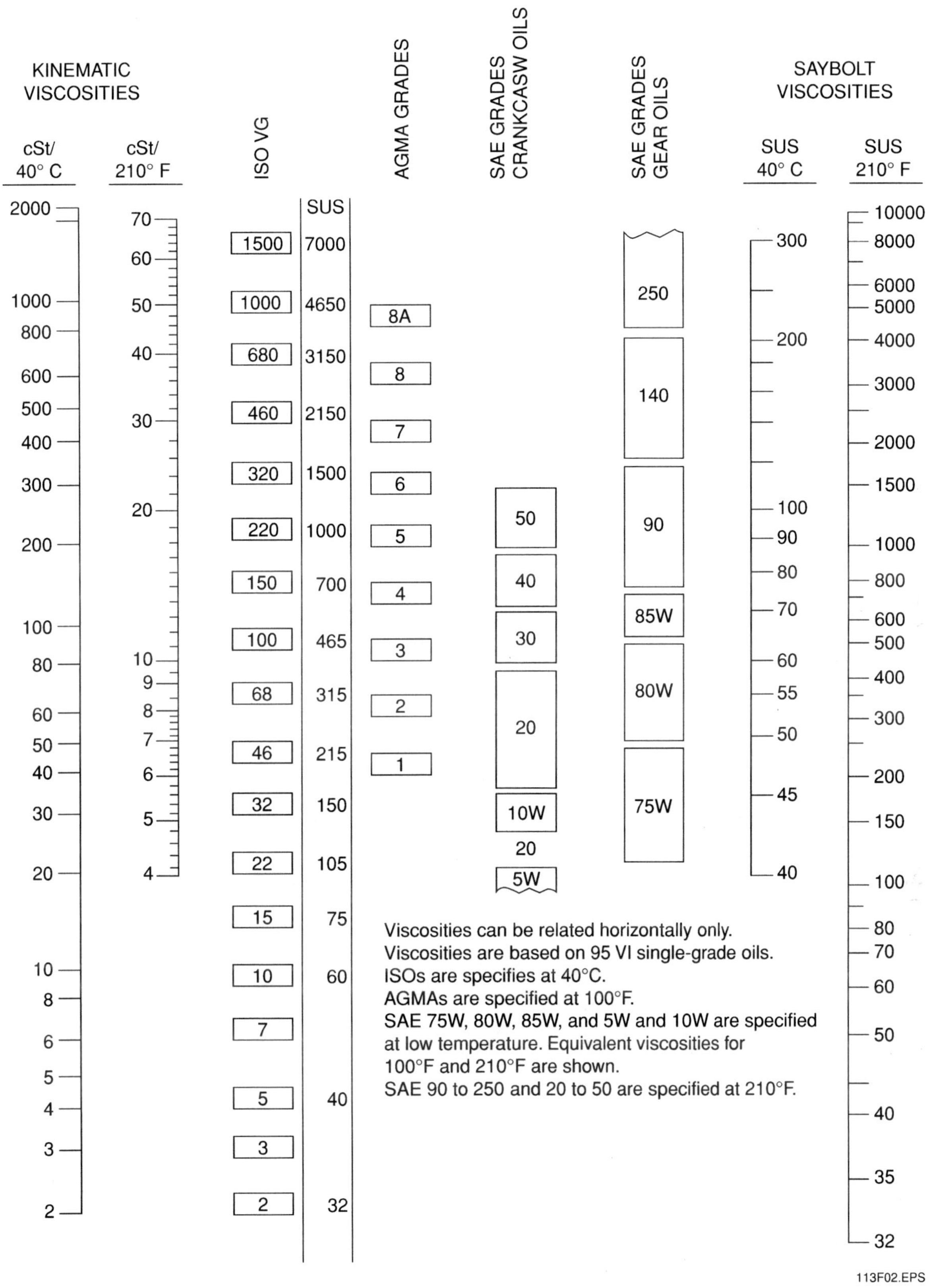

Figure 2 ◆ Viscosity equivalents.

3.2.6 Oxidation Resistance

Most lubricants are a hydrogen-carbon mixture known as a **hydrocarbon**. As hydrocarbons are exposed to air and heat, they take on oxygen from the air. This process is known as **oxidation**. Oxidation resistance is one of the most important qualities of a lubricant because all lubricants tend to oxidize over time. Oxidation breaks down the chemical structure of the lubricant and eventually destroys the lubricating qualities of the lubricant. The oxidation process is normally slow, but if the lubricant splashes around excessively or if it is exposed to high temperatures, the oxidation process speeds up.

3.2.7 Emulsion Resistance

Emulsion is a combination of substances that do not readily mix, such as water and oil. Since water and oil do not readily mix, emulsion is formed when the water is broken down into tiny droplets and carried around by the oil. This process is known as emulsification. Emulsion is undesirable in lubricants because it decreases the lubricating qualities of the lubricant and holds particles of dirt and foreign contaminants that can damage and cause excessive wear to bearings and other components in the system.

3.3.0 Properties of Grease

Grease is a solid or semisolid lubricant that is formed by adding a thickening agent, usually soap, to oil. Grease usually has at least three ingredients: oil, a thickening agent, and an additive. The most important ingredient is the lubricating oil. It is usually a type of mineral oil. The type of mineral used may vary in its viscosity or degree of refinement. Greases for high-temperature, low-speed applications are made with higher-viscosity oil; greases for low-temperature, high-speed applications are made with lower-viscosity oil.

The thickening agent is the second element in grease. The most common thickening agent is a soap that is made from fatty materials, either animal or vegetable, and some type of mineral or metal. Calcium, barium, and lithium are some types of metal that are used to make the soap. The function of the soap is to hold the oil and release it at a slow rate to provide the lubricating action. The soap itself may have some value as a lubricant, but the oil provides the lubricating action.

The third element in grease is some type of additive, although some greases may not have additives. Additives of different types are combined with grease to allow the grease to be used for more purposes. Additives help keep the grease from undergoing chemical changes, give the grease new characteristics, and protect metal parts from **corrosion**.

Grease produces the same lubricating action of oil, only in a different way. Oil forms a film that keeps surfaces apart to reduce friction and heat. Grease is not a liquid and cannot form a liquid film when it is initially applied. Grease forms the lubricating film when the surfaces begin to move and exert pressure on the grease. When pressure is applied, grease releases some of its oil, allowing the lubricating action to begin. When the motion is stopped and the pressure is reduced, grease tends to solidify again. The following properties of grease must be considered when selecting a grease for an application:

- Consistency
- **Dropping point**
- Pumpability
- Water resistance
- Stability

3.3.1 Consistency

The most important physical characteristic of a grease is its hardness or softness, which is known as the consistency of the grease. Consistency is determined by measuring the distance in tenths of a millimeter that a standard metal cone penetrates the grease under a standard load. This test is known as a penetration test.

The consistency of a grease varies with temperature, and generally a grease softens as the temperature increases. The American National Lubricating Grease Institute (NLGI) has developed a widely used classification system of grease consistencies and designated the consistencies of grease by **NLGI consistency numbers**. *Table 2* shows the NLGI grease classification system.

Table 2 NLGI Grease Classification System

NLGI Number	Penetration at 25°C (in mm)	Description of Firmness
000	445 to 475	Semifluid
00	400 to 430	Semifluid
0	355 to 385	Semifluid
1	310 to 340	Very Soft
2	265 to 295	Soft
3	220 to 250	Can be used in grease cup
4	175 to 205	Can be used in grease cup
5	130 to 160	Can be used in block-type grease cup
6	85 to 115	Block or cake

113T02.EPS

3.3.2 Dropping Point

As stated earlier, a grease becomes softer as the temperature increases. Eventually, the temperature at which the grease is soft enough for a drop to fall away or flow from the bulk of the grease is reached. This temperature is known as the dropping point. The dropping point is normally considered to be the highest temperature at which the grease can be used, but you must not always assume this. A grease may still be a satisfactory lubricant above its dropping point, but it will behave more like an oil than a grease above this point. Some greases will not give a dropping point because chemical decomposition of the grease will begin at a lower temperature than the temperature that breaks down the thickener in the grease.

3.3.3 Pumpability

The pumpability of a grease needs to be considered if the grease is to be pumped through a system from a centralized greasing system. The grease must be soft enough to flow through the system without clogging the pipes or components in the system.

3.3.4 Water Resistance

Water resistance of a grease is an important property of a grease that is to be used with machinery that can be damaged by water. Many greases break down or dissolve when they come in contact with water. Greases containing calcium or lithium soap as thickeners resist water very well, as do other greases that contain thickeners that are insoluble in water and retain the oil within the structure of the thickener.

3.3.5 Stability

The stability of a grease is the ability of the grease to retain its chemical and physical properties over time. In many applications, it is crucial that a stable grease be used to protect the component it is used with. Bearing applications are especially sensitive to instability. If the grease in a bearing breaks down, the grease will leak from the bearing and leave the bearing dry, causing it to seize. A stable grease will not break down due to temperature, air, water, or reactive chemicals, such as oxygen in the air.

3.4.0 Selecting Lubricants

Often, the best method for selecting a lubricant is to use the simplest lubrication technique that will perform the job satisfactorily. This usually consists of inserting a small amount of lubricant into a component during the initial assembly. This lubricant will never be replaced or replenished. This technique will not work if the loads or speeds are high or if the service life is long and continuous. It then becomes necessary to choose the lubricant with great care. Engineers or manufacturers determine the lubricant to be used in each situation. Some plants color-code lubricant containers and lubricant cartons.

The two main factors that must be considered when selecting lubricants are speed and load. Other factors include temperature, other substances in the environment, and special situations. One of the most frequently occurring lubrication practices in a plant involves lubricating bearings, and selecting bearing lubricants involves considering all of these factors.

3.4.1 Load and Speed

The two most important factors that determine the lubricant to use are load and speed. Load is the amount of pressure that is exerted on the lubricant when the system is operating. The greater the load, the greater the possibility that the molecules in the lubricant will break up. High-viscosity lubricants, greases, and solid lubricants are recommended for greater loads. If the speed is high, the lubricated surfaces tend to wear faster because the amount of heat caused by friction tends to be higher. Low-viscosity oils provide lower friction and better heat transfer at high operating speeds.

3.4.2 Other Selection Factors

Other selection factors that determine the type of lubricant to use include temperature, other substances in the environment, and special situations. Temperature variations change the viscosity of lubricants, as discussed earlier in this module. Low temperatures cause oils and greases to thicken, while high temperatures thin lubricants.

Substances in the environment of a given application also affect lubricating qualities. In a highly corrosive area, lubricants must be able to withstand any chemicals in the air that may tend to break down the lubricant. Water in a system requires a lubricant to be emulsion resistant.

There are also situations that require special consideration when selecting lubricants. For example, watches and instrument mechanisms require low friction and therefore need a very low-viscosity oil, while open gears and chains need a tacky oil or grease that has adhesive qualities because the lubricant may be thrown off the moving parts. Knowing what type of lubricant to use with a particular application comes with experience and knowing as much as possible about lubricants and their properties. Many times, considerations of product requirements are as important as the mechanical requirements of the machinery itself. An example is lubrication in plants dealing with foodstuffs; the lubricants are controlled by US Department of Agriculture standards. The machinery itself would require lubricants that would not endanger the consumer if small quantities got into the products. The lubricants would also have to be non-reactive and effective in the presence of sterilizing and cleaning products.

3.4.3 Bearing Lubrication

The lubrication of bearings is one of the most important and most frequent lubricating duties that you will perform in your job because of the large number of bearings found on most job sites. When selecting lubricants for bearings, you must consider the expected operating temperature, the bearing load and speed, and shaft and bearing clearances.

Boundary lubrication and fluid film lubrication both occur during the operation of machine bearings. Equipment design usually determines whether an oil or a grease is used for bearing lubrication. When oil is used, the most important factor to consider is the viscosity of the oil at the operating temperature of the bearing. Bearings in equipment used outdoors or in unheated areas require lubricants that have a low pour point. This is to ensure that the bearing will run freely at low temperatures when the equipment is started. Bearings in equipment to be used in high-temperature areas must have a heavier lubricant that will not thin at the higher temperatures. Greases are used to lubricate bearings in low-speed applications that do not generate much heat. Grease is frequently used when the running speed of a plain bearing does not exceed 200 to 300 rpms. Grease tends to stay in place and does not run out of the bearing as freely as oil does, and it seals out contaminants to keep them from entering the bearing housing.

The most common problem with lubricating bearings is overlubrication. Too much lubrication can cause an increase in operating temperature and a decrease in viscosity. If the lubricant becomes too thin, it cannot carry the load inside the bearing, and the bearing will fail and have to be replaced. Too much lubricant being forced into the bearing will burst the seals on many bearings, causing the lubricant to escape from the bearing and allowing contaminants to enter.

3.5.0 Additives

An additive is a substance that is added to a lubricant to improve the qualities of the lubricant. The following are the most common types of additives in use:

- Pour point depressants
- Oxidation **inhibitors**
- Viscosity index improvers
- Antifoam agents
- Rust inhibitors
- Extreme-pressure (EP) additives
- Detergents/**dispersants**
- Emulsifying agents
- Demulsifiers
- Antiwear agents
- Tackiness agents

3.5.1 Pour Point Depressants

Pour point depressants are additives that lower the temperature at which lubricants thicken. The pour point can be lowered by as much as 30°C with the addition of 1 percent or less of added depressants.

3.5.2 Oxidation Inhibitors

When oxygen is allowed to combine with an oil, oxidation occurs. Oxidation can cause acids to form, which can cause the oil to thicken. Oxidation inhibitors, also called antioxidants, help prevent acids, varnish, and sludge from forming. Zinc, barium, and calcium dithiophosphates are used to prevent oxidation in severe services, such as in internal combustion engines. These additives are most useful in systems whose operating temperature is less than 120°C.

3.5.3 Viscosity Index Improvers

Viscosity index improvers are used to thicken a light oil to a higher viscosity without changing the original viscosity-temperature coefficient. This enables a lower-viscosity oil to work at higher temperatures without breaking down. For example, a 10W-base stock oil can be thickened to a 10W-50 product. Caution must be used in high shear rate situations, such as in gearboxes, because the viscosity index improvers slowly lose their thickening power.

3.5.4 Antifoam Agents

Antifoam agents, such as methyl silicone polymers, prevent bubbles or foam from forming in oil used in a circulating system. Moderate foaming interrupts the boundary layer of lubrication and may cause early wear and breakdown. Severe foaming can cause the oil to overflow and cause catastrophic failure of the machine. Antifoam agents are not completely soluble in oil and work by forming minute droplets of low surface tension that help break foam bubbles. A very small amount can defoam oil in engines, turbines, gear sets, and aircraft applications.

3.5.5 Rust Inhibitors

Rust inhibitors are surface-active substances that adsorb as a film on steel and iron surfaces to protect the surfaces from being attacked by moisture. Rust that forms inside moving parts can start with pitting and metal removal and then progress to complete failure of the device if left unchecked. Moisture enters closed systems by condensation, leaks, or contamination within the lubricant. Under moderate conditions, you may use some acids, such as organic amines. Under severe conditions, such as seawater exposure, outdoor use, or long-term storage, lubricants need inhibitors such as organic phosphates and sodium and calcium sulfonates that adhere more strongly.

3.5.6 Extreme-Pressure (EP) Additives

Extreme-pressure (EP) additives, also called film-strength additives, strengthen the oil used to protect metal surfaces from very heavy loads. Automotive hypoid gears, slideways in machine tools, and heavy metal-cutting operations are some applications for EP lubricants. Additives such as chlorine, active sulfur, sulfur phosphate, and synthetic polymers are used in extreme conditions in which severe metal-to-metal contact and high temperatures occur. In the high spots where metal contact and loading is extreme, these EP additives form low shear-strength layers of iron chloride and iron sulfide. These layers prevent destructive welding, excessive metal transfer, and severe surface breakdown. They also bear some of the compressive load and cushion the shock of start-up or heavy load engagement.

3.5.7 Detergents/Dispersants

Detergents added to lubricants are designed to prevent or reduce high-temperature deposits of sludge, varnish, carbon, and lead compounds on internal combustion engine parts. They work by absorbing the insoluble particles and suspending them until the lubricant can be replaced. Dispersants do the same thing for engines or other applications with relatively low operating temperatures.

3.5.8 Emulsifying Agents

Emulsifying agents are mainly used to keep water from collecting and are used in components in closed systems in which water separation and draining is not possible. Emulsifying agents form an oil film around water particles that leak or condense into a system. The emulsifying agents then break up the water particles into tiny droplets that remain suspended in the oil. This prevents the water from contacting the metal surfaces and causing corrosion. Since they mix with water, emulsifying agents also allow for clean-up with hot water.

3.5.9 Demulsifiers

Demulsifiers are additives that prevent water from breaking up into smaller droplets. The demulsifiers separate the water from the lubricant to allow the water to be drained or removed easily.

3.5.10 Antiwear Agents

Antiwear agents produce a film on a surface by either chemical or physical **adsorption** to minimize friction and wear under boundary lubrication conditions. Some of these compounds contain oxygen, sulfur, organic chlorine, phosphorus, and combinations of these elements. These additives reduce friction and provide a low shear-strength layer like the EP additives but are designed for less severe applications.

3.5.11 Tackiness Agents

Tackiness agents help lubricants stick to metal surfaces. They are especially useful when added to heavy oils used for lubricating large, exposed, slow-moving gears or slides. Oils and greases with tackiness agents hold up better in wet conditions, such as in outdoor and marine equipment.

3.6.0 Lubricating Oils

Petroleum lubricants are mostly a complex mixture of hydrocarbon molecules. Their physical properties, such as viscosity, temperature range, and performance, depend mostly on the relative amounts of **paraffinic**, aromatic, and alicyclic components.

Paraffinic oils have lower viscosity and density and higher freezing points. They also tend to oxidize unless inhibitors are added.

Aromatic oils tend to resist oxidation but form insoluble black sludge at high temperatures. The viscosity of aromatics can change rapidly with temperature changes. This change can cause high density and darkening of the oil.

Alicyclic oils have a low pour point, low oxidation stability, and other properties that fall between those of the aromatics and paraffinics. Most of the paraffinic oils are composed of both paraffinic and alicyclic structures. The following oils are widely used in industrial applications and are combinations of each of these types of components:

- Turbine
- Hydraulic
- Gear
- Steam cylinder

3.6.1 Turbine Oils

In gas, steam, and internal combustion turbines, the high speeds and very tight clearances require that the lubrication be highly specialized. The additives used in turbine-grade oils include rust and corrosion inhibitors, oxidation inhibitors, demulsifiers, and antifoam agents. The high quality specifications for turbines make these oils a standard by which other lubricants are measured. The abrasive, wet, or less-than-ideal conditions in which turbines are used, combined with the high speed and load that they experience, rule out any substitutes. When properly maintained, some steam turbine oils can run 25 years without change because they are in a closed system and therefore are easier to protect from contaminants.

3.6.2 Hydraulic Oils

Hydraulic oils are usually chosen based on their viscosity, viscosity index, and pour point characteristics. The velocity of an oil in a hydraulic system is very important for efficient pump operation and must be within strict velocity limits determined by the pump manufacturer. If the oil is too heavy, it may cause excessive friction that may damage the pump bearings and cause system failure. Hydraulic oils normally range from 150 to 700 SUS, or 32 to 150 cSt, at 40°C.

The viscosity index of a hydraulic oil is very important if the system operates at a wide variety of temperatures. The oil must be able to flow easily in cold conditions and must not get too thin in warmer conditions. In conditions in which the temperature changes, a high viscosity index is recommended. The pour point of a hydraulic oil must be known when the equipment is to be operated in low temperatures.

3.6.3 Gear Oils

Gears have teeth that meet only at certain points and then roll away from each other. These points of contact must be properly lubricated with an oil film that is heavy enough to cushion shock and prevent the gear teeth from being damaged. There is a large variety of lubricating oils to match the various sizes and types of gears. Gear oils may be straight mineral oils, or they may contain additives to improve film strength and the ability to carry the load or handle the pressure of the application. Gear-oil viscosity must be high enough to protect the gear teeth under normal operations. It must also be low enough to transfer heat out of the gearbox and lubricate the bearings.

3.6.4 Steam Cylinder Oils

Steam cylinder oils are special-purpose lubricants that must be low enough in viscosity to be **atomized** and high enough to maintain a lubricating film on the walls of hot cylinders. They must also be resistant to steam and water. Steam cylinder oils are very heavy, and their viscosity is measured at the standard temperature of 210°F because they are usually used in very hot applications. The following are the three types of steam cylinder oils:

- *Light* – 100 to 120 SUS
- *Medium* – 120 to 150 SUS
- *Heavy* – Over 150 SUS

4.0.0 ◆ LUBRICATION EQUIPMENT

Several types of equipment and methods are used to apply lubricants. Some machines and equipment require manual lubrication; others use mechanical lubricating devices. These devices automatically lubricate bearings after a certain number of revolutions or hours of operation. The following sections explain lubrication equipment and how lubricants are applied.

4.1.0 Manual Lubrication Equipment

Lubrication equipment can be portable or part of the machine or equipment that is lubricated. Some lubrication equipment is used to apply oil; others are used to apply grease. Manual lubrication equipment is hand-operated and is used to apply oil or grease. The operator must know how to refill and maintain this equipment. Some common types of manual lubrication equipment are the following:

- Lever guns
- Transfer pumps
- Gear lube dispensers
- Bucket pumps

4.1.1 Lever Guns

Lever guns are hand-operated and develop high pumping pressure with little effort. They are reloaded by suction, cartridge, or filler pumps. Lever guns are used to grease bearings on equipment and machinery. *Figure 3* shows a lever gun.

Filler pumps are used to fill lever guns quickly and easily. Filler pumps are attached to original refinery grease containers by clamps. The filler nipple of a lever gun is attached to the filler socket on the pump, and the filler pump handle is manually pumped until the lever gun is filled with lubricant. A follower plate must be in place to ensure that there are no air pockets. *Figure 4* shows two types of filler pumps.

> **CAUTION**
>
> Lever guns develop pressures in excess of 10,000 psi, which can damage seals. Operate lever guns with care.

4.1.2 Transfer Pumps

Transfer pumps are used to apply fluids or heavy lubricants quickly from original containers. Pumps used for heavy lubricants use a pressure system for easier operation. *Figure 5* shows manually operated transfer pumps.

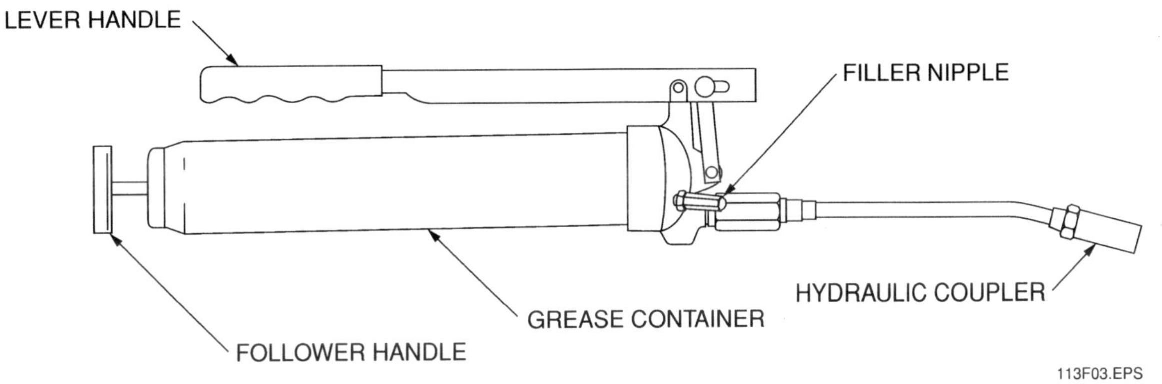

Figure 3 ◆ Lever gun.

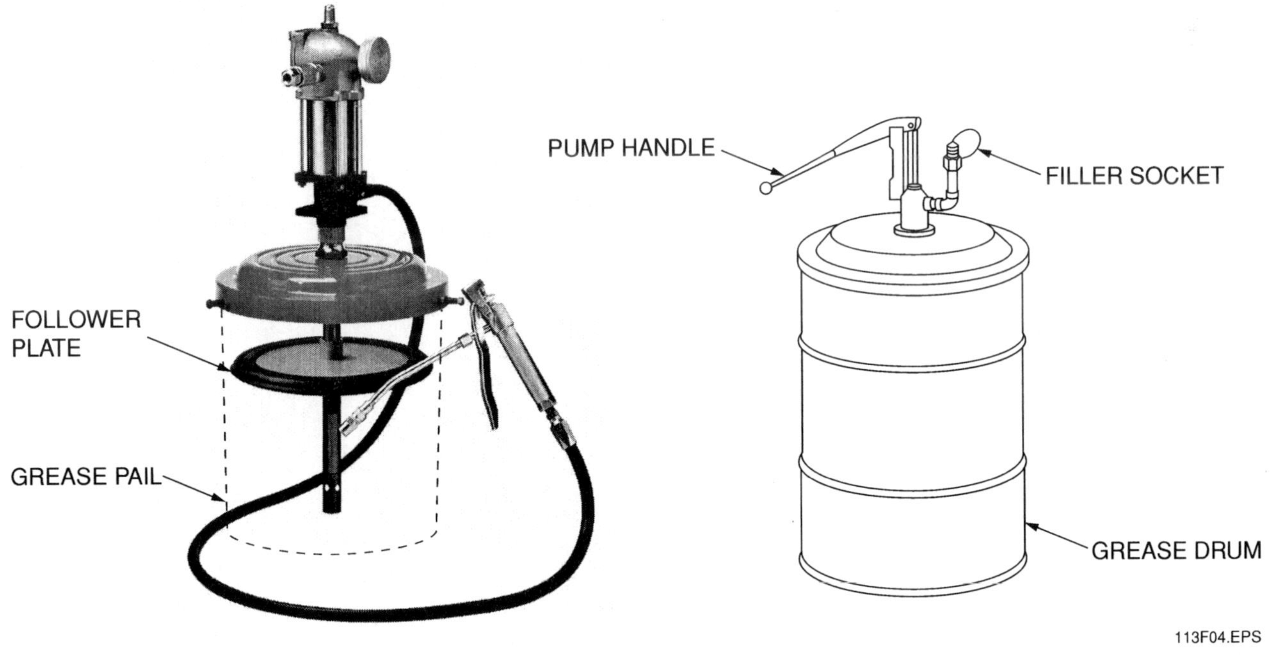

Figure 4 ◆ Filler pumps.

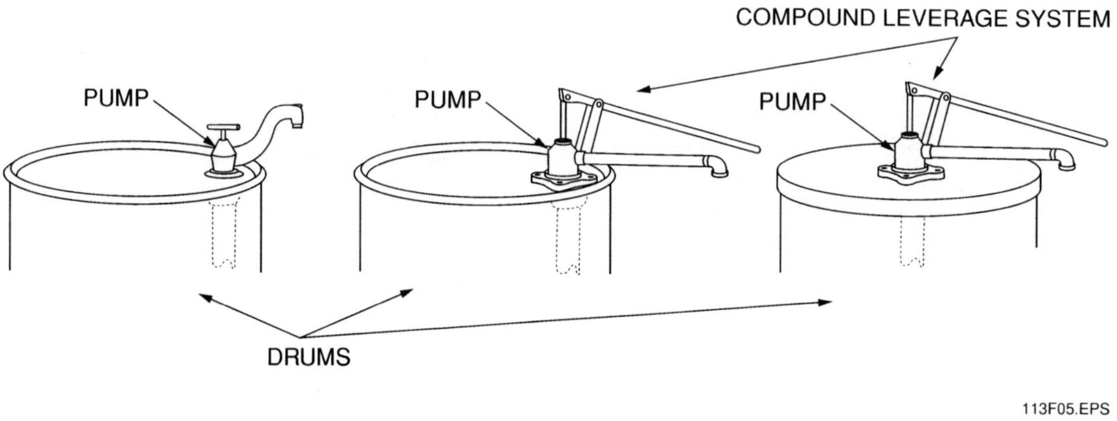

Figure 5 ◆ Manually operated transfer pumps.

4.1.3 Gear Lube Dispensers

Gear lube dispensers are used to fill transmissions and gearboxes with high-viscosity lubricants. They are attached directly to the original refinery lubricant container. Some models have a meter that indicates the amount of lubricant dispensed. *Figure 6* shows a metered gear lube dispenser.

4.1.4 Bucket Pumps

Bucket pumps are high-pressure, high-volume dispensers used to lubricate machinery and equipment. They are attached directly to the original refinery lubricant container and pump grease or heavy oil. *Figure 7* shows two types of bucket pumps.

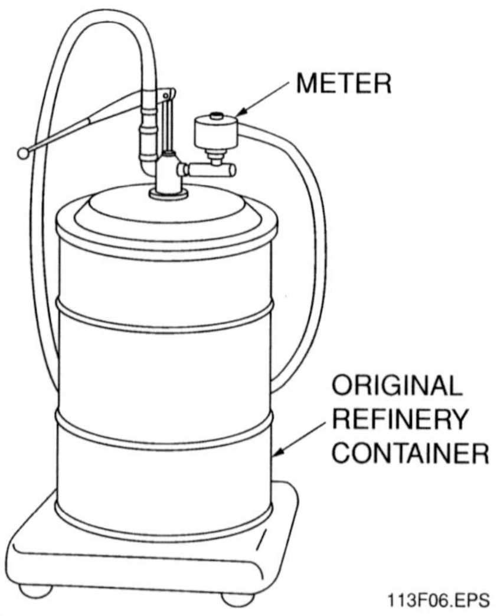

Figure 6 ◆ Gear lube dispenser.

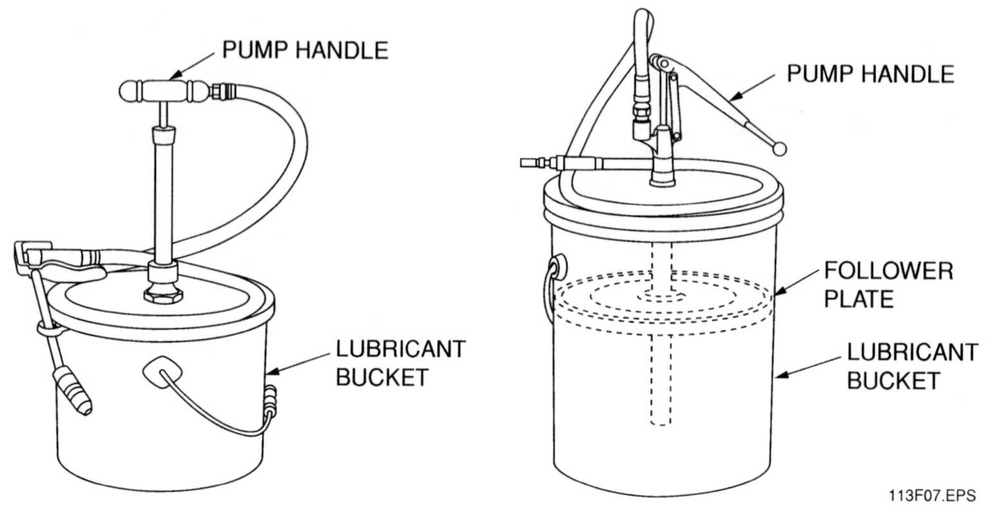

Figure 7 ◆ Bucket pumps.

4.2.0 Power-Operated Lubrication Equipment

Power-operated lubrication equipment uses direct air pressure or air motors to pump or apply lubricants. Power-operated lubrication equipment can be permanent or portable. Power-operated lubrication guns, lubrication trucks, automatic lubrication equipment, and internal lubricators are types of power-operated lubrication equipment. *Figure 8* shows a power-operated lubrication gun.

4.2.1 Lubrication Trucks

Lubrication trucks are mobile units used for lubricating and refueling heavy equipment and machinery. These trucks contain several types of grease, oil, and fuel.

4.2.2 Automatic Lubrication Equipment

Automatic lubrication systems *(Figure 9)* vary to reflect the characteristics of the machinery to be lubricated. For example, many large production plants have conveyor systems that transport the various parts and assemblies through the stages of production. The conveyors have automatic lubricant systems to supply one lubricant to the driving mechanism. At the same time, a high pressure lubricant is applied to the bearings of the carriages that hold and travel with the assemblies and parts. Still a third lubricant system may be applied to the bearings of the drive units that move the conveyors.

Automatic lubrication systems also reflect the requirements of the processes and materials in production. In mines and mills, spray systems *(Figure 10)* may be necessary to keep a constant film of lubricant on large gears or surfaces where extremely hot, heavy, and abrasive material may be transported. Rotary pumps are used to pump grease to all of the bearings of a system. In other

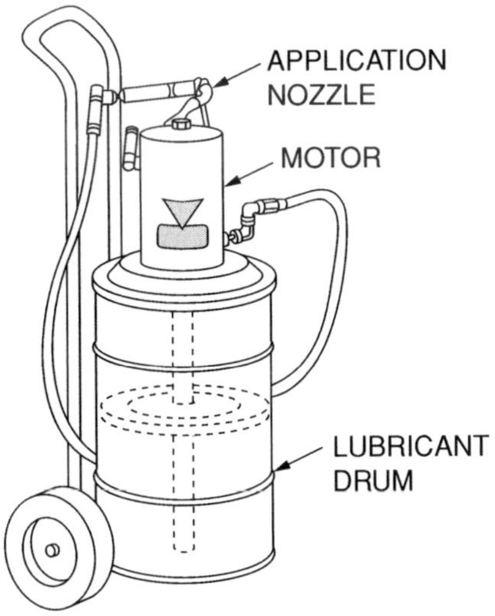

113F08.EPS

Figure 8 ◆ Power-operated lubrication gun.

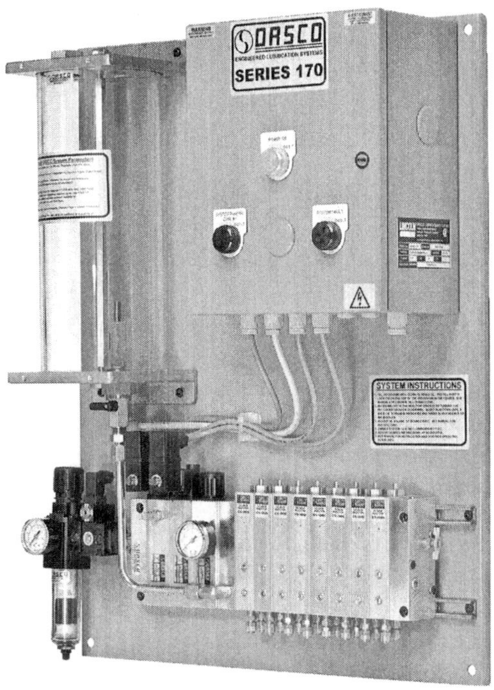

113F09.EPS

Figure 9 ◆ Automatic lubrication system.

113F10.EPS

Figure 10 ◆ Spray system.

cases, individual bearings are supplied with pressurized dispensing containers of lubricant (*Figure 11*), attached to the bearing housings. Such single-point automatic dispensers can be set to provide lubrication at a rate that will empty the container over a set period of months.

Automatic lubrication equipment is designed to provide a system for lubricating the bearings on a machine without shutting down the machine. An air-operated, gas-propelled, or electrically operated pump forces lubricants through lubrication lines to an injector or metering device. The metering device measures out the correct amount of lubricant needed for each bearing. This system is called a centralized lubrication system. *Figure 12* shows a centralized lubrication system.

Figure 11 ◆ Pressurized dispensing container.

4.2.3 Internal Lubricators

Internal lubricators are designed to provide automatic lubrication to all the parts in a machine. Oil lines, or galleys, are drilled or cast within the parts that make up the machine. Oil is pumped through the lines from a reservoir or sump to all the bearings. The bearings of an internal combustion engine, air compressor, or gearbox use this type of lubrication system.

Some internal lubrication systems use a splash, a circulating pump, or a combination splash-and-pump method for lubricating bearings and gears. The splash method is used in gear cases and simple crankcases. The gear or crankshaft rotates through the lubricant and carries it to other parts within the case as the gear rotates.

Rod bearing caps used with the splash method have scoops that direct the lubricant through a hole in the bearing cap. *Figure 13* shows circulating pump and splash internal lubrication systems.

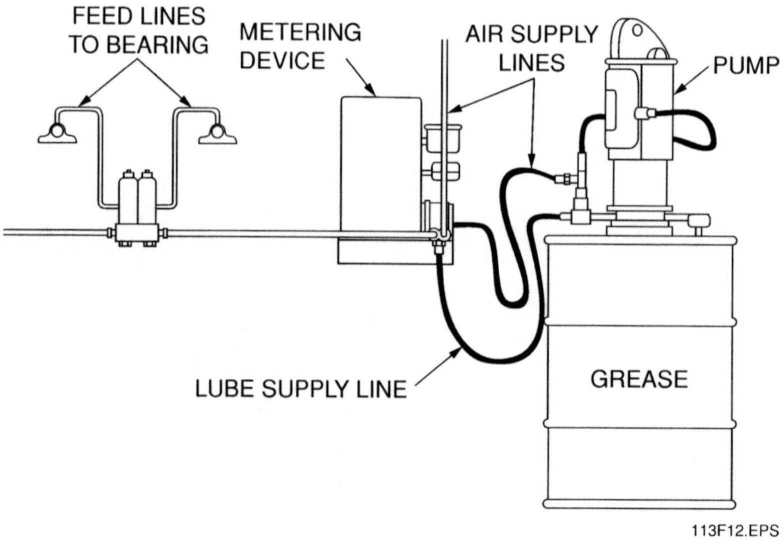

Figure 12 ◆ Centralized lubrication system.

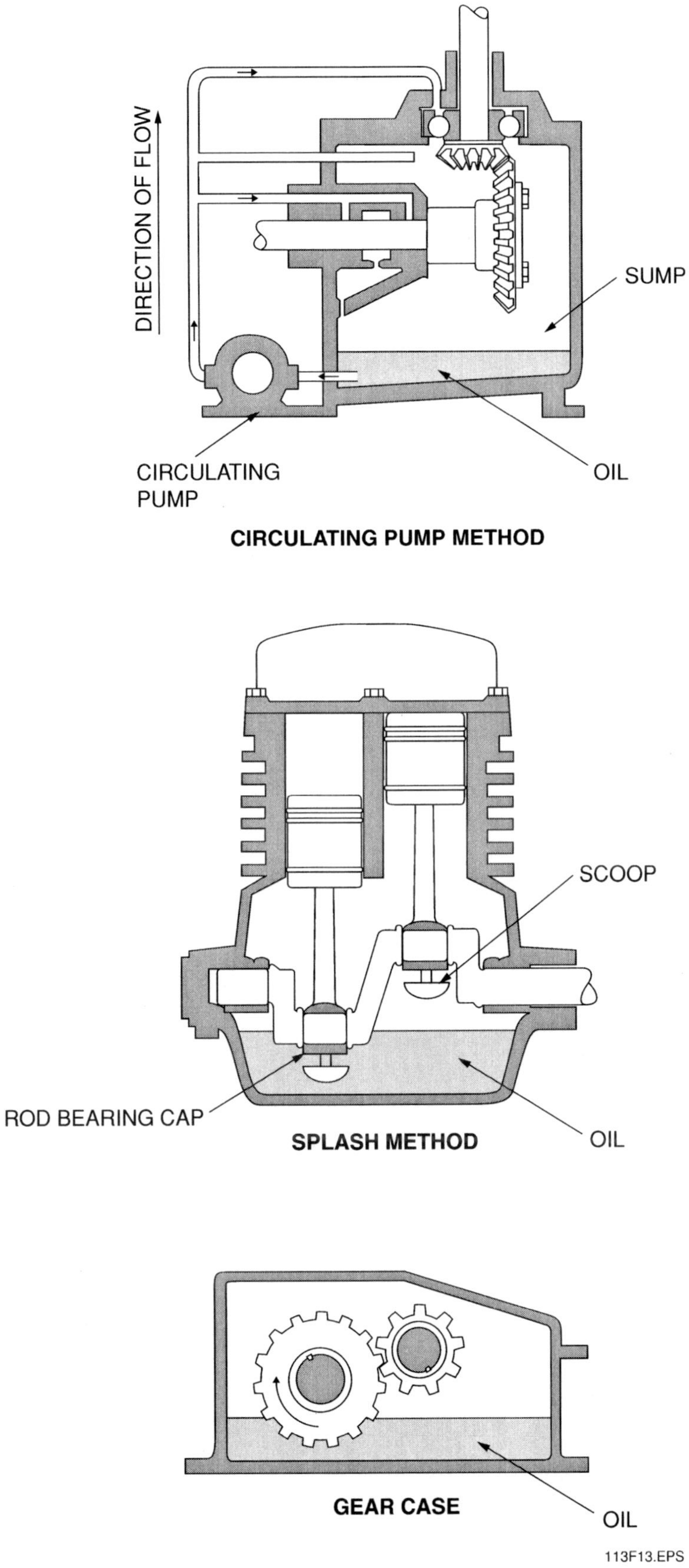

Figure 13 ◆ Internal lubrication systems.

4.3.0 Lubrication Fittings

There are many types of lubrication fittings that are used with lubrication equipment. Fittings can be straight, angled, standard thread, spin drive, or straight drive. Application fittings are divided into three main types: hydraulic, flush, and button-head. Hydraulic fittings are the most commonly used fittings. Flush fittings are used where space is limited and fittings might be damaged. Button-head fittings are used on earthmoving equipment, conveyors, and mining machinery where difficult conditions are encountered and large quantities of grease are required. *Figure 14* shows these different types of fittings.

There are other fittings that perform other functions, such as venting and relieving pressure. The following fittings perform other functions:

- Atmospheric vents
- Pressure-relief valves
- Couplers and nozzles
- Special-purpose fittings

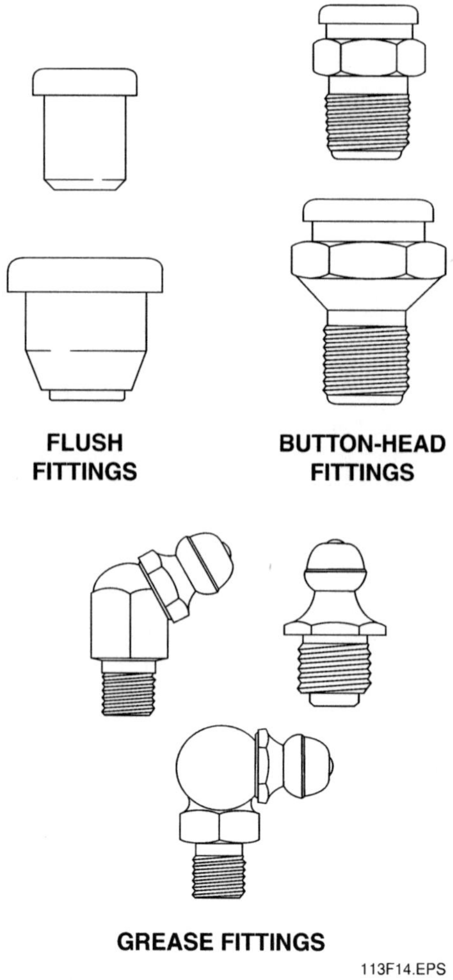

Figure 14 ◆ Fittings.

4.3.1 Atmospheric Vents

Atmospheric vents are used in hot-running gear cases or drive housings to prevent gas or lubricant pressure from building up. The gas or pressure escapes through a slotted surface, or vent, under the protective steel cap. Atmospheric vents are often installed in fill-hole locations and must be removed to fill the gearbox or check the lubricant level (*Figure 15*).

4.3.2 Pressure-Relief Valves

Pressure-relief valves (*Figure 16*) are used when oil or grease pressure must be controlled. They relieve pressure built up through the vent ports during equipment operation or when filling gear cases or bearing housings with pressure guns.

4.3.3 Couplers and Nozzles

Couplers and nozzles are used on hand-operated grease guns to make contact with different types of fittings. Hydraulic couplers are used for hydraulic fittings. Button-head couplers are used for button-head fittings. Button-head couplers are made in two sizes: standard $\frac{5}{8}$-inch diameter and large $\frac{7}{8}$-inch diameter. Needle nozzles are used for all flush fittings and hard-to-reach hydraulic fittings. *Figure 17* shows different types of couplers and nozzles.

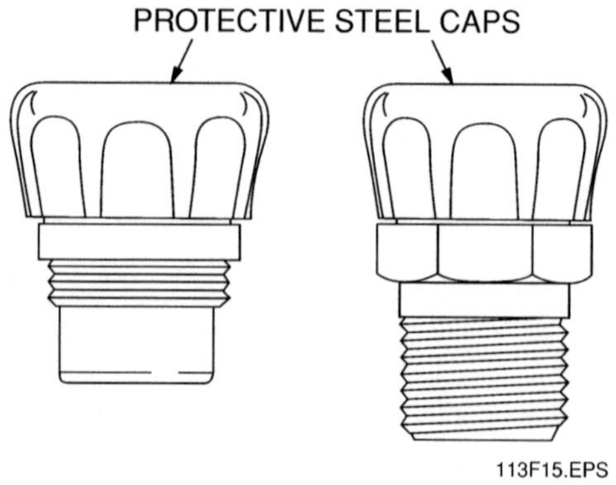

Figure 15 ◆ Atmospheric vents.

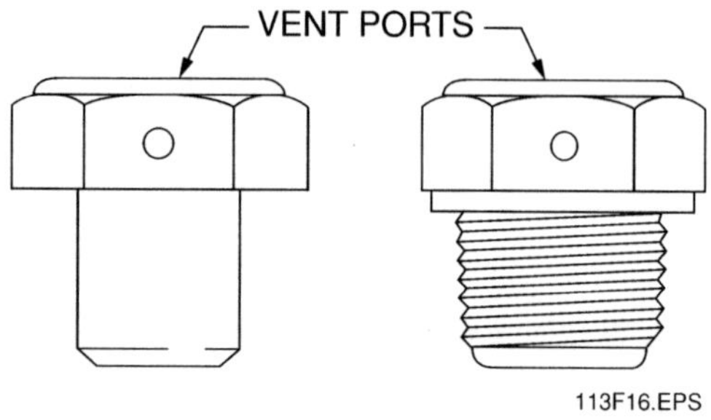

Figure 16 ◆ Pressure-relief valves.

4.3.4 *Special-Purpose Fittings*

Leak-proof fittings and safety vent fittings are used for special purposes. Leak-proof fittings have a special synthetic rubber seal used for light- or heavy-viscosity lubricants. They also have a ball check in the tip for added protection against dirt. Safety vent fittings have a vent slot cut vertically through the threads. This vent slot provides an air escape when filling a bearing fill and serves as a bearing fill indicator. *Figure 18* shows special-purpose fittings.

5.0.0 ◆ LUBRICATING METHODS

There are different methods for lubricating equipment. The method used depends on the equipment being lubricated. The choice of method involves considerations of size; that is, a large area of friction surface will require a large source and distribution system. If the equipment to be lubricated is cyclic; that is, if it does a repeated motion through the same space, a point application such as a sprayer is a good option. If part of the motion can take the component through an enclosure, an oil bath may be a good choice. If the equipment is only subject to small friction loads, hand oiling can solve the problem cheaply.

Other considerations include keeping the lubricant from being distributed where it is a problem. Oil on common walkways or stairs is a serious hazard, as is lubricant sprayed on electrical or high-temperature equipment.

5.1.0 Oiling Methods

Oil can be applied to bearings and moving parts manually or automatically, using different types of tools and equipment. The manual and automatic oiling methods are explained in the following sections.

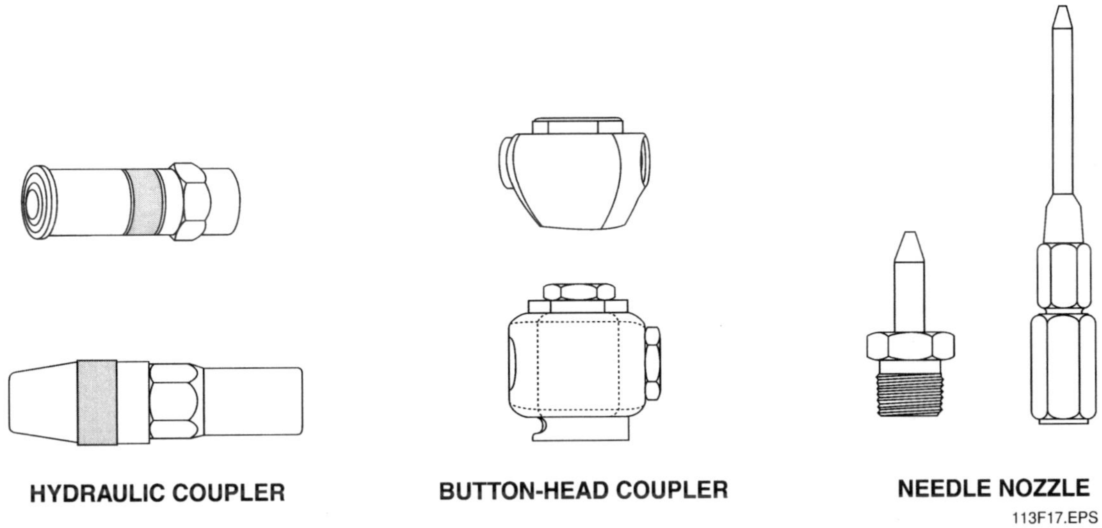

Figure 17 ◆ Couplers and nozzles.

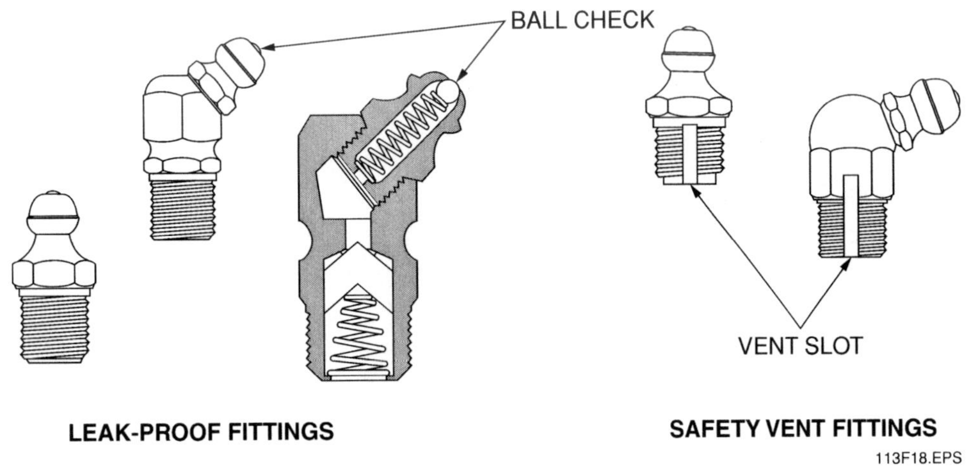

Figure 18 ◆ Special-purpose fittings.

5.1.1 Manual Oiling

Manual oiling means applying oils to bearings and moving parts using hand-operated oilers or sprays. Hand-operated oilers are used to apply a few drops of oil to small bearings or moving parts that do not need frequent oiling. Bearings on electric motors have small oil cups or reservoirs containing a wick of absorbent cotton or fiber material. These oil cups are filled regularly from a hand-operated oiler.

Oils are also applied manually. For wire rope and some types of roller chains, oils are applied with a spray can (*Figure 19*).

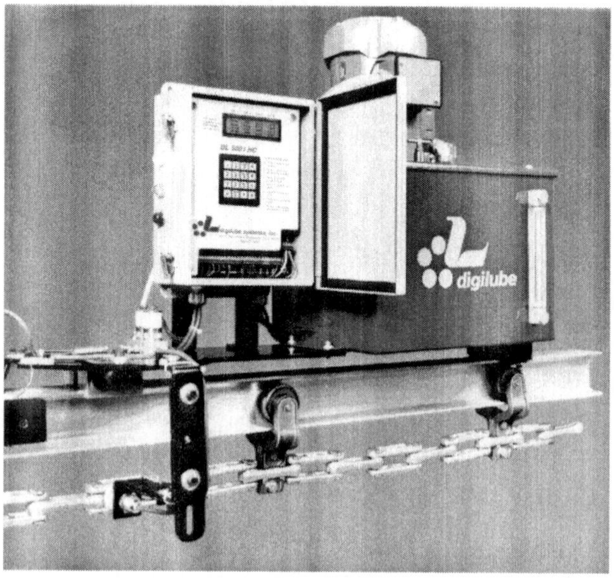

113F19.EPS

Figure 19 ◆ Spray lubrication.

> **WARNING!**
> The equipment must be locked out and tagged out before lubricating with a cloth or brush to prevent personal injury.

Oil cups and parts should be cleaned before oiling to prevent dust and dirt from entering the bearing with the oil. Excessive wear caused by dust and dirt entering the bearing may cause bearing failure.

5.1.2 Automatic Oiling

Automatic oiling means the oil is applied to the moving parts by a special system that is part of the machine or by a centralized lubrication system. These systems use a pump to supply oil to the lubrication points. The oil is filtered and circulated through the system from an oil sump or reservoir. The proper oil level must be maintained by regularly checking and adding oil.

Some automatic systems pump oil through a series of lines and metering devices to specific bearings or lubrication points. The pump is controlled by a timer or metering device on the machine. The parts are oiled at regular intervals, as controlled by the timer. The oil is forced through the bearing or part and not recirculated through the reservoir. The reservoir level must be maintained by manual filling. *Figure 20* shows an automatic oiling system.

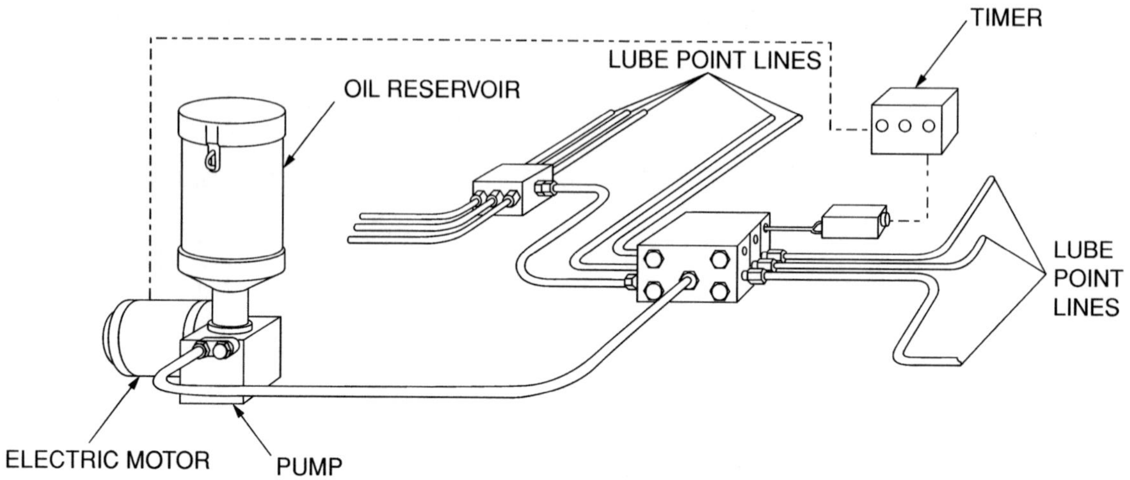

113F20.EPS

Figure 20 ◆ Automatic oiling system.

5.2.0 Greasing Methods

Grease can be applied to bearings and moving parts manually and automatically, using different types of tools and lubrication equipment. Manual and automatic greasing methods are explained in the following sections.

5.2.1 Manual Greasing

Manual greasing means applying grease to bearings and moving parts using hand-operated lever guns and bucket pumps. Grease can also be applied with a screw-type grease cup, a brush, a stick, or with your hands.

High pressures are required to pump the grease through a fitting and into the grease cavity around a bearing. Lever guns and bucket pumps can develop pressures up to 10,000 psi. Overpumping the handle can cause excessive grease pressure that could blow out the bearing seals. Some fittings and grease cavities are equipped with a vent groove or plug that is removed to relieve pressure and to allow old grease to be forced out.

5.2.2 Automatic Greasing

Automatic greasing means applying grease to a large number of machine bearings at the same time through a centralized greasing system. This system pumps grease from a reservoir through feed lines to an injector for each bearing. A metering device controls the amount of grease that each bearing receives.

A timer automatically starts and stops the pump at regular intervals. Each bearing is automatically supplied with grease when the pump is operated. The operator must keep the reservoir filled with grease and regularly inspect the system for leaks and broken pipes. *Figure 21* shows a centralized greasing system.

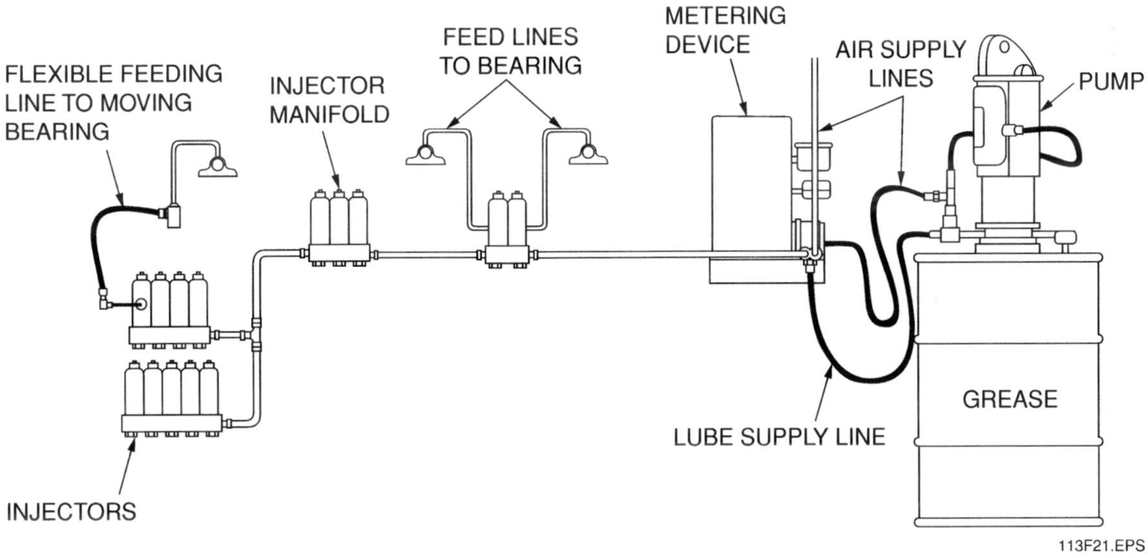

Figure 21 ◆ Centralized greasing system.

6.0.0 ◆ LUBRICATION CHARTS

Lubrication charts are provided by the manufacturer for each machine or piece of equipment. Each chart is designed for a specific machine and cannot be used for any other machine. Before lubricating any machine, you must read and follow the lubrication chart that comes with the operating manual for that machine. The correct chart must be used to prevent bearing damage. *Table 3* is a sample lubrication chart. This chart is an example only and must not be used for any machine or equipment on the job.

The chart heading identifies the sections of the lubrication chart. The chart sections identify the

Table 3 Sample Lubrication Chart

Index No.	Component	Number/Type Lube Points	Lube and Method	Interval (Hours)
1	Master leveling cylinder (base pin)	1 grease fitting	MPG – pressure gun	50
2	Master leveling cylinder (rod-end pin bushing)	1 grease fitting	MPG – pressure gun	50
3	Master leveling cylinder (rod-end pin)	1 grease fitting	MPG – pressure gun	50
4	Boom pivot shaft	2 grease fittings	MPG – pressure gun	10
5	Lift cylinder (rod-end pin)	1 grease fitting	MPG – pressure gun	10
6	Power track links	N/A	SAE 10 – oil can	100
7	Speed reducer (turntable)	Fill plug	EPGL (SAE 90)	500
8	Slave leveling cylinder (base pin)	1 grease fitting	MPG – pressure gun	50
9	Lower mid and fly boom	N/A	Dry lubricant spray	50
10	Boom side wear pads	N/A	Dry lubricant spray	200
11	Slave leveling cylinder (rod-end pin)	1 grease fitting	MPG – pressure gun	10
12	Platform door hinges	2 grease fittings	MPG – pressure gun	100
13	Platform door latch	N/A	SAE 10 – oil can	100
14	Platform attach pin	1 grease fitting	MPG – pressure gun	10
15	Fly section drive chain no. 50	N/A	MPG – brush/oil can	25
16	Drive hub	Fill plug	EPGL (SAE 90)	500
17	Swing bearing gear	N/A	MPG – brush	100
18	Swing bearing	4 grease fittings	MPG – pressure gun	100
19	Speed-reducer pinion	N/A	MPG – brush	100
20	Lift cylinder (base pin)	1 grease fitting	MPG – pressure gun	50
21	Wheel bearings	N/A	EPLG (SAE 90)	500
22	Front spindle	2 grease fittings	MPG – pressure gun	50
23	Tie rod (spindle – attach end)	1 grease fitting	MPG – pressure gun	50
24	Steer cylinder	1 grease fitting	MPG – pressure gun	50
25	Steer/towing hitch pivot pin	2 grease fittings	MPG – pressure gun	50
26	Tie rod (hitch – attach end)	2 grease fittings	MPG – pressure gun	50
27	Engine crankcase	Fill cap	EO (SAE 90)	50
28	Engine oil filter	N/A	Replaceable cartridge	50
29	Engine air filter	Reservoir	EQ (SAE 30)	50
30	Door access panels	N/A	SAE 10 – oil can	200

Notes:
1. Be sure to lubricate like items on each side of the machine.
2. Recommended lubricating intervals are based on normal use. If the machine is subjected to severe operating conditions, the user must adjust lubricating requirements accordingly.
3. Lubricating intervals are calculated on 50 hours of machine operation per week.
4. The lube and method column lists the type of lubricant required and the method of application. Lubricant types are shown by abbreviations. The following abbreviations are commonly used:
   - MPG – multipurpose grease
   - EPGL – extreme-pressure gear lube
   - EO – engine crankcase oil

113T03.EPS

index numbers, the components or parts to be lubricated, the number and type of lube points, the lubricant to be used, the method for lubricating, and the schedule for applying the lubricant. Each lube point or part that requires lubrication is shown and numbered on an illustration. These numbers are listed on the lube chart as index numbers. *Figure 22* shows index numbers on an illustration.

The component column of the lubrication chart lists the names of the parts identified by index numbers in the illustration.

The number/type lube points column lists the number and type of lube point found on the part indicated by the index number. The types of lube points are referred to as grease fittings or fill plugs. Lube points requiring lubrication that do not have fittings are shown as N/A.

The interval column lists how often a lubricant must be applied to each lube point. The interval is usually shown in hours of operation. Sometimes the words daily and as required are used. The notes section explains special precautions and recommendations.

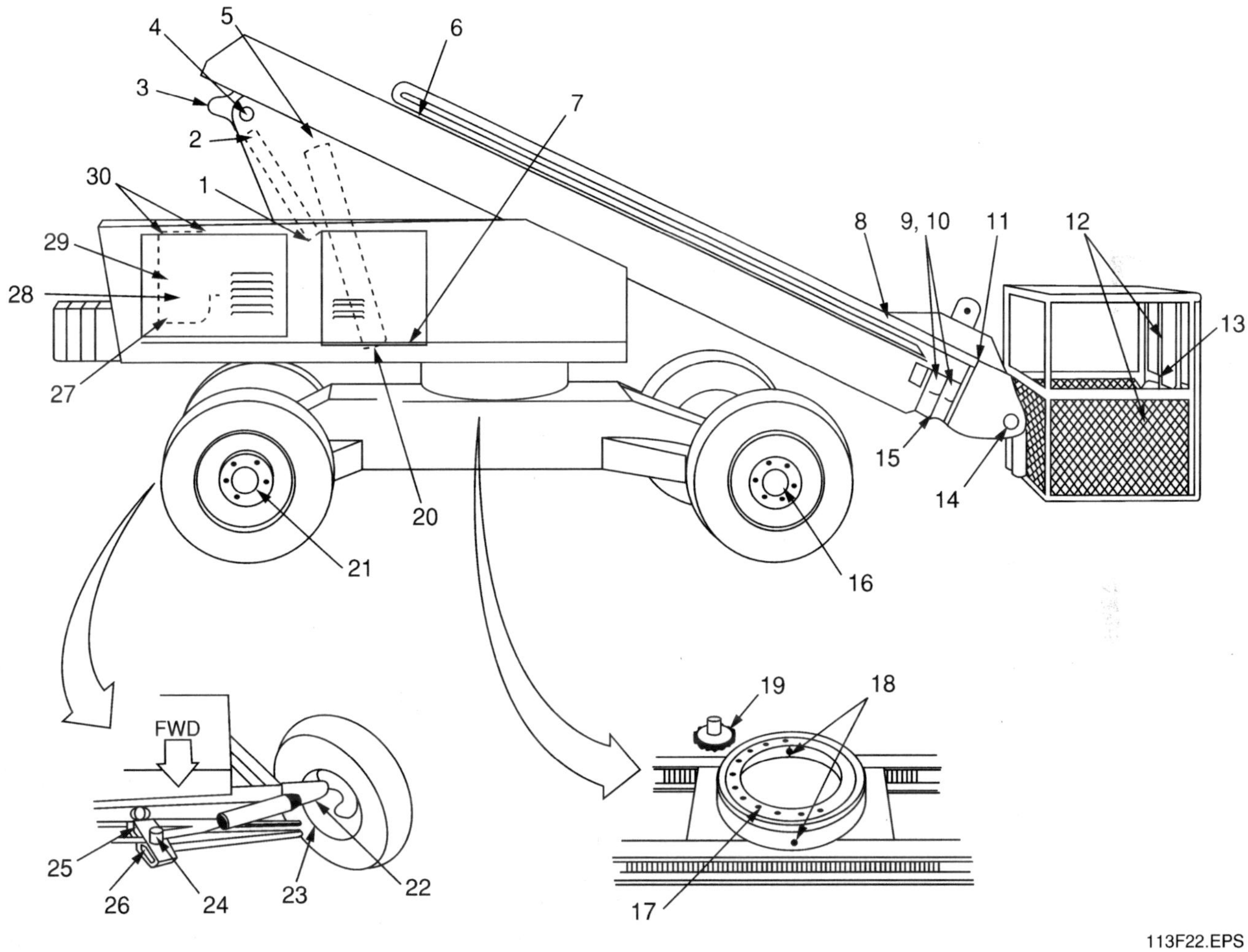

Figure 22 ◆ Index numbers on illustration.

Review Questions

1. The resistance to movement between two surfaces is called _____.
   a. lubrication
   b. friction
   c. planing
   d. momentum

2. The federal agency that works to minimize workplace hazards is called _____.
   a. EPA
   b. FBI
   c. OSHA
   d. ASTM

3. The entity responsible for supplying MSDSs to you is _____.
   a. OSHA
   b. EPA
   c. your employer
   d. the state government

4. It is *not* required to give the chemical identity of a chemical on an MSDS if it is _____.
   a. a trade secret
   b. not important
   c. too long a word
   d. short

5. The emergency first aid instructions are found in the _____ section of an MSDS.
   a. First-Aid Measures
   b. Identity and Emergency Information
   c. Fire and Explosion Hazard
   d. Physical Data

6. The clean-up procedures for a chemical are described in the _____ section of an MSDS.
   a. Hazardous Ingredients
   b. Identity and Emergency Information
   c. Exposure Controls, Personal Protection
   d. Accidental Release Measures

7. The three classifications of lubrications are _____.
   a. solid, liquid, and gas
   b. artificial, synthetic, and real
   c. liquid, semisolid, and solid
   d. oil, grease, and other

8. The condition when mating surfaces are partially lubricated is called _____.
   a. dry friction
   b. mixed film lubrication
   c. boundary lubrication
   d. fluid film lubrication

9. When the load is entirely supported by pressure within the separating lubricant film, it is called _____.
   a. dry friction
   b. mixed film lubrication
   c. boundary lubrication
   d. fluid film lubrication

10. The measure of the internal friction of a fluid is its _____.
    a. friction
    b. viscosity
    c. expense
    d. emulsion

11. Viscosities of automotive oils and gear-lubricating oils are expressed in _____ ratings.
    a. EPA
    b. SAE
    c. OSHA
    d. NLGI

12. Viscosity ratings for industrial lubricants are stated in _____.
    a. SUS
    b. SAE
    c. ANSI
    d. SAS

Review Questions

13. The lowest temperature at which a lubricant will flow freely is called its _____.
    a. flash point
    b. fire point
    c. drop point
    d. pour point

14. When hydrocarbons take in oxygen from the air, this is known as _____.
    a. viscosity
    b. hydrogenation
    c. oxidation
    d. vapor pressure

15. When pressure is applied to grease, it releases some of its _____.
    a. oxygen
    b. lithium
    c. cadmium
    d. oil

Summary

Any piece of equipment, machine, tool, or other mechanism that has rotating or moving parts experiences friction. Friction is the rubbing of one part against another. Lubrication is required wherever friction is a problem and must be controlled. Lubrication is the application of any substance that reduces friction by creating a slippery film between two surfaces. Lubricants may be liquid, semisolid, or solid. Every mechanic must understand lubrication and know how to perform lubrication activities safely and efficiently.

Notes

Trade Terms Introduced in This Module

Adsorption: The collection of a substance, either gas or liquid, in condensed form on a surface or to substances other than itself.

Atomize: To reduce a liquid to a fine spray.

Bungs: Holes in the top of a lubricant container that are usually kept closed by a plug.

Carbon dioxide (CO_2): A colorless, odorless gas produced by complete combustion of a hydrocarbon fuel-air mixture. CO_2 is a component of the exhaust from an internal combustion engine.

Centistoke (cSt): A unit of kinematic viscosity as measured in a capillary tube viscometer under constant pressure and temperature. One stoke equals 100 centistokes.

Corrosion: The removal of mass from the surface of metals by chemicals that weaken and degrade the material properties.

Decomposition: The breaking down into the component parts by chemical reaction.

Dispersant: An oil additive that maintains potential deposits finely dispersed and suspended in the oil so they will not settle out and collect on machine surfaces.

Disposal: The act of throwing out or away.

Dropping point: The temperature at which a grease passes from a semisolid to a liquid state under specified test conditions.

Dunnage: Wood blocks used as supports to keep materials off the ground.

Emulsion: The combination of two liquids that are not usually mixable and in which one liquid is finely dispersed in the other.

Environmental Protection Agency (EPA): A federal agency charged with administering and enforcing federal regulations governing air and water pollution control.

Fire point: The lowest temperature at which the tested sample ignites and sustains combustion for a specified time.

Flash point: The lowest temperature at which the tested sample vapors ignite.

Friction: The resistive force created when two surfaces slide or rub together. Rolling friction occurs in ball bearings. Fluid friction is measured by viscosity.

Grease: A physical mixture of base oil and thickeners, usually with other additives, having a solid or semifluid consistency.

Hydrocarbons: Compounds containing only carbon and hydrogen. Petroleum consists mainly of hydrocarbons.

Inhibitor: A compound added to a lubricant to prevent negative and destructive reactions.

NLGI consistency number: A simplified system for rating the consistency (hardness) of a lubricating grease as measured by the *ASTM Worked Penetration* (60 strokes) at 77°F.

Oil: A general term for a water-insoluble, viscous liquid that possesses lubricating properties.

Oxidation: The process of combining with oxygen. All petroleum products react with oxygen to some degree, and this increases with temperature increases.

Paraffin: Hydrocarbons belonging to the series starting with methane. Paraffins are saturated with respect to hydrogen. Paraffin wax is an example of a high molecular weight paraffin.

Polymerization: The joining of chemical molecules to form long chains called polymers. This joining reaction can be hazardous because some chemicals release great amounts of heat when they polymerize.

Pour point: The lowest temperature at which a lubricant will pour or flow under specified conditions of standardized *ASTM Test D 97.*

Saybolt universal seconds (SUS, SSU): A unit of measure of lubricating oil viscosity used in the U.S.

Viscosity: The viscosity of an oil is its stiffness or internal friction. Higher viscosity is thicker or heavy, and a lower viscosity is thinner or light.

Viscosity index (VI): An empirical number indicating the effect of change of temperature on the viscosity of oil. A low VI indicates a relatively large change of viscosity with temperature; a high VI indicates a relatively small change.

Appendix

Material Safety Data Sheet

MATERIAL SAFETY DATA SHEET
TIMKEN CONSTRUCTION AND OFF-HIGHWAY GREASE

1. CHEMICAL PRODUCT AND COMPANY IDENTIFICATION

TRADE NAME TIMKEN CONSTRUCTION AND OFF-HIGHWAY GREASE

PART No. GR219

PRODUCT USE Lubricating Grease

SUPPLIER The Timken Corporation
1835 Dueber Ave.
P.O. Box 6930
Canton,OH 44706-0930 USA
Tel: (330) 438-3000

CONTACT PERSON David Pierman

2. COMPOSITION, INFORMATION ON INGREDIENTS

INGREDIENT NAME	CAS No.	WEIGHT
LUBRICANT BASE OIL (PETROLEUM), HIGHLY REFINED**(2)	Mixture	60-80 %
CALCIUM SULFONATE COMPLEX THICKENER	Proprietary	10-30 %
*MOLYBDENUM SULFIDE (MoS2)	1317-33-5	3-7 %
*BENZENESULFONIC ACID, DODECYL-, CALCIUM SALT	26264-06-2	1-3 %

* This chemical(s) is hazardous according to OSHA/WHIMIS criteria

COMPOSITION COMMENTS Refer to section eight for exposure limits on ingredients.
Chemical ingredients not regulated by OSHA, SARA, state or federal agencies are treated confidentially.
**(2) The base oil for this product can be a mixture of any of the following highly refined petroleum streams:
CAS 64741-88-4; CAS 64741-89-5; CAS 64741-96-4; CAS 64741-97-5; CAS 64742-01-4; CAS 64742-52-5; CAS 64742-53-6; CAS 64742-54-7; CAS 64742-55-8; CAS 64742-56-9; CAS 64742-57-0; CAS 64742-62-7; CAS 64742-63-8; CAS 64742-65-0; CAS 72623-83-7; CAS 72623-85-9; CAS 72623-86-0; CAS 72623-87-1.
Carcinogenicity: The petroleum base oils contained in this product have been highly refined by a variety of processes including solvent extraction, hydrotreating, and dewaxing to remove aromatics and improve performance characteristics. None of the oils used are listed as a carcinogen by NTP, IARC, or OSHA.

3. HAZARDS IDENTIFICATION

EMERGENCY OVERVIEW	Not regarded as a health hazard under current legislation.
INHALATION	Inhalation hazard at room temperature is unlikely due to the low volatility of this product. Heating can generate vapors that may cause respiratory irritation, nausea and headaches.
INGESTION	May cause stomach pain or vomiting.
SKIN	Prolonged or repeated contact leads to drying of skin.
EYES	May be slightly irritating to eyes.
SENSITIZATION	No known information.
CARCINOGENICITY	IARC: Not listed as a Group 1, 2A, or 2B agent. OSHA: Not regulated. NTP: Not listed.
TERATOGENICITY	No known information.
MUTAGENICITY	No known information.
HEALTH WARNINGS	INHALATION. Heating can generate vapors that may cause respiratory irritation, nausea and headaches. Inhalation hazard at room temperature is unlikely due to the low volatility of this product. SKIN CONTACT. Repeated or prolonged contact can result in drying of the skin. EYE CONTACT. Slightly irritating. INGESTION. Can cause stomach ache and vomiting.
ROUTE OF ENTRY	Inhalation. Skin and/or eye contact. Ingestion.

4. FIRST AID MEASURES

INHALATION	Move the exposed person to fresh air at once. For breathing difficulties oxygen may be necessary. Get medical attention if any discomfort continues.
EYES	Rinse with water. Contact physician if discomfort continues.
SKIN	Remove contaminated clothing. Wash skin thoroughly with soap and water. Get medical attention if any discomfort continues. INJECTION INJURY WARNING: If product is injected into or under the skin, or into any part of the body, regardless of the appearance of the wound or its size, the individual should be evaluated immediately by a physician as a surgical emergency. Even though initial symptoms from high pressure injection may be minimal or absent, early surgical treatment within the first few hours may significantly reduce the ultimate extent of injury.
INGESTION	DO NOT INDUCE VOMITING! Get medical attention immediately!

5. FIRE FIGHTING MEASURES

FLASH POINT (°C)	246 (475°F) Cd OC (Cleveland open cup).
FLAMMABILITY LIMIT - LOWER(%)	N/D

FLAMMABILITY LIMIT - UPPER(%) N/D

EXTINGUISHING MEDIA Water spray, fog or mist. Foam. Carbon dioxide (CO_2). Dry chemicals, sand, dolomite etc.

SPECIAL FIRE FIGHTING PROCEDURES Use water to keep fire exposed containers cool and disperse vapors. Water spray may be used to flush spills away from exposures and dilute spills to non-flammable mixtures. Avoid water in straight hose stream; will scatter and spread fire. Keep run-off water out of sewers and water sources. Dike for water control.

UNUSUAL FIRE & EXPLOSION HAZARDS Volume and pressure increases strongly when heated. Risk of container explosion in fire.

HAZARDOUS COMBUSTION PRODUCTS Acrid smoke/fumes. Oxides of: Carbon. Sulfur.

PROTECTIVE MEASURES IN CASE OF FIRE Self-contained breathing equipment and chemical resistant clothing recommended.

6. ACCIDENTAL RELEASE MEASURES

PERSONAL PRECAUTIONS Minimize skin contact.

PRECAUTIONS TO PROTECT THE ENVIRONMENT Keep product out of sewers and watercourses by diking or impounding. Advise authorities if product has entered or may enter sewers, watercourses or extensive land areas. Assure conformity with applicable government regulations.

SPILL CLEAN-UP PROCEDURES Provide good ventilation. Use appropriate protective clothing. Carefully collect spilled material in closed containers and leave for disposal according to local regulations. Do not let washing down water contaminate ponds or waterways. Rinse area with water.

7. HANDLING AND STORAGE

HANDLING PRECAUTIONS Keep away from heat, sparks and open flame. Ventilate well, avoid breathing vapors. Use approved respirator if air contamination is above accepted level. Containers should be kept tightly closed. Avoid spilling, skin and eye contact. Eye wash and emergency shower must be available at the work place.

STORAGE PRECAUTIONS Keep away from heat, sparks and open flame. Store separated from: Acids. Oxidizing materials.

STORAGE CRITERIA Chemical storage.

8. EXPOSURE CONTROLS, PERSONAL PROTECTION

COMPONENT	STD	TWA	STEL	TWA	STEL
LUBRICANT BASE OIL (PETROLEUM), HIGHLY REFINED**(2)	OSHA			5 mg/m3 **(1)	
	ACGIH			5 mg/m3 **(1)	10 mg/m3 **(1)
MOLYBDENUM SULFIDE (MoS_2)	OSHA			N/E	
	ACGIH	3 mg/m3 (resp) as Mo		10 mg/m3 (ihl) as Mo	

INGREDIENT COMMENTS **(1) For respirable oil mist.

PROTECTIVE EQUIPMENT

ENGINEERING CONTROLS	Use engineering controls to reduce air contamination to permissible exposure level.
VENTILATION	No specific ventilation requirements noted, but forced ventilation may still be required if air contamination exceeds acceptable level.
RESPIRATORS	No specific recommendation made, but respiratory protection may still be required under exceptional circumstances when excessive air contamination exists.
PROTECTIVE GLOVES	Chemical resistant gloves required for prolonged or repeated contact. Use protective gloves made of: Neoprene, nitrile, polyethylene or PVC.
EYE PROTECTION	Use eye protection.
PROTECTIVE CLOTHING	Wear appropriate clothing to prevent repeated or prolonged skin contact.
HYGIENIC WORK PRACTICES	Wash at the end of each work shift and before eating, smoking and using the toilet.

9. PHYSICAL AND CHEMICAL PROPERTIES

APPEARANCE/PHYSICAL STATE	Grease.		
COLOR	Grey.		
ODOR	Mild (or faint). Petroleum.		
SOLUBILITY DESCRIPTION	Insoluble in water.		
DENSITY	0.96	**Temperature (°C)**	15.6 (60°F)
VAPOR DENSITY (air=1)	> 5		
VAPOR PRESSURE	< 0.01 mmHg	**Temperature (°C)**	20 (68°F)
EVAPORATION RATE	< 1	**Reference**	BuAc=1
pH-VALUE, CONC. SOLUTION	N/A		

10. STABILITY AND REACTIVITY

STABILITY	Normally stable.
CONDITIONS TO AVOID	Avoid contact with acids and oxidizing substances.
HAZARDOUS POLYMERIZATION	Will not polymerize.

113A01D.EPS

HAZARDOUS DECOMPOSITION PRODUCTS Oxides of: Carbon. Sulfur.

11. TOXICOLOGICAL INFORMATION

TOXICOLOGICAL INFORMATION No experimental toxicological data on the preparation as such is available.

COMPONENT **LUBRICANT BASE OIL (PETROLEUM), HIGHLY REFINED**(2)**

TOXICOLOGICAL DATA Carcinogenicity. IP 346: <3%

TOXIC DOSE - LD 50 N/A.

TOXIC CONC. - LC 50 N/A.

12. ECOLOGICAL INFORMATION

ECOLOGICAL INFORMATION No data on possible environmental effects have been found.

13. DISPOSAL CONSIDERATIONS

DISPOSAL METHODS Spilled material, unused contents and empty containers must be disposed of in accordance with local, state and federal regulations.

14. TRANSPORT INFORMATION

DOT HAZARD CLASS Not regulated.

IDENTIFICATION No. N/A

TDGR CLASS Not Regulated. Non réglementé.

SEA TRANSPORT NOTES Not regulated per IMDG.

AIR TRANSPORT NOTES Not regulated per IATA.

15. REGULATORY INFORMATION

US FEDERAL REGULATIONS: COMPONENT	SARA 302	CERCLA	SARA 313
CALCIUM SULFONATE COMPLEX THICKENER	No	No	No
MOLYBDENUM SULFIDE (MoS2)	No	No	No
BENZENESULFONIC ACID, DODECYL-, CALCIUM SALT	No	1 000 lbs	No
LUBRICANT BASE OIL (PETROLEUM), HIGHLY REFINED**(2)	No	No	No

SARA HAZARD CATEGORIES None

113A01E.EPS

US STATE REGULATIONS: BY COMPONENT	CA	FL	MA	MN	NJ	PA	RI
BENZENESULFONIC ACID, DODECYL-, CALCIUM SALT						EH	

INVENTORIES: COMPONENT	CAN	US	EU	AUS	JAP	KOR	CHN	PHLP
LUBRICANT BASE OIL (PETROLEUM), HIGHLY REFINED**(2)	DSL	Yes	EINECS	Yes	Yes	Yes	Yes	Yes
CALCIUM SULFONATE COMPLEX THICKENER	DSL	Yes	EINECS	Yes	Yes	Yes	Yes	Yes
MOLYBDENUM SULFIDE (MoS_2)	DSL	Yes	EINECS	Yes	Yes	Yes	Yes	Yes
BENZENESULFONIC ACID, DODECYL-, CALCIUM SALT	DSL	Yes	EINECS	Yes	Yes	Yes	Yes	Yes

16. OTHER INFORMATION

NFPA-HMIS: HEALTH	Irritation, minor residual injury (1) - HMIS/NFPA
NFPA-HMIS: FLAMMABILITY	Burns only if pre-heated (1) - HMIS/NFPA
NFPA-HMIS: REACTIVITY	Normally Stable (0) - HMIS/NFPA
HMIS PERS. PROTECT. INDEX	B - Safety Eyewear and Gloves
PREPARED BY	James W. Hermann
DATE	2005-04-08
PRINTING DATE:	2005-04-08
DISCLAIMER	While the information and recommendations set forth herein are believed to be accurate as of the date thereof, The Timken Corporation makes no warranty with respect thereto and disclaims all liability from reliance therein.

*** Information revised since previous MSDS version**

Resources & Acknowledgments

Additional Resources

This module is intended to be a thorough resource for task training. The following reference works are suggested for further study. These are optional materials for continued education rather than for task training.

Shell Lubricants Handbook, available through Shell Lubricant Sales Offices and Suppliers, published yearly.

Chevron Salesfax Digest, available through Chevron Lubricant Sales Offices and Suppliers, published yearly.

Mobil Brief Product Descriptions, available through Mobil Lubricant Sales Offices and Suppliers, published yearly.

Figure Credits

Lincoln Industrial Corp., 113F04 (line art), 113F06, 113F07, 113F09, 113F10, 113F14–113F18

Lube Technology, 113F04 (photo)

H-T-L Perma USA, 113F11

Digilube Systems, Inc., 113F19

The Timken Company, Appendix

CONTREN® LEARNING SERIES — USER UPDATE

NCCER makes every effort to keep these textbooks up-to-date and free of technical errors. We appreciate your help in this process. If you have an idea for improving this textbook, or if you find an error, a typographical mistake, or an inaccuracy in NCCER's Contren® textbooks, please write us, using this form or a photocopy. Be sure to include the exact module number, page number, a detailed description, and the correction, if applicable. Your input will be brought to the attention of the Technical Review Committee. Thank you for your assistance.

Instructors – If you found that additional materials were necessary in order to teach this module effectively, please let us know so that we may include them in the Equipment/Materials list in the Annotated Instructor's Guide.

Write: Product Development and Revision
National Center for Construction Education and Research
3600 NW 43rd St. Bldg G, Gainesville, FL 32606

Fax: 352-334-0932

E-mail: curriculum@nccer.org

Craft Module Name

Copyright Date Module Number Page Number(s)

Description

(Optional) Correction

(Optional) Your Name and Address

SMAW – Equipment and Setup

29107-09

29107-09 SMAW – Equipment and Setup

WELDING LEVEL ONE

29112-09 SMAW – Open V-Groove Welds

29111-09 SMAW – Groove Welds with Backing

29110-09 Joint Fit-Up and Alignment

29109-09 SMAW – Beads and Fillet Welds

29108-09 Shielded Metal Arc Electrodes

29107-09 SMAW – Equipment and Setup

29106-09 Weld Quality

29105-09 Base Metal Preparation

29104-09 Air Carbon Arc Cutting and Gouging

29103-09 Plasma Arc Cutting

29102-09 Oxyfuel Cutting

29101-09 Welding Safety

Core Curriculum: Introductory Craft Skills

This course map shows all of the modules in *Welding Level One.* The suggested training order begins at the bottom and proceeds up. Skill levels increase as you advance on the course map. The local Training Program Sponsor may adjust the training order.

Objectives

When you have completed this module, you will be able to do the following:

1. Identify and explain shielded metal arc welding (SMAW) safety.
2. Explain welding electrical current.
3. Identify welding power supplies and their characteristics.
4. Explain how to set up welding power supplies.
5. Set up a machine for welding.
6. Identify tools used for weld cleaning.

Trade Terms

Alternating current (AC)
Amperage
Arc
Direct current (DC)
Electrode
Polarity
Primary current
Step-down transformer
Voltage

Prerequisites

Before you begin this module, it is recommended that you successfully complete the following: *Core Curriculum* and *Welding Level One*, Modules 29101-09 through 29106-09.

Contents

Topics to be presented in this module include:

1.0.0 INTRODUCTION

Shielded metal **arc** welding (SMAW) is often referred to as stick welding. This welding process can use either **alternating current (AC)** or **direct current (DC)**, with adjustments made on the welding machine to regulate the level of current. The **electrodes** (rods) used are typically coated with a clay-like mixture of fluorides, carbonates, oxides, metal alloys, and binding materials to provide a gas shield for the weld puddle. This coating hardens and covers the weld during its cooling cycle. The coating protects the weld from the atmosphere, which can cause weld degradation; then it is removed to expose the finished, shiny weld beneath. *Figure 1* shows a welder using a mobile SMAW welding machine outdoors.

This module provides an overview of the different types of equipment used for SMAW, safety concerns, characteristics of the currents associated with SMAW, and some of the accessories necessary to properly connect the SMAW equipment. After completion of this module, you will have a better understanding of SMAW equipment and how to properly connect the equipment for welding.

107F01.EPS

Figure 1 A mobile SMAW machine being used outdoors.

2.0.0 SMAW SAFETY

Welding is a dangerous profession, not so much because of the welding itself, but because of the environments in which the work is often performed. Any welder must consider a number of safety issues associated with welding. Primary safety concerns involve the following:

- Moving welding equipment
- Electrical hazards
- Lifting hazards
- Working at heights
- Welding hazards

2.1.0 Moving Welding Equipment

Welding in the shop usually means that you are using a welding machine that is either set up on a stationary platform in or near a welding booth, or is on a portable cart that can be rolled around the shop floor. When welding is performed in the field, the welding machines may be mounted on the back of a truck, on a skid, or on a mobile wagon or trailer. There will be times when the welding machines must be lifted, by a crane or some other rigging device, to some elevated location. Whenever a welding machine is moved, it becomes a potential hazard. Machines can fall off their carts or be dropped. Even a 40-pound welding machine can seriously injure a foot or hand when dropped from its cart. Consider the damage a mobile welding machine could do if it got away from its tow vehicle, or if the brakes or chocks allowed it to roll down an incline. Be extremely aware of what is happening to a welding machine as it is being moved.

2.2.0 Electrical Hazards

Welding machines use electrical power. Most operate on 110V, 120V, 240V, or 480V levels. Regardless of the **voltage** level, be aware of any electrical power source associated with your welding machine. Electrical shock or possible death can result if you encounter the electrical power supplied to a welding machine. Always know where the main electrical breaker is located for the electrical outlet supplying electricity to your machine. Whenever possible, turn that breaker off before connecting or disconnecting the welding machine's plug and the outlet.

Another concern with electricity is moisture, especially water. Electricity, water, and the human body are a dangerous mix. Make sure that all your welding cables are in good condition and that your workpiece clamp is properly installed.

2.3.0 Lifting Hazards

Welders sometimes forget the basics of lifting as they transport their welding machines, cables, and even the parts to be welded. Use rigging or lifting devices whenever possible. Get help if you must lift a heavy object by hand. Always remember to use tag lines when hoisting heavy objects that tend to spin or swing.

2.4.0 Working at Heights

Welders may have to be suspended from the side of a structure to weld something. Remember to wear all the approved fall protection devices when performing such work. Make sure to inspect your fall protection equipment items each time you use them, and be certain that they are current and up-to-date.

2.5.0 Welding Hazards

Although this module addresses the preparation of SMAW equipment, there may be a time when you must test-operate a welding machine before using it. When that situation occurs, make sure that you are wearing all the required PPE and that the work area meets all welding safety requirements.

3.0.0 Welding Current

When the term *current* is used as a noun, it means a flow or a stream. In electrical terms, current is the flow of an electrical charge from one point to another. One of those points is earth ground. The other point is a source of stored electrical energy. The battery in your car is stored energy. Electrical charges (electricity) will not flow unless they find a path connected to earth ground or something that is grounded. Humans standing on the earth are grounded. To your car's battery, your car appears to be grounded. Buildings built on or around steel beams are grounded because their beams are anchored into the ground. Structures that have not been built with beams anchored in the ground must be grounded through wires attached to rods driven into the ground. Machines found in facilities must be grounded. The point of this discussion is to establish the fact that the electrical power used in welding machines is useless unless it is brought close to, or is in contact with, a grounded object. In welding, there must be current flow between the workpiece and the welding machine.

The arc that is produced during SMAW is an electrical current that jumps across the gap from the tip of the electrode to the surface of the workpiece, as long as the workpiece clamp from the SMAW lead is attached to the workpiece (*Figure 2*). The current originates in the transformer or generator of the welding machine, is routed through one welding lead and then through the workpiece, and finally returns to the transformer or generator through the other welding lead. If the electrode tip is touched and held against the workpiece instead of allowing for a gap between the two, the electrode will pass the electricity directly to the workpiece without creating an arc. Maximum available current from the welding machine will be passed through the circuit. No welding will take place in this situation because an arc across the gap is needed to heat the workpiece and the electrode to a molten state.

Hot Tip

Striking an Arc

Two general methods of striking the initial arc are scratching the electrode on the workpiece and tapping the electrode on the workpiece.

The scratching method is generally used by most beginners and even by many experienced welders. However, it is usually not acceptable for code or finished-product welding unless the burn trails from scratching are within the area covered by the weld. Some of the newer inverter power sources do not activate the open-circuit voltage until after an electrode touches a surface. With these machines, a scratching motion must be accomplished in order to activate the required open-circuit voltage and start the arc.

107F02.EPS

Figure 2 Striking an arc.

WARNING!

Never remove your welding helmet if the electrode is stuck or frozen to the workpiece. You must free the electrode with your helmet in place because the electrode will flash when it breaks free. First remove the holder, then remove the electrode with pliers.

3.1.0 Types of Welding Current

SMAW power supplies are designed to produce an output that has a nearly constant current. The currents may be either AC or DC. The type of current used depends on the type of welding to be done and/or the welding equipment available.

3.1.1 AC Welding Current

Alternating current in welding machines is derived from either a transformer-type welding machine or an alternator-type machine. The transformer-type machine changes high-voltage, low-current commercial AC power to the low-voltage, high-current power required for welding. An alternator-type machine uses an electric or fuel-driven motor to turn a rotor inside a number of electromagnets to produce the current needed for welding.

AC current alternates between positive and negative values only because the voltage changes from positive to negative values. In *Figure 3*, which is a meter display of 120VAC, the top half of the cycle of the voltage path is considered positive. The current follows the voltage. In the bottom half of the cycle, the voltage is considered negative, with the current continuing to follow the voltage. In one complete cycle, the wave crosses the axis or zero point twice. The peaks or extreme high or low points on the wave indicate the maximum level of the voltage, either positive (on the top) or negative (on the bottom). The number of cycles completed in one second is called the frequency. Frequency varies depending on the number of poles (magnetic field coils) installed in the generator during the generation stages of the power and on the speed of the rotor rotation. The standard frequency in the United States is set at 60 cycles per second.

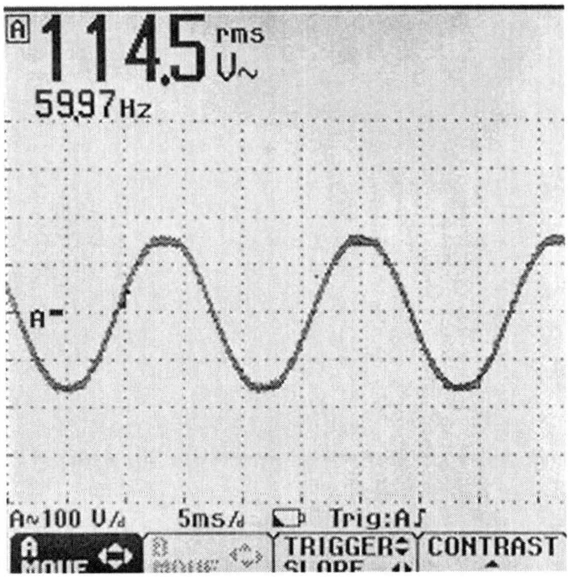

107F03.EPS

Figure 3 AC current waveform.

Think About It

Voltage Level

One cycle of our standard 60 cycles per second AC power occurs in 0.0167 second, with the corresponding rapid changes in voltage levels unnoticeable during the operation of machinery or in the arc of a welder. What is the voltage level of the AC power at the point where the waveform crosses the center axis?

3.1.2 DC Welding Current

Direct current is electrical current that has no frequency. It does not travel from positive to negative and negative to positive within the same operation. The waveform for DC power resembles a straight line; therefore, it is not really a wave. Remember, frequency is associated with AC power only because frequency refers to the change in direction during one cycle, and DC power does not change direction during its flow. Direct current in welding is derived either from a transformer-rectifier type of system or from a generator. In a transformer-rectifier type of system, AC voltage first is reduced (transformed) to a lower voltage level through the application of a transformer, then it is converted (rectified) into DC voltage. Most gasoline-powered SMAW welding machines produce DC current through a generator. SMAW units that plug into 60-cycle AC power generally produce DC current through the application of a transformer and a rectifier.

3.2.0 Polarity

Polarity only applies to DC current. Polarity in welding is determined by the way the welding leads are connected either to the welding machine or to the workpiece. The universally accepted theory is that DC current always travels from the negative pole to the positive pole, with the minus-marked terminal on the welding machine indicating the negative pole and the plus-marked terminal indicating the positive pole. In welding or in any other DC circuit, you cannot change that process. You can, however, change the path the current takes in going from the negative pole to the positive pole. In welding, this can be done simply by changing the welding lead connections at the machine or, in newer machines, by changing a switch position on the machine.

Most, if not all, welding leads are configured with one end of one lead holding an electrode or rod and the other lead equipped on one end with a clamping device to attach to the workpiece. Both of the other ends of these two cables typically are equipped with ring-type or plug-in connectors for attaching the leads to the welding machine. When the electrode-holder lead is connected to the plus or positive terminal on the machine and the lead with the clamping device is connected to the minus or negative terminal on the welding machine, direct current electrode positive (DCEP) polarity is established. If these leads are manually switched at the welding machine terminals or through the operation of the manual polarity switch located on the machine, the electrode-holder lead becomes the negative lead and the workpiece clamping device lead becomes the positive lead. This is referred to as direct current electrode negative (DCEN) polarity. *Figure 4* shows labeled diagrams of DCEP and DCEN hookups.

CAUTION

On machines that have polarity switches, do not switch the polarity during operation. This may damage the machine.

Hot Tip

Properties of an Arc

An arc produces a temperature in excess of 6,000°F at the tip of an electrode. This extreme heat is what melts the base metal and electrode, producing a pool of molten metal often referred to as a crater. As the crater solidifies behind the weld, it makes the bond.

3.3.0 Characteristics of Welding Current

The voltage in a welding machine, whether the machine is connected to commercial power or driven by an engine, is reduced or stepped down by a transformer called a **step-down transformer**. The step-down transformer reduces the voltage level and raises the current level. The higher current allows the arc to occur with a voltage potential that carries less of a risk of electrocution. However, electrical shock in welding remains a hazard that can be reduced or eliminated only by using proper personal protective equipment and applying safe welding practices.

3.3.1 Voltage

Voltage is the measure of the electromotive force or pressure that causes current to flow in a circuit. Two types of voltage are associated with welding current: open-circuit voltage and operating voltage. Open-circuit voltage is the voltage present when the welding machine is on but no arc is present. For SMAW, open-circuit voltage is usually between 50 and 80 volts.

Operating voltage, or arc voltage, is the voltage measured after the arc has been struck. This voltage is lower than open-circuit voltage and usually measures between 18 and 45 volts. The higher, open-circuit voltage is required to establish the arc because the air gap between the electrode and the work has high resistance to current flow. Once the arc has been established, less voltage is needed, and the welding machine automatically compensates by lowering the voltage.

This characteristic of an SMAW machine is called variable-voltage, constant-current power and is shown as a volt-ampere output curve in *Figure 5*. Operating voltage is dependent on the current range selected. However, SMAW voltage can be altered somewhat by a change in the arc gap.

3.3.2 Amperage

Amperage is a measurement of the electric current flowing in a circuit. The unit of measurement for current is the ampere (amp, or A). The number of amps produced by the welding machine determines the amount of heat available to melt the work and the electrode. The current is increased or decreased according to the size of electrode being used and the position in which the welding is being performed.

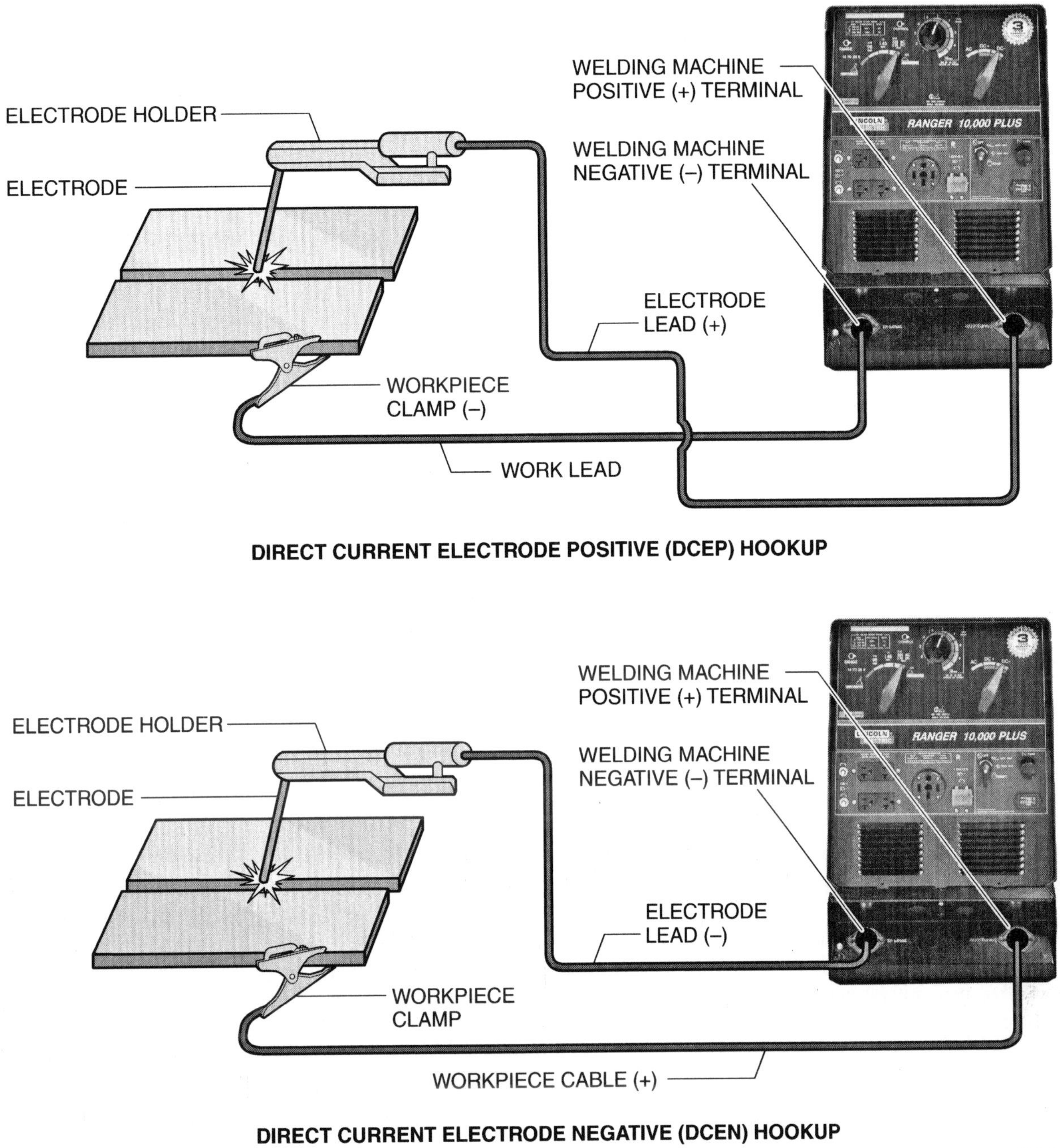

107F04.EPS

Figure 4 DCEP and DCEN hookups.

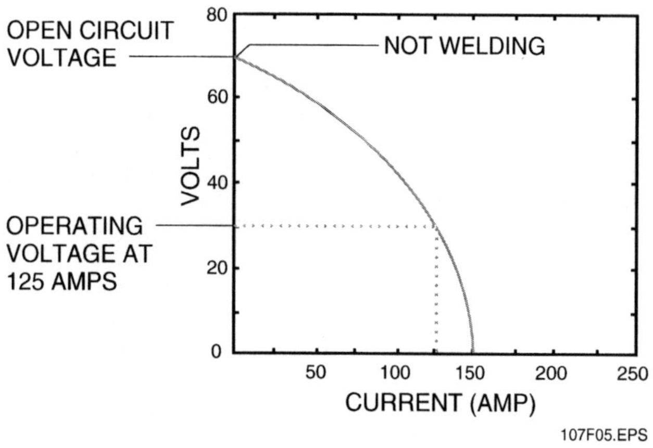

Figure 5 Volt-ampere output curve.

4.0.0 SMAW MACHINES

Welding machines, like any other tools, have specific classifications, different special types, and rated capacities. A lathe, for example, might be rated by its throw, speeds, and special toolings. SMAW welding machines are broken down by classifications, types, and ratings.

4.1.0 SMAW Machine Classifications

SMAW welding machines are classified by the type of welding current they produce: AC, DC, or AC/DC.

An SMAW welding machine classified as an AC machine will produce only AC welding voltage and current, and a DC machine will produce only DC welding voltage and current. If a welding machine produces both AC and DC welding voltage and current, it is classified as an AC/DC welding machine.

Welding machines that produce only DC welding voltage and current can be further classified by the characteristics of the welding current they produce. If the welding current varies between the higher open-circuit voltage and the lower operating voltage, it is classified as a variable-voltage constant-current power source. This type of welding machine is used for SMAW. If open-circuit and operating voltage are nearly the same, the machine is classified as a constant-voltage or constant-potential DC welding machine. This type of machine is used for GMAW and FCAW. *Figure 6* shows the difference in outputs between constant-current and constant-voltage welding machines.

4.2.0 SMAW Machine Types

Several different types of basic SMAW welding machines are available to produce the current necessary for welding. They include transformers, transformer rectifiers, inverter power sources, and engine-driven generators.

Transformers, transformer rectifiers, and electric motor generators all require electrical power from commercial power lines to operate. This high-voltage, low-amperage current from commercial power lines coming into the welding machine is called the **primary current**. The primary current first flows from the power lines to an electrical circuit breaker or disconnecting fuse box. From the disconnecting device, the current flows to a receptacle. The disconnecting device can be used to shut off the flow of current to the receptacle. The welding machine is plugged into the receptacle. The primary current required for welding machines can be 120V or 240V single-phase or 480V three-phase.

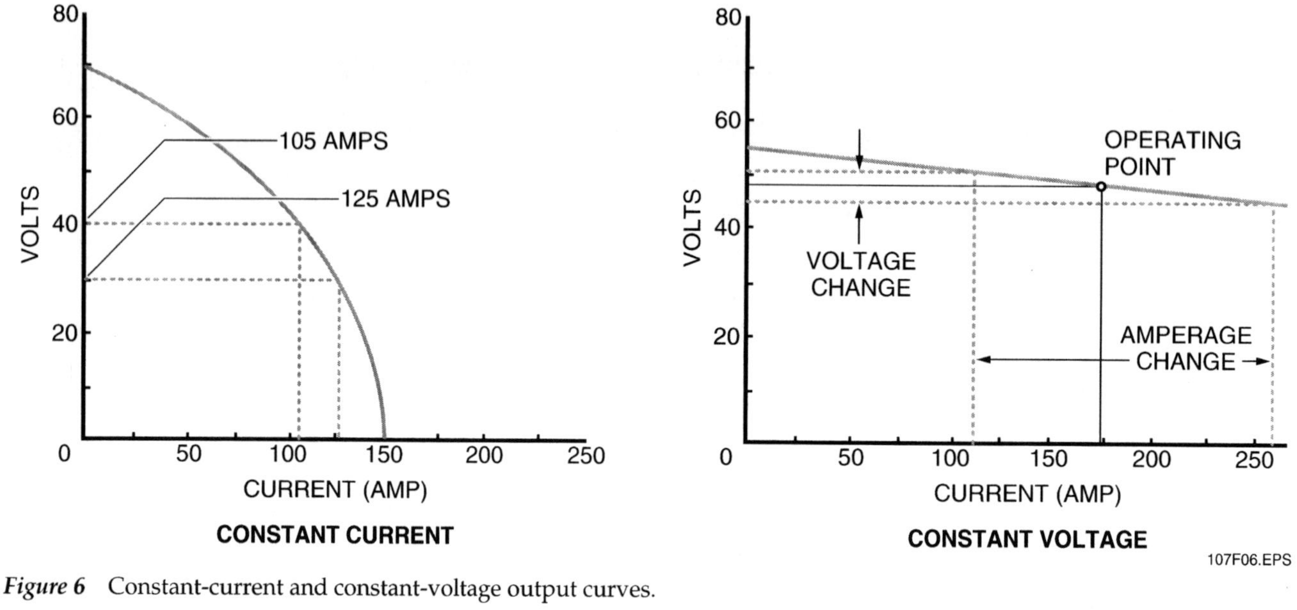

Figure 6 Constant-current and constant-voltage output curves.

WARNING! Coming into contact with the primary current of a welding machine can cause electrocution. Ensure that welding machines are properly grounded.

Figure 7 shows a typical welding machine circuit.

4.2.1 Transformer Welding Machines

Transformer welding machines without rectifiers produce AC welding current. They use a voltage step-down transformer, which converts high-voltage, low-amperage current from commercial power lines to low-voltage, high-amperage welding current. The primary power required for a transformer welder can be 240V single-phase or 480V three-phase. Special light-duty transformer welding machines used for sheet metal work are designed to be plugged into a 120V outlet. However, most light-duty transformer welding machines operate more efficiently on 240V primary power. Heavy-duty industrial transformer welders require 480V three-phase primary power.

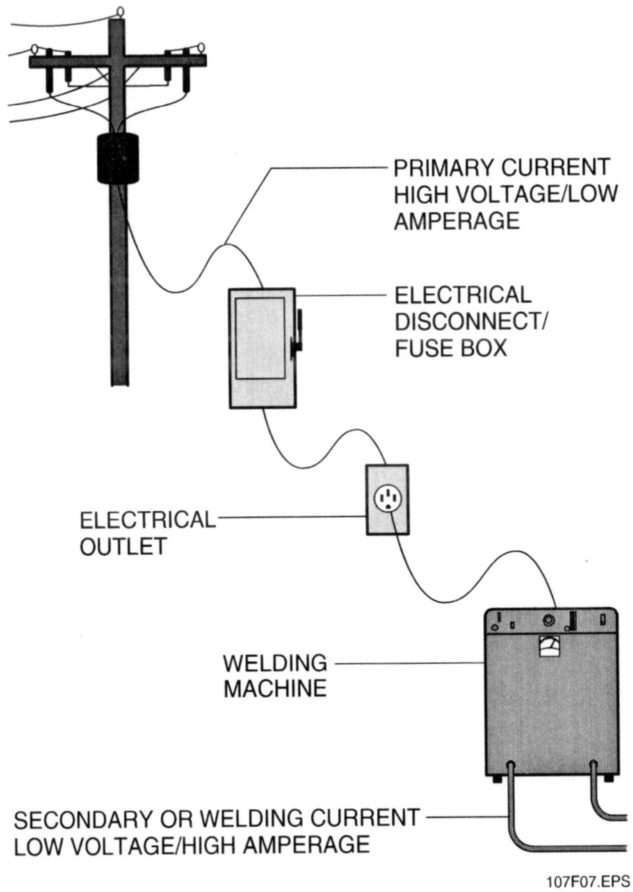

Figure 7 Welding machine primary circuit.

Transformer welders are not as common as other types of welding machines on the job site, but they are used for special jobs. A transformer welder has an ON/OFF switch, an amperage control, and terminals for connecting the electrode lead and the workpiece lead. *Figure 8* shows the components of a typical transformer welding machine.

4.2.2 Transformer-Rectifier Welding Machines

A rectifier is a device that converts AC current to DC current. A transformer-rectifier welding machine uses a transformer to reduce the primary voltage to welding voltage and then rectifies the current to change it from AC to DC. Transformer-rectifier welding machines can be designed to produce AC and DC welding current or DC current only. Transformer rectifiers that produce both AC and DC welding current are usually intended for lighter duty than those that produce DC current only. Depending on their size, transformer-rectifier welding machines may require 240V single-phase power, 240V three-phase power, or 480V three-phase power.

WARNING! Coming into contact with the primary circuit of a welding machine can cause electrocution. Ensure that you take proper safety precautions to avoid contact.

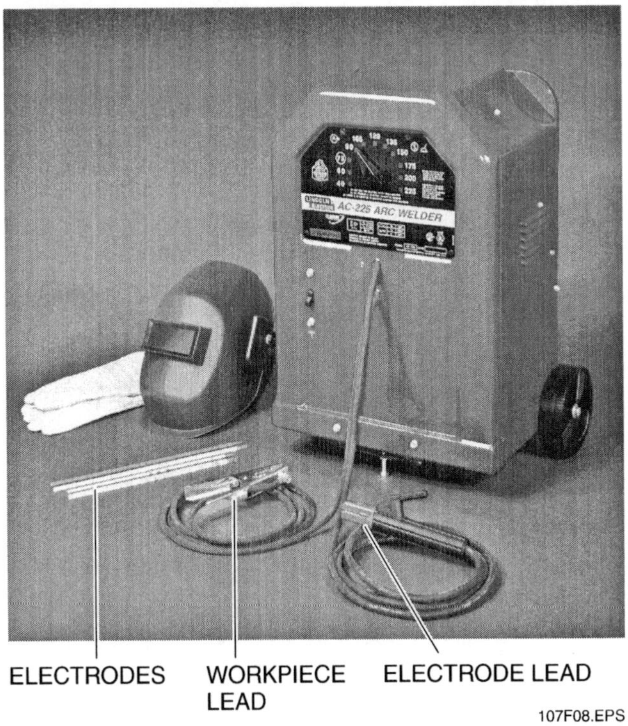

Figure 8 A transformer welding machine.

Transformer rectifiers can also be designed to produce variable-voltage constant-current or constant-voltage (constant-potential) DC welding current. Some are even designed to provide both variable and constant voltage; those are referred to as multi-process power sources since they can be used for any welding process. Refer to the manufacturer's documentation to determine how the machine is marked.

Transformer rectifiers have an ON/OFF switch and an amperage control. The welding cables (electrode lead and workpiece lead) are connected to terminals. They often have selector switches to select DCEP or DCEN or to select AC welding current if it is available. If there is no selector switch, the cables must be manually changed on the machine terminals to select the type of current desired. *Figure 9* shows the various parts of a typical transformer-rectifier welding machine.

Multiple transformer rectifiers are available grouped into a single cabinet, called a pack. When the pack contains six welding machines, it is called a six-pack; when it contains eight welding machines, it is called an eight-pack (*Figure 10*).

The pack in *Figure 10* has eight individual welding machines. They receive their electrical power from a power distribution panel at the end of the rack. The panel has one switch that allows all the machines to be turned off at once. Each individual welding machine has its own amperage control and welding cables. In addition, each welding machine has a receptacle for a remote current control. When the remote current control is plugged in, it can be used to adjust the amperage from a remote location without going back to the pack. The pack unit also has 120V receptacles for power tools and lights.

107F09.EPS

Figure 9 A transformer-rectifier welding machine.

> **Hot Tip**
>
> **AC/DC Machines**
>
> Transformer-rectifier welding machines that can provide either AC or DC current do not utilize the rectifier circuit when welding with AC current. In this situation, the welding machine functions like a transformer welding machine.

Packs are mounted on skids and have lifting eyes so they can be moved easily. They require only one primary power connection, which is normally 480V three-phase. Packs are common on construction sites and existing facilities during rebuilds when many welding machines are required.

4.2.3 Inverter Power Sources

Inverter power sources increase the frequency of the incoming primary power. This provides a smaller, lighter power source with a faster response time and much more waveform control for pulse welding. Inverter power sources are used where limited space and portability are important factors. The controls on these machines vary according to size and application. *Figure 11* shows a close-up view of an inverter power source.

The inverter power source shown in *Figure 11* is designed for outdoor use. It is used either as a single, stand-alone unit or in a rack with multiple inverter power sources. It supports both GTAW (gas tungsten arc welding) and SMAW processes, and can be used for gouging. With up to 275 amps of output power, it allows the welder to choose from a greater variety of stick electrodes, especially the E6010 and E7018 electrodes.

4.2.4 Engine-Driven Generator Machines

Welding machines can also be powered by gasoline or diesel engines. The engine can be connected to a generator. Engine-driven generators produce DC welding current.

To produce welding current, the generator must turn at a required number of revolutions per minute. Engines that power generators have governors to control the engine speed. Most governors have a welding speed switch. The switch can be set to idle the engine when no welding is

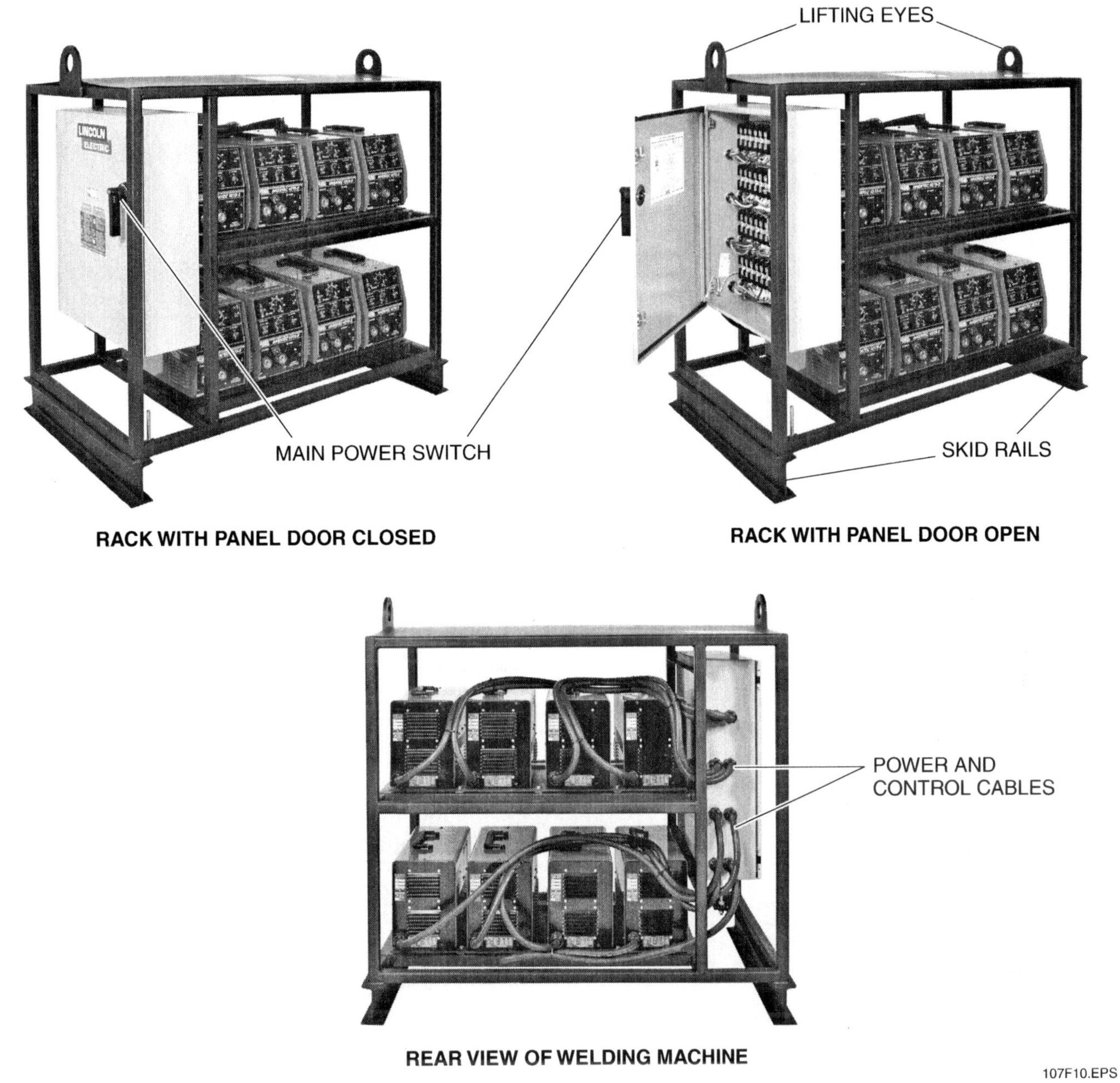

Figure 10 Eight-pack welding machine.

taking place. When the electrode is touched to the base metal, the governor will automatically increase the speed of the engine to the required rpm for welding. After about 15 seconds of no welding, the engine will automatically return to idle. This saves fuel and reduces wear on the generator. The switch can also be set to enable the engine to run continuously at welding speed. *Figure 12* shows an example of an engine-driven generator welding machine.

Engine-driven generators often have an auxiliary power unit to produce 120VAC current for lighting, power tools, and other electrical equipment. When 120VAC current is required, the engine-driven generator must run continuously at welding speed.

Engine-driven generators have engine controls and welding current controls. The engine controls vary with the type and size but normally include the following items:

- Starter
- Voltage gauge
- Temperature gauge
- Fuel gauge
- Hour meter

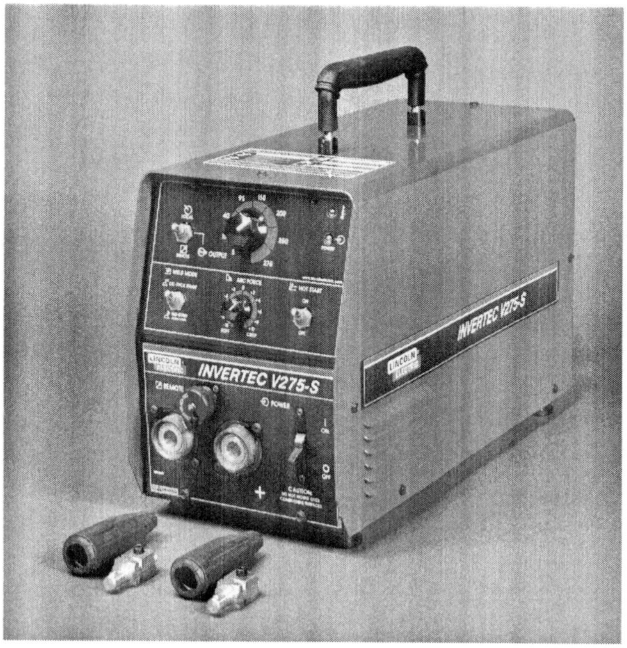

107F11.EPS

Figure 11 An inverter power source.

107F12.EPS

Figure 12 Engine-driven welding machine.

Engine-driven generators have an amperage control. To change polarity, there may be a polarity switch, or you may have to manually change the welding cables at the welding current terminals.

Many engine-driven generators are mounted on trailers. This makes them portable so they can be used in the field where electricity is not available for other types of welding machines. The disadvantage of engine-driven generators is that they are costly to purchase, operate, and maintain because of the added cost of the drive unit.

GOING GREEN

Welding Machine Maintenance

When performing maintenance on fuel-powered welding machine engines, make sure to prevent fuel and oil spills. Such spills, even small ones, can harm the environment. Always clean up and properly dispose of any spilled materials.

Hot Tip

Engines

The size and type of engine used depends on the size of the welding machine. Single-cylinder engines are used to power small generators. Four- and six-cylinder engines are used to power larger generators.

4.3.0 SMAW Machine Ratings

The size of a welding machine is determined by the amperage output of the machine at a given duty cycle. The duty cycle of a welding machine is based on a ten-minute period. It is the percentage of ten minutes that the machine can continuously produce its rated amperage without overheating. For example, a machine with a rated output of 300A at 60 percent duty cycle can deliver 300A of welding current for six minutes out of every ten (60 percent of ten minutes) without overheating.

The duty cycle of a welding machine will be 10 percent, 20 percent, 30 percent, 40 percent, 60 percent, or 100 percent. A welding machine having a duty cycle of from 10 to 40 percent is considered a light-duty to medium-duty machine. Most industrial, heavy-duty machines for manual welding have a 60 percent duty cycle. Machines designed for automatic welding operations have a 100 percent duty cycle.

With the exception of 100 percent duty cycle machines, the maximum amperage that a welding machine will produce is always higher than its rated capacity. A welding machine rated 300A at 60 percent duty cycle will generally put out a maximum of 375A to 400A. However, since the duty cycle is a function of its rated capacity, the duty cycle will decrease as the amperage is raised over 300A. Welding at 375A with a welding machine rated at 300A with a 60 percent duty cycle will lower the duty cycle to about 30 percent. If welding continues for more than three

> **Hot Tip**
>
> **Welding Machine Circuit Breakers**
>
> Most welding machines have a heat-activated circuit breaker that will shut off the machine automatically when it overheats. The machine cannot be turned on again until it has cooled.

out of ten minutes, the machine will probably overheat.

If the amperage is set below the rated amperage, the duty cycle increases. Setting the amperage at 200A for a welding machine rated at 300A with a 60 percent duty cycle will increase the duty cycle to 100 percent. *Figure 13* shows the relationship between amperage and duty cycle.

4.4.0 Welding Cable

Cables that carry welding current are designed for maximum strength and flexibility. The conductors inside the cable are made of fine strands of copper wire. The copper strands are covered with layers of rubber reinforced with nylon or dacron cord. *Figure 14* shows a cutaway section of welding cable.

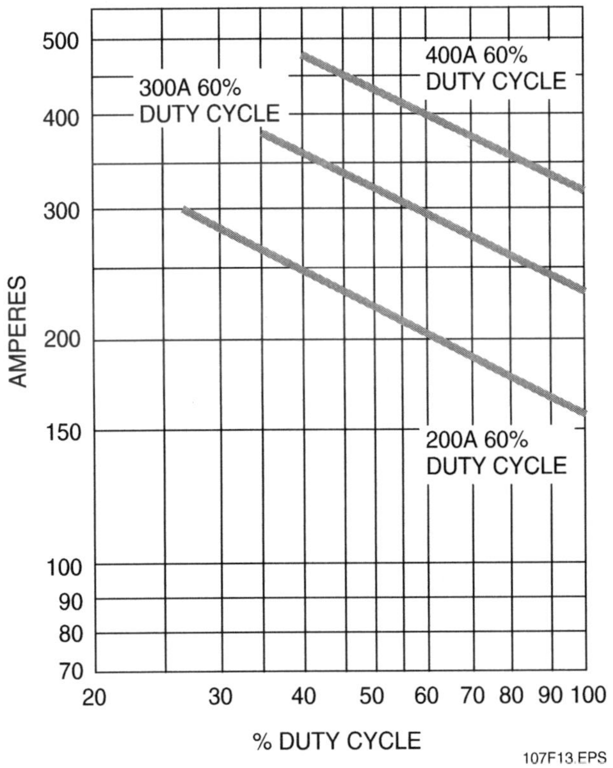

Figure 13 Amperage and duty cycle chart.

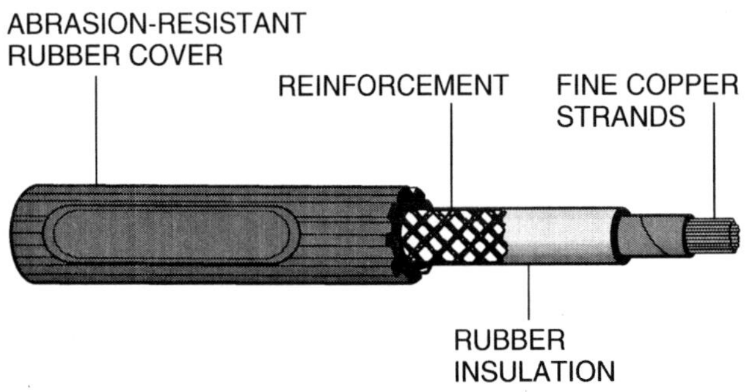

Figure 14 Welding cable.

The size of a welding cable is based on the number of copper strands it contains. Large-diameter cable has more copper strands and can carry more welding current. Typically the smallest welding cable size is No. 8 and the largest is No. 4/0 (four aught).

When selecting welding cable size, the amperage load and the distance the current will travel must be taken into account. The longer the distance the current has to travel, the larger the cable must be to reduce voltage drop and heating caused by electrical resistance. When selecting welding cable, use the rated capacity of the welding machine for the cable amperage requirement. To determine the distance, measure both the electrode and ground leads and add the two lengths together. To identify the welding cable size required, refer to a recommended welding cable size table furnished by most welding cable manufacturers. *Table 1* shows typical welding cable sizes for various welding currents and cable distances.

4.5.0 SMAW Cable Connectors

Welding cables must be equipped with the proper end connections. End connections used on welding cables include lugs, quick disconnects, ground clamps, and electrode holders.

> **CAUTION**
>
> If the end connection is not tightly secured to the cable, the connection will overheat. An overheated connection will cause variations in the welding current and permanent damage to the connector and/or cable. Repair any connectors that overheat.

4.5.1 Lugs and Quick Disconnects

Lugs are used at the end of the welding cable to connect the cable to the welding machine current terminals. Lugs come in various sizes to match the welding cable size and are mechanically crimped

Table 1 Welding Cable Sizes

Machine Size In Amperes	Duty Cycle (%)	Recommended Cable Sizes for Manual Welding — Copper Cable Sizes for Combined Lengths of Electrodes Plus Ground Cable				
		Up to 50 Feet	50–100 Feet	150 Feet	150–200 Feet	200–250 Feet
100	20	#8	#4	#3	#2	#1
180	20	#5	#4	#3	#2	#1
180	30	#4	#4	#3	#2	#1
200	50	#3	#3	#2	#1	#1/0
200	60	#2	#2	#2	#1	#1/0
225	20	#4	#3	#2	#1	#1/0
250	30	#3	#3	#2	#1	#1/0
300	60	#1/0	#1/0	#1/0	#2/0	#3/0
400	60	#2/0	#2/0	#2/0	#3/0	#4/0
500	60	#2/0	#2/0	#3/0	#3/0	#4/0
600	60	#3/0	#3/0	#3/0	#4/0	* * *
650	60	#3/0	#3/0	#4/0	* *	* * *

* * Use Double Strand of #2/0

* * * Use Double Strand of #3/0

onto the welding cable. Lugs are normally ringed to prevent the welding lead from pulling loose from the stud connector on the machine.

Quick disconnects are also mechanically connected to the cable ends. They are insulated and serve as cable extensions for splicing two lengths of cable together. Quick disconnects are connected or disconnected with a half twist. When using quick disconnects, make sure that they are tightly connected to prevent overheating or arcing in the connector. They must be kept clean of dirt, sand, mud, and other contaminants to ensure a good connection. *Figure 15* shows typical lugs and quick disconnects used as welding cable connectors.

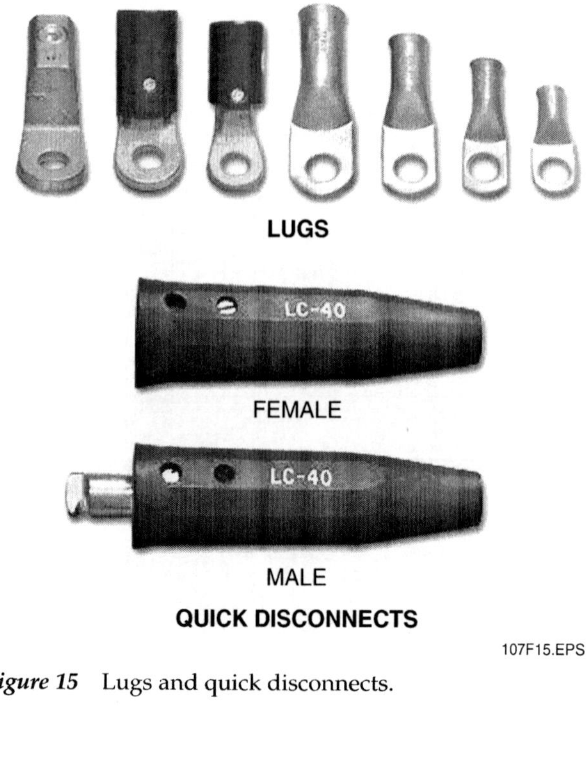

Figure 15 Lugs and quick disconnects.

4.5.2 Workpiece Clamps

Workpiece clamps establish the connection between the end of the workpiece and the workpiece lead. Workpiece clamps are mechanically connected to the welding cable and come in a variety of shapes and sizes. The size of a workpiece clamp is the rated amperage that it can carry without overheating. When selecting a workpiece clamp, be sure that it is rated at least as high as the rated capacity of the power source on which it will be used. *Figure 16* shows several examples of workpiece clamps.

4.5.3 Electrode Holders

Electrode holders grasp the electrode and provide the electrical contact between the electrode and the end of the welding cable. Electrode holders are also mechanically connected to the welding cable and come in a variety of styles and sizes. The size of an electrode holder is the rated amperage it will carry without overheating. The physical size of an electrode holder is also proportional to its amperage capacity. A small 200A electrode holder will be light and easy to handle but will not hold large electrodes that require more than 200A. When selecting an electrode holder, select the size for the amperage and electrodes being used. *Figure 17* shows two styles of electrode holders.

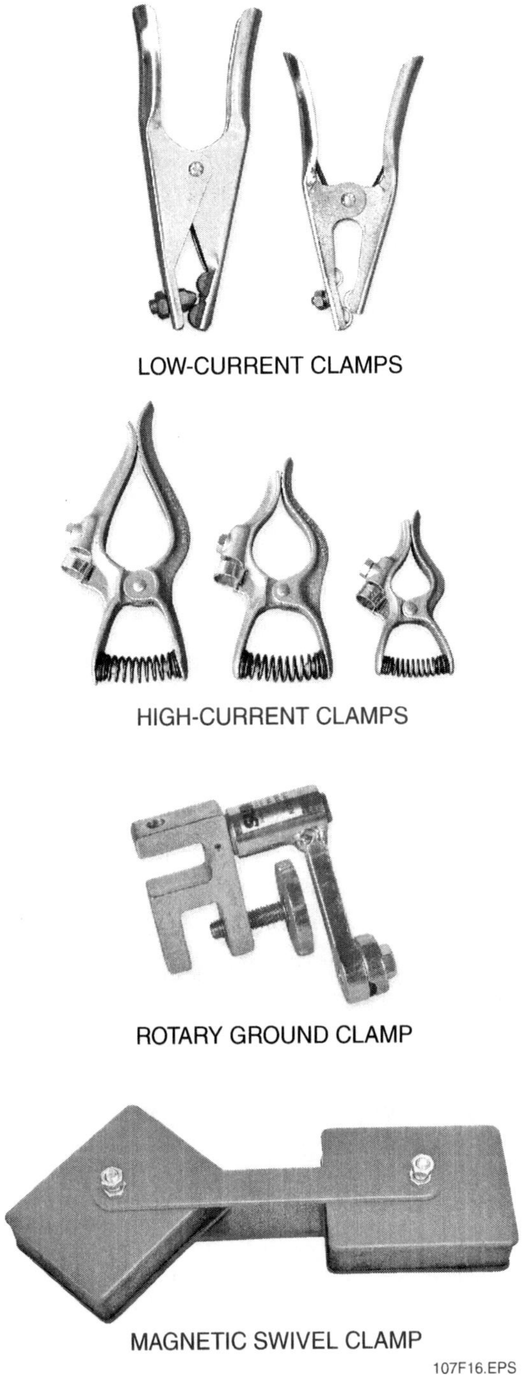

Figure 16 Workpiece clamps.

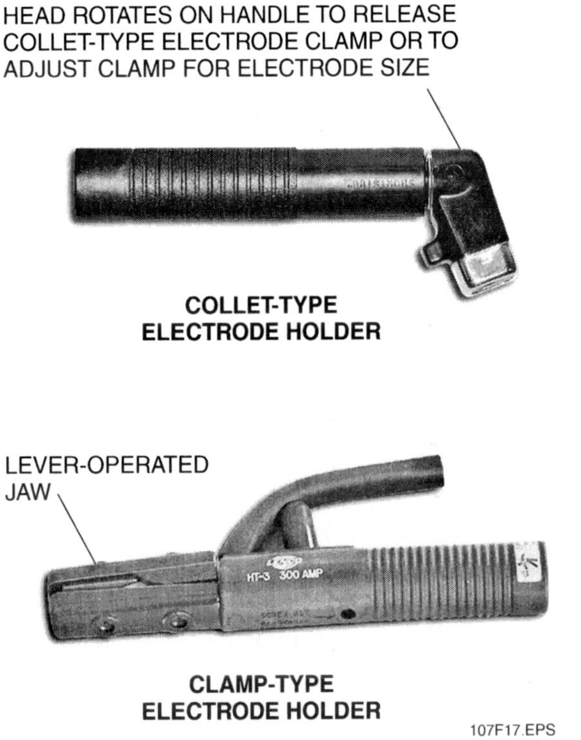

Figure 17 Electrode holders.

5.0.0 SMAW Equipment Setup

In order to safely and efficiently weld using the SMAW process, the equipment must be properly selected and set up. Accomplishing proper SMAW equipment setup involves doing the following procedures:

- Selecting the proper SMAW equipment
- Choosing the machine's location
- Moving a welding machine
- Stringing welding cable
- Locating the workpiece clamp

5.1.0 Selecting the Proper SMAW Equipment

In selecting any tool for a particular job, there are a number of considerations to be examined. To select the proper SMAW machine, the following factors must be considered:

- *The welding process* – As previously shown, SMAW requires a variable-voltage power source with a constant current. Other processes may require constant-voltage power sources, but this module focuses on SMAW.
- *The type of welding current (AC or DC)* – Most SMAW processes use DC current.
- *The maximum amperage required* – Consider the composition of the material, its thickness, and other material properties to determine the amperage requirements of the machine.
- *The primary power requirements* – If AC power with the proper receptacle for the welding machine is not available, then a portable or machine-driven unit must be used.

5.2.0 Welding Machine Location

The welding machine should be located near the work to be performed. Select a site where the machine will not be in the way but will be protected from welding or cutting sparks. There should be good air circulation to keep the machine cool, and the environment should be free from explosive or corrosive fumes and as free as possible from dust and dirt. Welding machines have internal cooling fans that will pull these materials into the machine if they are present. The site should be free of standing water or water leaks. If an engine-driven generator is used, locate it so it can be easily refueled and serviced.

WARNING!

Never run an engine-driven generator indoors because it can cause carbon monoxide poisoning.

There should also be easy access to the site so the machine can be started, stopped, and adjusted as needed. If the machine is to be plugged into an outlet, be sure the outlet has been properly installed by a licensed electrician to ensure it is grounded. Also, be certain to identify the location of the electrical disconnect for the outlet before plugging in the welding machine.

5.3.0 Moving a Welding Machine

Large, engine-driven generators are mounted on a trailer frame and can be easily moved by a pickup truck or tractor using a trailer hitch. Other types of welding machines may have a skid base or may be mounted on steel or rubber wheels. Be careful when moving welding machines that are mounted on wheels. Some machines are top-heavy and may fall over in a tight turn or if the floor or ground is uneven or soft.

WARNING!

If a welding machine starts to fall over, do not attempt to hold it. Welding machines are very heavy and you can be severely injured if a welding machine falls on you.

Many welding machines have lifting eyes. Lifting eyes are usually provided to move machines mounted on skids. Before lifting any welding machine, refer to the equipment specifications for the weight. Be sure the lifting device and tackle are rated at more than the weight of the machine. When lifting a welding machine, always use a shackle. Never attempt to lift a machine by placing the lifting hook in the machine eye. Before lifting or moving a welding machine, make sure the welding cables are coiled and secure. *Figure 18* shows how to lift a welding machine.

Hot Tip

Typical SMAW Setup

A typical SMAW setup, using elements previously described, is shown here for an engine-driven welding machine. However, any type of SMAW welding machine could be substituted. For DC welding machines, the electrode lead would be connected to the machine as DCEN or DCEP as required.

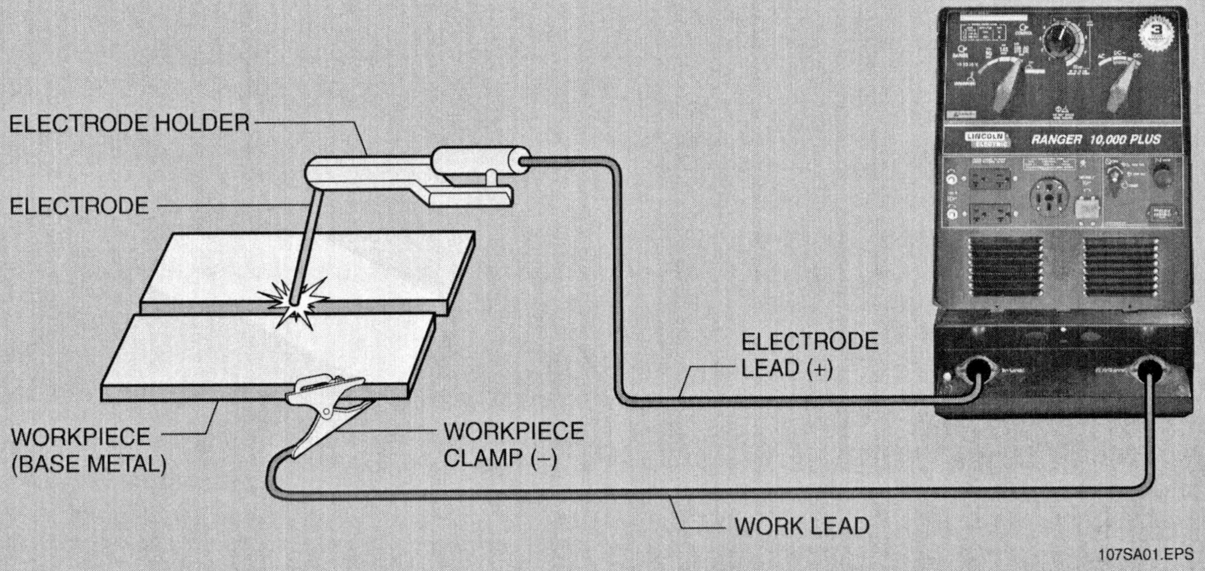

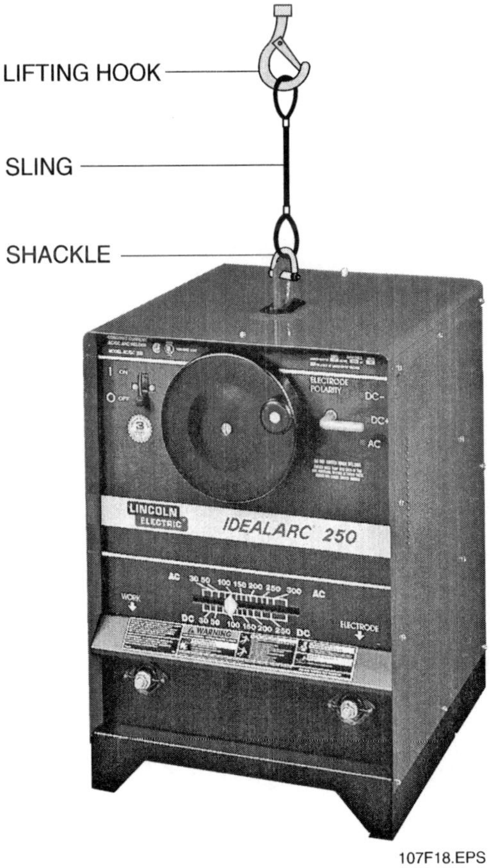

Figure 18 Lifting a welding machine.

Hot Tip

Moving Welding Machines on Wheels

It is not uncommon on industrial construction projects to move all welding machines using cranes or cherry pickers (small hydraulic boom equipment), even those on trailers or machines with wheels, because of the limited access for vehicular traffic in these facilities.

5.4.0 Stringing Welding Cable

Before stringing welding cable, inspect the cable for damage. Cuts or breaks in the insulation must be repaired according to procedures. Breaks in the welding cable insulation could arc to equipment or to any metal surface the welding contacts. This will damage the surface and cause additional damage to the welding cable. If the welding cable cannot be repaired, it should be discarded.

Welding cables should be long enough to reach the work area but not so long that they must always be coiled. Welding cables must be strung to prevent tripping hazards or damage from traffic. Keep welding cables out of walkways and aisles. If welding cables must cross a walkway or aisle, string the cables overhead. If the cables cannot be strung overhead, use boards or ramps to prevent workers from tripping and to protect the cables from both foot and vehicular traffic. *Figure 19* shows how to protect welding cables with a ramp.

When stringing welding cables, take care not to damage equipment. Do not string welding cable over instrumentation wires and tubing. These can be easily damaged by the weight or movement of the cables. If there are moving parts, keep the welding cables well away from nip points. If the equipment suddenly starts, the welding cables could be pulled into the equipment. When there is new construction or rebuild work taking place, equipment such as cable trays, conduit, piping, and pipe hangers may not be permanently attached. The weight of welding cables could cause these items to collapse. Check such equipment before stringing welding cables.

When stringing welding cable overhead or between floors, use rope to tie off the cables. The weight of the welding cables could damage connections or pull an individual holding onto the electrode holder off balance. Rope is relatively nonconductive and nonabrasive, and it will support the weight of the welding cable without damaging it.

5.5.0 Locating the Workpiece Clamp

The workpiece clamp must be properly located to prevent damage to surrounding equipment. If the welding current travels through any type of bearing, seal, valve, or contacting surface, it could cause severe damage from arcing, which would require replacement of the item. Carefully check the area to be welded and position the workpiece clamp so the welding current will not pass through any contacting surface. If in doubt, ask your supervisor for assistance before proceeding.

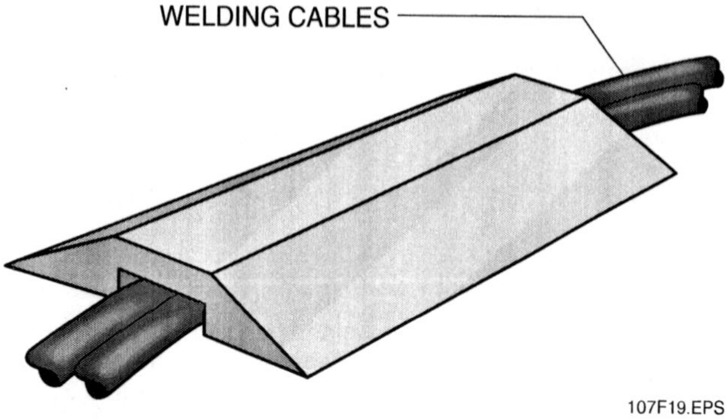

Figure 19 Protecting welding cables.

CAUTION

Welding current can severely damage bearings, seals, valves, and contacting surfaces. Position the workpiece clamp to prevent welding current from passing through them.

Welding current passing through electrical or electronic equipment will cause severe damage. Disconnect the workpiece lead at the battery to protect the electrical system before welding on any type of mobile equipment. If welding will be performed near the battery, remove the battery. Batteries produce hydrogen gas, which is extremely explosive. A welding spark could cause the battery to explode.

WARNING!

Do not weld near batteries. A welding spark could cause a battery to explode, showering the area with battery acid.

Before welding on or near electronic or electrical equipment or cabinets, contact an electrician. The electrician will isolate the system to protect it.

CAUTION

The slightest spark of welding current can destroy electronic or electrical equipment. Have an electrician check the equipment and, if necessary, isolate the system before welding.

Workpiece clamps must never be connected to pipes carrying flammable or corrosive materials. The welding current could cause overheating or sparks, resulting in an explosion or fire.

The workpiece clamp must make good electrical contact when it is connected. Dirt and paint will inhibit the connection and cause arcing, resulting in overheating of the workpiece clamp. Dirt and paint also affect the welding current and can cause defects in the weld. Clean the surface before connecting the workpiece clamp. If the workpiece clamp is damaged and does not close securely onto the surface, change it.

Case History

Trailer Mud Flaps

A welder was given the job of reattaching a mud flap to the back of a refrigerated trailer. The loaded trailer was parked right next to the welding machine, and it looked to be a very simple job. Because it was so close, the employee reached over and connected the workpiece clamp to the front of the trailer and then pulled the cable with the electrode holder to the back end. After cleaning and positioning the bracket, the mud flap was welded to the trailer. The entire job took less than 15 minutes to complete.

When the customer returned to pick up the trailer, he noticed that the refrigeration unit was off and there was no power at the display panel. A service technician determined that the microprocessor on the refrigeration unit had failed and needed to be replaced. After further investigation, it was determined that the pressure and temperature sensors on the refrigeration unit had provided a path for the welding current to get to the microprocessor. The current was too high for the microprocessor and burned out some of its components. As a result, the welding shop was required to pay more than $1,000 in parts and labor to replace the microprocessor.

The Bottom Line: If the welder had isolated the electronic components from the welding circuit, or at least had located the workpiece clamp close to the weld joint, then the damage would not have occurred.

6.0.0 Starting SMAW Welding Machines

Starting an electrically powered welder is a simple matter of flipping a switch and making adjustments. The source of energy is constantly available. Starting a motor-driven welding machine is a bit more complicated than starting an electrically powered machine.

6.1.0 Energizing Electrically Powered Welding Machines

Electrically powered welding machines are plugged into an electrical outlet. The electrical requirements (primary current) will be on the equipment specification tag displayed prominently on the machine. Most machines will require single-phase 240V current or three-phase 480V current. Machines requiring single-phase 240V power will have a three-prong plug. Machines requiring three-phase 480V power will have a four-prong plug. *Figure 20* shows grounded electrical plugs for welding machines.

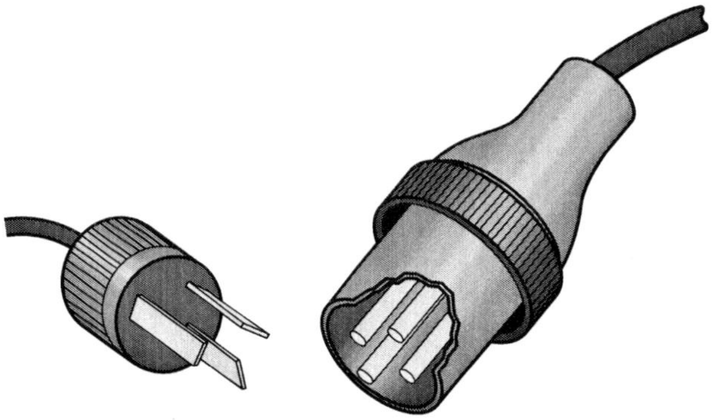

Figure 20 Grounded electrical plugs.

If a welding machine does not have a power plug, an electrician must connect the machine. The electrician will add a plug or will hard-wire the machine directly into a disconnecting device.

WARNING!

Never use a welding machine until you identify the electrical disconnect. In the event of an emergency, you must be able to quickly turn off the power to the welding machine at the disconnect to prevent injury or even death.

6.2.0 Starting Engine-Driven Welding Machines

Before welding can take place with an engine-driven welding machine, the engine must be checked and then started. As with a car engine, the engine powering the welding machine must also have routine maintenance performed.

6.2.1 Prestart Checks

Many facilities will have prestart checklists that must be completed and signed prior to starting and operating an engine-driven welding machine. If your site has such a checklist, you must complete and sign it. If your site does not have a prestart checklist, perform the following checks before starting the engine:

- Check the oil level using the engine oil dipstick. If the oil is low, add the proper amount of appropriate grade oil. Do not overfill.
- Check the coolant level in the radiator if the engine is liquid-cooled. If the coolant level is low, add coolant. Do not remove the radiator cap on a hot engine.

CAUTION

Do not add plain water to radiators that contain antifreeze. Antifreeze not only protects radiators from freezing in cold weather, it also has rust inhibitors and additives to aid in cooling. If the antifreeze is diluted, it will not function properly. If the weather turns cold, the system may freeze, causing damage to the radiator, engine block, and water pump.

- Check the fuel level. The unit may have a fuel gauge or a dipstick. If the fuel is low, add the correct fuel, diesel or gasoline, to the fuel tank. The type of fuel required should be marked on the fuel tank. If not, contact your supervisor to verify the fuel required, and have the tank marked.

CAUTION

Adding gasoline to a diesel engine or diesel to a gasoline engine will cause severe engine problems and can cause a fire hazard. Always be sure to add the correct type of fuel to the fuel tank.

- Check the battery water level, unless the battery is sealed. Add water if the battery water level is low.

WARNING!

Battery cell vapors can be explosive. Use extreme care when working with or around a serviceable battery. Avoid exposing the battery to sparks or arcs of any type. Many serious accidents are reported annually due to batteries exploding in workers' faces, caused by accidental arcing or sparking.

- Look at the electrode holder and make sure it is not touching the workpiece. If the electrode holder is touching the workpiece, it will arc and overheat the welding system when the welding machine is started.
- Open the fuel shutoff valve if the equipment has one. If there is a fuel shutoff valve, it will be located in the fuel line between the fuel tank and the carburetor.
- Record the hours from the hour meter if the equipment has one. An hour meter records the total number of hours the engine runs. This information is used to determine when the engine needs to be serviced. The hours will be displayed on a gauge similar to an odometer.

NOTE

Clean the engine and equipment after each use. Refer to the manufacturer's instructions.

6.2.2 Starting the Engine

Most engines will have an ON/OFF ignition switch and a starter. They may be combined into a key switch similar to the ignition switch on a car. To start the engine, turn on the ignition switch and press the starter. Release the starter when the engine starts. The engine speed will be controlled by the governor. If the governor switch is set for idle, the engine will slow to an idle after a few seconds. If the governor is set to welding speed, the engine will continue to run at welding speed.

Small engine-driven welding machines may have an ON/OFF switch and a pull rope. These machines are started by turning on the ignition switch and pulling the cord, which is similar to starting a lawn mower. These engines do not have a battery.

Engine-driven welding machines should be started about five to ten minutes before they are needed for welding. This will allow the engine to warm up before a welding load is placed on it.

WARNING!

When you start the welding machine, always make sure the welding electrode and/or electrode holder are secure and not touching any nearby worker or any conductive surface. Engine cowlings, or covers, should be closed and latched prior to starting the machine to protect workers from accidental contact with moving parts.

6.2.3 Stopping the Engine

If no welding is required for 30 or more minutes, stop the engine by turning off the ignition switch. If you are finished with the welding machine for the day, also close the fuel valve if there is one. Allow the machine to cool before storing leads inside the engine compartment.

6.2.4 Preventive Maintenance

Engine-driven welding machines require regular preventive maintenance to keep the equipment operating properly. Most sites will have a preventive maintenance schedule based on the hours that the engine operates. In severe conditions, such as a

GOING GREEN

Fuel-Powered Machine Maintenance

Proper maintenance of fuel-powered welding machine engines helps to protect the environment. If a fuel-powered engine is smoking or leaking fluids, have it turned in for maintenance.

very dusty environment or cold weather, maintenance may have to be performed more frequently.

CAUTION

To prevent equipment damage, perform preventive maintenance as recommended by the site procedures or the manufacturer's maintenance schedule in the equipment manual.

Responsibility for performing preventive maintenance will vary from site to site. Check with your supervisor to determine who is responsible for performing preventive maintenance.

When performing preventive maintenance, follow the manufacturer's guidelines in the equipment manual. Typical tasks to be performed as part of a preventive maintenance schedule include the following:

- Changing the oil, gas filter, and air filter
- Checking/changing the antifreeze
- Greasing the undercarriage
- Repacking the wheel bearings
- Replacing worn drive belts
- Maintaining/replacing the spark plugs
- Maintaining the battery
- Cleaning surfaces

7.0.0 Tools for Cleaning Welds

Even though some of the hand tools used to clean welds were presented in the tools section, it is worthwhile to mention some of them again and explain their functions. Hand tools such as files, chipping hammers, and wire brushes, as well as small power tools such as slag chippers and scalers, are common tools used by welders to prepare surfaces before welding and to clean welds after they are completed.

WARNING!

Always wear eye and hand protection when preparing and cleaning weldment surfaces. Remember that finished welds remain extremely hot for extended periods and can cause serious burns. Chipping slag and other coatings from finished welds is very hazardous and requires proper eye protection, including goggles or face shields.

7.1.0 Hand Tools

It is extremely important that the base metals are thoroughly cleaned and prepared before welding. Dirty or corroded surfaces are very difficult, if not impossible, to weld properly. Also, before any additional weld beads can be applied over previous beads, the previous pass must be cleaned of any slag that was deposited along with the weld bead. Many mechanical methods have been developed to assist in these cleaning and removing processes, including sandblasting and ultrasonic cleaning. However, many welders prefer simple manual procedures using tools such as files, wire brushes, and chipping hammers. *Figure 21* shows some of the popular hand tools used by welders to prepare base metals prior to welding and to remove slag after the bead has been laid.

7.2.0 Pneumatic Cleaning and Slag Removal Tools

Weld slag chippers and needle scalers *(Figure 22)* are pneumatically powered. They are used by welders to clean surfaces and to remove slag from cuts and welds. Weld slag chippers and needle scalers are also excellent for removing paint or hardened dirt, but they are not very effective for removing surface corrosion. Weld slag chippers have a single chisel, and needle scalers have 18 to 20 blunt steel needles approximately 10" long. Most weld slag chippers can be converted to needle scalers with a needle scaler attachment.

107F21.EPS

Figure 21 Typical hand tools used in welding.

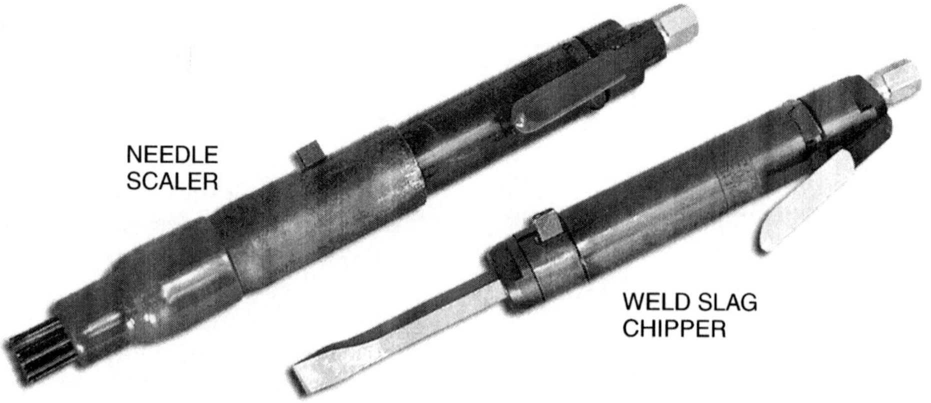

107F22.EPS

Figure 22 Weld slag chipper and pneumatic needle scaler.

Review Questions

1. SMAW is often referred to as _____.
   a. AC welding
   b. stick welding
   c. gap welding
   d. resistive welding

2. When working at heights, make sure to inspect your fall protection equipment items each time you use them, and make sure that they are current and up-to-date.
   a. True
   b. False

3. During SMAW welding, current jumps across the _____ from the tip of the electrode to the surface of the workpiece.
   a. workpiece clamp
   b. gap
   c. electrode holder
   d. molten metal

4. In AC current, the number of cycles completed in one second is called the _____.
   a. polarity
   b. travel
   c. frequency
   d. current value

5. Electrical current that has no frequency is known as _____.
   a. DC current
   b. AC current
   c. AC voltage
   d. 120VAC

6. _____ only applies to DC current.

a. Frequency
b. Polarity
c. Amperage
d. Current flow

7. The voltage in a welding machine is reduced or stepped down by a(n) _____.

a. limiter
b. generator
c. alternator
d. transformer

8. The voltage that is present when the welding machine is on but no arc is present is _____ voltage.

a. open-circuit
b. operating
c. zero
d. minimum

9. Most light-duty transformer welding machines operate more efficiently on _____ primary power.

a. 480V three-phase
b. 120V single-phase
c. 240V
d. DC power

10. A(n) _____ converts alternating current to direct current.

a. transformer
b. rectifier
c. alternator
d. generator

11. Governors on engine-driven generator machines directly control the _____ of the machine.

a. temperature
b. speed
c. rectifier
d. primary power

12. The duty cycle of a welding machine is based on a _____ period.

a. 10-minute
b. 15-minute
c. 20-minute
d. 30-minute

13. Welding cable lugs are generally _____ to prevent the lead from pulling loose from the welding machine terminal.

a. slotted
b. soldered in place
c. ringed
d. insulated

14. To determine the amperage requirements of a welding machine, consider the _____.

a. type of material being welded, its thickness, and its properties
b. environment in which the welding must be performed
c. weight of the base metal being welded
d. type of machine being used

15. Machines requiring single-phase 240V power will have a _____-prong plug.

a. two
b. three
c. four
d. five

SUMMARY

Shielded metal arc welding is often referred to as stick welding. SMAW remains one of the most widely used forms of welding in the industry because of its simplicity and relatively low cost. However, it is not as foolproof as some of the newer technology now used in the welding industry. Whatever type of welding is being used, proper selection and setup of equipment and accessories are necessary to achieve a reliable weld while maintaining a safe working environment.

Included in this module are the welding currents associated with SMAW, the various types of SMAW machines, and some of the accessories needed to connect SMAW equipment. The steps in selecting and setting up SMAW equipment are also covered. SMAW is the basis on which all other welding processes have been developed. Understanding the SMAW process is recommended for achieving success in the other processes.

Trade Terms Quiz

Fill in the blank with the correct key term that you learned from your study of this module.

1. ____________ is electrical current received from conventional power lines used to energize welding machines.
2. The terminal point to which electricity is brought in the welding operation and from which the arc is produced to do the welding is called the ____________.
3. ____________ is electrical current that reverses its flow at set intervals.
4. ____________ is the measurement of electromotive force or pressure that causes current to flow in a circuit.
5. ____________ is electrical current that flow in one direction only.
6. An electrical device that uses two wire coils of different sizes to convert a high-voltage, low-current power source into a lower voltage, higher current output is known by the term ____________.
7. A measurement of the rate of flow of electric current is called ____________.
8. The direction of flow of electrical current in a direct current welding circuit is called ____________.
9. The flow of electricity through an air gap or gaseous space is called a(n) ____________.

Trade Terms

Alternating current (AC)
Amperage
Arc
Direct current (DC)
Electrode
Polarity
Primary current
Step-down transformer
Voltage

Cornerstone of Craftsmanship

Bennett B. (Ben) Grimmett, P.E., M.B.A.

Principal Engineer Babcock & Wilcox Companies – Nuclear Operations Group
Lynchburg, VA

Born in Amherst, Ohio, Ben Grimmett was one of four brothers. His father was a maintenance mechanic at Ford Motor Company who performed welding both on the job and at home. Though he only had a sixth grade education, his father was, according to Ben, "one of the cleverest individuals I have ever known." Ben started out his welding career by taking vocational welding and fabrication programs at the Lorain County Vocational School. With his father as a role model of hard work and ingenuity, Ben began to further his career through additional education. He went on to earn a B.S. in welding engineering from The Ohio State University and an M.B.A. from Franklin University. Ben is now employed as a Principal Engineer at Babcock & Wilcox Companies – Nuclear Operations Group in Lynchburg, Virginia.

How did you become interested in the welding industry?

As a kid, I remember watching my dad build a swing set out of old pipes and chains. As he was welding the pieces together, I remember watching through the welding hood and being amazed by the process and the associated phenomena—the noises, the sparks, the bright lights. I was quickly hooked on it, even at that young age. It was fascinating to a 10-year-old boy. Since then, welding has always been and will remain an extremely interesting science and technology to me. Welding is truly one of the most under-appreciated technologies in the world.

I was also fortunate enough to have an older engineer take an active interest in my career when I was at Terra Technical College. Jim Harris, who works as a senior staff engineer at Ashland Chemical, became my mentor and helped guide me at a time in my life when I needed someone to help me find direction. Jim continues to be an advisor, confidante, and one of my best friends.

What do you think it takes to be a success in your trade?

Make solutions, not excuses. If you make a mistake, accept it as a learning experience. Nobody wants to hear an excuse. What people want to hear is how you are going to rectify the situation.

If you are not experiencing some failure in life, then you simply aren't doing anything. You need to be able to make decisions, accept responsibility for those decisions, and be willing to see them through to completion. Nothing in the world is more annoying than an engineer who does not make decisions.

Find a mentor. A mentor gives you someone to discuss your goals in life with and who can offer advice on how to achieve them. Mentoring is a two-way relationship. A mentor takes an active, vested interest in seeing you succeed, is never jealous, and takes great satisfaction in sharing your success.

What are some of the things you do in your job?

In my job, one of the interesting and challenging things I do is create solutions to unique problems and situations. I evaluate and develop methods, testing procedures, training materials, and programs. I also develop cost estimates, proposals, and scopes-of-work, which includes a lot of project management.

What do you like most about your job?

I like the variety, the depth of projects, and the opportunity to work with some very talented people.

What would you say to someone entering the trade today?

Be able to roll up your sleeves and talk to the person on the floor, then be able to turn around and talk to the CEO of the company—in language both the floorperson and the senior manager understand. Do not be afraid to challenge yourself and take some risks to get noticed. I have had some absolutely stunning fiascos in my life, but each one provided a unique learning opportunity and a chance to evaluate my principles.

Always think long term and big picture, but be able to see the trees and the forest so that you can communicate effectively. Also, be able to sit back and laugh at yourself.

Trade Terms Introduced in This Module

Alternating current (AC): Electrical current that reverses its flow at set intervals.

Amperage: A measurement of the rate of flow of electric current.

Arc: The flow of electricity through an air gap or gaseous space.

Direct current (DC): Electrical current that flows in one direction only.

Electrode: Terminal point to which electricity is brought in the welding operation and from which the arc is produced to do the welding.

Polarity: The direction of flow of electrical current in a direct current welding circuit.

Primary current: Electrical current received from conventional power lines used to energize welding machines.

Step-down transformer: An electrical device that uses two wire coils of different sizes to convert a high-voltage, low-current power source into a lower voltage, higher current output.

Voltage: A measurement of electromotive force or pressure that causes current to flow in a circuit.

Additional Resources

This module is intended to present thorough resources for task training. The following reference works are suggested for further study. These are optional materials for continued education rather than for task training.

OSHA Standard 1926.351, Arc Welding and Cutting.

The Procedure Handbook of Arc Welding, 2000. Cleveland, OH: The Lincoln Electric Company.

Stick Electrode Welding Guide, 2004. Cleveland, OH: The Lincoln Electric Company.

Stick Electrode Product Catalog, 2008. Cleveland, OH: The Lincoln Electric Company. www.lincolnelectric.com

CONTREN® LEARNING SERIES – USER UPDATE

NCCER makes every effort to keep these textbooks up-to-date and free of technical errors. We appreciate your help in this process. If you have an idea for improving this textbook, or if you find an error, a typographical mistake, or an inaccuracy in NCCER's Contren® textbooks, please write us, using this form or a photocopy. Be sure to include the exact module number, page number, a detailed description, and the correction, if applicable. Your input will be brought to the attention of the Technical Review Committee. Thank you for your assistance.

Instructors – If you found that additional materials were necessary in order to teach this module effectively, please let us know so that we may include them in the Equipment/Materials list in the Annotated Instructor's Guide.

Write: Product Development and Revision
National Center for Construction Education and Research
3600 NW 43rd St., Bldg. G, Gainesville, FL 32606

Fax: 352-334-0932

E-mail: curriculum@nccer.org

Craft ______ Module Name ______

Copyright Date ______ Module Number ______ Page Number(s) ______

Description ______

______

______

______

(Optional) Correction ______

______

______

(Optional) Your Name and Address ______

______

______

Glossary

Abrasive: A rough material used for sanding, grinding, sharpening, or cutting.

Actuator: The part of a regulating valve that converts electrical or fluid energy to mechanical energy to position the valve.

Adjacent side: The side of a right triangle that is next to the reference angle.

Adsorption: The collection of a substance, either gas or liquid, in condensed form on a surface or to substances other than itself.

Aerial lift: A mobile work platform designed to transport and raise personnel, tools, and materials to overhead work areas.

Align: To line up two or more parts.

Alternating current (AC): Electrical current that reverses its flow at set intervals.

American Society for Testing Materials International (ASTM): Founded in 1898, a scientific and technical organization, formed for the development of standards on the characteristics and performance of materials, products, systems, and services; publishes specifications and standards relating to fasteners.

Amperage: A measurement of the rate of flow of electric current.

Ampere: A unit of electrical current, also known as amp.

Amperes: The unit of measure for current.

Angle valve: A type of globe valve in which the piping connections are at right angles.

Anneal: To soften a metal by heat treatment.

Anti-two-blocking devices: Devices that provide warnings and prevent two-blocking from occurring. Two-blocking occurs when the lower load block or hook comes into contact with the upper load block, boom point, or boom point machinery. The likely result is failure of the rope or release of the load or hook block.

Apex: The point at which the lines of a figure converge.

Arc: The flow of electricity through an air gap or gaseous space.

Architectural drawings: Working drawings consisting of plans, elevations, details, and other information necessary for the construction of a building. Architectural drawings usually include:

- A site (plot) plan indicating the location of the building on the property
- Floor plans showing the walls and partitions for each floor or level
- Elevations of all exterior faces of the building
- Several vertical cross sections to indicate clearly the various floor levels and details of the footings, foundations, walls, floors, ceilings, and roof construction
- Large-scale detail drawings showing such construction details as may be required

Arithmetic numbers: Numbers that have definite numerical values, such as 4, 6.3, and ⅝.

Atomize: To reduce a liquid to a fine spray.

Backfire: A loud snap or pop as a torch flame is extinguished.

Ball valve: A type of plug valve with a spherical disc.

Bevel: An angle cut or ground on the end of a piece of solid material.

Bird caging: A deformation of wire rope that causes the strands or lays to separate and balloon outward like the vertical bars of a bird cage.

Blueprint: An exact copy or reproduction of an original drawing.

Body: The main part of the valve. It contains the disc, seat, and valve ports. The body of the valve is directly connected to the piping by threaded, welded, or flanged ends.

Bonnet: The part of a valve containing the valve stem and packing.

Bungs: Holes in the top of a lubricant container that are usually kept closed by a plug.

Burr: A sharp, ragged edge produced by cutting sheet metal.

Butterfly valve: A quarter-turn valve with a plate-like disc that stops flow when the outside area of the disc seals against the inside of the valve body.

Capacity: The volume of liquid handled by a pump per unit of time, expressed as gallons per minute (gpm) or cubic feet per minute (cfm).

Carbon dioxide (CO_2): A colorless, odorless gas produced by complete combustion of a hydrocarbon fuel-air mixture. CO_2 is a component of the exhaust from an internal combustion engine.

Carburizing flame: A flame burning with an excess amount of fuel; also called a reducing flame.

Centistoke (cSt): A unit of kinematic viscosity as measured in a capillary tube viscometer under constant pressure and temperature. One stoke equals 100 centistokes.

Chamfer: An angle cut or ground only on the edge of a piece of material.

Check valve: A valve that allows flow in one direction only.

Cherry picker: A truck-mounted or trailer-mounted lift designed for both indoor and outdoor use.

Chuck: The part of a machine that holds a piece of work tightly in the machine. A chuck is normally used only when the pipe or cutter will be rotated.

Circle: A continuous curved line that encloses a space, with every point on the line the same distance from the center of the circle.

Circuit breaker: A device designed to protect circuits from overloads and able to be reset after tripping.

Circuit: A complete path of electrical components and conductors.

Circumference: The distance around a circle.

Clearance: The amount of space between the threads of bolts and their nuts.

Component: A single part in a system.

Compressed: Pressed or squeezed together.

Compressor: A motor-driven machine used to supply compressed air for pneumatic tools.

Concentric circles: Circles having a common center point.

Construction drawings: Drawings that show in a clear, concise manner exactly what is required of the mechanic or builder.

Contamination: To make impure by contact or mixture.

Continuity: An uninterrupted electrical path for current flow.

Control valve: A globe valve automatically controlled to regulate flow through the valve.

Corrosion: The removal of mass from the surface of metals by chemicals that weaken and degrade the material properties.

Creep: The loss of thickness of a gasket, which results in bolt torque loss and leakage.

Cubic: The designation of a given unit representing volume.

Cylinder: A shape created by a circle moving in a straight line through space perpendicular to the surface of the circle.

Decomposition: The breaking down into the component parts by chemical reaction.

Deformation: A change in the shape of a material or component due to an applied force or temperature.

Detail drawings: An enlarged, detailed view taken from an area of a drawing and shown in a separate view.

Diameter: The width of a circle. The measurement from side to side.

Die: A tool used to make male threads on a pipe or a bolt.

Dimensions: Sizes or measurements printed on a drawing.

Direct current (DC): Electrical current that flows in one direction only.

Disc: Part of a valve used to control the flow of system fluid.

Discharge: The exit side of the pump from which the fluid leaves the pump.

Dispersant: An oil additive that maintains potential deposits finely dispersed and suspended in the oil so they will not settle out and collect on machine surfaces.

Disposal: The act of throwing out or away.

Drag lines: The lines on the kerf that result from the travel of the cutting oxygen stream into, through, and out of the metal.

Dropping point: The temperature at which a grease passes from a semisolid to a liquid state under specified test conditions.

Dross: The material (oxidized and molten metal) that is expelled from the kerf when cutting using a thermal process.

Dunnage: Wood blocks used as supports to keep materials off the ground.

Elastomeric: Elastic or rubber-like; flexible, pliable.

Electrical drawings: A means of conveying a large amount of exact, detailed information in an abbreviated language. Consists of lines, symbols, dimensions, and notations to accurately convey an engineer's designs to electricians and electronic systems technicians who install the electrical system on a job.

Electrode: Terminal point to which electricity is brought in the welding operation and from which the arc is produced to do the welding.

Elevation drawings: An architectural drawing showing height, but not depth; usually the front, rear, and sides of a building or object.

Emulsion: The combination of two liquids that are not usually mixable and in which one liquid is finely dispersed in the other.

Environmental Protection Agency (EPA): A federal agency charged with administering and enforcing federal regulations governing air and water pollution control.

Equalizer beam: A beam used to distribute weight on multi-crane lifts.

Ethylene propylene dieneterpolymer (EPDM): A gasket material and general-purpose polymer that is heat, ozone, and weather resistant. EPDM is not oil resistant.

Exploded diagram: A diagram that shows all the separated parts of the machine and where they connect together, with lines showing where each piece connects to the assembly.

Exponent: A number or symbol placed to the right and above another number, symbol, or expression, denoting the power to which the latter is to be raised.

Extrusion: A gasket protruding out of a flange.

Factors: The numbers that can be multiplied together to produce a given product.

Ferrous metals: Metals containing iron.

Fire point: The lowest temperature at which the tested sample ignites and sustains combustion for a specified time.

Fixed block: The upper block of a block and tackle. The block that is attached to the support.

Flanges: Projecting rims or collars on pipes used to hold them in place, give them strength, or attach them to something else.

Flash point: The lowest temperature at which the tested sample vapors ignite.

Flashback: The flame burning back into the tip, torch, hose, or regulator, causing a high-pitched whistling or hissing sound.

Floor plan: A drawing of a building as if a horizontal cut were made through a building at about window level, and the top portion removed. The floor plan is what would appear if the remaining structure were viewed from above.

Foot pounds (ft lbs): The normal method used for measuring the amount of torque being applied to bolts or nuts.

Forklift: A machine designed to move bulk items around the job site.

Formula: An equation that states a rule.

Frequency: The number of cycles completed each second by a given AC voltage; usually expressed in hertz; one hertz = one cycle per second.

Friction: The resistive force created when two surfaces slide or rub together. Rolling friction occurs in ball bearings. Fluid friction is measured by viscosity.

Fuse holder: A device used to hold a fuse that may be part of a circuit or may simply hold a spare fuse.

Galling: An uneven wear pattern between trim and seat that causes friction between the moving parts; surface damage on mating, moving metal parts that is caused by friction

Gate valve: A valve with a straight-through flow design that exhibits very little resistance to flow. It is normally used for open/shut applications.

Generator: A machine designed to generate electricity.

Globe valve: A valve in which flow is always parallel to the stem as it goes past the seat.

Governor: A device used to provide automatic control of speed or power for an internal combustion engine.

Grease: A physical mixture of base oil and thickeners, usually with other additives, having a solid or semifluid consistency.

Ground fault interrupter (GFI): A device that detects grounds on a power line and interrupts power flow to the attached outlet(s).

Hauling line: The line of a lifting device that is pulled by hand to raise the load.

Head: The pressure at the base of a column of fluid of a specific height. For water, 100 feet of head equals 43 psi. The pressure at any point in a liquid can be thought of as being caused by a vertical column of the liquid which, due to its weight, exerts a pressure equal to the pressure at the point in question. The height of this column is called the static head and is expressed in terms of feet of liquid.

Hertz: One cycle per second.

Horsepower (hp): A unit of power equal to 745.7 watts or 33,000 foot-pounds per minute.

Hydraulic: Operated by fluid pressure.

Hydrocarbons: Compounds containing only carbon and hydrogen. Petroleum consists mainly of hydrocarbons.

Hypotenuse: The longest side of a right triangle. It is always located opposite the right angle.

Impervious: Cannot be penetrated.

Inch pounds (in lbs): A method of measuring the amount of torque applied to small bolts or nuts that require measurement in smaller increments than foot pounds.

Inert: Unreactive. Exhibiting no chemical activity.

Inhibitor: A compound added to a lubricant to prevent negative and destructive reactions.

Kerf: The edge of the cut.

Key: A machined metal part that fits into a keyway and prevents parts such as gears or pulleys from rotating on a shaft.

Keyway: A machined slot in a shaft and on parts such as gears and pulleys that accepts a key.

Kinetic energy: Energy of motion.

Kinking: Bending a rope so severely that the bend is permanent and individual wires or fibers are damaged.

Lift: The actual travel of the disc away from the closed position when a valve is relieving.

Linear flow: Flow in which the output is directly proportional to the input.

Literal numbers: Letters that represent arithmetic numbers, such as x, y, and h. Also known as algebraic numbers.

Net positive suction head (NPSH): The amount of suction head required to prevent vaporization of the pumped liquid.

Neutral flame: A flame burning with correct proportions of fuel gas and oxygen.

NLGI consistency number: A simplified system for rating the consistency (hardness) of a lubricating grease as measured by the *ASTM Worked Penetration* (60 strokes) at 77°F.

Nominal size: A means of expressing the size of a bolt or screw. It is the approximate diameter of a bolt or screw.

Ohms: The unit of electrical resistance.

Oil: A general term for a water-insoluble, viscous liquid that possesses lubricating properties.

Opposite side: The side of a right triangle that is directly across from the reference angle.

Oxidation: The process of combining with oxygen. All petroleum products react with oxygen to some degree, and this increases with temperature increases.

Oxidizing flame: A flame burning with an excess amount of oxygen.

Ozone: A form of oxygen, usually created when electricity passes through the oxygen.

Packing: Material used to make a dynamic seal, preventing system fluid leakage around a valve stem.

Paraffin: Hydrocarbons belonging to the series starting with methane. Paraffins are saturated with respect to hydrogen. Paraffin wax is an example of a high molecular weight paraffin.

Parallel: A circuit wired in parallel has multiple paths for current to flow.

Parts of line: The number of ropes between the upper and lower blocks of a block and tackle. These lines carry the load. Parts of line are also called falls.

Perpendicular: At a right angle to the plane of a line or surface.

Phonographic: When referring to the facing of a pipe flange, serrated grooves cut into the facing, resembling those on a phonograph record.

Pi: A number that represents the ratio of the circumference to the diameter of a circle. Pi is approximately 3.1416 and is represented by the Greek letter π.

Pierce: To penetrate through metal plate with an oxyfuel cutting torch.

Pipe fitting: A unit attached to a piping system.

Pitch: The number of threads per inch on bolts, screws, and threaded rods.

Plan view: A drawing made as though the viewer were looking straight down (from above) on an object.

Plastic flow: The flowing of gasket material under stress.

Plug valve: A quarter-turn valve with a ported disc.

Plug: The moving part of a valve trim (plug and seat) that either opens or restricts the flow through a valve in accordance with its position relative to the valve seat, which is the stationary part of a valve trim.

Pneumatic: Run by or using compressed air.

Polarity: In DC circuits, the direction of current flow is from negative to positive; it is critical that the connections be identically arranged.

Polarity: The direction of flow of electrical current in a direct current welding circuit.

Polymerization: The joining of chemical molecules to form long chains called polymers. This joining reaction can be hazardous because some chemicals release great amounts of heat when they polymerize.

Positioner: A field-based device that takes a signal from a control system and ensures that the control device is at the setting required by the control system.

Pour point: The lowest temperature at which a lubricant will pour or flow under specified conditions of standardized *ASTM Test D 97*.

Powered industrial trucks: See forklift.

Primary current: Electrical current received from conventional power lines used to energize welding machines.

Programmable logic controller: A computerized control device that operates a particular machine, based on input from sensors and instructions from the operator.

Proportional control: A control that increases speed in proportion to the movement of the control.

Pyramid: A shape with a multi-sided base, and sides that converge at a point.

Radius: A straight line from the center of a circle to a point on the edge of the circle.

Rectangular: Description of a shape having parallel sides and four right angles.

Relief valve: A valve that automatically opens when a preset amount of pressure is exerted on the valve disc.

Resilience: The ability of a gasket to return to its original shape after being compressed.

Resistance: The ability of a gasket to withstand the effects of chemicals and other substances without damage or change.

Revolutions per minute (rpm): The number of complete revolutions an object will make in one minute.

Run: The horizontal distance from one pipe to another.

Saybolt universal seconds (SUS, SSU): A unit of measure of lubricating oil viscosity used in the U.S.

Scale: On a drawing, the size relationship between an object's actual size and the size it is drawn. Scale also refers to the measuring tool used to determine this relationship.

Schedule: A systematic method of presenting equipment lists on a drawing in tabular form.

Schematic diagram: A detailed diagram showing complicated circuits, such as control circuits.

Seat: The part of a valve against which the disc presses to stop flow through the valve.

Sectional view: A cutaway drawing that shows the inside of an object or building.

Series: A circuit wired in series has only one path for current to flow.

Set: The vertical distance from the line of flow of a pipe and the line of flow of the pipe to which it is attached.

Sheave: A grooved wheel used as a belt pulley.

Shims: A thin strip of wood or metal used to align parts.

Site plan: A drawing showing the location of a building or buildings on the building site. Such drawings frequently show topographical lines, electrical and communication lines, water and sewer lines, sidewalks, driveways, and similar information.

Sling angle: The angle formed by the legs of a sling with respect to the horizontal when tension is put upon the load.

Soapstone: Soft, white stone used to mark metal.

Society of Automotive Engineers (SAE): An organization that publishes specifications and standards relating to fasteners.

Solid: A figure enclosing a volume.

Sphere: A shape whose surface is everywhere the same distance from a central point.

Static: Having no motion; at rest.

Step-down transformer: An electrical device that uses two wire coils of different sizes to convert a high-voltage, low-current power source into a lower voltage, higher current output.

Straight blade duplex connector: An electrical connector or style of outlet.

Suction: The inlet side of a pump where the fluid is supplied to the pump.

Tetrafluoroethylene (TFE) (Teflon®): Used for many high-pressure, high-temperature applications.

Thermal transients: Short-lived temperature spikes.

Thread classes: Threads are distinguished by three classifications according to the amount of tolerance the threads provide between the bolt and nut.

Thread identification: Standard symbols used to identify threads.

Thread standards: An established set of standards for machining threads.

Throttling: The regulation of flow through a valve.

Tolerance: The difference between the allowed maximum and minimum limits of size; the amount of difference allowed from a standard.

Torque: A twisting force used to apply a clamping force to a mechanical joint; the turning force applied to a fastener.

Travel: The diagonal distance from one pipe to another.

Twist-lock connector: A type of electrical connector.

Unified National Coarse (UNC) thread: A standard type of coarse thread.

Unified National Extra Fine (UNEF) thread: A standard type of extra-fine thread.

Unified National Fine (UNF) thread: A standard type of fine thread.

Valve body: The part of a valve containing the passages for fluid flow, valve seat, and inlet and outlet connections.

Valve stem: The part of a valve that raises, lowers, or turns the valve disc.

Valve trim: The combination of the valve plug and the valve seat.

Velocity: Speed, expressed in distance over time, that is, ft/sec or ft/min.

Viscosity: The viscosity of an oil is its stiffness or internal friction. Higher viscosity is thicker or heavy, and a lower viscosity is thinner or light.

Viscosity index (VI): An empirical number indicating the effect of change of temperature on the viscosity of oil. A low VI indicates a relatively large change of viscosity with temperature; a high VI indicates a relatively small change.

Voltage: A measurement of electromotive force or pressure that causes current to flow in a circuit

Volume: The amount of space occupied by an object.

Watt: A unit of power that is equal to one joule per second.

Wedge: The disc in a gate valve.

Wire drawing: The erosion of a valve seat under high velocity flow through which thin, wire-like gullies are eroded away.

Written specifications: A written description of what is required by the owner, architect, and engineer in the way of materials and workmanship. Together with working drawings, the specifications form the basis of the contract requirements for construction.

Yoke bushing: The bearing between the valve stem and the valve yoke.

Index

Index

C

D

E

F

H

Q

R

U

V